# Android 物联网开发从入门到实战

孙光宇　张玲玲　编著

清华大学出版社
北　京

## 内容简介

本书内容分为5篇，共计17章，循序渐进地讲解了Android物联网开发的基本知识。本书从获取源码和搭建应用开发环境开始讲起，依次讲解了基础知识篇、数据传输篇、信息识别篇、传感器应用篇和技术提高篇这5大部分内容。在讲解每一个知识时，都遵循了理论联系实际的讲解方式，从内核分析到接口API实现，再到实战演练，最后到综合实例演练，彻底剖析了物联网项目开发的完整实现流程。本书几乎涵盖了当下Android物联网开发的绝大多数内容，讲解方法通俗易懂并且详细，不但适合应用高手们学习，也特别便于初学者学习和理解。

本书适合Android驱动开发者、Linux开发人员、Android物联网开发人员、Android爱好者、Android源码分析人员、Android应用开发人员、Android传感器开发人员、Android智能家居开发人员、Android可穿戴设备人员的学习，也可以作为相关培训学校和大专院校相关专业的教学用书。



图书在版编目（CIP）数据

Android物联网开发从入门到实战/孙光宇，张玲玲编著．—北京：清华大学出版社，2015（2021.8重印）
ISBN 978-7-302-40084-4

I. ①A… II. ①孙… ②张… III. ①移动终端-应用程序-程序设计 ②互联网络-应用 ③智能技术-应用 IV. ①TN929.53 ②TP393.4 ③TP18

中国版本图书馆CIP数据核字（2015）第089629号

**责任编辑：**朱英彪
**封面设计：**刘　超
**版式设计：**牛瑞瑞
**责任校对：**王　云
**责任印制：**丛怀宇

**出版发行：**清华大学出版社
**网　　址：**http://www.tup.com.cn，http://www.wqbook.com
**地　　址：**北京清华大学学研大厦A座　　**邮　　编：**100084
**社 总 机：**010-62770175　　**邮　　购：**010-62786544
**投稿与读者服务：**010-62776969，c-service@tup.tsinghua.edu.cn
**质量反馈：**010-62772015，zhiliang@tup.tsinghua.edu.cn

**印 装 者：**北京富博印刷有限公司
**经　　销：**全国新华书店
**开　　本：**203mm×260mm　**印　　张：**42　**字　　数：**1183千字
（附DVD光盘1张）
**版　　次：**2015年7月第1版　**印　　次：**2021年8月第6次印刷
**定　　价：**99.80元

产品编号：061534-02

# 前　　言

2007 年 11 月 5 日，谷歌公司宣布基于 Linux 平台的开源手机操作系统 Android 诞生，该平台号称是首个为移动终端打造的真正开放和完整的移动软件平台。本书将和广大读者一起领略这款系统的神奇之处。

## 市场占有率高居第一

截至 2014 年 9 月，Android 在手机市场上的占有率从 2013 年的 68.8%上升到 80%。而 iOS 则从 2013 年的 19.4%下降到 15.5%，WP 系统从原来的 2.7%小幅上升到 3.6%。

从数据上来看，Android 平台占据了市场的主导地位，继续充当老大的角色。Android 市场的占有率增加幅度较大，WP 市场小幅增长，但 iOS 却有所下降。就目前来看，智能手机的市场已经饱和，大多数人都在各个平台间转换。而就在这样一个市场上，Android 的占有率还增长了 10%左右，确实不易。

## 为开发人员提供了成长的“沃土”

（1）保证开发人员可以迅速转型到 Android 应用开发

Android 应用程序是通过 Java 语言开发的，只要具备 Java 开发基础，就能很快上手并掌握。作为单独的 Android 应用开发，对 Java 编程门槛的要求并不高，即使没有编程经验的门外汉，也可以在突击学习 Java 之后学习 Android。另外，Android 完全支持 2D、3D 和数据库，并且和浏览器实现了集成。所以通过 Android 平台，程序员可以迅速、高效地开发出绚丽多彩的应用，例如常见的工具软件、管理软件、互联网软件和游戏等。

（2）定期召开奖金丰厚的 Android 开发大赛

为了吸引更多的用户使用 Android 开发，已经成功举办了奖金为数千万美元的开发者竞赛，鼓励开发人员创建出创意十足、十分有用的软件。这种大赛对于开发人员来说，不但能练习自己的开发水平，并且高额的奖金也是学员们学习的动力。

（3）开发人员可以利用自己的作品赚钱

为了能让 Android 平台吸引更多的关注，谷歌提供了一个专门下载 Android 应用的门店：Android Market，地址是 https://play.google.com/store。在这个门店里允许开发人员发布应用程序，也允许 Android 用户下载获取自己喜欢的程序。作为开发者，需要申请开发者账号，申请后才能将自己的程序上传到 Android Market，并且可以对自己的软件进行定价。只要你的软件程序足够吸引人，你就可以获得很好的金钱回报。这样实现了程序员学习和赚钱的两不误，所以吸引了更多开发人员加入到 Android 大军中来。

## 本书的内容

Android 系统从诞生到现在的短短几年时间，凭借其操作易用性和开发的简洁性，赢得了广大用户和开发者的支持，成为市场占有率第一的智能设备系统。本书内容分为 5 篇，共计 17 章，循序渐进地讲解了开发 Android 外设项目的基本知识。本书从获取源码和搭建应用开发环境开始讲起，依次讲解了基础知识篇、数据传输篇、信息识别篇、传感器应用篇和技术提高篇 5 部分内容。在讲解每一个知识时，都遵循了理论联系实际的讲解方式，从内核分析到接口实现，再到实战演练，最后到综合实例演练，彻底剖析了物联网开发的完整实现流程。本书几乎涵盖了当下 Android 物联网开发的绝大多数内容，讲解方法通俗易懂并且详细，不但适合应用高手们学习，也特别便于初学者学习并理解。

本书适合 Android 驱动开发者、Linux 开发人员、Android 物联网开发人员、Android 爱好者、Android 源码分析人员、Android 应用开发人员、Android 传感器开发人员、Android 智能家居开发人员、Android 可穿戴设备人员的学习，也可以作为相关培训学校和大专院校相关专业的教学用书。

## 本书的版本

Android 系统自 2008 年 9 月发布第一个版本 1.1 以来，截至 2014 年 10 月发布最新版本 5.0，一共存在十多个版本。由此可见，Android 系统升级频率较快，一年之中最少有两个新版本诞生。如果过于追求新版本，会造成力不从心的结果。所以在此建议广大读者："不必追求最新的版本，我们只需关注最流行的版本即可"。据官方统计，截至 2014 年 10 月 25 日，占据前 3 位的版本分别是 Android 4.3、Android 4.4 和 Android 4.2，其实这 3 个版本的区别并不是很大，只是在某领域的细节上进行了更新。为了及时体验 Android 系统的最新功能，在本书中使用的版本是目前（本书成稿时）最新的 Android 5.0。

## 本书特色

本书内容十分丰富，并且讲解细致。我们的目标是通过一本图书，提供多本图书的价值，读者可以根据自己的需要有选择地阅读。在内容的编写上，本书具有以下特色。

（1）内容全面，讲解细致

本书几乎涵盖了 Android 物联网开发所需要的主要知识点，详细讲解了每一个典型物联网项目的实现过程和具体移植方法。每一个知识点都力求用详实和易懂的语言展现在读者面前。

（2）遵循合理的主线进行讲解

为了使广大读者彻底弄清楚 Android 物联网开发的各个知识点，在讲解每一个知识点时，从 Linux 内核开始讲起，依次剖析了底层架构、API 接口连接和具体应用的知识。遵循了从底层到顶层，实现了 Android 物联网开发大揭秘的目标。

（3）章节独立，自由阅读

本书中的每一章内容都可以独自成书，读者既可以按照本书编排的章节顺序进行学习，也可以根据自己的需求对某一章节进行针对性的学习，并且和传统古板的计算机书籍相比，阅读本书会带来很大的快乐。

（4）实例典型，实用性强

本书讲解了现实中最典型的 Android 物联网项目的实现方法和架构技巧，这些经典应用都是在商业项目中最需要的部分。读者可以直接将本书中的知识灵活运用，应用到自己的项目中，实现无缝对接。

（5）附配资源丰富

本书配有丰富的学习资源，除源代码、PPT 之外，还实录了 84 个高清学习视频，既有实用的知识点讲解视频，又有详细的实例开发视频，全面、深入、细致地解析 Android 物联网开发的方方面面。除此以外，本书额外赠送了 38 个 Android 应用开发学习视频，以及 15 个 Android 应用开发综合案例，包括仿小米录音机、音乐播放器、跟踪定位系统、仿陌陌交友系统、手势音乐播放器、智能家居系统、湿度测试仪、象棋游戏、抢滩登陆游戏、九宫格数独游戏、健康饮食系统、仓库管理系统、个人财务系统、仿去哪儿酒店预定系统、仿开心网客户端等。通过这些附配资源，读者的学习过程会更加方便、快捷。

## 读者对象

☑ 初学 Android 编程的自学者。
☑ Linux 开发人员。
☑ 大中专院校的老师和学生。
☑ 毕业设计的学生。
☑ Android 编程爱好者。
☑ 相关培训机构的老师和学员。
☑ 从事 Android 开发的程序员。

参与本书编写的人员还有周秀、付松柏、邓才兵、钟世礼、谭贞军、张加春、王教明、万春潮、郭慧玲、侯恩静、程娟、王文忠、陈强、何子夜、李天祥、周锐、朱桂英、张元亮、张韶青、秦丹枫。本团队在编写的过程中，得到了清华大学出版社工作人员的大力支持，正是各位编辑的求实、耐心和效率，才使得本书在这么短的时间内出版。另外也十分感谢我们的家人，在我们写作的时候给予的巨大支持。

由于编者水平有限，如有纰漏和不尽如人意之处，诚请读者提出意见或建议，以便修订并使之更臻完善。另外，我们提供了售后支持网站（http://www.chubanbook.com/）和 QQ 群（192153124），读者朋友如有疑问可以在此提出，一定会得到满意的答复。

编　者

# 目　　录

## 第 1 篇　基础知识篇


第 1 章　Android 系统介绍 ............................. 2

1.1　纵览智能设备系统 ............................. 2

1.1.1　Symbian（塞班） ............................. 2

1.1.2　Android（安卓） ............................. 2

1.1.3　iOS（苹果系统） ............................. 3

1.1.4　Windows Phone（微软系统） ............................. 4

1.1.5　BlackBerry OS（黑莓） ............................. 4

1.2　分析 Android 成功的秘诀 ............................. 5

1.2.1　强有力的业界支持 ............................. 5

1.2.2　研发阵容强大 ............................. 6

1.2.3　为开发人员“精心定制” ............................. 6

1.2.4　开源 ............................. 6

1.3　剖析 Android 系统架构 ............................. 7

1.3.1　底层操作系统层（OS） ............................. 7

1.3.2　各种库（Libraries）和 Android 运行环境（RunTime） ............................. 8

1.3.3　Application Framework（应用程序框架） ............................. 9

1.3.4　顶层应用程序（Application） ............................. 9

1.4　核心组件 ............................. 9

1.4.1　Activity 界面 ............................. 9

1.4.2　Intent 和 Intent Filters ............................. 10

1.4.3　Service 服务 ............................. 10

1.4.4　Broadcast Receiver 发送广播 ............................. 11

1.4.5　用 Content Provider 存储数据 ............................. 11

1.5　进程和线程 ............................. 12

1.5.1　什么是进程 ............................. 12

1.5.2　什么是线程 ............................. 12

1.6　物联网技术改变未来 ............................. 13

1.6.1　什么是物联网 ............................. 13

1.6.2　发展历程 ............................. 13

1.6.3　Android 正在成为物联网标准操作系统 ...... 14

1.7　智能设备凶猛来袭 ............................. 15

1.7.1　常见的 Android 智能设备 ............................. 15

1.7.2　新兴热点——可穿戴设备 ............................. 17

1.7.3　可穿戴设备的发展前景分析 ............................. 19

1.7.4　Android 对穿戴设备的支持——Android Wear ............................. 120

第 2 章　搭建 Android 开发环境 ............................. 21

2.1　在 Linux 系统获取 Android 源码 ............................. 21

2.2　在 Windows 平台获取 Android 源码 ...... 22

2.3　编译源码 ............................. 24

2.3.1　搭建编译环境 ............................. 25

2.3.2　开始编译 ............................. 26

2.3.3　在模拟器中运行 ............................. 27

2.3.4　常见的错误分析 ............................. 27

2.3.5　实践演练——演示两种编译 Android 程序的方法 ............................. 28

2.4　编译 Android Kernel ............................. 32

2.4.1　获取 Goldfish 内核代码 ............................. 34

2.4.2　获取 MSM 内核代码 ............................. 34

2.4.3　获取 OMAP 内核代码 ............................. 34

2.4.4　编译 Android 的 Linux 内核 ............................. 34

2.5　搭建 Android 应用开发环境 ............................. 36

2.5.1　安装 JDK ............................. 37

2.5.2　获取并安装 Eclipse 和 Android SDK ............................. 39

2.5.3　快速安装 SDK ............................. 42

2.5.4　安装 ADT ............................. 43

2.5.5　验证设置 ............................. 45

2.5.6　创建 Android 虚拟设备（AVD） ............................. 46

2.5.7　启动 AVD 模拟器 ............................. 48

## 第 2 篇　数据传输篇


第 3 章　基本数据通信 ................................ 52

3.1　HTTP 通信 ................................ 52

3.1.1　Android 中的 HTTP ................................ 52

3.1.2　使用 Apache 接口................................ 53

3.1.3　在 Android 中使用 java.net ................................ 57

3.1.4　使用 Android 网络接口................................ 59

3.1.5　实战演练——在屏幕中传递 HTTP 参数..... 60

3.1.6　实战演练——在户外运动过程中访问 HTTP 地图 ................................ 64

3.2　使用 Socket 实现数据通信................................ 67

3.2.1　基于 Socket 的 Java 网络编程 ................................ 67

3.2.2　使用 TCP 协议传输数据................................ 68

3.3　下载数据 ................................ 72

3.3.1　实战演练——下载远程服务器中的图片..... 72

3.3.2　实战演练——下载网络中的 JSON 信息 ..... 73

3.3.3　实战演练——下载并播放网络 MP3........... 78

3.4　上传数据 ................................ 84

3.4.1　实战演练——上传文件到远程服务器......... 84

3.4.2　实战演练——使用 GET 方式上传数据 ....... 87

3.4.3　实战演练——使用 POST 方式上传数据..... 91

3.5　处理 XML 数据 ................................ 94

3.5.1　XML 的概述................................ 95

3.5.2　XML 的语法................................ 95

3.5.3　获取 XML 文档................................ 96

3.5.4　SAX 常用的接口和类................................ 98

3.5.5　实战演练——使用 SAX 解析 XML 数据....100

3.5.6　实战演练——使用 DOM 解析 XML 数据 ................................103

第 4 章　蓝牙技术详解 ................................ 106

4.1　短距离无线通信技术概览 ................... 106

4.1.1　ZigBee——低功耗、自组网......................106

4.1.2　WiFi——大带宽支持家庭互联 ..................107

4.1.3　蓝牙——4.0 进入低功耗时代......................107

4.1.4　NFC——必将逐渐远离历史舞台...............108

4.2　低功耗蓝牙基础 ................................ 108

4.2.1　低功耗蓝牙的架构................................109

4.2.2　低功耗蓝牙分类 ................................ 109

4.2.3　低功耗蓝牙的集成方式 ................................ 110

4.2.4　低功耗蓝牙的特点 ................................ 111

4.2.5　BLE 和传统蓝牙 BR/EDR 技术的对比...... 111

4.3　蓝牙 4.0 BLE 基础................................ 112

4.3.1　低功耗是最大特点 ................................ 112

4.3.2　推动了可穿戴设备的兴起 ................................ 112

4.4　蓝牙规范 ................................ 113

4.4.1　Bluetooth 常用规范 ................................ 114

4.4.2　蓝牙协议体系结构 ................................ 114

4.4.3　低功耗（BLE）蓝牙协议................................ 116

4.4.4　基于 GATT 的协议/服务................................ 116

4.4.5　双模协议栈和单模协议栈 ................................ 117

4.5　低功耗蓝牙协议栈详解 ....................... 118

4.5.1　什么是蓝牙协议栈 ................................ 118

4.5.2　蓝牙协议体系中的协议 ................................ 119

4.6　TI 公司的低功耗蓝牙 ........................ 121

4.6.1　获取蓝牙协议栈................................ 121

4.6.2　BLE 蓝牙协议栈结构................................ 123

4.6.3　BLE 低功耗蓝牙系统架构................................ 124

4.6.4　硬件抽象层 HAL 和 BLE 低功耗蓝牙协议栈................................ 125

4.7　实战演练——使用蓝牙控制电风扇..... 129

第 5 章　Android 蓝牙系统详解.................... 142

5.1　Android 系统中的蓝牙模块 ................ 142

5.2　分析蓝牙模块的源码 .......................... 143

5.2.1　初始化蓝牙芯片 ................................ 144

5.2.2　蓝牙服务................................ 144

5.2.3　管理蓝牙电源 ................................ 145

5.3　和蓝牙相关的类 ................................ 145

5.3.1　BluetoothSocket 类 ................................ 145

5.3.2　BluetoothServerSocket 类 ................................ 147

5.3.3　BluetoothAdapter 类................................ 147

5.3.4　BluetoothClass.Service 类................................ 154

5.3.5　BluetoothClass.Device 类................................ 154

5.4　Android BlueDroid 架构详解 ............... 155

5.4.1 Android 系统中 BlueDroid 的架构 ............155
5.4.2 Application Framework 层分析 ............155
5.4.3 分析 Bluetooth System Service 层 ............163
5.4.4 JNI 层详解 ............163
5.4.5 HAL 层详解 ............168
5.5 Android 中的低功耗蓝牙协议栈 ............168
5.5.1 低功耗蓝牙协议栈基础 ............169
5.5.2 低功耗蓝牙 API 详解 ............169

## 第 3 篇　信息识别篇

第 6 章　语音识别技术详解 ............216
6.1 语音识别技术基础 ............216
6.1.1 语音识别的发展历史 ............216
6.1.2 语音识别技术的发展历程 ............217
6.2 Text-To-Speech 技术 ............217
6.2.1 Text-To-Speech 基础 ............217
6.2.2 Text-To-Speech 的实现流程 ............218
6.2.3 实战演练——使用 Text-To-Speech 实现语音识别 ............221
6.2.4 实战演练——借助开源项目实现中文语音识别 ............222
6.3 Voice Recognition 技术详解 ............227
6.3.1 Voice Recognition 技术基础 ............228
6.3.2 实战演练——使用 Voice Recognition 实现语音识别 ............230
6.4 实战演练——为设备中所有的 APP 实现语音提醒功能 ............233

第 7 章　手势识别实战 ............261
7.1 Android 中的事件监听机制 ............261
7.1.1 Android 系统中的监听事件 ............261
7.1.2 Android 事件监听器的回调方法 ............262
7.1.3 Android 事件处理的两种模型 ............263
7.1.4 基于自定义监听器的事件处理流程 ............264
7.2 手势识别技术介绍 ............266
7.2.1 手势识别类 GestureDetector ............266
7.2.2 手势检测器类 GestureDetector ............267
7.2.3 手势识别处理事件和方法 ............270
7.3 实战演练——通过点击的方式移动图片 ............271
7.4 实战演练——实现各种手势识别 ............274
7.4.1 布局文件 main.xml ............274
7.4.2 隐藏屏幕顶部的电池等图标和标题内容 ............274
7.4.3 监听触摸屏幕中各种常用的手势 ............275
7.4.4 根据监听到的用户手势创建视图 ............276
7.5 实战演练——实现手势翻页效果 ............278
7.5.1 布局文件 main.xml ............278
7.5.2 监听手势 ............278

第 8 章　在物联网设备中处理多媒体数据 ............284
8.1 Android 多媒体系统架构基础 ............284
8.1.1 OpenMax 框架介绍 ............285
8.1.2 OpenCore 框架介绍 ............286
8.1.3 StageFright 框架介绍 ............287
8.2 Graphics 类详解 ............288
8.2.1 Graphics 类基础 ............288
8.2.2 使用 Graphics 类 ............288
8.3 二维动画 ............293
8.3.1 类 Drawable ............293
8.3.2 实现 Tween 动画效果 ............294
8.3.3 实现 Frame 动画效果 ............296
8.4 OpenGL ES 详解 ............301
8.4.1 OpenGL ES 基础 ............301
8.4.2 Android 用到 OpenGL ES ............301
8.4.3 OpenGL ES 的基本操作 ............302
8.4.4 绘制图形 ............308
8.5 音频开发 ............312
8.5.1 音频接口类 ............312
8.5.2 AudioManager 控制铃声 ............313
8.6 录音详解 ............321
8.6.1 使用 MediaRecorder 接口录制音频 ............321
8.6.2 使用 AudioRecord 接口录音 ............324
8.7 在物联网设备中播放音乐 ............327

8.7.1 使用 AudioTrack 播放音频......327
8.7.2 使用 MediaPlayer 播放音频......328
8.7.3 使用 SoundPool 播放音频......328
8.8 为物联网设备实现振动功能......329
8.9 实战闹钟功能......330
第 9 章 GPS 地图定位......331
9.1 位置服务......331
9.1.1 类 location 详解......331
9.1.2 实现定位服务功能......332
9.1.3 实战演练——在 Android 设备中实现 GPS 定位......334
9.2 随时更新位置信息......336
9.2.1 库 Maps 中的类......337
9.2.2 使用 LocationManager 监听位置......337
9.2.3 实战演练——监听当前设备的坐标和海拔......339
9.3 在设备中使用地图......348
9.3.1 添加 Google Map 密钥......348
9.3.2 使用 Map API 密钥......351
9.3.3 实战演练——在 Android 设备中使用谷歌地图实现定位......353
9.4 接近警报......357
9.4.1 类 Geocoder 基础......357
9.4.2 Geocoder 的公共构造器和公共方法......359
9.4.3 实战演练——接近某个位置时实现自动提醒......361

## 第 4 篇 传感器应用篇

第 10 章 Android 传感器系统架构详解......376
10.1 Android 传感器系统概述......376
10.2 Java 层详解......377
10.3 Frameworks 层详解......383
10.3.1 监听传感器的变化......383
10.3.2 注册监听......384
10.4 JNI 层详解......395
10.4.1 实现本地函数......396
10.4.2 处理客户端数据......400
10.4.3 处理服务端数据......403
10.4.4 封装 HAL 层的代码......417
10.4.5 消息队列处理......421
10.5 HAL 层详解......424
10.6 Android 传感器应用开发基础......432
10.6.1 查看包含的传感器......432
10.6.2 模拟器测试工具——SensorSimulator......434
10.6.3 实战演练——检测当前设备支持的传感器......437
第 11 章 光线传感器和磁场传感器......439
11.1 光线传感器详解......439
11.1.1 光线传感器介绍......439
11.1.2 使用光线传感器的方法......440
11.1.3 实战演练——获取设备中光线传感器的值......441
11.1.4 实战演练——显示设备中光线传感器的强度......443
11.2 磁场传感器详解......445
11.2.1 什么是磁场传感器......445
11.2.2 磁场传感器的分类......446
11.2.3 Android 系统中的磁场传感器......446
11.2.4 实战演练——获取磁场传感器的 3 个分量......447
11.2.5 实战演练——演示常用传感器的基本用法......448
第 12 章 加速度传感器、方向传感器和陀螺仪传感器......458
12.1 加速度传感器详解......458
12.1.1 加速度传感器的分类......458
12.1.2 加速度传感器的主要应用领域......459
12.1.3 线性加速度传感器的原理......460
12.1.4 Android 系统中的加速度传感器......461
12.1.5 实战演练——获取 X、Y、Z 轴的加速度值......462
12.1.6 实战演练——实现仿微信“摇一摇”效果......464

12.2 方向传感器详解 ........................................ 471
12.2.1 方向传感器基础 ........................................ 471
12.2.2 Android 中的方向传感器 ........................................ 472
12.2.3 实战演练——测试当前设备的 3 个方向值 ........................................ 473
12.2.4 实战演练——开发一个指南针程序 ........................................ 475
12.3 陀螺仪传感器详解 ........................................ 477
12.3.1 陀螺仪传感器基础 ........................................ 477
12.3.2 Android 中的陀螺仪传感器 ........................................ 478
12.4 实战演练——联合使用加速度传感器和陀螺仪传感器 ........................................ 481
12.4.1 系统介绍界面 ........................................ 481
12.4.2 系统主界面 ........................................ 484
第 13 章 旋转向量传感器、距离传感器和气压传感器 ........................................ 504
13.1 旋转向量传感器详解 ........................................ 504
13.1.1 Android 中的旋转向量传感器 ........................................ 504
13.1.2 实战演练——确定设备当前的方向 ........................................ 505
13.2 距离传感器详解 ........................................ 516
13.2.1 距离传感器介绍 ........................................ 516
13.2.2 Android 系统中的距离传感器 ........................................ 517
13.2.3 实战演练——实现自动锁屏功能 ........................................ 519
13.3 气压传感器详解 ........................................ 525
13.3.1 气压传感器基础 ........................................ 526
13.3.2 气压传感器在智能手机中的应用 ........................................ 526
13.3.3 实战演练——开发一个 Android 气压计 ........................................ 526
第 14 章 温度传感器和湿度传感器 ........................................ 536
14.1 温度传感器详解 ........................................ 536
14.1.1 温度传感器介绍 ........................................ 536
14.1.2 Android 系统中的温度传感器 ........................................ 537
14.1.3 实战演练——开发一个 Android 温度计 ........................................ 539
14.1.4 实战演练——测试电池的温度 ........................................ 541
14.2 湿度传感器详解 ........................................ 553
14.2.1 Android 系统中的湿度传感器 ........................................ 553
14.2.2 实战演练——获取远程湿度传感器的数据 ........................................ 554

## 第 5 篇 技术提高篇

第 15 章 条形码解析技术详解 ........................................ 562
15.1 Android 拍照系统结构基础 ........................................ 562
15.2 底层程序详解 ........................................ 564
15.2.1 V4L2 API ........................................ 564
15.2.2 操作 V4L2 的流程 ........................................ 565
15.2.3 V4L2 驱动框架 ........................................ 567
15.2.4 实现 Video 核心层 ........................................ 568
15.3 拍照系统的硬件抽象层 ........................................ 571
15.3.1 Andorid 2.1 及其以前的版本 ........................................ 571
15.3.2 Andorid 2.2 及其以后的版本 ........................................ 572
15.3.3 实现 Camera 硬件抽象层 ........................................ 574
15.4 拍照系统的 Java 部分 ........................................ 575
15.5 开发拍照应用程序 ........................................ 581
15.5.1 通过 Intent 调用系统的照相机 Activity ........................................ 581
15.5.2 调用 Camera API 拍照 ........................................ 582
15.5.3 总结 Camera 拍照的流程 ........................................ 583
15.6 解析二维码 ........................................ 585
15.6.1 QR Code 码的特点 ........................................ 585
15.6.2 实战演练——使用 Android 相机解析二维码 ........................................ 585
第 16 章 NFC 近场通信技术详解 ........................................ 594
16.1 近场通信技术基础 ........................................ 594
16.1.1 NFC 技术的特点 ........................................ 594
16.1.2 NFC 的工作模式 ........................................ 594
16.1.3 NFC 和蓝牙的对比 ........................................ 595
16.2 射频识别技术详解 ........................................ 595
16.2.1 RFID 技术简介 ........................................ 596
16.2.2 RFID 技术的组成 ........................................ 596
16.2.3 RFID 技术的特点 ........................................ 597
16.2.4 RFID 技术的工作原理 ........................................ 597
16.3 Android 系统中的 NFC ........................................ 598
16.3.1 分析 Java 层 ........................................ 599

16.3.2 分析 JNI 部分 ........................................614
16.3.3 分析底层........................................619
16.4 在 Android 系统编写 NFC APP 的方法 ........................................619
16.5 实战演练——使用 NFC 发送消息 ........................................622

第 17 章 Google Now 和 Android Wear 详解 ........................................627
17.1 Google Now 介绍 ........................................627
17.1.1 搜索引擎的升级——Google Now ........................................627
17.1.2 Google Now 的用法 ........................................628
17.2 Android Wear 详解 ........................................629
17.2.1 什么是 Android Wear ........................................629
17.2.2 搭建 Android Wear 开发环境 ........................................630
17.3 开发 Android Wear 程序 ........................................634
17.3.1 创建通知........................................634
17.3.2 创建声音........................................637
17.3.3 给通知添加页面........................................639
17.3.4 通知堆........................................640
17.3.5 通知语法介绍 ........................................641
17.4 实战演练——开发一个 Android Wear 程序 ........................................642
17.5 实战演练——实现手机和 Android Wear 的交互........................................649
仿小米录音机 ........................................DVD
一个音乐播放器 ........................................DVD
跟踪定位系统 ........................................DVD
仿陌陌交友系统 ........................................DVD
手势音乐播放器 ........................................DVD
智能家居系统 ........................................DVD
湿度测试仪 ........................................DVD
象棋游戏 ........................................DVD
iPad 抢滩登陆 ........................................DVD
OpenSudoku 九宫格数独游戏 ........................................DVD
健康饮食 ........................................DVD
仓库管理系统 ........................................DVD
个人财务系统 ........................................DVD
高仿去哪儿酒店预定 ........................................DVD
仿开心网客户端 ........................................DVD

# 第 1 篇

## 基础知识篇

- 第 1 章　Android 系统介绍
- 第 2 章　搭建 Android 开发环境

# 第 1 章 Android 系统介绍

2007 年，Google 公司推出了一款无与伦比的移动智能设备系统——Android，这是一种建立在 Linux 基础之上的为手机、平板等移动设备提供的软件解决方案。截至 2013 年，根据知名 IDC 公司的统计，Android 系统在世界智能手机发货量中占据 75%的份额，已经成为了当今最受欢迎的智能设备系统之一。本章将引领读者一起来了解 Android 系统的发展历程和背景，充分体验这款操作系统的成功之处。

## 1.1 纵览智能设备系统

**知识点讲解：光盘:视频\知识点\第 1 章\纵览智能设备系统.avi**

在当今市面中有很多智能手机系统，在 Android 推出之前，智能手机系统领域塞班、苹果、微软互不相让，呈三足鼎立之势。除此之外，还有占份额较小的 PDA、黑莓等。本节将一一介绍这些智能手机系统。

### 1.1.1 Symbian（塞班）

Symbian 作为昔日智能手机的王者，在 2005—2010 年曾一度盛行，街上大大小小拿的很多都是诺基亚的 Symbian 手机，N70—N73—N78—N97，诺基亚 N 系列曾经被称为“N=无限大”的手机。对硬件的水平要求低，操作简单，省电，软件众多是 Symbian 系统手机的重要特点。

在国内软件开发市场内，基本每一个软件都会有对应的塞班手机版本。而塞班开发之初的目标是要保证在较低资源的设备上能长时间稳定可靠地运行，这导致了塞班的应用程序开发有着较为陡峭的学习曲线，开发成本较高。但是程序的运行效率很高。例如 5800 的 128MB 的 RAM，后台可以同时运行十几个程序而保持操作流畅(多任务功能是特别强大的)，即使几天不关机它的剩余内存也能保持稳定。

虽然在 Android、iOS 的围攻之下，诺基亚推出了塞班^3 系统，甚至依然为其更新（Symbian Anna，Symbian Belle），从外在的用户界面到内在的功能特性都有了显著提升，例如可自由定制的全新窗体部件、更多主屏、全新下拉式菜单等。

由于对新兴的社交网络和 Web 2.0 内容支持欠佳，塞班占智能手机的市场份额日益萎缩。2010 年末，其市场占有量已被 Android 超过。自 2009 年底开始，包括摩托罗拉、三星电子、LG、索尼爱立信等各大厂商纷纷宣布终止塞班平台的研发，转而投入 Android 领域。2011 年初，诺基亚宣布将与微软成立战略联盟，推出基于 Windows Phone 的智能手机，从而在事实上放弃了经营多年的塞班，塞班退市已成定局。

### 1.1.2 Android（安卓）

Android 一词最早出现于法国作家利尔亚当（Auguste Villiers de l'Isle-Adam）在 1886 年发表的科幻

小说《未来夏娃》（*L'ève future*）中。他将外表像人的机器起名为 Android。

从 2008 年 HTC 和 Google 联手推出第一台 Android 手机 G1 开始，到 2014 年 10 月 15 日（美国太平洋时间），Google 公司发布全新 Android 操作系统 Android 5.0 为止，Android 系统经过了多个版本的发展。从 2011 年第一季度开始，Android 在全球的市场份额首次超过塞班系统，跃居全球第一。2014 年 8 月 15 日消息，根据 IDC 发布的 2014 年第二季度智能手机市场的最新数据显示，苹果 iOS 和谷歌 Android 两大系统平台继续领跑。Android 阵营增长则更惊人，达到了 33.3%，出货量达到了 2.553 亿台。Android 系统的市场份额得到了提高，从 2013 年第二季度的 79.6%增长到了 2014 年第二季度的 84.7%。具体信息如图 1-1 所示。

Top Five Smartphone Operating Systems, Worldwide Shipments, and Market Share, 2014Q2 (Units in Millions) - IDC/AppleInsider

| Operating System | Q2 2014 Shipment Volume | Q2 2014 Market Share | Q2 2013 Shipment Volume | Q2 2013 Market Share | Year-Over-Year Growth |
|---|---|---|---|---|---|
| Android | 255.3 | 84.7% | 191.5 | 79.6% | 33.3% |
| iOS | 35.2 | 11.7% | 31.2 | 13.0% | 12.7% |
| Windows Phone | 7.4 | 2.5% | 8.2 | 3.4% | -9.4% |
| BlackBerry | 1.5 | 0.5% | 6.7 | 2.8% | -78.0% |
| Others | 1.9 | 0.6% | 2.9 | 1.2% | -32.2% |
| Total | 301.3 | 100.0% | 240.5 | 100.0% | 25.3% |

图 1-1　2014 年 8 月智能手机平台调查表

由此可见，Android 系统的市场占有率位居第一，并且毫无压力。Android 机型数量庞大，简单易用，相当自由的系统能让厂商和客户轻松地定制各种各样的 ROM，定制各种桌面部件和主题风格。简单而华丽的界面得到广大客户的认可，对手机进行刷机也是不少 Android 用户所津津乐道的事情。

可惜 Android 版本数量较多，市面上同时存在着 1.6、2.0、2.1、2.2、2.3、4.4.2 等各种版本的 Android 系统手机，应用软件和各版本系统的兼容性对程序开发人员是一种很大的挑战。同时由于开发门槛低，导致应用数量虽然很多，但是应用质量参差不齐，甚至出现不少恶意软件，导致一些用户受到损失。同时 Android 没有对各厂商在硬件上进行限制，导致一些用户在低端机型上体验不佳。另一方面，因为 Android 的应用主要使用 Java 语言开发，其运行效率和硬件消耗一直是其他手机用户所诟病的地方。

### 1.1.3　iOS（苹果系统）

iOS 作为苹果移动设备 iPhone 和 iPad 的操作系统，在 App Store 的推动之下，成为了世界上引领潮流的操作系统之一。原本这个系统名为“iPhone OS”，直到 2010 年 6 月 7 日 WWDC 大会上宣布改名为“iOS”。iOS 用户操作界面的最大特性是可以使用多点触控的方式完成所有操作。控制方法包括滑动、轻触开关及按键。与系统交互包括滑动（Swiping）、轻按（Tapping）、挤压（Pinching，通常用于缩小）及反向挤压（Reverse Pinching or unpinching，通常用于放大）。此外，通过其自带的加速器，可以令其旋转设备时改变其 y 轴以令屏幕改变方向，这样的设计令 iPhone 更便于使用。

- ☑ 最早 iPhone OS 1.0：内置于 iPhone 一代手机中，借助 iPhone 流畅的触摸屏幕，iPhone OS 给用户带来了极为优秀的使用体验，相比当时的手机可以用惊艳来形容。
- ☑ iPhone OS 2.0：随着 iPhone 3G 的发布，App Store 诞生。App Store 为第三方软件的提供者提供了一个方便而又高效的软件销售平台，在软件开发者与最终用户之间架起了一座沟通与销售的桥梁，从而极大地丰富了 iPhone 手机应用功能。
- ☑ iPhone OS 3.0：iPhone 3GS 开始支持复制粘贴。

- ☑ iOS 4：在 iPhone 4 推出时，苹果决定将原来 iPhone OS 系统重新定名为“iOS”，并发布新一代操作系统“iOS 4”。在这个版本中，开始正式支持多任务功能，通过双击 HOME 键实现切换。
- ☑ iOS 5：加入了 Siri 语音操作助手功能，用户可以与手机实现语言上的人机交互，该功能可以实现对用户的语音识别，完成一些较为复杂的操作，使用 Siri 来实现查询天气、进行导航、询问时间、设定闹钟、查询股票甚至发送短信等功能，方便了用户的使用。

从最初的 iPhone OS，演变至最新的 iOS 系统，iOS 成为了苹果新的移动设备操作系统，横跨 iPod Touch、iPad、iPhone，成为苹果最强大的操作系统。甚至新一代的 Mac OS X Lion 也借鉴了 iOS 系统的一些设计，可以说 iOS 是苹果的又一个成功的操作系统，能给用户带来极佳的使用体验。

优秀系统设计以及严格的 App Store，iOS 作为应用数量最多的移动设备操作系统，加上强大的硬件支持以及 iOS 5 内置的 Siri 语音助手，无疑使得用户体验得到更大的提升，感受科技带来的好处。

### 1.1.4　Windows Phone（微软系统）

早在 2004 年时，微软就开始以“Photon”的计划代号开始研发 Windows Mobile 的一个重要版本更新。直到 2008 年，在 iOS 和 Android 的巨大冲击之下，微软重新组织了 Windows Mobile 的小组，并继续开发一个新的移动操作系统。

Windows Phone，简称 WP，是微软发布的一款手机操作系统，它将微软旗下的 Xbox Live 游戏、Xbox Music 音乐与独特的视频体验集成至手机中。微软公司于 2010 年 10 月 11 日晚上 9 点 30 分正式发布了智能手机操作系统 Windows Phone，并将其使用接口称为“Modern”接口。2011 年 2 月，“诺基亚”与微软达成全球战略同盟并深度合作共同研发。2011 年 9 月 27 日，微软发布 Windows Phone 7.5。2012 年 6 月 21 日，微软正式发布 Windows Phone 8，采用和 Windows 8 相同的 Windows NT 内核，同时也针对市场的 Windows Phone 7.5 发布 Windows Phone 7.8。现有 Windows Phone 7 手机都将无法升级至 Windows Phone 8。

Windows Phone 具有桌面定制、图标拖曳、滑动控制等一系列前卫的操作体验。其主屏幕通过提供类似仪表盘的体验来显示新的电子邮件、短信、未接来电、日历约会等，对重要信息保持时刻更新。它还包括一个增强的触摸屏界面，更方便手指操作；以及一个最新版本的 IE Mobile 浏览器——该浏览器在一项由微软赞助的第三方调查研究中，和参与调研的其他浏览器和手机相比，可以执行指定任务的比例超过 48%。很容易看出微软在用户操作体验上所做出的努力，而史蒂夫·鲍尔默也表示：“全新的 Windows 手机把网络、个人电脑和手机的优势集于一身，让人们可以随时随地享受到想要的体验”。

Windows Phone 力图打破人们与信息和应用之间的隔阂，提供适用于人们包括工作和娱乐在内完整生活的方方面面，最优秀的端到端体验。

### 1.1.5　BlackBerry OS（黑莓）

BlackBerry 系统，即黑莓系统，是加拿大 Research In Motion（RIM）公司推出的一种无线手持邮件解决终端设备的操作系统，由 RIM 自主开发。它和其他手机终端使用的 Symbian、Windows Mobile、IOS 等操作系统有所不同，BlackBerry 系统的加密性能更强、更安全。

安装有 BlackBerry 系统的黑莓机，指的不单单只是一台手机，而是由 RIM 公司所推出，包含服务器（邮件设定）、软件（操作接口）以及终端（手机）大类别的 Push Mail 实时电子邮件服务。

“黑莓”移动邮件设备基于双向寻呼技术。该设备与 RIM 公司的服务器相结合，依赖于特定的服

务器软件和终端，兼容现有的无线数据链路，实现了遍及北美、随时随地收发电子邮件的梦想。这种装置并不以奇妙的图片和彩色屏幕夺人耳目，甚至不带发声器。“9・11”事件之后，由于 BlackBerry 及时传递了灾难现场的信息，而在美国掀起了拥有一部 BlackBerry 终端的热潮。

黑莓赖以成功的最重要原则——针对高级白领和企业人士，提供企业移动办公的一体化解决方案。企业有大量的信息需要及时处理，出差在外时，也需要一个无线的可移动的办公设备。企业只要装一个移动网关，一个软件系统，用手机的平台实现无缝链接，无论何时何地，员工都可以用手机进行办公。它最大的方便之处是提供了邮件的推送功能：即由邮件服务器主动将收到的邮件推送到用户的手持设备上，而不需要用户频繁地连接网络查看是否有新邮件。

黑莓系统稳定性非常优秀，其独特定位也深得商务人士所青睐。可是也因此在大众市场上得不到优势，国内用户和应用资源也较少。

**背景说明：**

（1）2010 年 9 月，诺基亚宣布将从 2011 年 4 月起从 Symbian 基金会（Symbian Foundation）手中收回 Symbian 操作系统控制权。由此看来，诺基亚在 2008 年全资收购塞班公司之后希望继续扩大塞班影响力的愿望并没有实现。

（2）在苹果和 Android 的强大市场攻势下，诺基亚在 2011 年 2 月 11 日宣布与微软达成广泛战略合作关系，并将 Windows Phone 作为其主要的智能手机操作系统。这家芬兰手机巨头试图通过结盟扭转颓势。

（3）2011 年 8 月 15 日，谷歌和摩托罗拉移动公司共同宣布，谷歌将以每股 40.00 美元现金收购摩托罗拉移动，总额约 125 亿美元，相比摩托罗拉移动股份的收盘价溢价了 63%，双方董事会都已全票通过该交易。谷歌 CEO 拉里・佩奇表示，摩托罗拉移动完全专注于 Android 系统，收购摩托罗拉移动之后，将增强整个 Android 生态系统。佩奇同时表示，Android 将继续开源，收购的一个目的是为了获得专利。

（4）2013 年 9 月 3 日，微软公司宣布将以 37.9 亿欧元的价格收购诺基亚的设备和服务部门，同时还将以 16.5 亿欧元的价格收购诺基亚的相关技术专利，本次交易总额达到 54.4 亿欧元，其中有 3.2 万名员工将从诺基亚转入微软，整笔交易预计将于 2014 年第一季度完成。

（5）2013 年 9 月 24 日消息，黑莓表示已经与由 Fairfax Financial Holdings 主导的财团达成交易，准备以 47 亿美元出售，但是后来没有任何爆炸性消息发布。

# 1.2　分析 Android 成功的秘诀

**知识点讲解：光盘:视频\知识点\第 1 章\分析 Android 成功的秘诀.avi**

从 2007 年诞生，到 2014 年占据市场 80%的份额，为什么 Android 系统能够在这么短的时间内成为移动智能设备市场占有率的第一名？本节将从 4 个方面来为读者解答这个问题。

## 1.2.1　强有力的业界支持

Android 系统基于 Linux 内核，是一款开源的手机操作系统。正是因为如此，在 Android 刚刚崭露头角之后，各大手机厂商和电信部门纷纷加入到了 Android 联盟当中。Android 联盟由业界内的世界级企业组成，主要成员包括中国移动、摩托罗拉、高通、T-Mobile、三星、LG、HTC 等在内的 30 多家技术和无线应用的领军企业。Android 通过与运营商、设备制造商、开发商和其他有关各方结成深层次的合作伙伴关系，希望借助建立标准化、开放式的移动电话软件平台，在移动产业内形成一个开放式的生态系统。

### 1.2.2 研发阵容强大

Android 的研发队伍阵容强大，包括摩托罗拉、Google、HTC（宏达电子）、PHILIPS、T-Mobile、高通、魅族、三星、LG 以及中国移动在内的 34 家企业，这些一个个响亮的名字都在业界内堪称大佬。他们都将基于该平台开发手机的新型业务，应用之间的通用性和互联性将在最大程度上得到保持。无论是从硬件到软件，还是到电信服务商，Android 从一开始便成为了业界内的宠儿，被当作重点新秀而培养。这样 Android 系统在强大的开发团队的培育和呵护下，最终顺利地功成名就，成为了一方霸主。

### 1.2.3 为开发人员“精心定制”

Google 公司一直视程序员为前进动力的源泉，为了提高程序员们的开发积极性，不但为开发人员提供了一流的开发装备和软件服务，而且还提出了振奋人心的奖励机制。

（1）保证开发人员可以迅速转型为 Android 应用开发人员

Android 应用程序是通过 Java 语言开发的，只要具备 Java 开发基础，就能很快地上手并掌握。作为单独的 Android 应用开发，对 Java 编程门槛的要求并不高，即使没有编程经验的门外汉，也可以在突击学习 Java 之后不影响学习 Android。另外，Android 完全支持 2D、3D 和数据库，并且和浏览器实现了集成。所以通过 Android 平台，程序员可以迅速、高效地开发出绚丽多彩的应用，例如常见的工具软件、管理软件、互联网应用和游戏等。

（2）定期召开奖金丰厚的 Android 开发大赛

为了吸引更多的用户使用 Android 开发，已经成功举办了奖金为数千万美元的开发者竞赛。鼓励开发人员创建出创意十足、十分有用的软件。对于开发人员来说，这种大赛不但能练习自己的开发水平，并且高额的奖金也是学员们学习的动力。

（3）开发人员可以利用自己的作品赚钱

为了能让 Android 平台吸引更多的关注，谷歌提供了一个专门下载 Android 应用的门店：Android Market，地址是 https://play.google.com/store。在这个门店里允许开发人员发布应用程序，也允许 Android 用户下载获取自己喜欢的程序。作为开发者，需要申请开发者账号，申请后才能将自己的程序上传到 Android Market，并且可以对自己的软件进行定价。只要你的软件程序足够吸引人，你就可以获得很好的金钱回报。这样实现了程序员学习和赚钱的两不误，所以吸引了更多开发人员加入到 Android 大军中来。

### 1.2.4 开源

Android 是一款开源的系统，开源意味着对开发人员和手机厂商来说是完全无偿免费使用的。正是因为这一原因，所以吸引了全世界各地无数程序员的热情。于是很多手机厂商都纷纷采用 Android 作为自己产品的系统，这当然也包括很多山寨厂商。因为免费所以降低了成本，因此提高了利润。而对

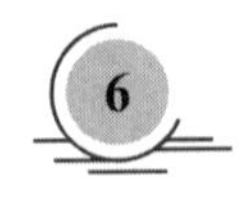

于开发人员来说，因为 Android 深受众多移动设备产品厂商所青睐，所以这方面的人才也变得愈发珍贵。

# 1.3　剖析 Android 系统架构

**知识点讲解：光盘:视频\知识点\第 1 章\剖析 Android 系统架构.avi**

Android 系统是一个移动设备的开发平台，其软件层次结构包括操作系统（OS）、中间件（MiddleWare）和应用程序（Application）。根据 Android 的软件框图，其软件层次结构自下而上分为以下 4 层。

（1）操作系统层（OS）。

（2）各种库（Libraries）和 Android 运行环境（RunTime）。

（3）应用程序框架（Application Framework）。

（4）应用程序（Application）。

上述各个层的具体结构如图 1-2 所示。

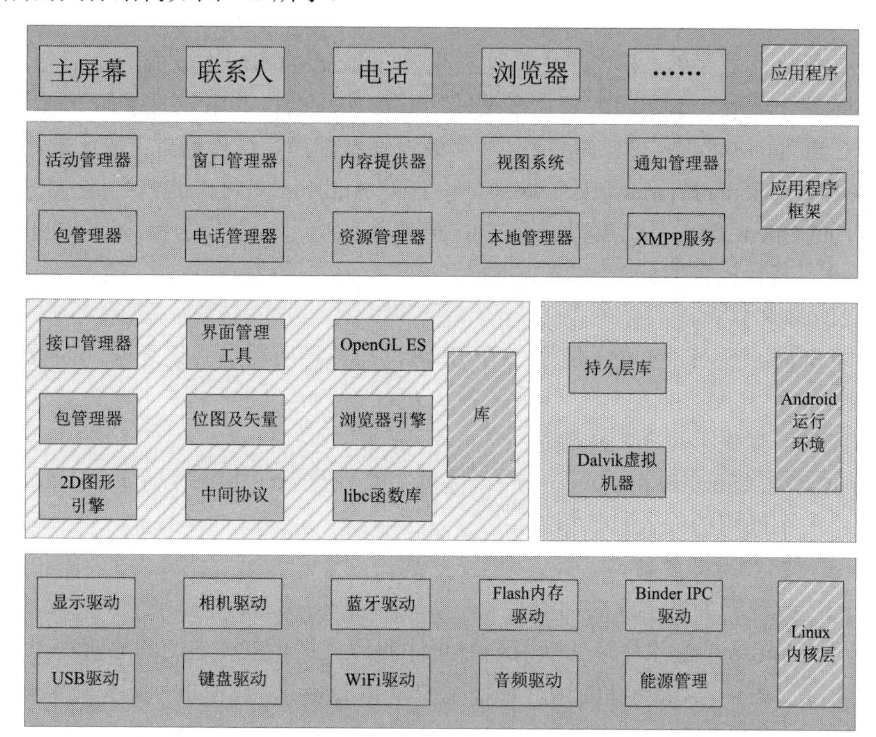


图 1-2　Android 操作系统的组件结构图

本节将详细介绍 Android 操作系统的基本组件结构方面的知识。

## 1.3.1　底层操作系统层（OS）

因为 Android 源于 Linux，使用了 Linux 内核，所以 Android 使用 Linux 2.6 作为操作系统。Linux 2.6 是一种标准的技术，Linux 也是一个开放的操作系统。Android 对操作系统的使用包括核心和驱动程序两部分，Android 的 Linux 核心为标准的 Linux 2.6 内核，Android 更多的是需要一些与移动设备相关的驱动程序。主要的驱动如下。

☑ 显示驱动（Display Driver）：是常用的基于 Linux 的帧缓冲（Frame Buffer）驱动。
☑ Flash 内存驱动（Flash Memory Driver）：是基于 MTD 的 Flash 驱动程序。
☑ 照相机驱动（Camera Driver）：常用基于 Linux 的 V4L（Video for Linux）驱动。
☑ 音频驱动（Audio Driver）：常用基于 ALSA（Advanced Linux Sound Architecture，高级 Linux 声音体系）驱动。
☑ WiFi 驱动（Camera Driver）：基于 IEEE 802.11 标准的驱动程序。
☑ 键盘驱动（KeyBoard Driver）：作为输入设备的键盘驱动。
☑ 蓝牙驱动（Bluetooth Driver）：基于 IEEE 802.15.1 标准的无线传输技术。
☑ Binder IPC 驱动：Andoid 一个特殊的驱动程序，具有单独的设备节点，提供进程间通信的功能。
☑ Power Management（能源管理）：用于管理电池电量等信息。

### 1.3.2 各种库（Libraries）和 Android 运行环境（RunTime）

本层次对应一般嵌入式系统，相当于中间件层次。Android 的本层次分成两个部分，一个是各种库，另一个是 Android 运行环境。本层的内容大多是使用 C++实现的，其中包含了如下所示的各种库。

☑ C 库：C 语言的标准库，也是系统中一个最为底层的库，C 库通过 Linux 的系统调用来实现。
☑ 多媒体框架（Media Framework）：这部分内容是 Android 多媒体的核心部分，基于 PacketVideo（即PV）的OpenCORE，从功能上本库一共分为两大部分，一部分是音频、视频的回放（PlayBack），另一部分则是音视频的记录（Recorder）。
☑ SGL：2D 图像引擎。
☑ SSL：即 Secure Socket Layer 位于 TCP/IP 协议与各种应用层协议之间，为数据通信提供安全支持。
☑ OpenGL ES：提供了对 3D 的支持。
☑ 界面管理工具（Surface Management）：提供了管理显示子系统等功能。
☑ SQLite：一个通用的嵌入式数据库。
☑ WebKit：网络浏览器的核心。
☑ FreeType：位图和矢量字体的功能。

在一般情况下，Android 的各种库是以系统中间件的形式提供的，它们的显著特点是与移动设备的平台的应用密切相关。另外，Android 的运行环境主要是指 Dalvik（虚拟机）技术。Dalvik 和一般的 Java 虚拟机（Java VM）是有区别的。

☑ Java 虚拟机：执行的是 Java 标准的字节码（Bytecode）。在最新的 Android 5.0 版本中，将使用 ART 为默认的运行环境，Java 虚拟机只是作为一个备选项而即将被淘汰。
☑ Dalvik：执行的是 Dalvik 可执行格式（.dex）中的执行文件。在执行的过程中，每一个应用程序即一个进程（Linux 的一个 Process）。

二者最大的区别在于 Java 虚拟机是基于栈的虚拟机（Stack-based），而 Dalvik 是基于寄存器的虚拟机（Register-based）。显然，后者最大的好处在于可以根据硬件实现更大的优化，这更适合移动设备的特点。

### 1.3.3　Application Framework（应用程序框架）

在整个 Android 系统中，和应用开发最相关的是 Application Framework，在这一层，Android 为应用程序层的开发者提供了各种功能强大的 APIs，这实际上是一个应用程序的框架。由于上层的应用程序是以 Java 构建的，在本层提供了程序中所需要的各种控件，例如 Views（视图组件）、List（列表）、Grid（栅格）、Text Box（文本框）、Button（按钮），甚至还有一个嵌入式的 Web 浏览器。

一个基本的 Andoid 应用程序可以利用应用程序框架中的以下 5 个部分。

☑　Activity：活动。

☑　Broadcast Intent Receiver：广播意图接收者。

☑　Service：服务。

☑　Content Provider：内容提供者。

☑　Intent and Intent Filter：意图和意图过滤器。

### 1.3.4　顶层应用程序（Application）

Android 的应用程序主要是用户界面（User Interface）方面的，本层通常使用 Java 语言编写，其中还可以包含各种被放置在 res 目录中的资源文件。Java 程序和相关资源在经过编译后，会生成一个 APK 包。Android 本身提供了主屏幕（Home）、联系人（Contact）、电话（Phone）和浏览器（Browers）等众多的核心应用。同时应用程序的开发者还可以使用应用程序框架层的 API 实现自己的程序。这也是 Android 开源的巨大潜力的体现。

## 1.4　核　心　组　件

**知识点讲解：光盘:视频\知识点\第 1 章\核心组件.avi**

在分析 Android 4.4 的源码之前，很有必要了解一下 Android 应用程序的核心组件功能。一个典型的 Android 应用程序通常由 5 个组件组成，这 5 个组件构成了 Android 的核心功能。本节将详细讲解这 5 大组件的基本知识。

### 1.4.1　Activity 界面

Activities 是这 5 个组件中最常用的一个组件。程序中 Activity 通常的表现形式是一个单独的界面（screen）。每个 Activity 都是一个单独的类，它扩展实现了 Activity 基础类。这个类显示为一个由 Views 组成的用户界面，并响应事件。大多数程序有多个 Activity。例如，一个文本信息程序有这么几个界面：显示联系人列表界面、写信息界面、查看信息界面或者设置界面等。每个界面都是一个 Activity。切换到另一个界面就是载入一个新的 Activity。某些情况下，一个 Activity 可能会给前一个 Activity 返回值——例如，一个让用户选择相片的 Activity 会把选择到的相片返回给其调用者。

打开一个新界面后，前一个界面就被暂停，并放入历史栈中（界面切换历史栈）。使用者可以回溯前面已经打开的存放在历史栈中的界面，也可以从历史栈中删除没有界面价值的界面。Android 在历史栈中保留程序运行产生的所有界面：从第一个界面，到最后一个。

## 1.4.2 Intent 和 Intent Filters

Android 通过一个专门的 Intent 类来进行界面的切换。Intent 描述了程序想做什么（Intent 意为意图、目的、意向）。Intent 类还有一个相关类 Intent Filter。Intent 请求来做什么事情，Intent Filters 则描述了一个 Activity（或下文的 Intent Receiver）能处理什么意图。显示某人联系信息的 Activity 使用了一个 Intent Filter，就是说它知道如何处理应用到此人数据的 View（视图）操作。Activities 在文件 AndroidManifest.xml 中使用 Intent Filters。

通过解析 Intents 可以实现 Activity 的切换，我们可以使用 startActivity(myIntent)启用新的 Activity。系统会考察所有安装程序的 Intent Filters，然后找到与 myIntent 匹配最好的 Intent Filters 所对应的 Activity。这个新 Activity 能够接收 Intent 传来的消息，并因此被启用。解析 Intents 的过程发生在 startActivity 被实时调用时，这样做有如下两个好处。

（1）Activities 仅发出一个 Intent 请求，便能重用其他组件的功能。

（2）Activities 可以随时被替换为有等价 Intent Filter 的新 Activity。

## 1.4.3 Service 服务

Service 是一个没有 UI 且长驻系统的代码，最常见的例子是媒体播放器从播放列表中播放歌曲。在媒体播放器程序中，可能有一个或多个 Activities 让用户选择播放的歌曲。然而在后台播放歌曲时无须 Activity 干涉，因为用户希望在音乐播放的同时能够切换到其他界面。既然这样，媒体播放器 Activity 需要通过 Context.startService()启动一个 Service，这个 Service 在后台运行以保持继续播放音乐。在媒体播放器被关闭之前，系统会保持音乐后台播放 Service 的正常运行。可以用 Context.bindService()方法连接到一个 Service 上（如果 Service 未运行的话，连接后还会启动它），连接后就可以通过一个 Service 提供的接口与 Service 进行通话。对音乐 Service 来说，提供了暂停和重放等功能。

### 1. 如何使用服务

在 Android 系统中，有如下两种使用 Service 服务的方法。

（1）通过调用 Context.startService()启动服务，调用 Context.stopService()结束服务，startService()可以传递参数给 Service。

（2）通过调用 Context.bindService()启动，调用 Context.unbindService()结束，还可以通过 Service Connection 访问 Service。二者可以混合使用，例如可以先 startService()再 unbindService()。

### 2. Service 的生命周期

在使用 startService()方法启动服务后，即使调用 startService()的进程结束了，Service 还仍然存在，一直到有进程调用 stopService()或 Service 自己灭亡（stopSelf()）为止。

在 bindService()后，Service 就和调用 bindService()的进程同生共死，也就是说当调用 bindService()的进程死了，那么它绑定的 Service 也要跟着被结束，当然期间也可以调用 unbindService()让 Service

结束。

当混合使用上述两种方式时，例如你 startService()（启动服务）了，我 bindService()（绑定服务）了，那么只有你 stopService()（停止服务）了，而且我也 unbindService()（解除服务绑定）了，这个 Service 才会被结束。

#### 3. 进程生命周期

在 Android 系统中，会尝试保留那些启动了的或者绑定了的服务进程，具体规则如下所示。

（1）如果该服务正在进程的 onCreate()、onStart()或者 onDestroy()这些方法中执行时，那么主进程将会成为一个前台进程，以确保此代码不会被停止。

（2）如果服务已经开始，那么它的主进程的重要性会低于所有的可见进程，但是会高于不可见进程。由于只有少数几个进程是用户可见的，所以只要不是内存特别低，该服务就不会停止。

（3）如果有多个客户端绑定了服务，只要客户端中的一个对于用户是可见的，就可以认为该服务可见。

### 1.4.4 Broadcast Receiver 发送广播

在 Android 系统中，Broadcast Receiver 是一个广播接收器组件。广播接收器是一个专注于接收广播通知信息，并做出对应处理的组件。很多广播是源自于系统代码的，例如，通知时区改变、电池电量低、拍摄了一张照片或者用户改变了语言选项。应用程序也可以进行广播——例如，通知其他应用程序一些数据下载完成并处于可用状态。应用程序可以拥有任意数量的广播接收器以对所有它感兴趣的通知信息予以响应，所有的接收器均继承自 BroadcastReceiver 基类。

在 Android 系统中，Broadcast Receiver 广播接收器没有用户界面。然而，它们可以启动一个 Activity 来响应它们收到的信息，或者用 NotificationManager 来通知用户。通知可以用很多种方式来吸引用户的注意力——闪动背灯、振动、播放声音等。一般来说是在状态栏上放一个持久的图标，用户可以打开它并获取消息。

Android 中的广播事件有两种，一种是系统广播事件，例如 ACTION_BOOT_COMPLETED（系统启动完成后触发）、ACTION_TIME_CHANGED（系统时间改变时触发）和 ACTION_BATTERY_LOW（电量低时触发）等。另外一种是我们自定义的广播事件。

在 Android 系统中，广播事件的基本流程如下所示。

（1）注册广播事件：注册方式有两种，一种是静态注册，即在 AndroidManifest.xml 文件中定义，注册的广播接收器必须要继承 Broadcast Receiver；另一种是动态注册，是在程序中使用 Context.Register Receiver 注册，注册的广播接收器相当于一个匿名类。两种方式都需要 Intent Filter。

（2）发送广播事件：通过 Context.sendBroadcast 来发送，由 Intent 来传递注册时用到的 Action。

（3）接收广播事件：当发送的广播被接收器监听到后，会调用它的 onReceive()方法，并将包含消息的 Intent 对象传给它。onReceive 中代码的执行时间不要超过 5s，否则 Android 会弹出超时对话框。

### 1.4.5 用 Content Provider 存储数据

在 Android 系统中，应用程序会把数据存放在 SQLite 数据库格式文件中，或者存放在其他有效设备中。如果想让其他程序能够使用我们程序中的数据，就需要用到 Content Provider。Content Provider

是一个实现了一系列标准方法的类，这个类使得其他程序能存储、读取某种 Content Provider 可处理的数据。

# 1.5　进程和线程

**知识点讲解：光盘:视频\知识点\第 1 章\进程和线程.avi**

Android 系统中也有进程和线程，代表当前系统中正在运行的程序。当第一次运行某个组件时，Android 会启动一个进程。在默认情况下，所有的组件和程序运行在这个进程和线程中，也可以安排组件在其他的进程或者线程中运行。本节将简要讲解 Android 进程和线程的基本知识。

## 1.5.1　什么是进程

组件运行的进程由 manifest file 控制。组件的节点一般都包含一个 process 属性，例如<activity>、<service>、<receiver>和<provider>节点。属性 process 可以设置组件运行的进程，可以配置组件在一个独立进程中运行，或者多个组件在同一个进程中运行，甚至可以多个程序在一个进程中运行，当然前提是这些程序共享一个 User ID 并给定同样的权限。另外<application>节点也包含了 process 属性，用来设置程序中所有组件的默认进程。

当更加常用的进程无法获取足够内存时，Android 会智能地关闭不常用的进程。当下次启动程序时会重新启动这些进程。当决定哪个进程需要被关闭时，Android 会考虑哪个对用户更加有用。例如 Android 会倾向于关闭一个长期不显示在界面的进程来支持一个经常显示在界面的进程。是否关闭一个进程决定于组件在进程中的状态。

## 1.5.2　什么是线程

如果用户界面需要很快对用户进行响应，就需要将一些费时的操作，如网络连接、下载或者非常占用服务器时间的操作等放到其他线程。也就是说，即使为组件分配了不同的进程，有时也需要再分配线程。

线程是通过 Java 的标准对象 Thread 来创建的，在 Android 中提供了如下管理线程的方法。

（1）Looper 在线程中运行一个消息循环。

（2）Handler 传递一个消息。

（3）HandlerThread 创建一个带有消息循环的线程。

（4）Android 让一个应用程序在单独的线程中，指导它创建自己的线程。

（5）应用程序组件（Activity、Service、Broadcast Receiver）都在理想的主线程中实例化。

（6）当被系统调用时，没有一个组件应该执行长时间或是阻塞操作（例如网络呼叫或是计算循环），这将中断所有在该进程的其他组件。

（7）可以创建一个新的线程来执行长期操作。

# 1.6　物联网技术改变未来

知识点讲解：光盘:视频\知识点\第 1 章\物联网技术改变未来.avi

物联网是新一代信息技术的重要组成部分，其英文名称是 the Internet of Things。顾名思义，物联网就是物物相连的互联网。本节将详细讲解物联网技术的基础知识。

## 1.6.1　什么是物联网

物联网有两层含义，第一，物联网的核心和基础仍然是互联网，是在互联网基础上延伸和扩展的网络；第二，其用户端延伸和扩展到了任何物品与物品之间进行信息交换和通信。物联网就是“物物相连的互联网”。物联网通过智能感知、识别技术与普适计算，广泛应用于网络的融合中，也因此被称为继计算机、互联网之后世界信息产业发展的第三次浪潮。物联网是互联网的应用拓展，与其说物联网是网络，不如说物联网是业务和应用。因此，应用创新是物联网发展的核心，以用户体验为核心的创新 2.0 是物联网发展的灵魂。

由此可见，物联网利用局部网络或互联网等通信技术把传感器、控制器、机器、人员和物等通过新的方式联系在一起，形成人与物、物与物相联，实现信息化、远程管理控制和智能化的网络。物联网是互联网的延伸，它包括互联网及互联网上所有的资源，兼容互联网所有的应用，但物联网中所有的元素（所有的设备、资源及通信等）都是个性化和私有化的。

## 1.6.2　发展历程

物联网的实践应用最早可以追溯到 1990 年施乐公司的网络可乐贩售机——Networked Coke Machine。

1991 年，美国麻省理工学院（MIT）的 Kevin Ash-ton 教授首次提出物联网的概念。

1995 年，比尔·盖茨在《未来之路》一书中也曾提及物联网，但未引起广泛重视。

1999 年，美国麻省理工学院建立了“自动识别中心（Auto-ID）”，提出“万物皆可通过网络互联”，阐明了物联网的基本含义。早期的物联网是依托射频识别（RFID）技术的物流网络，随着技术和应用的发展，物联网的内涵已经发生了较大变化。

2003 年，美国《技术评论》提出传感网络技术将是未来改变人们生活的十大技术之首。

2004 年，日本总务省（MIC）提出 u-Japan 计划，该战略力求实现人与人、物与物、人与物之间的连接，希望将日本建设成一个随时、随地、任何物体、任何人均可连接的泛网络社会。

2005 年 11 月 17 日，在突尼斯举行的信息社会世界峰会（WSIS）上，国际电信联盟（ITU）发布了《ITU 互联网报告 2005：物联网》，引用了“物联网”的概念。物联网的定义和范围已经发生了变化，覆盖范围有了较大的拓展，不再只是指基于 RFID 技术的物联网。

2006 年，韩国确立了 u-Korea 计划，该计划旨在建立无所不在的社会（ubiquitous society），在民众的生活环境里建设智能型网络（如 IPv6、BcN、USN）和各种新型应用（如 DMB、Telematics、RFID），让民众可以随时随地享有科技智慧服务。2009 年，韩国通信委员会出台了《物联网基础设施构建基本规划》，将物联网确定为新增长动力，提出到 2012 年实现“通过构建世界最先进的物联网基础设施，打造未来广播通信融合领域超一流信息通信技术强国”的目标。

2008 年后，为了促进科技发展，寻找经济新的增长点，各国政府开始重视下一代的技术规划，将目光放在了物联网上。

2009 年，欧盟执委会发表了欧洲物联网行动计划，描绘了物联网技术的应用前景，提出欧盟政府要加强对物联网的管理，促进物联网的发展。

2009 年 1 月 28 日，美国总统与美国工商业领袖举行了一次“圆桌会议”，作为仅有的两名代表之一，IBM 首席执行官彭明盛首次提出“智慧地球”这一概念，建议新政府投资新一代的智慧型基础设施。当年，美国将新能源和物联网列为振兴经济的两大重点。

2009 年 2 月 24 日，2009 IBM 论坛上，IBM 大中华区首席执行官公布了名为“智慧的地球”的最新策略。此概念一经提出，即得到美国各界的高度关注，甚至有分析认为 IBM 公司的这一构想极有可能上升至美国的国家战略，并在世界范围内引起轰动。

而今天，“智慧地球”战略被不少美国人认为与当年的“信息高速公路”有许多相似之处，同样被他们认为是振兴经济、确立竞争优势的关键战略。该战略能否掀起如当年互联网革命一样的科技和经济浪潮，不仅为美国关注，更为世界所关注。

2009 年 8 月，无锡市率先建立了“感知中国”研究中心，中国科学院、运营商、多所大学在无锡建立了物联网研究院，无锡市江南大学还建立了全国首家实体物联网工厂学院。从此物联网被正式列为国家五大新兴战略性产业之一，写入“政府工作报告”，物联网在中国受到了全社会极大的关注，其受关注程度是美国、欧盟以及其他各国不可比拟的。

截至 2010 年，发改委、工信部等部委正在会同有关部门，在新一代信息技术方面开展研究，以形成支持新一代信息技术的一些新政策措施，从而推动我国经济的发展。

物联网作为一个新经济增长点的战略新兴产业，具有良好的市场效益，《2013—2017 年中国物联网行业应用领域市场需求与投资预测分析报告》数据表明，2010 年物联网在安防、交通、电力和物流领域的市场规模分别为 600 亿元、300 亿元、280 亿元和 150 亿元。2011 年中国物联网产业市场规模达到 2600 多亿元。

### 1.6.3 Android 正在成为物联网标准操作系统

Linux 基金会执行董事吉姆 • 泽林（Jim Zemlin）认为，Android 已经在智能设备市场所向披靡。iOS 只在苹果的 iPhone 和 iPad 上运行，市场深度不够。相反 Android 充斥于各种形状和尺寸及品牌的硬件设备上。不同类型的屏幕、移动芯片和传感器都可以与 Android 完美配合，任何人都可以对 Android 进行优化，使它胜任各种工作。泽林以中国上汽集团举例，上汽集团只靠着 6 个软件开发人员和 Android 就完成了内置的车载信息娱乐系统。

美国宇航局艾姆斯研究中心的年轻工程师则利用 Android 系统开发了宇宙飞船的大脑，硬件大小不过一颗棒球。卫星通常需要花费数百万美元进行建造和发射，而这款控制系统的造价不过 1.5 万美元，之所以造价如此低就是因为采用了开源 Android 系统。

同样，Xively 公司的副总裁菲利普也展示了他最骄傲的物联网作品：一款基于 Android 的农业灌溉系统。这款系统利用一个小型防水芯片建立调节水量的网络。菲利普表示，“有了 Android，你可以开发一些低功耗的产品，它很容易进行用户界面和触摸控制的开发，同样很容易处理数据传输。”

Android 的崛起对于微软来说并不是好事。微软的嵌入式系统目前运行于福特汽车、NCR 收款机

等产品，但是微软的嵌入式系统并不廉价。正像微软在智能手机和平板电脑端的作为，微软在物联网嵌入式操作的表现并不理想。菲利普表示，他了解到大量的公司想要做智能计步器、网络连接的 LED 照明和其他可以与 iPhones 和 iPads 互联的设备。但最后最有可能的就是，这些设备将会选择 Android 或更简单的系统。至于 iOS，苹果似乎完全没有让其运行在非苹果产品上的想法。

安迪·鲁宾（Andy Rubin），谷歌 Android 的长期领导者，目前成立了一个位于加利福尼亚州洛斯阿尔托斯的孵化器，在那里他与朋友进行包括 Android 在内的有趣开发项目。鲁宾说，谷歌已经收到了大量的对于 Android 进行物联网化的请求。物联网还在起步，但是似乎 Android 已经做好了准备。事实终将证明，Android 会在物联网中取得和在移动设备中一样的强势地位。

# 1.7　智能设备凶猛来袭

**知识点讲解：光盘:视频\知识点\第 1 章\智能设备凶猛来袭.avi**

因为 Android 系统的免费和开源，也因为系统本身强大的功能性，使得 Android 系统不仅被用于手机设备上，而且也被广泛用于其他智能设备中，例如当前的新兴热点——可穿戴设备。本节将简要介绍除了手机产品之外，常见的搭载 Android 系统的智能设备。

## 1.7.1　常见的 Android 智能设备

（1）智能电视

Android 智能电视是指搭载了安卓操作系统的电视，使得电视智能化，能让电视机实现网页浏览、视频电影观看、聊天办公游戏等与平板电脑和智能手机一样的功能。其凭借安卓系统让电视实现智能化的提升，数十万款安卓市场的应用、游戏等内容可随意安装。例如海尔的 MOOKA 模卡 U42H7030 便是一款搭载 Android 4.2 系统的智能电视，如图 1-3 所示。

图 1-3　搭载 Android 4.2 系统的智能电视

（2）机顶盒

Android 机顶盒是指像智能手机一样，具有全开放式平台，搭载了安卓操作系统，可以由用户自行安装和卸载软件、游戏等第三方服务商提供的程序，通过此类程序来不断对电视的功能进行扩充，并可以通过网线、无线网络来实现上网冲浪的新一代机顶盒的总称。

通过使用 Android 机顶盒，可以让电视具有上网、看网络视频、玩游戏、看电子书、听音乐等功能，使电视成为一个低成本的平板电脑，Android 机顶盒不仅仅是一个高清播放器，更具有一种全新的人机交互模式，既区别于电脑，又有别于触摸屏，Android 机顶盒配备红外感应条，遥控器一般采用空中飞鼠，这样就可以方便地实现触摸屏上的各种单点操作，可以方便地在电视上玩愤怒的小鸟、植物大战僵尸等经典游戏。例如乐视公司的 LeTV 机顶盒便是基于 Android 打造的，如图 1-4 所示。

（3）游戏机

Android 游戏机就像 Android 智能手表一样，在 2013 年出现了爆炸式增长。在 CES 展会上，NVIDIA 的 Project Shield 掌上游戏主机以绝对震撼的姿态亮相，之后又有 Ouya 和 Gamestick 相继推出。不久前，Mad Catz 也发布了一款 Andriod 游戏机。

（4）智能手表

智能手表，是将手表内置智能化系统、搭载智能手机系统而连接于网络而实现多功能，能同步手机中的电话、短信、邮件、照片、音乐等。2013 年 3 月媒体报道，苹果、三星、谷歌等科技巨头都将在 2013 年晚些时候发布智能手表。美国市场研究公司 Current Analysis 分析师艾维·格林加特（Avi Greengart）认为 2013 年可能会成为智能手表元年。例如三星的 Galaxy Gear 便是一款搭载 Android 系统的智能手表设备，如图 1-5 所示。

图 1-4 基于 Android 的 LeTV 机顶盒

图 1-5 搭载 Android 系统的 Galaxy Gear

（5）智能家居

智能家居是以住宅为平台，利用综合布线技术、网络通信技术、智能家居-系统设计方案安全防范技术、自动控制技术、音视频技术将家居生活有关的设施集成，构建高效的住宅设施与家庭日程事务的管理系统，提升家居安全性、便利性、舒适性、艺术性，并实现环保节能的居住环境。

智能家居是在互联网的影响之下的物联化体现。智能家居通过物联网技术将家中的各种设备（如音视频设备、照明系统、窗帘控制、空调控制、安防系统、数字影院系统、网络家电以及三表抄送等）连接到一起，提供家电控制、照明控制、窗帘控制、电话远程控制、室内外遥控、防盗报警、环境监测、暖通控制、红外转发以及可编程定时控制等多种功能和手段。与普通家居相比，智能家居不仅具有传统的居住功能，还兼备建筑、网络通信、信息家电、设备自动化，是集系统、结构、服务、管理为一体的高效、舒适、安全、便利、环保的居住环境，可以提供全方位的信息交互功能，帮助家庭与外部保持信息交流畅通，优化人们的生活方式，帮助人们有效安排时间，增强家居生活的安全性，甚至为各种能源费用节约资金。

例如乐得威公司的 GW-9311 智能主机产品便是一款 Android 智能家居产品，如图 1-6 所示。

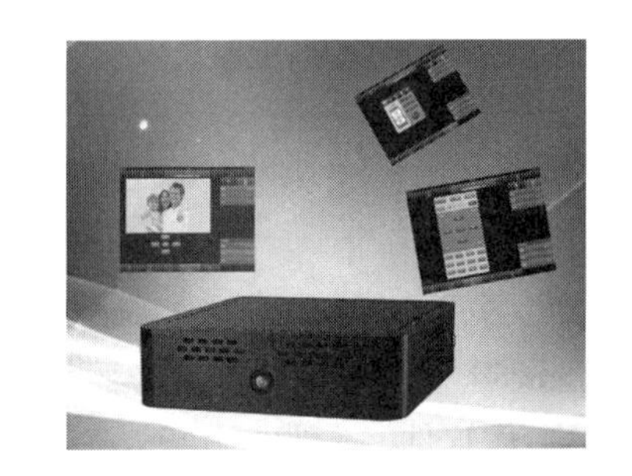

图 1-6 乐得威公司的 GW-9311 智能主机

上述智能设备只是冰山一角，随着物联网和云服务的普及和发展，将有更多的智能设备诞生。到那个时候，Android

系统更是如鱼得水，将拥有一个更美好的未来。

### 1.7.2　新兴热点——可穿戴设备

在最近两年来，随着 Android 和 iOS 系统的发展，可穿戴设备逐渐展现在广大用户的面前。谷歌眼镜、苹果手表等新颖而又时尚的设备吸引了广大用户的眼球，相信在未来这些设备必将引领时尚的潮流，成为科技界的主流产品之一。自从谷歌推出 Google 眼镜产品之后，可穿戴计算设备便成为了当今科技界的火热话题之一。在 CES 2013 和 CES 2014（国际电子展）上，也有不少公司推出了眼镜、腕带等各种可穿戴计算设备，可穿戴计算也越来越火热。

#### 1. 发展背景

穿戴设备看似是一个新兴事物，但实际上其发展历史可以上溯到 20 世纪 80 年代。多伦多大学 Steve Mann 教授被人称为“可穿戴计算之父”，公认的第一个赛博格（Cyborg）——这是一个特殊的群体，他们很像科幻小说中的一些角色，利用机器设备来增强自己的感觉，从而加强对环境的掌控。

Steve Mann 自 20 世纪 80 年代就开始尝试制作类似于 Google Glass 这样可以架在自己的鼻梁上，以第一人称的角度来记录周遭事物的眼镜。Mann 最初设计的设备是戴在头盔上的，而经过多年的实验和反复改进，他的头戴式智能眼镜变得越来越轻巧。后来 Mann 成功开发出令智能眼镜小型化，并与电脑和网络相连的技术 EyeTab，这比 Google 眼镜要早 13 年。

以 Google 眼镜为代表的这种穿戴设备，和智能手机最大的不同是把用户的眼睛和手从设备上解放出来了，所以不需要从兜里掏出一个东西，也不需要低下头去看它，它永远在你前面。所以我们有理由相信，可穿戴计算设备（不一定是 Google 眼镜）一定可以给人们带来更大的自由，并且在将来一定会成为潮流趋势。

#### 2. 发展现状介绍

可穿戴计算设备将成为继智能手机、平板电脑之后的又一个潮流。当前可穿戴技术正在处于一种过渡时期，一些看似疯狂的想法逐渐在摸索中变得更加成熟。在 CES 2014 电子消费展上，我们已经看到类似 Pebble Steel 这样拥有更精致设计的智能手表，还有很多其他运动腕带、智能眼镜等产品参与展出，下面一起来回顾一下这些出色的可穿戴设备。

（1）Google Project Glass

谷歌眼镜（Google Project Glass）是由谷歌公司于 2012 年 4 月发布的一款“拓展现实”眼镜，如图 1-7 所示。谷歌眼镜具有和智能手机一样的功能，可以通过声音控制拍照、视频通话和辨明方向以及上网冲浪、处理文字信息和电子邮件等。

图 1-7　谷歌眼镜

2013 年 4 月 10 日，美国科技博客 Gizmodo 发布了一张图片，揭示了谷歌智能眼镜的工作原理。谷歌眼镜承载着可穿戴设备的开端，它极具想象空间，前途不可限量。但现在看来，它暂时只是一个手机伴侣。基础通信、文字输入还依赖手机。

2013 年 11 月 12 日，发布谷歌眼镜的一系列新功能，包括搜索歌曲、扫描已保存播放列表，以及收听高保真音乐等。美国东部时间 2014 年 4 月 15 日早上 9

点，Google Glass 正式开放网上订购。

（2）苹果智能手表

苹果智能手表，是苹果正在秘密研发的智能手表产品。苹果手表可能将采用 1.5 至 2 英寸显示屏，并将采用指纹识别技术。苹果成立了一支 100 人左右的开发团队，专门开发这款设备。2013 年 5 月，凯基证券分析师郭明池（Ming-ChiKuo）在一份报告中指出，苹果手表的零部件尚未成熟，苹果手表最早 2014 年下半年投产。而苹果 CEO 蒂姆-库克（TimCook）曾表示，苹果期待 2013 年秋季和整个 2014 年将推出令人兴奋的新产品。苹果手表的预期效果如图 1-8 所示。

（3）MetaWatch

MetaWatch 是一款智能手表，其设计师此前设计了奢侈品手机 Vertu，所以非常擅长将时尚元素与电子产品有机结合，如图 1-9 所示。

图 1-8　苹果手表

图 1-9　MetaWatch

由此可以看到，MetaWatch 的金属表盘充满质感，皮革腕带也显得十分高级，无论是搭配西装还是 T 恤，都十分合适。MetaWatch 支持 10 至 15 米防水，支持 iOS 设备，用户可以从 App Store 中下载应用程序，实现更多信息的通知功能。

（4）Garmin Vivofit

健身腕带 Garmin Vivofit——知名 GPS 厂商 Garmin 此前推出过运动手表产品，此次更是推出了 Vivofit 健身腕带，全面进入到运动监测设备市场。这款运动腕带的设计充满活力，拥有多种配色款式，不仅能够实现全面的运动数据监测，还支持心率监测，与手机端的应用搭配，可实现出色的健身运动监测功能，如图 1-10 所示。

（5）英特尔智能手表及手镯

芯片巨头英特尔在 CES 2014 上也宣布进入可穿戴设备领域，将陆续推出智能手表、手镯等产品。有意思的是，英特尔的智能手镯将与时尚百货 Barneys 合作推出，或许能够让可穿戴设备在时尚领域更进一步，如图 1-11 所示。

图 1-10　Garmin Vivofit

图 1-11　英特尔智能手表及手镯

### 1.7.3 可穿戴设备的发展前景分析

可穿戴设备是延续性地穿戴在人体上，具备先进的电路系统、无线联网及独立处理能力的终端设备，其具备最重要的两个特点是可长期穿戴和智能化。在智能手机和平板的发展进入停滞期后，以智能眼镜、手表等为代表的智能可穿戴设备成为谷歌、苹果等巨头下一部竞争的主战场。著名科技媒体 Android Authority 的撰稿人奈特·斯旺纳（Nate Swanner）曾经撰文对 2014 年进行了展望，他总结道：穿戴设备将会蓬勃发展。消费者有望买到谷歌眼镜，而且市场上会出现很多智能手表。如果你有意购买一款这样的设备，请务必保持谨慎乐观的心态。毕竟，你在这类产品爆炸式发展初期买下的设备，确实有可能在未来几年里登上“有史以来最糟糕产品”的榜单。

（1）智能手机推动力

智能手机是可穿戴设备爆发的核心驱动力之一，智能手机现在已经不只是拿来打电话，或者是上网，其内建的处理器与操作系统具有强大的运算能力，使它成为远程的计算机引擎，而同时智能手机具有广泛的用户基础，预计未来，手机可能作为智能控制中心和计算系统，使可穿戴设备、平板、笔记本、电视等所有终端保持互联，而每个人的身体及可穿戴设备将变成微网络，身上配戴的各式与智能手机联接的装饰，提供各类功能并与智能手机、云端进行数据计算和交互。根据权威统计数据证明，中国移动互联网用户已超过桌面互联网用户。

（2）跨国公司推动力

随着智能手机渗透率快速提升，便携性要求出现、硬件配置提升、传感器及电池改善，可穿戴设备的便携、云端互联等性能优势将越来越明显，预计可穿戴设备将继智能手机之后成为下一个爆发性增长点。尤其是苹果、谷歌、微软、亚马逊和 Facebook 5 大平台及相应开发者都进入可穿戴设备领域时，后台数据及前端检测传输更加完善时，可穿戴设备将会变成主流。按照 ShareThis 2013 年 6 月最新数据，消费者在移动设备上点击和分享内容的行为是桌面电脑的 2 倍，随着社交网络越发重要，可供分享的数据暴增。以社交分享平台中份额最高的 Facebook 和 Google 来研究，按照 Searchmetrics 数据，目前 Facebook 的分享量以每个月 10%的速度增长，Google 分享量目前以每月 19%的速度增长，截至 2013 年 4 月，Facebook 和 Google 的数据分享量相比 2012 年初增长分别是 202%及 788%，数据分享爆发时代到来。即时的数据分享和社交网络的需求将导致用户对各类移动终端需求不断提升，可穿戴设备在数据分享和社交领域具备放量基础。

（3）用户推动力

用户对健身、医疗及健康监测等需求也在持续抬升，未来可穿戴设备作为新一代智能终端，将成为新的移动平台市场及生态圈，硬件终端不仅是营收增长点，也将成为黏住客户的产品形态，进而围绕消费者形成可穿戴设备、手机、平板、笔记本、电视、汽车等终端互联互通的一体化智能方案，因此原软硬件、互联网等各类厂商均参与到推出硬件终端产品的环节中来。

在可穿戴设备领域应用中，目前较受欢迎的应用是娱乐和社交，而较快进入商用的功能是健身、医疗及健康监测。娱乐和社交领域的典型产品包括 Google/百度智能眼镜、Sony/三星/果壳智能手表等，医疗和健康领域，可穿戴式设备主要包括脉搏血氧仪、葡萄糖监测、心电图（ECG）、助听器、药物输送等类型产品，目前主要功能进行一体化整合成为趋势，如三星智能手表同时也具备健康监测功能。

据数据显示，2012 年中国可穿戴便携移动医疗设备市场销售规模达到 4.2 亿元，预计到 2015 年这一市场规模将超过 10 亿元，到 2017 年中国可穿戴便携移动医疗设备市场销售规模将接近 50 亿元，市场年复合增长达到 60%。

HIS 预计全球范围内与健康相关的可穿戴设备 App 应用装机量（或下载量）会从 2012 年的 1.56 亿上升至 2017 年的 2.48 亿，随着开发者加入及生态环境改善，可穿戴设备将放量增长。

第三方机构 EndpointTechnologiesAssociates 预计，若未来 5 年可穿戴设备市场占比达 4000 万个，便可能为开发商带来 4 亿美元的商机，而程序内广告（in-APPadvertising）可能大幅提升营收。

### 1.7.4 Android 对穿戴设备的支持——Android Wear

自从 Bluetooth Smart 低耗能技术推出后，穿戴设备的开发创造了良好的条件。苹果是从 iOS 5（iPhone 4S 以及以上版本的手机）开始支持 Bluetooth Smart，而 Google 直到 Android 4.3 才开始支持。Google 在 Android 4.3 中添加了 Bluetooth Smart，在操作系统层面建立一个统一标准。也就是说，从 Android 4.3 开始，将完全支持新蓝牙传输技术，这样就为可穿戴设备开发铺平了道路。而在 Android 4.4 中，新增加了地磁旋转矢量、脚步探测器和计步器 3 个传感器类别，这些功能很可能是面向谣传的谷歌 Android 智能手表、谷歌眼镜以及非谷歌出厂的设备。随着更多的厂商在产品中加入运动传感器，追踪人们运动的 Android 手机应用也将从该新功能中获益。

北京时间 2014 年 3 月 19 日，谷歌正式公布了可穿戴设备操作系统 Android Wear。Android Wear 是 Android 的一个修改版，基于 Google Now 语音识别技术，针对可穿戴计算设备设计，最初将被用在智能手表中。谷歌同时表示，LG、华硕、HTC、摩托罗拉移动和三星将是 Android Wear 的硬件合作伙伴，而博通、Imagination、英特尔、联发科和高通将是芯片合作伙伴。Fossil Group 将于 2014 年晚些时候推出采用 Android Wear 的智能手表。LG 和谷歌将在谷歌 I/O 开发者大会上发布智能手表，而 LG 将是推出谷歌智能手表的首家合作伙伴。

Android Wear 与谷歌眼镜类似，将基于 Google Now 和语音命令。通过 OK Google 的语音指令，用户可以提问或发送文字消息。谷歌表示，Android Wear 的设计是为了提供相关性更好的信息，以及来自社交媒体应用的通知、消息应用的提示，以及购物、新闻和拍照应用的通知等。这一修改版 Android 系统将专注于健康和运动追踪功能。FitBit Force 和耐克 FuelBand 等产品推动了这类功能的发展。谷歌还希望，Android Wear 将成为联系用户与其他设备，包括电视机和计算机的纽带。业内人士希望，谷歌的进入将有助于提升智能手表的设计美学。尽管可穿戴计算设备目前非常热门，但相关产品的销售情况并不火爆。三星第一代 Galaxy Gear 和索尼 SmartWatch 仍是小众产品，这两款产品的尺寸过大，并不时尚。对可穿戴计算设备的其他不满还包括电池续航时间过短，以及缺少某些实用功能等。

# 第2章　搭建Android开发环境

Android作为一项新兴技术，在进行开发前首先要搭建一个对应的开发环境。Android开发包括底层开发和应用开发，底层开发大多数是指和硬件相关的开发，并且是基于Linux环境的，例如开发驱动程序。应用开发是指开发能在Android系统上运行的程序，例如游戏、地图等。本书的重点是物联网开发技术，这要求开发人员既需要掌握底层开发的知识，也需要上层应用开发的知识。所以本章将详细讲解搭建Android底层和应用开发环境的知识，为读者步入本书后面知识的学习打下基础。

## 2.1　在Linux系统获取Android源码

**知识点讲解：光盘:视频\知识点\第2章\在Linux系统获取Android源码.avi**

在Linux系统中，通常使用Ubuntu来下载和编译Android源码。由于Android的源码内容很多，Google采用了git的版本控制工具，并对不同的模块设置不同的git服务器，我们可以用repo自动化脚本来下载Android源码，下面介绍如何一步一步地获取Android源码的过程。

（1）下载repo

在用户目录下，创建bin文件夹，用于存放repo，并把该路径设置到环境变量中去，命令如下：

```
$ mkdir ~/bin
$ PATH=~/bin:$PATH
```

下载repo的脚本，用于执行repo，命令如下：

```
$ curl https://dl-ssl.google.com/dl/googlesource/git-repo/repo > ~/bin/repo
```

设置可执行权限，命令如下：

```
$ chmod a+x ~/bin/repo
```

（2）初始化一个repo的客户端

在用户目录下，创建一个空目录，用于存放Android源码，命令如下：

```
$ mkdir AndroidCode
$ cd AndroidCode
```

进入到AndroidCode目录，并运行repo下载源码，下载主线分支的代码，主线分支包括最新修改的bug，以及并未正式发出版本的最新源码，命令如下：

```
$ repo init -u https://android.googlesource.com/platform/manifest
```

下载其他分支，正式发布的版本，可以通过添加-b参数来下载，命令如下：

```
$ repo init -u https://android.googlesource.com/platform/manifest -b
                                                android-4.4_r1
```

在下载过程中会需要填写Name和Email，填写完毕之后，选择Y进行确认，最后提示repo初始化完成，这时可以开始同步Android源码了，同步过程很漫长，需要耐心等待，执行下面命令开始同步代码：

```
$ repo sync
```

经过上述步骤后，便开始下载并同步 Android 源码，界面效果如图 2-1 所示。

```
Checking out files: 100% (2497/2497), done. out files:  31% (790/2497)
Checking out files: 100% (1654/1654), done. out files:  39% (649/1654)
Checking out files: 100% (3471/3471), done.
Checking out files: 100% (24607/24607), done.ut files:  21% (5269/24607)
Checking out files: 100% (2431/2431), done. out files:  32% (800/2431)
Checking out files: 100% (18696/18696), done.
Checking out files: 100% (4276/4276), done. out files:  48% (2058/4276)
Checking out files: 100% (2216/2216), done. out files:  49% (1093/2216)
Checking out files: 100% (857/857), done.ng out files:  13% (115/857)
Checking out files: 100% (1141/1141), done. out files:   2% (28/1141)
Checking out files: 100% (431/431), done.ng out files:  10% (46/431)
Checking out files: 100% (175/175), done.ng out files:  10% (19/175)
Checking out files: 100% (135/135), done.
Checking out files: 100% (378/378), done.
Checking out files: 100% (433/433), done.
Checking out files: 100% (2407/2407), done. out files:  30% (724/2407)
Checking out files: 100% (2489/2489), done.
Checking out files: 100% (2493/2493), done.
Checking out files: 100% (177/177), done.ng out files:  15% (27/177)
Checking out files: 100% (137/137), done.
Checking out files: 100% (40775/40775), done.ut files:   5% (2199/40775)
Checking out files: 100% (93/93), done.
Checking out files: 100% (450/450), done.
Checking out files: 100% (5265/5265), done. out files:  41% (2167/5265)
Syncing work tree: 100% (329/329), done.
```

图 2-1　下载同步

## 2.2　在 Windows 平台获取 Android 源码

**知识点讲解：光盘:视频\知识点\第 2 章\在 Windows 平台获取 Android 源码.avi**

在 Windows 平台获取 Android 源码的原理和 Linux 相同，但是需要预先在 Windows 平台上面搭建一个 Linux 模拟环境，笔者使用的是 cygwin 工具。cygwin 的作用是构建一套在 Windows 上的 Linux 模拟环境，下载 cygwin 工具的地址为 http://cygwin.com/install.html。

下载成功后会得到一个名为 setup.exe 的可执行文件，通过此文件可以更新和下载最新的工具版本，具体流程如下：

（1）启动 cygwin，如图 2-2 所示。

（2）单击“下一步”按钮，选中第一个单选按钮：从网络下载安装，如图 2-3 所示。

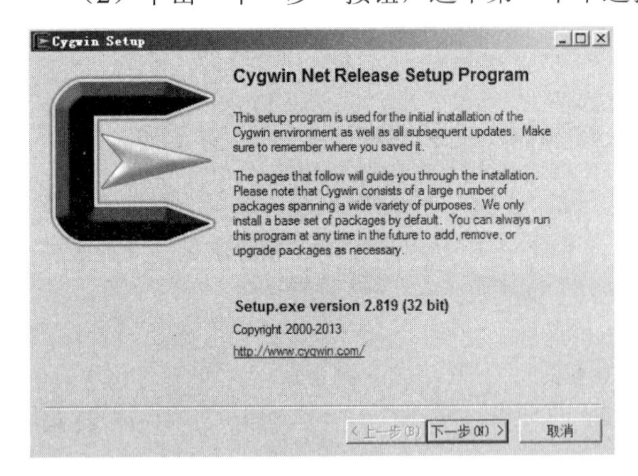


图 2-2　启动 cygwin

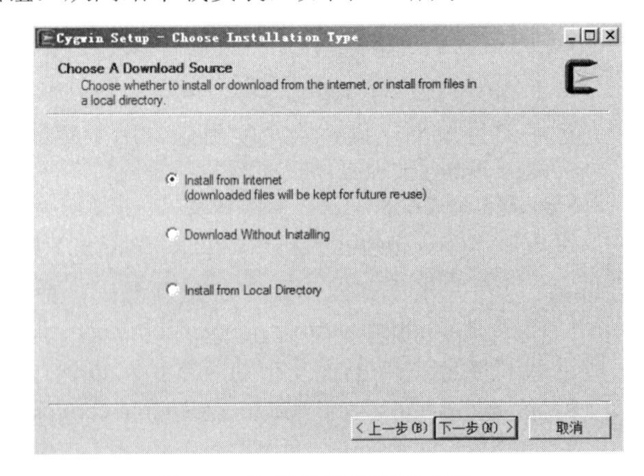


图 2-3　选择从网络下载安装

（3）单击“下一步”按钮，选择安装根目录，如图 2-4 所示。

（4）单击“下一步”按钮，选择临时文件目录，如图 2-5 所示。

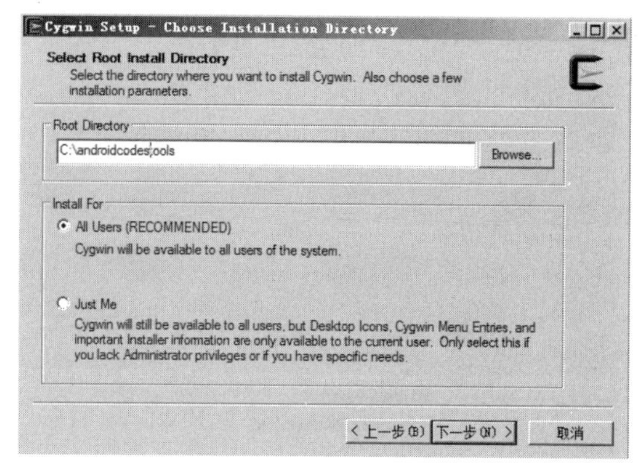


图 2-4　选择安装根目录

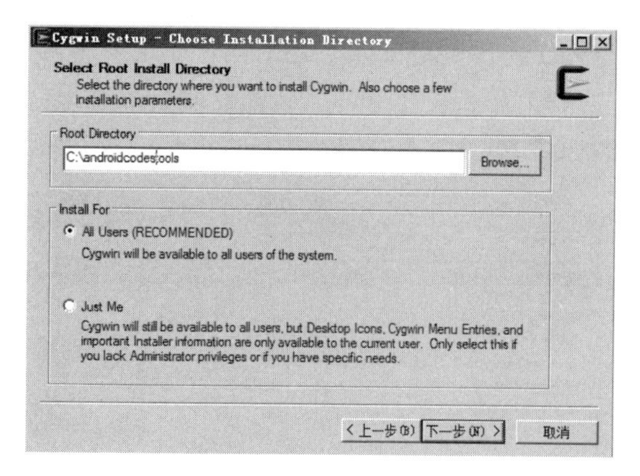


图 2-5　选择临时文件目录

（5）单击“下一步”按钮，设置网络代理。如果所在网络需要代理，则在这一步进行设置，如果不用代理，则选择直接下载，如图 2-6 所示。

（6）单击“下一步”按钮，选择下载站点。一般选择离我们比较近的站点，速度会比较快，这里选择的是台湾站点，如图 2-7 所示。

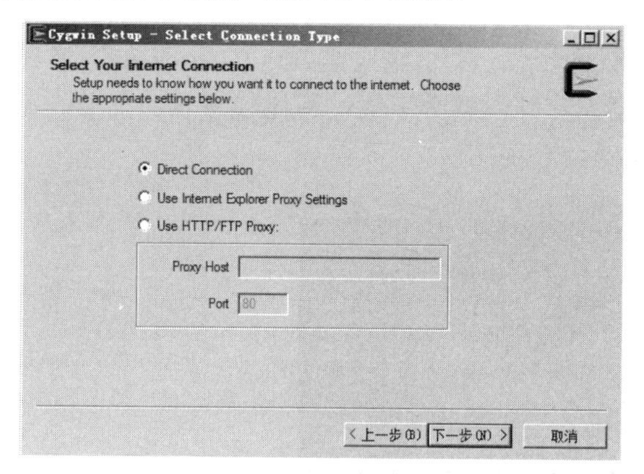


图 2-6　设置网络代理

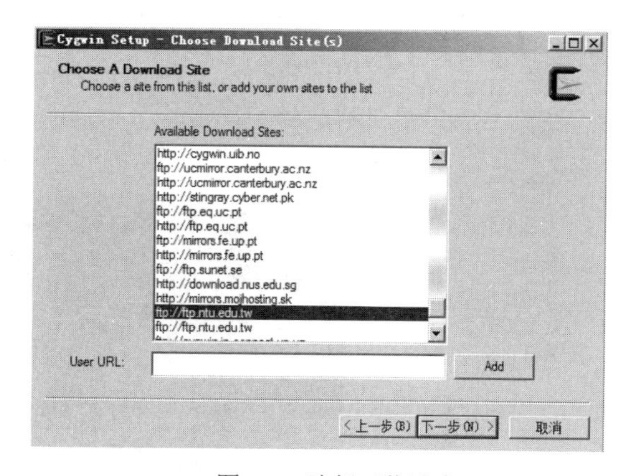


图 2-7　选择下载站点

（7）单击“下一步”按钮，开始更新工具列表，如图 2-8 所示。

（8）单击“下一步”按钮，选择需要下载的工具包。在此需要依次下载 curl、git、python 这些工具，如图 2-9 所示。

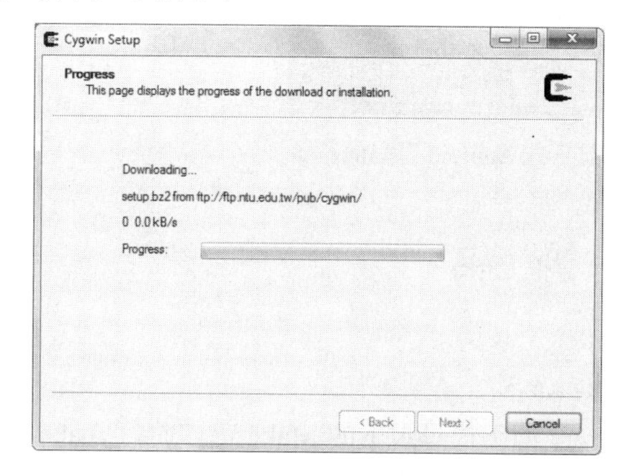


图 2-8　更新工具列表

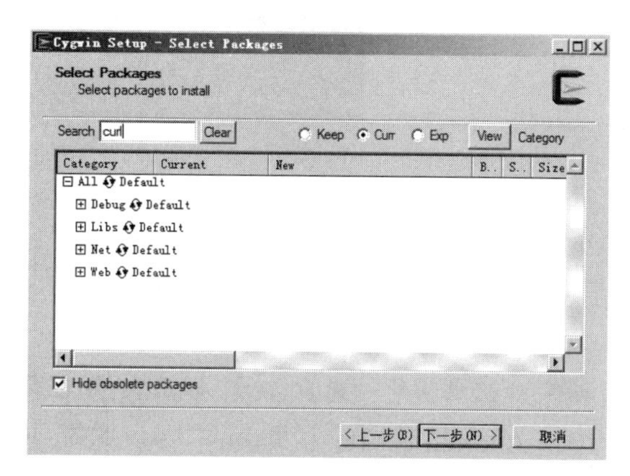


图 2-9　依次下载工具

为了确保能够安装上述工具，一定要用鼠标双击变为 Install 形式，如图 2-10 所示。

（9）单击“下一步”按钮，系统显示下载进度条，如图 2-11 所示。

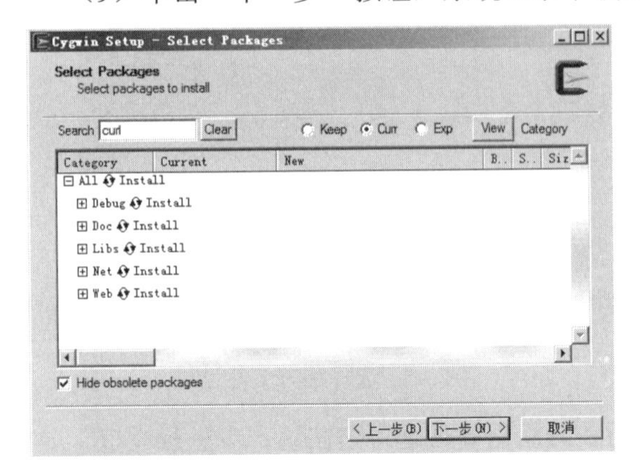

图 2-10　务必设置为 Install 形式

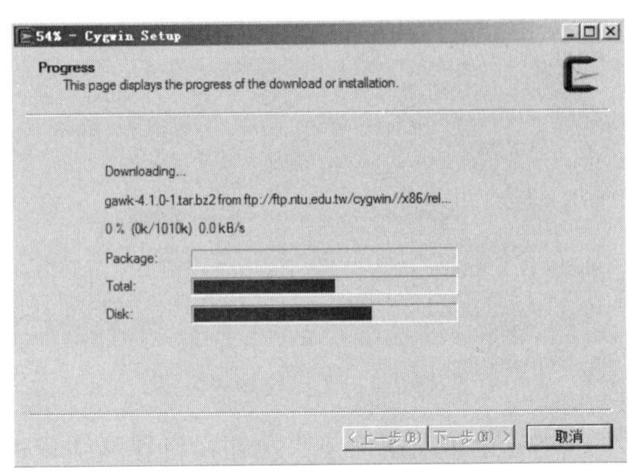


图 2-11　下载进度条

如果下载安装成功会出现提示信息，单击“完成”按钮即完成安装。当安装好 cygwin 后，打开 cygwin，会模拟出一个 Linux 的工作环境，然后按照 Linux 平台的源码下载方法即可下载 Android 源码。

建议读者在下载 Android 源码时，严格按照官方提供的步骤进行，地址是 http://source.android.com/source/downloading.html，这一点对初学者来说尤为重要。另外，整个下载过程比较漫长，需要大家耐心等待。如图 2-12 所示是笔者下载 Android 4.4 时的机器命令截图。

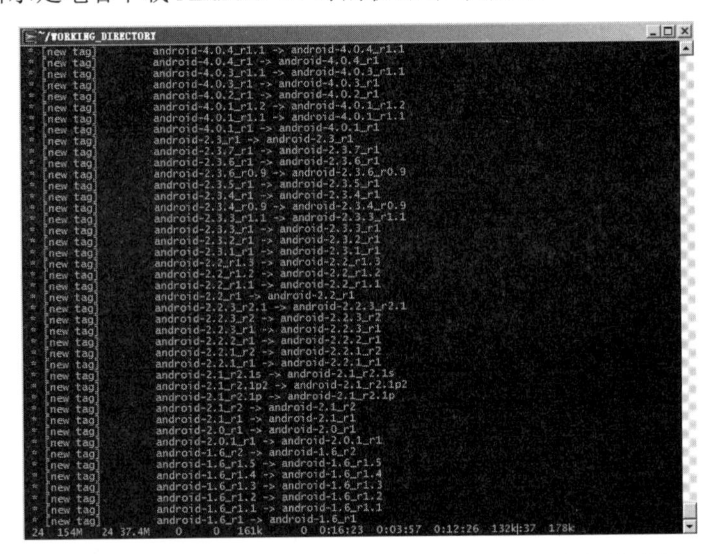

图 2-12　在 Windows 中用 cygwin 工具下载 Android 源码的截图

# 2.3　编译源码

**知识点讲解：光盘:视频\知识点\第 2 章\编译源码.avi**

编译 Android 源码的方法非常简单，只需使用 Android 源码根目录下的 Makefile，执行 make 命令即可轻松实现。当然在编译 Android 源码之前，首先要确定已经完成同步工作。进入 Android 源码目录

使用 make 命令进行编译，使用此命令的格式如下：

```
$: cd ~/Android 4.3（这里的 Android 4.3 就是我们下载源码的保存目录）
$: make
```

编译 Android 源码可以得到“~/project/android/cupcake/out”目录，笔者的截图界面如图 2-13 所示。

```
注意：external/doclava/src/com/google/doclava/Stubs.java 使用了未经检查或不安全
的操作。
注意：要了解详细信息，请使用 -Xlint:unchecked 重新编译。
Notice file: external/doclava/NOTICE -- out/host/linux-x86/obj/NOTICE_FILES/src/
/framework/doclava.jar.txt
Install: out/host/linux-x86/framework/doclava.jar
target Java: core (out/target/common/obj/JAVA_LIBRARIES/core_intermediates/class
es)
注意：某些输入文件使用或覆盖了已过时的 API。
注意：要了解详细信息，请使用 -Xlint:deprecation 重新编译。
注意：某些输入文件使用了未经检查或不安全的操作。
注意：要了解详细信息，请使用 -Xlint:unchecked 重新编译。
Copying: out/target/common/obj/JAVA_LIBRARIES/core_intermediates/classes-jarjar.
jar
Copying: out/target/common/obj/JAVA_LIBRARIES/core_intermediates/emma_out/lib/cl
asses-jarjar.jar
Copying: out/target/common/obj/JAVA_LIBRARIES/core_intermediates/classes.jar
target Java: conscrypt (out/target/common/obj/JAVA_LIBRARIES/conscrypt_intermedi
ates/classes)
注意：某些输入文件使用或覆盖了已过时的 API。
注意：要了解详细信息，请使用 -Xlint:deprecation 重新编译。
注意：某些输入文件使用了未经检查或不安全的操作。
注意：要了解详细信息，请使用 -Xlint:unchecked 重新编译。
host Prebuilt: jarjar (out/host/common/obj/JAVA_LIBRARIES/jarjar_intermediates/javalib.jar)
Install: out/host/linux-x86/framework/jarjar.jar
JarJar: out/target/common/obj/JAVA_LIBRARIES/conscrypt_intermediates/classes-jarjar.jar
Copying: out/target/common/obj/JAVA_LIBRARIES/conscrypt_intermediates/emma_out/lib/classes-jarjar.jar
Copying: out/target/common/obj/JAVA_LIBRARIES/conscrypt_intermediates/classes.jar
target Java: bouncycastle (out/target/common/obj/JAVA_LIBRARIES/bouncycastle_intermediates/classes)
注意：某些输入文件使用或覆盖了已过时的 API。
注意：要了解详细信息，请使用 -Xlint:deprecation 重新编译。
注意：某些输入文件使用了未经检查或不安全的操作。
注意：要了解详细信息，请使用 -Xlint:unchecked 重新编译。
JarJar: out/target/common/obj/JAVA_LIBRARIES/bouncycastle_intermediates/classes-jarjar.jar
Copying: out/target/common/obj/JAVA_LIBRARIES/bouncycastle_intermediates/emma_out/lib/classes-jarjar.jar
Copying: out/target/common/obj/JAVA_LIBRARIES/bouncycastle_intermediates/classes.jar
target Java: ext (out/target/common/obj/JAVA_LIBRARIES/ext_intermediates/classes)
```

图 2-13　编译过程的界面截图

整个编译过程也是非常漫长，需要读者耐心等待。

## 2.3.1　搭建编译环境

在编译 Android 源码之前，需要先进行环境搭建工作。下面以 Ubuntu 系统为例讲解搭建编译环境以及编译 Android 源码的方法，具体流程如下：

（1）安装 JDK，编译 Android 4.3 的源码需要 JDK 1.6，下载 jdk-6u22-linux-i586.bin 后进行安装，对应命令如下：

```
$ cd /usr
$ mkdir java
$ cd java
$ sudo cp jdk-6u22-linux-i586.bin 所在目录 ./
$ sudo chmod 755 jdk-6u22-linux-i586.bin
$ sudo sh jdk-6u22-linux-i586.bin
```

（2）设置 JDK 环境变量，将如下环境变量添加到主文件夹目录下的.bashrc 文件中，然后用 source 命令使其生效，加入的环境变量代码如下：

```
export JAVA_HOME=/usr/java/jdk1.6.0_23
export JRE_HOME=$JAVA_HOME/jre
export CLASSPATH=.:$JAVA_HOME/lib:$JRE_HOME/lib:$CLASSPATH
export PATH=$PATH:$JAVA_HOME/bin:$JAVA_HOME/bin/tools.jar:$JRE_HOME/bin
export ANDROID_JAVA_HOME=$JAVA_HOME
```

（3）安装需要的包，读者可以根据编译过程中的提示进行选择，可能需要的包的安装命令如下：

```
$ sudo apt-get install git-core bison zlib1g-dev flex libx12-dev gperf sudo aptitude install git-core gnupg flex
bison gperf libsdl-dev libesd0-dev libwxgtk2.6-dev build-essential zip curl libncurses5-dev zlib1g-dev
```

## 2.3.2 开始编译

当所依赖的包安装完成之后，就可以开始编译 Android 源码了，具体步骤如下。

（1）进行编译初始化工作，在终端中执行下面的命令：

```
source build/envsetup.sh
```

或：

```
. build/envsetup.sh
```

执行后将会输出：

```
source build/envsetup.sh
including device/asus/grouper/vendorsetup.sh
including device/asus/tilapia/vendorsetup.sh
including device/generic/armv7-a-neon/vendorsetup.sh
including device/generic/armv7-a/vendorsetup.sh
including device/generic/mips/vendorsetup.sh
including device/generic/x86/vendorsetup.sh
including device/samsung/maguro/vendorsetup.sh
including device/samsung/manta/vendorsetup.sh
including device/samsung/toroplus/vendorsetup.sh
including device/samsung/toro/vendorsetup.sh
including device/ti/panda/vendorsetup.sh
including sdk/bash_completion/adb.bash
```

（2）选择编译目标，命令如下：

```
lunch full-eng
```

执行后会输出如下所示的提示信息：

```
============================================
PLATFORM_VERSION_CODENAME=REL
PLATFORM_VERSION=4.3
TARGET_PRODUCT=full
TARGET_BUILD_VARIANT=eng
TARGET_BUILD_TYPE=release
TARGET_BUILD_APPS=
TARGET_ARCH=arm
TARGET_ARCH_VARIANT=armv7-a
HOST_ARCH=x86
HOST_OS=linux
HOST_OS_EXTRA=Linux-2.6.32-45-generic-x86_64-with-Ubuntu-10.04-lucid
HOST_BUILD_TYPE=release
BUILD_ID=JOP40C
OUT_DIR=out
============================================
```

（3）开始编译代码，在终端中执行下面的命令：

```
make -j4
```

其中“-j4”表示用 4 个线程进行编译。整个编译进度根据不同机器的配置而需要不同的时间。例如笔者电脑为 intel i5-2300 四核 2.8、4GB 内存，经过近 4 小时才编译完成。当出现下面的信息时表示编译完成：

```
target Java: ContactsTests (out/target/common/obj/APPS/ContactsTests_intermediates/classes)
target Dex: Contacts
```

```
Done!
Install: out/target/product/generic/system/app/Browser.odex
Install: out/target/product/generic/system/app/Browser.apk
Note: Some input files use or override a deprecated API.
Note: Recompile with -Xlint:deprecation for details.
Copying: out/target/common/obj/APPS/Contacts_intermediates/noproguard.classes.dex
target Package: Contacts (out/target/product/generic/obj/APPS/Contacts_intermediates/package.apk)
 'out/target/common/obj/APPS/Contacts_intermediates/classes.dex' as 'classes.dex'...
Processing target/product/generic/obj/APPS/Contacts_intermediates/package.apk
Done!
Install: out/target/product/generic/system/app/Contacts.odex
Install: out/target/product/generic/system/app/Contacts.apk
build/tools/generate-notice-files.py out/target/product/generic/obj/NOTICE.txt out/target/product/generic/obj/ NOTICE.
html "Notices for files contained in the filesystem images in this directory:" out/target/product/generic/obj/
NOTICE_FILES/src
Combining NOTICE files into HTML
Combining NOTICE files into text
Installed file list: out/target/product/generic/installed-files.txt
Target system fs image: out/target/product/generic/obj/PACKAGING/systemimage_intermediates/system.img
Running:   mkyaffs2image -f out/target/product/generic/system out/target/product/generic/obj/ PACKAGING/
systemimage_intermediates/system.img
Install system fs image: out/target/product/generic/system.img
DroidDoc took 5331 sec. to write docs to out/target/common/docs/doc-comment-check
```

## 2.3.3　在模拟器中运行

在模拟器中运行的步骤比较简单，只需在终端中执行下面的命令即可：

```
emulator
```

运行成功后的效果如图 2-14 所示。

图 2-14　在模拟器中的编译执行效果

## 2.3.4　常见的错误分析

虽然编译方法非常简单，但是作为初学者来说很容易出错，下面列出了其中常见的编译错误类型。

（1）缺少必要的软件

进入到 Android 目录下，使用 make 命令编译，可能会出现如下错误提示。

```
host C: libneo_cgi <= external/clearsilver/cgi/cgi.c
external/clearsilver/cgi/cgi.c:22:18: error: zlib.h: No such file or directory
```

上述错误是因为缺少 zlib1g-dev，需要使用 apt-get 命令从软件仓库中安装 zlib1g-dev，具体命令如下所示。

```
sudo apt-get install zlib1g-dev
```

同理需要安装下面的软件，否则也会出现上述类似的错误。

```
sudo apt-get install flex
sudo apt-get install bison
sudo apt-get install gperf
sudo apt-get install libsdl-dev
sudo apt-get install libesd0-dev
sudo apt-get install libncurses5-dev
sudo apt-get install libx12-dev
```

（2）没有安装 Java 环境 JDK

当安装所有上述软件后，运行 make 命令再次编译 Android 源码。如果在之前忘记安装 Java 环境 JDK，则此时会出现很多 Java 文件无法编译的错误，如果打开 Android 的源码，可以看到在目录 android/dalvik/libcore/dom/src/test/java/org/w3c/domts 中有很多 Java 源文件。

这充分说明在编译 Android 之前必须先安装 Java 环境 JDK，安装流程如下：

☑ 从 Oracle 官方网站下载 jdk-6u16-linux-i586.bin 文件，然后安装。

在 Ubuntu 8.04 中，/etc/profile 文件是全局的环境变量配置文件，它适用于所有的 shell。在登录 Linux 系统时应该先启动/etc/profile 文件，然后再启动用户目录下的~/.bash_profile、~/.bash_login 或~/.profile 文件中的其中一个，执行的顺序和上面的排序一样。如果~/.bash_profile 文件存在，则还会执行~/.bashrc 文件。在此只需要把 JDK 的目录放到/etc/profile 目录下即可。

```
JAVA_HOME=/usr/local/src/jdk1.6.0_16
PATH=$PATH:$JAVA_HOME/bin:/usr/local/src/android-sdk-linux_x86-1.1_r1/tools:~/bin
```

☑ 重新启动机器，输入 java –version 命令，输出下面的信息则表示配置成功。

```
ava version "1.6.0_16"
Java(TM) SE Runtime Environment (build 1.6.0_16-b01)
Java HotSpot(TM) Client VM (build 13.2-b01, mixed mode, sharing)
```

当成功编译 Android 源码后，在终端会输出如下提示。

```
Target system fs image: out/target/product/generic/obj/PACKAGING/systemimage_unopt_intermediates/ system.img
Install system fs image: out/target/product/generic/system.img
Target ram disk: out/target/product/generic/ramdisk.img
Target userdata fs image: out/target/product/generic/userdata.img
Installed file list: out/target/product/generic/installed-files.txt
root@dfsun2009-desktop:/bin/android#
```

### 2.3.5 实践演练——演示两种编译 Android 程序的方法

Android 编译环境本身比较复杂，并且不像普通的编译环境那样只有顶层目录下才有 Makefile 文件，而其他的每个 component 都使用统一标准的 Android.mk 文件。不过这并不是我们熟悉的 Makefile，而是经过 Android 自身编译系统的很多处理。所以说要真正理清其中的联系还比较复杂，不过这种方式的好处在于，编写一个新的 Android.mk 给 Android 增加一个新的 Component 会变得比较简单。为了使读者更加深入地理解在 Linux 环境下编译 Android 程序的方法，下面将分别演示两种编译 Android 程序的方法。

### 1. 编译 Native C（本地 C 程序）的 helloworld 模块

编译 Java 程序可以直接采用 Eclipse 的集成环境来完成，实现方法非常简单，在这里就不再重复了。接下来将主要针对 C/C++进行说明，通过一个例子来讲解在 Android 中增加一个 C 程序的 Hello World 的方法。

（1）在$(YOUR_ANDROID)/development 目录下创建一个名为 hello 的目录，并用$(YOUR_ NDROID)指向 Android 源代码所在的目录。

```
- # mkdir $(YOUR_ANDROID)/development/hello
```

（2）在目录$(YOUR_ANDROID)/development/hello/下编写一个名为 hello.c 的 C 语言文件，文件 hello.c 的实现代码如下：

```
#include <stdio.h>
int main()
{
    printf("Hello World!\n");//输出 Hello World
return 0;
}
```

（3）在目录$(YOUR_ANDROID)/development/hello/下编写 Android.mk 文件。这是 Android Makefile 的标准命名，不能更改。文件 Android.mk 的格式和内容可以参考其他已有的 Android.mk 文件的写法，针对 helloworld 程序的 Android.mk 文件内容如下：

```
LOCAL_PATH:= $(call my-dir)
include $(CLEAR_VARS)
LOCAL_SRC_FILES:= \
    hello.c
LOCAL_MODULE := helloworld
include $(BUILD_EXECUTABLE)
```

上述各个内容的具体说明如下。

- ☑ LOCAL_SRC_FILES：用来指定源文件。
- ☑ LOCAL_MODULE：指定要编译的模块的名字，在下一步编译时将会用到。
- ☑ include $(BUILD_EXECUTABLE)：表示要编译成一个可执行文件，如果想编译成动态库则可用 BUILD_SHARED_LIBRARY，这些具体用法可以在$(YOUR_ANDROID)/build/core/config.mk 中查到。

（4）回到 Android 源代码顶层目录进行编译。

```
# cd $(YOUR_ANDROID) && make helloworld
```

在此需要注意，make helloworld 中的目标名 helloworld 就是上面 Android.mk 文件中由 LOCAL_MODULE 指定的模块名。最终的编译结果如下：

```
target thumb C: helloworld <= development/hello/hello.c
target Executable: helloworld (out/target/product/generic/obj/EXECUTABLES/helloworld_intermediates/LINKED/
helloworld)
target Non-prelinked: helloworld (out/target/product/generic/symbols/system/bin/helloworld)
target Strip: helloworld (out/target/product/generic/obj/EXECUTABLES/helloworld_intermediates/helloworld)
Install: out/target/product/generic/system/bin/helloworld
```

（5）如果和上述编译结果相同，则编译后的可执行文件存放在目录 out/target/product/generic/system/bin/helloworld 中。

这样通过 adb push 将它传送到模拟器上，再通过 adb shell 登录到模拟器终端后就可以执行了。

### 2. 手工编译 C 模块

在前面讲解了通过标准的 Android.mk 文件来编译 C 模块的具体流程，其实我们可以直接运用 gcc 命令行来编译 C 程序，这样可以更好地了解 Android 编译环境的细节。具体流程如下所示。

（1）在 Android 编译环境中，提供了 showcommands 选项来显示编译命令行，可以通过打开这个选项来查看一些编译时的细节。

（2）在具体操作之前需要使用如下命令把前面的 helloworld 模块清除。

```
# make clean-helloworld
```

上面的 make clean-$(LOCAL_MODULE)命令是 Android 编译环境提供的 make clean 的方式。

（3）使用 showcommands 选项重新编译 helloworld，具体命令如下：

```
# make helloworld showcommands
build/core/product_config.mk:229: WARNING: adding test OTA key
target thumb C: helloworld <= development/hello/hello.c
prebuilt/linux-x86/toolchain/arm-eabi-4.3.1/bin/arm-eabi-gcc -I system/core/include -I hardware/libhardware/
include -I hardware/ril/include -I dalvik/libnativehelper/include -I frameworks/base/include -I external/
skia/include -I out/target/product/generic/obj/include -I bionic/libc/arch-arm/include -I bionic/libc/include
-I bionic/libstdc++/include -I bionic/libc/kernel/common -I bionic/libc/kernel/arch-arm -I bionic/libm/ include -I
bionic/libm/include/arch/arm -I bionic/libthread_db/include -I development/hello -I out/target/product/generic/obj/
EXECUTABLES/helloworld_intermediates -c -fno-exceptions -Wno-multichar -march=armv5te -mtune=xscale
-msoft-float -fpic -mthumb-interwork -ffunction-sections -funwind-tables -fstack-protector -D__ARM_ARCH_5__
-D__ARM_ARCH_5T__ -D__ARM_ARCH_5E__ -D__ARM_ARCH_5TE__ -include system/core/include/arch/
linux-arm/AndroidConfig.h -DANDROID -fmessage-length=0 -W -Wall -Wno-unused -DSK_RELEASE
-DNDEBUG -O2 -g -Wstrict-aliasing=2 -finline-functions -fno-inline-functions-called-once -fgcse-after-reload
-frerun-cse-after-loop -frename-registers -DNDEBUG -UDEBUG -mthumb -Os -fomit- frame-pointer
-fno-strict-aliasing -finline-limit=64 -MD -o out/target/product/generic/obj/EXECUTABLES/
helloworld_intermediates/hello.o development/hello/hello.c

target Executable: helloworld (out/target/product/generic/obj/EXECUTABLES/helloworld_intermediates/LINKED/
helloworld)

prebuilt/linux-x86/toolchain/arm-eabi-4.3.1/bin/arm-eabi-g++ -nostdlib -Bdynamic -Wl,-T,build/core/armelf.x
-Wl,-dynamic-linker,/system/bin/linker -Wl,--gc-sections -Wl,-z,nocopyreloc -o out/target/product/generic/obj/
EXECUTABLES/helloworld_intermediates/LINKED/helloworld -Lout/target/product/generic/obj/lib -Wl,-rpath-link=
out/ target/product/generic/obj/lib -lc -lstdc++ -lm out/target/product/generic/obj/lib/crtbegin_dynamic.o out/
target/ product/generic/obj/EXECUTABLES/helloworld_intermediates/hello.o -Wl,--no-undefined prebuilt/linux-x86/
toolchain/arm-eabi-4.3.1/bin/../lib/gcc/arm-eabi/4.3.1/interwork/libgcc.a out/target/product/generic/obj/lib/crtend_
android.o

target Non-prelinked: helloworld
(out/target/product/generic/symbols/system/bin/helloworld)

out/host/linux-x86/bin/acp -fpt out/target/product/generic/obj/EXECUTABLES/helloworld_intermediates/LINKED/
helloworld out/target/product/generic/symbols/system/bin/helloworld

target Strip: helloworld (out/target/product/generic/obj/EXECUTABLES/helloworld_intermediates/helloworld)

out/host/linux-x86/bin/soslim  --strip  --shady  --quiet  out/target/product/generic/symbols/system/bin/helloworld
```

```
--outfile out/target/product/generic/obj/EXECUTABLES/helloworld_intermediates/helloworld

Install: out/target/product/generic/system/bin/helloworld

out/host/linux-x86/bin/acp -fpt out/target/product/generic/obj/EXECUTABLES/helloworld_intermediates/helloworld out/target/ product/generic/system/bin/helloworld
```

从上述命令行可以看到，Android 编译环境所用的交叉编译工具链如下：

```
prebuilt/linux-x86/toolchain/arm-eabi-4.3.1/bin/arm-eabi-gcc
```

其中参数-I 和-L 分别指定了所用的 C 库头文件和动态库文件路径分别是 bionic/libc/include 和 out/target/product/generic/obj/lib，其他还包括很多编译选项以及-D 所定义的预编译宏。

（4）利用上面的编译命令来手工编译 helloworld 程序，首先手工删除上次编译得到的 helloworld 程序。

```
# rm out/target/product/generic/obj/EXECUTABLES/helloworld_intermediates/hello.o
# rm out/target/product/generic/system/bin/helloworld
```

然后再用 gcc 编译以生成目标文件。

```
# prebuilt/linux-x86/toolchain/arm-eabi-4.3.1/bin/arm-eabi-gcc -I bionic/libc/arch-arm/include -I bionic/libc/include -I bionic/libc/kernel/common -I bionic/libc/kernel/arch-arm -c -fno-exceptions -Wno-multichar -march=armv5te -mtune=xscale -msoft-float -fpic -mthumb-interwork -ffunction-sections -funwind-tables -fstack-protector -D__ARM_ARCH_5__ -D__ARM_ARCH_5T__ -D__ARM_ARCH_5E__ -D__ARM_ARCH_5TE__ -include system/core/include/arch/linux-arm/AndroidConfig.h -DANDROID -fmessage-length=0 -W -Wall -Wno-unused -DSK_RELEASE -DNDEBUG -O2 -g -Wstrict-aliasing=2 -finline-functions -fno-inline-functions-called-once -fgcse-after-reload -frerun-cse-after-loop -frename-registers -DNDEBUG -UDEBUG -mthumb -Os -fomit-frame-pointer -fno-strict-aliasing -finline-limit=64 -MD -o out/target/product/generic/ obj/EXECUTABLES/helloworld_intermediates/hello.o development/hello/hello.c
```

如果此时与 Android.mk 编译参数进行比较，会发现上面主要减少了不必要的参数“-I”。

（5）生成可执行文件。

```
# prebuilt/linux-x86/toolchain/arm-eabi-4.3.1/bin/arm-eabi-gcc -nostdlib -Bdynamic -Wl,-T,build/core/armelf.x -Wl,-dynamic-linker,/system/bin/linker -Wl,--gc-sections -Wl,-z,nocopyreloc -o out/target/product/ generic/obj/EXECUTABLES/helloworld_intermediates/LINKED/helloworld -Lout/target/product/generic/obj/lib -Wl,-rpath-link=out/ target/product/generic/obj/lib -lc -lm out/target/product/generic/obj/EXECUTABLES/helloworld_ intermediates/hello.o out/target/product/generic/obj/lib/crtbegin_dynamic.o -Wl,--no-undefined./prebuilt/linux-x86/ toolchain/arm-eabi-4.3.1/bin/../lib/gcc/arm-eabi/4.3.1/interwork/libgcc.a out/target/product/generic/obj/lib/crtend_android.o
```

在此需要特别注意的是参数-Wl,-dynamic-linker,/system/bin/linker，它指定了 Android 专用的动态链接器是/system/bin/linker，而不是平常使用的 ld.so。

（6）使用命令 file 和 readelf 来查看生成的可执行程序。

```
# file out/target/product/generic/obj/EXECUTABLES/helloworld_intermediates/LINKED/helloworld
out/target/product/generic/obj/EXECUTABLES/helloworld_intermediates/LINKED/helloworld: ELF 32-bit LSB executable, ARM, version 1 (SYSV), dynamically linked (uses shared libs), not stripped
# readelf -d out/target/product/generic/obj/EXECUTABLES/helloworld_intermediates/LINKED/helloworld |grep NEEDED
 0x00000001 (NEEDED)                     Shared library: [libc.so]
 0x00000001 (NEEDED)                     Shared library: [libm.so]
```

这就是 ARM 格式的动态链接可执行文件，在运行时需要 libc.so 和 libm.so。当提示 not stripped 时表示它还没被 STRIP（剥离）。嵌入式系统中为节省空间通常将编译完成的可执行文件或动态库进行剥离，即去掉其中多余的符号表信息。在前面 make helloworld showcommands 命令的最后也可以看到，Android 编译环境中使用了 out/host/linux-x86/bin/soslim 工具进行 STRIP。

# 2.4 编译 Android Kernel

**知识点讲解：光盘:视频\知识点\第 2 章\编译 Android Kernel.avi**

编译 Android Kernel 代码就是编译 Android 内核代码，在进行具体编译工作之前需要先了解在 Android 开源系统中包含的如下 3 部分代码。

☑ 仿真器公共代码：对应的工程名是 kernel/common.get。

☑ MSM 平台的内核代码：对应的工程名是 kernel/msm.get。

☑ OMAP 平台的内核代码：对应的工程名是 kernel/omap.get。

本节将详细讲解编译上述 Android Kernel 的基本知识。

## 2.4.1 获取 Goldfish 内核代码

Goldfish 是一种虚拟的 ARM 处理器，通常在 Android 的仿真环境中使用。在 Linux 的内核中，Goldfish 作为 ARM 体系结构的一种“机器”。在 Android 的发展过程中，Goldfish 内核的版本也从 Linux 2.6.25 升级到了 Linux 3.4，此处理器的 Linux 内核和标准的 Linux 内核有以下 3 个方面的差别。

☑ Goldfish 机器的移植。

☑ Goldfish 一些虚拟设备的驱动程序。

☑ Android 中特有的驱动程序和组件。

Goldfish 处理器有两个版本，分别是 ARMv5 和 ARMv7，在一般情况下，只需使用 ARMv5 版本即可。在 Android 开源工程的代码仓库中，使用 git 工具得到 Goldfish 内核代码的命令如下：

```
$ git clone git://android.git.kernel.org/kernel/common.git
```

在其 Linux 源代码的根目录中，配置和编译 Goldfish 内核的过程如下：

```
$make ARCH=arm goldfish_defconfig .config
$make ARCH=arm CROSS_COMPILE={path}/arm-none-linux-gnueabi-
```

其中，CROSS_COMPILE 的 path 值用于指定交叉编译工具的路径。

编译结果如下：

```
LD vmlinux
SYSMAP system.map
SYSMAP .tmp_system.map
OBJCOPY arch/arm/boot/Image
Kernel: arch/arm/boot/Image is ready
AS arch/arm/boot/compressed/head.o
GZIP arch/arm/boot/compressed/piggy.gz
AS arch/arm/boot/compressed/piggy.o
CC arch/arm/boot/compressed/misc.o
LD arch/arm/boot/compressed/vmlinux
  OBJCONPY arch/arm/boot/zImage
  Kernel: arch/arm/boot/zImage is ready
```

☑ vmlinux：是 Linux 进行编译和连接之后生成的 Elf 格式的文件。

☑ Image：是未经过压缩的二进制文件。

☑　piggy：是一个解压缩程序。

☑　zImage：是解压缩程序和压缩内核的组合。

在 Android 源代码的根目录中，vmlinux 和 zImage 分别对应 Android 代码 prebuilt 中的预编译的 arm 内核。使用 zImage 可以替换 prebuilt 中的 prebuilt/android-arm/目录下的 goldfish_defconfig，此文件的主要片段如下：

```
CONFIG_ARM=y
#
# System Type
#
CONFIG_ARCH_GOLDFISH=y
#
# Goldfish options
#
CONFIG_MACH_GOLDFISH=y
# CONFIG_MACH_GOLDFISH_ARMV7 is not set
```

因为 Goldfish 是 ARM 处理器，所以 CONFIG_ARM 宏需要被使能，CONFIG_ARCH_GOLDFISH 和 CONFIG_MACH_GOLDFISH 宏是 Goldfish 处理器这类机器使用的配置宏。

在 goldfish_defconfig 中，与 Android 系统相关的宏如下：

```
#
# android
#
CONFIG_ANDROID=y
CONFIG_ANDROID_BUNDER_IPC=y #binder ipc 驱动程序
CONFIG_ANDROID_LOGGER=y #log 记录器驱动程序
# CONFIG_ANDROID_RAM_CONSOLE is not set
CONFIG_ANDROID_TIMED_OUTPUT=y #定时输出驱动程序框架
CONFIG_ANDROID_LOW_MEMORY_KILLER=y
CONFIG_ANDROID_PMEM=y #物理内存驱动程序
CONFIG_ASHMEM=y #匿名共享内存驱动程序
CONFIG_RTC_INTF_ALARM=y
CONFIG_HAS_WAKELOCK=y 电源管理相关的部分 wakelock 和 earlysuspend
CONFIG_HAS_EARLYSUSPEND=y
CONFIG_WAKELOCK=y
CONFIG_WAKELOCK_STAT=y
CONFIG_USER_WAKELOCK=y
CONFIG_EARLYSUSPEND=y
```

goldfish_defconfig 配置文件中，另外有一个宏是处理器虚拟设备的“驱动程序”，其内容如下所示。

```
CONFIG_MTD_GOLDFISH_NAND=y
CONFIG_KEYBOARD_GOLDFISH_EVENTS=y
CONFIG_GOLDFISH_TTY=y
CONFIG_BATTERY_GOLDFISH=y
CONFIG_FB_GOLDFISH=y
CONFIG_MMC_GOLDFISH=y
CONFIG_RTC_DRV_GOLDFISH=y
```

在 Goldfish 处理器的各个配置选项中，体系结构和 Goldfish 的虚拟驱动程序基于标准 Linux 的内容的驱动程序框架，但是这些设备在不同的硬件平台的移植方式不同；Android 专用的驱动程序是 Android 中特有的内容，非 Linux 标准，但是和硬件平台无关。

和原Linux内核相比，Android内核增加了Android的相关驱动，对应的目录是kernel/drivers/android。主要分为以下几类驱动。

☑ Android IPC 系统：Binder(binder.c)。
☑ Android 日志系统：Logger(logger.c)。
☑ Android 电源管理：Power(power.c)。
☑ Android 闹钟管理：Alarm(alarm.c)。
☑ Android 内存控制台：Ram_console(ram_console.c)。
☑ Android 时钟控制的 gpio：Timed_gpio(timed_gpio.c)。

对于本书讲解的驱动程序开发来说，我们比较关心的是 Goldfish 平台下相关的驱动文件，具体说明如下。

（1）字符输出设备：kernel/drivers/char/goldfish_tty.c。
（2）图像显示设备（Frame Buffer）：kernel/drivers/video/goldfishfb.c。
（3）键盘输入设备文件：kernel/drivers/input/keyboard/goldfish_events.c。
（4）RTC 设备（Real Time Clock）文件：kernel/drivers/rtc/rtc-goldfish.c。
（5）USB Device 设备文件：kernel/drivers/usb/gadget/android_adb.c。
（6）SD 卡设备文件：kernel/drivers/mmc/host/goldfish.c。
（7）FLASH 设备文件：kernel/drivers/mtd/devices/goldfish_nand.c、kernel/drivers/mtd/devices/goldfish_nand_reg.h。
（8）LED 设备文件：kernel/drivers/leds/ledtrig-sleep.c。
（9）电源设备：kernel/drivers/power/goldfish_battery.c。
（10）音频设备：kernel/arch/arm/mach-goldfish/audio.c。
（11）电源管理：kernel/arch/arm/mach-goldfish/pm.c。
（12）时钟管理：kernel/arch/arm/mach-goldfish/timer.c。

## 2.4.2 获取 MSM 内核代码

在当前市面中，谷歌的手机产品 G1 是基于 MSM 内核的，MSM 是高通公司的应用处理器，在 Android 代码库中公开了对应的 MSM 的源代码。在 Android 开源工程的代码仓库中，使用 git 工具得到 MSM 内核代码的命令如下：

```
$ git clone git://android.git.kernel.org/kernel/msm.git
```

## 2.4.3 获取 OMAP 内核代码

OMAP 是德州仪器公司的应用处理器，Android 使用的是 OMAP3 系列的处理器。在 Android 代码库中公开了对应的 MSM 的源代码，使用 git 工具得到 MSM 内核代码的命令如下：

```
$ git clone git://android.git.kernel.org/kernel/omap.git
```

## 2.4.4 编译 Android 的 Linux 内核

了解了上述 3 类 Android 内核后，下面开始讲解编译 Android 内核的方法。在此假设以 Ubuntu 8.10

为例，完整编译 Android 内核的流程如下。

（1）构建交叉编译环境

Android 的默认硬件处理器是 ARM，因此我们需要在自己的机器上构建交叉编译环境。交叉编译器 GNU Toolchain for ARM Processors 的下载地址是 http://www.codesourcery.com/gnu_toolchains/arm/download.html。

单击 GNU/Linux 对应的链接，再单击 Download Sourcery CodeBench Lite 5.1 2012.02-117 链接后直接下载，如图 2-15 所示。

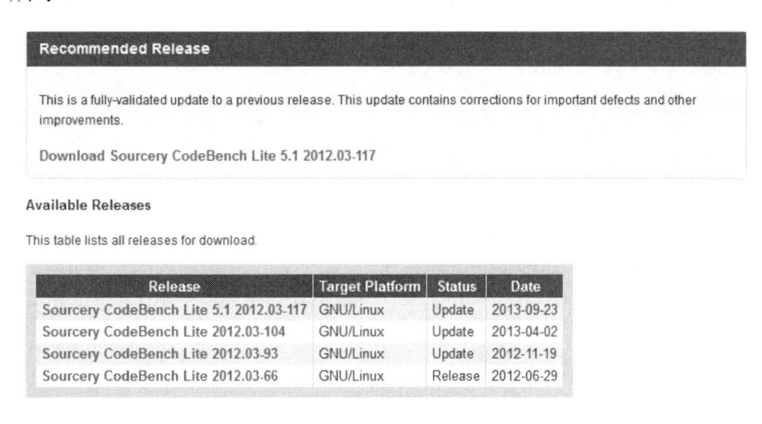


图 2-15　下载交叉编译器

把 arm-2008q2-72-arm-none-linux-gnueabi-i686-pc-linux-gnu.tar.bz2 解压到一个目录下，例如~/programes/，并加入 PATH 环境变量：

```
vim ~/.bashrc
```

然后添加：

```
ARM_TOOLCHIAN=~/programes/arm-2008q3/bin/
export PATH=${PATH}:${ARM_TOOLCHIAN};
```

保存后并执行：source ~/.bashrc。

（2）获取内核源码

源码地址是 http://code.google.com/p/android/downloads/list。

选择的内核版本要与选用的模拟器版本尽量一致。下载并解压后得到 kernel.git 文件夹：

```
tar -xvf ~/download/linux-3.2.5-android-4.3_r1.tar.gz
```

（3）获取内核编译配置信息文件

编译内核时需要使用 configure，通常 configure 有很多选项，我们往往不知道需要哪些选项。在运行 Android 模拟器时，有一个文件/proc/config.gz，这是当前内核的配置信息文件，把 config.gz 获取并解压到 kernel.git/下，然后改名为.config。命令如下：

```
cd kernel.git/
emulator &
adb pull /proc/config.gz
gunzip config.gz
mv config .config
```

（4）修改 Makefile

修改第 195 行的代码：

```
CROSS_COMPILE = arm-none-linux-gnueabi-
```

将 CROSS_COMPILE 值改为 arm-none-linux-gnueabi-，这是我们安装的交叉编译工具链的前缀，修改此处意在告诉 make 在编译时要使用该工具链。然后注释掉第 562 和 563 行的如下代码：

```
#LDFLAGS_BUILD_ID = $(patsubst -Wl$(comma)%,%,/
# $(call ld-option, -Wl$(comma)--build-id,))
```

必须将上述代码中的 build-id 值注释掉，因为目前版本的 Android 内核不支持该选项。

（5）编译

使用 make 进行编译，并同时生成 zImage。

```
LD        arch/arm/boot/compressed/vmlinux
OBJCOPY arch/arm/boot/zImage
Kernel: arch/arm/boot/zImage is ready
```

这样生成 zImage 大小为 1.23MB，android-sdk-linux_x86-4.3_r1/tools/lib/images/kernel-qemu 是 1.24MB。

（6）使用模拟器加载内核测试

其命令如下：

```
cd android/out/cupcake/out/target/product/generic
emulator -image system.img -data userdata.img -ramdisk ramdisk.img -kernel ~/project/android/kernel.git/arch/
arm/boot/zImage &
```

至此，模拟器加载成功。

## 2.5 搭建 Android 应用开发环境

**知识点讲解：光盘:视频\知识点\第 2 章\搭建 Android 应用开发环境.avi**

Android SDK 是开发 Android 应用程序所必须具备的工具，在搭建之前需要先确定基于 Android 应用软件所需要开发环境的要求，具体如表 2-1 所示。

表 2-1 开发系统所需求参数

| 项 目 | 版 本 要 求 | 说 明 | 备 注 |
|---|---|---|---|
| 操作系统 | Windows XP 以上或 Vista Mac OS X 10.4.8+Linux Ubuntu Drapper 以上 | 根据自己的电脑自行选择 | 选择自己最熟悉的操作系统 |
| 软件开发包 | Android SDK | 选择最新版本的 SDK | 截止到目前，最新手机版本是 5.0 |
| IDE | Eclipse IDE+ADT | Eclipse3.3(Europa)，3.4(Ganymede)ADT(Android Development Tools)开发插件 | 选择 for Java Developer |
| 其他 | JDK Apache Ant | Java SE Development Kit 5 或 6Linux 和 Mac 上使用 Apache Ant 1.6.5+，Windows 上使用 1.7+版本 | （单独的 JRE 是不可以的，必须要有 JDK），不兼容 Gnu Java 编译器（gcj） |

Android 工具是由多个开发包组成的，具体说明如下。

☑ JDK：可以到网址 http://java.sun.com/javase/downloads/index.jsp 处下载。

☑ Eclipse（Europa）：可以到网址 http://www.eclipse.org/downloads/下载 Eclipse IDE for Java Developers。

☑ Android SDK：可以到网址 http://developer.android.com 下载。

☑ 还有对应的开发插件。

## 2.5.1　安装 JDK

JDK（Java Development Kit）是整个 Java 的核心，包括了 Java 运行环境、Java 工具和 Java 基础的类库。JDK 是学好 Java 的第一步，是开发和运行 Java 环境的基础，当用户要对 Java 程序进行编译时，必须先获得对应操作系统的 JDK，否则将无法编译 Java 程序。在安装 JDK 之前需要先获得 JDK，获得 JDK 的操作流程如下：

（1）登录 Oracle 官方网站，网址为 http://developers.sun.com/downloads/，如图 2-16 所示。

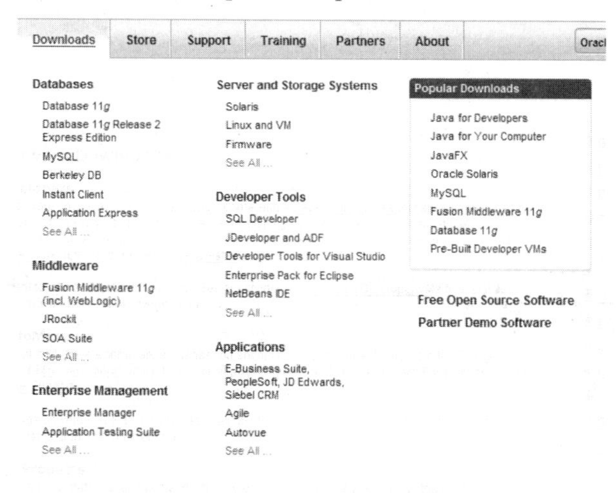


图 2-16　Oracle 官方下载页面

（2）在图 2-16 中可以看到有很多版本，在此选择当前最新的版本 Java 7，下载页面如图 2-17 所示。

（3）在图 2-17 中单击 JDK 下方的 Download 按钮，在弹出的新界面中选择将要下载的 JDK，笔者在此选择的是 Windows x86 版本，如图 2-18 所示。

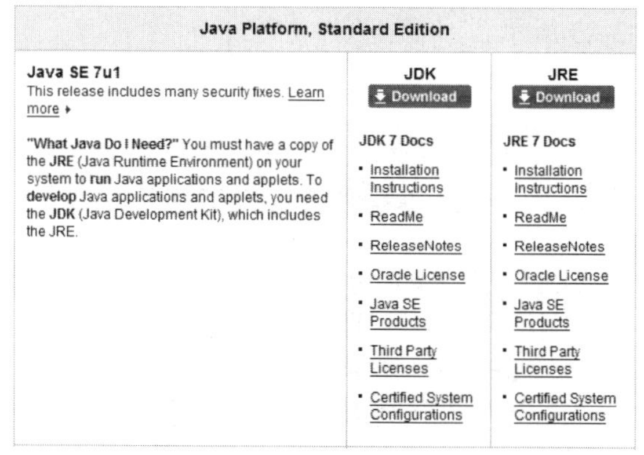


图 2-17　JDK 下载页面

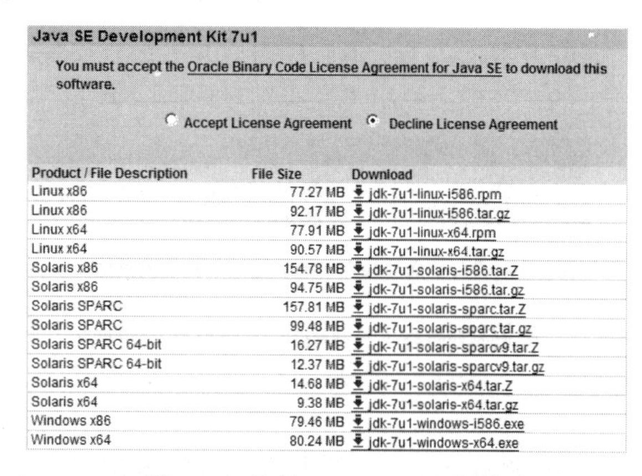


图 2-18　选择 Windows x86 版本

（4）下载完成后双击下载的.exe 文件开始进行安装，将弹出“安装向导”对话框，在此单击“下一步”按钮，如图 2-19 所示。

（5）弹出“安装路径”对话框，在此选择文件的安装路径，如图 2-20 所示。

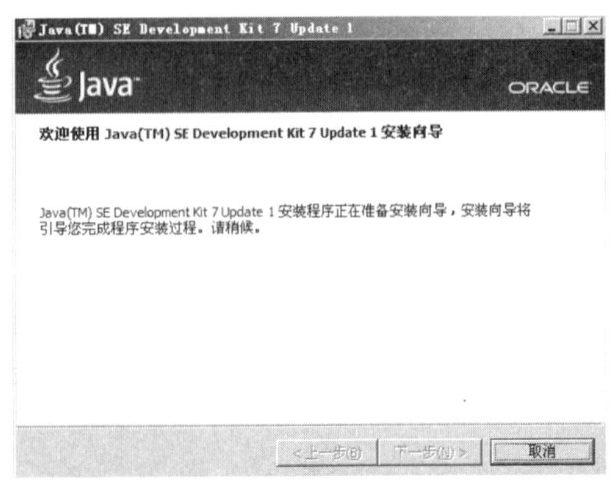


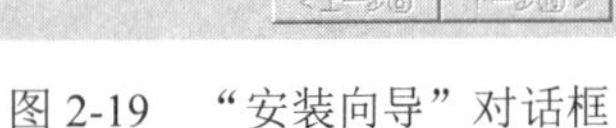
图 2-19　“安装向导”对话框

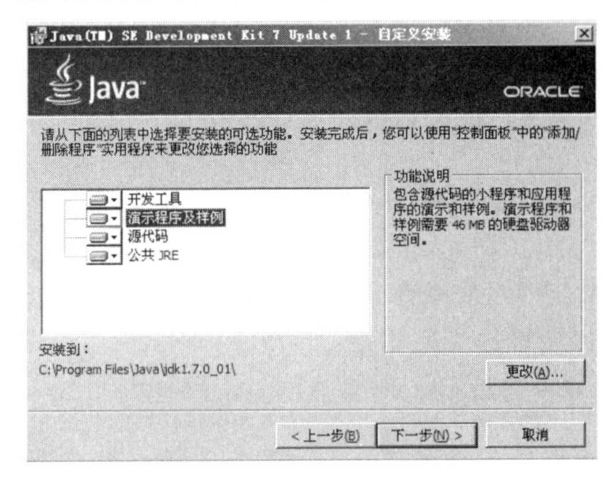


图 2-20　“安装路径”对话框

（6）在此设置安装路径是“E:\jdk1.7.0_01\”，然后单击“下一步”按钮开始在安装路径解压缩下载的文件，如图 2-21 所示。

（7）完成后弹出“目标文件夹”对话框，在此选择要安装的位置，如图 2-22 所示。

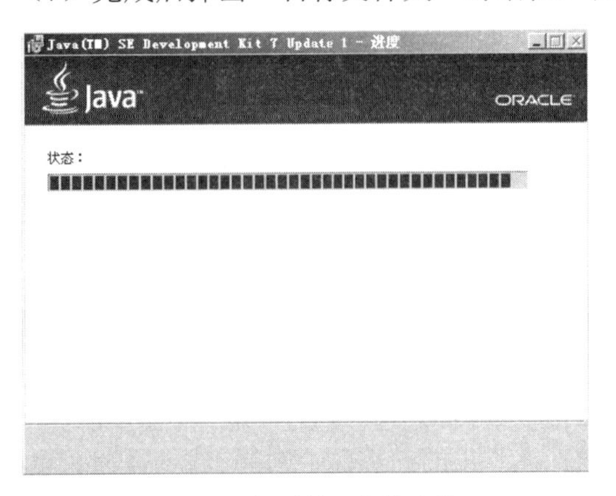


图 2-21　解压缩下载的文件

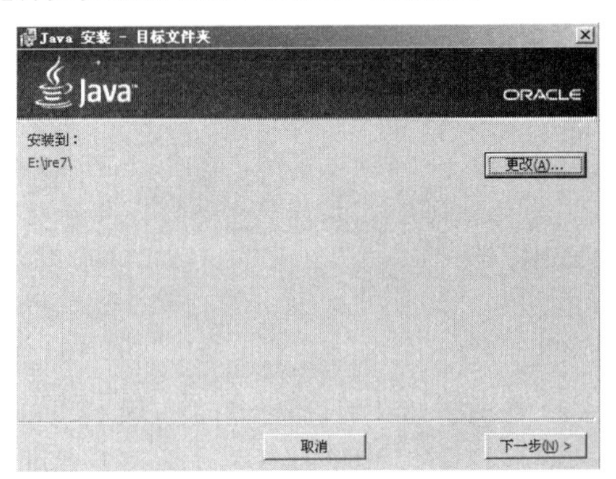


图 2-22　“目标文件夹”对话框

（8）单击“下一步”按钮后开始正式安装，如图 2-23 所示。

（9）完成后弹出“完成”对话框，单击“完成”按钮后完成整个安装过程，如图 2-24 所示。

图 2-23　继续安装

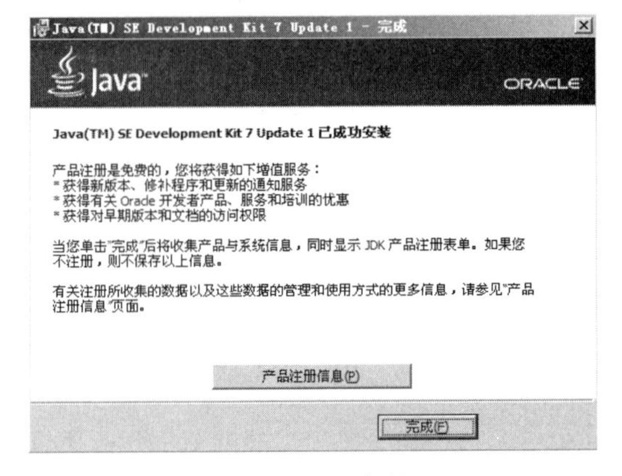


图 2-24　完成安装

**注意：** 完成安装后可以检测是否安装成功，方法是选择“开始”|“运行”命令，在弹出的对话框中输入“cmd”并按 Enter 键，在打开的 CMD 窗口中输入“java–version”，如果显示如图 2-25 所示的提示信息，则说明安装成功。

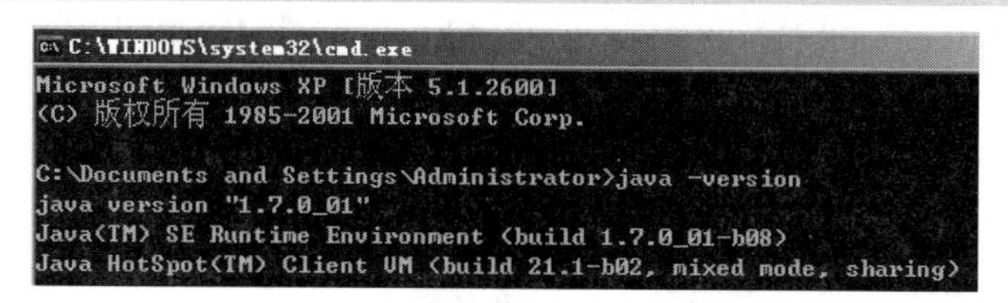


图 2-25　CMD 窗口

如果检测到没有安装成功，需要将其目录的绝对路径添加到系统的 PATH 中。具体做法如下：

（1）右击“我的电脑”图标，并选择“属性”命令，在弹出的对话框中选择“高级”选项卡，单击“环境变量”按钮，然后单击“系统变量”栏中的“新建”按钮，在弹出对话框的“变量名”文本框中输入“JAVA_HOME”，在“变量值”文本框中输入刚才的目录，如设置为“C:\Program Files\Java\jdk1.7.0_01”，如图 2-26 所示。

（2）再次新建一个变量名为 classpath，其变量值如下所示。

```
.;%JAVA_HOME%/lib/rt.jar;%JAVA_HOME%/lib/tools.jar
```

单击“确定”按钮找到 PATH 的变量，双击或单击编辑，在变量值最前面添加如下值。

```
%JAVA_HOME%/bin;
```

具体如图 2-27 所示。

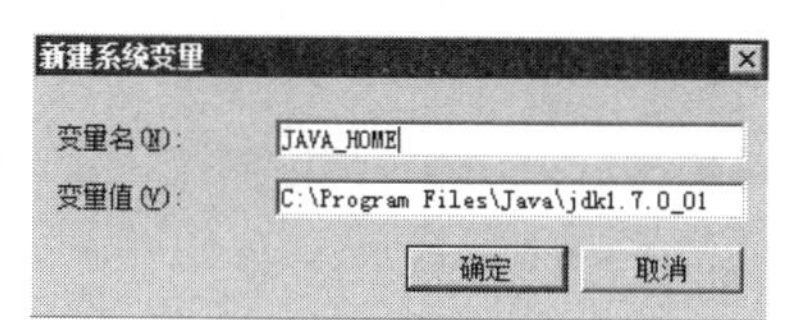


图 2-26　设置系统变量

图 2-27　设置系统变量

（3）再选择“开始”|“运行”命令，在弹出的对话框中输入“cmd”并按 Enter 键，在打开的 CMD 窗口中输入“java–version”，如果显示如图 2-28 所示的提示信息，则说明安装成功。

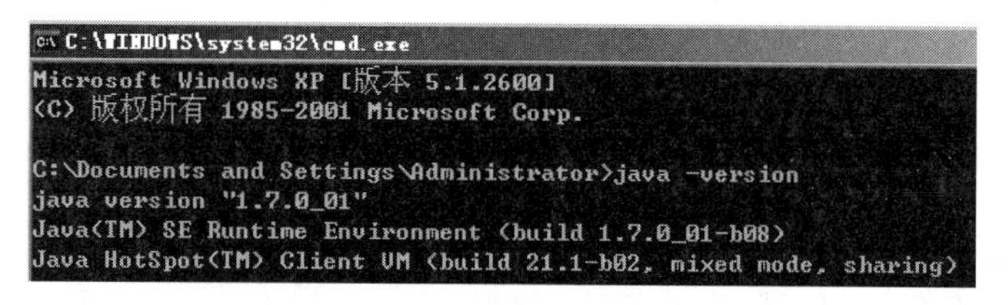


图 2-28　CMD 窗口

**注意：** 上述变量设置中，是按照笔者本人的安装路径设置的，笔者安装的 JDK 的路径是 C:\Program Files\Java\jdk1.7.0_01。

## 2.5.2　获取并安装 Eclipse 和 Android SDK

在安装好 JDK 后，接下来需要安装 Eclipse 和 Android SDK。Eclipse 是进行 Android 应用开发的一

个集成工具，而 Android SDK 是开发 Android 应用程序锁必须具备的框架。在 Android 官方公布的最新版本中，已经将 Eclipse 和 Android SDK 这两个工具进行了集成，一次下载即可同时获得这两个工具。

获取并安装 Eclipse 和 Android SDK 的具体步骤如下：

（1）登录 Android 的官方网站 http://developer.android.com/index.html，如图 2-29 所示。

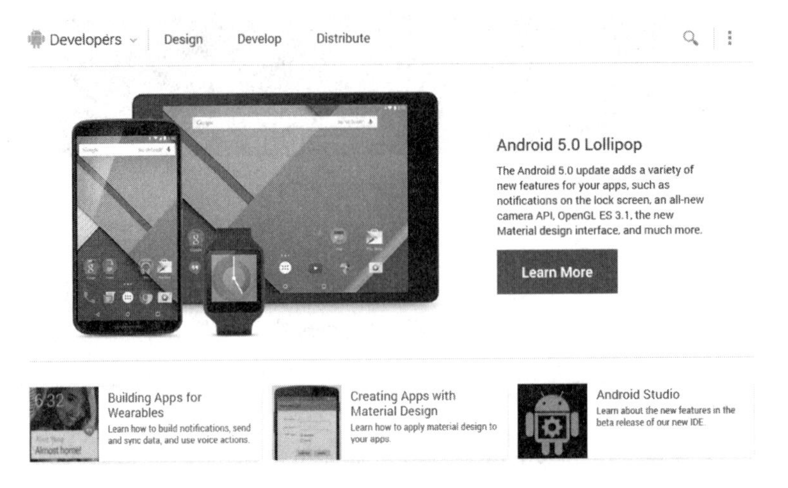


图 2-29　Android 的官方网站

（2）单击中部的 Get the SDK 链接，如图 2-30 所示。

图 2-30　单击 Get the SDK 链接

（3）在弹出的新页面中单击 Download the SDK 按钮，如图 2-31 所示。

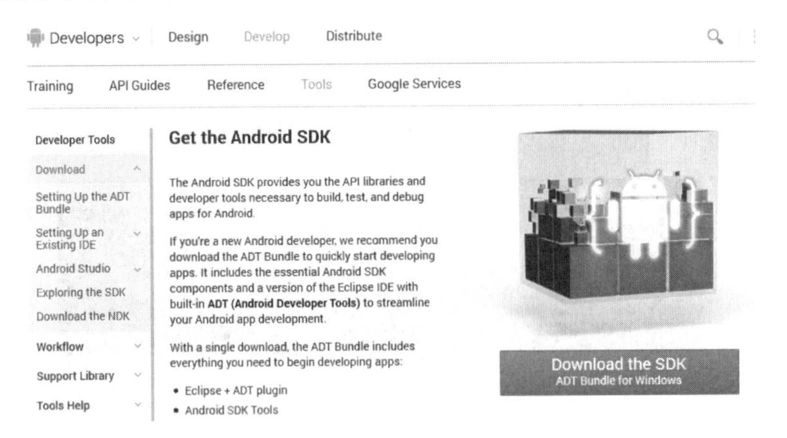


图 2-31　单击 Download the SDK 按钮

（4）在弹出的 Get the Android SDK 界面中选中 I have read and agree with the above terms and conditions 复选框，然后在下面的单选按钮中选择系统的位数。例如笔者的机器是 32 位的，所以选中 32-bit 单选按钮，如图 2-32 所示。

（5）单击图 2-32 中的 Download the SDK ADT Bundle for Windows 按钮后开始下载工作，下载的目标文件是一个压缩包，如图 2-33 所示。

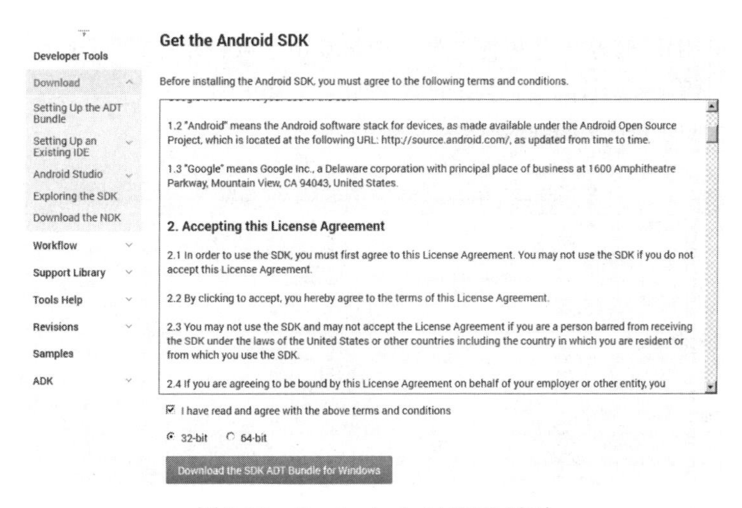


图 2-32　Get the Android SDK 界面

（6）将下载得到的压缩包进行解压，解压后的目录结构如图 2-34 所示。

图 2-33　开始下载目标文件压缩包　　　　图 2-34　解压后的目录结构

由此可见，Android 官方已经将 Eclipse 和 Android SDK 实现了集成。双击 eclipse 目录中的 eclipse.exe 可以打开 Eclipse，界面效果如图 2-35 所示。

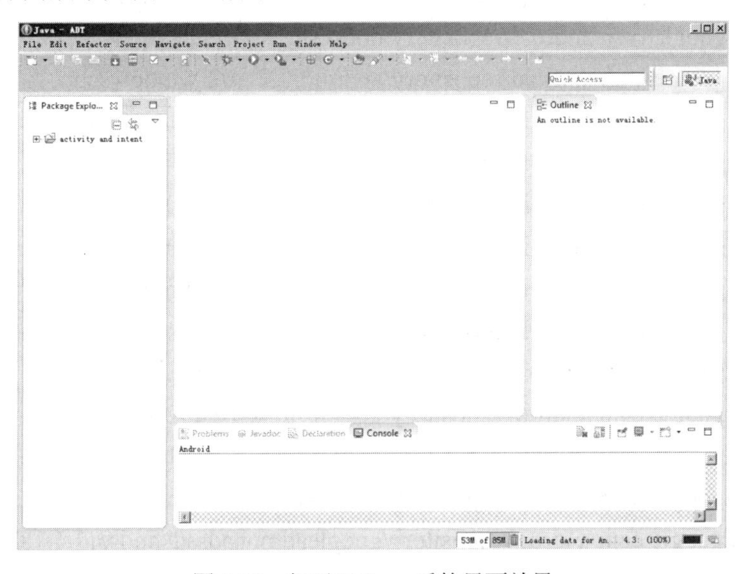


图 2-35　打开 Eclipse 后的界面效果

（7）打开 Android SDK 的方法有两种，第一种是双击下载目录中的 SDK Manager.exe 文件，第二种是在 Eclipse 工具栏中单击图标。打开后的效果如图 2-36 所示。

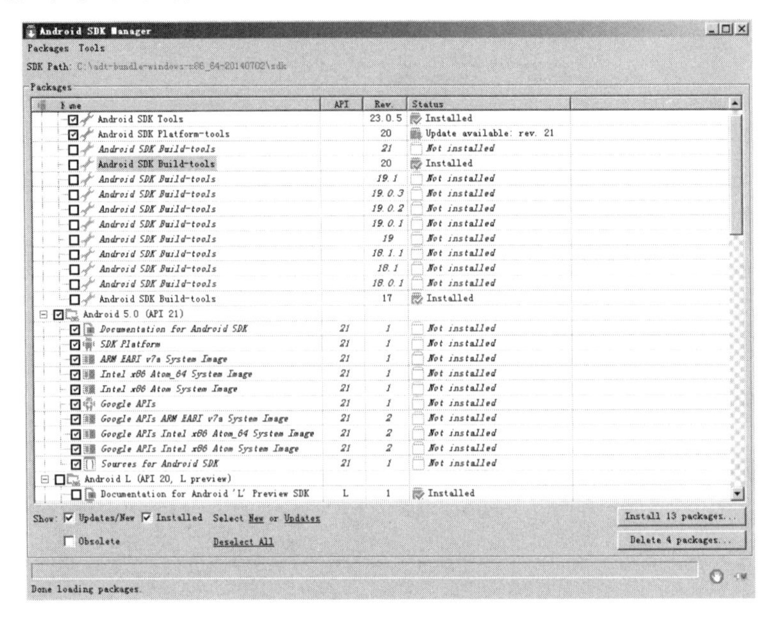


图 2-36　打开 Android SDK 后的界面效果

## 2.5.3　快速安装 SDK

通过 Android SDK Manager 在线安装的速度非常慢，而且有时容易挂掉。其实我们可以先从网络中寻找到 SDK 资源，用迅雷等下载工具下载后，将其放到指定目录后就可以完成安装。具体方法是先下载 android-sdk-windows，其可以更新，然后在 android-sdk-windows 下双击 setup.exe，在更新的过程中会发现安装 Android SDK 的速度是 1Kb/s，此时打开迅雷，分别输入下面的地址：

- ☑ https://dl-ssl.google.com/android/repository/platform-tools_r05-windows.zip
- ☑ https://dl-ssl.google.com/android/repository/docs-3.1_r02-linux.zip
- ☑ https://dl-ssl.google.com/android/repository/android-2.2_r02-windows.zip
- ☑ https://dl-ssl.google.com/android/repository/android-2.3.3_r02-linux.zip
- ☑ https://dl-ssl.google.com/android/repository/android-2.1_r02-windows.zip
- ☑ https://dl-ssl.google.com/android/repository/samples-2.3.3_r02-linux.zip
- ☑ https://dl-ssl.google.com/android/repository/samples-2.2_r02-linux.zip
- ☑ https://dl-ssl.google.com/android/repository/samples-2.1_r02-linux.zip
- ☑ https://dl-ssl.google.com/android/repository/compatibility_r02.zip
- ☑ https://dl-ssl.google.com/android/repository/tools_r12-windows.zip
- ☑ https://dl-ssl.google.com/android/repository/google_apis-10_r02.zip
- ☑ https://dl-ssl.google.com/android/repository/android-2.3.1_r02-linux.zip
- ☑ https://dl-ssl.google.com/android/repository/usb_driver_r04-windows.zip
- ☑ https://dl-ssl.google.com/android/repository/googleadmobadssdkandroid-4.1.0.zip
- ☑ https://dl-ssl.google.com/android/repository/market_licensing-r01.zip

☑　https://dl-ssl.google.com/android/repository/market_billing_r01.zip
☑　https://dl-ssl.google.com/android/repository/google_apis-8_r02.zip
☑　https://dl-ssl.google.com/android/repository/google_apis-7_r01.zip
☑　https://dl-ssl.google.com/android/repository/google_apis-9_r02.zip
……

可以继续根据自己的开发要求选择不同版本的 API。

下载完后将它们复制到 android-sdk-windows/Temp 目录下，然后再运行 setup.exe，选中需要的 API 选项，完成安装。记得保留原始文件，因为放在 temp 目录下的文件装好后立刻会消失。

## 2.5.4　安装 ADT

Android 为 Eclipse 定制了一个专用插件 Android Development Tools（ADT），此插件为用户提供了一个强大的开发 Android 应用程序的综合环境。ADT 扩展了 Eclipse 的功能，可以让用户快速地建立 Android 项目，创建应用程序界面。要安装 Android Development Tools plug-in，需要首先打开 Eclipse IDE，然后进行如下操作：

（1）打开 Eclipse 后，依次选择菜单栏中的 Help | Install New Software 命令，如图 2-37 所示。

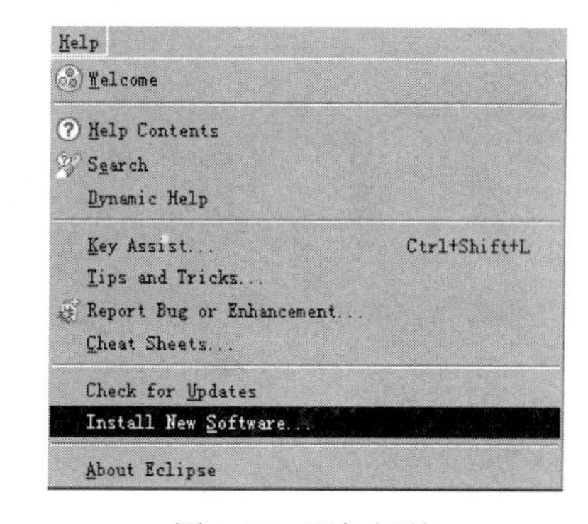


图 2-37　添加插件

（2）在弹出的对话框中单击 Add 按钮，如图 2-38 所示。

（3）在弹出的 Add Site 对话框中分别输入名字和地址，名字可以自己命名，例如“123”，但是在 Location 中必须输入插件的网络地址 http://dl-ssl.google.com/Android/eclipse/，如图 2-39 所示。

（4）单击 OK 按钮，此时在 Install 窗口将会显示系统中可用的插件，如图 2-40 所示。

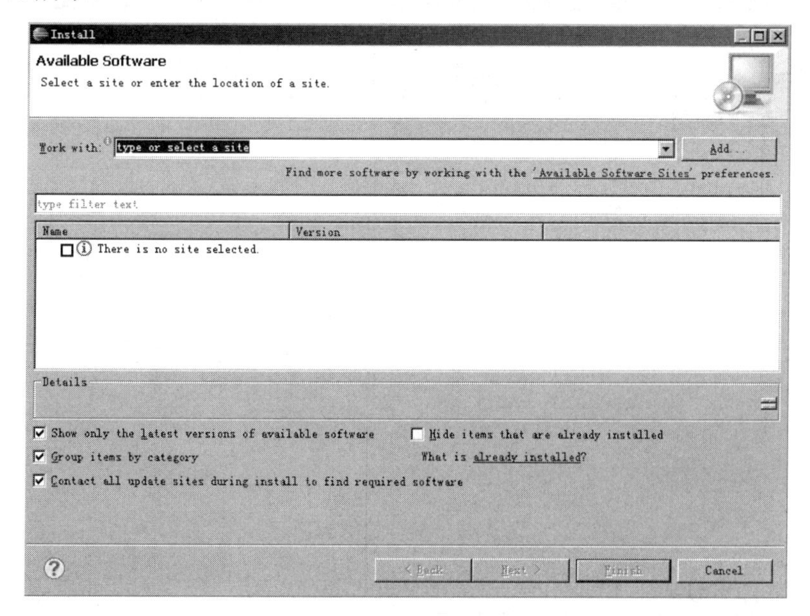


图 2-38　添加插件

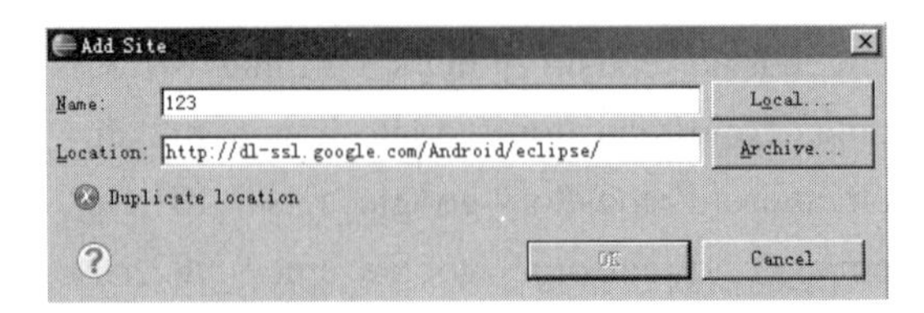


图 2-39 设置地址

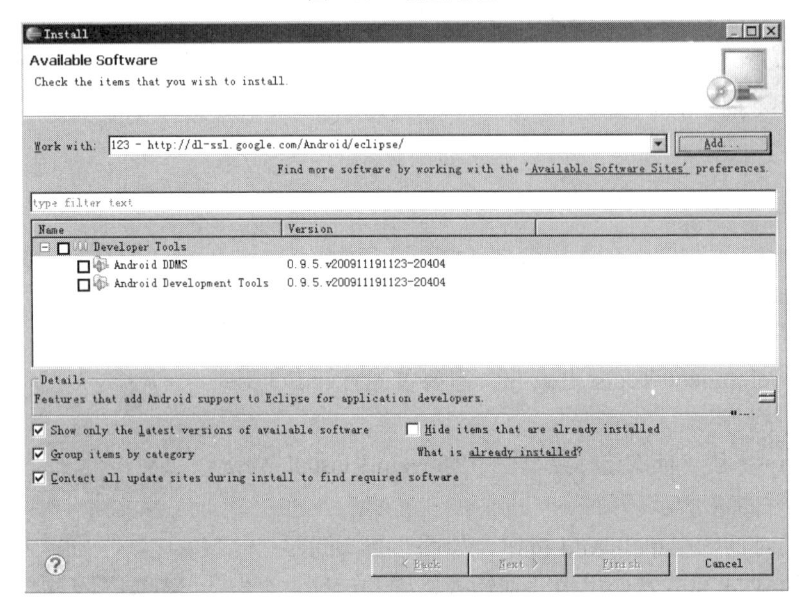


图 2-40 插件列表

（5）选中 Android DDMS 和 Android Development Tools 复选框，然后单击 Next 按钮进入安装界面，如图 2-41 所示。

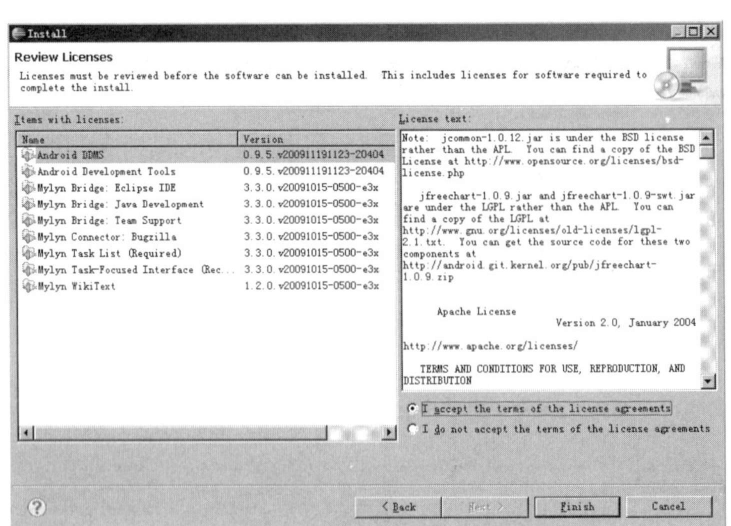


图 2-41 插件安装界面

（6）选中 I accept the terms of the license agreements 单选按钮，单击 Finish 按钮，开始进行安装，如图 2-42 所示。

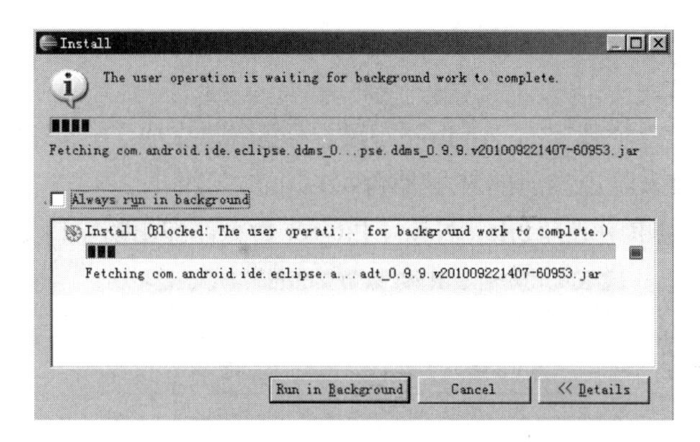


图 2-42　开始安装

**注意**：在步骤（6）中，可能会发生计算插件占用资源情况，过程有点慢。完成后会提示重启 Eclipse 来加载插件，等重启后就可以用了。并且不同版本的 Eclipse 安装插件的方法和步骤是不同的，但是都大同小异，读者可以根据操作提示自行解决。

## 2.5.5　验证设置

经过本章前面内容的介绍，已经讲解了搭建安装 Android 基本环境的知识。在完成安装之后，还需要一些具体验证和设置工作。本节将详细讲解验证和设置 Android 开发环境的基本知识。

### 1．设定 Android SDK Home

当完成上述插件装备工作后，此时还不能使用 Eclipse 创建 Android 项目，还需要在 Eclipse 中设置 Android SDK 的主目录。

（1）打开 Eclipse，在菜单中选择 Window | Preferences 命令，如图 2-43 所示。

（2）在弹出的界面左侧选择 Android 节点后，在右侧设定 Android SDK 所在目录为 SDK Location，单击 OK 按钮完成设置，如图 2-44 所示。

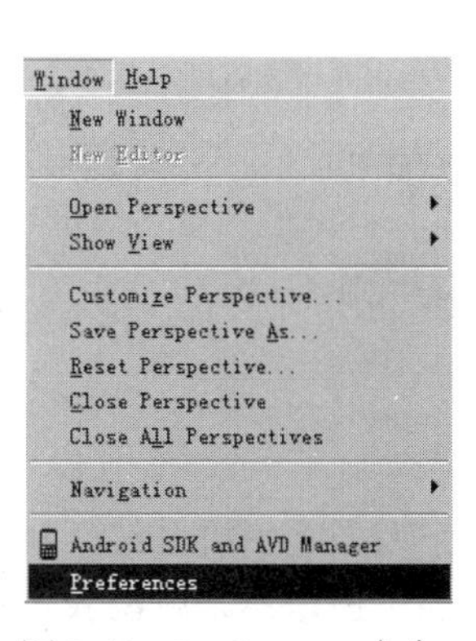


图 2-43　Preferences 命令

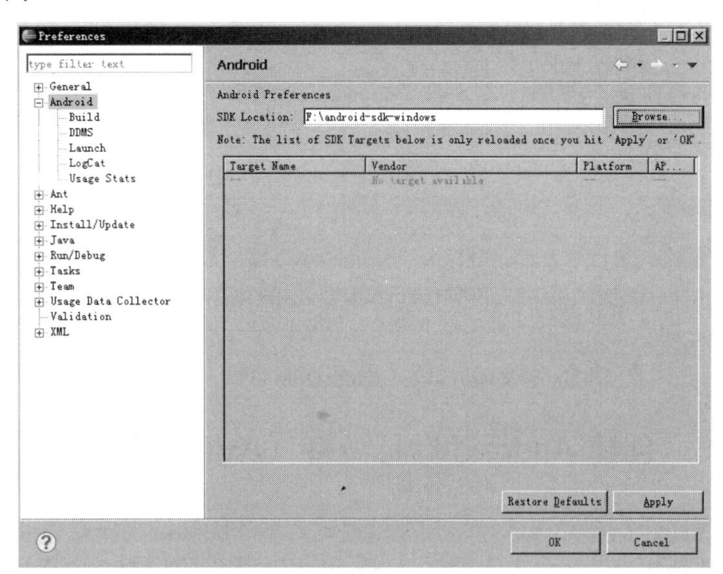


图 2-44　选择 Android 节点

### 2. 验证开发环境

经过前面步骤的讲解，一个基本的 Android 开发环境搭建完成了。都说实践是检验真理的唯一标准，下面通过新建一个项目来验证当前的环境是否可以正常工作。

（1）打开 Eclipse，在菜单中选择 File | New | Project 命令，弹出的窗口如图 2-45 所示。

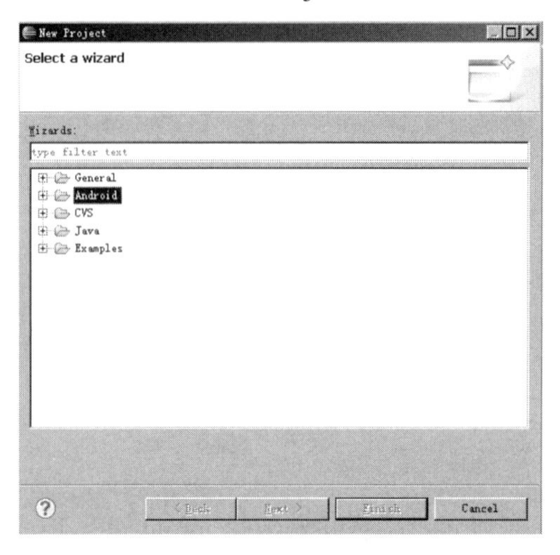


图 2-45　新建项目

（2）在图 2-45 中选择 Android 节点，单击 Next 按钮后打开 New Android Application 窗口，在对应的文本框中输入必要的信息，如图 2-46 所示。

（3）单击 Finish 按钮后 Eclipse 会自动完成项目的创建工作，最后会看到如图 2-47 所示的项目结构。

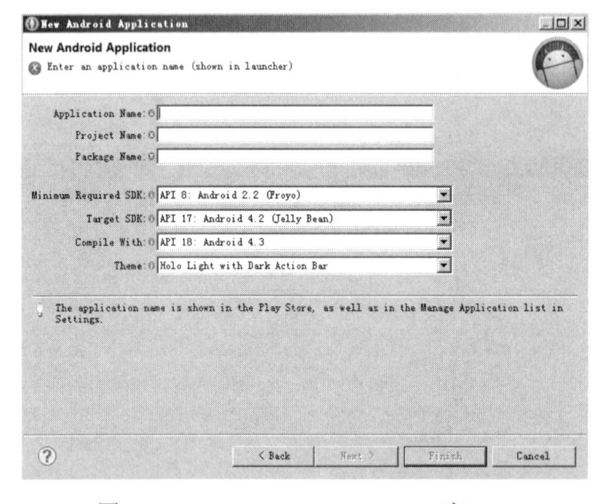


图 2-46　New Android Application 窗口

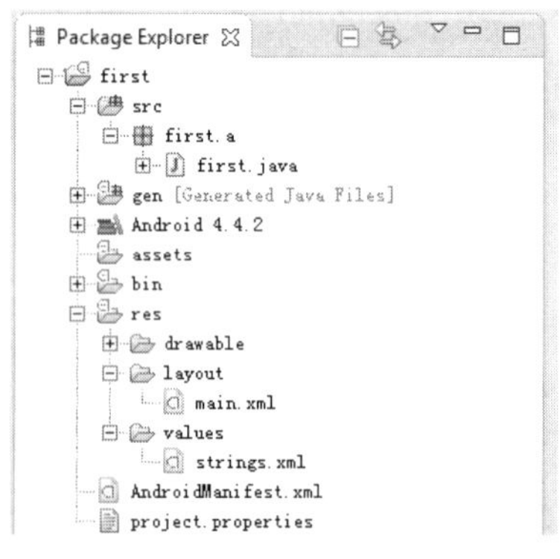


图 2-47　项目结构

## 2.5.6　创建 Android 虚拟设备（AVD）

程序开发需要调试，只有经过调试之后才能知道程序是否正确运行。作为一款手机系统，我们怎么样才能在电脑平台之上调试 Android 程序呢？不用担心，谷歌为我们提供了模拟器来解决我们担心的问题。所谓模拟器，就是指在电脑上模拟安卓系统，可以用这个模拟器来调试并运行开发的 Android

程序。开发人员不需要一个真实的 Android 手机，只通过电脑即可模拟运行一个手机，即可开发出应用在手机上面的程序。

AVD 全称为 Android 虚拟设备（Android Virtual Device），每个 AVD 模拟了一套虚拟设备来运行 Android 平台，这个平台至少要有自己的内核、系统图像和数据分区，还可以有自己的 SD 卡和用户数据以及外观显示等。创建 AVD 的基本步骤如下：

（1）单击 Eclipse 菜单中的图标，如图 2-48 所示。

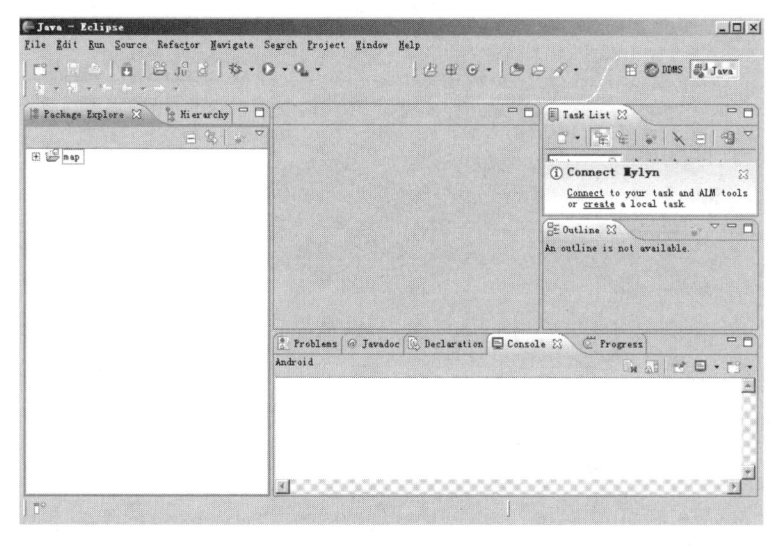


图 2-48　Eclipse

（2）在弹出的 Android Virtual Device（AVD）Manager 界面中选择 Android Virtual Devices 选项卡，如图 2-49 所示。

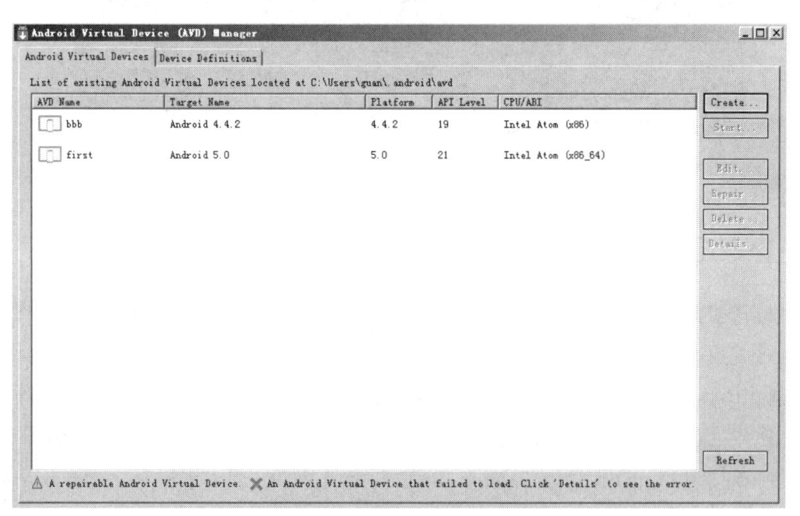


图 2-49　Android Virtual Device（AVD）Manager 界面

在 Android Virtual Devices 选项卡中列出了当前已经安装的 AVD 版本，我们可以通过右侧的按钮来创建、删除或修改 AVD。主要按钮的具体说明如下。

☑ Create...：创建一个新的 AVD，单击此按钮在弹出的界面中可以创建一个新 AVD，如图 2-50 所示。

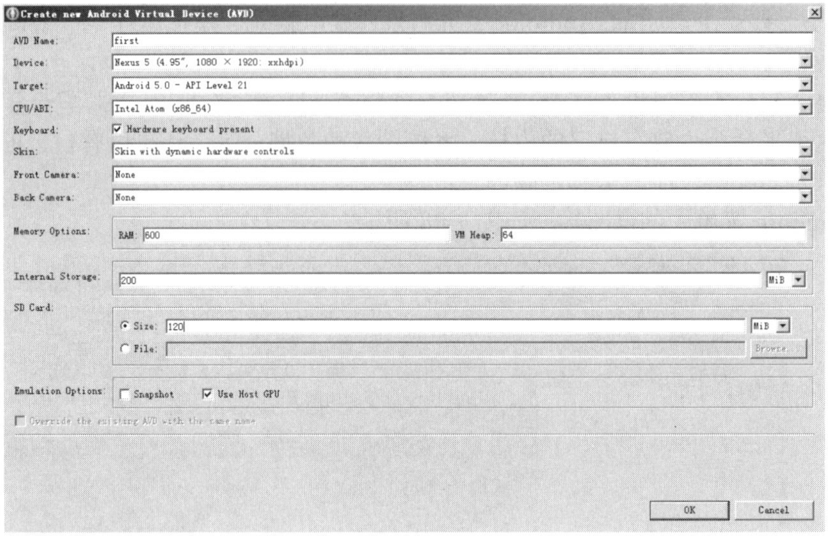

图 2-50　新建 AVD 界面

☑ Edit...：修改已经存在的 AVD。

☑ Delete...：删除已经存在的 AVD。

☑ Start...：启动一个 AVD 模拟器。

**注意：** 可以在 CMD 中创建或删除 AVD，例如可以按照如下 CMD 命令创建一个 AVD。

android create avd --name <your_avd_name> --target <targetID>

其中，your_avd_name 是需要创建的 AVD 的名字，在 CMD 界面中如图 2-51 所示。

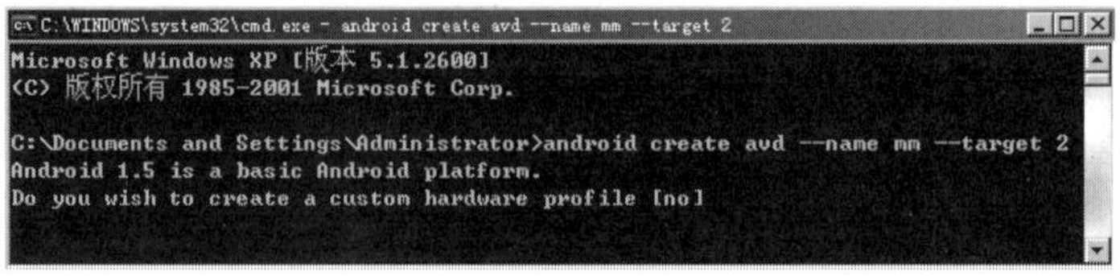

图 2-51　CMD 界面

## 2.5.7　启动 AVD 模拟器

对于 Android 程序的开发者来说，模拟器的推出给开发者在开发上和测试上带来了很大的便利。无论在 Windows 下还是 Linux 下，Android 模拟器都可以顺利运行。并且官方提供了 Eclipse 插件，可以将模拟器集成到 Eclipse 的 IDE 环境。Android SDK 中包含的模拟器的功能非常齐全，电话本、通话等功能都可正常使用（当然用户没办法真的从这里打电话），甚至其内置的浏览器和 Maps 都可以联网。用户可以使用键盘输入，单击模拟器按键输入，甚至还可以使用鼠标单击、拖动屏幕进行操作。模拟器在电脑上模拟运行的效果如图 2-52 所示。

### 1．模拟器和真机究竟有何区别

当然 Android 模拟器不能完全替代真机，具体来说有如下差异。

☑ 模拟器不支持呼叫和接听实际来电，但可以通过控制台模拟电话呼叫（呼入和呼出）。

☑ 模拟器不支持 USB 连接。

图 2-52　模拟器

- ☑ 模拟器不支持相机/视频捕捉。
- ☑ 模拟器不支持音频输入（捕捉），但支持输出（重放）。
- ☑ 模拟器不支持扩展耳机。
- ☑ 模拟器不能确定连接状态。
- ☑ 模拟器不能确定电池电量水平和交流充电状态。
- ☑ 模拟器不能确定 SD 卡的插入/弹出。
- ☑ 模拟器不支持蓝牙。

### 2. 启动 AVD 模拟器的基本流程

在调试时需要启动 AVD 模拟器，启动 AVD 模拟器的基本流程如下：

（1）选择图 2-49 列表中名为 first 的 AVD，单击 Start... 按钮后弹出 Launch Options 界面，如图 2-53 所示。

（2）单击 Launch 按钮后将会运行名为 mm 的模拟器，运行界面效果如图 2-54 所示。

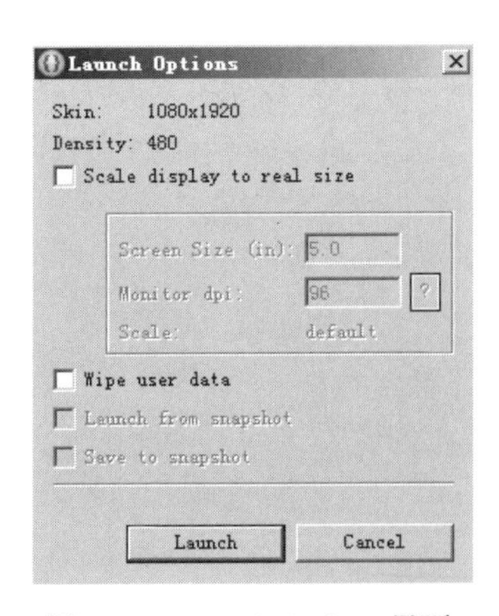


图 2-53　Launch Options 界面

图 2-54　模拟运行成功

# 第 2 篇

## 数据传输篇

- 第 3 章　基本数据通信
- 第 4 章　蓝牙技术详解
- 第 5 章　Android 蓝牙系统详解

# 第 3 章　基本数据通信

在 Android 物联网开发应用中，经常需要网络、蓝牙和红外等设备进行通信传输，以实现不同设备之间的通信功能。例如可以将我们的设备和远程服务器相连接，获取和身体健康相关的指数等信息，也可以在车载设备中查看当前车内的温度信息和湿度信息。本章将详细介绍在 Android 系统中实现基本数据通信的方法，为读者步入本书后面知识的学习打下基础。

## 3.1　HTTP 通信

**知识点讲解：光盘:视频\知识点\第 3 章\HTTP 通信.avi**

超文本传输协议（HyperText Transfer Protocol，HTTP）是互联网上应用最为广泛的一种网络协议。所有的 WWW 文件都必须遵守这个标准。设计 HTTP 最初的目的是为了提供一种发布和接收 HTML 页面的方法。HTTP 是一个客户端和服务器端请求和应答的标准（TCP）。客户端是终端用户，服务器端是网站。通过使用 Web 浏览器、网络爬虫或者其他的工具，客户端发起一个到服务器上指定端口（默认端口为 80）的 HTTP 请求。我们称这个客户端为用户代理（User Agent）。应答的服务器上存储着（一些）资源，例如 HTML 文件和图像。我们称这个应答服务器为源服务器（Origin Server）。在用户代理和源服务器中间可能存在多个中间层，例如代理、网关，或者隧道（Tunnels）。尽管 TCP/IP 协议是互联网上最流行的应用，HTTP 协议并没有规定必须使用它和（基于）它支持的层。事实上，HTTP 可以在任何其他互联网协议上，或者在其他网络上实现。HTTP 只假定（其下层协议提供）可靠的传输，任何能够提供这种保证的协议都可以被其使用。本节首先简要介绍 HTTP 技术的相关基本理论知识，为读者步入本书后面知识的学习打下基础。

### 3.1.1　Android 中的 HTTP

在 Android 系统中，为 HTTP 协议提供了如下 3 种通信接口。

☑　标准 Java 接口：java.net。

☑　Apache 接口：org.apache.http。

☑　Android 网络接口：android.net.http。

网络编程在无线应用程序开发过程中起到了重要的作用。在 Android 系统中包括 Apache HttpClient 库，这个库是执行 Android 中的网络操作的首选方法。除此之外，Android 还可允许通过标准的 Java 联网 API（java.net 包）来访问网络。即便使用 java.net 包，也是在内部使用该 Apache 库。

为了访问互联网，需要设置应用程序获取 android.permission.INTERNET 权限的许可。

在 Android 系统中，存在如下与网络连接相关的包。

（1）java.net

提供联网相关的类，包括流和数据报套接字、互联网协议以及通用的 HTTP 处理。此为多用途的联网资源。经验丰富的 Java 开发人员可立即使用此惯用的包来创建应用程序。

（2）java.io

尽管未明确联网，但其仍然非常重要。此包中的各种类通过其他 Java 包中提供的套接字和链接来使用。它们也可用来与本地文件进行交互（与网络进行交互时经常发生）。

（3）java.nio

包含表示具体数据类型的缓冲的各种类。便于基于 Java 语言的两个端点之间的网络通信。

（4）org.apache.*

表示可为进行 HTTP 通信提供精细控制和功能的各种包。可将 Apache 识别为普通的开源 Web 服务器。

（5）android.net

包括核心 java.net.*类之外的各种附加的网络接入套接字。此包包括 URL 类，其通常在传统联网之外的 Android 应用程序开发中使用。

（6）android.net.http

包含可操作 SSL 证书的各种类。

（7）android.net.wifi

包含可管理 Android 平台中 WiFi（802.11 无线以太网）所有方面的各种类。目前市场上大多数移动设备均配备有 WiFi 能力，尤其随着 Android 在对制造商（如诺基亚和 LG）的翻盖手机的研发方面取得了进展。

（8）android.telephony.gsm

包含管理和发送短信（文本）消息所要求的各种类。随着时间的推移，可能将引入一种附加的包，以提供有关非 GSM 网络（如 CDMA 或类似 android.telephony.cdma）的类似功能。

## 3.1.2 使用 Apache 接口

因为在 Android 平台中，使用得最多的是 Apache 接口。所以本节详细介绍使用 Apache 接口（org.apache.http）实现网络连接的基本知识。在 Apache HttpClient 库中，提供了如下对网络连接有用的包。

☑ org.apache.http.HttpResponse
☑ org.apache.http.client.HttpClient
☑ org.apache.http.client.methods.HttpGet
☑ org.apache.http.impl.client.DefaultHttpClient
☑ HttpClient httpclient=new DefaultHttpClient()

如果想从服务器检索此信息，则需要使用 HttpGet 类的构造器，例如下面的代码。

```
HttpGet request=new HttpGet("http://innovator.samsungmobile.com");
```

然后用 HttpClient 类的 execute 方法中的 HttpGet 对象来检索 HttpResponse 对象，例如下面的代码。

```
HttpResponse response = client.execute(request);
```

接着读取已检索的响应，例如下面的代码。

```
BufferedReader rd = new BufferedReader
                              (new InputStreamReader(response.getEntity().getContent()));
    String line = "";
```

```
    while ((line = rd.readLine()) != null) {
        Log.d("output: ",line);
    }
```

在 Android 系统中，可以采用 HttpPost 和 HttpGet 来封装 post 请求和 get 请求，再使用 HttpClient 的 excute 方法发送 post 或者 get 请求并返回服务器的响应数据。使用 Apache 联网的基本流程如下所示。

（1）设置连接和读取超时时间，并新建 HttpClient 对象，例如下面的代码。

```
//设置连接超时时间和数据读取超时时间
HttpParams httpParams = new BasicHttpParams();
HttpConnectionParams.setConnectionTimeout(httpParams,
        KeySource.CONNECTION_TIMEOUT_INT);
HttpConnectionParams.setSoTimeout(httpParams,
        KeySource.SO_TIMEOUT_INT);
//新建 HttpClient 对象
HttpClient httpClient = new DefaultHttpClient(httpParams)
```

（2）实现 Get 请求，例如下面的代码。

```
//获取请求
HttpGet get = new HttpGet(url);
if (headers != null)
{
    Set<String> setHead = headers.keySet();
    Iterator<String> iteratorHead = setHead.iterator();
    while (iteratorHead.hasNext())
    {
        String headerName = iteratorHead.next();
        String headerValue = (String) headers.get(headerName);
        MyLog.d(headerName, headerValue);
        get.setHeader(headerName, headerValue);
    }
}
    response = httpClient.execute(get);
```

（3）实现 Post 发送请求处理，例如下面的代码。

```
HttpPost post = new HttpPost(KeySource.HOST_URL_STR);
Map<String, String> headers = heads;
Set<String> setHead = headers.keySet();
Iterator<String> iteratorHead = setHead.iterator();
while (iteratorHead.hasNext())
{
    String headName = iteratorHead.next();
    String headValue = (String) headers.get(headName);
    post.setHeader(headName, headValue);
}
/**
* 通常的 HTTP 实体需要在执行上下文时动态生成
* HttpClient 提供使用 EntityTemplate 实体类和 ContentProducer 接口支持动态实体
* 内容制作是通过写需求的内容到一个输出流，每次请求时都会产生
* 因此，通过 EntityTemplate 创建实体通常是独立的，重复性好
*/
ContentProducer cp = new ContentProducer()
    {
```

```
            public void writeTo(OutputStream outstream)
                    throws IOException
            {
                Writer writer = new OutputStreamWriter(outstream,"UTF-8");
                writer.write(requestBody);
                writer.flush();
                writer.close();
            }
        };
        HttpEntity entity = new EntityTemplate(cp);
        post.setEntity(entity);
}
response = httpClient.execute(post);
```

（4）通过 Response 响应请求，例如下面的代码。

```
if (response.getStatusLine().getStatusCode() == 200)
        {
            /**
             * 因为直接调用 toString 可能会导致某些中文字符出现乱码的情况。所以此处使用 toByteArray，
             * 如果需要转成 String 对象，可以先调用 EntityUtils.toByteArray 方法将消息实体转成 byte 数组，
             * 再由 new String(byte[] bArray)转换成字符串
             */
             byte[] bResultXml = EntityUtils.toByteArray(response.getEntity());
             if (bResultXml != null)
             {
                 String    strXml = new String(bResultXml, "utf-8");
             }
        }
```

这样使用 Apache 实现联网处理数据交互的过程就完成了，无论多么复杂的项目，都需要遵循上面的流程。

Apache 的核心功能是 HttpClient，和网络有关的功能几乎都需要用 HttpClient 来实现。在 Android 开发中经常会用到网络连接功能与服务器进行数据的交互，为此 Android 的 SDK 提供了 Apache 的 HttpClient 来方便我们使用各种 Http 服务。可以把 HttpClient 想象成一个浏览器，通过它的 API 我们可以很方便地发出 GET 请求和 POST 请求。

例如，只需要以下几行代码就能发出一个简单的 GET 请求并打印响应结果。

```
try {
        //创建一个默认的 HttpClient
        HttpClient httpclient = new DefaultHttpClient();
        //创建一个 GET 请求
        HttpGet request = new HttpGet("www.google.com");
        //发送 GET 请求，并将响应内容转换成字符串
        String response = httpclient.execute(request, new BasicResponseHandler());
        Log.v("response text", response);
    } catch (ClientProtocolException e) {
        e.printStackTrace();
    } catch (IOException e) {
        e.printStackTrace();
    }
```

肯定有读者禁不住要问为什么上述代码要使用单例 HttpClient 呢？这只是一段演示代码，实际的项

目中的请求与响应处理会复杂一些，并且还要考虑到代码的容错性，但这并不是本篇的重点。读者重点注意代码的第 3 行：

```
HttpClient httpclient = new DefaultHttpClient();
```

在发出 HTTP 请求前先创建了一个 HttpClient 对象，而在实际项目中，我们很可能在多处需要进行 HTTP 通信，这时不需要为每个请求都创建一个新的 HttpClient。因为之前已经提到，HttpClient 就像一个小型的浏览器，对于整个应用，只需要一个 HttpClient 就够了。由此可以得出，使用简单的单例就可以实现，例如下面的代码。

```
public class CustomerHttpClient {
    private static HttpClient customerHttpClient;
    private CustomerHttpClient() {
    }

    public static HttpClient getHttpClient() {
        if(null == customerHttpClient) {
            customerHttpClient = new DefaultHttpClient();
        }
        return customerHttpClient;
    }
}
```

但是如果同时有多个请求需要处理呢？答案是使用多线程。假如现在应用程序使用同一个 HttpClient 来管理所有的 Http 请求，一旦出现并发请求，那么一定会出现多线程的问题。这就好像我们的浏览器只有一个标签页却有多个用户，A 要上 Google，B 要上 baidu，这时浏览器就会忙不过来。幸运的是，HttpClient 提供了创建线程安全对象的 API，能够帮助我们很快地得到线程安全的“浏览器”。例如下面的代码很好地解决了多线程问题。

```
public class CustomerHttpClient {
    private static final String CHARSET = HTTP.UTF_8;
    private static HttpClient customerHttpClient;
    private CustomerHttpClient() {
    }
    public static synchronized HttpClient getHttpClient() {
        if (null == customerHttpClient) {
            HttpParams params = new BasicHttpParams();
            //设置一些基本参数
            HttpProtocolParams.setVersion(params, HttpVersion.HTTP_1_1);
            HttpProtocolParams.setContentCharset(params,CHARSET);
            HttpProtocolParams.setUseExpectContinue(params, true);
            HttpProtocolParams
                    .setUserAgent(
                            params,
                            "Mozilla/3.0(Linux;U;Android
2.2.1;en-us;Nexus One Build.FRG83) "
                                    + "AppleWebKit/553.1(KHTML,like Gecko) Version/4.0 Mobile Safari/533.1");
            //超时设置
            /*从连接池中取连接的超时时间*/
            ConnManagerParams.setTimeout(params, 1000);
            /*连接超时*/
            HttpConnectionParams.setConnectionTimeout(params, 2000);
```

```
            /*请求超时*/
            HttpConnectionParams.setSoTimeout(params, 4000);
            //设置 HttpClient 支持 HTTP 和 HTTPS 两种模式
            SchemeRegistry schReg = new SchemeRegistry();
            schReg.register(new Scheme("http", PlainSocketFactory.getSocketFactory(), 80));
            schReg.register(new Scheme("https", SSLSocketFactory.getSocketFactory(), 443));
            //使用线程安全的连接管理来创建 HttpClient
            ClientConnectionManager conMgr = new ThreadSafeClientConnManager(params, schReg);
            customerHttpClient = new DefaultHttpClient(conMgr, params);
        }
        return customerHttpClient;
    }
}
```

在上面的代码中，通过 getHttpClient 方法为 HttpClient 配置了一些基本参数和超时设置，然后使用 ThreadSafeClientConnManager 来创建线程安全的 HttpClient。

### 3.1.3　在 Android 中使用 java.net

本节将通过具体代码来演示在 Android 中使用 java.net 的基本流程。

（1）在文件 AndroidManifest.xml 中添加 android.permission.INTERNET 权限，这样才允许应用程序访问网络。具体实现代码如下：

```
<?xml version="1.0" encoding="utf-8"?>
<manifest xmlns:android="http://schemas.android.com/apk/res/android"
    package="com.net"
    android:versionCode="1"
    android:versionName="1.0">
  <uses-sdk android:minSdkVersion="8" />
  <uses-permission android:name="android.permission.INTERNET"></uses-permission>
    <application android:icon="@drawable/icon" android:label="@string/app_name">
     <activity android:name=".NetworkingProject" android:label="@string/app_name">
            <intent-filter>
                <action android:name="android.intent.action.MAIN" />
                <category android:name="android.intent.category.LAUNCHER" />
            </intent-filter>
        </activity>
    </application>
</manifest>
```

（2）编写布局文件 main.xml，主要代码如下：

```
<?xml version="1.0" encoding="utf-8"?>
<LinearLayout xmlns:android="http://schemas.android.com/apk/res/android"
    android:orientation="vertical"
    android:layout_width="fill_parent"
    android:layout_height="fill_parent"
    >
    <TextView
        android:text="Enter URL"
        android:id="@+id/textView1"
        android:layout_width="wrap_content"
```

```
            android:layout_height="wrap_content">
        </TextView>
        <EditText
            android:id="@+id/editText1"
            android:layout_width="match_parent"
            android:text="http://innovator.samsungmobile.com"
            android:layout_height="wrap_content">
        </EditText>
        <Button
            android:text="Click Here"
            android:id="@+id/button1"
            android:layout_width="wrap_content"
            android:layout_height="wrap_content">
         </Button>
        <EditText
            android:id="@+id/editText2"
            android:layout_width="match_parent"
            android:layout_height="fill_parent">
        </EditText>
</LinearLayout>
```

（3）编写主程序文件 NetworkingProject.java，功能也是创建一个可以查看网页 HTML 代码的 Java 程序。具体实现代码如下：

```
package com.net;
import java.io.BufferedReader;
import java.io.InputStreamReader;
import java.net.URL;
import java.net.URLConnection;
import android.app.Activity;
import android.os.Bundle;
import android.view.View;
import android.view.View.OnClickListener;
import android.widget.Button;
import android.widget.TextView;
public class NetworkingProject extends Activity {
Button bt;
 TextView textView1;
 TextView textView2;
@Override
 public void onCreate(Bundle savedInstanceState) {
    super.onCreate(savedInstanceState);
    setContentView(R.layout.main);
    bt = (Button) findViewById(R.id.button1);
    textView1 = (TextView) findViewById(R.id.editText1);
    textView2 = (TextView) findViewById(R.id.editText2);
     bt.setOnClickListener(new OnClickListener() {
         @Override
          public void onClick(View v) {
    textView2.setText("");
            try {
                    URL url = new URL(textView1.getText().toString());
```

```
            URLConnection conn = url.openConnection();
            BufferedReader rd = new BufferedReader(new
                        InputStreamReader(conn.getInputStream()));
    String line = "";
    while ((line = rd.readLine()) != null) {
      textView2.append(line);
    }
            } catch (Exception exe) {
    exe.printStackTrace();
            }
    }
  });
  }
}
```

执行上述代码后，可以在手机浏览器中查看输入网址网页的 HTML 代码，如图 3-1 所示。

图 3-1　执行效果

## 3.1.4　使用 Android 网络接口

在 Android 平台中，我们可以使用 Android 网络接口 android.net.http 来处理 HTTP 请求。android.net.http 是 android.net 中的一个包，在里面主要包含处理 SSL 证书的类。在 android.net.http 中存在如下 4 个类。

- ☑ AndroidHttpClient
- ☑ SslCertificate
- ☑ SslCertificate.DName
- ☑ SslError

其中 AndroidHttpClient 就是用来处理 HTTP 请求的。

android.net.*实际上是通过对 Apache 的 HttpClient 的封装来实现的一个 HTTP 编程接口，同时还提供了 HTTP 请求队列管理，以及 HTTP 连接池管理，以提高并发请求情况下（如转载网页时）的处理效率，除此之外还有网络状态监视等接口。

下面是一个通过 AndroidHttpClient 访问服务器的最简例子。

```
import import android.net.http.AndroidHttpClient;
      try {
            AndroidHttpClient client = AndroidHttpClient.newInstance("your_user_agent");
            //创建 HttpGet 方法，该方法会自动处理 URL 地址的重定向
            HttpGet httpGet = new HttpGet ("http://www.test_test.com/");
            HttpResponse response = client.execute(httpGet);
            if (response.getStatusLine().getStatusCode() != HttpStatus.SC_OK) {
                  //错误处理
            }
            //关闭连接
            client.close();
      } catch (Exception ee) {
      }
```

另外当我们的应用需要同时从不同的主机获取数目不等的数据，并且仅关心数据的完整性而不关

心其先后顺序时，也可以使用这部分的接口。典型用例就是 android.webkit 在转载网页和下载网页资源时，具体可参考 android.webkit.*中的相关类来实现。

### 3.1.5 实战演练——在屏幕中传递 HTTP 参数

经过前面的学习，了解到 HTTP 是一种网络传输协议，现实中的大多数网页都是通过 HTTP://WWW.的形式实现显示的。在具体应用时，一些需要的数据都是通过其参数传递的。和网络 HTTP 有关的是 HTTP protocol，在 Android SDK 中，集成了 Apache 的 HttpClient 模块。通过这些模块，可以方便地编写出和 HTTP 有关的程序。在 Android SDK 中通常使用 HttpClient 4.0。在下面的实例中插入了两个按钮：

☑ 一个按钮用于以 POST 方式获取网站数据。

☑ 一个按钮用于以 GET 方式获取数据，并以 TextView 对象来显示由服务器端返回网页内容来显示连接结果。在之前需要先建立和 HTTP 的连接，连接之后才能获取 Web Server 返回的结果。

| 题　目 | 目　的 | 源 码 路 径 |
| --- | --- | --- |
| 实例 3-1 | 在物联网设备屏幕中传递 HTTP 参数 | 光盘:\daima\3\httpexample |

（1）编写布局文件

编写布局文件 main.xml，主要代码如下：

```
<?xml version="1.0" encoding="utf-8"?>
<LinearLayout
  xmlns:android="http://schemas.android.com/apk/res/android"
  android:background="@drawable/white"
  android:orientation="vertical"
  android:layout_width="fill_parent"
  android:layout_height="fill_parent"
  >
  <TextView
    android:id="@+id/myTextView1"
    android:layout_width="fill_parent"
    android:layout_height="wrap_content"
    android:text="@string/title"/>
  <Button
    android:id="@+id/myButton1"
    android:layout_width="wrap_content"
    android:layout_height="wrap_content"
    android:text="@string/str_button1" />
    <Button
    android:id="@+id/myButton2"
    android:layout_width="wrap_content"
    android:layout_height="wrap_content"
    android:text="@string/str_button2" />
</LinearLayout>
```

（2）编写程序文件

编写文件 httpSHI.java，其具体实现流程如下。

☑ 引用 apache.http 相关类实现 HTTP 联机，然后引用 java.io 与 java.util 相关类来读写档案。具体实现代码如下：

```
/*引用 apache.http 相关类来建立 HTTP 联机*/
import org.apache.http.HttpResponse;
import org.apache.http.NameValuePair;
import org.apache.http.client.ClientProtocolException;
import org.apache.http.client.entity.UrlEncodedFormEntity;
import org.apache.http.client.methods.HttpGet;
import org.apache.http.client.methods.HttpPost;
import org.apache.http.impl.client.DefaultHttpClient;
import org.apache.http.message.BasicNameValuePair;
import org.apache.http.protocol.HTTP;
import org.apache.http.util.EntityUtils;
/*必须引用 java.io 与 java.util 相关类来读写档案*/
import irdc.httpSHI.R;
import java.io.IOException;
import java.util.ArrayList;
import java.util.List;
import java.util.regex.Matcher;
import java.util.regex.Pattern;

import android.app.Activity;
import android.os.Bundle;
import android.view.View;
import android.widget.Button;
import android.widget.TextView;
```

☑ 使用 OnClickListener 来聆听单击第一个按钮事件，声明网址字符串并使用建立 Post 方式联机，最后通过 mTextView1.setText 输出提示字符。具体实现代码如下：

```
/*设定 OnClickListener 来聆听 OnClick 事件*/
mButton1.setOnClickListener(new Button.OnClickListener()
{
  /*覆写 onClick 事件*/
  @Override
  public void onClick(View v)
  {
    /*声明网址字符串*/
    String uriAPI = "http://www.dubblogs.cc:8751/Android/Test/API/Post/index.php";
    /*建立 HTTP Post 联机*/
    HttpPost httpRequest = new HttpPost(uriAPI);
    /*
     * Post 运行传送变量必须用 NameValuePair[]数组存储
    */
    List <NameValuePair> params = new ArrayList <NameValuePair>();
    params.add(new BasicNameValuePair("str", "I am Post String"));
    try
    {
      httpRequest.setEntity(new UrlEncodedFormEntity(params, HTTP.UTF_8));
      /*取得 HTTP 输出*/
      HttpResponse httpResponse = new DefaultHttpClient().execute(httpRequest);
      /*如果状态码为 200 */
```

```
        if(httpResponse.getStatusLine().getStatusCode() == 200)
        {
          /*获取应答字符串*/
          String strResult = EntityUtils.toString(httpResponse.getEntity());
          mTextView1.setText(strResult);
        }
        else
        {
          mTextView1.setText("Error Response: "+httpResponse.getStatusLine().toString());
        }
      }
      catch (ClientProtocolException e)
      {
        mTextView1.setText(e.getMessage().toString());
        e.printStackTrace();
      }
      catch (IOException e)
      {
        mTextView1.setText(e.getMessage().toString());
        e.printStackTrace();
      }
      catch (Exception e)
      {
        mTextView1.setText(e.getMessage().toString());
        e.printStackTrace();
      }

    }
});
```

☑ 使用 OnClickListener 来聆听单击第二个按钮的事件，声明网址字符串并建立 Get 方式的联机功能，分别实现发出 HTTP 获取请求、获取应答字符串和删除冗余字符操作，最后通过 mTextView1.setText 输出提示字符。具体实现代码如下：

```
mButton2.setOnClickListener(new Button.OnClickListener()
{
  @Override
  public void onClick(View v)
  {
    /*声明网址字符串*/
    String uriAPI = "http://www.XXXX.cc:8751/index.php?str=I+am+Get+String";
    /*建立 HTTP Get 联机*/
    HttpGet httpRequest = new HttpGet(uriAPI);
    try
    {
      /*发出 HTTP 获取请求*/
      HttpResponse httpResponse = new DefaultHttpClient().execute(httpRequest);
      /*若状态码为 200 ok*/
      if(httpResponse.getStatusLine().getStatusCode() == 200)
      {
        /*获取应答字符串*/
```

```
            String strResult = EntityUtils.toString(httpResponse.getEntity());
            /*删除冗余字符*/
            strResult = eregi_replace("(\r\n|\r|\n|\n\r)","",strResult);
            mTextView1.setText(strResult);
          }
          else
          {
            mTextView1.setText("Error Response: "+httpResponse.getStatusLine().toString());
          }
        }
        catch (ClientProtocolException e)
        {
          mTextView1.setText(e.getMessage().toString());
          e.printStackTrace();
        }
        catch (IOException e)
        {
          mTextView1.setText(e.getMessage().toString());
          e.printStackTrace();
        }
        catch (Exception e)
        {
          mTextView1.setText(e.getMessage().toString());
          e.printStackTrace();
        }
      }
    });
}
```

☑ 定义替换字符串函数 eregi_replace 来替换一些非法字符，具体实现代码如下：

```
  /*字符串替换函数*/
  public String eregi_replace(String strFrom, String strTo, String strTarget)
  {
    String strPattern = "(?i)"+strFrom;
    Pattern p = Pattern.compile(strPattern);
    Matcher m = p.matcher(strTarget);
    if(m.find())
    {
      return strTarget.replaceAll(strFrom, strTo);
    }
    else
    {
      return strTarget;
    }
  }
}
```

（3）声明网络连接权限

在文件 AndroidManifest.xml 中声明网络连接权限，具体实现代码如下：

```
<uses-permission android:name="android.permission.INTERNET"></uses-permission>
```

执行后的效果如图 3-2 所示，单击图中的按钮能够以不同方式获取 HTTP 参数。

图 3-2　单击“使用 POST 方式”按钮后的效果

### 3.1.6　实战演练——在户外运动过程中访问 HTTP 地图

在使用智能物联网设备的过程中，经常需要远程服务器的信息。例如作为一名户外驴友成员，为了确保自己行走的路程准确无误，可以随时访问服务器中的地图来进行对照。在下面的实例中，以一名户外驴友使用物联网设备为背景，讲解了在 Android 物联网设备中使用 HTTP 访问服务器中地图的过程。

在下面的实例中首先创建了 HttpGet 和 HttpPost 对象，并将要请求的 URL 对象构造方法传入 HttpGet、HttpPost 对象中。然后通过 HttpClient 接口的实现类 DefaultClent 的 excute(HttpUriRequest request)方法实现连接处理。因为已经知道 HttpGet 和 HttpPost 类都实现了 HttpUriRequest 接口，所以可以将前面创建好的 HttpGet 或者 HttpPost 对象传入以得到 HttpResponse 对象。最后通过 HttpResponse 获取返回的 HTTP 资源信息，然后再做提取工作。

| 题　　目 | 目　　的 | 源 码 路 径 |
|---|---|---|
| 实例 3-2 | 通过 Apache HTTP 访问 HTTP 资源 | 光盘:\daima\3\httpEX |

本实例的具体实现流程如下所示。

（1）编写布局文件 main.xml，在界面中分别插入 3 个 Button 按钮和两个 EditText 控件，主要代码如下：

```
<LinearLayout android:orientation="horizontal"
    android:layout_width="fill_parent" android:layout_height="wrap_content">
    <TextView android:layout_width="wrap_content"
        android:layout_height="wrap_content" android:text="url:" />
    <EditText android:id="@+id/urlText" android:layout_width="fill_parent"
        android:layout_height="wrap_content"
        android:text="" />
</LinearLayout>
<LinearLayout android:orientation="horizontal"
    android:layout_width="fill_parent" android:layout_height="wrap_content"
    android:gravity="right">
    <Button android:id="@+id/getBtn" android:text="GET 请求"
        android:layout_width="wrap_content" android:layout_height="wrap_content" />
    <Button android:id="@+id/postBtn" android:text="POST 请求"
        android:layout_width="wrap_content" android:layout_height="wrap_content" />
</LinearLayout>
<TextView android:id="@+id/resultView" android:layout_width="fill_parent"
    android:layout_height="wrap_content" />
<LinearLayout android:orientation="horizontal"
```

```
            android:layout_width="fill_parent" android:layout_height="wrap_content">
            <TextView android:layout_width="wrap_content"
                android:layout_height="wrap_content" android:text="图片 url:" />

            <EditText android:id="@+id/imageurlText" android:layout_width="fill_parent"
                android:layout_height="wrap_content" android:text="" />
        </LinearLayout>
        <Button android:id="@+id/imgBtn" android:text="获取图片"
            android:layout_width="wrap_content" android:layout_height="wrap_content"
            android:layout_gravity="right" />
        <ImageView android:id="@+id/imgeView01"
            android:layout_height="wrap_content" android:layout_width="fill_parent" />
    </LinearLayout>
```

（2）编写核心文件 HTTPDemoActivity.java，根据 EditText 控件中输入的数据来访问远程 HTTP 资源，并将得到的信息转换成为一个输出流并返回。在整个实现过程中需要通过 url 创建 HttpGet 对象，并通过 DefaultClient 的 execute 方法返回一个 HttpResponse 对象。文件 HTTPDemoActivity.java 的主要实现代码如下：

```
private String request(String method, String url) {
    HttpResponse httpResponse = null;
    StringBuffer result = new StringBuffer();
    try {
        if (method.equals("GET")) {
            //1.通过 url 创建 HttpGet 对象
            HttpGet httpGet = new HttpGet(url);
            //2.通过 DefaultClient 的 execute 方法执行返回一个 HttpResponse 对象
            HttpClient httpClient = new DefaultHttpClient();
            httpResponse = httpClient.execute(httpGet);
            //3.取得相关信息
            //取得 HttpEntiy
            HttpEntity httpEntity = httpResponse.getEntity();
            //得到一些数据
            //通过 EntityUtils 并指定编码方式取到返回的数据
            result.append(EntityUtils.toString(httpEntity, "utf-8"));
            //得到 StatusLine 接口对象
            StatusLine statusLine = httpResponse.getStatusLine();

            //得到协议
            ;
            result.append("协议:" + statusLine.getProtocolVersion() + "\r\n");
            int statusCode = statusLine.getStatusCode();

            result.append("状态码:" + statusCode + "\r\n");

        } else if (method.equals("POST")) {

            //1.通过 url 创建 HttpGet 对象
            HttpPost httpPost = new HttpPost(url);
            //2.通过 DefaultClient 的 execute 方法执行返回一个 HttpResponse 对象
```

```
                HttpClient httpClient = new DefaultHttpClient();
                httpResponse = httpClient.execute(httpPost);
                //3.取得相关信息
                //取得 HttpEntiy
                HttpEntity httpEntity = httpResponse.getEntity();
                //得到一些数据
                //通过 EntityUtils 并指定编码方式取到返回的数据
                result.append(EntityUtils.toString(httpEntity, "utf-8"));
                StatusLine statusLine = httpResponse.getStatusLine();
                statusLine.getProtocolVersion();
                int statusCode = statusLine.getStatusCode();

                result.append("状态码:" + statusCode + "\r\n");

            }
    } catch (Exception e) {
        Toast.makeText(HTTPDemoActivity.this, "网络连接异常", Toast.LENGTH_LONG).show();
    }
    return result.toString();
}

public void getImage(String url) {
    try {
        //1.通过 url 创建 HttpGet 对象
        HttpGet httpGet = new HttpGet(url);
        //2.通过 DefaultClient 的 execute 方法执行返回一个 HttpResponse 对象
        HttpClient httpClient = new DefaultHttpClient();
        HttpResponse httpResponse = httpClient.execute(httpGet);
        //3.取得相关信息
        //取得 HttpEntiy
        HttpEntity httpEntity = httpResponse.getEntity();
        //4.通过 HttpEntiy.getContent 得到一个输入流
        InputStream inputStream = httpEntity.getContent();
        System.out.println(inputStream.available());

        //通过传入的流再通过 Bitmap 工厂创建一个 Bitmap
        Bitmap bitmap = BitmapFactory.decodeStream(inputStream);
        //设置 imageView
        imageView.setImageBitmap(bitmap);
    } catch (Exception e) {
        Toast.makeText(HTTPDemoActivity.this, "网络连接异常", Toast.LENGTH_LONG)
        .show();
    }
}
```

（3）在设置文件 AndroidManifest.xml 中添加访问网络资源的权限，具体实现代码如下所示。

```
<uses-permission android:name="android.permission.INTERNET"/>
```

（4）设置一个 Java 服务器环境，在里面添加服务器资源供前面的 Android 客户端来访问。将光盘中源码的 Servers 部分复制到本地 Java 服务器的 Tomcat 中。最终客户端在模拟器中的执行效果如图 3-3 所示。

图 3-3　获取服务器中的地图

# 3.2　使用 Socket 实现数据通信

**知识点讲解：光盘:视频\知识点\第 3 章\使用 Socket 实现数据通信.avi**

在物联网设备的数据传输应用中，通常可以通过 TCP、IP 或 UDP 这 3 种协议实现数据传输。在传输数据的过程中，需要通过一个双向的通信连接实现数据的交互。在这个传输过程中，通常将这个双向链路的一端称为 Socket，一个 Socket 通常由一个 IP 地址和一个端口号来确定。由此可见，在整个数据传输过程中，Socket 的作用是巨大的。在 Java 编程应用中，Socket 是 Java 网络编程的核心。因为 Java 是 Andord 应用开发的主流语言，所以本节将详细讲解在 Android 系统中使用 Socket 实现通信的基本知识，为读者步入本书后面知识的学习打下基础。

## 3.2.1　基于 Socket 的 Java 网络编程

网络上的两个程序通过一个双向的通信连接实现数据的交换，这个双向链路的一端称为一个 Socket。Socket 通常用来实现客户方和服务方的连接。Socket 是 TCP/IP 协议的一个十分流行的编程界面，一个 Socket 由一个 IP 地址和一个端口号唯一确定。但是，Socket 所支持的协议种类也不只 TCP/IP 一种，因此两者之间是没有必然联系的。在 Java 环境下，Socket 编程主要是指基于 TCP/IP 协议的网络编程。

### 1. Socket 通信的过程

Server 端 Listen（监听）某个端口是否有连接请求，Client 端向 Server 端发出 Connect（连接）请求，Server 端向 Client 端发回 Accept（接受）消息。这样一个连接就建立起来了。Server 端和 Client 端都可以通过 Send、Write 等方法与对方通信。

在 Java 网络编程应用中，对于一个功能齐全的 Socket 来说，其工作过程包含如下所示的基本步骤。

（1）创建 Socket。

（2）打开连接到 Socket 的输入/输出流。

（3）按照一定的协议对 Socket 进行读/写操作。

（4）关闭 Socket。（在实际应用中，并未使用到显示的 close，虽然很多文章都推荐如此，不过在笔者的程序中，可能因为程序本身比较简单，要求不高，所以并未造成什么影响。）

2. 创建 Socket

在 Java 网络编程应用中，在包 java.net 中提供了 Socket 和 ServerSocket 两个类，分别用来表示双向连接的客户端和服务端。这是两个封装得非常好的类，其中包含了如下所示的构造方法。

- ☑ Socket(InetAddress address, int port)
- ☑ Socket(InetAddress address, int port, boolean stream)
- ☑ Socket(String host, int prot)
- ☑ Socket(String host, int prot, boolean stream)
- ☑ Socket(SocketImpl impl)
- ☑ Socket(String host, int port, InetAddress localAddr, int localPort)
- ☑ Socket(InetAddress address, int port, InetAddress localAddr, int localPort)
- ☑ ServerSocket(int port)
- ☑ ServerSocket(int port, int backlog)
- ☑ ServerSocket(int port, int backlog, InetAddress bindAddr)

在上述构造方法中，参数 address、host 和 port 分别是双向连接中另一方的 IP 地址、主机名和端口号，stream 指明 socket 是流 socket 还是数据报 socket，localPort 表示本地主机的端口号，localAddr 和 bindAddr 是本地机器的地址（ServerSocket 的主机地址），impl 是 socket 的父类，既可以用来创建 serverSocket，又可以用来创建 Socket。count 则表示服务端所能支持的最大连接数。例如：

```
Socket client = new Socket("123.0.01.", 80);
ServerSocket server = new ServerSocket(80);
```

**注意：** 每一个端口提供一种特定的服务，只有给出正确的端口，才能获得相应的服务。0～1023 的端口号为系统所保留，例如 http 服务的端口号为 80，telnet 服务的端口号为 21，ftp 服务的端口号为 23，所以在选择端口号时，最好选择一个大于 1023 的数以防止发生冲突。另外，在创建 Socket 时如果发生错误，将产生 IOException，在程序中必须对之作出处理。所以在创建 Socket 或 ServerSocket 时必须捕获或抛出例外。

### 3.2.2 使用 TCP 协议传输数据

TCP/IP 通信协议是一种可靠的网络协议，能够在通信的两端各建立一个 Socket，从而在通信的两端之间形成网络虚拟链路。一旦建立了虚拟的网络链路，两端的程序就可以通过虚拟链路进行通信。Java 语言对 TCP 网络通信提供了良好的封装，通过 Socket 对象代表两端的通信端口，并通过 Socket 产生的 I/O 流进行网络通信。

（1）使用 ServerSocket

在 Java 程序中，使用类 ServerSocket 接受其他通信实体的连接请求。对象 ServerSocket 的功能是监听来自客户端的 Socket 连接，如果没有连接则会一直处于等待状态。在类 ServerSocket 中包含了如下监听客户端连接请求的方法。

Socket accept()：如果接收到一个客户端 Socket 的连接请求，该方法将返回一个与客户端 Socket 对应的 Socket，否则该方法将一直处于等待状态，线程也被阻塞。

为了创建 ServerSocket 对象，ServerSocket 类提供了如下构造器。

☑ ServerSocket(int port)：用指定的端口 port 创建一个 ServerSocket，该端口应该有一个有效的端口整数值 0～65535。

☑ ServerSocket(int port,int backlog)：增加一个用来改变连接队列长度的参数 backlog。

☑ ServerSocket(int port,int backlog,InetAddress localAddr)：在机器存在多个 IP 地址的情况下，允许通过 localAddr 这个参数来指定将 ServerSocket 绑定到指定的 IP 地址。

当使用 ServerSocket 后，需要使用 ServerSocket 中的方法 close()关闭该 ServerSocket。在通常情况下，因为服务器不会只接收一个客户端请求，而是会不断地接收来自客户端的所有请求，所以可以通过循环来不断地调用 ServerSocket 中的方法 accept()。例如下面的代码。

```
//创建一个 ServerSocket，用于监听客户端 Socket 的连接请求
ServerSocket ss = new ServerSocket(30000);
//采用循环不断接收来自客户端的请求
while (true)
{
//每当接收到客户端 Socket 的请求，服务器端也对应产生一个 Socket
Socket s = ss.accept();
//下面就可以使用 Socket 进行通信了
...
}
```

在上述代码中，创建的 ServerSocket 没有指定 IP 地址，该 ServerSocket 会绑定到本机默认的 IP 地址。在代码中使用 30000 作为该 ServerSocket 的端口号，通常推荐使用 10000 以上的端口，主要是为了避免与其他应用程序的通用端口冲突。

（2）使用 Socket

在客户端可以使用 Socket 的构造器实现和指定服务器的连接，在 Socket 中可以使用如下两个构造器。

☑ Socket(InetAddress/String remoteAddress,int port)：创建连接到指定远程主机、远程端口的 Socket，该构造器没有指定本地地址、本地端口，默认使用本地主机的默认 IP 地址，默认使用系统动态指定的 IP 地址。

☑ Socket(InetAddress/String remoteAddress, int port, InetAddress localAddr, int localPort)：创建连接到指定远程主机、远程端口的 Socket，并指定本地 IP 地址和本地端口号，适用于本地主机有多个 IP 地址的情形。

在使用上述构造器指定远程主机时，既可使用 InetAddress 来指定，也可以使用 String 对象指定，在 Java 中通常使用 String 对象指定远程 IP，例如 192.163.2.23。当本地主机只有一个 IP 地址时，建议使用第一个方法，因为这样更简单。例如下面的代码。

```
//创建连接到本机、30000 端口的 Socket
Socket s = new Socket("123.0.0.1" , 30000);
```

当程序执行上述代码后会连接到指定服务器，让服务器端的 ServerSocket 的方法 accept()向下执行，于是服务器端和客户端就产生一对互相连接的 Socket。上述代码连接到“远程主机”的 IP 地址是 123.0.0.1，此 IP 地址总是代表本级的 IP 地址。因为笔者示例程序的服务器端、客户端都是在本机运行，所以 Socket 连接到远程主机的 IP 地址使用 123.0.0.1。

当客户端、服务器端产生对应的 Socket 之后，程序无须再区分服务器端和客户端，而是通过各自的 Socket 进行通信。在 Socket 中提供如下两个方法获取输入流和输出流。

☑ InputStream getInputStream()：返回该 Socket 对象对应的输入流，让程序通过该输入流从 Socket 中取出数据。

☑ OutputStream getOutputStream()：返回该 Socket 对象对应的输出流，让程序通过该输出流向 Socket 中输出数据。

（3）TCP 中的多线程

在实际应用中，通过 ServerSocket 通常只能实现 Server 和 Client 之间的简单通信操作，当服务器接收到客户端连接之后，服务器向客户端输出一个字符串，而客户端也只是读取服务器的字符串后就退出了。在实际应用中，客户端可能需要和服务器端保持长时间通信，即服务器需要不断地读取客户端数据，并向客户端写入数据，客户端也需要不断地读取服务器数据，并向服务器写入数据。

当使用 readLine()方法读取数据时，如果在该方法成功返回之前线程被阻塞，则程序无法继续执行。所以此服务器很有必要为每个 Socket 单独启动一条线程，每条线程负责与一个客户端进行通信。另外，因为客户端读取服务器数据的线程同样会被阻塞，所以系统应该单独启动一条线程，该线程专门负责读取服务器数据。

假设要开发一个聊天室程序，在服务器端应该包含多条线程，其中每个 Socket 对应一条线程，该线程负责读取 Socket 对应输入流的数据（从客户端发送过来的数据），并将读到的数据向每个 Socket 输出流发送一遍（将一个客户端发送的数据“广播”给其他客户端），因此需要在服务器端使用 List 来保存所有的 Socket。在具体实现时，为服务器提供了如下两个类。

☑ 创建 ServerSocket 监听的主类。

☑ 处理每个 Socket 通信的线程类。

（4）实现非阻塞 Socket 通信

在 Java 应用程序中，可以使用 NIO API 来开发高性能网络服务器。当程序执行输入/输出操作后，在这些操作返回之前会一直阻塞该线程，服务器必须为每个客户端都提供一条独立线程进行处理。这说明前面的程序是基于阻塞式 API 的，当服务器需要同时处理大量客户端时，这种做法会降低性能。

在 Java 应用程序中可以用 NIO API 让服务器使用一个或有限几个线程来同时处理连接到服务器上的所有客户端。在 Java 的 NIO 中，为非阻塞式的 Socket 通信提供了下面的特殊类。

☑ Selector：是 SelectableChannel 对象的多路复用器，所有希望采用非阻塞方式进行通信的 Channel 都应该注册到 Selector 对象。可通过调用此类的静态 open()方法来创建 Selector 实例，该方法将使用系统默认的 Selector 来返回新的 Selector。Selector 可以同时监控多个 SelectableChannel 的 I/O 状况，是非阻塞 I/O 的核心。一个 Selector 实例有如下 3 个 SelectionKey 的集合。

  ➢ 所有 SelectionKey 集合：代表了注册在该 Selector 上的 Channel，这个集合可以通过 keys()方法返回。

  ➢ 被选择的 SelectionKey 集合：代表了所有可通过 select()方法监测到、需要进行 I/O 处理的 Channel，这个集合可以通过 selectedKeys()返回。

  ➢ 被取消的 SelectionKey 集合：代表了所有被取消注册关系的 Channel，在下一次执行 select()方法时，这些 Channel 对应的 SelectionKey 会被彻底删除，程序通常无须直接访问该集合。

除此之外，Selector 还提供了如下和 select()相关的方法。

  ➢ int select()：监控所有注册的 Channel，当它们中间有需要处理的 I/O 操作时，该方法返回，并将对应的 SelectionKey 加入被选择的 SelectionKey 集合中，该方法返回这些 Channel

的数量。

- int select(long timeout)：可以设置超时时长的 select()操作。
- int selectNow()：执行一个立即返回的 select()操作，相对于无参数的 select()方法而言，该方法不会阻塞线程。
- Selector wakeup()：使一个还未返回的 select()方法立刻返回。

☑ SelectableChannel：它代表可以支持非阻塞 I/O 操作的 Channel 对象，可以将其注册到 Selector 上，这种注册的关系由 SelectionKey 实例表示。在 Selector 对象中，可以使用 select()方法设置允许应用程序同时监控多个 I/O Channel。Java 程序可调用 SelectableChannel 中的 register()方法将其注册到指定 Selector 上，当该 Selector 上某些 SelectableChannel 上有需要处理的 I/O 操作时，程序可以调用 Selector 实例的 select()方法获取它们的数量，并通过 selectedKeys()方法返回它们对应的 SelectKey 集合。这个集合的作用巨大，因为通过该集合就可以获取所有需要处理 I/O 操作的 SelectableChannel 集。

对象 SelectableChannel 支持阻塞和非阻塞两种模式，其中所有 channel 默认都是阻塞模式，我们必须使用非阻塞式模式才可以利用非阻塞 I/O 操作。

在 SelectableChannel 中提供了如下两个方法来设置和返回该 Channel 的模式状态。

☑ SelectableChannel configureBlocking(boolean block)：设置是否采用阻塞模式。

☑ boolean isBlocking()：返回该 Channel 是否是阻塞模式。

不同的 SelectableChannel 所支持的操作不一样，例如 ServerSocketChannel 代表一个 ServerSocket，它就只支持 OP_ACCEPT 操作。在 SelectableChannel 中提供了如下方法来返回它支持的所有操作。

int validOps()：返回一个 bit mask，表示这个 channel 上支持的 I/O 操作。

除此之外，SelectableChannel 还提供了如下方法获取它的注册状态。

- boolean isRegistered()：返回该 Channel 是否已注册在一个或多个 Selector 上。
- SelectionKey keyFor(Selector sel)：返回该 Channel 和 sel Selector 之间的注册关系，如果不存在注册关系，则返回 null。
- SelectionKey：该对象代表 SelectableChannel 和 Selector 之间的注册关系。
- ServerSocketChannel：支持非阻塞操作，对应于 java.net.ServerSocket 这个类，提供了 TCP 协议 I/O 接口，只支持 OP_ACCEPT 操作。该类也提供了 accept()方法，功能相当于 ServerSocket 提供的 accept()方法。
- SocketChannel：支持非阻塞操作，对应于 java.net.Socket 这个类，提供了 TCP 协议 I/O 接口，支持 OP_CONNECT、OP_READ 和 OP_WRITE 操作。这个类还实现了 ByteChannel 接口、ScatteringByteChannel 接口和 GatheringByteChannel 接口，所以可以直接通过 SocketChannel 来读写 ByteBuffer 对象。

服务器上所有 Channel 都需要向 Selector 注册，包括 ServerSocketChannel 和 SocketChannel。该 Selector 则负责监视这些 Socket 的 I/O 状态，当其中任意一个或多个 Channel 具有可用的 I/O 操作时，该 Selector 的 select()方法将会返回大于 0 的整数，该整数值就表示该 Selector 上有多少个 Channel 具有可用的 I/O 操作，并提供了 selectedKeys()方法来返回这些 Channel 对应的 SelectionKey 集合。正是通过 Selector 才使得服务器端只需要不断地调用 Selector 实例的 select()方法，就可以知道当前所有 Channel 是否有需要处理的 I/O 操作。当 Selector 上注册的所有 Channel 都没有需要处理的 I/O 操作时，将会阻塞 select()方法，此时调用该方法的线程被阻塞。

我们继续以聊天室为例，讲解非阻塞 Socket 通信在 Java 应用项目中的实现过程。我们的目标是，在服务器端使用循环不断获取 Selector 的 select()方法返回值，当该返回值大于 0 时就处理该 Selector 上被选择 SelectionKey 所对应的 Channel。在具体实现时，服务器端使用 ServerSocketChannel 来监听客户端的连接请求，程序先调用它的 socket()方法获得关联 ServerSocket 对象，再用该 ServerSocket 对象绑定到指定监听 IP 和端口。最后在服务器端调用 Selector 的 select()方法来监听所有 Channel 上的 I/O 操作。

# 3.3 下载数据

**知识点讲解：光盘:视频\知识点\第 3 章\下载数据.avi**

下载是指通过网络进行文件传输，把互联网或其他电子计算机上的信息保存到本地电脑上的一种网络活动。下载可以显式或隐式地进行，只要是获得本地电脑上所没有的信息的活动，都可以认为是下载，如在线观看。在 Android 物联网开发过程中，下载功能是十分常见的一个应用。本节将详细讲解在 Android 物联网中实现远程下载数据的基本知识，为读者步入本书后面知识的学习打下基础。

## 3.3.1 实战演练——下载远程服务器中的图片

| 题　　目 | 目　　的 | 源 码 路 径 |
|---|---|---|
| 实例 3-3 | 在 Android 设备中下载网络服务器中的图片 | 光盘:\daima\3\GetPictureEX |

实例文件 GetAPictureFromInternetActivity.java 的主要实现代码如下：

```
public class GetAPictureFromInternetActivity extends Activity {
    private EditText pathText;
    private ImageView imageView;

    @Override
    public void onCreate(Bundle savedInstanceState) {
        super.onCreate(savedInstanceState);
        setContentView(R.layout.main);
        pathText = (EditText) this.findViewById(R.id.path);
        imageView = (ImageView) this.findViewById(R.id.imageView);
    }

    public void showimage(View v){
     String path = pathText.getText().toString();
          try {
               Bitmap bitmap = ImageService.getImage(path);
               imageView.setImageBitmap(bitmap);
          } catch (Exception e) {
               e.printStackTrace();
               Toast.makeText(getApplicationContext(), R.string.error, 1).show();
          }
    }
}
```

执行后的效果如图 3-4 所示。

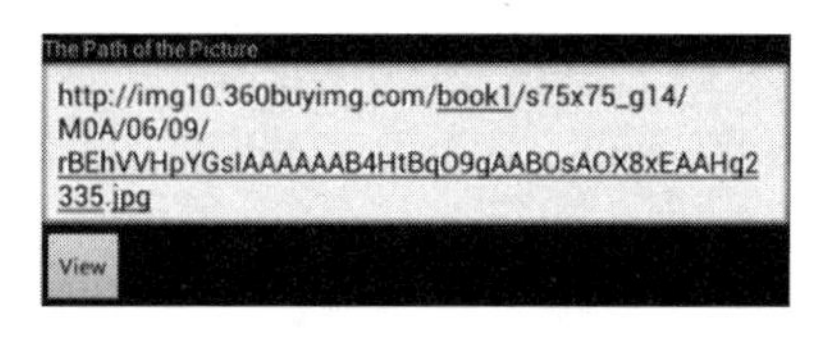


图 3-4　执行效果

## 3.3.2　实战演练——下载网络中的 JSON 信息

JSON 是 JavaScript Object Notation 的缩写，是一种轻量级的数据交换格式。JSON 是基于 JavaScript（Standard ECMA-262 3rd Edition-December 1999）的一个子集，采用完全独立于语言的文本格式，但是也使用了类似于 C 语言家族的习惯（包括 C、C++、C#、Java、JavaScript、Perl、Python 等）。这些特性使 JSON 成为理想的数据交换语言，易于阅读和编写，同时也易于机器解析和生成。

下面将通过一个具体实例的实现过程，详细讲解在 Android 物联网设备中远程下载服务器中的 JSON 数据的方法。

| 题　　目 | 目　　的 | 源 码 路 径 |
|---|---|---|
| 实例 3-4 | 获取远程服务器端的 JSON 数据 | 光盘:\daima\3\jsonEX |

本实例的具体实现流程如下所示。

（1）使用 Eclipse 新建一个 JavaEE 工程作为服务器端，设置工程名为“ServerForJSON”。自动生成工程文件后，打开文件 web.xml 进行配置，配置后的代码如下：

```
<?xml version="1.0" encoding="UTF-8"?>
<web-app id="WebApp_ID" version="2.4" xmlns="http://java.sun.com/xml/ns/j2ee"xmlns:xsi= "http://www.w3.org/ 2001/XMLSchema-instance" xsi:schemaLocation="http://java.sun.com/xml/ns/j2ee http://java.sun.com/xml/ns/j2ee/web-app_2_4.xsd">
	<display-name>ServerForJSON</display-name>
	<servlet>
		<display-name>NewsListServlet</display-name>
		<servlet-name>NewsListServlet</servlet-name>
		<servlet-class>com.guan.server.xml.NewsListServlet</servlet-class>
	</servlet>
	<servlet-mapping>
		<servlet-name>NewsListServlet</servlet-name>
		<url-pattern>/NewsListServlet</url-pattern>
	</servlet-mapping>

	<welcome-file-list>
		<welcome-file>index.html</welcome-file>
		<welcome-file>index.jsp</welcome-file>
	</welcome-file-list>
</web-app>
```

（2）编写业务接口 Bean 的实现文件 NewsService.java，具体实现代码如下：

```
public interface NewsService {
	/**
	* 获取最新的资讯信息
```

```
    */
    public List<News> getLastNews();
}
```

设置业务 Bean 的名称为 NewsServiceBean，实现文件 NewsServiceBean.java 的具体实现代码如下：

```
package com.guan.server.service.implement;

import java.util.ArrayList;
import java.util.List;

import com.guan.server.domain.News;
import com.guan.server.service.NewsService;


public class NewsServiceBean implements NewsService {
    /**
    *获取最新的视频资讯
    */
    public List<News> getLastNews(){
        List<News> newes = new ArrayList<News>();
        newes.add(new News(10, "aaa", 20));
        newes.add(new News(45, "bbb", 10));
        newes.add(new News(89, "Android is good", 50));
        return newes;
    }
}
```

（3）创建一个名为 News 的实现类，实现文件 News.java 的具体实现代码如下：

```
package com.guan.server.domain;
public class News {
    private Integer id;
    private String title;
    private Integer timelength;
    public News(Integer id, String title, Integer timelength) {
        this.id = id;
        this.title = title;
        this.timelength = timelength;
    }
    public Integer getId() {
        return id;
    }
    public void setId(Integer id) {
        this.id = id;
    }
    public String getTitle() {
        return title;
    }
    public void setTitle(String title) {
        this.title = title;
    }
    public Integer getTimelength() {
        return timelength;
```

```
    }
    public void setTimelength(Integer timelength) {
        this.timelength = timelength;
    }
}
```

（4）编写文件 NewsListServlet，具体实现代码如下：

```
public class NewsListServlet extends HttpServlet {
    private static final long serialVersionUID = 1L;
    private NewsService newsService = new NewsServiceBean();
    protected void doGet(HttpServletRequest request, HttpServletResponse response) throws ServletException,
IOException {
        doPost(request, response);
    }
    protected void doPost(HttpServletRequest request, HttpServletResponse response) throws ServletException,
IOException {
        List<News> newes = newsService.getLastNews();//获取最新的视频资讯
            //    [{id:20,title:"xxx",timelength:90},{id:10,title:"xbx",timelength:20}]
            StringBuilder json = new StringBuilder();
            json.append('[');
            for(News news : newes){
                json.append('{');
                json.append("id:").append(news.getId()).append(",");
                json.append("title:\"").append(news.getTitle()).append("\",");
                json.append("timelength:").append(news.getTimelength());
                json.append("},");
            }
            json.deleteCharAt(json.length() - 1);
            json.append(']');
            request.setAttribute("json", json.toString());
            request.getRequestDispatcher("/WEB-INF/page/jsonnewslist.jsp").forward(request, response);
    }
}
```

（5）新建一个 JSP 文件 jsonnewslist.jsp，在其中引入 JSON 功能，具体实现代码如下：

```
<%@ page language="java" contentType="text/plain; charset=UTF-8" pageEncoding="UTF-8"%>${json}
```

（6）使用 Eclipse 新建一个名为 GetNewsInJSONFromInternet 的 Android 工程文件，在文件 AndroidManifest.xml 中声明对网络权限的应用，具体实现代码如下：

```
<?xml version="1.0" encoding="utf-8"?>
<manifest xmlns:android="http://schemas.android.com/apk/res/android"
      package="com.guan.internet.json"
      android:versionCode="1"
      android:versionName="1.0">
    <application android:icon="@drawable/icon" android:label="@string/app_name">
        <activity android:name="com.guan.internet.json.MainActivity"
                  android:label="@string/app_name">
            <intent-filter>
                <action android:name="android.intent.action.MAIN" />
                <category android:name="android.intent.category.LAUNCHER" />
            </intent-filter>
        </activity>
    </application>
```

```
    <uses-sdk android:minSdkVersion="8" />
<!-- 访问 internet 权限 -->
<uses-permission android:name="android.permission.INTERNET"/>
</manifest>
```

（7）编写主界面布局文件 main.xml，具体实现代码如下：

```
<?xml version="1.0" encoding="utf-8"?>
<LinearLayout xmlns:android="http://schemas.android.com/apk/res/android"
    android:orientation="vertical"
    android:layout_width="fill_parent"
    android:layout_height="fill_parent"
    >
<ListView
    android:layout_width="fill_parent"
    android:layout_height="wrap_content"
    android:id="@+id/listView"
    />
</LinearLayout>
```

在上述代码中，通过 ListView 控件列表显示获取的 JSON 数据。其中 ListView 的 Item 显示的数据为 item.xml，具体实现代码如下：

```
<?xml version="1.0" encoding="utf-8"?>
<LinearLayout
  xmlns:android="http://schemas.android.com/apk/res/android"
  android:orientation="horizontal"
  android:layout_width="fill_parent"
  android:layout_height="wrap_content">

  <TextView
   android:layout_width="200dp"
   android:layout_height="wrap_content"
   android:id="@+id/title"
  />
    <TextView
   android:layout_width="fill_parent"
   android:layout_height="wrap_content"
   android:id="@+id/timelength"
  />
</LinearLayout>
```

（8）编写文件 MainActivity.java，功能是获取 JSON 数据并显示数据，具体实现代码如下：

```
public class MainActivity extends Activity {
    @Override
    public void onCreate(Bundle savedInstanceState) {
        super.onCreate(savedInstanceState);
        setContentView(R.layout.main);
        ListView listView = (ListView) this.findViewById(R.id.listView);

        String length = this.getResources().getString(R.string.length);
        try {
              List<News> newes = NewsService.getJSONLastNews();
              List<HashMap<String, Object>> data = new ArrayList<HashMap<String,Object>>();
```

```
            for(News news : newes){
                HashMap<String, Object> item = new HashMap<String, Object>();
                item.put("id", news.getId());
                item.put("title", news.getTitle());
                item.put("timelength", length+ news.getTimelength());
                data.add(item);
            }
            SimpleAdapter adapter = new SimpleAdapter(this, data, R.layout.item,
                    new String[]{"title", "timelength"}, new int[]{R.id.title, R.id.timelength});
            listView.setAdapter(adapter);
        } catch (Exception e) {
            e.printStackTrace();
        }
    }
}
```

（9）编写文件 NewsService.java，定义方法 getJSONLastNews()请求前面搭建的 JavaEE 服务器，当获取 JSON 输入流后解析 JSON 的数据，并返回集合中的数据。文件 NewsService.java 的具体实现代码如下：

```
public class NewsService {
    /**
    *获取最新视频资讯
    */
    public static List<News> getJSONLastNews() throws Exception{
        String path = "http://192.163.1.100:8080/ServerForJSON/NewsListServlet";
        HttpURLConnection conn = (HttpURLConnection) new URL(path).openConnection();
        conn.setConnectTimeout(5000);
        conn.setRequestMethod("GET");
        if(conn.getResponseCode() == 200){
            InputStream json = conn.getInputStream();
            return parseJSON(json);
        }
        return null;
    }
    private static List<News> parseJSON(InputStream jsonStream) throws Exception{
        List<News> list = new ArrayList<News>();
        byte[] data = StreamTool.read(jsonStream);
        String json = new String(data);
        JSONArray jsonArray = new JSONArray(json);
        for(int i = 0; i < jsonArray.length() ; i++){
            JSONObject jsonObject = jsonArray.getJSONObject(i);
            int id = jsonObject.getInt("id");
            String title = jsonObject.getString("title");
            int timelength = jsonObject.getInt("timelength");
            list.add(new News(id, title, timelength));
        }
        return list;
    }
}
```

至此，整个实例介绍完毕，执行后将成功获取服务器端 JSON 的数据。

### 3.3.3 实战演练——下载并播放网络 MP3

为了节约手机的存储空间，在听音乐时可以用从网络中下载的方式播放 MP3。在本实例中，首先插入 4 个按钮，分别用于播放、暂停、重新播放和停止。执行后，通过 Runnable 发起运行线程，在线程中通过网络传输方式远程下载指定的 MP3 文件。下载完毕后，临时保存到 SD 卡中，这样可以通过 4 个按钮对其进行控制。当关闭程序后，会自动删除 SD 卡中的临时性文件。

| 题　目 | 目　的 | 源 码 路 径 |
| --- | --- | --- |
| 实例 3-5 | 播放网络中的 MP3 | 光盘:\daima\3\mpEX |

编写主程序文件 mp.java，具体实现流程如下所示。

（1）定义方法 currentFilePath 用于记录当前正在播放 MP3 的 URL 地址，定义 currentTempFilePath 表示当前播放 MP3 的路径。具体实现代码如下：

```
/*记录当前正在播放 MP3 的地址 URL*/
private String currentFilePath = "";
/*当前播放 MP3 的路径*/
private String currentTempFilePath = "";
private String strVideoURL = "";
```

（2）使用 strVideoURL 设置要播放 MP3 文件的网址，并设置透明度。具体实现代码如下：

```
public void onCreate(Bundle savedInstanceState)
{
  super.onCreate(savedInstanceState);
  setContentView(R.layout.main);
  /*MP3 文件不会被下载到 local*/
  strVideoURL = "http://www.lrn.cn/zywh/xyyy/yyxs/200805/W020080505536315331313.mp3";
  mTextView01 = (TextView)findViewById(R.id.myTextView1);
  /*设置透明度*/
  getWindow().setFormat(PixelFormat.TRANSPARENT);
  mPlay = (ImageButton)findViewById(R.id.play);
  mReset = (ImageButton)findViewById(R.id.reset);
  mPause = (ImageButton)findViewById(R.id.pause);
  mStop = (ImageButton)findViewById(R.id.stop);
```

（3）编写单击"播放"按钮所触发的处理事件，具体实现代码如下：

```
/*播放按钮*/
mPlay.setOnClickListener(new ImageButton.OnClickListener()
{
  public void onClick(View view)
  {
    /*调用播放影片 Function*/
    playVideo(strVideoURL);
    mTextView01.setText
    (
      getResources().getText(R.string.str_play).toString()+
      "\n"+ strVideoURL
    );
  }
});
```

（4）编写单击“重播”按钮所触发的处理事件，具体实现代码如下：

```
/* 重新播放 */
mReset.setOnClickListener(new ImageButton.OnClickListener()
{
  public void onClick(View view)
  {
    if(bIsReleased == false)
    {
      if (mMediaPlayer01 != null)
      {
        mMediaPlayer01.seekTo(0);
        mTextView01.setText(R.string.str_play);
      }
    }
  }
});
```

（5）编写单击“暂停”按钮所触发的处理事件，具体实现代码如下：

```
/*暂停播放*/
mPause.setOnClickListener(new ImageButton.OnClickListener()
{
  public void onClick(View view)
  {
    if (mMediaPlayer01 != null)
    {
      if(bIsReleased == false)
      {
        if(bIsPaused==false)
        {
          mMediaPlayer01.pause();
          bIsPaused = true;
          mTextView01.setText(R.string.str_pause);
        }
        else if(bIsPaused==true)
        {
          mMediaPlayer01.start();
          bIsPaused = false;
          mTextView01.setText(R.string.str_play);
        }
      }
    }
  }
});
```

（6）编写单击“停止”按钮所触发的处理事件，具体实现代码如下：

```
/*停止*/
mStop.setOnClickListener(new ImageButton.OnClickListener()
{
  public void onClick(View view)
  {
    try
    {
```

```
            if (mMediaPlayer01 != null)
            {
              if(bIsReleased==false)
              {
                mMediaPlayer01.seekTo(0);
                mMediaPlayer01.pause();
                //mMediaPlayer01.stop();
                //mMediaPlayer01.release();
                //bIsReleased = true;
                mTextView01.setText(R.string.str_stop);
              }
            }
          }
          catch(Exception e)
          {
            mTextView01.setText(e.toString());
            Log.e(TAG, e.toString());
            e.printStackTrace();
          }
        }
      });
    }
```

（7）定义方法 playVideo(final String strPath)来播放指定的 MP3，其播放的是存储卡中暂时保存的 MP3 文件，具体实现代码如下：

```
private void playVideo(final String strPath)
{
  try
  {
    if (strPath.equals(currentFilePath)&& mMediaPlayer01 != null)
    {
      mMediaPlayer01.start();
      return;
    }
    currentFilePath = strPath;
    mMediaPlayer01 = new MediaPlayer();
    mMediaPlayer01.setAudioStreamType(2);
```

（8）编写 setOnErrorListener 来监听错误处理，具体实现代码如下：

```
/*错误事件*/
mMediaPlayer01.setOnErrorListener(new MediaPlayer.OnErrorListener()
{
  @Override
  public boolean onError(MediaPlayer mp, int what, int extra)
  {
    Log.i(TAG, "Error on Listener, what: " + what + "extra: " + extra);
    return false;
  }
});
```

（9）编写 setOnBufferingUpdateListener 来监听 MediaPlayer 缓冲区的更新，具体实现代码如下：

```
/*捕捉使用 MediaPlayer 缓冲区的更新事件*/
mMediaPlayer01.setOnBufferingUpdateListener(new MediaPlayer.OnBufferingUpdateListener()
```

```
{
  @Override
  public void onBufferingUpdate(MediaPlayer mp, int percent)
  {
    Log.i(TAG, "Update buffer: " + Integer.toString(percent)+ "%");
  }
});
```

（10）编写 setOnCompletionListener 来监听播放完毕所触发的事件，具体实现代码如下：

```
/*播放完毕所触发的事件*/
mMediaPlayer01.setOnCompletionListener(new MediaPlayer.OnCompletionListener()
{
  @Override
  public void onCompletion(MediaPlayer mp)
  {
    //delFile(currentTempFilePath);
    Log.i(TAG,"mMediaPlayer01 Listener Completed");
  }
});
```

（11）编写 setOnPreparedListener 来监听开始阶段的事件，具体实现代码如下：

```
/*开始阶段的监听 Listener*/
mMediaPlayer01.setOnPreparedListener(new MediaPlayer.OnPreparedListener()
{
  @Override
  public void onPrepared(MediaPlayer mp)
  {
    Log.i(TAG,"Prepared Listener");
  }
});
```

（12）将文件存到 SD 卡后，通过方法 mMediaPlayer01.start()播放 MP3。具体实现代码如下：

```
  /*用 Runnable 来确保文件在存储完毕后才开始 start()*/
  Runnable r = new Runnable()
  {
    public void run()
    {
      try
      {
        /*setDataSource 将文件存到 SD 卡*/
        setDataSource(strPath);
        /*因为线程顺利进行，所以在 setDataSource 后运行 prepare()*/
        mMediaPlayer01.prepare();
        Log.i(TAG, "Duration: " + mMediaPlayer01.getDuration());

        /*开始播放 MP3*/
        mMediaPlayer01.start();
        bIsReleased = false;
      }
      catch (Exception e)
      {
        Log.e(TAG, e.getMessage(), e);
      }
```

```
        }
    };
    new Thread(r).start();
}
```

（13）如果有异常则输出提示，具体实现代码如下：

```
    catch(Exception e)
    {
      if (mMediaPlayer01 != null)
      {
        /*线程发生异常则停止播放*/
        mMediaPlayer01.stop();
        mMediaPlayer01.release();
      }
      e.printStackTrace();
    }
}
```

（14）定义函数 setDataSource 用于存储 URL 的 MP3 文件到存储卡。首先判断传入的地址是否为 URL，然后创建 URL 对象和临时文件。具体实现代码如下：

```
/*定义函数用于存储 URL 的 MP3 文件到存储卡*/
private void setDataSource(String strPath) throws Exception
{
  /*判断传入的地址是否为 URL*/
  if (!URLUtil.isNetworkUrl(strPath))
  {
    mMediaPlayer01.setDataSource(strPath);
  }
  else
  {
    if(bIsReleased == false)
    {
      /*创建 URL 对象*/
      URL myURL = new URL(strPath);
      URLConnection conn = myURL.openConnection();
      conn.connect();

      /*获取 URLConnection 的 InputStream*/
      InputStream is = conn.getInputStream();
      if (is == null)
      {
        throw new RuntimeException("stream is null");
      }
      /*创建临时文件*/
      File myTempFile = File.createTempFile("yinyue", "."+getFileExtension(strPath));
      currentTempFilePath = myTempFile.getAbsolutePath();
      FileOutputStream fos = new FileOutputStream(myTempFile);
      byte buf[] = new byte[128];
      do
      {
        int numread = is.read(buf);
        if (numread <= 0)
        {
```

```
                break;
            }
            fos.write(buf, 0, numread);
        }while (true);

        /*直到 fos 存储完毕，调用 MediaPlayer.setDataSource */
        mMediaPlayer01.setDataSource(currentTempFilePath);
        try
        {
            is.close();
        }
        catch (Exception ex)
        {
            Log.e(TAG, "error: " + ex.getMessage(), ex);
        }
      }
    }
}
```

（15）定义方法 getFileExtension(String strFileName)来获取音乐文件的扩展名，如果无法顺利获取扩展名，则默认为.dat。具体实现代码如下：

```
/*获取音乐文件扩展名自定义函数*/
private String getFileExtension(String strFileName)
{
    File myFile = new File(strFileName);
    String strFileExtension=myFile.getName();
    strFileExtension=(strFileExtension.substring(strFileExtension.lastIndexOf(".")+1)).toLowerCase();
    if(strFileExtension=="")
    {
        /*如果无法顺利获取扩展名则默认为.dat*/
        strFileExtension = "dat";
    }
    return strFileExtension;
}
```

（16）定义方法 delFile(String strFileName)来设置当离开程序时删除临时音乐文件，具体实现代码如下：

```
    /*离开程序时需要调用自定义函数删除临时音乐文件*/
    private void delFile(String strFileName)
    {
        File myFile = new File(strFileName);
        if(myFile.exists())
        {
            myFile.delete();
        }
    }

    @Override
    protected void onPause()
    {

        /*删除临时文件*/
```

```
    try
    {
      delFile(currentTempFilePath);
    }
    catch(Exception e)
    {
      e.printStackTrace();
    }
    super.onPause();
  }
}
```

执行后可以通过播放、暂停、重新播放和停止 4 个按钮来控制播放指定的 MP3 音乐，如图 3-5 所示。

图 3-5 执行效果

## 3.4 上 传 数 据

**知识点讲解：光盘:视频\知识点\第 3 章\上传数据.avi**

“上传”和“下载”是相对的，上传是指将信息从个人计算机（本地计算机）传递到中央计算机（远程计算机）系统上，让网络上的人都能看到。本节将详细讲解在 Android 物联网设备中上传数据的基本知识，为步入本书后面知识的学习打下基础。

### 3.4.1 实战演练——上传文件到远程服务器

在使用物联网设备时，可以使用拍照装置进行拍照。为了节省设备的容量，可以将照片上传到远程服务器。例如在下面的实例中，演示了在 Android 物联网设备中将文件上传到远程服务器的方法。

| 题　目 | 目　的 | 源 码 路 径 |
| --- | --- | --- |
| 实例 3-6 | 上传物联网设备中的文件到远程服务器 | 光盘:\daima\3\chuanEX |

编写主程序文件 chuan.java，具体实现流程如下。

（1）分别声明变量 newName、uploadFile 和 actionUrl，具体实现代码如下：

```
public class chuan extends Activity
{
  /*变量声明
  *newName：上传后在服务器上的文件名称
  *uploadFile：要上传的文件路径
  *actionUrl：服务器上对应的程序路径*/
  private String newName="image.jpg";
  private String uploadFile="/data/data/irdc.example9/image.jpg";
  private String actionUrl="http://127.127.0.1/upload/upload.jsp";
```

```
  private TextView mText1;
  private TextView mText2;
  private Button mButton;
```

（2）通过 mText1 对象获取文件路径，根据 mText2 设置上传网址，单击按钮后调用上传方法 uploadFile()。具体实现代码如下：

```
public void onCreate(Bundle savedInstanceState)
{
  super.onCreate(savedInstanceState);
  setContentView(R.layout.main);
  mText1 = (TextView) findViewById(R.id.myText2);
  mText1.setText("文件路径：\n"+uploadFile);
  mText2 = (TextView) findViewById(R.id.myText3);
  mText2.setText("上传网址：\n"+actionUrl);
  /*设置 mButton 的 onClick 事件处理*/
  mButton = (Button) findViewById(R.id.myButton);
  mButton.setOnClickListener(new View.OnClickListener()
  {
    public void onClick(View v)
    {
      uploadFile();
    }
  });
}
```

（3）定义方法 uploadFile()，将文件上传至 Server，具体实现代码如下：

```
/*上传文件至 Server 的方法*/
private void uploadFile()
{
  String end = "\r\n";
  String twoHyphens = "--";
  String boundary = "*****";
  try
  {
    URL url =new URL(actionUrl);
    HttpURLConnection con=(HttpURLConnection)url.openConnection();
    /*允许 Input、Output，不使用 Cache*/
    con.setDoInput(true);
    con.setDoOutput(true);
    con.setUseCaches(false);
    /* 设置传送的 method=POST */
    con.setRequestMethod("POST");
    /*setRequestProperty*/
    con.setRequestProperty("Connection", "Keep-Alive");
    con.setRequestProperty("Charset", "UTF-8");
    con.setRequestProperty("Content-Type","multipart/form-data;boundary="+boundary);
    /*设置 DataOutputStream*/
    DataOutputStream ds =
      new DataOutputStream(con.getOutputStream());
    ds.writeBytes(twoHyphens + boundary + end);
    ds.writeBytes("Content-Disposition: form-data; " +
                  "name=\"file1\";filename=\"" +
```

```
                    newName +"\"" + end);
    ds.writeBytes(end);
    /*取得文件的 FileInputStream*/
    FileInputStream fStream = new FileInputStream(uploadFile);
    /*设置每次写入 1024bytes*/
    int bufferSize = 1024;
    byte[] buffer = new byte[bufferSize];
    int length = -1;
    /*从文件读取数据至缓冲区*/
    while((length = fStream.read(buffer)) != -1)
    {
      /*将资料写入 DataOutputStream 中*/
      ds.write(buffer, 0, length);
    }
    ds.writeBytes(end);
    ds.writeBytes(twoHyphens + boundary + twoHyphens + end);
    fStream.close();
    ds.flush();
    /*取得 Response 内容*/
    InputStream is = con.getInputStream();
    int ch;
    StringBuffer b =new StringBuffer();
    while( ( ch = is.read() ) != -1 )
    {
      b.append( (char)ch );
    }
    /*将 Response 显示在 Dialog 对话框中*/
    showDialog(b.toString().trim());
    /*关闭 DataOutputStream*/
    ds.close();
  }
  catch(Exception e)
  {
    showDialog(""+e);
  }
}
```

（4）定义方法 showDialog(String mess)来显示提示对话框，具体实现代码如下：

```
  /*显示 Dialog 的方法*/
  private void showDialog(String mess)
  {
    new AlertDialog.Builder(example9.this).setTitle("Message")
     .setMessage(mess)
     .setNegativeButton("确定",new DialogInterface.OnClickListener()
     {
       public void onClick(DialogInterface dialog, int which)
       {
       }
     })
     .show();
  }
}
```

执行后单击“上传”按钮可以将指定的文件上传到服务器，如图 3-6 所示。

图 3-6　执行效果

## 3.4.2　实战演练——使用 GET 方式上传数据

在 Android 系统中可以通过 GET 方式或 POST 方式上传数据，本节将通过一个具体实例的实现过程，介绍在 Android 物联网设备中采用 GET 方式向服务器传递数据的基本方法。

| 题　目 | 目　的 | 源 码 路 径 |
| --- | --- | --- |
| 实例 3-7 | 在物联网设备中采用 GET 方式向服务器传递数据 | 光盘:\daima\3\getEX |

（1）创建一个名为 ServletForGETMethod 的 Servlet，功能是接收并处理通过 GET 方式上传的数据。文件 ServletForGETMethod.java 的具体实现代码如下：

```
@WebServlet("/ServletForGETMethod")
public class ServletForGETMethod extends HttpServlet {
	private static final long serialVersionUID = 1L;
	protected void doGet(HttpServletRequest request, HttpServletResponse response) throws ServletException,
IOException {
			String name= request.getParameter("name");
//			String name= new String(request.getParameter("name").getBytes("ISO8859-1"),"UTF-8");
			String age= request.getParameter("age");
			System.out.println("name: " + name );
			System.out.println("age: " + age );


	}
}
```

在上述代码中，为了避免出现中文乱码的问题，特意实现了 ISO8859-1 和 UTF-8 转换处理。下面的代码很好地解决了乱码问题。

```
<%@ page language="java" import="java.util.*" pageEncoding="UTF-8"%>
<%
String zh_value=new String(request.getParameter("zh_value").getBytes("ISO-8859-1"),"UTF-8")
%>
```

由此可见，在使用 get 方式传递数据时，需要使用如下所示的代码声明当前页的字符集。

```
pageEncoding="UTF-8" //声明当前页的字符集
```

（2）在配置文件 web.xml 中配置 ServletForGETMethod，具体实现代码如下：

```
<?xml version="1.0" encoding="UTF-8"?>
<web-app xmlns:xsi="http://www.w3.org/2001/XMLSchema-instance" xmlns=http://java.sun.com/xml/ns/javaee
xmlns:web="http://java.sun.com/xml/ns/javaee/web-app_2_5.xsd" xsi:schemaLocation="http://java.sun.com/xml/
ns/javaee http://java.sun.com/xml/ns/javaee/web-app_3_0.xsd" id="WebApp_ID" version="3.0">
  <display-name>ServerForGETMethod</display-name>
```

```
  <servlet>
    <display-name>ServletForGETMethod</display-name>
    <servlet-name>ServletForGETMethod</servlet-name>
    <servlet-class>com.guan.internet.servlet.ServletForGETMethod</servlet-class>
  </servlet>
  <servlet-mapping>
    <servlet-name>ServletForGETMethod</servlet-name>
    <url-pattern>/ServletForGETMethod</url-pattern>
  </servlet-mapping>
  <welcome-file-list>
    <welcome-file>index.html</welcome-file>
    <welcome-file>index.htm</welcome-file>
    <welcome-file>index.jsp</welcome-file>
    <welcome-file>default.html</welcome-file>
    <welcome-file>default.htm</welcome-file>
    <welcome-file>default.jsp</welcome-file>
  </welcome-file-list>
</web-app>
```

（3）打开 Eclipse，新建一个名为 UserInformation 的 Android 工程。然后编写界面布局文件 main.xml，具体实现代码如下：

```
<?xml version="1.0" encoding="utf-8"?>
<LinearLayout xmlns:android="http://schemas.android.com/apk/res/android"
    android:layout_width="fill_parent"
    android:layout_height="fill_parent"
    android:orientation="vertical" >
    <TextView
    android:layout_width="fill_parent"
    android:layout_height="wrap_content"
    android:text="@string/title"
    />
    <EditText
     android:layout_width="fill_parent"
     android:layout_height="wrap_content"
     android:id="@+id/title"
    />

    <TextView
    android:layout_width="fill_parent"
    android:layout_height="wrap_content"
    android:text="@string/length"
    />
    <EditText
     android:layout_width="fill_parent"
     android:layout_height="wrap_content"
     android:numeric="integer"
     android:id="@+id/length"
    />
    <Button
    android:layout_width="wrap_content"
    android:layout_height="wrap_content"
    android:text="@string/button"
    android:onClick="save"
```

```
    />
</LinearLayout>
```

（4）编写文件 UserInformationActivity.java，具体实现代码如下：

```
public class UserInformationActivity extends Activity {
    private EditText titleText;
        private EditText lengthText;

        @Override
        public void onCreate(Bundle savedInstanceState) {
            super.onCreate(savedInstanceState);
            setContentView(R.layout.main);

            titleText = (EditText) this.findViewById(R.id.title);
            lengthText = (EditText) this.findViewById(R.id.length);
        }

        public void save(View v){
         String title = titleText.getText().toString();
         String length = lengthText.getText().toString();
         try {
             boolean result = false;

             result = UserInformationService.save(title, length);

             if(result){
                     Toast.makeText(this, R.string.success, 1).show();
                 }else{
                     Toast.makeText(this, R.string.fail, 1).show();
                 }
             } catch (Exception e) {
                 e.printStackTrace();
                 Toast.makeText(this, R.string.fail, 1).show();
             }
        }
}
```

（5）编写业务类的实现文件 UserInformationService.java，主要实现代码如下：

```
public class UserInformationService {
    public static boolean save(String title, String length) throws Exception{
        String path = "http://192.163.1.100:8080/ServerForGETMethod/ServletForGETMethod";
        Map<String, String> params = new HashMap<String, String>();
        params.put("name", title);
        params.put("age", length);
        return sendGETRequest(path, params, "UTF-8");
    }
    /**
    *发送 GET 请求
    *@param path 请求路径
    *@param params 请求参数
    */
    private static boolean sendGETRequest(String path, Map<String, String> params, String encoding) throws
```

```
Exception{
        http://192.173.1.100:8080/ServerForGETMethod/ServletForGETMethod?title=xxxx&length=90
        StringBuilder sb = new StringBuilder(path);
        if(params!=null && !params.isEmpty()){
            sb.append("?");
            for(Map.Entry<String, String> entry : params.entrySet()){
                sb.append(entry.getKey()).append("=");
                sb.append(URLEncoder.encode(entry.getValue(), encoding));
                sb.append("&");
            }
            sb.deleteCharAt(sb.length() - 1);
        }
        HttpURLConnection conn = (HttpURLConnection) new URL(sb.toString()).openConnection();
        conn.setConnectTimeout(5000);
        conn.setRequestMethod("GET");
        if(conn.getResponseCode() == 200){
            return true;
        }
        return false;
    }
}
```

（6）在配置文件 AndroidManifest.xml 中声明网络访问权限，主要实现代码如下：

```
<uses-sdk android:minSdkVersion="18" />
    <application
        android:icon="@drawable/ic_launcher"
        android:label="@string/app_name" >
        <activity
            android:label="@string/app_name"
            android:name="com.guan.internet.userInformation.get.UserInformationActivity" >
            <intent-filter >
                <action android:name="android.intent.action.MAIN" />
                <category android:name="android.intent.category.LAUNCHER" />
            </intent-filter>
        </activity>
    </application>
     <uses-permission android:name="android.permission.INTERNET"/>
</manifest>
```

至此，整个实例讲解完毕，执行后的效果如图 3-7 所示。输入用户名和年龄后单击 save 按钮，会将输入的数据上传至服务器。

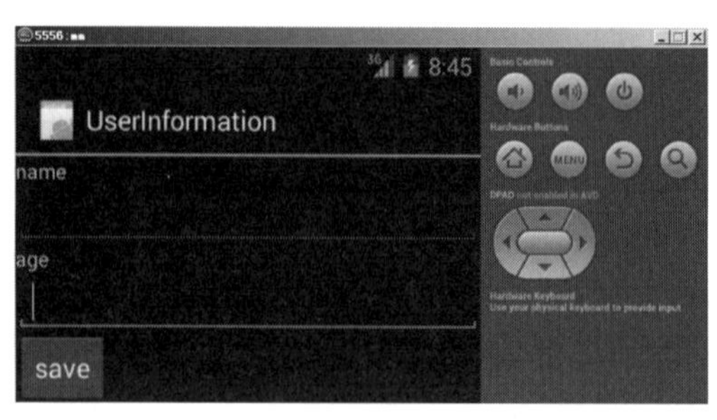


图 3-7　执行效果

### 3.4.3　实战演练——使用 POST 方式上传数据

在 Android 物联网设备中，采用 POST 方式向服务器传递数据的基本步骤如下：

（1）利用 Map 集合获取数据并进行数据处理。

（2）新建一个 StringBuilder 对象，得到 POST 传给服务器的数据。

（3）新建一个 HttpURLConnection 的 URL 对象，打开连接并传递服务器的 path，设置超时和允许对外连接数据。

（4）设置连接的 setRequestProperty 属性，并得到连接输出流。

```
outputStream =connection.getOutputStream();
```

（5）把得到的数据写入输出流中并刷新。

下面将通过一个具体实例的实现过程，介绍在 Android 物联网设备中采用 POST 方式向服务器传递数据的方法。

| 题　目 | 目　的 | 源 码 路 径 |
|---|---|---|
| 实例 3-8 | 在物联网设备中采用 POST 方式向服务器传递数据 | 光盘:\daima\3\postEX |

（1）创建一个名为 ServletForPOSTMethod 的 Servlet，功能是接收并处理通过 POST 方式上传的数据。实现文件 ServletForPOSTMethod.java 的具体实现代码如下：

```
@WebServlet("/ServletForPOSTMethod")
public class ServletForPOSTMethod extends HttpServlet {
	private static final long serialVersionUID = 1L;
	protected void doPost(HttpServletRequest request, HttpServletResponse response) throws ServletException,
IOException {
		String name= request.getParameter("name");
		String age= request.getParameter("age");
		System.out.println("name from POST method: " + name );
		System.out.println("age from POST method: " + age );
	}
}
```

（2）在配置文件 web.xml 中配置 ServletForGETMethod，具体实现代码如下：

```
<?xml version="1.0" encoding="UTF-8"?>
<web-app xmlns:xsi="http://www.w3.org/2001/XMLSchema-instance" xmlns=http://java.sun.com/xml/ns/javaee
xmlns:web="http://java.sun.com/xml/ns/javaee/web-app_2_5.xsd"
xsi:schemaLocation="http://java.sun.com/xml/ns/javaee  http://java.sun.com/xml/ns/javaee/web-app_3_0.xsd" id=
"WebApp_ID" version="3.0">
  <display-name>ServerForPOSTMethod</display-name>
  <welcome-file-list>
    <welcome-file>index.html</welcome-file>
    <welcome-file>index.htm</welcome-file>
    <welcome-file>index.jsp</welcome-file>
    <welcome-file>default.html</welcome-file>
    <welcome-file>default.htm</welcome-file>
    <welcome-file>default.jsp</welcome-file>
  </welcome-file-list>
</web-app>
```

（3）打开 Eclipse，新建一个名为 POST 的 Android 工程。然后编写界面布局文件 main.xml，具体实现代码如下：

```
<?xml version="1.0" encoding="utf-8"?>
<LinearLayout xmlns:android="http://schemas.android.com/apk/res/android"
    android:layout_width="fill_parent"
    android:layout_height="fill_parent"
    android:orientation="vertical" >
    <TextView
    android:layout_width="fill_parent"
    android:layout_height="wrap_content"
    android:text="@string/title"
    />
    <EditText
     android:layout_width="fill_parent"
     android:layout_height="wrap_content"
     android:id="@+id/title"
    />
    <TextView
    android:layout_width="fill_parent"
    android:layout_height="wrap_content"
    android:text="@string/length"
    />
    <EditText
     android:layout_width="fill_parent"
     android:layout_height="wrap_content"
     android:numeric="integer"
     android:id="@+id/length"
    />
    <Button
    android:layout_width="wrap_content"
    android:layout_height="wrap_content"
    android:text="@string/button"
    android:onClick="save"
    />
</LinearLayout>
```

（4）编写文件 UploadUserInformationByPOSTActivity.java，具体实现代码如下：

```
public class UploadUserInformationByPOSTActivity extends Activity {
     private EditText titleText;
    private EditText lengthText;
    @Override
    public void onCreate(Bundle savedInstanceState) {
        super.onCreate(savedInstanceState);
        setContentView(R.layout.main);

        titleText = (EditText) this.findViewById(R.id.title);
        lengthText = (EditText) this.findViewById(R.id.length);
    }

    public void save(View v){
     String title = titleText.getText().toString();
```

```
        String length = lengthText.getText().toString();
        try {
            boolean result = false;

            result = UploadUserInformationByPostService.save(title, length);

            if(result){
                    Toast.makeText(this, R.string.success, 1).show();
                }else{
                    Toast.makeText(this, R.string.fail, 1).show();
                }
            } catch (Exception e) {
                e.printStackTrace();
                Toast.makeText(this, R.string.fail, 1).show();
            }
        }
}
```

（5）编写业务类的实现文件 UploadUserInformationByPostService.java，主要实现代码如下：

```
public class UploadUserInformationByPostService {
    public static boolean save(String title, String length) throws Exception{
        String path = "http://192.163.1.100:8080/ServerForPOSTMethod/ServletForPOSTMethod";
        Map<String, String> params = new HashMap<String, String>();
        params.put("name", title);
        params.put("age", length);
        return sendPOSTRequest(path, params, "UTF-8");
    }

    /**
    *发送 POST 请求
    *@param path 请求路径
    *@param params 请求参数
    */
    private static boolean sendPOSTRequest(String path, Map<String, String> params, String encoding)
throws Exception{
        //    title=liming&length=30
        StringBuilder sb = new StringBuilder();
        if(params!=null && !params.isEmpty()){
            for(Map.Entry<String, String> entry : params.entrySet()){
                sb.append(entry.getKey()).append("=");
                sb.append(URLEncoder.encode(entry.getValue(), encoding));
                sb.append("&");
            }
            sb.deleteCharAt(sb.length() - 1);
        }
        byte[] data = sb.toString().getBytes();

        HttpURLConnection conn = (HttpURLConnection) new URL(path).openConnection();
        conn.setConnectTimeout(5000);
        conn.setRequestMethod("POST");
        conn.setDoOutput(true);//允许对外传输数据
        conn.setRequestProperty("Content-Type", "application/x-www-form-urlencoded");
        conn.setRequestProperty("Content-Length", data.length+"");
```

```
            OutputStream outStream = conn.getOutputStream();
            outStream.write(data);
            outStream.flush();
            if(conn.getResponseCode() == 200){
                return true;
            }
            return false;
    }
}
```

（6）编写配置文件 AndroidManifest.xml，声明网络访问权限，主要实现代码如下：

```
<manifest xmlns:android="http://schemas.android.com/apk/res/android"
    package="com.guan.internet.userInformation.post"
    android:versionCode="1"
    android:versionName="1.0" >
    <uses-sdk android:minSdkVersion="8" />
    <application
        android:icon="@drawable/ic_launcher"
        android:label="@string/app_name" >
        <activity
            android:label="@string/app_name"
            android:name="com.guan.internet.userInformation.post.UploadUserInformationByPOSTActivity" >
            <intent-filter >
                <action android:name="android.intent.action.MAIN" />
                <category android:name="android.intent.category.LAUNCHER" />
            </intent-filter>
        </activity>
    </application>
<uses-permission android:name="android.permission.INTERNET"/>
</manifest>
```

至此，整个实例讲解完毕，执行后的效果如图 3-8 所示。输入用户名和年龄后单击 save 按钮，会将输入的数据上传至服务器。

图 3-8　执行效果

# 3.5　处理 XML 数据

**知识点讲解：光盘:视频\知识点\第 3 章\处理 XML 数据.avi**

XML（eXtensible Markup Language）即可扩展标记语言，与 HTML 一样，都是 SGML（Standard Generalized Markup Language，标准通用标记语言）。XML 是 Internet 环境中跨平台的，依赖于内容的

技术，是当前处理结构化文档信息的有力工具。扩展标记语言 XML 是一种简单的数据存储语言，使用一系列简单的标记描述数据，而这些标记可以用方便的方式建立，虽然 XML 占用的空间要比二进制数据多，但是 XML 极其简单，易于掌握和使用。本节将详细讲解在 Android 智能设备中处理 XML 数据的基本知识，为读者步入本书后面知识的学习打下基础。

### 3.5.1　XML 的概述

XML 与 Access、Oracle 和 SQL Server 等数据库不同，数据库提供了更强有力的数据存储和分析能力，例如数据索引、排序、查找、相关一致性等，XML 仅是展示数据。事实上 XML 与其他数据表现形式最大的不同是它极其简单，这是一个看上去有点琐细的优点，但正是这点使 XML 与众不同。

XML 的简单使其易于在任何应用程序中读写数据，这使 XML 很快成为数据交换的唯一公共语言，虽然不同的应用软件也支持其他的数据交换格式，但不久之后它们都将支持 XML，那就意味着程序可以更容易地与 Windows、Mac OS、Linux 以及其他平台下产生的信息结合，然后可以很容易地加载 XML 数据到程序中并分析它，并以 XML 格式输出结果。

为了使得 SGML 显得用户友好，XML 重新定义了 SGML 的一些内部值和参数，去掉了大量的很少用到的功能，这些繁杂的功能使得 SGML 在设计网站时显得复杂化。XML 保留了 SGML 的结构化功能，这样就使得网站设计者可以定义自己的文档类型，XML 同时也推出一种新型文档类型，使得开发者也可以不必定义文档类型。

因为 XML 是 W3C 制定的，XML 的标准化工作由 W3C 的 XML 工作组负责，该小组成员由来自各个地方和行业的专家组成，他们通过 Email 交流对 XML 标准的意见，并提出自己的看法（www.w3.org/TR/WD-xml）。因为 XML 是一个公共格式，可以无须担心 XML 技术会成为少数公司的盈利工具，XML 不是一个依附于特定浏览器的语言。

### 3.5.2　XML 的语法

上面虽然讲解了 XML 的特点，但是初学者仍然不明白 XML 是用来做什么的，其实 XML 什么也不做，它只是用来存储数据的，对 HTML 语言进行扩展，它和 HTML 分工很明显，XML 是用来存储数据，而 HTML 是用来如何表现数据的，下面通过一段程序代码进行讲解，其代码（3-2.xml）如下：

```
<?xml version="1.0" encoding="utf-8"?>
<book>
<person>
<first>Kiran</first>
<last>Pai</last>
<age>22</age>
</person>
<person>
<first>Bill</first>
<last>Gates</last>
<age>46</age>
</person>
<person>
<first>Steve</first>
<last>Jobs</last>
```

```
<age>40</age>
</person>
</book>
```

上面的语法还可以写成汉语，如下面（3-3.xml）的代码：

```
<?xml version="1.0" encoding="utf-8"?>
        <项目>
              <名>天上星</名>
              <电子邮件>tianshangxing@hotmail.com</电子邮件>
              <住宅>何国何市何区何街道何番号</住宅>
              <电话>83-021-742745674</电话>
              <一言>XML 学习</一言>
</项目>
```

从上面两段代码可以看出，XML 的标记完全自由定义，不受约束，它只是用来存储信息，除了第一行固定以外，其他的只需前后标签一致，末标签不能省略，下面将 XML 语法格式总结如下。

☑ 在第一行必须对 XML 进行声明，也即声明 XML 的版本。

☑ 它的标记和 HTML 一样是成双成对出现的。

☑ XML 对标记的大小写十分敏感。

☑ XML 标记是用户自行定义，但是每一个标记必须有结束标记。

### 3.5.3 获取 XML 文档

如何获取 XML 文档十分简单，下面通过一个简单的 Java 代码获取 3.5.2 节讲解的 3-2.xml 中信息，其代码如下：

```
import java.io.File;
import org.w3c.dom.Document;
import org.w3c.dom.*;
import javax.xml.parsers.DocumentBuilderFactory;
import javax.xml.parsers.DocumentBuilder;
import org.xml.sax.SAXException;
import org.xml.sax.SAXParseException;
public class ReadAndPrintXMLFile{
public static void main (String argv []){
try {
    DocumentBuilderFactory docBuilderFactory
= DocumentBuilderFactory.newInstance();
                DocumentBuilder docBuilder
= docBuilderFactory.newDocumentBuilder();
                Document doc = docBuilder.parse (new File("3-2.xml"));
                doc.getDocumentElement ().normalize ();
                System.out.println ("Root element of the doc is "
  + doc.getDocumentElement().getNodeName());
                NodeList listOfPersons = doc.getElementsByTagName("person");
                int totalPersons = listOfPersons.getLength();
                System.out.println("Total no of people : " + totalPersons);
                for(int s=0; s<listOfPersons.getLength() ; s++){
                     Node firstPersonNode = listOfPersons.item(s);
                     if(firstPersonNode.getNodeType() == Node.ELEMENT_NODE){
```

```
                    Element firstPersonElement = (Element)firstPersonNode;
                   NodeList firstNameList =
 firstPersonElement.getElementsByTagName("first");
                    Element firstNameElement
= (Element)firstNameList.item(0);
                    NodeList textFNList = firstNameElement.getChildNodes();
                    System.out.println("First Name : " +
                          ((Node)textFNList.item(0)).getNodeValue().trim());
                  NodeList lastNameList
= firstPersonElement.getElementsByTagName("last");
                    Element lastNameElement = (Element)lastNameList.item(0);
                    NodeList textLNList = lastNameElement.getChildNodes();
                    System.out.println("Last Name : " +
                           ((Node)textLNList.item(0)).getNodeValue().trim());
                    NodeList ageList
 = firstPersonElement.getElementsByTagName("age");
                    Element ageElement = (Element)ageList.item(0);
                    NodeList textAgeList = ageElement.getChildNodes();
                    System.out.println("Age : " +
 ((Node)textAgeList.item(0)).getNodeValue().trim());
                }    }  }
        catch (SAXParseException err)
{
               System.out.println ("** Parsing error" + ", line "
                                          + err.getLineNumber () + ", uri " + err.getSystemId ());
               System.out.println(" " + err.getMessage ());     }
        catch (SAXException e) {
                Exception x = e.getException ();
                ((x == null) ? e : x).printStackTrace ();
        }
        catch (Throwable t) {
                t.printStackTrace ();
 }
          }
}
```

用户在 Java API 中还可以找到更多操作 XML 文档的方法，执行上述代码后得到如图 3-9 所示的结果。

**注意：** XML 文档其实比 HTML 文档更简单，XML 主要用来存储信息，不负责显示在页面。获取 XML 文档的方法有很多，也并不是只有 Java 语言，还有许多语言都可以调用，如 C#、PHP 和 ASP 等，也包括 HTML 语言。

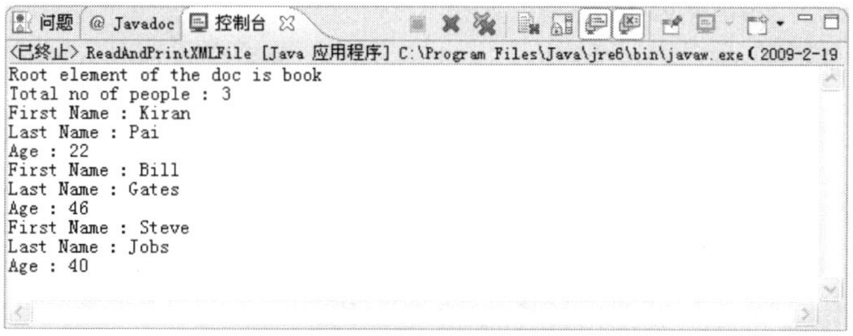

```
问题 @ Javadoc 控制台 ☒
<已终止> ReadAndPrintXMLFile [Java 应用程序] C:\Program Files\Java\jre6\bin\javaw.exe ( 2009-2-19
Root element of the doc is book
Total no of people : 3
First Name : Kiran
Last Name : Pai
Age : 22
First Name : Bill
Last Name : Gates
Age : 46
First Name : Steve
Last Name : Jobs
Age : 40
```

图 3-9　获取 XML 文档

### 3.5.4 SAX 常用的接口和类

SAX，全称 Simple API for XML，既是一种接口，也是一个软件包。SAX 最初是由 David Megginson 采用 Java 语言开发的，之后 SAX 很快在 Java 开发者中流行起来。San 现在负责管理其原始 API 的开发工作，这是一种公开的、开放源代码的软件。不同于其他大多数 XML 标准的是，SAX 没有语言开发商必须遵守的标准 SAX 参考版本。因此，SAX 的不同实现可能采用区别很大的接口。

在现实开发应用中，SAX 将其事件分为如下所示的接口。

☑ ContentHandler：定义与文档本身关联的事件（例如，开始和结束标记）。大多数应用程序都注册这些事件。

☑ DTDHandler：定义与 DTD 关联的事件。然而，它不定义足够的事件来完整地报告 DTD。如果需要对 DTD 进行语法分析，请使用可选的 DeclHandler。DeclHandler 是 SAX 的扩展，并且不是所有的语法分析器都支持它。

☑ EntityResolver：定义与装入实体关联的事件。只有少数几个应用程序注册这些事件。

☑ ErrorHandler：定义错误事件。许多应用程序注册这些事件以便用它们自己的方式报错。

为简化工作，SAX 在 DefaultHandler 类中提供了这些接口的默认实现。在大多数情况下，为应用程序扩展 DefaultHandler 并覆盖相关的方法要比直接实现一个接口更容易。

#### 1．XMLReader

如果为注册事件处理器并启动语法分析器，应用程序应该使用 XMLReader 接口，实现方法是使用 XMLReader 中的 parse()方法来启动，具体语法格式如下：

```
parser.parse(args[0]);
```

XMLReader 中的主要方法如下。

☑ parse()：对 XML 文档进行语法分析。parse()有两个版本：一个接收文件名或 URL，另一个接收 InputSource 对象。

☑ setContentHandler()、setDTDHandler()、setEntityResolver()和 setErrorHandler()：让应用程序注册事件处理器。

☑ setFeature()和 setProperty()：控制语法分析器如何工作。它们采用一个特性或功能标识（一个类似于名称空间的 URI 和值）。功能采用 Boolean 值，而特性采用"对象"。

最常用的 XMLReaderFactory 功能如下所示。

☑ http://xml.org/sax/features/namespaces：所有 SAX 语法分析器都能识别它。如果将它设置为 true（默认值），则在调用 ContentHandler 的方法时，语法分析器将识别出名称空间并解析前缀。

☑ http://xml.org/sax/features/validation：它是可选的。如果将它设置为 true，则验证语法分析器将验证该文档。非验证语法分析器忽略该功能。

#### 2．XMLReaderFactory

XMLReaderFactory 用于创建语法分析器对象，它定义了 createXMLReader()的如下两个版本。

☑ 一个采用语法分析器的类名作为参数。

☑ 一个从 org.xml.sax.driver 系统特性中获得类名称。

对于 Xerces，类是 org.apache.xerces.parsers.SAXParser。应该使用 XMLReaderFactory，因为它易于

切换至另一种 SAX 语法分析器。实际上，只需要更改一行然后重新编译。

```
XMLReaderparser=XMLReaderFactory.createXMLReader(
"org.apache.xerces.parsers.SAXParser");
```

为获得更大的灵活性，应用程序可以从命令行读取类名或使用不带参数的 createXMLReader()。因此可以不重新编译就更改语法分析器。

### 3．InputSource

InputSource 控制语法分析器如何读取文件，包括 XML 文档和实体。在大多数情况下，文档是从 URL 装入的。但有特殊需求的应用程序可以覆盖 InputSource。例如，这可以用来从数据库中装入文档。

### 4．ContentHandler

ContentHandler 是最常用的 SAX 接口，因为它定义 XML 文档的事件。在 ContentHandler 中声明了如下所示的事件。

（1）startDocument()/endDocument()：通知应用程序文档的开始或结束。

（2）startElement()/endElement()：通知应用程序标记的开始或结束。属性作为 Attributes 参数传递。即使只有一个标记，“空”元素（例如，<imghref="logo.gif"/>）也生成 startElement()和 endElement()。

（3）startPrefixMapping()/endPrefixMapping()：通知应用程序名称空间作用域。几乎不需要该信息，因为当 http://xml.org/sax/features/namespaces 为 true 时，语法分析器已经解析了名称空间。

（4）当语法分析器在元素中发现文本（已经过语法分析的字符数据）时，characters()/ignorable Whitespace()会通知应用程序。要知道，语法分析器负责将文本分配到几个事件（更好地管理其缓冲区）。ignorableWhitespace 事件用于由 XML 标准定义的可忽略空格。

（5）processingInstruction()：将处理指令通知应用程序。

（6）skippedEntity()：通知应用程序已经跳过了一个实体（即当语法分析器未在 DTD/schema 中发现实体声明时）。

（7）setDocumentLocator()：将 Locator 对象传递到应用程序；请注意，不需要 SAX 语法分析器提供 Locator，但是如果它提供了，则必须在任何其他事件之前激活该事件。

### 5．属性

在 startElement()事件中，应用程序在 Attributes 参数中接收属性列表。

```
Stringattribute=attributes.getValue("","price");
```

Attributes 定义下列方法。

☑ getValue(i)/getValue(qName)/getValue(uri,localName)：返回第 i 个属性值或给定名称的属性值。

☑ getLength()：返回属性数目。

☑ getQName(i)/getLocalName(i)/getURI(i)：返回限定名（带前缀）、本地名（不带前缀）和第 i 个属性的名称空间 URI。

☑ getType(i)/getType(qName)/getType(uri,localName)：返回第 i 个属性的类型或者给定名称的属性类型。类型为字符串，即在 DTD 所使用的 CDATA、ID、IDREF、IDREFS、NMTOKEN、NMTOKENS、ENTITY、ENTITIES 或 NOTATION。

**注意：** Attributes 参数仅在 startElement()事件期间可用。如果在事件之间需要它，则用 AttributesImpl 复制一个。

### 6．定位器

Locator 为应用程序提供行和列的位置。不需要语法分析器来提供 Locator 对象。Locator 定义下列方法。

- ☑ getColumnNumber()：返回当前事件结束时所在的那一列。在 endElement()事件中，它将返回结束标记所在的最后一列。
- ☑ getLineNumber()：返回当前事件结束时所在的行。在 endElement()事件中，它将返回结束标记所在的行。
- ☑ getPublicId()：返回当前文档事件的公共标识。
- ☑ getSystemId()：返回当前文档事件的系统标识。

### 7．DTDHandler

DTDHandler 声明两个与 DTD 语法分析器相关的事件，具体如下所示。

- ☑ notationDecl()：通知应用程序已经声明了一个标记。
- ☑ nparsedEntityDecl()：通知应用程序已经发现了一个未经过语法分析的实体声明。

### 8．EntityResolver

EntityResolver 接口仅定义一个事件 resolveEntity()，它返回 InputSource。因为 SAX 语法分析器已经可以解析大多数 URL，所以很少应用程序实现 EntityResolver。例外情况是目录文件，它将公共标识解析成系统标识。如果在应用程序中需要目录文件，请下载 NormanWalsh 的目录软件包（请参阅参考资料）。

### 9．ErrorHandler

ErrorHandler 接口定义错误事件。处理这些事件的应用程序可以提供定制错误处理。安装了定制错误处理器后，语法分析器不再抛出异常。抛出异常是事件处理器的责任。接口定义了错误的 3 个级别或严重性对应的 3 个方法。

- ☑ warning()：警示那些不是由 XML 规范定义的错误。例如，当没有 XML 声明时，某些语法分析器发出警告。它不是错误（因为声明是可选的），但是它可能值得注意。
- ☑ error()：警示那些由 XML 规范定义的错误。
- ☑ fatalError()：警示那些由 XML 规范定义的致命错误。

### 10．SAXException

SAX 定义的大多数方法都可以抛出 SAXException。当对 XML 文档进行语法分析时，SAXException 会抛出一个错误，这里的错误可以是语法分析错误，也可以是事件处理器中的错误要报告来自事件处理器的其他异常，可以将异常封装在 SAXException 中。

## 3.5.5 实战演练——使用 SAX 解析 XML 数据

Android 是最常用的智能手机平台，XML 是数据交换的标准媒介。在 Android 系统中可以使用标准的 XML 生成器、解析器、转换器 API，对 XML 进行解析和转换。本实例的功能是在物联网中使用 SAX 解析 XML 数据。

| 题　目 | 目　的 | 源 码 路 径 |
| --- | --- | --- |
| 实例 3-9 | 在物联网中使用 SAX 解析 XML 数据 | 光盘:\daima\3\XML_ParserEX |

（1）编写布局文件 main.xml，具体实现代码如下：

```
<?xml version="1.0" encoding="utf-8"?>
<LinearLayout xmlns:android="http://schemas.android.com/apk/res/android"
    android:layout_width="fill_parent"
    android:layout_height="fill_parent"
    android:orientation="vertical" >
    <TextView
        android:layout_width="fill_parent"
        android:layout_height="wrap_content"
        android:text="@string/hello" />

</LinearLayout>
```

（2）解析功能的核心文件是 SAXForHandler.java，主要实现代码如下：

```
public class SAXForHandler extends DefaultHandler {
    private static final String TAG = "SAXForHandler";
    private List<Person> persons;
    private String perTag ;//通过此变量，记录前一个标签的名称
    Person person;//记录当前 Person

    public List<Person> getPersons() {
        return persons;
    }

    //适合在此事件中触发初始化行为
    public void startDocument() throws SAXException {
        persons = new ArrayList<Person>();
        Log.i(TAG , "***startDocument()***");
    }

    public void startElement(String uri, String localName, String qName,
            Attributes attributes) throws SAXException {
        if("person".equals(localName)){
            for ( int i = 0; i < attributes.getLength(); i++ ) {
                Log.i(TAG ,"attributeName:" + attributes.getLocalName(i)
                        + "_attribute_Value:" + attributes.getValue(i));
                person = new Person();
                person.setId(Integer.valueOf(attributes.getValue(i)));
            }
        }
        perTag = localName;
        Log.i(TAG , qName+"***startElement()***");
    }

    public void characters(char[] ch, int start, int length) throws SAXException {
        String data = new String(ch, start, length).trim();
       if(!"".equals(data.trim())){
```

```
                Log.i(TAG ,"content: " + data.trim());
            }
            if("name".equals(perTag)){
                    person.setName(data);
            }else if("age".equals(perTag)){
                    person.setAge(new Short(data));
            }
    }

    public void endElement(String uri, String localName, String qName)
            throws SAXException {
        Log.i(TAG , qName+"***endElement()***");
        if("person".equals(localName)){
            persons.add(person);
            person = null;
        }
        perTag = null;
    }

    public void endDocument() throws SAXException {
        Log.i(TAG , "***endDocument()***");
    }
}
```

（3）单元测试文件 PersonServiceTest.java 的具体代码如下：

```
public void testSAXGetPersons() throws Throwable{
    InputStream inputStream = this.getClass().getClassLoader().
            getResourceAsStream("wang.xml");
    SAXForHandler saxForHandler = new SAXForHandler();
    SAXParserFactory spf = SAXParserFactory.newInstance();
    SAXParser saxParser = spf.newSAXParser();
    saxParser.parse(inputStream, saxForHandler);
    List<Person> persons = saxForHandler.getPersons();
    inputStream.close();
    for(Person person:persons){
        Log.i(TAG, person.toString());
    }
}
```

此时使用 Eclipse 启动 Android 模拟器，执行后的效果如图 3-10 所示。

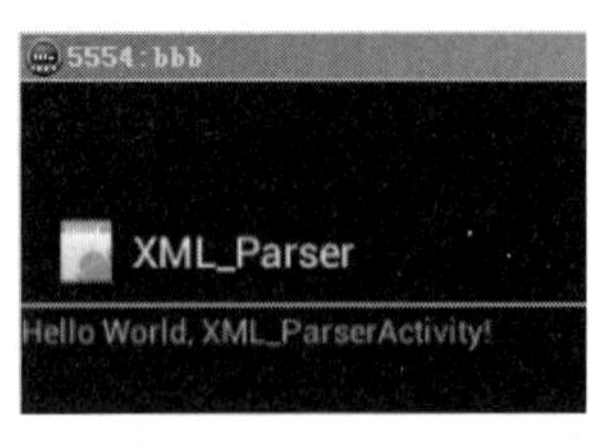


图 3-10　执行效果

（4）开始具体测试，在 Eclipse 中导入本实例项目，在 Outline 面板中右击 testSAXGetPersons():void，如图 3-11 所示，在弹出的快捷菜单中选择 Run As | Android JUnit Test 命令，如图 3-12 所示。

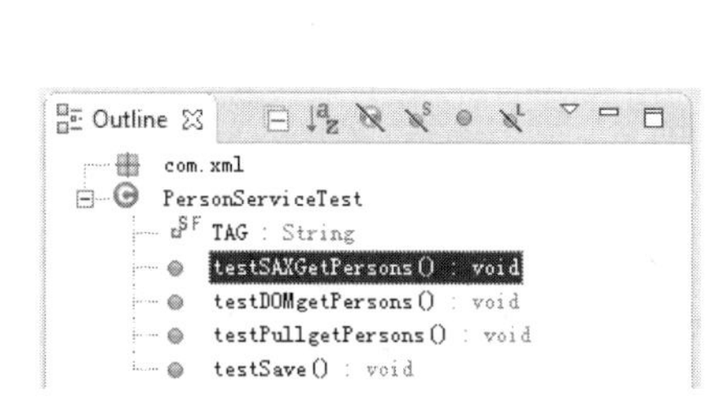


图 3-11　右击 testSAXGetPersons:void

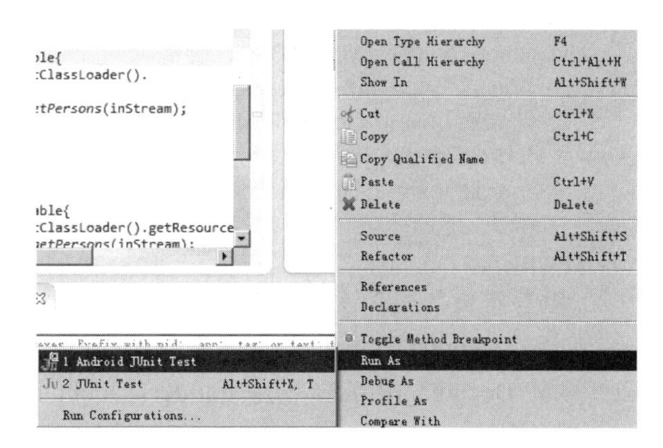


图 3-12　选择 Android JUnit Test 命令

此时将在 Logcat 中显示测试的解析结果，如图 3-13 所示。

Declaration　Console　LogCat

Search for messages. Accepts Java regexes. Prefix with pid:, app:, tag: or text: to limit scope.　verbose

| L... | Time | PID | TID | Application | Tag | Text |
| --- | --- | --- | --- | --- | --- | --- |
| I | 10-20 10:54:12.385 | 1349 | 1362 | com.mmm.xml | SAXForHandler | age***endElement()*** |
| I | 10-20 10:54:12.395 | 1349 | 1362 | com.mmm.xml | SAXForHandler | person***endElement()** |
| I | 10-20 10:54:12.395 | 1349 | 1362 | com.mmm.xml | SAXForHandler | persons***endElement()* |
| I | 10-20 10:54:12.405 | 1349 | 1362 | com.mmm.xml | SAXForHandler | ***endDocument()*** |
| I | 10-20 10:54:12.405 | 1349 | 1362 | com.mmm.xml | PersonServi... | id=23,name=NEwii,age=21 |
| I | 10-20 10:54:12.405 | 1349 | 1362 | com.mmm.xml | PersonServi... | id=20,name=WiiWill,age= |
| I | 10-20 10:54:12.405 | 1349 | 1362 | com.mmm.xml | TestRunner | finished: testSAXGetPer |
| I | 10-20 10:54:12.405 | 1349 | 1362 | com.mmm.xml | TestRunner | passed: testSAXGetPerso |
| D | 10-20 10:54:12.456 | 1339 | 1339 |  | AndroidRuntime | Shutting down VM |

图 3-13　解析结果

**注意：**如果 Android 下的 Eclipse 界面中没有 Logcat 面板，只需依次选择 Eclipse 菜单栏中的 Window | show view | other | Android 命令，然后选择 Logcat 后即可在 Eclipse 界面看到 Logcat 面板。

### 3.5.6　实战演练——使用 DOM 解析 XML 数据

DOM 是 Document Object Model 的简称，被译为文件对象模型，是 W3C 组织推荐的处理可扩展置标语言的标准编程接口。Document Object Model 的历史可以追溯至 20 世纪 90 年代后期微软与 Netscape 的“浏览器大战”，双方为了在 JavaScript 与 Jscript 间一决生死，于是大规模地赋予浏览器强大的功能。微软在网页技术上加入了不少专属事物，即有 VBScript、ActiveX 以及微软的 DHTML 格式等，使不少网页使用非微软平台及浏览器无法正常显示。本实例将演示在 Android 物联网中使用 DOM 技术来解析并生成 XML 的方法。

| 题　　目 | 目　　的 | 源 码 路 径 |
| --- | --- | --- |
| 实例 3-10 | 在物联网中使用 DOM 解析 XML 数据 | 光盘:\daima\3\XML_ParserEX |

本实例的具体实现流程如下。

（1）编写布局文件 main.xml，具体实现代码如下：

```
<?xml version="1.0" encoding="utf-8"?>
<LinearLayout xmlns:android="http://schemas.android.com/apk/res/android"
    android:layout_width="fill_parent"
    android:layout_height="fill_parent"
```

```
    android:orientation="vertical" >
    <TextView
        android:layout_width="fill_parent"
        android:layout_height="wrap_content"
        android:text="@string/hello" />

</LinearLayout>
```

（2）编写解析功能的核心文件 DOMPersonService.java，具体实现流程如下：

☑ 创建 DocumentBuilderFactory 对象 factory，并调用 newInstance()创建新实例。

☑ 创建 DocumentBuilder 对象 builder，DocumentBuilder 将实现具体的解析工作以创建 Document 对象。

☑ 解析目标 XML 文件以创建 Document 对象。

文件 DOMPersonService.java 的具体实现代码如下：

```
public class DOMPersonService {
    public static List<Person> getPersons(InputStream inStream) throws Exception{
        List<Person> persons = new ArrayList<Person>();
        DocumentBuilderFactory factory = DocumentBuilderFactory.newInstance();
        DocumentBuilder builder = factory.newDocumentBuilder();
        Document document = builder.parse(inStream);
        Element root = document.getDocumentElement();
        NodeList personNodes = root.getElementsByTagName("person");
        for(int i=0; i < personNodes.getLength() ; i++){
            Element personElement = (Element)personNodes.item(i);
            int id = new Integer(personElement.getAttribute("id"));
            Person person = new Person();
            person.setId(id);
            NodeList childNodes = personElement.getChildNodes();
            for(int y=0; y < childNodes.getLength() ; y++){
                if(childNodes.item(y).getNodeType()==Node.ELEMENT_NODE){
                    if("name".equals(childNodes.item(y).getNodeName())){
                        String name = childNodes.item(y).getFirstChild().getNodeValue();
                        person.setName(name);
                    }else if("age".equals(childNodes.item(y).getNodeName())){
                        String age = childNodes.item(y).getFirstChild().getNodeValue();
                        person.setAge(new Short(age));
                    }
                }
            }
            persons.add(person);
        }
        inStream.close();
        return persons;
    }
}
```

（3）编写单元测试文件 PersonServiceTest.java，具体代码如下：

```
public void testDOMgetPersons() throws Throwable{
    InputStream inStream = this.getClass().getClassLoader().
            getResourceAsStream("wang.xml");
    List<Person> persons = DOMPersonService.getPersons(inStream);
```

```
    for(Person person : persons){
        Log.i(TAG, person.toString());
    }
}
```

（4）开始具体测试，在 Eclipse 中导入本实例项目，在 Outline 面板中右击 testDOMgetPersons():void，如图 3-14 所示，在弹出的快捷菜单中选择 Run As | Android JUnit Test 命令，如图 3-15 所示。

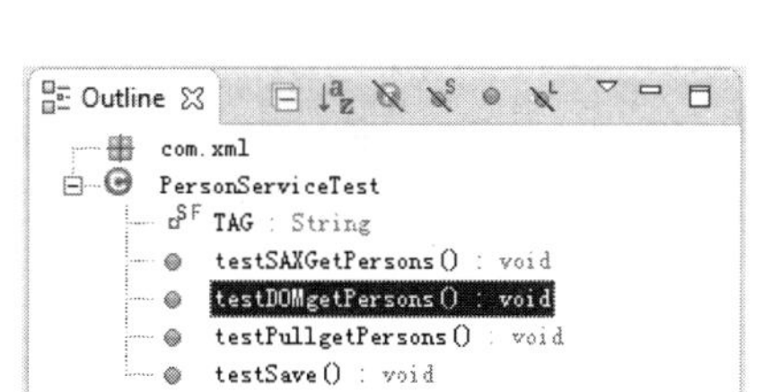


图 3-14　右击 testDOMgetPersons():void

图 3-15　选择 Android JUnit Test 命令

此时将在 Logcat 中显示测试的解析结果，如图 3-16 所示。

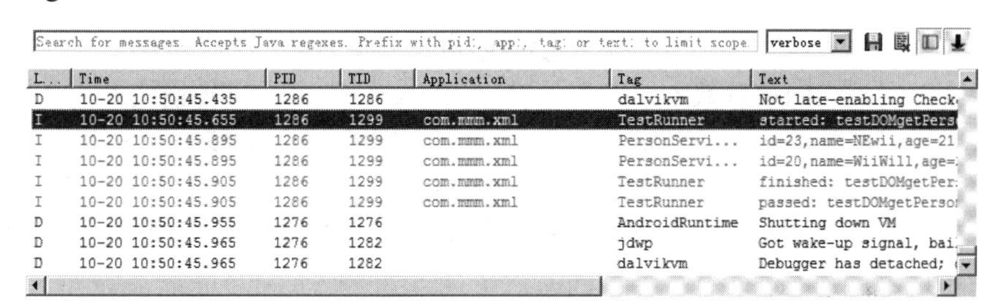


图 3-16　解析结果

# 第4章 蓝牙技术详解

蓝牙这一名称来自于10世纪的一位丹麦国王 Harald Blatand，Blatand 在英文里的意思可以被解释为 Bluetooth。因为国王喜欢吃蓝莓，牙龈每天都是蓝色的所以叫蓝牙。蓝牙的创始人是瑞典爱立信公司，爱立信早在1994年就已进行研发。1997年，爱立信与其他设备生产商联系，并激发了他们对该项技术的浓厚兴趣。1998年2月，5个跨国大公司，包括爱立信、诺基亚、IBM、东芝及 Intel 组成了一个特殊兴趣小组（SIG），他们共同的目标是建立一个全球性的小范围无线通信技术，也就是现在的蓝牙。在 Android 设备中，蓝牙技术是常用的一种近距离数据传输技术。本章将详细讲解蓝牙技术的基本知识。

## 4.1 短距离无线通信技术概览

**知识点讲解：光盘:视频\知识点\第4章\短距离无线通信技术概览.avi**

在物联网中物与网相连的最后数米，发挥关键作用的是短距离无线传输技术。目前有多种短距离无线传输技术可以应用在物联网中，在我国，除已经得到大规模应用的 RFID 之外，还有 WiFi、ZigBee、蓝牙等比较成熟的技术，以及基于这些技术发展而来的新技术。这些技术各具特点，因对其传输速度、距离、耗电量等方面的要求不同，形成了各自不同的物联网应用场景。本节将简要介绍当今实现短距离无线通信的常用技术。

### 4.1.1 ZigBee——低功耗、自组网

ZigBee 以其鲜明的技术特点在物联网中受到了高度关注，该技术使用的频段分别为 2.4GHz、868MHz（欧洲）及 915MHz（美国）。其主要的技术特点如下：一是数据传输速率低，只有 10Kbps～250Kbps；二是功耗低，低传输速率带来了仅为1毫瓦的低发射功率。据估算，ZigBee 设备仅靠两节5号电池就可以维持长达6个月到两年左右的使用时间，这是 ZigBee 的一个独特优势；三是成本低，因为 ZigBee 传输速率低、协议简单；四是网络容量大，每个 ZigBee 网络最多可以支持255个设备，一个区域内可以同时存在最多100个 ZigBee 网络，网络组成灵活。ZigBee 芯片的主要企业有德州仪器、飞思卡尔等。市场调研机构 ABI Research 的一份数据显示，2005—2012年，ZigBee 市场的年均复合增长率为63%。

"ZigBee 是从家庭自动化开始的，在瑞典哥德堡就是从智能电表开始，然后进一步用到燃气表、水表、热力表等家庭各种计量表。"在2011年中国无线世界暨物联网大会上 ZigBee 联盟大中华区代表黄家瑞说："ZigBee 在智能电表里不仅仅是远程抄表工具，它是一个终端，也是一个网关，这些网关结合在一起，整个小区就变成了智能电网小区，智能电表可以搜集家里所有家电的用电信息。"

目前，ZigBee 正在完善其网关标准，2011年7月底发布了第10个标准 ZigBee Gateway（ZigBee

网关）。ZigBee Gateway 提供了一种简单、高成本效益的互联网连接方式，使服务提供商、企业和个人消费者有机会运行这些设备并将 ZigBee 网络连接至互联网。ZigBee Gateway 是 ZigBee Network Devices（ZigBee 网络设备）这一新类别范畴的首个标准，这将使 ZigBee 发展进一步提速。

### 4.1.2　WiFi——大带宽支持家庭互联

WiFi 是以太网的一种无线扩展技术，如果有多个用户同时通过一个热点接入，带宽将被这些用户共享，WiFi 的速率会降低，处于 2.4GHz 频段的 WiFi 信号受墙壁阻隔的影响较小。WiFi 的传输速率随着技术的演进还在不断提高，我国电信运营商在构建无线城市中采用的 WiFi 技术部分已经升级到 802.11n，最高速率从 802.11g 标准的 11Mbps 提高到 50Mbps 以上。在 WiFi 产业链中，最大的芯片企业是博通。

“在过去几年里整个 WiFi 技术和产品发货量达到 20 亿个，整个 WiFi 产品销售每年都是以两位数的速度持续增长。”WiFi 联盟董事 Myron Hattig 说：“在 2011 年我们还会销售 10 亿个产品。”

在笔记本电脑和手机上已经得到广泛应用的 WiFi 正在向消费电子产品渗透，Myron Hattig 说：“除了手机外，已经有 25%的消费类电子设备使用 WiFi，在打印机、洗衣机上都在使用 WiFi，家用电器生产商协会将 WiFi 作为一个更高级别的智能电器沟通技术。WiFi 可以将设备与设备相连，从而使整个家庭的家用电器、电子设备相连。”

最大 WiFi 芯片制造商博通正在推动 WiFi Direct 标准的商用，以支持这种设备到设备的直连。特别是在家庭互联中，相片、视频等大数据量的业务在手机、平板电脑、电视等设备中的直连应用前景广阔。Myron Hattig 告诉记者：“直连技术可将平板电脑的内容展示在电视上，相关产品会在 2012 年发布。”

基于 WiFi 上发展起来的 WIGIG 也是未来家庭互联市场有力的竞争技术。该技术可工作在 40GHz～60GHz 的超高频段，其传输速度可以达到 1Gbps 以上，不能穿过墙壁。目前英特尔、高通等芯片企业在支持 WIGIG 发展，该技术还在完善中，如需要进一步降低功耗等。

### 4.1.3　蓝牙——4.0 进入低功耗时代

使用“蓝牙”技术可以有效地简化移动通信终端设备之间的通信，也能够成功地简化设备与互联网之间的通信，从而使数据传输变得更加迅速高效，为无线通信拓宽道路。蓝牙采用分散式网络结构以及快跳频和短包技术，支持点对点及点对多点通信，工作在全球通用的 2.4GHz ISM（即工业、科学、医学）频段。蓝牙技术的数据传输速率为 1Mbps，采用时分双工传输方案实现全双工传输。

蓝牙是一种 Bluetooth 传输无线技术，许多行业的制造商都积极地在其产品中实施此技术，以减少使用零乱的电线，实现无缝连接、流传输立体声，传输数据或进行语音通信。Bluetooth 技术在 2.4GHz 波段运行，该波段是一种无须申请许可证的工业、科技、医学（ISM）无线电波段。正因如此，使用 Bluetooth 技术不需要支付任何费用。但必须向手机提供商注册使用 GSM 或 CDMA，除了设备费用外，不需要为使用 Bluetooth 技术再支付任何费用。

Bluetooth 技术得到了空前广泛的应用，集成该技术的产品从手机、汽车到医疗设备，使用该技术的用户从消费者、工业市场到企业等，不一而足。低功耗、小体积以及低成本的芯片解决方案使得 Bluetooth 技术甚至可以应用于极微小的设备中。

Bluetooth 技术是一项即时技术，它不要求固定的基础设施，且易于安装和设置，不需要电缆即可实现连接。新用户使用亦不费力，只需拥有 Bluetooth 品牌产品，检查可用的配置文件，将其连接至使用同一配置文件的另一 Bluetooth 设备即可。后续的 PIN 码流程就如同用户在 ATM 机器上操作一样简单。外出时，可以随身带上用户的个人局域网（PAN），甚至可以与其他网络连接。

蓝牙可以在包括移动电话、PDA、无线耳机、笔记本电脑、相关外设等众多设备之间进行无线信息交换。蓝牙采用分散式网络结构以及快跳频和短包技术，支持点对点及点对多点通信，工作在全球通用的 2.4GHz 频段，其数据速率为 1Mbps。

2010 年 7 月，以低功耗为特点的蓝牙 4.0 标准推出，蓝牙大中华区技术市场经理吕荣良将其看作蓝牙第二波发展高潮的标志，他表示："蓝牙可以跨领域应用，主要有 4 个生态系统，分别是智能手机与笔记本电脑等终端市场、消费电子市场、汽车前装市场和健身运动器材市场。"

NFC 和 UWB 曾经是十分受关注的短距离无线接入技术，但其发展已经日渐势微。业内专家认为，无线频谱的规划和利用在短距离通信中日益重要。短距离通信技术目前主要采用 2.4GHz 的开放频谱，但随着物联网的发展和大量短距离通信技术的应用，频谱需求会快速增长，视频、图像等大数据量的通信正在寻求更高频段的解决方案。

### 4.1.4 NFC——必将逐渐远离历史舞台

NFC 是近场通信（Near Field Communication）的缩写，此技术由非接触式射频识别（RFID）演变而来，由飞利浦半导体（现恩智浦半导体）、诺基亚和索尼共同研制开发，其基础是 RFID 及互连技术。NFC 是一种短距高频的无线电技术，在 13.56MHz 频率运行于 20 厘米距离内。其传输速度有 106Kbit/秒、212Kbit/秒或者 424Kbit/秒 3 种。目前近场通信已通过成为 ISO/IEC IS 18092 国际标准、ECMA-340 标准与 ETSI TS 102 190 标准。NFC 采用主动和被动两种读取模式。

NFC 近场通信技术是由非接触式射频识别（RFID）及互联互通技术整合演变而来的，在单一芯片上结合感应式读卡器、感应式卡片和点对点的功能，能在短距离内与兼容设备进行识别和数据交换。工作频率为 13.56MHz，但是使用这种手机支付方案的用户必须更换特制的手机。目前这项技术在日韩被广泛应用。手机用户凭着配置了支付功能的手机就可以行遍全国：他们的手机可以用作机场登机验证、大厦的门禁钥匙、交通一卡通、信用卡、支付卡等。

NFC 和蓝牙（Bluetooth）都是短程通信技术，而且都被集成到移动电话。但 NFC 不需要复杂的设置程序。NFC 也可以简化蓝牙连接。NFC 略胜蓝牙的地方在于设置程序较短，但无法达到低功率蓝牙（Bluetooth Low Energy）的速度。在两台 NFC 设备相互连接的设备识别过程中，使用 NFC 来替代人工设置会使创建连接的速度大大加快，会少于十分之一秒。

## 4.2 低功耗蓝牙基础

**知识点讲解：光盘:视频\知识点\第 4 章\低功耗蓝牙基础.avi**

BLE 是 Bluetooth Low Energy 的缩写，意为低功耗蓝牙，是对传统蓝牙 BR/EDR 技术的补充。尽管 BLE 和传统蓝牙都被称为蓝牙标准，并且都共享射频，但 BLE 是一个完全不一样的技术。BLE 不具备和传统蓝牙 BR/EDR 的兼容性，是专为小数据率、离散传输的应用而设计的。本节将详细讲解低

功耗蓝牙技术的基本知识。

### 4.2.1 低功耗蓝牙的架构

BLE 协议架构总体上分成 3 层，从下到上分别是控制器（Controller）、主机（Host）和应用端（Apps）。三者可以在同一芯片内实现，也可以分不同芯片实现，控制器（Controller）是处理射频数据解析、接收和发送，主机（Host）是控制不同设备之间如何进行数据交换，应用端（Apps）实现具体应用。

（1）控制器（Controller）

Controller 实现射频相关的模拟和数字部分，完成最基本的数据发送和接收，Controller 对外接口是天线，对内接口是主机控制器接口 HCI（Host Controller Interface）；控制器包含物理层 PHY（Physical Layer）、链路层 LL（Linker Layer）、直接测试模式 DTM（Direct Test Mode）以及主机控制器接口 HCI。

☑ 物理层 PHY

GFSK 信号调制，2402～2480MHz，40 个 channel，每两个 channel 间隔 2MHz（经典蓝牙协议是 1MHz），数据传输速率是 1Mbps。

☑ 直接测试模式 DTM

为射频物理层测试接口，射频数据分析之用。

☑ 链路层 LL

基于物理层 PHY 之上，实现数据通道分发、状态切换、数据包校验和加密等；链路层 LL 分两种通道：广播通道（Advertising Channels）和数据通道（Data Channels）；广播通道有 3 个，即 37ch（2402MHz）、38ch（2426MHz）和 39ch（2480MHz），每次广播都会往这 3 个通道同时发送（并不会在这 3 个通道之间跳频），为防止某个通道被其他设备阻塞，以至于设备无法配对或广播数据，只所以定 3 个广播通道是一种权衡，少了可能会被阻塞，多了会加大功耗，另外 3 个广播通道刚好避开了 WiFi 的 1ch、6ch 和 11ch，所以在 BLE 广播时，不至于被 WiFi 影响（如果要干扰 BLE 广播数据，一个最简单的办法就是，同时阻塞 3 个广播通道）；当 BLE 匹配之后，链路层 LL 由广播通道切换到数据通道，数据通道有 37 个，数据传输时会在这 37 个通道间切换，切换规则在设备间匹配时约定。

（2）主机（Host）/控制器（Controller）接口 HCI

HCI 作为一种接口，存在于主机和控制器中，控制器通过 HCI 发送数据和事件给主机，主机通过 HCI 发送命令和数据给控制器。HCI 逻辑上定义一系列的命令、事件；物理上有 UART、SDIO 和 USB，实际可能包含里面的任意一种或几种。

### 4.2.2 低功耗蓝牙分类

BLE 通常应用在传感器和智能手机或者平板的通信中。到目前为止，只有很少的智能机和平板支持 BLE，如 iPhone 4S 以后的苹果手机、Motorola Razr 和 the new iPad 及其以后的 iPad。安卓手机也逐渐支持 BLE，安卓的 BLE 标准在 2013 年 7 月 24 日刚发布。智能机和平板会带双模蓝牙的基带和协议栈，协议栈中包括 GATT 及以下的所有部分，但是没有 GATT 之上的具体协议。所以，这些具体的协议需要在应用程序中实现，实现时需要基于各个 GATT API 集。这样有利于在智能机端简单地实现具体协议，也可以在智能机端简单地开发出一套基于 GATT 的私有协议。

在现实应用中，低功耗蓝牙分为单模（Bluetooth Smart）和双模（Bluetooth Smart Ready）两种设备。BLE 和蓝牙 BR/EDR 有所区分，这样可以让我们用 3 种方式将蓝牙技术集成到具体设备中。因为不再是所有现有的蓝牙设备可以和另一个蓝牙设备进行互联，所以准确描述产品中蓝牙的版本是非常重要的。下面将详细讲解单模蓝牙和双模蓝牙的基本知识。

（1）单模蓝牙

单模蓝牙设备被称为 Bluetooth Smart 设备，并且有专用的 Logo，如图 4-1 所示。

在现实应用中，手表、运动传感器等小型设备通常是基于低功耗单模蓝牙的。为了实现极低的功耗效果，在硬件和软件上都进行了优化，这样的设备只能支持 BLE。单模蓝牙芯片往往是一个带有单模蓝牙协议栈的产品，这个协议栈通常是芯片商免费提供的。

（2）双模蓝牙

双模蓝牙设备被称为 Bluetooth Smart Ready 设备，并且有专用的 Logo，如图 4-2 所示。

图 4-1　Bluetooth Smart 设备

图 4-2　Bluetooth Smart Ready 设备

双模设备支持蓝牙 BR/EDR 和 BLE。在双模设备中，BR/EDR 和 BLE 技术使用同一个射频前端和天线。典型的双模设备有智能手机、平板电脑、PC 和 Gateway。这些设备可以接收到通过 BLE 或者蓝牙 BR/EDR 设备发送过来的数据，这些设备往往都有足够的供电能力。双模设备和 BLE 设备通信的功耗低于双模设备和蓝牙 BR/EDR 设备通信的功耗。在使用双模解决方案时，需要用一个外部处理器才可以实现蓝牙协议栈。

### 4.2.3　低功耗蓝牙的集成方式

尽管有单模和双模方案的区别，但是在设备中集成蓝牙技术的方式有多种，其中最为常用的方式有模块和芯片。

（1）模块

在现实应用中，最简单和快速的方式是使用一个嵌入式模块。此类模块包含了天线、嵌入了协议栈并提供多种不同的接口：UART、USB、SPI 和 $I^2C$，可以通过这些接口和处理器连接。模块会提供一种简单的接口来控制蓝牙的功能。很多的模块公司都会提供带 CE、FCC 和 IC 认证的产品。这样的模块可以只是蓝牙 BR/EDR 的、双模式的或者单模式的。

如果是蓝牙 BR/EDR 和双模的方案，还可以采用 HCI 模块。HCI 模块不带蓝牙协议栈，其他的和上述的模块是一样的。所以，这样的模块会更便宜。HCI 模块只是提供了硬件接口，在这样的方案中，蓝牙协议栈需要第三方提供。这样的第三方协议栈要求能在主设备的处理器中运行，如斯图曼提供的 BlueCode+SR。使用 HCI 模块需要将软件移植到最终的硬件中。

从理论上讲，提供单模的 HCI 模块也是可以的。然而，所有的芯片公司都已经将 GATT 集成到他们的芯片中，所以市面上不会有 HCI 单模模块出现。

（2）芯片

通过芯片来集成 BLE 是从物料角度最低成本的方式，但是，这需要很多的前期工作和花费大量的时间。虽然在软件上只需要将协议栈移植到目标平台即可，但硬件方面则需要对 RF 的 layout 和天线的设计非常有经验。这些公司提供的 BLE 芯片有 Broadcom、CSR、EM Microelectronic、Nordic 和 TI。

### 4.2.4　低功耗蓝牙的特点

在实际应用过程中，BLE 的低功耗并不是通过优化空中的无线射频传输实现的，而是通过改变协议的设计来实现的。为了实现极低的功耗效果，通常 BLE 协议设计为：在不必要射频的时候，彻底将空中射频关断。

与传统蓝牙 BR/EDR 相比，BLE 通过如下 3 大特性实现低功耗效果。

☑　缩短无线开启时间。

☑　快速建立连接。

☑　降低收发峰值功耗（具体由芯片决定）。

缩短无线开启时间的第一个技巧是只用 3 个“广告”信道，第二个技巧是通过优化协议栈来降低工作周期。一个在广告的设备可以自动和一个在搜索的设备快速建立连接，所以可以在 3 毫秒内完成连接的建立和数据的传输。

在现实应用中，低功耗设计可能会带来一些牺牲，例如音频数据无法通过 BLE 来进行传输。尽管如此，BLE 仍然是一种非常出色的技术，依然会支持跳频（37 个数据信道），并且采用了一种改进的 GFSK 调制方法来提高链路的稳定性。BLE 也是非常安全的技术，因为在芯片级提供了 128bit AES 加密。

单模设备可以作为 Master 或者 Slave，但是不能同时充当两种角色。这意味着 BLE 只能建立简单的星状拓扑，不能实现散射网。在 BLE 的无线电规范中，定义了低功耗蓝牙的最高数据率为 305Kbps，但这只是理论数据。在实际应用中，数据的吞吐量取决于上层协议栈。而 UART 的速度、处理器的能力和主设备都会影响数据吞吐能力。

高的数据吞吐能力的 BLE 只有通过私有方案或者基于 ATT notification 才能实现。事实上，如果是高数据率或高数据量的应用，蓝牙 BR/EDR 通常显得更加省电。

### 4.2.5　BLE 和传统蓝牙 BR/EDR 技术的对比

BLE 和传统蓝牙 BR/EDR 技术的对比如表 4-1 所示。

表 4-1　BLE 和传统蓝牙 BR/EDR 技术的对比

| 对　　比 | Bluetooth BR/EDR | Bluetooth Low Energy |
|---|---|---|
| Frequency | 2400～2483.5MHz | 2400～2483.5MHz |
| Deep Sleep | ~80μA | <5μA |
| Idle | ~8mA | ~1mA |
| Peak Current | 4～40mA | 10～30mA |
| Range | 500m (Class 1) / 50m (Class 2) | 100m |
| Min. Output Power | 0dBm (Class 1) / −6dBm (Class 2) | −20dBm |
| Max. Output Power | +20dBm (Class 1) / +4dBm (Class 2) | +10dBm |
| Receiver Sensitivity | ≥−70dBm | ≥−70dBm |
| Encryption | 64bit / 128bit | AES-128bit |
| Connection Time | 100ms | 3ms |
| Frequency Hopping | Yes | Yes |
| Advertising Channel | 32 | 3 |
| Data Channel | 79 | 37 |
| Voice Capable | Yes | No |

# 4.3 蓝牙 4.0 BLE 基础

**知识点讲解：光盘:视频\知识点\第 4 章\蓝牙 4.0 BLE 基础.avi**

蓝牙 4.0 也被称为 Bluetooth Smart，而 BLE 是 Bluetooth Low Energy 的缩写，属于蓝牙低功耗协议，Android 4.3 以上版本及苹果手机等都支持蓝牙 4.0 BLE，主要面向传感器应用市场提供短时间小数据传输，例如健康领域的手机监测血压、体育领域的手机计步器等。本节将详细讲解蓝牙 4.0 BLE 的基础知识。

## 4.3.1 低功耗是最大特点

蓝牙 4.0 是 2012 年最新蓝牙版本，是 3.0 的升级版本；较 3.0 版本更省电、成本低、3 毫秒低延迟、超长有效连接距离、AES-128 加密等，通常被用在蓝牙耳机、蓝牙音箱等设备上。

蓝牙 4.0 最重要的特性是省电，极低的运行和待机功耗可以使一粒纽扣电池连续工作数年之久。此外，低成本和跨厂商互操作性，3 毫秒低延迟、AES-128 加密等诸多特色，可以用于计步器、心律监视器、智能仪表、传感器物联网等众多领域，大大扩展了蓝牙技术的应用范围。

蓝牙 4.0 已经走向了商用，在最新款的 Xperia Z、Galaxy S3、Galaxy S4、Note2、Note3、SurfaceRT、iPhone 5S、iPhone 5、iPhone 4S、魅族 MX3、Moto Droid Razr、HTC One X、小米手机 2、The New iPad、iPad 4、 MacBook Air、Macbook Pro，以及台商 ACER AS3951 系列/Getway NV57 系列、ASUS UX21/31 三星 NOTE 系列上都已应用了蓝牙 4.0 技术。很多品牌已推出蓝牙 4.0 版本周边设备，同时支持蓝牙 4.0 和 NFC 的 WOOWI HERO、jabra MOTION 和 WOOWI 泡我等，支持蓝牙 4.0 的音箱有 Big jambox、Braven 等产品。

蓝牙技术联盟（Bluetooth SIG）2010 年 7 月 7 日宣布，正式采纳蓝牙 4.0 核心规范（Bluetooth Core Specification Version 4.0），并启动对应的认证计划。会员厂商可以提交其产品进行测试，通过后将获得蓝牙 4.0 标准认证。该技术拥有极低的运行和待机功耗，使用一粒纽扣电池甚至可连续工作数年之久。

蓝牙 4.0 BLE 的主要特性如下：

☑ 超低的峰值、平均和待机模式功耗。
☑ 使用标准纽扣电池可运行一年乃至数年。
☑ 低成本。
☑ 不同厂商设备交互性。
☑ 无线覆盖范围增强。
☑ 完全向下兼容。
☑ 低延迟（APT-X）。

## 4.3.2 推动了可穿戴设备的兴起

到目前为止，当大家谈到可穿戴设备时都要提到一个参数：支持蓝牙还是用无线网络与智能手机相连。这是衡量可穿戴设备是否能与智能手机上的软件顺利“对话”的主要依据。其实在过去的一段

时间内，大家已经习惯了“Bluetooth X.0 版”的说法，其实从 Bluetooth 4.0 开始，这项技术被 Bluetooth SIG（Special Interest Group，负责推动蓝牙技术标准的开发和将其授权给制造商的非营利组织）改名为 Bluetooth Smart 或 Bluetooth Smart Ready。Bluetooth SIG 首席营销官 Suke Jawanda 对 PingWest 说：“未来 Bluetooth SIG 也将继续淡化 X.0 的概念，将更加强调 Bluetooth Smart，原因是 X.0 是说给极客听的，而 Bluetooth SIG 希望普通消费者也能听懂。”

可穿戴设备与智能手机之间的数据传输方式对蓝牙技术的要求也与以往不同。Suke Jawanda 用自己手腕上的 Fitbit Flex 举例，“过去当我们谈到蓝牙技术和数据传输，主要考虑的是类似 Spotify 这种在一个较长的时间段里输送数据的需求，现在像 Fitbit Flex 是先收集数据，再断续地在某些‘时刻’里将数据传送到用户的手机上，两种数据传输方式不同。Bluetooth Smart 的低耗能技术就可以满足这一需求。”

Suke Jawanda 向 PingWest 解释说：“现在不少设备制造商都在强调自己支持蓝牙低耗能技术（Bluetooth Low Energy），其实它只是 Bluetooth Smart 其中的一个功能。支持 Bluetooth Smart 的设备都支持蓝牙低耗能技术”。

虽然特意强调低耗能技术是给消费者造成一种“省电省流量”的印象，但是换个角度来看，每个产品说明自己支持 Bluetooth Smart 的背后就是可穿戴设备为什么在此时流行的重要原因之一——Bluetooth Smart 对操作系统和硬件设备的支持情况，决定了可穿戴设备能否以较低的成本与软件进行数据传输，接下来才是解决软件获得数据之后怎么处理的问题。

根据 Suke Jawanda 的介绍，Bluetooth Smart 对可穿戴设备的支持分成硬件和软件。以 Fitbit 为例，如果 Fitbit 开发一款新的产品，支持 Bluetooth Smart，同时要求软件，也就是从 iOS 或者 Android——操作系统层面要支持 Bluetooth Smart，苹果是从 iOS 5（iPhone 4S 以及以上版本的手机）开始支持 Bluetooth Smart，而 Google 直到 Android 4.3 才开始支持。

当然 Bluetooth Smart 并不是唯一推动穿戴设备发展的原因，有些可穿戴设备可以用其他方式传输数据，例如无线网络，但是人们不可能一直在无线网络环境下生活。

那除了可穿戴设备外，汽车除了用蓝牙接打电话，还能做些什么？Suke Jawanda 说：“现在我们知道汽车能做到的是通过蓝牙进行语音操作、接打电话，未来我们想象的是用蓝牙技术可以不再用钥匙，你的手机就可以作为车钥匙；另一个是利用更多的传感器收集数据，让车与车之间‘对话’，例如你的车可以知道前后 3 辆车的时速，当他们减速时你的车能提醒你前方的车在减速可能是遇到什么情况等。但这里最大的问题是汽车行业技术滞后，例如你现在看到的一个汽车领域的新技术，真正应用到生产、被推广恐怕是 2～3 年后的事情，而且人买一辆车的期待是要用 10～15 年的，也就是你买了一辆车之后 10 年内可能都体验不到汽车领域的新技术了，这个问题现在还没有很好的解决方案。”。

**注意：** 本节的内容引用自“ZOL 网的科技频道：http://news.zol.com.cn/article/179109.html”。

# 4.4 蓝牙规范

**知识点讲解：光盘:视频\知识点\第 4 章\蓝牙规范.avi**

蓝牙规范即 Bluetooth Profile，Bluetooth SIG 定义了许多 Profile。Profile 的目的是要确保 Bluetooth 设备间的互通性（Interoperability），但是 Bluetooth 产品无须实现所有的 Bluetooth 规范 Profile。本节将详细讲解蓝牙规范的基本知识，为读者步入本书后面知识的学习打下基础。

### 4.4.1 Bluetooth 常用规范

在 Bluetooth 系统中，定义了如下所示的常用规范。

（1）蓝牙立体声音讯传输协议 A2DP

蓝牙立体声音讯传输协议（Advance Audio Distribution Profile），功能是播放立体声。

（2）基本图像规范

基本图像规范（Basic Imaging Profile）的功能是在装置之间传送图像，可以将其再细分为如下所示的类别。

- ☑ Image Push
- ☑ Image Pull
- ☑ Advanced Image Printing
- ☑ Automatic Archive
- ☑ Remote Camera
- ☑ Remote Display

（3）基本打印规范

基本打印规范（Basic Printing Profile）可以将文件、电子邮件传至打印机打印，主要包含如下所示的分类。

- ☑ 无线电话规范（Cordless Telephony Profile）：设置了蓝牙无线电话之间沟通的规范。
- ☑ 内通信规范（Intercom Profile）：是另类的 TCS（Telephone Control protocol Specification）基底规范，两个 Bluetooth 通信设备间沟通的规范。
- ☑ 拨号网络规范：Baseband、LMP、L2CAP、SDP、RFCOMM 协定所需要的传输需求。
- ☑ 传真规范（Fax Profile）：能传输传真的资料。
- ☑ 人机界面规范（Human Interface Device Profile）：可以支援鼠标和键盘功能。
- ☑ 头戴式通话器规范（Headset Profile）：能够将声音传送到蓝牙耳机设备。
- ☑ 序列埠规范（Serial Port Profile）：用来取代有线的 RS-232Cable 。
- ☑ SIM 卡存取规范（SIM Access Profile）：用于存取手机内的 SIM 卡。
- ☑ 同步规范（Synchronization Profile）：建立在 serial port profile、generic access profile 与 generic access profile 之上。
- ☑ 档案传输规范（File Transfer Profile）：Bluetooth 可以利用 OBEX 通信协定来传送档案。
- ☑ 泛用存取规范（Generic Access Profile）：用来建立连线。
- ☑ 泛用物件交换规范（Generic Object Exchange Profile）：使用 OBEX 进行物件交换。
- ☑ 物件交换规范（Object Push Profile）：Bluetooth 利用 OBEX 通信协定在两个设备间交换资料。
- ☑ 个人局域网路规范（Personal Area Networking Profile）：可以支持蓝牙网络第三层协定。
- ☑ 电话簿存取规范（Phone Book Access Profile）：可以在装置之间互换电话簿。
- ☑ 影像分享规范（Video Distribution Profile）：可以使用 H.263 编码算法来分享影像信息。

### 4.4.2 蓝牙协议体系结构

整个蓝牙协议体系结构可分为底层硬件模块、中间协议层和高端应用层 3 大部分。链路管理层

（LMP）、基带层（BBP）和蓝牙无线电信道构成蓝牙的底层模块。BBP 层负责跳频和蓝牙数据及信息帧的传输。LMP 层负责连接的建立和拆除以及链路的安全和控制，它们为上层软件模块提供了不同的访问入口，但是两个模块接口之间的消息和数据传递必须通过蓝牙主机控制器接口的解释才能进行。也就是说，中间协议层包括逻辑链路控制与适配协议（L2CAP）、服务发现协议（SDP）、串口仿真协议（RFCOMM）和电话控制协议规范（TCS）。L2CAP 完成数据拆装、服务质量控制、协议复用和组提取等功能，是其他上层协议实现的基础，因此也是蓝牙协议栈的核心部分。SDP 为上层应用程序提供一种机制来发现网络中可用的服务及其特性。在蓝牙协议栈的最上部是高端应用层，它对应于各种应用模型的剖面，是剖面的一部分，目前定义了 13 种剖面。

（1）蓝牙低层模块

蓝牙的低层模块是蓝牙技术的核心，是任何蓝牙设备都必须包括的部分。蓝牙工作在 2.4GHz 的 ISM 频段。采用了蓝牙结束的设备能够提供高达 720kbit/s 的数据交换速率。

蓝牙支持电路交换和分组交换两种技术，分别定义了两种链路类型，即面向连接的同步链路（SCO）和面向无连接的异步链路（ACL）。为了在很低的功率状态下也能使蓝牙设备处于连接状态，蓝牙规定了 3 种节能状态，即停等（Park）状态、保持（Hold）状态和呼吸（Sniff）状态。这几种工作模式按照节能效率以升序排依次是 Sniff 模式、Hold 模式、Park 模式。

蓝牙采用 3 种纠错方案，分别是 1/3 前向纠错（FEC）、2/3 前向纠错和自动重发（ARQ）。前向纠错的目的是减少重发的可能性，但同时也增加了额外开销。然而在一个合理的无错误率环境中，多余的投标会减少输出，故分组定义的本身也保持灵活的方式，因此，在软件中可定义是否采用 FEC。一般而言，在信道的噪声干扰比较大时蓝牙系统会使用前向纠错方案，以保证通信质量：对于 SCO 链路，使用 1/3 前向纠错；对于 ACL 链路，使用 2/3 前向纠错。在无编号的自动请求重发方案中，一个时隙传送的数据必须在下一个时隙得到收到的确认。只有数据在接收端通过了报头错误检测和循环冗余校验（CRC）后认为无错时，才向发送端发回确认消息，否则返回一个错误消息。

蓝牙系统的移动性和开放性使得安全问题变得及其重要。虽然蓝牙系统所采用的调频技术已经提供了一定的安全保障，但是蓝牙系统仍然需要链路层和应用层的安全管理。在链路层中，蓝牙系统提供了认证、加密和密钥管理等功能。每个用户都有一个个人标识码（PIN）,它会被译成 128bit 的链路密钥（Link Key）来进行单双向认证。一旦认证完毕，链路就会以不同长度的密码（Encryphon Key）来加密（此密码以 shit 为单位增减，最大的长度为 128bit），链路层安全机制提供了大量的认证方案和一个灵活的加密方案（即允许协商密码的长度）。当来自不同国家的设备互相通信时，这种机制是极其重要的，因为某些国家会指定最大密码长度。蓝牙系统会选取微微网中各个设备的最小的最大允许密码长度。例如，美国允许 128bit 的密码长度，而西班牙仅允许 48bit，这样当两国的设备互通时，将选择 48bit 来加密。蓝牙系统也支持高层协议栈的不同应用体内的特殊的安全机制。例如两台计算机在进行商业卡信息交流时，一台计算机就只能访问另一台计算机的该项业务，而无权访问其他业务。蓝牙安全机制依赖 PIN 在设备间建立信任关系，一旦这种关系建立起来了，这些 PIN 就可以存储在设备中以便将来更快捷地连接。

（2）软件模块

L2CAP 是数据链路层的一部分，位于基带协议之上。L2CAP 向上层提供面向连接的和无连接的数据服务，它的功能包括协议的复用能力、分组的分割和重新组装（Segmentation And Reassembly）以及提取（Group Abstraction）。L2CAP 允许高层协议和应用发送和接受高达 64KB 的数据分组。

SDP 为应用提供了一个发现可用协议和决定这些可用协议的特性的方法。蓝牙环境下的服务发现与传统的网络环境下的服务发现有很大的不同，在蓝牙环境下，移动的 RF 环境变化很大，因此业务的参数也是不断变换的。SDP 将强调蓝牙环境的独特的特性。蓝牙使用基于客户/服务器机制定义了根据蓝牙服务类型和属性发现服务的方法，还提供了服务浏览的方法。

RFCOMM 是射频通信协议，它可以仿真串行电缆接口协议，符合 ETSI0710 串口仿真协议。通过 RFCOMM，蓝牙可以在无线环境下实现对高层协议，如 PPP、TCP/IP、WAP 等的支持。另外，RFCOMM 可以支持 AT 命令集，从而可以实现移动电话机和传真机及调制解调器之间的无线连接。

蓝牙对语音的支持是它与 WLAN 相区别的一个重要标志。蓝牙电话控制规范是一个基于 ITU-T 建议 Q.931 的采用面向比特的协议，它定义了用于蓝牙设备间建立语音和数据呼叫的呼叫控制信令以及用于处理蓝牙 TCS 设备的移动性管理过程。

### 4.4.3 低功耗（BLE）蓝牙协议

BLE 不再支持传统蓝牙 BR/EDR 的协议，例如传统蓝牙中的 SPP 协议在 BLE 中就不复存在。在 BLE 应用中，所有的协议或者服务都是基于 GATT（Generic Attribute Profile）的。尽管有些传统蓝牙中的协议，如 HID 被移植到了 BLE 中，但是在 BLE 的应用中必须区分协议和服务。其中服务描述自身有什么特点和形式，并且描述清楚如何应用这些特点以及需要什么安全机制。而应用协议定义了其使用的服务，说明是传感器端还是接收端，定义 GATT 的角色（Server/Client）和 GAP 的角色（Peripheral/Central）。

和蓝牙 BR/EDR 协议相比，因为所有的功能都是集成在 GATT 终端，这些基于其上的应用协议只是对 GATT 提供的功能的使用，所以基于 GATT 的应用协议非常简单。

### 4.4.4 基于 GATT 的协议/服务

截止到 2013 年 7 月，现有的基于 GATT 的协议/服务如表 4-2 所示。

表 4-2　基于 GATT 的协议/服务

| GATT-Based Specifications（Qualifiable） | | Adopted Version |
|---|---|---|
| ANP | Alert Notification Profile | 1.0 |
| ANS | Alert Notification Service | 1.0 |
| BAS | Battery Service | 1.0 |
| BLP | Blood Pressure Profile | 1.0 |
| BLS | Blood Pressure Service | 1.0 |
| CPP | Cycling Power Profile | 1.0 |
| CPS | Cycling Power Service | 1.0 |
| CSCP | Cycling Speed and Cadence Profile | 1.0 |
| CSCS | Cycling Speed and Cadence Service | 1.0 |
| CTS | Current Time Service | 1.0 |
| DIS | Device Information Service | 1.1 |
| FMP | Find Me Profile | 1.0 |
| GLP | Glucose Profile | 1.0 |
| HIDS | HID Service | 1.0 |

续表

| GATT-Based Specifications（Qualifiable） | | Adopted Version |
|---|---|---|
| HOGP | HID over GATT Profile | 1.0 |
| HTP | Health Thermometer Profile | 1.0 |
| HTS | Health Thermometer Service | 1.0 |
| HRP | Heart Rate Profile | 1.0 |
| HRS | Heart Rate Service | 1.0 |
| IAS | Immediate Alert Service | 1.0 |
| LLS | Link Loss Service | 1.0 |
| LNP | Location and Navigation Profile | 1.0 |
| LNS | Location and Navigation Service | 1.0 |
| NDCS | Next DST Change Service | 1.0 |
| PASP | Phone Alert Status Profile | 1.0 |
| PASS | Phone Alert Status Service | 1.0 |
| PXP | Proximity Profile | 1.0 |
| RSCP | Running Speed and Cadence Profile | 1.0 |
| RSCS | Running Speed and Cadence Service | 1.0 |
| RTUS | Reference Time Update Service | 1.0 |
| ScPP | Scan Parameters Profile | 1.0 |
| ScPS | Scan Parameters Service | 1.0 |
| TIP | Time Profile | 1.0 |
| TPS | Tx Power Service | 1.0 |

## 4.4.5　双模协议栈和单模协议栈

图4-3展示了斯图曼双模协议栈BlueCode+SR的具体架构，在此架构图中包含了SPP、HDP和GATT所需要的所有部分。

图 4-4 展示了单模协议栈的一种典型协议栈设计。

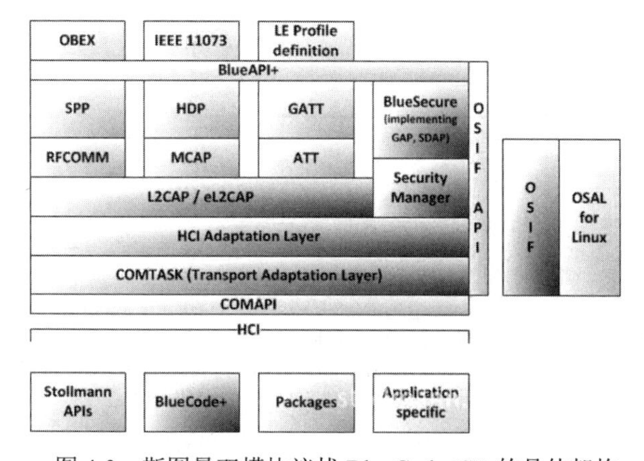


图 4-3　斯图曼双模协议栈 BlueCode+SR 的具体架构

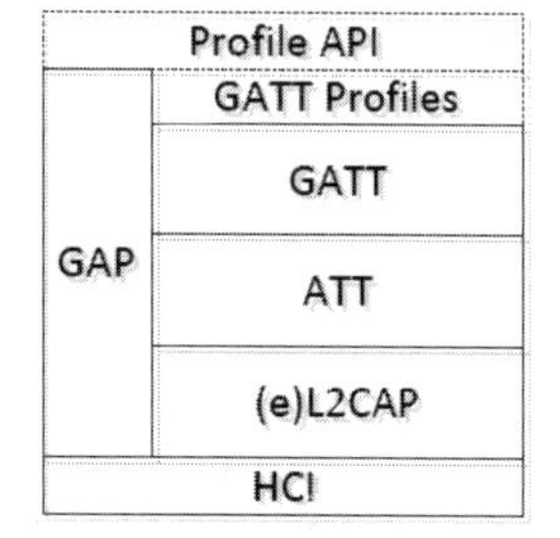


图 4-4　单模协议栈的一种设计

在单模协议栈中一般不会包含具体协议，所以需要在具体的应用程序中实现每一个具体应用对应

的协议。这和传统蓝牙有非常大的区别，传统蓝牙会在协议栈中实现每个具体应用相关的协议，如 SPP、HDP 等。和双模协议栈相比，BLE 无须一个主处理器来实现它的协议栈，所以极低功耗的集成成为可能。大多数的单模芯片或者模块都是自带协议栈的。

因为 BLE 单模产品（芯片或者模块）中的协议栈只是实现了 GATT 层，所以通常需要将具体应用对应的协议集成到该单模产品中。甚至芯片商都开始提供带有具体协议和 Sample Code 的 SDK。但是，仍然没有真正能拿到手的解决方案。

# 4.5 低功耗蓝牙协议栈详解

**知识点讲解：光盘:视频\知识点\第 4 章\低功耗蓝牙协议栈详解.avi**

在大家的印象中，提到协议栈时都会想到开放式系统互联（OSI）协议栈，OSI 协议栈定义了厂商们如何才能生产可以与其他厂商的产品一起工作的产品。协议栈是指一组协议的集合，例如把大象装到冰箱里需要 3 步，每步就是一个协议，3 步组成一个协议栈。本节将详细讲解低功耗蓝牙协议栈的基本知识，为读者步入本书后面知识的学习打下基础。

## 4.5.1 什么是蓝牙协议栈

蓝牙协议栈就是 SIG（Special Intersted Group）定义的一组协议的规范，目标是允许遵循规范的蓝牙应用能够进行相互间操作，图 4-5 展示了完整蓝牙协议栈和部分 Profile。

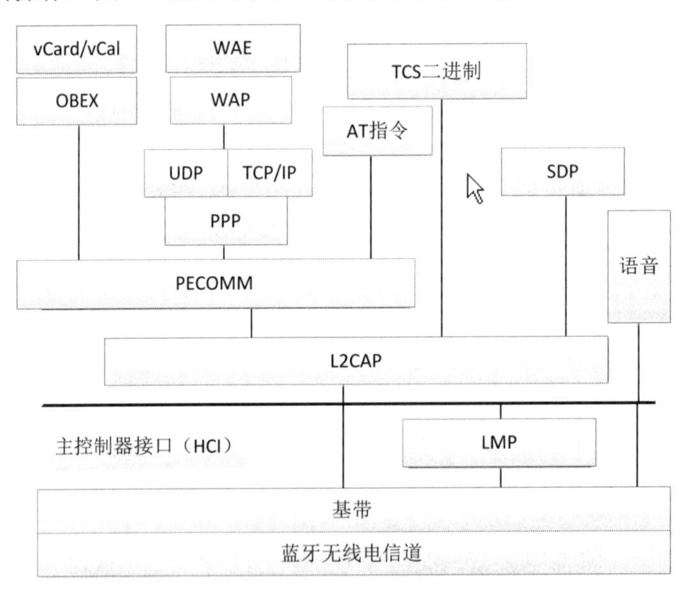


图 4-5 完整蓝牙协议栈和部分 Profile

在蓝牙系统中，Profile 是配置文件，其定义了可能的应用，蓝牙配置文件表达了一般行为，蓝牙设备可以通过这些行为与其他设备进行通信。在蓝牙系统中定义了广泛的配置文件，描述了许多不同类型的使用案例。按照蓝牙规格中提供的指导，开发商可以创建应用程序以与其他符合蓝牙规格的设备协同工作。到目前为止，在蓝牙系统中一共有 20 多个 Profile，在 www.bluetooth.com 中有各个 Profile

的详细说明文档。在这些众多的协议栈中，其中已经实现了的协议栈如下。

- ☑ Widcomm：第一个 Windows 上的协议栈，由 Widcomm 公司开发，也就是现在的 Broadcom。
- ☑ Microsoft Windows Stack：Windows XP SP2 中包括了这个内建的协议栈，开发者也可以调用其 API 开发第三方软件。
- ☑ Toshiba Stack：这也是基于 Windows 的，不支持第三方开发，但它把协议栈授权给一些 laptop 商，例如 Sony。它支持的 Profile 有 SPP、DUN、FAX、LAP、OPP、FTP、HID、HCRP、PAN、BIP、HSP、HFP、A2DP、AVRCP 和 GAVDP。
- ☑ BlueSoleil：著名的 IVT 公司的产品，该产品可以用于桌面和嵌入式，也支持第三方开发，例如 DUN、FAX、HFP、HSP、LAP、OBEX、OPP、PAN SPP、AV、BIP、FTP、GAP、HID、SDAP 和 SYNC。
- ☑ Blues：是 Linux 官方协议栈，该协议栈的上层用 Socket 封装，便于开发者使用，通过 DBUS 与其他应用程序通信。
- ☑ Affix：是 NOKIA 公司的协议栈，在 Symbian 系统上运行。
- ☑ BlueDragon：是东软公司产品，在 2002 年 6 月就通过了蓝牙的认证，支持的 Profile 有 SDP、Serial-DevB、AVCTP、AVRCP-Controller、AVRCP-Target、Headset-AG、Headset-HS、OPP-Client、OPP-Server、CT-GW、CT-Term、Intercom、FT-Server、FT-Client、GAP、SDAP、Serial-DevA、AVDTP、GAVDP、A2DP-Source 和 A2DP-Sink。
- ☑ BlueMagic：这是美国 Open Interface 公司 for portable embedded device 的协议栈，iPhone（apple）、nav-u（sony）等很多电子产品都用该商业的协议栈，BlueMagic 3.0 是第一个通过 Bluetooth 协议栈 1.1 认证的协议栈。
- ☑ BCHS-Bluecore Host Software：这是蓝牙芯片 CSR 的协议栈，同时也提供了一些上层应用的 Profile 的库，当然也是为嵌入式产品提供的服务，支持的 Profile 有 A2DP、AVRCP、PBAP、BIP、BPP、CTP、DUN、FAX、FM API、FTP GAP、GAVDP、GOEP、HCRP、Headset、HF1.5、HID、ICP、JSR82、LAP Message Access Profile、OPP、PAN、SAP、SDAP、SPP、SYNC 和 SYNC ML。
- ☑ Windows CE：微软给 Windows CE 开发的协议栈，但是 Windows CE 本身也支持其他的协议栈。
- ☑ BlueLet：是 IVT 公司 for embedded product 的轻量级协议栈。

## 4.5.2　蓝牙协议体系中的协议

在蓝牙协议体系中的协议中，按 SIG 的关注程度分为如下 4 层。

- ☑ 核心协议：BaseBand、LMP、L2CAP 和 SDP。
- ☑ 电缆替代协议：RFCOMM。
- ☑ 电话传送控制协议：TCS-Binary 和 AT 命令集。
- ☑ 选用协议：PPP、UDP/TCP/IP、OBEX、WAP、vCard、vCal、IrMC 和 WAE。

除上述协议层外，规范还定义了主机控制器接口（HCI），它为基带控制器、连接管理器、硬件状态和控制寄存器提供命令接口。在图 4-5 中，HCI 位于 L2CAP 的下层，但 HCI 也可位于 L2CAP 上层。

蓝牙核心协议由 SIG 制定的蓝牙专用协议组成，绝大部分蓝牙设备都需要核心协议（加上无线部分），而其他协议则根据应用的需要而定。总之，电缆替代协议、电话控制协议和被采用的协议在核心协议基础上构成了面向应用的协议。

在现实应用中，常用蓝牙核心协议类型如下所示。

（1）基带协议

基带和链路控制层确保微微网内各蓝牙设备单元之间由射频构成的物理连接。蓝牙的射频系统是一个跳频系统，其任一分组在指定时隙、指定频率上发送。它使用查询和分页进程同步不同设备间的发送频率和时钟，为基带数据分组提供了两种物理连接方式，即面向连接（SCO）和无连接（ACL），而且在同一射频上可实现多路数据传送。ACL 适用于数据分组，SCO 适用于话音以及话音与数据的组合，所有的话音和数据分组都附有不同级别的前向纠错（FEC）或循环冗余校验（CRC），而且可进行加密。此外，对于不同数据类型（包括连接管理信息和控制信息）都分配一个特殊通道。

可使用各种用户模式在蓝牙设备间传送话音，面向连接的话音分组只需经过基带传输，而不到达 L2CAP。话音模式在蓝牙系统内相对简单，只需开通话音连接即可传送话音。

（2）连接管理协议（LMP）

该协议负责各蓝牙设备间连接的建立。它通过连接的发起、交换、核实，进行身份认证和加密，通过协商确定基带数据分组大小。它还控制无线设备的电源模式和工作周期，以及微微网内设备单元的连接状态。

（3）逻辑链路控制和适配协议（L2CAP）

该协议是基带的上层协议，可以认为它与 LMP 并行工作，它们的区别在于，当业务数据不经过 LMP 时，L2CAP 为上层提供服务。L2CAP 向上层提供面向连接的和无连接的数据服务，它采用了多路技术、分割和重组技术、群提取技术。L2CAP 允许高层协议以 64K 字节长度收发数据分组。虽然基带协议提供了 SCO 和 ACL 两种连接类型，但 L2CAP 只支持 ACL。

（4）服务发现协议（SDP）

发现服务在蓝牙技术框架中起着至关重要的作用，它是所有用户模式的基础。使用 SDP 可以查询到设备信息和服务类型，从而在蓝牙设备间建立相应的连接。

（5）电缆替代协议（RFCOMM）

RFCOMM 是基于 ETSI-07.10 规范的串行线仿真协议。它在蓝牙基带协议上仿真 RS-232 控制和数据信号，为使用串行线传送机制的上层协议（如 OBEX）提供服务。

（6）电话控制协议

☑ 二元电话控制协议（TCS-Binary 或 TCSBIN）：是面向比特的协议，它定义了蓝牙设备间建立语音和数据呼叫的控制信令，定义了处理蓝牙 TCS 设备群的移动管理进程。基于 ITU TQ.931 建议的 TCSBinary 被指定为蓝牙的二元电话控制协议规范。

☑ AT 命令集电话控制协议：SIG 定义了控制多用户模式下移动电话和调制解调器的 AT 命令集，该 AT 命令集基于 ITU TV.250 协议和 GSM07.07 指令集，它还可以用于传真业务。

（7）选用协议

☑ 点对点协议（PPP）：在蓝牙技术中，PPP 位于 RFCOMM 上层，完成点对点的连接。

☑ TCP/UDP/IP：该协议是由互联网工程任务组制定，广泛应用于互联网通信的协议。在蓝牙设备中，使用这些协议是为了与互联网相连接的设备进行通信。

☑ 对象交换协议（OBEX）：IrOBEX（简写为 OBEX）是由红外数据协会（IrDA）制定的会话层

协议，它采用简单的和自发的方式交换目标。OBEX 是一种类似于 HTTP 的协议，它假设传输层是可靠的，采用客户机/服务器模式，独立于传输机制和传输应用程序接口（API）。

例如电子名片交换格式（vCard）、电子日历及日程交换格式（vCal）都是开放性规范，它们都没有定义传输机制，而只是定义了数据传输格式。SIG 采用 vCard/vCal 规范，是为了进一步促进个人信息交换。

☑ 无线应用协议（WAP）：该协议是由无线应用协议论坛制定的，它融合了各种广域无线网络技术，其目的是将互联网内容和电话传送的业务传送到数字蜂窝电话和其他无线终端上。

# 4.6　TI 公司的低功耗蓝牙

**知识点讲解：光盘:视频\知识点\第 4 章\TI 公司的低功耗蓝牙.avi**

BLE 低功耗蓝牙协议有很多个版本，不同的厂商提供的低功耗蓝牙协议会有所区别。因为目前市场的表现情况，本节将详细讲解使用最为频繁的 TI（德州仪器）公司 BLE 低功耗蓝牙协议的基本知识。

## 4.6.1　获取蓝牙协议栈

TI（德州仪器）公司提供了多个版本的 BLE 低功耗蓝牙协议栈，读者可以登录其官方网站来下载，如图 4-6 所示。

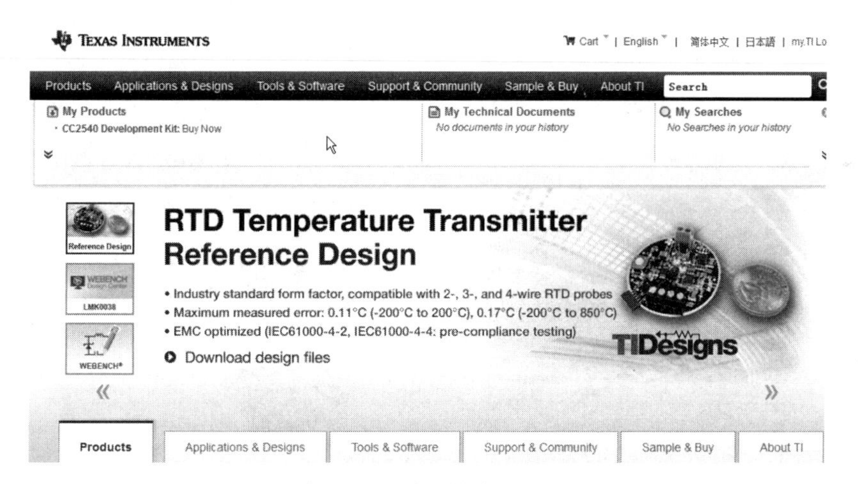


图 4-6　TI 公司的官方站点

笔者下载的版本是 BLE-CC254x-1.3.exe，双击此文件后可以进行安装工作，具体安装过程如下：

（1）进入“解压缩”界面，单击 Next 按钮，如图 4-7 所示。

（2）进入同意安装协议界面，选中 I accept the agreement 单选按钮，单击 Next 按钮，如图 4-8 所示。

（3）进入选择安装路径界面，通过 Browse 按钮选择安装路径，如图 4-9 所示。

（4）进入准备安装界面，单击 Install 按钮开始安装，如图 4-10 所示。

（5）进入安装进度界面，此过程需要耐心等待，如图 4-11 所示。

（6）进入安装完成界面，整个安装过程结束，如图 4-12 所示。

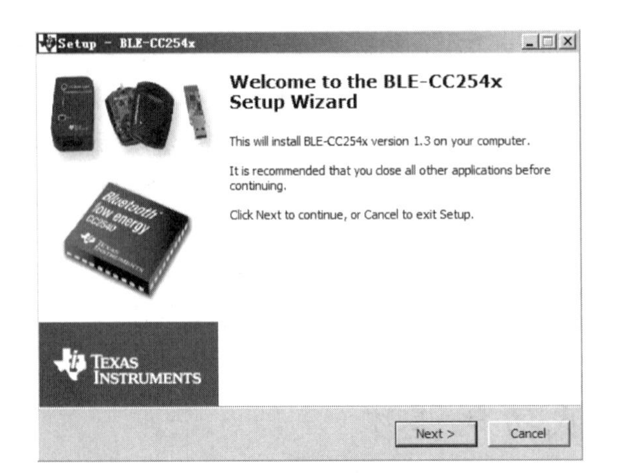


图 4-7 “解压缩”界面

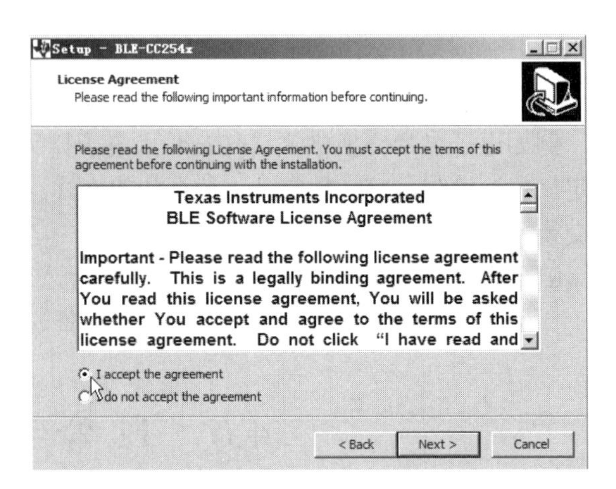


图 4-8 同意安装协议界面

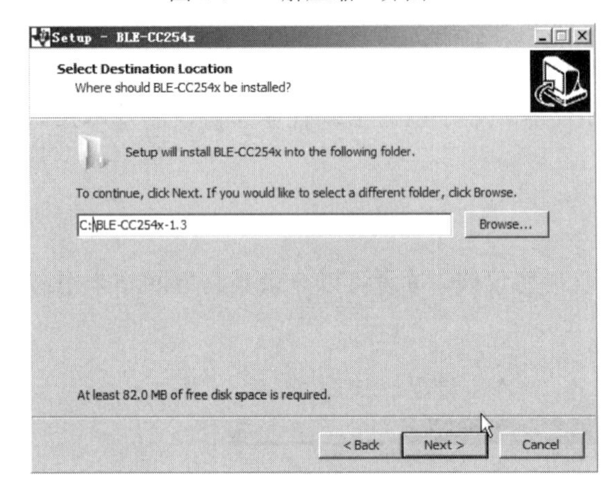


图 4-9 选择安装路径界面

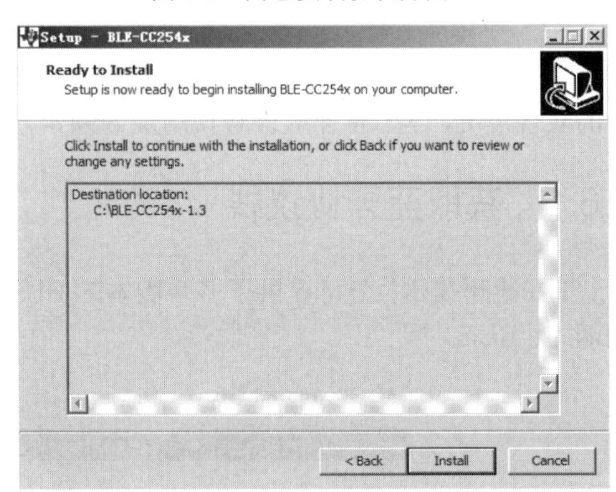


图 4-10 准备安装界面

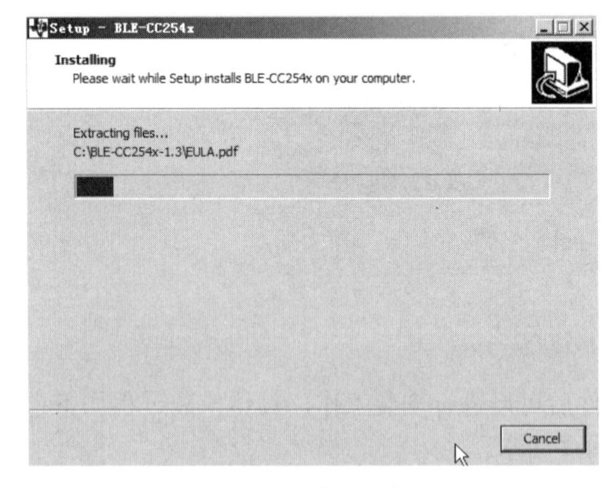


图 4-11 安装进度界面

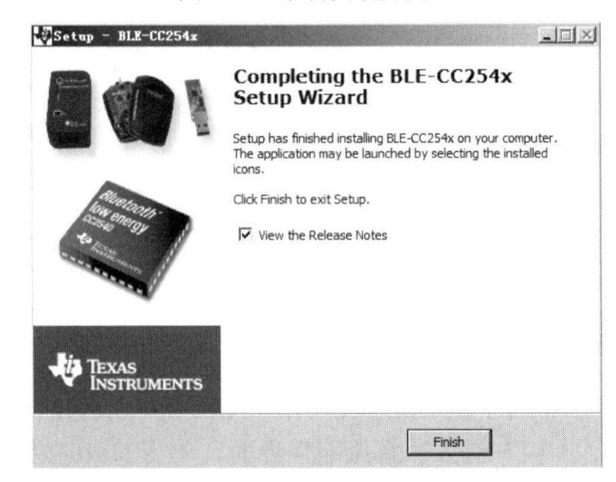


图 4-12 安装完成界面

安装完成后，需要使用 IAR 集成开发环境打开工程文件。例如 TI 公司在 Projects\ble\SimpleBLE Peripheral\目录下提供了实例工程，通过使用 IAR 工具打开.eww 文件的方式可以浏览整个工程。

**注意：** 读者可以自行下载并安装 IAR 集成开发环境。

## 4.6.2　BLE 蓝牙协议栈结构

**注意：** 在后面的内容中，笔者参考了 TI 公司的官方资料《CC2540Bluetooth Low Energy Software Developer's Guide (Rev. B)》，部分图片也是直接引用自上述参考文档。

BLE 蓝牙协议栈分为两个部分，分别是控制器和主机。对于 4.0 以前的蓝牙，这两部分是分开的。所有 profile（剧本，用来定义设备或组件的角色）和应用都建构在 GAP 或 GATT 之上。BLE 蓝牙协议栈的结构如图 4-13 所示。

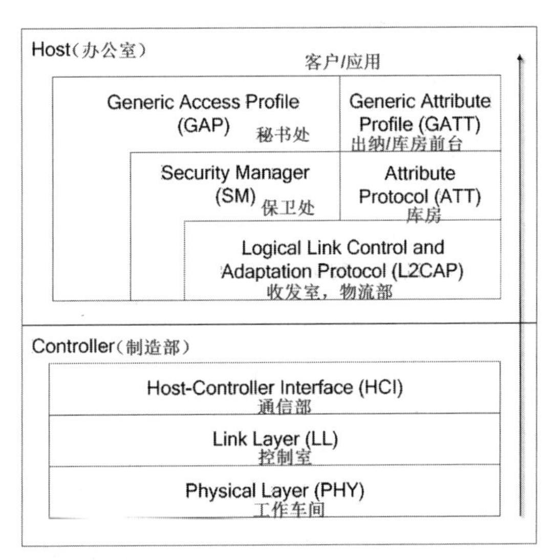


图 4-13　BLE 蓝牙协议栈的结构

在 BLE 蓝牙协议栈的结构中，从上到下的具体说明如下。

☑ PHY 层：工作车间，1Mbps 自适应跳频 GFSK（高斯频移键控），运行在免许可证使用的 2.4GHz 频段中。

☑ LL 层：为 RF 控制器，控制室，控制设备处于准备（Standby）、广播、监听/扫描（Scan）、初始化、连接这 5 种状态的一种。5 种状态切换描述为：未连接时，设备广播信息（向周围邻居讲“我来了”），另外一个设备一直监听或按需扫描，两个设备连接初始化（搬几把椅子到院子），设备连接上了（开聊）。发起聊天的设备为主设备，接收聊天的设备为从设备，同一次聊天只能有一个意见领袖，即主设备和从设备不能切换。

☑ HCI 层：为接口层，通信部，向上为主机提供软件应用程序接口（API），对外为外部硬件控制接口，可以通过串口、SPI、USB 来实现设备控制。

☑ L2CAP 层：物流部，负责行李打包和拆封处，提供数据封装服务。

☑ SM 层：保卫处，提供配对和密匙分发，实现安全连接和数据交换。

☑ ATT 层：库房，负责数据检索。

☑ GATT 层：出纳/库房前台，出纳负责处理向上与应用打交道，而库房前台负责向下把检索任务子进程交给 ATT 库房去做，其关键工作是为检索工作提供合适的 profile 结构，而 profile 由检索关键词（Characteristics）组成。

☑ GAP 层：秘书处，对上级提供应用程序接口，对下级管理各级职能部门，尤其是指示 LL 层控制室 5 种状态切换，指导保卫处做好机要工作。

蓝牙为了实现同多个设备相连或实现多功能的目标，也实现了功能扩充，这就产生了调度问题。因为，虽然软件和协议栈可扩充，但终究最底层的执行部门只有一个。为了实现多事件和多任务切换，需要把事件和任务对应的应用，以及其相关的提供支撑“办公室”和“工厂”的对象打包起来，并起一个名字 OSAL 操作系统抽象层，类似于集团公司以下的子公司。

如果实现软件和硬件的低耦合，使软件不经改动或很少改动即可应用在另外的硬件上，这样就方便硬件改造、升级、迁移后软件的移植。HAL 硬件抽象层正是用来抽象各种硬件的资源，告知给软件。其作用类似于嵌入式系统设备驱动的定义硬件资源的“.h”头文件，其角色类似于现代工厂的设备管理部。

## 4.6.3 BLE 低功耗蓝牙系统架构

BLE 低功耗蓝牙系统架构如图 4-14 所示。

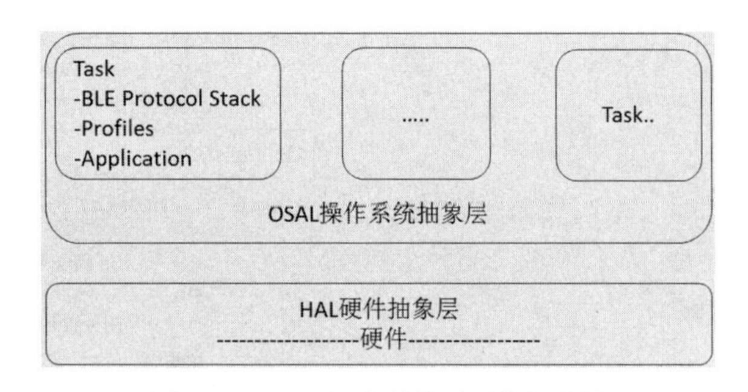


图 4-14 BLE 低功耗蓝牙系统架构图

由此而见，BLE 低功耗蓝牙软件主要由两部分组成，分别是 OSAL 操作系统抽象层和 HAL 硬件抽象层，多个 Task 任务和事件在 OSAL 管理下工作，而每个任务和事件又包括 3 个部分，分别是 BLE 协议栈、profiles 和应用程序。

（1）OSAL 操作系统抽象层

OSAL 作为调度核心，BLE 协议栈、profile 定义和所有的应用都围绕它来实现。OSAL 不是传统大家使用的操作系统，而是一个允许软件建立和执行事件的循环。软件功能是由任务事件来实现的，创建的任务事件需要完成如下所示的工作。

☑ 创建 task identifier 任务 ID。

☑ 编写任务初始化进程，并需要添加到 OSAL 初始化进程中，这就是说系统启动后不能动态添加功能。

☑ 编写任务处理程序。

☑ 提供消息服务。

BLE 协议栈的各层都是以 OSAL 任务方式实现，由于 LL 控制室的时间要求最为迫切，所以其任务优先级最高。为了实现任务管理，OSAL 通过消息处理、存储管理和计时器定时等附加服务实现。

（2）系统启动流程

为了使用 OSAL，在 main 函数的最后要启动一个名为 osal_start_system 的进程，该进程会调用由特定应用决定的启动函数 osalInitTasks 来启动系统。osalInitTasks 逐个调用 BLE 协议栈各层的启动进程来初始化协议栈。随后设置一个任务的 8bit 任务 ID（Task ID），跳入循环等待执行任务，系统启动完成。

（3）任务事件与事件处理

☑ 进程优先级和任务 ID

 ➢ 任务优先级决定于任务 ID，任务 ID 越小，优先级越高。

➢ BLE 协议栈各层的任务优先级比应用程序的高。
➢ 初始化协议栈后，越早调入的任务，任务 ID 越高，优先级越低，即系统倾向于处理新到的任务。

☑ 事件变量和旗语

每个事件任务由对应的 16bit 事件变量来标示，事件状态由旗号（Taskflag）来标示。如果事件处理程序已经完成，但其旗号并没有移除，OSAL 会认为事情还没有完成而继续在该程序中不返回。例如，在 SimpleBLEPeripheral 实例工程中，当事件 START_DEVICE_EVT 发生，其处理函数 SimpleBLEPeripheral_ProcessEvent 就运行，结束后返回 16bit 事件变量，并清除旗语 SBP_START_DEVICE_EVT。

☑ 事件处理表单

每当 OSAL 事件检测到了有任务事件，其相应的处理进程将被添加到由处理进程指针构成的事件处理表单中，该表单名为 taskArr（Taskarray）。taskArr 中各个事件进程的顺序和 osalInitTasks 初始化函数中任务 ID 的顺序是对应的。

☑ 事件调度的方法

有两种事件调度的方法，最简单的方法是使用 osal_set_event 函数（函数原型在 OSAL.h 文件中），在这个函数中，用户可以像定义函数参数一样设置任务 ID 和事件旗语。第二种方法是使用 osal_start_timerEx 函数（函数原型在 OSAL_Timers.h 文件中），使用方法同 osal_set_event 函数，而第三个以毫秒为单位的参数 osal_start_timerEx 则指示该事件处理必须要在这个限定时间内，通过定时器来为事件处理计时。

（4）存储管理

存储管理类似于 Linux 嵌入式系统内存分配 C 函数 mem_alloc，OSAL 利用 osal_mem_alloc 提供基本的存储管理，但 osal_mem_alloc 只有一个用于定义 byte 数的参数。对应的内存释放函数为 osal_mem_free。

（5）进程间通信——通过消息机制实现

不同的子系统通过 OSAL 的消息机制通信。消息即为数据，数据种类和长度都不限定。消息收发的过程描述如下所示。

在接收信息时调用函数 osal_msg_allocate 创建消息占用内存空间（已经包含了 osal_mem_alloc 函数功能），需要为该函数指定空间大小，该函数返回内存空间地址指针，利用该指针就可以把所需数据复制到该空间。

在发送数据时调用函数 osal_msg_send，需为该函数指定发送目标任务，OSAL 通过旗语 SYS_EVENT_MSG 告知目标任务，目标任务的处理函数调用 osal_msg_receive 来接收发来的数据。建议每个 OSAL 任务都有一个消息处理函数，每当任务收到一个消息后，通过消息的种类来确定需要本任务做相应处理。消息接收并处理完成，调用函数 osal_msg_deallocate 来释放内存（已经包含了 osal_mem_free 函数功能）。

### 4.6.4　硬件抽象层 HAL 和 BLE 低功耗蓝牙协议栈

当新的硬件平台做好后，只需修改 HAL，而不需修改 HAL 之上的协议栈的其他组件和应用程序。

（1）BLE 库文件

TI 蓝牙协议栈是以单独一个库文件提供的，并没有提供源代码，因此不做深入说明。对于 TI 的

BLE 实例应用来说，这个单独库文件完全够用，因为已经列出了所有的库文件。

（2）GAP 秘书处

☑ 角色（即服务/功能）

在 TI 实例中，GAP 运行在如下一种角色中。

☑ Broadcaster：广播员——我在，但只可远观，不可连接。

☑ Observer：观察员——看看谁在，但我只远观，不连接。

☑ Peripheral：外设（从机）——我在，谁要我就跟谁走，协议栈单层连接。

☑ Central：核心（主机）——看看谁在，并且愿意跟我走我就带她/他走，协议栈单层或多层连接，目前最多支持 3 个同时连接。

虽然指标显示 BLE 可以同时扮演多个角色，但是在 TI 提供的 BLE 实例应用中默认只支持外设角色。每一种角色都由一个剧本来定义。

☑ 连接

在主从机连接过程中，一个典型的低功耗蓝牙系统同时包含外设和核心（主机），两者的连接过程是：外设角色向外发送自己的信息（设备地址、名字等），主机收到外设广播信息后，发送扫描请求给外设，外设响应主机的请求，连接建立完成。

连接参数主要有通信间隙、外设鄙视、最大耐心等待时间等，具体说明如下。

☑ 通信间隙：蓝牙通信是间断的、跳频的，每次连接都可能选择不同的子频带。跳频的好处是避免频道拥塞，间断连接的好处是节省功耗，通信间隙就是指两次连接之间的时间间隔。这个间隔以 1.25ms 为基本单位，最小 6 单位，最大 3200 单位，间隙越小通信越及时，间隙越大功耗越低。

☑ 外设鄙视：外设与主机建立连接以后，主机总会定期发送问候信息到外设，外设不接收，这些主机发送的信息就浮云般飘过。可以忽略的连接事件个数从 0～499 个，最多不超过 32 秒。有效连接间隙= 连接间隙×（1+外设鄙视）。

☑ 最大耐心等待时间：指的是为了创建一个连接，主机允许的最大等候时间，在这个时间内，不停地尝试连接。范围是 10～3200 个通信间隙基本单位（1.25ms）。

以上 3 个参数大小设置优劣是显而易见的，连接参数的设置请阅读本小节后面的内容。

假如主机采用从机并不舒坦的参数来请求连接，有如主从机已经连接了，但从机有想法了，要改参数条约。通过“连接参数更新请求（Connection Parameter Update Request）”来解决问题，交由 L2CAP“收发室物流处”处理。

在实现加密处理时可以利用配对实现，利用密匙来加密授权连接。典型的过程是：外设向主机请求口令一个（passkey）以便进行配对，待主机发送了正确的口令之后，连接通信通过主从机互换密码来校验。由于蓝牙通信是间断通信，如果一个应用需要经常通信，而每次通信都要重新申请连接，那将是劳神费力的，为此 GAP 安全卫士 SM 提供了一种长期签证(Long-termset of Keys)，叫做绑定(Bonding)，这样每次建立连接通关流程就简便并快捷。

（3）出纳 GATT

GATT 负责两个设备间通信的数据交互。共有两种角色：出纳员（GATTClient）和银行（GATTServer），银行提供资金，出纳从银行存取款。银行可以同时面对多个出纳员。这两种角色和主从机等角色是无关的。

GATT 把工作拆分成几部分来实现：读关键词和描述符，用来去库房查找提取数据，并写/读关键

词和描述符。

GATT 银行（GATTServer）的业务部门（API）主要提供两个主要的功能：一是服务功能，注册或销毁服务，并作为回调函数；二是管理功能，添加或删除 GATT 银行业务。

一个角色定义的剧本可以同时定义多个角色，每个角色的服务、关键词、关键值和描述符都以句柄（Attributes）形式保存在角色提供的服务上。所有的服务都是一个 gattAttribute_t 类型的 array，在文件 gatt.h.中定义。

（4）调用 GAP 和 GATT 的一般过程

调用 GAP 和 GATT 的一般过程如下：

☑ API 调用。

☑ 协议栈响应并返回。

☑ 协议栈发送一个 OSAL 消息（数据）去调用相应任务事件。

☑ 调用任务去接收和处理消息。

☑ 消息清除。

以设备初始化为 GAP 外设角色来举例说明，外设角色由其剧本来决定，实例程序在文件 peripheral.c 内。

☑ 调用 API 函数 GAP_DeviceInit。

☑ GAP 检查后开始初始化，返回值为 SUCCESS (0x00)，并通知 BLE 工作。

☑ BLE 协议栈发送 OSAL 消息给外设角色剧本（Peripheral Roleprofile），消息内容包括要干什么（Eventvalue）GAP_MSG_EVENT 和指标是什么（Opcodevalue，参数）。

☑ 角色剧本的服务任务就收到了事件请求 SYS_EVENT_MSG，表示有消息来了。

☑ 角色剧本接收消息，并拆看到底是什么事，接着把消息数据转换（Cast）成具体要干的事情，并完成相应的工作（这里为 gapDeviceInitDoneEvent_t）。

☑ 角色剧本清除消息并返回。例如 GATT 客户端设备想从 GATT 服务器端读取数据，即 GATT 出纳想从 GATT 银行那边取钱。

☑ 应用程序调用 GATT 子进程 API 函数 GATT_ReadCharValue，传递的参数为连接句柄、关键词句柄和自身任务的 ID。

☑ GATT 答应了这个请求，返回值为 SUCCESS (0x00)，向下告知 BLE 有活干了。

☑ BLE 协议栈在下次建立蓝牙连接时，发送取钱的指令给银行，当银行通知，我们正好有柜员空闲，于是把钱取出来交给 BLE。

☑ BLE 接着就把取到的钱包成消息（OSAL message），通过出纳 GATT 返回给应用程序。消息内包含 GATT_MSG_EVENT 和修改了的 ATT_READ_RSP。

☑ 应用程序接收到从 OSAL 来的 SYS_EVENT_MSG 事件，表示钱可能到了。

☑ 应用程序接收消息，拆包检查，并拿走需要的钱。

☑ 最后应用程序把包装袋销毁。

（5）GAP 角色剧本

在 TI 的 BLE 实例应用中提供了 3 种 GAP 角色剧本，分别是保卫处角色和几种 GATT 出纳/库管示例程序服务角色。

☑ GAP 外设剧本

其 API 函数在 peripheral.h 中定义，包括如下所示的信息。

- ➢ GAPROLE_ADVERT_ENABLED：广播使能。
- ➢ GAPROLE_ADVERT_DATA：包含在广播中的信息。
- ➢ GAPROLE_SCAN_RSP_DATA：用于回复主机扫描请求的信息。
- ➢ GAPROLE_ADVERT_OFF_TIME：表示外设关闭广播持续时间，该值为零表示无限期关闭广播直到下一次广播使能信号到来。
- ➢ GAPROLE_PARAM_UPDATE_ENABLE：使能自动更新连接参数，可以让外设连接失败时自动调整连接参数以便重新连接。
- ➢ GAPROLE_MIN_CONN_INTERVAL：设置最小连接间隙，默认值为 80 个单位（每单位 1.25ms）。
- ➢ GAPROLE_MIN_CONN_INTERVAL：设置最大连接间隙，默认值为 3200 个单位。
- ➢ GAPROLE_SLAVE_LATENCY：是一个外设参数，默认值为零。表示处于连接后，从机可以做出不响应连接请求的距离数目，即跳过几个交互的连接。
- ➢ GAPROLE_TIMEOUT_MULTIPLIER：最大耐心等待时间，默认值为 1000 个单位。

函数 GAPRole_StartDevice 用来初始化 GAP 外设角色，其唯一的参数是 gapRolesCBs_t，这个参数是一个包含两个函数指针的结构体，这两个函数是 pfnStateChange 和 pfnRssiRead，前者标示状态，后者标示 RSSI 已经被读走了。

☑ 多角色同时扮演

在此以设备同时为外设和广播员两种角色，方法是去除前文外设的定义剧本 peripheral.c 和 peripheral.h，添加新的剧本 peripheralBroadcaster.c 和 peripheralBroadcaster.h；定义处理器值 PLUS_BROADCASTER。

☑ GAP 主机剧本

与外设剧本相似，主机剧本的 API 函数在 central.h 中定义，包括 GAPCentralRole_GetParameter 和 GAPCentralRole_SetParameter 以及其他。如 GAPROLE_PARAM_UPDATE_ENABLE 连接参数自动更新使能的功能，和外设角色一样。

GAPCentralRole_StartDevice 函数用来初始化 GAP 主机角色，其唯一的参数是 gapCentralRolesCBs_t，这个参数是一个包含两个函数指针的结构体，这两个函数是 eventCB 和 rssiCB，每次 GAP 时间发生，前者都会被调用，后者标示 RSSI 已经被读走。

☑ GAP 绑定管理器剧本

GAP 绑定管理器剧本用于保持长期的连接。同时支持外设配置和主机配置。当建立了配对连接后，如果绑定使能，绑定管理器就维护这个连接。主要参数如下所示。

- ➢ GAPBOND_PAIRING_MODE
- ➢ GAPBOND_MITM_PROTECTION
- ➢ GAPBOND_IO_CAPABILITIES
- ➢ GAPBOND_IO_CAP_DISPLAY_ONLY
- ➢ GAPBOND_BONDING_ENABLED

函数 GAPBondMgr_Register 用来初始化 GAP 主机角色，其唯一的参数是 gapBondCBs_t，这个参数是一个包含两个函数指针的结构体，这两个函数是 pairStateCB 和 passcodeCB，前者返回状态，后者用于配对时产生 6 位数字口令（passcode）。

☑ 编写一个剧本来创建（定义）新的角色（功能、服务）

以 SimpleGATT Profile 为剧本名称，包含 simpleGATTProfile.c 和 simpleGATTProfile.h 两个文件。

主要包含如下所示的 API 函数。

- SimpleProfile_AddService：用于初始化的进程，作用是添加服务句柄（serviceattributes）到句柄组（attributetable）内，寄存器读取和回写。
- SimpleProfile_SetParameter：设置剧本关键词。
- SimpleProfile_GetParameter：获取设置剧本关键词。
- SimpleProfile_RegisterAppCBs：注册 simpleProfile 回调函数。
- simpleProfile_ReadAttrCB：读 simpleProfile 回调函数。
- simpleProfile_WriteAttrCB：写 simpleProfile 回调函数。
- simpleProfile_HandleConnStatusCB：连接 simpleProfile 状态函数。

此实例剧本共有如下所示 5 个关键词：

- ☑ SIMPLEPROFILE_CHAR1
- ☑ SIMPLEPROFILE_CHAR2
- ☑ SIMPLEPROFILE_CHAR3
- ☑ SIMPLEPROFILE_CHAR4
- ☑ SIMPLEPROFILE_CHAR5

为节省本书的篇幅，TI 公司的低功耗蓝牙协议栈的基本知识介绍完毕。有关此协议栈的具体知识，请读者登录其官方站点查看帮助文档。

## 4.7　实战演练——使用蓝牙控制电风扇

**知识点讲解：光盘:视频\知识点\第 4 章\使用蓝牙控制电风扇.avi**

本节将通过一个具体实例的实现过程，详细讲解使用蓝牙技术传输数据的过程。本实例的功能是，使用蓝牙将 Android 应用程序和风扇建立连接，在 Android 应用程序中可以设置一个湿度值，并根据设置的湿度值来控制风扇的转速。

| 题　目 | 目　的 | 源 码 路 径 |
| --- | --- | --- |
| 实例 4-1 | 使用蓝牙控制电风扇的转动 | 光盘:\daima\4\lanEX |

本实例的功能是使用蓝牙控制电风扇的转动，在开始之前使用蓝牙对 Android 应用程序和电风扇建立连接，Android 应用程序是建立在 Arduino 开发板之上的，当然也可以建立在任何 Android 开发板中，也包括 Android 可穿戴设备。蓝牙、开发板、风扇的连接如图 4-15 所示。

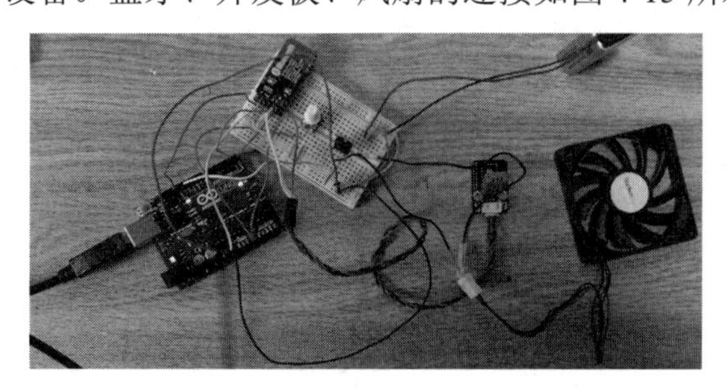

图 4-15　硬件设备的连接

本实例的具体实现流程如下所示。

（1）为了降低开发成本，在本实例使用的是 DHT 公司的低端传感器产品。在使用之前，需要将文件 DHT.cpp 和 DHT.h 包含在 Arduino 开发板的库文件夹的工程中。其中文件 DHT.h 是一个头文件，功能是设置需要多少时间进行转换跟踪，具体实现代码如下：

```
#ifndef DHT_H
#define DHT_H
#if ARDUINO >= 100
 #include "Arduino.h"
#else
 #include "WProgram.h"
#endif

/* DHT library

MIT license
written by Adafruit Industries
*/

#define MAXTIMINGS 85

#define DHT11 11
#define DHT22 22
#define DHT21 21
#define AM2301 21

class DHT {
 private:
  uint8_t data[6];
  uint8_t _pin, _type, _count;
  boolean read(void);
  unsigned long _lastreadtime;
  boolean firstreading;

 public:
  DHT(uint8_t pin, uint8_t type, uint8_t count=6);
  void begin(void);
  float readTemperature(bool S=false);
  float convertCtoF(float);
  float readHumidity(void);

};
#endif
```

文件 DHT.cpp 的功能是提供 DHT 的温度传感器和湿度传感器测试，具体实现代码如下：

```
#include "DHT.h"

DHT::DHT(uint8_t pin, uint8_t type, uint8_t count) {
  _pin = pin;
  _type = type;
  _count = count;
```

```
  firstreading = true;
}

void DHT::begin(void) {
  //建立指针
  pinMode(_pin, INPUT);
  digitalWrite(_pin, HIGH);
  _lastreadtime = 0;
}

float DHT::readTemperature(bool S) {
  float f;

  if (read()) {
    switch (_type) {
    case DHT11:
      f = data[2];
      if(S)
          f = convertCtoF(f);

      return f;
    case DHT22:
    case DHT21:
      f = data[2] & 0x7F;
      f *= 256;
      f += data[3];
      f /= 10;
      if (data[2] & 0x80)
     f *= -1;
      if(S)
     f = convertCtoF(f);

      return f;
    }
  }
  Serial.print("Read fail");
  return NAN;
}

float DHT::convertCtoF(float c) {
     return c * 9 / 5 + 32;
}

float DHT::readHumidity(void) {
  float f;
  if (read()) {
    switch (_type) {
    case DHT11:
      f = data[0];
      return f;
    case DHT22:
```

```
    case DHT21:
      f = data[0];
      f *= 256;
      f += data[1];
      f /= 10;
      return f;
    }
  }
  Serial.print("Read fail");
  return NAN;
}


boolean DHT::read(void) {
  uint8_t laststate = HIGH;
  uint8_t counter = 0;
  uint8_t j = 0, i;
  unsigned long currenttime;

  //提升指针，并设置等待 250 毫秒
  digitalWrite(_pin, HIGH);
  delay(250);

  currenttime = millis();
  if (currenttime < _lastreadtime) {
    //此时翻转
    _lastreadtime = 0;
  }
  if (!firstreading && ((currenttime - _lastreadtime) < 2000)) {
    return true; //return last correct measurement
    //delay(2000 - (currenttime - _lastreadtime));
  }
  firstreading = false;
  /*
    Serial.print("Currtime: "); Serial.print(currenttime);
    Serial.print(" Lasttime: "); Serial.print(_lastreadtime);
  */
  _lastreadtime = millis();

  data[0] = data[1] = data[2] = data[3] = data[4] = 0;

  //降低 20 毫秒
  pinMode(_pin, OUTPUT);
  digitalWrite(_pin, LOW);
  delay(20);
  cli();
  digitalWrite(_pin, HIGH);
  delayMicroseconds(40);
  pinMode(_pin, INPUT);

  //读取时间
```

```
  for ( i=0; i< MAXTIMINGS; i++) {
    counter = 0;
    while (digitalRead(_pin) == laststate) {
      counter++;
      delayMicroseconds(1);
      if (counter == 255) {
        break;
      }
    }
    laststate = digitalRead(_pin);

    if (counter == 255) break;

    //忽略前 3 个转换
    if ((i >= 4) && (i%2 == 0)) {
      data[j/8] <<= 1;
      if (counter > _count)
        data[j/8] |= 1;
      j++;
    }

  }

  sei();

  /*
  Serial.println(j, DEC);
  Serial.print(data[0], HEX); Serial.print(", ");
  Serial.print(data[1], HEX); Serial.print(", ");
  Serial.print(data[2], HEX); Serial.print(", ");
  Serial.print(data[3], HEX); Serial.print(", ");
  Serial.print(data[4], HEX); Serial.print(" =? ");
  Serial.println(data[0] + data[1] + data[2] + data[3], HEX);
  */

  //读取到 40 位便进行校验并比较
  if ((j >= 40) &&
      (data[4] == ((data[0] + data[1] + data[2] + data[3]) & 0xFF)) ) {
    return true;
  }


  return false;

}
```

编写测试程序 DHTtester.pde 来获取 DHT 传感器温度和湿度值，具体实现代码如下：

```
DHT dht(DHTPIN, DHTTYPE);

void setup() {
  Serial.begin(9600);
```

```
  Serial.println("DHTxx test!");

  dht.begin();
}

void loop() {
  float h = dht.readHumidity();
  float t = dht.readTemperature();

  if (isnan(t) || isnan(h)) {
    Serial.println("Failed to read from DHT");
  } else {
    Serial.print("Humidity: ");
    Serial.print(h);
    Serial.print(" %\t");
    Serial.print("Temperature: ");
    Serial.print(t);
    Serial.println(" *C");
  }
}
```

（2）开始编写 Android 应用程序，首先编写界面布局文件 activity_control.xml，功能是设置蓝牙启动测试按钮，具体实现代码如下：

```
<RelativeLayout
     xmlns:android="http://schemas.android.com/apk/res/android"
     xmlns:tools="http://schemas.android.com/tools"
          android:layout_width="match_parent"
          android:layout_height="match_parent"
          >

     <LinearLayout
          android:layout_width="match_parent"
          android:layout_height="match_parent"
          android:orientation="vertical"
          android:paddingBottom="@dimen/activity_vertical_margin"
          android:paddingLeft="@dimen/activity_horizontal_margin"
          android:paddingRight="@dimen/activity_horizontal_margin"
          android:paddingTop="@dimen/activity_vertical_margin"
          tools:context=".ControlActivity" >

          <TextView
               android:id="@+id/textView1"
               android:layout_width="wrap_content"
               android:layout_height="wrap_content"
               android:layout_alignParentLeft="true"
               android:layout_alignParentTop="true"
               android:layout_marginLeft="17dp"
               android:layout_marginTop="15dp"
               android:text="" />

          <Button
```

```
        android:id="@+id/connect"
        android:layout_width="wrap_content"
        android:layout_height="wrap_content"
        android:layout_alignLeft="@+id/textView1"
        android:layout_below="@+id/textView1"
        android:layout_marginTop="10dp"
        android:text="Connect"
        android:layout_gravity="center" />

    <ImageView
         android:id="@+id/power_button"
         android:layout_width="130dp"
         android:layout_height="130dp"
         android:contentDescription="@string/power_control"
         android:src="@drawable/humidigator_off"
          android:layout_gravity="center_horizontal"
          android:layout_weight="7" />
    <TextView
        android:id="@+id/current_data"
        android:layout_width="wrap_content"
        android:layout_height="36dp"
        android:layout_marginTop="69dp"
        android:layout_gravity="center_horizontal"
        android:gravity="center_horizontal"
        android:textSize="16sp"
        android:textColor="@color/white"
        android:text="Press 'Connect' to connect to Humidlgator" />

    <LinearLayout
         android:orientation="vertical"
         android:layout_width="match_parent"
         android:layout_height="wrap_content"
         android:gravity="center"
         android:layout_weight="12"
         android:layout_marginBottom="20dp"
          >

         <LinearLayout
              android:layout_width="match_parent"
              android:layout_height="wrap_content"
              android:orientation="horizontal"
              android:gravity="center"
              android:id="@+id/setHumLayout"
              >

              <TextView
                android:layout_width="wrap_content"
                android:layout_height="wrap_content"
                android:gravity="center"
                android:text="Set Humidity:"
                android:textSize="30dp"
```

```
                android:textColor="@color/white"
                android:textStyle="normal"
                android:layout_marginRight="15dp"
               />

            <TextView
               android:id="@+id/humidity"
                   android:layout_width="wrap_content"
                android:layout_height="wrap_content"
                android:text="50"
                android:textSize="30dp"
                android:layout_marginTop="3dp"
                android:textColor="@color/white"
                android:layout_gravity="center_vertical" />
            <TextView
               android:gravity="center"
               android:layout_width="wrap_content"
               android:layout_height="match_parent"
               android:text="%"
               android:textSize="30dp"
               android:layout_marginTop="3dp"
               android:textColor="@color/white"
               android:layout_gravity="center_vertical"
            />

           </LinearLayout>

            <SeekBar
               android:id="@+id/humidity_seek"
            android:layout_width="300dp"
              android:layout_height="wrap_content"
              android:layout_gravity="center"
              android:progress="50"
                 />
        </LinearLayout>
    </LinearLayout>
</RelativeLayout>
```

然后编写主 Android 应用程序文件 ControlActivity.java，功能是根据用户触摸屏幕操作来开启关闭蓝牙测试开关，将用户设置的值通过蓝牙传递给 DHT 传感器程序以控制电风扇的转动。文件 ControlActivity.java 的具体实现代码如下：

```
public class ControlActivity extends Activity implements OnClickListener, OnSeekBarChangeListener{

    Button Connect;
    ToggleButton OnOff;
    TextView Result;
    private String dataToSend;

    boolean powerOn = false;

    private static final String TAG = "Ian";
```

```
 private BluetoothAdapter mBluetoothAdapter = null;
 private BluetoothSocket btSocket = null;
 private OutputStream outStream = null;
 private static String address = "20:13:08:28:03:16";
 private static final UUID MY_UUID = UUID
            .fromString("00001101-0000-1000-8000-00805F9B34FB");
 private InputStream inStream = null;
Handler handler = new Handler();
byte delimiter = 10;
boolean stopWorker = false;
int readBufferPosition = 0;
byte[] readBuffer = new byte[1024];

private TextView setHum;
private SeekBar humSeek;
private LinearLayout setHumLayout;

ImageView powerButton;

 @Override
 protected void onCreate(Bundle savedInstanceState) {
      super.onCreate(savedInstanceState);
      setContentView(R.layout.activity_control);

      CheckBt();
      BluetoothDevice device = mBluetoothAdapter.getRemoteDevice(address);
      Log.e("Ian", device.toString());

      Connect = (Button) findViewById(R.id.connect);
      powerButton = (ImageView) findViewById(R.id.power_button);
      Result = (TextView) findViewById(R.id.current_data);
      setHum = (TextView) findViewById(R.id.humidity);
      humSeek = (SeekBar) findViewById(R.id.humidity_seek);
      setHumLayout = (LinearLayout) findViewById(R.id.setHumLayout);

      Connect.setOnClickListener(this);
      powerButton.setOnClickListener(this);
      humSeek.setOnSeekBarChangeListener(this);

      powerButton.setVisibility(View.INVISIBLE);
      setHumLayout.setVisibility(View.INVISIBLE);
      humSeek.setVisibility(View.INVISIBLE);

 }

private void CheckBt() {
      mBluetoothAdapter = BluetoothAdapter.getDefaultAdapter();

      if (!mBluetoothAdapter.isEnabled()) {
           Toast.makeText(getApplicationContext(), "Bluetooth Disabled !",
```

```
                    Toast.LENGTH_SHORT).show();
        }

        if (mBluetoothAdapter == null) {
            Toast.makeText(getApplicationContext(),
                    "Bluetooth null !", Toast.LENGTH_SHORT)
                    .show();
        }
    }

        public void Connect() {
            Log.d(TAG, address);
            BluetoothDevice device = mBluetoothAdapter.getRemoteDevice(address);
            Log.d(TAG, "Connecting to ... " + device);
            mBluetoothAdapter.cancelDiscovery();
            try {
                btSocket = device.createRfcommSocketToServiceRecord(MY_UUID);
                btSocket.connect();
                Log.d(TAG, "Connection made.");
            } catch (IOException e) {
                try {
                    btSocket.close();
                } catch (IOException e2) {
                    Log.d(TAG, "Unable to end the connection");
                }
                Log.d(TAG, "Socket creation failed");
            }

            beginListenForData();
        }

    private void writeData(String data) {
        try {
            outStream = btSocket.getOutputStream();
        } catch (IOException e) {
            Log.d(TAG, "Bug BEFORE Sending stuff", e);
        }

        String message = data;
        byte[] msgBuffer = message.getBytes();

        try {
            outStream.write(msgBuffer);
        } catch (IOException e) {
            Log.d(TAG, "Bug while sending stuff", e);
        }
    }

    @Override
    protected void onDestroy() {
        super.onDestroy();
```

```
            try {
                btSocket.close();
            } catch (IOException e) {
            }
    }

    public void beginListenForData()    {
        try {
                inStream = btSocket.getInputStream();
            } catch (IOException e) {
            }

        Thread workerThread = new Thread(new Runnable()
        {
            public void run()
            {
                while(!Thread.currentThread().isInterrupted() && !stopWorker)
                {
                    try
                    {
                        int bytesAvailable = inStream.available();
                        if(bytesAvailable > 0)
                        {
                            byte[] packetBytes = new byte[bytesAvailable];
                            inStream.read(packetBytes);
                            for(int i=0;i<bytesAvailable;i++)
                            {
                                byte b = packetBytes[i];
                                if(b == delimiter)
                                {
                                    byte[] encodedBytes = new byte[readBufferPosition];
                                    System.arraycopy(readBuffer, 0, encodedBytes, 0, encodedBytes.length);
                                    final String data = new String(encodedBytes, "US-ASCII");
                                    readBufferPosition = 0;
                                    handler.post(new Runnable()
                                    {
                                        public void run()
                                        {

                                            Result.setText(data);
                                            Result.setTextSize(25);

                                        }
                                    });
                                }
                                else
                                {
                                    readBuffer[readBufferPosition++] = b;
                                }
                            }
```

```
                        }
                    }
                    catch (IOException ex)
                    {
                        stopWorker = true;
                    }
                }
            }
        });

        workerThread.start();
    }

@Override
public boolean onCreateOptionsMenu(Menu menu) {
    getMenuInflater().inflate(R.menu.control, menu);
    return true;

}

@Override
public void onClick(View control) {
    switch (control.getId()) {
    case R.id.connect:
            Connect();
            powerButton.setVisibility(View.VISIBLE);
            setHumLayout.setVisibility(View.VISIBLE);
            humSeek.setVisibility(View.VISIBLE);

            Connect.setVisibility(View.INVISIBLE);
        break;
    case R.id.power_button:
        if (!powerOn){
            powerOn = true;
            powerButton.setImageResource(R.drawable.humidigator_on);
            dataToSend = "1";
            writeData(dataToSend);
        }
        else if (powerOn){
            powerOn = false;
            powerButton.setImageResource(R.drawable.humidigator_off);
            dataToSend = "0";
            writeData(dataToSend);
        }

        break;
    }
}

@Override
public void onProgressChanged(SeekBar seekBar, int progress,
```

```
                boolean fromUser) {
            setHum.setText(Integer.toString(humSeek.getProgress()));
    }
    @Override
    public void onStartTrackingTouch(SeekBar seekBar) {

    }
    @Override
    public void onStopTrackingTouch(SeekBar seekBar) {

    }
}
```

至此，整个实例介绍完毕，执行后的效果如图 4-16 所示。

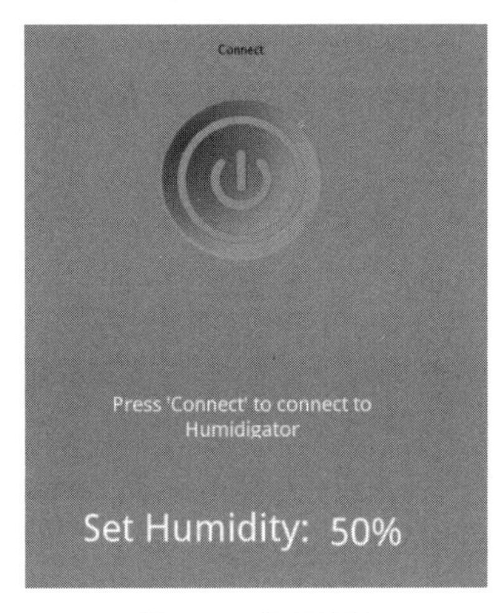


图 4-16　执行效果

本实例的目的是演示通过蓝牙和传感器设备控制硬件的过程，这说明蓝牙技术在近距离传输中的作用，所以必然被可穿戴设备广泛接收并支持。

# 第 5 章　Android 蓝牙系统详解

Android 系统包含了对蓝牙网络协议栈的支持，这使得蓝牙设备能够无线连接其他蓝牙设备交换数据。Android 的应用程序框架提供了访问蓝牙功能的 APIs。这些 APIs 让应用程序能够无线连接其他蓝牙设备，实现点对点，或点对多点的无线交互功能。本章将首先讲解 Android 系统中蓝牙模块的基本知识，为读者步入本书后面知识的学习打下基础。

## 5.1　Android 系统中的蓝牙模块

**知识点讲解：光盘:视频\知识点\第 5 章\Android 系统中的蓝牙模块.avi**

Android 平台的蓝牙系统是基于 BlueZ 实现的，是通过 Linux 中一套完整的蓝牙协议栈开源实现的。当前 BlueZ 被广泛应用于各种 Linux 版本中，并被芯片公司移植到各种芯片平台上所使用。在 Linux 2.6 内核中已经包含了完整的 BlueZ 协议栈，在 Android 系统中已经移植并嵌入进了 BlueZ 的用户空间实现，并且随着硬件技术的发展而不断更新。

蓝牙（Bluetooth）技术实际上是一种短距离无线电技术。在 Android 系统的蓝牙模块中，除了使用 Kernel 支持外，还需要用户空间的 BlueZ 支持。

Android 平台中蓝牙模块的基本层次结构如图 5-1 所示。

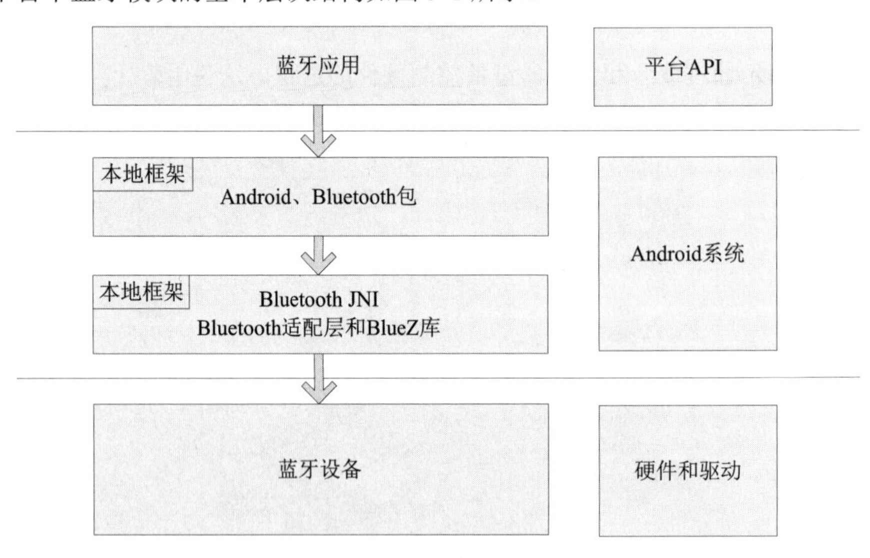


图 5-1　蓝牙系统的层次结构

Android 平台中蓝牙系统从上到下主要包括 Java 框架中的 BlueTooth 类、Android 适配库、BlueZ 库、驱动程序和协议，其系统结构如图 5-2 所示。

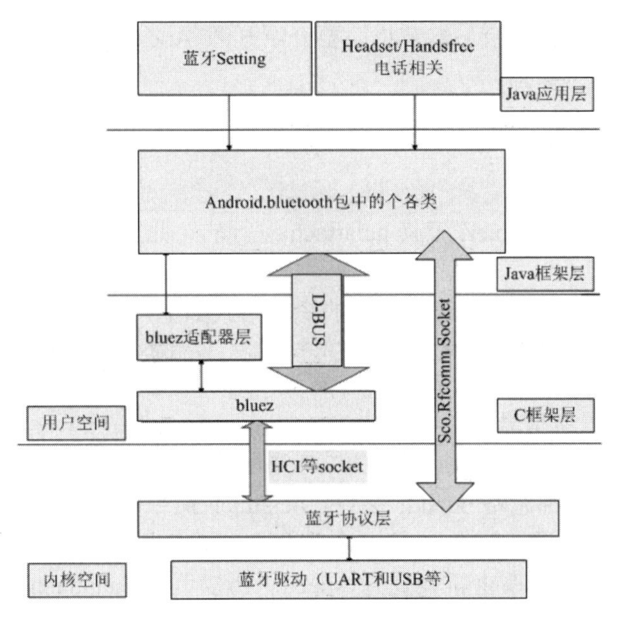


图 5-2　蓝牙系统结构

在图 5-2 中各个层次结构的具体说明如下。

（1）BlueZ 库

Android 蓝牙设备管理的库的路径是 external/bluez/。

可以分别生成库 libbluetooth.so、libbluedroid.so 和 hcidump 等众多相关工具和库。BlueZ 库提供了对用户空间蓝牙的支持，在里面包含了主机控制协议 HCI 以及其他众多内核实现协议的接口，并且实现了所有蓝牙应用模式 Profile。

（2）蓝牙的 JNI 部分

此部分的代码路径是 frameworks/base/core/jni/。

（3）Java 框架层

Java 框架层的实现代码保存在如下路径中。

☑　frameworks/base/core/java/android/bluetooth：蓝牙部分对应应用程序的 API。

☑　frameworks/base/core/java/android/Server：蓝牙的服务部分。

蓝牙的服务部分负责管理并使用底层本地服务，并封装成系统服务。而在 android.bluetooth 部分中包含了各个蓝牙平台的 API 部分，以供应用程序层所使用。

（4）BlueTooth 的适配库

BlueTooth 适配库的代码路径是 system/bluetooth/。

该层用于生成库 libbluedroid.so 以及相关工具和库，能够实现对蓝牙设备的管理，例如蓝牙设备的电源管理。

## 5.2　分析蓝牙模块的源码

**知识点讲解：光盘:视频\知识点\第 5 章\分析蓝牙模块的源码.avi**

要想掌握蓝牙系统的开发原理，需要首先分析 Android 中的蓝牙源码并了解其核心构造，只有这样才能对蓝牙应用开发做到游刃有余。本节将简要介绍开源 Android 中蓝牙模块相关的代码，为读者步入本书后面知识的学习打下基础。

## 5.2.1 初始化蓝牙芯片

初始化蓝牙芯片工作是通过 BlueZ 工具 hciattach 进行的，此工具在 external/bluetooth/tools 目录的文件中实现。

hciattach 命令主要用来初始化蓝牙设备，它的命令格式如下：

```
hciattach [-n] [-p] [-b] [-t timeout] [-s initial_speed] <tty> <type | id> [speed] [flow|noflow] [bdaddr]
```

在上述格式中，最重要的参数就是 type 和 speed，type 决定了要初始化的设备的型号，可以使用 hciattach-l 来列出所支持的设备型号。

并不是所有的参数对所有的设备都是适用的，有些设备会忽略一些参数设置，例如，查看 hciattach 的代码就可以看到，多数设备都忽略 bdaddr 参数。hciattach 命令内部的工作步骤是：首先打开制定的 tty 设备，然后做一些通用的设置，如 flow 等，接着设置波特率为 initial_speed，然后再根据 type 调用各自的初始化代码，最后将波特率重新设置为 speed。所以调用 hciattach 时，要根据实际情况设置 initial_speed 和 speed。

对于 type BCSP 来说，它的初始化代码只做了一件事，就是完成 BCSP 协议的同步操作，它并不对蓝牙芯片做任何的 pskey 设置。

## 5.2.2 蓝牙服务

在蓝牙服务方面一般不要我们自己定义，只需要使用初始化脚本文件 init.rc 中的默认内容即可。例如下面的代码。

```
service bluetoothd /system/bin/logwrapper /system/bin/bluetoothd -d -n
    socket bluetooth stream 660 bluetooth bluetooth
    socket dbus_bluetooth stream 660 bluetooth bluetooth
    # init.rc does not yet support applying capabilities, so run as root and
    # let bluetoothd drop uid to bluetooth with the right linux capabilities
    group bluetooth net_bt_admin misc
    disabled

# baudrate change 115200 to 1152000(Bluetooth)
service changebaudrate /system/bin/logwrapper /system/xbin/bccmd_115200 -t bcsp -d /dev/s3c2410_serial1
psset -r 0x1be 0x126e
    user bluetooth
    group bluetooth net_bt_admin
    disabled
    oneshot

#service hciattach /system/bin/logwrapper /system/bin/hciattach -n -s 1152000 /dev/s3c2410_serial1 bcsp
1152000
service hciattach /system/bin/logwrapper /system/bin/hciattach -n -s 115200 /dev/s3c2410_serial1 bcsp 115200
    user bluetooth
    group bluetooth net_bt_admin misc
    disabled

service hfag /system/bin/sdptool add --channel=10 HFAG
    user bluetooth
```

```
    group bluetooth net_bt_admin
    disabled
    oneshot

service hsag /system/bin/sdptool add --channel=11 HSAG
    user bluetooth
    group bluetooth net_bt_admin
    disabled
    oneshot

service opush /system/bin/sdptool add --channel=12 OPUSH
    user bluetooth
    group bluetooth net_bt_admin
    disabled
    oneshot

service pbap /system/bin/sdptool add --channel=19 PBAP
    user bluetooth
    group bluetooth net_bt_admin
    disabled
    oneshot
```

在上述代码中，每一个 service 后面都列出了一种 Android 服务。

### 5.2.3　管理蓝牙电源

在 Android 系统的软件目录 system/bluetooth/中实现了 libbluedroid。

可以调用 rfkill 接口来控制电源管理，如果已经实现了 rfkill 接口，则无须再进行配置。如果在文件 init.rc 中已经实现了 hciattach 服务，则说明在 libbluedroid 中已经实现对其调用以操作蓝牙的初始化。

## 5.3　和蓝牙相关的类

**知识点讲解：光盘:视频\知识点\第 5 章\和蓝牙相关的类.avi**

经过本章前面内容的学习，已经了解了 Android 系统中蓝牙的基本知识。根据对上述从底层到应用的学习，了解了蓝牙的工作原理和机制。本节将详细讲解在 Android 系统中和蓝牙相关的类，为读者步入本书后面知识的学习打好基础。

### 5.3.1　BluetoothSocket 类

#### 1．BluetoothSocket 类基础

类 BluetoothSocket 的定义格式如下：

```
public static class Gallery.LayoutParams extends ViewGroup.LayoutParams
```

类 BluetoothSocket 的定义结构如下：

```
java.lang.Object
android.view. ViewGroup.LayoutParams
android.widget.Gallery.LayoutParams
```

Android 的蓝牙系统和 Socket 套接字密切相关，蓝牙端的监听接口和 TCP 的端口类似，都是使用

了 Socket 和 ServerSocket 类。在服务器端，使用 BluetoothServerSocket 类来创建一个监听服务端口。当一个连接被 BluetoothServerSocket 所接受，它会返回一个新的 BluetoothSocket 来管理该连接。在客户端，使用一个单独的 BluetoothSocket 类去初始化一个外接连接和管理该连接。

最通常使用的蓝牙端口是 RFCOMM，它是被 Android API 支持的类型。RFCOMM 是一个面向连接，通过蓝牙模块进行的数据流传输方式，它也被称为串行端口规范（Serial Port Profile，SPP）。

为了创建一个 BluetoothSocket 去连接到一个已知设备，使用方法 BluetoothDevice.createRfcommSocketToServiceRecord()。然后调用 connect()方法去尝试一个面向远程设备的连接。这个调用将被阻塞指导一个连接已经建立或者该链接失效。

为了创建一个 BluetoothSocket 作为服务端（或者“主机”），每当该端口连接成功后，无论它初始化为客户端，或者被接收作为服务端，都通过方法 getInputStream()和 getOutputStream()来打开 I/O 流，从而获得各自的 InputStream 和 OutputStream 对象

BluetoothSocket 类的线程是安全的，因为 close()方法总会马上放弃外界操作并关闭服务端口。

### 2．BluetoothSocket 类的公共方法

（1）public void close()

功能：马上关闭该端口并且释放所有相关的资源。在其他线程的该端口中引起阻塞，从而使系统马上抛出一个 I/O 异常。

异常：IOExccption。

（2）public void connect()

功能：尝试连接到远程设备。该方法将阻塞，指导一个连接建立或者失效。如果该方法没有返回异常值，则该端口现在已经建立。当设备查找正在进行时，创建对远程蓝牙设备的新连接不可被尝试。设备查找在蓝牙适配器上是一个重量级过程，并且肯定会降低一个设备的连接。使用 cancelDiscovery()方法会取消一个外界的查询，因为这个查询并不由活动所管理，而是作为一个系统服务来运行，所以即使它不能直接请求一个查询，应用程序也总会调用 cancelDiscovery()方法。使用方法 close()可以用来放弃从另一线程而来的调用。

异常：IOException，表示一个错误，例如连接失败。

（3）public InputStream getInputStream()

功能：通过连接的端口获得输入数据流。即使该端口未连接，该输入数据流也会返回。不过在该数据流上的操作将抛出异常，直到相关的连接已经建立。

返回值：输入流。

异常：IOException。

（4）public OutputStream getOutputStream()

功能：通过连接的端口获得输出数据流。即使该端口未连接，该输出数据流也会返回。不过在该数据流上的操作将抛出异常，直到相关的连接已经建立。

返回值：输出流。

异常：IOException。

（5）public BluetoothDevice getRemoteDevice()

功能：获得该端口正在连接或者已经连接的远程设备。

返回值：远程设备。

## 5.3.2　BluetoothServerSocket 类

### 1．BluetoothServerSocket 类基础

类 BluetoothServerSocket 的格式如下：

```
public final class BluetoothServerSocket extends Object implements Closeable
```

类 BluetoothServerSocket 的结构如下：

```
java.lang.Object
android.bluetooth.BluetoothServerSocket
```

### 2．BluetoothServerSocket 类的公共方法

（1）public BluetoothSocketaccept(int timeout)

功能：阻塞直到超时时间内的连接建立。在一个成功建立的连接上返回一个已连接的 BluetoothSocket 类。每当该调用返回时，它可以在此调用去接收以后新来的连接。close()方法可以用来放弃从另一线程来的调用。

参数 timeout：表示阻塞超时时间。

返回值：已连接的 BluetoothSocket。

异常：IOException，表示出现错误，例如该调用被放弃或超时。

（2）public BluetoothSocket accept()

功能：阻塞直到一个连接已经建立。在一个成功建立的连接上返回一个已连接的 BluetoothSocket 类。每当该调用返回时，它可以在此调用去接收以后新来的连接。使用 close()方法可以用来放弃从另一线程来的调用。

返回值：已连接的 BluetoothSocket。

异常：IOException，表示出现错误，例如该调用被放弃或者超时。

（3）public void close()

功能：马上关闭端口，并释放所有相关的资源。在其他线程的该端口中引起阻塞，从而使系统马上抛出一个 I/O 异常。关闭 BluetoothServerSocket 不会关闭接受自 accept()的任意 BluetoothSocket。

异常：IOException。

## 5.3.3　BluetoothAdapter 类

### 1．BluetoothAdapter 类基础

类 BluetoothAdapter 的格式如下：

```
public final class BluetoothAdapter extends Object
```

类 BluetoothAdapter 的结构如下：

```
java.lang.Object
android.bluetooth.BluetoothAdapter
```

BluetoothAdapter 代表本地的蓝牙适配器设备，通过此类可以让用户执行基本的蓝牙任务。例如初始化设备的搜索，查询可匹配的设备集，使用一个已知的 MAC 地址来初始化一个 BluetoothDevice 类，创建一个 BluetoothServerSocket 类以监听其他设备对本机的连接请求等。

为了得到这个代表本地蓝牙适配器的 BluetoothAdapter 类，需要调用静态方法 getDefaultAdapter()，

这是所有蓝牙动作使用的第一步。当拥有本地适配器以后，用户可以获得一系列的 BluetoothDevice 对象，这些对象代表所有拥有 getBondedDevice()方法的已经匹配的设备；用 startDiscovery()方法来开始设备的搜寻；或者创建一个 BluetoothServerSocket 类，通过 listenUsingRfcommWithServiceRecord(String, UUID)方法来监听新来的连接请求。

**注意：** 大部分方法需要 BLUETOOTH 权限，一些方法同时需要 BLUETOOTH_ADMIN 权限。

### 2. BluetoothAdapter 类的常量

（1）String ACTION_DISCOVERY_FINISHED

广播事件：本地蓝牙适配器已经完成设备的搜寻过程。需要 BLUETOOTH 权限接收。

常量值：android.bluetooth.adapter.action.DISCOVERY_FINISHED。

（2）String ACTION_DISCOVERY_STARTED

广播事件：本地蓝牙适配器已经开始对远程设备的搜寻过程。它通常牵涉到一个大概需时 12 秒的查询扫描过程，紧跟着是一个对每个获取到自身蓝牙名称的新设备的页面扫描。用户会发现一个把 ACTION_FOUND 常量通知为远程蓝牙设备的注册。设备查找是一个重量级过程。当查找正在进行时，用户不能尝试对新的远程蓝牙设备进行连接，同时存在的连接将获得有限制的带宽以及高等待时间。用户可用 cancelDiscovery()类来取消正在执行的查找进程。需要 BLUETOOTH 权限接收。

常量值：android.bluetooth.adapter.action.DISCOVERY_STARTED。

（3）String ACTION_LOCAL_NAME_CHANGED

广播活动：本地蓝牙适配器已经更改了它的蓝牙名称。该名称对远程蓝牙设备是可见的，它总是包含了一个带有名称的 EXTRA_LOCAL_NAME 附加域。需要 BLUETOOTH 权限接收。

常量值：android.bluetooth.adapter.action.LOCAL_NAME_CHANGED。

（4）String ACTION_REQUEST_DISCOVERABLE

Activity 活动：显示一个请求被搜寻模式的系统活动。如果蓝牙模块当前未打开，该活动也将请求用户打开蓝牙模块。被搜寻模式和 SCAN_MODE_CONNECTABLE_DISCOVERABLE 等价。当远程设备执行查找进程时，它允许其发现该蓝牙适配器。从隐私安全考虑，Android 不会将被搜寻模式设置为默认状态。该意图的发送者可以选择性地运用 EXTRA_DISCOVERABLE_DURATION 这个附加域去请求发现设备的持续时间。普遍来说，对于每一请求，默认的持续时间为 120 秒，最大值则可达到 300 秒。

Android 运用 onActivityResult(int, int, Intent)回收方法来传递该活动结果的通知。被搜寻的时间（以秒为单位）将通过 resultCode 值来显示，如果用户拒绝被搜寻，或者设备产生了错误，则通过 RESULT_CANCELED 值来显示。

每当扫描模式变化时，应用程序可以通过 ACTION_SCAN_MODE_CHANGED 值来监听全局的消息通知。例如，当设备停止被搜寻以后，该消息可以被系统通知给应用程序。需要 BLUETOOTH 权限。

常量值：android.bluetooth.adapter.action.REQUEST_DISCOVERABLE。

（5）String ACTION_REQUEST_ENABLE

Activity 活动：显示一个允许用户打开蓝牙模块的系统活动。当蓝牙模块完成打开工作，或者当用户决定不打开蓝牙模块时，系统活动将返回该值。Android 运用 onActivityResult(int, int, Intent)回收方法来传递该活动结果的通知。如果蓝牙模块被打开，将通过 resultCode 值 RESULT_OK 来显示；如果用户拒绝该请求，或者设备产生了错误，则通过 RESULT_CANCELED 值来显示。每当蓝牙模块被打开或者关闭，应用程序可以通过 ACTION_STATE_CHANGED 值来监听全局的消息通知。需要 BLUETOOTH 权限。

常量值：android.bluetooth.adapter.action.REQUEST_ENABLE。

（6）String ACTION_SCAN_MODE_CHANGED

广播活动：指明蓝牙扫描模块或者本地适配器已经发生变化。它总是包含 EXTRA_SCAN_MODE 和 EXTRA_PREVIOUS_SCAN_MODE。这两个附加域各自包含了新的和旧的扫描模式。需要 BLUETOOTH 权限。

常量值：android.bluetooth.adapter.action.SCAN_MODE_CHANGED。

（7）String ACTION_STATE_CHANGED

广播活动：本来的蓝牙适配器的状态已经改变，例如蓝牙模块已经被打开或者关闭。它总是包含 EXTRA_STATE 和 EXTRA_PREVIOUS_STATE。这两个附加域各自包含了新的和旧的状态。需要 BLUETOOTH 权限接收。

常量值：android.bluetooth.adapter.action.STATE_CHANGED。

（8）int ERROR

功能：标记该类的错误值。确保和该类中的任意其他整数常量不相等。它为需要一个标记错误值的函数提供了便利。例如：

```
Intent.getIntExtra(BluetoothAdapter.EXTRA_STATE, BluetoothAdapter.ERROR)
```

常量值：-2147483648(0x80000000)。

（9）String EXTRA_DISCOVERABLE_DURATION

功能：试图在 ACTION_REQUEST_DISCOVERABLE 常量中作为一个可选的整型附加域，来为短时间内的设备发现请求一个特定的持续时间。默认值为 120 秒，超过 300 秒的请求将被限制。这些值是可以变化的。

常量值：android.bluetooth.adapter.extra.DISCOVERABLE_DURATION。

（10）String EXTRA_LOCAL_NAME

功能：试图在 ACTION_LOCAL_NAME_CHANGED 常量中作为一个字符串附加域，来请求本地蓝牙的名称。

常量值：android.bluetooth.adapter.extra.LOCAL_NAME。

（11）String EXTRA_PREVIOUS_SCAN_MODE

功能：试图在 ACTION_SCAN_MODE_CHANGED 常量中作为一个整型附加域，来请求以前的扫描模式。可以取的值如下：

☑ SCAN_MODE_NONE

☑ SCAN_MODE_CONNECTABLE

☑ SCAN_MODE_CONNECTABLE_DISCOVERABLE

常量值：android.bluetooth.adapter.extra.PREVIOUS_SCAN_MODE。

（12）String EXTRA_PREVIOUS_STATE

功能：试图在 ACTION_STATE_CHANGED 常量中作为一个整型附加域，来请求以前的供电状态。可以取的值如下：

☑ STATE_OFF

☑ STATE_TURNING_ON

☑ STATE_ON

☑ STATE_TURNING_OFF

常量值：android.bluetooth.adapter.extra.PREVIOUS_STATE。

（13）String EXTRA_SCAN_MODE

功能：试图在 ACTION_SCAN_MODE_CHANGED 常量中作为一个整型附加域，来请求当前的扫描模式，可以取的值如下：

☑ SCAN_MODE_NONE

☑ SCAN_MODE_CONNECTABLE

☑ SCAN_MODE_CONNECTABLE_DISCOVERABLE

常量值：android.bluetooth.adapter.extra.SCAN_MODE。

（14）String EXTRA_STATE

功能：试图在 ACTION_STATE_CHANGED 常量中作为一个整型附加域，来请求当前的供电状态。可以取的值如下：

☑ STATE_OFF

☑ STATE_TURNING_ON

☑ STATE_ON

☑ STATE_TURNING_OFF

常量值：android.bluetooth.adapter.extra.STATE。

（15）int SCAN_MODE_CONNECTABLE

功能：指明在本地蓝牙适配器中，查询扫描功能失效，但页面扫描功能有效。因此该设备不能被远程蓝牙设备发现，但如果以前曾经发现过该设备，则远程设备可以对其进行连接。

常量值：21(0x00000015)。

（16）int SCAN_MODE_CONNECTABLE_DISCOVERABLE

功能：指明在本地蓝牙适配器中，查询扫描功能和页面扫描功能都有效。因此该设备既可以被远程蓝牙设备发现，也可以被其连接。

常量值：23(0x00000017)。

（17）int SCAN_MODE_NONE

功能：指明在本地蓝牙适配器中，查询扫描功能和页面扫描功能都失效。因此该设备既不可以被远程蓝牙设备发现，也不可以被其连接。

常量值：20(0x00000014)。

（18）int STATE_OFF

功能：指明本地蓝牙适配器模块已经关闭。

常量值：10(0x0000000a)。

（19）int STATE_ON

功能：指明本地蓝牙适配器模块已经打开，并且准备被使用。

（20）int STATE_TURNING_OFF

功能：指明本地蓝牙适配器模块正在关闭。本地客户端可以立刻尝试友好地断开任意外部连接。

常量值：13(0x0000000d)。

（21）int STATE_TURNING_ON

功能：指明本地蓝牙适配器模块正在打开。然而本地客户在尝试使用这个适配器之前需要为 STATE_ON 状态而等待。

常量值：11(0x0000000b)。

### 3．BluetoothAdapter 类的公共方法

（1）public boolean cancelDiscovery()

功能：取消当前的设备发现查找进程，需要 BLUETOOTH_ADMIN 权限。因为对蓝牙适配器而言，查找是一个重量级的过程，因此这个方法必须在尝试连接到远程设备前使用 connect()方法进行调用。发现的过程不会由活动来管理，但是它会作为一个系统服务来运行，因此即使它不能直接请求这样的一个查询动作，也必须取消该搜索进程。如果蓝牙状态不是 STATE_ON，这个 API 将返回 false。蓝牙打开后，等待 ACTION_STATE_CHANGED 更新成 STATE_ON。

返回值：成功则返回 true，有错误则返回 false。

（2）public static boolean checkBluetoothAddress(String address)

功能：验证皆如“00:43:A8:23:10:F0”之类的蓝牙地址，字母必须为大写才有效。

参数 address：字符串形式的蓝牙模块地址。

返回值：地址正确则返回 true，否则返回 false。

（3）public boolean disable()

功能：关闭本地蓝牙适配器，不能在没有明确关闭蓝牙的用户动作中使用。这个方法友好地停止所有的蓝牙连接，停止蓝牙系统服务，以及对所有基础蓝牙硬件进行断电。没有用户的直接同意，蓝牙永远不能被禁止。这个 disable()方法只提供了一个应用，该应用包含了一个改变系统设置的用户界面（例如“电源控制”应用）。

这是一个异步调用方法：该方法将马上获得返回值，用户要通过监听 ACTION_STATE_CHANGED 值来获取随后的适配器状态改变的通知。如果该调用返回 true 值，则该适配器状态会立刻从 STATE_ON 转向 STATE_TURNING_OFF，稍后则会转为 STATE_OFF 或者 STATE_ON。如果该调用返回 false，那么系统已经有一个保护蓝牙适配器被关闭的问题，比如该适配器已经被关闭了。

需要 BLUETOOTH_ADMIN 权限。

返回值：如果蓝牙适配器的停止进程已经开启则返回 true，如果产生错误则返回 false。

（4）public boolean enable()

功能：打开本地蓝牙适配器，不能在没有明确打开蓝牙的用户动作中使用。该方法将为基础的蓝牙硬件供电，并且启动所有的蓝牙系统服务。没有用户的直接同意，蓝牙永远不能被禁止。如果用户为了创建无线连接而打开了蓝牙模块，则其需要 ACTION_REQUEST_ENABLE 值，该值将提出一个请求用户允许以打开蓝牙模块的会话。这个 enable()值只提供了一个应用，该应用包含了一个改变系统设置的用户界面（例如“电源控制”应用）。

这是一个异步调用方法：该方法将马上获得返回值，用户要通过监听 ACTION_STATE_CHANGED 值来获取随后的适配器状态改变的通知。如果该调用返回 true 值，则该适配器状态会立刻从 STATE_OFF 转向 STATE_TURNING_ON，稍后则会转为 STATE_OFF 或者 STATE_ON。如果该调用返回 false，那么说明系统已经有一个保护蓝牙适配器被打开的问题，例如飞行模式，或者该适配器已经被打开。需要 BLUETOOTH_ADMIN 权限。

返回值：如果蓝牙适配器的打开进程已经开启则返回 true，如果产生错误则返回 false。

（5）public String getAddress()

功能：返回本地蓝牙适配器的硬件地址，例如 00:11:22:AA:BB:CC。

需要 BLUETOOTH 权限。

返回值：字符串形式的蓝牙模块地址。

（6）public Set<BluetoothDevice>getBondedDevices()

功能：返回已经匹配到本地适配器的 BluetoothDevice 类的对象集合。如果蓝牙状态不是 STATE_ON，这个 API 将返回 false。蓝牙打开后，等待 ACTION_STATE_CHANGED 更新成 STATE_ON。需要 BLUETOOTH 权限。

返回值：未被修改的 BluetoothDevice 类的对象集合，如果有错误则返回 null。

（7）public static synchronized BluetoothAdapter getDefaultAdapter()

功能：获取对默认本地蓝牙适配器的操作权限。目前 Android 只支持一个蓝牙适配器，但是 API 可以被扩展为支持多个适配器。该方法总是返回默认的适配器。

返回值：返回默认的本地适配器，如果蓝牙适配器在该硬件平台上不能被支持，则返回 null。

（8）public String getName()

功能：获取本地蓝牙适配器的蓝牙名称，这个名称对于外界蓝牙设备而言是可见的。需要 BLUETOOTH 权限。

返回值：该蓝牙适配器名称，如果有错误则返回 null。

（9）public BluetoothDevice getRemoteDevice(String address)

功能：为给予的蓝牙硬件地址获取一个 BluetoothDevice 对象。合法的蓝牙硬件地址必须为大写，格式类似于"00:11:22:33:AA:BB"。checkBluetoothAddress(String)方法可以用来验证蓝牙地址的正确性。BluetoothDevice 类对于合法的硬件地址总会产生返回值，即使这个适配器从未见过该设备。

参数 address：合法的蓝牙 MAC 地址。

异常：IllegalArgumentException，如果地址不合法。

（10）public int getScanMode()

功能：获取本地蓝牙适配器的当前蓝牙扫描模式，蓝牙扫描模式决定本地适配器可连接并且/或者可被远程蓝牙设备所连接。需要 BLUETOOTH 权限，可能的取值如下：

☑ SCAN_MODE_NONE

☑ SCAN_MODE_CONNECTABLE

☑ SCAN_MODE_CONNECTABLE_DISCOVERABLE

如果蓝牙状态不是 STATE_ON，则这个 API 将返回 false。蓝牙打开后，等待 ACTION_STATE_CHANGED 更新成 STATE_ON。

返回值：扫描模式。

（11）public int getState()

功能：获取本地蓝牙适配器的当前状态，需要 BLUETOOTH 类。可能的取值如下：

☑ STATE_OFF

☑ STATE_TURNING_ON

☑ STATE_ON

☑ STATE_TURNING_OFF

返回值：蓝牙适配器的当前状态。

（12）public boolean isDiscovering()

功能：如果当前蓝牙适配器正处于设备发现查找进程中，则返回真值。设备查找是一个重量级过程。当查找正在进行时，用户不能尝试对新的远程蓝牙设备进行连接，同时存在的连接将获得有限制

的带宽以及高等待时间。用户可用 cencelDiscovery()类来取消正在执行的查找进程。

应用程序也可以为 ACTION_DISCOVERY_STARTED 或者 ACTION_DISCOVERY_FINISHED 进行注册，从而当查找开始或者完成时，可以获得通知。

如果蓝牙状态不是 STATE_ON，这个 API 将返回 false。蓝牙打开后，等待 ACTION_STATE_CHANGED 更新成 STATE_ON。需要 BLUETOOTH 权限。

返回值：如果正在查找，则返回 true。

（13）public boolean isEnabled()

功能：如果蓝牙正处于打开状态并可用，则返回真值，与 getBluetoothState()==STATE_ON 等价。需要 BLUETOOTH 权限。

返回值：如果本地适配器已经打开，则返回 true。

（14）public BluetoothServerSocket listenUsingRfcommWithServiceRecord(String name, UUID uuid)

功能：创建一个正在监听的安全的带有服务记录的无线射频通信（RFCOMM）蓝牙端口。一个对该端口进行连接的远程设备将被认证，对该端口的通信将被加密。使用 accpet()方法可以获取从监听 BluetoothServerSocket 处新来的连接。该系统分配一个未被使用的无线射频通信通道来进行监听。

该系统也将注册一个服务探索协议（SDP）记录，该记录带有一个包含了特定的通用唯一识别码（Universally Unique Identifier，UUID）、服务器名称和自动分配通道的本地 SDP 服务。远程蓝牙设备可以用相同的 UUID 来查询自己的 SDP 服务器，并搜寻连接到了哪个通道上。如果该端口已经关闭，或者如果该应用程序异常退出，则这个 SDP 记录会被移除。使用 createRfcommSocketToServiceRecord (UUID)可以从另一使用相同 UUID 的设备来连接到这个端口。需要 BLUETOOTH 权限。

参数：

☑ name：SDP 记录下的服务器名。

☑ uuid：SDP 记录下的 UUID。

返回值：一个正在监听的无线射频通信蓝牙服务端口。

异常：IOException，表示产生错误，例如蓝牙设备不可用，或者许可无效，或者通道被占用。

（15）public boolean setName(String name)

功能：设置蓝牙或者本地蓝牙适配器的昵称，这个名字对于外界蓝牙设备而言是可见的。合法的蓝牙名称最多拥有 248 位 UTF-8 字符，但是很多外界设备只能显示前 40 个字符，有些可能只限制前 20 个字符。

如果蓝牙状态不是 STATE_ON，这个 API 将返回 false。蓝牙打开后，等待 ACTION_STATE_CHANGED 更新成 STATE_ON。需要 BLUETOOTH_ADMIN 权限。

参数 name：一个合法的蓝牙名称。

返回值：如果该名称已被设定，则返回 true，否则返回 false。

（16）public boolean startDiscovery()

功能：开始对远程设备进行查找的进程，它通常牵涉到一个大概需时 12 秒的查询扫描过程，紧跟着是一个对每个获取到自身蓝牙名称的新设备的页面扫描。这是一个异步调用方法：该方法将马上获得返回值，注册 ACTION_DISCOVERY_STARTED 和 ACTION_DISCOVERY_FINISHED 准确地确定该探索是处于开始阶段或者完成阶段。注册 ACTION_FOUND 以活动远程蓝牙设备已找到的通知。

设备查找是一个重量级过程。当查找正在进行时，用户不能尝试对新的远程蓝牙设备进行连接，同时存在的连接将获得有限制的带宽以及高等待时间。用户可用 cencelDiscovery()类来取消正在执行的查找进程。发现的过程不会由活动来进行管理，但是它会作为一个系统服务来运行，因此即使它不能直接请求这样的一个查询动作，也必须取消该搜索进程。设备搜寻只寻找已经被连接的远程设备。许多蓝牙设备默认不会被搜寻到，并且需要进入到一个特殊的模式当中。

如果蓝牙状态不是 STATE_ON，这个 API 将返回 false。蓝牙打开后，等待 ACTION_STATE_CHANGED 更新成 STATE_ON。需要 BLUETOOTH_ADMIN 权限。

返回值：成功返回 true，错误返回 false。

### 5.3.4 BluetoothClass.Service 类

类 BluetoothClass.Service 的格式如下：

```
public static final class BluetoothClass.Service extends Object
```

类 BluetoothClass.Service 的结构如下：

```
java.lang.Object
android.bluetooth.BluetoothClass.Service
```

类 BluetoothClass.Service 用于定义所有的服务类常量，任意 BluetoothClass 由 0 或多个服务类编码组成。在类 BluetoothClass.Service 中包含如下所示的常量：

- ☑ int AUDIO
- ☑ int CAPTURE
- ☑ int INFORMATION
- ☑ int LIMITED_DISCOVERABILITY
- ☑ int NETWORKING
- ☑ int OBJECT_TRANSFER
- ☑ int POSITIONING
- ☑ int RENDER
- ☑ int TELEPHONY

### 5.3.5 BluetoothClass.Device 类

类 BluetoothClass.Device 的格式如下：

```
public final class BluetoothClass.Device extends Object
```

类 BluetoothClass.Device 的结构如下：

```
java.lang.Object
android.bluetooth.BluetoothClass.Device
```

类 BluetoothClass.Device 用于定义所有的设备类的常量，每个 BluetoothClass 有一个带有主要和较小部分的设备类进行编码。里面的常量代表主要和较小的设备类部分（完整的设备类）的组合。BluetoothClass.Device.Major 的常量只能代表主要设备类。

BluetoothClass.Device 有一个内部类，此内部类定义了所有的主要设备类常量。内部类的定义格式如下：

```
class BluetoothClass.Device.Major
```

**注意：** 至此，Android 中的蓝牙类介绍完毕。在调用这些类时，首先要确保 API Level 至少为版本 5 以上，并且还需添加相应的权限，例如，使用通信需要在文件 androidmanifest.xml 中加入<uses-permission android:name="android.permission.BLUETOOTH"/>权限，而在开关蓝牙时需要加入 android.permission.BLUETOOTH_ADMIN 权限。

# 5.4　Android BlueDroid 架构详解

**知识点讲解：光盘:视频\知识点\第 5 章\Android BlueDroid 架构详解.avi**

了解了 Android 系统中低功耗蓝牙协议栈 BlueDroid 的基本知识后，本节将详细讲解 Android 源码中低功耗协议栈 BlueDroid 的具体架构知识，为读者步入本书后面知识的学习打下基础。

## 5.4.1　Android 系统中 BlueDroid 的架构

在 Android 新系统中，采用 BlueDroid 作为默认的协议栈，BlueDroid 分为如下所示的两个部分。

☑ Bluetooth Embedded System（BTE）：实现了 BT 的核心功能。

☑ Bluetooth Application Layer (BTA)：用于和 Android framework 层进行交互。

在 Android 新系统中，BT 系统服务通过 JNI 与 BT stack 进行交互，并且通过 Binder IPC 通信与应用交互，这个系统服务同时也提供给 RD 获取不同的 BT profiles。图 5-3 展示了 BT stack 一个大体的结构。

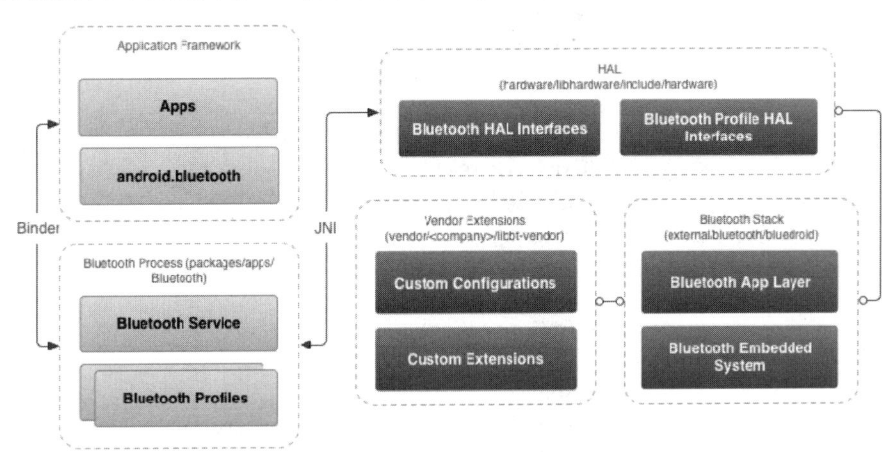


图 5-3　BT stack 的结构

## 5.4.2　Application Framework 层分析

在 Application Framework 层中，功能是利用 android.bluetooth APIS 和 Bluetooth Hardware 层进行交互，也就是通过 Binder IPC 机制调用 Bluetooth 进程。Application Framework 层的代码位于目录 framework/base/core/java/android.bluetooth/中。

在文件 framework/base/core/java/android/bluetooth/BluetoothA2dp.java 中定义了 connect(Bluetoothevice) 方法，功能是调用 Binder IPC 通信机制调用文件 packages/apps/Bluetooth/src/com/android/bluetooth/a2dp/A2dpService.java 中的一个内部私有类。

文件 BluetoothA2dp.java 的具体实现代码如下：

```
public final class BluetoothA2dp implements BluetoothProfile {
    private static final String TAG = "BluetoothA2dp";
    private static final boolean DBG = true;
    private static final boolean VDBG = false;
```

```
    @SdkConstant(SdkConstantType.BROADCAST_INTENT_ACTION)
    public static final String ACTION_CONNECTION_STATE_CHANGED =
        "android.bluetooth.a2dp.profile.action.CONNECTION_STATE_CHANGED";
    @SdkConstant(SdkConstantType.BROADCAST_INTENT_ACTION)
    public static final String ACTION_PLAYING_STATE_CHANGED =
        "android.bluetooth.a2dp.profile.action.PLAYING_STATE_CHANGED";
    public static final int STATE_PLAYING = 10;
    public static final int STATE_NOT_PLAYING = 11;

    private Context mContext;
    private ServiceListener mServiceListener;
    private IBluetoothA2dp mService;
    private BluetoothAdapter mAdapter;

    final private IBluetoothStateChangeCallback mBluetoothStateChangeCallback =
            new IBluetoothStateChangeCallback.Stub() {
                public void onBluetoothStateChange(boolean up) {
                    if (DBG) Log.d(TAG, "onBluetoothStateChange: up=" + up);
                    if (!up) {
                        if (VDBG) Log.d(TAG,"Unbinding service...");
                        synchronized (mConnection) {
                            try {
                                mService = null;
                                mContext.unbindService(mConnection);
                            } catch (Exception re) {
                                Log.e(TAG,"",re);
                            }
                        }
                    } else {
                        synchronized (mConnection) {
                            try {
                                if (mService == null) {
                                    if (VDBG) Log.d(TAG,"Binding service...");
                                    if (!mContext.bindService(new Intent(IBluetoothA2dp.class.getName()),
mConnection, 0)) {
                                        Log.e(TAG, "Could not bind to Bluetooth A2DP Service");
                                    }
                                }
                            } catch (Exception re) {
                                Log.e(TAG,"",re);
                            }
                        }
                    }
                }
        };
BluetoothA2dp(Context context, ServiceListener l) {
        mContext = context;
        mServiceListener = l;
        mAdapter = BluetoothAdapter.getDefaultAdapter();
        IBluetoothManager mgr = mAdapter.getBluetoothManager();
        if (mgr != null) {
```

```
            try {
                mgr.registerStateChangeCallback(mBluetoothStateChangeCallback);
            } catch (RemoteException e) {
                Log.e(TAG,"",e);
            }
        }

        if (!context.bindService(new Intent(IBluetoothA2dp.class.getName()), mConnection, 0)) {
            Log.e(TAG, "Could not bind to Bluetooth A2DP Service");
        }
    }

void close() {
        mServiceListener = null;
        IBluetoothManager mgr = mAdapter.getBluetoothManager();
        if (mgr != null) {
            try {
                mgr.unregisterStateChangeCallback(mBluetoothStateChangeCallback);
            } catch (Exception e) {
                Log.e(TAG,"",e);
            }
        }

        synchronized (mConnection) {
            if (mService != null) {
                try {
                    mService = null;
                    mContext.unbindService(mConnection);
                } catch (Exception re) {
                    Log.e(TAG,"",re);
                }
            }
        }
    }

    public void finalize() {
        close();
    }
    public boolean connect(BluetoothDevice device) {
        if (DBG) log("connect(" + device + ")");
        if (mService != null && isEnabled() &&
            isValidDevice(device)) {
            try {
                return mService.connect(device);
            } catch (RemoteException e) {
                Log.e(TAG, "Stack:" + Log.getStackTraceString(new Throwable()));
                return false;
            }
        }
        if (mService == null) Log.w(TAG, "Proxy not attached to service");
        return false;
```

```
}
public boolean disconnect(BluetoothDevice device) {
    if (DBG) log("disconnect(" + device + ")");
    if (mService != null && isEnabled() &&
        isValidDevice(device)) {
        try {
            return mService.disconnect(device);
        } catch (RemoteException e) {
            Log.e(TAG, "Stack:" + Log.getStackTraceString(new Throwable()));
            return false;
        }
    }
    if (mService == null) Log.w(TAG, "Proxy not attached to service");
    return false;
}

/**
*{@inheritDoc}
*/
public List<BluetoothDevice> getConnectedDevices() {
    if (VDBG) log("getConnectedDevices()");
    if (mService != null && isEnabled()) {
        try {
            return mService.getConnectedDevices();
        } catch (RemoteException e) {
            Log.e(TAG, "Stack:" + Log.getStackTraceString(new Throwable()));
            return new ArrayList<BluetoothDevice>();
        }
    }
    if (mService == null) Log.w(TAG, "Proxy not attached to service");
    return new ArrayList<BluetoothDevice>();
}
public List<BluetoothDevice> getDevicesMatchingConnectionStates(int[] states) {
    if (VDBG) log("getDevicesMatchingStates()");
    if (mService != null && isEnabled()) {
        try {
            return mService.getDevicesMatchingConnectionStates(states);
        } catch (RemoteException e) {
            Log.e(TAG, "Stack:" + Log.getStackTraceString(new Throwable()));
            return new ArrayList<BluetoothDevice>();
        }
    }
    if (mService == null) Log.w(TAG, "Proxy not attached to service");
    return new ArrayList<BluetoothDevice>();
}
public int getConnectionState(BluetoothDevice device) {
    if (VDBG) log("getState(" + device + ")");
    if (mService != null && isEnabled()
        && isValidDevice(device)) {
        try {
            return mService.getConnectionState(device);
```

```
            } catch (RemoteException e) {
                Log.e(TAG, "Stack:" + Log.getStackTraceString(new Throwable()));
                return BluetoothProfile.STATE_DISCONNECTED;
            }
        }
        if (mService == null) Log.w(TAG, "Proxy not attached to service");
        return BluetoothProfile.STATE_DISCONNECTED;
    }
    public boolean setPriority(BluetoothDevice device, int priority) {
        if (DBG) log("setPriority(" + device + ", " + priority + ")");
        if (mService != null && isEnabled()
            && isValidDevice(device)) {
            if (priority != BluetoothProfile.PRIORITY_OFF &&
                priority != BluetoothProfile.PRIORITY_ON){
              return false;
            }
            try {
                return mService.setPriority(device, priority);
            } catch (RemoteException e) {
                Log.e(TAG, "Stack:" + Log.getStackTraceString(new Throwable()));
                return false;
            }
        }
        if (mService == null) Log.w(TAG, "Proxy not attached to service");
        return false;
    }
    public int getPriority(BluetoothDevice device) {
        if (VDBG) log("getPriority(" + device + ")");
        if (mService != null && isEnabled()
            && isValidDevice(device)) {
            try {
                return mService.getPriority(device);
            } catch (RemoteException e) {
                Log.e(TAG, "Stack:" + Log.getStackTraceString(new Throwable()));
                return BluetoothProfile.PRIORITY_OFF;
            }
        }
        if (mService == null) Log.w(TAG, "Proxy not attached to service");
        return BluetoothProfile.PRIORITY_OFF;
    }
    public boolean isA2dpPlaying(BluetoothDevice device) {
        if (mService != null && isEnabled()
            && isValidDevice(device)) {
            try {
                return mService.isA2dpPlaying(device);
            } catch (RemoteException e) {
                Log.e(TAG, "Stack:" + Log.getStackTraceString(new Throwable()));
                return false;
            }
        }
        if (mService == null) Log.w(TAG, "Proxy not attached to service");
```

```
        return false;
    }
    public boolean shouldSendVolumeKeys(BluetoothDevice device) {
        if (isEnabled() && isValidDevice(device)) {
            ParcelUuid[] uuids = device.getUuids();
            if (uuids == null) return false;

            for (ParcelUuid uuid: uuids) {
                if (BluetoothUuid.isAvrcpTarget(uuid)) {
                    return true;
                }
            }
        }
        return false;
    }
    public static String stateToString(int state) {
        switch (state) {
        case STATE_DISCONNECTED:
            return "disconnected";
        case STATE_CONNECTING:
            return "connecting";
        case STATE_CONNECTED:
            return "connected";
        case STATE_DISCONNECTING:
            return "disconnecting";
        case STATE_PLAYING:
            return "playing";
        case STATE_NOT_PLAYING:
          return "not playing";
        default:
            return "<unknown state " + state + ">";
        }
    }

    private ServiceConnection mConnection = new ServiceConnection() {
        public void onServiceConnected(ComponentName className, IBinder service) {
            if (DBG) Log.d(TAG, "Proxy object connected");
            mService = IBluetoothA2dp.Stub.asInterface(service);

            if (mServiceListener != null) {
                mServiceListener.onServiceConnected(BluetoothProfile.A2DP, BluetoothA2dp.this);
            }
        }
        public void onServiceDisconnected(ComponentName className) {
            if (DBG) Log.d(TAG, "Proxy object disconnected");
            mService = null;
            if (mServiceListener != null) {
                mServiceListener.onServiceDisconnected(BluetoothProfile.A2DP);
            }
        }
    };
```

在上述代码中，定义了 A2dpService 对象 service，并调用 getService()方法。A2dpService 是一个继承于类 ProfileService 的子类，而 ProfileService 是继承于类 Service 的子类。文件 A2dpService.java 的主要实现代码如下：

```
public class A2dpService extends ProfileService {
    private static final boolean DBG = false;
    private static final String TAG="A2dpService";

    private A2dpStateMachine mStateMachine;
    private Avrcp mAvrcp;
    private static A2dpService sAd2dpService;

    protected String getName() {
        return TAG;
    }

    protected IProfileServiceBinder initBinder() {
        return new BluetoothA2dpBinder(this);
    }

    protected boolean start() {
        mStateMachine = A2dpStateMachine.make(this, this);
        mAvrcp = Avrcp.make(this);
        setA2dpService(this);
        return true;
    }

    protected boolean stop() {
        mStateMachine.doQuit();
        mAvrcp.doQuit();
        return true;
    }

    protected boolean cleanup() {
        if (mStateMachine!= null) {
            mStateMachine.cleanup();
        }
        if (mAvrcp != null) {
            mAvrcp.cleanup();
            mAvrcp = null;
        }
        clearA2dpService();
        return true;
    }

    public static synchronized A2dpService getA2dpService(){
        if (sAd2dpService != null && sAd2dpService.isAvailable()) {
            if (DBG) Log.d(TAG, "getA2DPService(): returning " + sAd2dpService);
            return sAd2dpService;
        }
        if (DBG)  {
```

```
            if (sAd2dpService == null) {
                Log.d(TAG, "getA2dpService(): service is NULL");
            } else if (!(sAd2dpService.isAvailable())) {
                Log.d(TAG,"getA2dpService(): service is not available");
            }
        }
        return null;
    }

    private static synchronized void setA2dpService(A2dpService instance) {
        if (instance != null && instance.isAvailable()) {
            if (DBG) Log.d(TAG, "setA2dpService(): set to: " + sAd2dpService);
            sAd2dpService = instance;
        } else {
            if (DBG)   {
                if (sAd2dpService == null) {
                    Log.d(TAG, "setA2dpService(): service not available");
                } else if (!sAd2dpService.isAvailable()) {
                    Log.d(TAG,"setA2dpService(): service is cleaning up");
                }
            }
        }
    }

    private static synchronized void clearA2dpService() {
        sAd2dpService = null;
    }

    public boolean connect(BluetoothDevice device) {
        enforceCallingOrSelfPermission(BLUETOOTH_ADMIN_PERM,
                                       "Need BLUETOOTH ADMIN permission");

        if (getPriority(device) == BluetoothProfile.PRIORITY_OFF) {
            return false;
        }

        int connectionState = mStateMachine.getConnectionState(device);
        if (connectionState == BluetoothProfile.STATE_CONNECTED ||
            connectionState == BluetoothProfile.STATE_CONNECTING) {
            return false;
        }

        mStateMachine.sendMessage(A2dpStateMachine.CONNECT, device);
        return true;
    }

    boolean disconnect(BluetoothDevice device) {
        enforceCallingOrSelfPermission(BLUETOOTH_ADMIN_PERM,
                                       "Need BLUETOOTH ADMIN permission");
        int connectionState = mStateMachine.getConnectionState(device);
        if (connectionState != BluetoothProfile.STATE_CONNECTED &&
```

```
            connectionState != BluetoothProfile.STATE_CONNECTING) {
            return false;
        }

        mStateMachine.sendMessage(A2dpStateMachine.DISCONNECT, device);
        return true;
    }
```

由此可见，在接下来的通信过程中通过 Binder IPC 通信机制调用了文件 A2dpService.java 中的 connect(BluetoothDevice)方法。上述过程就是 Bluetooth Application Framework 与 Bluetooth Process 之间的调用过程。

## 5.4.3　分析 Bluetooth System Service 层

在 Android 系统中，Bluetooth System Service 位于 packages/apps/Bluetooth 目录下，将其打包成一个“Android App（Android 应用程序）”包，并且在 Android Framework 层中实现 BT Service 和各种 profile。BT App 接下来会通过 JNI 调用到 HAL 层。

在文件 A2dpService.java 中，connect()方法会发送一个 StateMachine.sendMessage(A2dpStateMachine.CONNECT,device)的 message（信息 0，这个 message 会被 A2dpStateMachine 对象的 processMessage (Message)方法接收到），对应代码如下：

```
case CONNECT:
                BluetoothDevice device = (BluetoothDevice) message.obj;
                broadcastConnectionState(device, BluetoothProfile.STATE_CONNECTING,
                                BluetoothProfile.STATE_DISCONNECTED);

                if (!connectA2dpNative(getByteAddress(device)) ) {
                    broadcastConnectionState(device, BluetoothProfile.STATE_DISCONNECTED,
                                BluetoothProfile.STATE_CONNECTING);
                    break;
                }

                synchronized (A2dpStateMachine.this) {
                    mTargetDevice = device;
                    transitionTo(mPending);
                }
                sendMessageDelayed(CONNECT_TIMEOUT, 30000);
                break;
```

在上述代码中，会通过“connectA2dpNative(getByteAddress(device);”代码行设置通过 JNI 调用到 Native（本地程序）：

```
private native boolean connectA2dpNative(byte[] address);
```

## 5.4.4　JNI 层详解

在 Android 系统中，和 Bluetooth 有关的 JNI 代码位于目录 packages/apps/bluetooth/jni 中。

JNI 层的代码会调用到 HAL 层，并且在确信一些 BT 操作被触发时从 HAL 获取一些回调，例如当 BT 设备被发现时。在 A2dp 连接的例子中，BT System Service 会通过 JNI 调用文件 com_android_bluetooth_ a2dp.cpp 中的方法，此文件的主要实现代码如下：

```
namespace android {
static jmethodID method_onConnectionStateChanged;
static jmethodID method_onAudioStateChanged;

static const btav_interface_t *sBluetoothA2dpInterface = NULL;
static jobject mCallbacksObj = NULL;
static JNIEnv *sCallbackEnv = NULL;

static bool checkCallbackThread() {
    //if (sCallbackEnv == NULL) {
    sCallbackEnv = getCallbackEnv();
    //}

    JNIEnv* env = AndroidRuntime::getJNIEnv();
    if (sCallbackEnv != env || sCallbackEnv == NULL) return false;
    return true;
}

static void bta2dp_connection_state_callback(btav_connection_state_t state, bt_bdaddr_t* bd_addr) {
    jbyteArray addr;

    ALOGI("%s", __FUNCTION__);

    if (!checkCallbackThread()) {                                          \
        ALOGE("Callback: '%s' is not called on the correct thread", __FUNCTION__); \
        return;                                                            \
    }
    addr = sCallbackEnv->NewByteArray(sizeof(bt_bdaddr_t));
    if (!addr) {
        ALOGE("Fail to new jbyteArray bd addr for connection state");
        checkAndClearExceptionFromCallback(sCallbackEnv, __FUNCTION__);
        return;
    }

    sCallbackEnv->SetByteArrayRegion(addr, 0, sizeof(bt_bdaddr_t), (jbyte*) bd_addr);
    sCallbackEnv->CallVoidMethod(mCallbacksObj, method_onConnectionStateChanged, (jint) state,addr);
    checkAndClearExceptionFromCallback(sCallbackEnv, __FUNCTION__);
    sCallbackEnv->DeleteLocalRef(addr);
}

static void bta2dp_audio_state_callback(btav_audio_state_t state, bt_bdaddr_t* bd_addr) {
    jbyteArray addr;

    ALOGI("%s", __FUNCTION__);

    if (!checkCallbackThread()) {                                          \
        ALOGE("Callback: '%s' is not called on the correct thread", __FUNCTION__); \
        return;                                                            \
    }
```

```
    addr = sCallbackEnv->NewByteArray(sizeof(bt_bdaddr_t));
    if (!addr) {
        ALOGE("Fail to new jbyteArray bd addr for connection state");
        checkAndClearExceptionFromCallback(sCallbackEnv, __FUNCTION__);
        return;
    }

    sCallbackEnv->SetByteArrayRegion(addr, 0, sizeof(bt_bdaddr_t), (jbyte*) bd_addr);
    sCallbackEnv->CallVoidMethod(mCallbacksObj, method_onAudioStateChanged, (jint) state,addr);
    checkAndClearExceptionFromCallback(sCallbackEnv, __FUNCTION__);
    sCallbackEnv->DeleteLocalRef(addr);
}

static btav_callbacks_t sBluetoothA2dpCallbacks = {
    sizeof(sBluetoothA2dpCallbacks),
    bta2dp_connection_state_callback,
    bta2dp_audio_state_callback
};

static void classInitNative(JNIEnv* env, jclass clazz) {
    int err;
    const bt_interface_t* btInf;
    bt_status_t status;

    method_onConnectionStateChanged =
        env->GetMethodID(clazz, "onConnectionStateChanged", "(I[B)V");

    method_onAudioStateChanged =
        env->GetMethodID(clazz, "onAudioStateChanged", "(I[B)V");
    /*
    if ( (btInf = getBluetoothInterface()) == NULL) {
        ALOGE("Bluetooth module is not loaded");
        return;
    }

    if ( (sBluetoothA2dpInterface = (btav_interface_t *)
          btInf->get_profile_interface(BT_PROFILE_ADVANCED_AUDIO_ID)) == NULL) {
        ALOGE("Failed to get Bluetooth A2DP Interface");
        return;
    }
    if ( (status = sBluetoothA2dpInterface->init(&sBluetoothA2dpCallbacks)) != BT_STATUS_SUCCESS) {
        ALOGE("Failed to initialize Bluetooth A2DP, status: %d", status);
        sBluetoothA2dpInterface = NULL;
        return;
    }*/

    ALOGI("%s: succeeds", __FUNCTION__);
}

static void initNative(JNIEnv *env, jobject object) {
    const bt_interface_t* btInf;
```

```
    bt_status_t status;

    if ( (btInf = getBluetoothInterface()) == NULL) {
        ALOGE("Bluetooth module is not loaded");
        return;
    }

    if (sBluetoothA2dpInterface !=NULL) {
         ALOGW("Cleaning up A2DP Interface before initializing...");
         sBluetoothA2dpInterface->cleanup();
         sBluetoothA2dpInterface = NULL;
    }

    if (mCallbacksObj != NULL) {
         ALOGW("Cleaning up A2DP callback object");
         env->DeleteGlobalRef(mCallbacksObj);
         mCallbacksObj = NULL;
    }

    if ( (sBluetoothA2dpInterface = (btav_interface_t *)
          btInf->get_profile_interface(BT_PROFILE_ADVANCED_AUDIO_ID)) == NULL) {
        ALOGE("Failed to get Bluetooth A2DP Interface");
        return;
    }

    if ( (status = sBluetoothA2dpInterface->init(&sBluetoothA2dpCallbacks)) != BT_STATUS_SUCCESS) {
        ALOGE("Failed to initialize Bluetooth A2DP, status: %d", status);
        sBluetoothA2dpInterface = NULL;
        return;
    }

    mCallbacksObj = env->NewGlobalRef(object);
}

static void cleanupNative(JNIEnv *env, jobject object) {
    const bt_interface_t* btInf;
    bt_status_t status;

    if ( (btInf = getBluetoothInterface()) == NULL) {
        ALOGE("Bluetooth module is not loaded");
        return;
    }

    if (sBluetoothA2dpInterface !=NULL) {
        sBluetoothA2dpInterface->cleanup();
        sBluetoothA2dpInterface = NULL;
    }

    if (mCallbacksObj != NULL) {
        env->DeleteGlobalRef(mCallbacksObj);
        mCallbacksObj = NULL;
```

```
    }
}

static jboolean connectA2dpNative(JNIEnv *env, jobject object, jbyteArray address) {
    jbyte *addr;
    bt_bdaddr_t * btAddr;
    bt_status_t status;

    ALOGI("%s: sBluetoothA2dpInterface: %p", __FUNCTION__, sBluetoothA2dpInterface);
    if (!sBluetoothA2dpInterface) return JNI_FALSE;

    addr = env->GetByteArrayElements(address, NULL);
    btAddr = (bt_bdaddr_t *) addr;
    if (!addr) {
        jniThrowIOException(env, EINVAL);
        return JNI_FALSE;
    }

    if ((status = sBluetoothA2dpInterface->connect((bt_bdaddr_t *)addr)) != BT_STATUS_SUCCESS) {
        ALOGE("Failed HF connection, status: %d", status);
    }
    env->ReleaseByteArrayElements(address, addr, 0);
    return (status == BT_STATUS_SUCCESS) ? JNI_TRUE : JNI_FALSE;
}

static jboolean disconnectA2dpNative(JNIEnv *env, jobject object, jbyteArray address) {
    jbyte *addr;
    bt_status_t status;

    if (!sBluetoothA2dpInterface) return JNI_FALSE;

    addr = env->GetByteArrayElements(address, NULL);
    if (!addr) {
        jniThrowIOException(env, EINVAL);
        return JNI_FALSE;
    }

    if ( (status = sBluetoothA2dpInterface->disconnect((bt_bdaddr_t *)addr)) != BT_STATUS_SUCCESS) {
        ALOGE("Failed HF disconnection, status: %d", status);
    }
    env->ReleaseByteArrayElements(address, addr, 0);
    return (status == BT_STATUS_SUCCESS) ? JNI_TRUE : JNI_FALSE;
}

static JNINativeMethod sMethods[] = {
    {"classInitNative", "()V", (void *) classInitNative},
    {"initNative", "()V", (void *) initNative},
    {"cleanupNative", "()V", (void *) cleanupNative},
    {"connectA2dpNative", "([B)Z", (void *) connectA2dpNative},
    {"disconnectA2dpNative", "([B)Z", (void *) disconnectA2dpNative},
};
```

```
int register_com_android_bluetooth_a2dp(JNIEnv* env)
{
    return jniRegisterNativeMethods(env, "com/android/bluetooth/a2dp/A2dpStateMachine",
                                    sMethods, NELEM(sMethods));
}

}
```

在上述代码中用到了结构体对象 sBluetoothA2dpInterface，此对象在方法 initNative(JNIEnv*env, jobject object)中定义获取，即如下所示的代码。

```
if ( (sBluetoothA2dpInterface = (btav_interface_t *)
      btInf->get_profile_interface(BT_PROFILE_ADVANCED_AUDIO_ID)) == NULL) {
    ALOGE("Failed to get Bluetooth A2DP Interface");
    return;
}
```

### 5.4.5 HAL 层详解

硬件抽象层用于定义 android.bluetooth APIs 和 BT process 调用的标准接口，BT HAL 的头文件位于文件 hardware/libhardware/include/hardware/bluetooth.h 和 hardware/libhardware/include/hardware/bt_*.h 中。

JNI 中的 sBluetoothA2dpIntcrfacc 是一个 btav_interface_t 结构体，位于 hardware/libhardware/include/hardware/bt_av.h 中，具体定义代码如下：

```
typedef struct {
    size_t size;
    bt_status_t (*init)( btav_callbacks_t* callbacks );
    bt_status_t (*connect)( bt_bdaddr_t *bd_addr );
    bt_status_t (*disconnect)( bt_bdaddr_t *bd_addr );
    void   (*cleanup)( void );
} btav_interface_t;
```

Android 系统新版本默认蓝牙协议栈 BlueDroid 在目录 external/bluetooth/bluedroid 下实现。

上述 stack 实现了通用的 BT HAL，并且也可以通过扩展和改变配置来自定义。例如 A2dp 的连接会调用到 external/bluetooth/bluedroid/btif/src/btif_av.c 的 connect()方法，此方法的具体实现代码如下：

```
static bt_status_t connect(bt_bdaddr_t *bd_addr)
{
    BTIF_TRACE_EVENT1("%s", __FUNCTION__);
    CHECK_BTAV_INIT();

    return btif_queue_connect(UUID_SERVCLASS_AUDIO_SOURCE, bd_addr, connect_int);
}
```

## 5.5 Android 中的低功耗蓝牙协议栈

**知识点讲解：光盘:视频\知识点\第 5 章\Android 中的低功耗蓝牙协议栈.avi**

从 Android 4.2 版本开始，Google 便更换了 Android 的蓝牙协议栈，从 BlueZ 换成 BlueDroid。从

Android 4.3 版本开始，提供了对蓝牙 4.0 BLE 的支持。本节将详细讲解 Android 系统中的蓝牙 4.0 BLE 的基本知识，为读者步入本书后面知识的学习打下基础。

## 5.5.1　低功耗蓝牙协议栈基础

为了确保 Android 系统可以更好地支持蓝牙 4.0 BLE，Broadcom 公司特意推出了适应于 Android 平台的开源低功耗蓝牙协议栈 BlueDroid，其开发文档和 API 是开源代码，在地址 https://github.com/briandbl/framework 中保存。

在上述开源代码中，低功耗蓝牙 API 支持 Android 平台上的低功耗蓝牙通信功能。通过使用 BlueDroid 协议栈，Android 应用程序可以枚举、发现并访问低功耗蓝牙的外部设备，并且实现了低功耗蓝牙规范。

从 Android 4.2 版本开始，低功耗蓝牙模块的整体结构如图 5-4 所示。

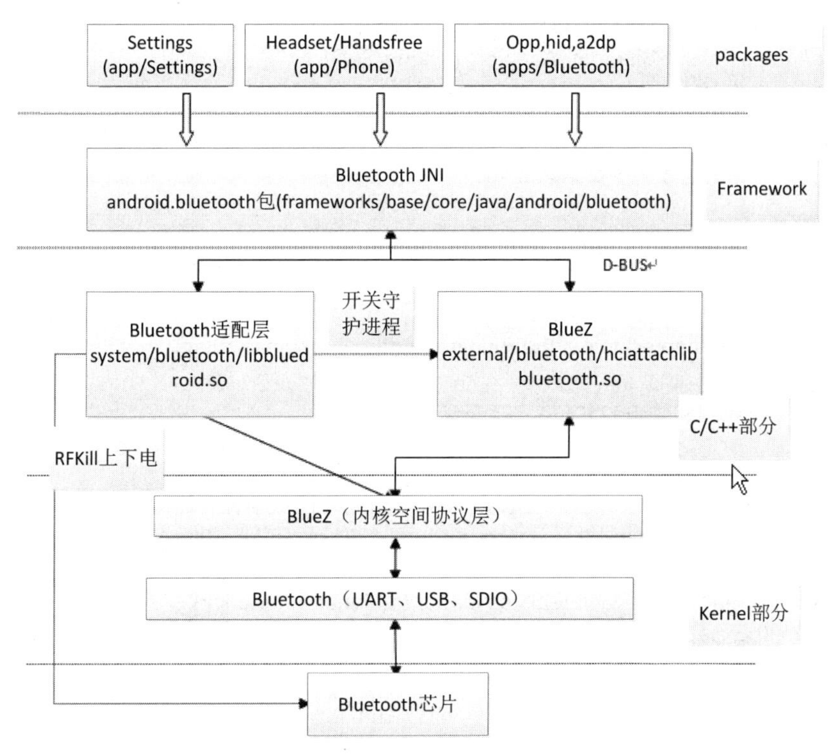


图 5-4　低功耗蓝牙模块的整体结构

**注意：** 虽然从 Android 4.2 版本开始，JNI 部分的代码在 packages 层中实现。但是为了便于读者从视觉上更加容易接受，所以将 JNI 部分绘制在了 Framework 层中。

## 5.5.2　低功耗蓝牙 API 详解

Broadcom 公司推出的低功耗蓝牙协议栈 BlueDroid 的开发文档和 API 是开源代码，被保存在地址 https://github.com/briandbl/framework 中。

下面将详细讲解 API 的基本功能和具体原理。

（1）本地蓝牙适配器设备

本功能不是由 Broadcom 公司提供的，而是由 Android SDK 提供的，源码位于目录 framework/base/core/java/android.bluetooth/BluetoothAdapter.java 中。

文件 BluetoothAdapter.java 实现了所有蓝牙交互的入口。通过使用类 BluetoothAdapter 可以实现如下所示的功能。

☑ 发现其他的蓝牙设备，查询匹配的设备集。

☑ 使用一个已知蓝牙地址来初始化蓝牙设备 BluetoothDevice。

☑ 创建一个能够监听其他设备通信的类 BluetoothSocket。

文件 BluetoothAdapter.java 的主要实现代码如下：

```
public static synchronized BluetoothAdapter getDefaultAdapter() {
    if (sAdapter == null) {
        IBinder b = ServiceManager.getService(BLUETOOTH_MANAGER_SERVICE);
        if (b != null) {
            IBluetoothManager managerService = IBluetoothManager.Stub.asInterface(b);
            sAdapter = new BluetoothAdapter(managerService);
        } else {
            Log.e(TAG, "Bluetooth binder is null");
        }
    }
    return sAdapter;
}

/**
*Use {@link #getDefaultAdapter} to get the BluetoothAdapter instance.
*/
BluetoothAdapter(IBluetoothManager managerService) {

    if (managerService == null) {
        throw new IllegalArgumentException("bluetooth manager service is null");
    }
    try {
        mService = managerService.registerAdapter(mManagerCallback);
    } catch (RemoteException e) {Log.e(TAG, "", e);}
    mManagerService = managerService;
    mLeScanClients = new HashMap<LeScanCallback, GattCallbackWrapper>();
}
public BluetoothDevice getRemoteDevice(byte[] address) {
    if (address == null || address.length != 6) {
        throw new IllegalArgumentException("Bluetooth address must have 6 bytes");
    }
    return new BluetoothDevice(String.format("%02X:%02X:%02X:%02X:%02X:%02X",
            address[0], address[1], address[2], address[3], address[4], address[5]));
}
public int getState() {
    try {
        synchronized(mManagerCallback) {
            if (mService != null)
            {
```

```
                int state=  mService.getState();
                if (VDBG) Log.d(TAG, "" + hashCode() + ": getState(). Returning " + state);
                return state;
            }
        }
    } catch (RemoteException e) {Log.e(TAG, "", e);}
    if (DBG) Log.d(TAG, "" + hashCode() + ": getState() :  mService = null. Returning STATE_OFF");
    return STATE_OFF;
}
public String getAddress() {
    try {
        return mManagerService.getAddress();
    } catch (RemoteException e) {Log.e(TAG, "", e);}
    return null;
}
public String getName() {
    try {
        return mManagerService.getName();
    } catch (RemoteException e) {Log.e(TAG, "", e);}
    return null;
}

public int getScanMode() {
    if (getState() != STATE_ON) return SCAN_MODE_NONE;
    try {
        synchronized(mManagerCallback) {
            if (mService != null) return mService.getScanMode();
        }
    } catch (RemoteException e) {Log.e(TAG, "", e);}
    return SCAN_MODE_NONE;
}
public boolean setScanMode(int mode, int duration) {
    if (getState() != STATE_ON) return false;
    try {
        synchronized(mManagerCallback) {
            if (mService != null) return mService.setScanMode(mode, duration);
        }
    } catch (RemoteException e) {Log.e(TAG, "", e);}
    return false;
}
```

（2）请求远程蓝牙设备

本功能也不是由Broadcom公司提供的，而是由Android SDK提供的，源码位于目录framework/base/core/java/android.bluetooth/BluetoothDevice.java中。

文件BluetoothDevice.java代表一个远程蓝牙设备，可以支持BLE低功耗设备、BR/EDR设备或Dual-mode类型的设备。通过使用类BluetoothDevice可以实现如下所示的功能。

☑ 请求获取远程蓝牙设备的连接。

☑ 查询获取远程蓝牙设备的名称、地址、类和链接状态。

文件BluetoothDevice.java的主要实现代码如下：

```
static IBluetooth getService() {
        synchronized (BluetoothDevice.class) {
            if (sService == null) {
                BluetoothAdapter adapter = BluetoothAdapter.getDefaultAdapter();
                sService = adapter.getBluetoothService(mStateChangeCallback);
            }
        }
        return sService;
    }

    static IBluetoothManagerCallback mStateChangeCallback = new IBluetoothManagerCallback.Stub() {

        public void onBluetoothServiceUp(IBluetooth bluetoothService)
                throws RemoteException {
            synchronized (BluetoothDevice.class) {
                sService = bluetoothService;
            }
        }

        public void onBluetoothServiceDown()
            throws RemoteException {
            synchronized (BluetoothDevice.class) {
                sService = null;
            }
        }
    };
    /*package*/ BluetoothDevice(String address) {
        getService();
        if (!BluetoothAdapter.checkBluetoothAddress(address)) {
            throw new IllegalArgumentException(address + " is not a valid Bluetooth address");
        }

        mAddress = address;
    }
    public static final Parcelable.Creator<BluetoothDevice> CREATOR =
            new Parcelable.Creator<BluetoothDevice>() {
        public BluetoothDevice createFromParcel(Parcel in) {
            return new BluetoothDevice(in.readString());
        }
        public BluetoothDevice[] newArray(int size) {
            return new BluetoothDevice[size];
        }
    };
    public boolean cancelBondProcess() {
        if (sService == null) {
            Log.e(TAG, "BT not enabled. Cannot cancel Remote Device bond");
            return false;
        }
        try {
            return sService.cancelBondProcess(this);
        } catch (RemoteException e) {Log.e(TAG, "", e);}
```

```
        return false;
    }
    public boolean removeBond() {
        if (sService == null) {
            Log.e(TAG, "BT not enabled. Cannot remove Remote Device bond");
            return false;
        }
        try {
            return sService.removeBond(this);
        } catch (RemoteException e) {Log.e(TAG, "", e);}
        return false;
    }
    public int getBondState() {
        if (sService == null) {
            Log.e(TAG, "BT not enabled. Cannot get bond state");
            return BOND_NONE;
        }
        try {
            return sService.getBondState(this);
        } catch (RemoteException e) {Log.e(TAG, "", e);}
        catch (NullPointerException npe) {
            Log.e(TAG, "NullPointerException for getBondState() of device ("+
                getAddress()+")", npe);
        }
        return BOND_NONE;
    }
    public BluetoothClass getBluetoothClass() {
        if (sService == null) {
            Log.e(TAG, "BT not enabled. Cannot get Bluetooth Class");
            return null;
        }
        try {
            int classInt = sService.getRemoteClass(this);
            if (classInt == BluetoothClass.ERROR) return null;
            return new BluetoothClass(classInt);
        } catch (RemoteException e) {Log.e(TAG, "", e);}
        return null;
    }
     public boolean fetchUuidsWithSdp() {
        try {
            return sService.fetchRemoteUuids(this);
        } catch (RemoteException e) {Log.e(TAG, "", e);}
            return false;
    }
    public boolean setPin(byte[] pin) {
        if (sService == null) {
            Log.e(TAG, "BT not enabled. Cannot set Remote Device pin");
            return false;
        }
        try {
            return sService.setPin(this, true, pin.length, pin);
```

```
        } catch (RemoteException e) {Log.e(TAG, "", e);}
        return false;
    }
    public boolean setPasskey(int passkey) {
        /*
        try {
            return sService.setPasskey(this, true, 4, passkey);
        } catch (RemoteException e) {Log.e(TAG, "", e);}*/
        return false;
    }
    public boolean setPairingConfirmation(boolean confirm) {
        if (sService == null) {
            Log.e(TAG, "BT not enabled. Cannot set pairing confirmation");
            return false;
        }
        try {
            return sService.setPairingConfirmation(this, confirm);
        } catch (RemoteException e) {Log.e(TAG, "", e);}
        return false;
    }
```

（3）实现客户端的低功耗蓝牙规范

在Broadcom公司提供的源码中，文件BleClientProfile.java的功能是实现客户端的低功耗蓝牙规范。在应用中要想访问远程设备中的低功耗蓝牙规范，就必须继承于类 BleClientProfile，并且需要提供要访问规范的必需参数和服务标识。通过 BleClientProfile 的派生类可以发起一个远程设备的连接，并且一个 BleClientProfile 类可能会包含多个 BleClientService 对象的实例。文件 BleClientProfile.java 的具体实现代码如下：

```
//下面是构造方法，功能是给当前规范的 UUID 和客户端应用上下文创建一个 BleClientProfile
    public BleClientProfile(Context context, BleGattID profileUuid)
    {
        Log.d(TAG, "new profile" + profileUuid.toString());

        this.mContext = context;
        this.mAppUuid = profileUuid;

        this.mConnectedDevices = new ArrayList<BluetoothDevice>();
        this.mConnectingDevices = new ArrayList<BluetoothDevice>();
        this.mDisconnectingDevices = new ArrayList<BluetoothDevice>();

        this.mClientIDToDeviceMap = new HashMap<Integer, BluetoothDevice>();
        this.mDeviceToClientIDMap = new HashMap<BluetoothDevice, Integer>();

        this.mCallback = new BleClientCallback();
        this.mSvcConn = new GattServiceConnection(context);
    }

    /**
    *初始化 BleClientProfile 对象
    */
    public void init(ArrayList<BleClientService> requiredServices,
```

```
          ArrayList<BleClientService> optionalServices)
{
    Log.d(TAG, "init (" + this.mAppUuid + ")");

    this.mRequiredServices = requiredServices;
    this.mOptionalServices = optionalServices;

    IBinder b = ServiceManager.getService(BleConstants.BLUETOOTH_LE_SERVICE);
    if (b == null) {
        throw new RuntimeException("Bluetooth Low Energy service not available");
    }
    this.mSvcConn.onServiceConnected(null, b);
}

/**
*清除和此规范有关的资源
*/
public synchronized void finish()
{
    if (this.mSvcConn != null) {
        this.mContext.unbindService(this.mSvcConn);
        this.mSvcConn = null;
    }
}

@Override

/**
*返回此规范是否已经成功注册到蓝牙协议栈中
*@see {@link #registerProfile()}
*/
public boolean isProfileRegistered()
{
    Log.d(TAG, "isProfileRegistered (" + this.mAppUuid + ")");
    return this.mClientIf != BleConstants.GATT_SERVICE_PRIMARY;
}

/**
*注册规范到蓝牙协议栈
*/
public int registerProfile()
{
    int ret = BleConstants.GATT_SUCCESS;
    Log.d(TAG, "registerProfile (" + this.mAppUuid + ")");

    if (this.mClientIf == BleConstants.GATT_SERVICE_PRIMARY)
    {
        try
        {
            this.mService.registerApp(this.mAppUuid, this.mCallback);
        } catch (RemoteException e) {
```

```
                Log.e(TAG, e.toString());
                ret = BleConstants.SERVICE_UNAVAILABLE;
            }
        }

        return ret;
    }

    /**
    *注销蓝牙协议栈中的规范
    */
    public void deregisterProfile()
    {
        Log.d(TAG, "deregisterProfile (" + this.mAppUuid + ")");

        if (this.mClientIf != BleConstants.GATT_SERVICE_PRIMARY)
            try {
                this.mService.unregisterApp(this.mClientIf);
            } catch (RemoteException e) {
                Log.e(TAG, "deregisterProfile() - " + e.toString());
            }
    }

    /**
    *设置一个活跃连接设备的加密等级
    */
    public void setEncryption(BluetoothDevice device, byte action)
    {
        try
        {
            this.mService.setEncryption(device.getAddress(), action);
        } catch (RemoteException e) {
            Log.e(TAG, e.toString());
        }
    }

    /**
    *当请求后台连接时，定义本地设备扫描远程低功耗设备的强度
    */
    public void setScanParameters(int scanInterval, int scanWindow)
    {
        try
        {
            this.mService.setScanParameters(scanInterval, scanWindow);
        } catch (RemoteException e) {
            Log.e(TAG, e.toString());
        }
    }

    /**
    *建立一个到远程设备的 GATT 连接
```

```
*/
public int connect(BluetoothDevice device)
{
    Log.d(TAG, "connect (" + this.mAppUuid + ")" + device.getAddress());

    int ret = BleConstants.GATT_SUCCESS;

    synchronized (this.mConnectingDevices) {
        this.mConnectingDevices.add(device);
    }

    synchronized (this.mDisconnectingDevices) {
        this.mDisconnectingDevices.remove(device);
    }
    try
    {
        this.mService.open(this.mClientIf, device.getAddress(), true);
    } catch (RemoteException e) {
        Log.e(TAG, e.toString());
        ret = BleConstants.GATT_ERROR;
    }

    return ret;
}

/**
*准备一个到远程蓝牙设备的后台连接
*/
public int connectBackground(BluetoothDevice device)
{
    Log.d(TAG,
            "connectBackground (" + this.mAppUuid + ")" + device.getAddress());

    int ret = BleConstants.GATT_SUCCESS;

    synchronized (this.mConnectingDevices) {
        this.mConnectingDevices.add(device);
    }

    synchronized (this.mDisconnectingDevices) {
        this.mDisconnectingDevices.remove(device);
    }
    try
    {
        this.mService.open(this.mClientIf, device.getAddress(), false);
    } catch (RemoteException e) {
        Log.e(TAG, e.toString());
        ret = BleConstants.GATT_ERROR;
    }

    return ret;
```

```
}

/**
*停止监听远程蓝牙设备试图发起的连接
*/
public int cancelBackgroundConnection(BluetoothDevice device)
{
    Log.d(TAG, "cancelBackgroundConnection (" + this.mAppUuid
            + ") - device " + device.getAddress());

    int ret = BleConstants.GATT_SUCCESS;
    try
    {
        this.mService.close(this.mClientIf, device.getAddress(), 0, false);
    } catch (RemoteException e) {
        Log.e(TAG, e.toString());
        ret = BleConstants.GATT_ERROR;
    }
    return ret;
}
/**
*断开一个到远程设备的 GATT 连接
*/
public int disconnect(BluetoothDevice device)
{
    Log.d(TAG,
            "disconnect (" + this.mAppUuid + ") - device " + device.getAddress());
    synchronized (this.mDisconnectingDevices) {
        this.mDisconnectingDevices.add(device);
    }
    int ret = BleConstants.GATT_SUCCESS;
    try
    {
        this.mService.close(this.mClientIf,
                device.getAddress(),
                ((Integer) this.mDeviceToClientIDMap.get(device)).intValue(),
                true);
    } catch (RemoteException e) {
        Log.e(TAG, e.toString());
        ret = BleConstants.GATT_ERROR;
    }
    return ret;
}
/**
*刷新当前客户端的规范
*/
public int refresh(BluetoothDevice device)
{
    Log.d(TAG,
            "refresh (" + this.mAppUuid + ") - address = " + device.getAddress());
    if (isDeviceDisconnecting(device)) {
```

```
            Log.d(TAG, "refresh (" + this.mAppUuid
                    + ") - Device unavailable!");
            return BleConstants.GATT_ERROR;
        }
        this.mRequiredServices.get(BleConstants.GATT_SERVICE_PRIMARY).refresh(device);
        return BleConstants.GATT_SUCCESS;
    }
    /**
    *刷新当前规范包含的特定服务
    */
    public int refreshService(BluetoothDevice device, BleClientService service)
    {
        Log.d(TAG, "refreshService (" + this.mAppUuid + ") address = s "
                + device.getAddress() + "service = " + service.getServiceId());
        return 0;
    }
    /**
    *在已经连接的设备列表中查找指定蓝牙设备的地址
    */
    public BluetoothDevice findConnectedDevice(String address)
    {
        BluetoothDevice ret = null;
        synchronized (this.mConnectedDevices) {
            for (int i = 0; i != this.mConnectedDevices.size(); i++) {
                BluetoothDevice d = (BluetoothDevice) this.mConnectedDevices.get(i);
                if (address.equalsIgnoreCase(d.getAddress())) {
                    ret = d;
                    break;
                }
            }
        }
        return ret;
    }
    /**
    *返回当前连接和等待连接中的所有远程设备集合
    */
    public BluetoothDevice[] getPendingConnections()
    {
        return (BluetoothDevice[]) this.mConnectingDevices.toArray(new BluetoothDevice[0]);
    }
    /**
    *设置一个蓝牙设备地址，在等待连接设备列表中查找一个远程设备
    */
    public BluetoothDevice findDeviceWaitingForConnection(String address)
    {
        BluetoothDevice ret = null;
        synchronized (this.mConnectingDevices) {
            for (int i = 0; i < this.mConnectingDevices.size(); i++) {
                BluetoothDevice d = (BluetoothDevice) this.mConnectingDevices.get(i);
                if (address.equalsIgnoreCase(d.getAddress())) {
                    ret = d;
```

```
                    break;
                }
            }
        }
        return ret;
    }
    void onServiceRefreshed(BleClientService s, BluetoothDevice device)
    {
        int i = this.mRequiredServices.indexOf(s);
        if (i + 1 < this.mRequiredServices.size()) {
            Log.d(TAG, "Refreshing next service");
            ((BleClientService) this.mRequiredServices.get(i + 1)).refresh(device);
        } else {
            onRefreshed(device);
        }
    }
    public void onInitialized(boolean success)
    {
        Log.d(TAG, "onInitialized");
        if (success)
            registerProfile();
    }
        public void onServiceConnected(ComponentName name, IBinder service)
        {
            Log.d(TAG, "Connected to GattService!");

            if (service != null)
                try {
                    BleClientProfile.this.mService = IBluetoothGatt.Stub.asInterface(service);

                    for (int i = 0; i < BleClientProfile.this.mRequiredServices.size(); i++) {
                        BleClientProfile.this.mRequiredServices.get(i)
                                .setProfile(BleClientProfile.this);
                    }

                    if (BleClientProfile.this.mOptionalServices != null) {
                        for (int i = 0; i < BleClientProfile.this.mOptionalServices.size(); i++) {
                            BleClientProfile.this.mOptionalServices
                                    .get(i).setProfile(BleClientProfile.this);
                        }
                    }

                    BleClientProfile.this.onInitialized(true);
                } catch (Throwable t) {
                    Log.e(TAG, "Unable to get Binder to GattService", t);
                    BleClientProfile.this.onInitialized(false);
                }
        }
```

（4）创建一个代表客户端角色设备上的低功耗蓝牙服务派生类

在 Broadcom 公司提供的源码中，文件 BleClientService.java 的功能是定义一个派生类，此派生类代表了客户端角色设备上的低功耗蓝牙服务。通过这个派生类可以允许应用程序读写低功耗蓝牙服务的特征，并在特征改变时注册通知。文件 BleClientService.java 的主要实现代码如下：

```
//定义代表客户端的低功耗服务
public abstract class BleClientService
{
    private static String TAG = "BleClientService";

    private BleClientProfile mProfile = null;
    private BleGattID mServiceId = null;
    private HashMap<BluetoothDevice, ArrayList<ServiceData>> mdeviceToDataMap =
            new HashMap<BluetoothDevice, ArrayList<ServiceData>>();
    private BleCharacteristicDataCallback mCallback =
            new BleCharacteristicDataCallback();
    private boolean mReadDescriptors = true;

    /**
    *创建一个新的低功耗蓝牙服务的 UUID
    *
    *@param serviceId
    */
    public BleClientService(BleGattID serviceId)
    {
        mServiceId = serviceId;
        if (mServiceId.getServiceType() == BleConstants.GATT_UNDEFINED)
            mServiceId.setServiceType(BleConstants.GATT_SERVICE_PRIMARY);
    }

    /**
    *返回服务的 UUID
    */
    public BleGattID getServiceId()
    {
        return mServiceId;
    }

    /**
    *写操作远程设备上的一个特性
    */
    public int writeCharacteristic(BluetoothDevice remoteDevice, int instanceId,
            BleCharacteristic characteristic)
    {
        Log.d(TAG, "writeCharacteristic");

        int ret = BleConstants.GATT_SUCCESS;
        int connID = BleConstants.GATT_INVALID_CONN_ID;

        if ((connID = mProfile.getConnIdForDevice(remoteDevice)) == BleConstants.GATT_INVALID_CONN_ID) {
            return BleConstants.GATT_INVALID_CONN_ID;
        }
        ServiceData s = getServiceData(remoteDevice, instanceId);
```

```
    if (s == null) {
        return ret;
    }
    s.writeIndex = s.characteristics.indexOf(characteristic);

    if ((s.characteristics != null) && (s.writeIndex >= BleConstants.GATT_SERVICE_PRIMARY)) {
        Log.d(TAG, "writeCharacteristic found characteristic in array:");
        Log.d(TAG,
                "Service = [instanceID = " + instanceId + " svcid = "
                        + mServiceId.toString() + " serviceType = "
                        + mServiceId.getServiceType());
        Log.d(TAG, "CharID = [instanceID = " + characteristic.getInstanceID()
                + " svcid = " + characteristic.getID().toString());
        BleGattID svcId = new BleGattID(instanceId, mServiceId.getUuid(),
                mServiceId.getServiceType());
        BleGattID cID = characteristic.getID();
        BluetoothGattCharID charID = new BluetoothGattCharID(svcId, cID);
        try
        {
            if (characteristic.isDirty()) {
                if (characteristic.getWriteType() == BleConstants.GATT_SUCCESS)
                    characteristic.setWriteType(2);
                characteristic.setDirty(false);
                mProfile.getGattService().writeCharValue(connID, charID,
                        characteristic.getWriteType(), characteristic.getAuthReq(),
                        characteristic.getValue());
            }
            else if (!characteristic.getDirtyDescQueue().isEmpty()) {
                ArrayList<BleDescriptor> descList =
                        characteristic.getDirtyDescQueue();
                BleDescriptor descObj = descList.get(0);

                Log.d(TAG, "writeCharacteristic - descriptor = "
                        + descObj.getID().toString());
                if (descObj.isDirty()) {
                    BluetoothGattCharDescrID descID = new BluetoothGattCharDescrID(
                            svcId, cID, descObj.getID());
                    descObj.setDirty(false);
                    mProfile.getGattService().writeCharDescrValue(connID,
                            descID, descObj.getWriteType(), descObj.getAuthReq(),
                            descObj.getValue());
                }

            }
            else
            {
                onWriteCharacteristicComplete(0, remoteDevice, characteristic);
            }
        } catch (RemoteException e) {
            ret = BleConstants.GATT_ERROR;
```

```
            }
        } else {
            onWriteCharacteristicComplete(0, remoteDevice, characteristic);
        }
        return ret;
    }

    /**
    *检索服务包含的所有特征
    */
    public ArrayList<BleCharacteristic> getAllCharacteristics(BluetoothDevice remoteDevice)
    {
        Log.d(TAG, "getAllCharacteristics");

        ServiceData s = getServiceData(remoteDevice, mServiceId.getInstanceID());
        if (null != s) {
            return s.characteristics;
        }
        return null;
    }

    /**
    *返回一个基于它的 ID 的服务特性
    */
    public BleCharacteristic getCharacteristic(BluetoothDevice remoteDevice,
            BleGattID characteristicID)
    {
        Log.d(TAG, "getCharacteristic charID = [" + characteristicID.toString()
                + "] instance ID = [" + characteristicID.getInstanceID() + "]");
        ServiceData s = getServiceData(remoteDevice, mServiceId.getInstanceID());
        if (s == null) {
            Log.d(TAG, "getCharacterisic - Service data not found");
            return null;
        }
        for (int i = 0; i < s.characteristics.size(); i++) {
            BleCharacteristic c = s.characteristics.get(i);
            if (c != null) {
                if (c.getID() != null) {
                    if ((c.getID().toString().equals(characteristicID.toString()))
                            && (c.getInstanceID() == characteristicID.getInstanceID()))
                    {
                        return c;
                    }
                }
                else
                    Log.d(TAG, "Error: Characteristic ID is null");
            }
            else {
                Log.d(TAG, "Error: Cannot retrieve characteristic");
            }
        }
```

```
        return null;
    }

    /**
    *返回所有服务实例的 IDs 列表
    */
    public int[] getAllServiceInstanceIds(BluetoothDevice remoteDevice)
    {
        Log.d(TAG, "getAllServiceInstanceIds");
        ArrayList<ServiceData> s = mdeviceToDataMap.get(remoteDevice);
        if (s != null) {
            int[] instanceIds = new int[s.size()];

            for (int i = 0; i < s.size(); i++) {
                instanceIds[i] = s.get(0).instanceID;
            }

            return instanceIds;
        }

        return null;
    }

    /**
    *刷新远程服务的所有实例的所有特性
    */
    public void refresh(BluetoothDevice remoteDevice)
    {
        Log.d(TAG, "Refresh (" + mServiceId.toString() + ")");

        ArrayList<ServiceData> s = mdeviceToDataMap.get(remoteDevice);
        if (s != null) {
            ServiceData sd = s.get(0);
            Log.e(TAG,
                    "refresh() - Service data found, reading first characteristic... (serviceType = "
                            + sd.serviceType + ")");
            readFirstCharacteristic(remoteDevice, new BleGattID(sd.instanceID,
                    getServiceId().getUuid(), sd.serviceType));
        } else {
            Log.e(TAG, "refresh() - Service data not found");
        }
    }

    /**
    *从远程设备上读取指定的特性
    *
    *@see {@link #onReadCharacteristicComplete(BluetoothDevice,BleCharacteristic)}
    */
    public int readCharacteristic(BluetoothDevice remoteDevice,
            BleCharacteristic characteristic)
    {
```

```
        int ret = BleConstants.GATT_SUCCESS;
        int connID = BleConstants.GATT_INVALID_CONN_ID;
        Log.d(TAG,
                "readCharacteristic - svc UUID = " + getServiceId().getUuid().toString()
                        + ", characteristic = " + characteristic.getID());

        BluetoothGattCharID charID = new BluetoothGattCharID(new BleGattID(
                characteristic.getInstanceID(), getServiceId().getUuid(), getServiceId()
                        .getServiceType()), characteristic.getID());

        if ((connID = mProfile.getConnIdForDevice(remoteDevice)) != BleConstants.GATT_INVALID_CONN_ID)
            readCharacteristicValue(connID, charID, characteristic.getAuthReq());
        else {
            ret = BleConstants.GATT_INVALID_CONN_ID;
        }
        return ret;
    }

    /**
    *设置一个远程设备特性的写操作已经完成的函数
    */
    public void onWriteCharacteristicComplete(int status, BluetoothDevice remoteDevice,
            BleCharacteristic characteristic)
    {
        Log.d(TAG, "onWriteCharacteristicComplete 1 status=" + status);
        if (status == BleConstants.GATT_INSUF_AUTHENTICATION) {
            Log.d(TAG,
                    "onWriteCharacteristicComplete rcv GATT_INSUF_AUTHENTICATION issue createBond");
            if (remoteDevice.createBond())
                Log.d(TAG, "onWriteCharacteristicComplete createBond request Accepted");
            else {
                Log.e(TAG, "onWriteCharacteristicComplete createBond request FAILED");
            }
        }
        else if (status == BleConstants.GATT_INSUF_ENCRYPTION) {
            Log.d(TAG,
                    "onWriteCharacteristicComplete rcv GATT_INSUF_ENCRYPTION check link can be encrypt
or not");
            if (remoteDevice.getBondState() == BluetoothDevice.BOND_BONDED) {
                Log.d(TAG,
                        "device bonded start to encrypt the link.   !!!! This case should not happen !!!!");
            } else {
                Log.d(TAG, "device is Not bonded start to pair");
                remoteDevice.createBond();
            }
        }
    }

    /**
    *设置一个远程特性已经改变的回调函数
    */
```

```
public void onCharacteristicChanged(BluetoothDevice remoteDevice,
        BleCharacteristic characteristic)
{
    Log.d(TAG, "onCharacteristicChanged");
}

/**
*设置服务已经完成一次刷新的回调函数
*/
public void onRefreshComplete(BluetoothDevice remoteDevice)
{
    Log.d(TAG, "onRefreshComplete");
}

/**
*当蓝牙服务需要连接认证时触发的回调函数
*/
public void onSetCharacteristicAuthRequirement(BluetoothDevice remoteDevice,
        BleCharacteristic characteristic, int instanceID)
{
    Log.d(TAG, "onSetCharacteristicAuthRequirement");
}

/**
*当一个设置的特性被更新，并且读取其值和描述符时会被调用
*/
public void onReadCharacteristicComplete(BluetoothDevice remoteDevice,
        BleCharacteristic characteristic)
{
    Log.d(TAG, "onReadCharacteristicComplete");
}

public void onReadCharacteristicComplete(int status, BluetoothDevice remoteDevice,
        BleCharacteristic characteristic)
{
    Log.d(TAG, "onReadCharacteristicComplete status=" + status);
    if (status == BleConstants.GATT_INSUF_AUTHENTICATION) {
        Log.d(TAG,
                "onReadCharacteristicComplete rcv GATT_INSUF_AUTHENTICATION issue createBond");
        remoteDevice.createBond();
        return;
    }
    if (status != BleConstants.GATT_INSUF_ENCRYPTION) {
        return;
    }

    Log.d(TAG,
            "onReadCharacteristicComplete rcv GATT_INSUF_ENCRYPTION check link can be encrypt
or not");
    if (remoteDevice.getBondState() == BluetoothDevice.BOND_BONDED) {
        Log.d(TAG,"device bonded start to encrypt the link.   !!!! This case should not happen !!!!");
```

```
        } else {
            Log.d(TAG, "device is Not bonded start to pair");
            remoteDevice.createBond();
        }
    }
    /**
    *注册此服务在服务器上的通知
    */
    public int registerForNotification(BluetoothDevice remoteDevice, int instanceID,BleGattID characteristicID)
      {
        int ret = BleConstants.GATT_SUCCESS;
        Log.d(TAG, "registerForNotification address: " + remoteDevice.getAddress());
        try {
            BleGattID svcId = new BleGattID(instanceID, getServiceId().getUuid(),
                    getServiceId().getServiceType());
            BluetoothGattCharID charId = new BluetoothGattCharID(svcId, characteristicID);
            mProfile.getGattService().registerForNotifications(
                    mProfile.getClientIf(), remoteDevice.getAddress(), charId);
        } catch (RemoteException e) {
            ret = BleConstants.GATT_ERROR;
        }
        return ret;
    }
    /**
    *取消从服务器的通知服务
    */
    public int unregisterNotification(BluetoothDevice remoteDevice, int instanceID,BleGattID characteristicID)
    {
        int ret = BleConstants.GATT_SUCCESS;
        Log.d(TAG, "unregisterNotification address: " + remoteDevice.getAddress());
        try {
            BleGattID svcId = new BleGattID(instanceID, getServiceId().getUuid(),
                    getServiceId().getServiceType());
            BluetoothGattCharID charId = new BluetoothGattCharID(svcId, characteristicID);

            mProfile.getGattService().deregisterForNotifications(
                    mProfile.getClientIf(), remoteDevice.getAddress(), charId);
        } catch (RemoteException e) {
            ret = BleConstants.GATT_ERROR;
        }
        return ret;
    }
    void setInstanceID(BluetoothDevice remoteDevice, int instanceId)
    {
        Log.d(TAG, "setInstanceID address = " + remoteDevice.getAddress());
        ServiceData sd = getServiceData(remoteDevice, instanceId);
        mServiceId.setInstanceId(instanceId);
        if (null == sd) {
            Log.d(TAG, "setInstanceID setting instance id (" + instanceId + ")");
            sd = new ServiceData();
            sd.instanceID = mServiceId.getInstanceID();
```

```
            sd.serviceType = mServiceId.getServiceType();
            ArrayList<ServiceData> s = mdeviceToDataMap.get(remoteDevice);
            if (null == s) {
                s = new ArrayList<ServiceData>();
            }
            s.add(sd);
            mdeviceToDataMap.put(remoteDevice, s);
        }
        sd.instanceID = mServiceId.getInstanceID();
        sd.serviceType = mServiceId.getServiceType();
        try
        {
            int connID = BleConstants.GATT_INVALID_CONN_ID;
            if ((connID = mProfile.getConnIdForDevice(remoteDevice)) != BleConstants.GATT_INVALID_
CONN_ID)
                mProfile.getGattService().registerServiceDataCallback(connID,
                        mServiceId, remoteDevice.getAddress(), mCallback);
        } catch (RemoteException e)
        {
            Log.d(TAG, e.toString());
        }
    }
//获得第一个特征描述符
public void onGetFirstCharacteristicDescriptor(int connID, int status,
                BluetoothGattID svcId, BluetoothGattID characteristicID,
                BluetoothGattID descriptorID)
        {
            Log.d(BleClientService.TAG, "onGetFirstCharacteristicDescriptor "
                    + characteristicID.toString() + " status = " + status);

            BluetoothDevice device = BleClientService.this.mProfile.getDeviceforConnId(connID);
            if ((device == null)
                    || (BleClientService.this.mProfile.isDeviceDisconnecting(device))) {
                Log.e(BleClientService.TAG,
                        "onGetFirstCharacteristicDescriptor() - Device is disconnecting...");
                return;
            }

            if (status != BleConstants.GATT_SUCCESS) {
                BleClientService.this.readNextCharacteristic(
                        BleClientService.this.mProfile.getDeviceforConnId(connID),
                        BleApiHelper.gatt2BleID(svcId),
                        BleApiHelper.gatt2BleID(characteristicID));
                return;
            }

            Log.d(BleClientService.TAG,
                    "characteristic ID = " + characteristicID.toString() + "instance ID = "
                            + characteristicID.getInstanceID());

            BleCharacteristic characteristic = findCharacteristic(connID,
```

```
                BleApiHelper.gatt2BleID(svcId), BleApiHelper.gatt2BleID(characteristicID));

        if (descriptorID.getUuidType() == BleConstants.GATT_UUID_TYPE_128) {
            String uuid128 = descriptorID.getUuid().toString();
            if (uuid128.equals("00002900-0000-1000-8000-00805f9b34fb"))
                characteristic.addDescriptor(new BleExtProperty());
            else if (uuid128.equals("00002902-0000-1000-8000-00805f9b34fb"))
                characteristic.addDescriptor(new BleClientConfig());
            else if (uuid128.equals("00002903-0000-1000-8000-00805f9b34fb"))
                characteristic.addDescriptor(new BleServerConfig());
            else if (uuid128.equals("00002904-0000-1000-8000-00805f9b34fb"))
                characteristic.addDescriptor(new BlePresentationFormat());
            else if (uuid128.equals("00002901-0000-1000-8000-00805f9b34fb"))
                characteristic.addDescriptor(new BleUserDescription());
            else
                characteristic.addDescriptor(new BleDescriptor(new BleGattID(
                        descriptorID.getUuid())));
        }
        else {
            switch (descriptorID.getUuid16()) {
                case BleConstants.GATT_UUID_CHAR_EXT_PROP16:
                    characteristic.addDescriptor(new BleExtProperty());
                    break;
                case BleConstants.GATT_UUID_CHAR_CLIENT_CONFIG16:
                    characteristic.addDescriptor(new BleClientConfig());
                    break;
                case BleConstants.GATT_UUID_CHAR_SRVR_CONFIG16:
                    characteristic.addDescriptor(new BleServerConfig());
                    break;
                case BleConstants.GATT_UUID_CHAR_PRESENT_FORMAT16:
                    characteristic.addDescriptor(new BlePresentationFormat());
                    break;
                case BleConstants.GATT_UUID_CHAR_DESCRIPTION16:
                    characteristic.addDescriptor(new BleUserDescription());
                    break;
                default:
                    characteristic.addDescriptor(new BleDescriptor(new BleGattID(
                            descriptorID.getUuid16())));
            }

        }

        BleClientService.this.readNextCharDescriptor(
                BleClientService.this.mProfile.getDeviceforConnId(connID),
                BleApiHelper.gatt2BleID(svcId), BleApiHelper.gatt2BleID(characteristicID),
                BleApiHelper.gatt2BleID(descriptorID));

    }
```

（5）定义服务器端的角色的低功耗规范

在 Broadcom 公司提供的源码中，文件 BleServerProfile.java 的功能是定义了服务器端的角色的低功耗规范，在创建一个新的低功耗规范之前，需要先继承于这个类，并提供标识要访问规范所必需的参数和服务。通常来说，一个 BleServerProfile 派生的类包含一个或多个 BleServerService 对象。在 BleServerProfile 派生的类中，包含低功耗规范中定义服务的 BleServerService 对象的集合。文件 BleServerProfile.java 的主要实现代码如下：

```
public abstract class BleServerProfile
{
    private static final boolean D = true;
    private static final String TAG = "BleServerProfile";
    private Context mCtxt = null;
    private BleGattID mAppid;
    ArrayList<BleServerService> mServiceArr = null;
    private HashMap<String, Integer> mConnMap = null;
    private HashMap<Integer, Integer> mMtuMap = null;
    private IBluetoothGatt mService;
    private int mSvcCreated = 0;
    private int mSvcStarted = 0;
    private byte mAppHandle = -1;
    private int mProfileStatus = 2;
    private GattServiceConnection mSvcConn;

    public BleServerProfile(Context ctxt, BleGattID appId,ArrayList<BleServerService> serviceArr)
    {
        mAppid = appId;
        mCtxt = ctxt;
        mServiceArr = serviceArr;
        mConnMap = new HashMap<String, Integer>();
        mMtuMap = new HashMap<Integer, Integer>();
        mSvcConn = new GattServiceConnection(null);
        Intent i = new Intent();
        i.setClassName("com.broadcom.bt.app.system", "com.broadcom.bt.app.system.GattService");
        mCtxt.bindService(i, mSvcConn, 1);

        throw new RuntimeException("Not implemented");
    }
/*取消和此规范相关的资源*/
    public synchronized void finish()
    {
        if (mSvcConn != null) {
            mCtxt.unbindService(mSvcConn);
            mSvcConn = null;
        }
    }

    public void finalize()
    {
        finish();
    }
/*初始化相关的服务*/
    void initProfile()
    {
```

```
        Log.i("BleServerProfile", "initProfile()");
        try {
            mService.registerServerProfileCallback(mAppid, new BleServerProfileCallback(this));
        } catch (Throwable t) {
            Log.e("BleServerProfile", "Unable to start profile", t);
        }
    }

    void notifyAction(int event)
    {
        if ((event == 0) && (++mSvcCreated == mServiceArr.size()))
        {
            Log.i("BleServerProfile",
                    "All services created successfully. Calling onInitialized");
            onInitialized(true);
        } else if ((event == 4) && (--mSvcCreated == 0))
        {
            Log.i("BleServerProfile",
                    "All services stopped successfully. Calling onStopped");
            onStopped();
        } else if ((event == 2) && (++mSvcStarted == mServiceArr.size()))
        {
            Log.i("BleServerProfile",
                    "All services started successfully. Calling onStarted");
            onStarted(true);
        } else if (event == 1) {
            Log.i("BleServerProfile",
                    "One of the services creation failed. Calling onInitialized");
            mProfileStatus = 2;
            onInitialized(false);
        } else if (event == 3) {
            Log.i("BleServerProfile",
                    "One of the services start failed. Calling onStarted");
            mProfileStatus = 2;
            onStarted(false);
        } else {
            Log.e("BleServerProfile", "Unknown action from a service");
        }
    }
/*启用和此规范有关的所有服务*/
    public void startProfile()
    {
        Log.i("BleServerProfile", "startProfile()");
        if (mService == null) {
            Log.i("BleServerProfile", "Remote service object is null.. Returning..");
            return;
        }

        for (int i = 0; i < mServiceArr.size(); i++) {
            if (!((BleServerService) mServiceArr.get(i)).isRegistered()) {
                Log.i("BleServerProfile",
```

```
                    "One of the services is not registered. Stopping all the services");
                stopProfile();
                return;
            }

            ((BleServerService) mServiceArr.get(i)).startService();
        }
    }
/*停止和此规范有关的所有服务*/
    public void stopProfile()
    {
        Log.i("BleServerProfile", "stopProfile()");
        for (int i = 0; i < mServiceArr.size(); i++)
            ((BleServerService) mServiceArr.get(i)).stopService();
    }
/*注销所有相关的服务*/
    public void finishProfile()
    {
        Log.i("BleServerProfile", "finishProfile()");
        for (int i = 0; i < mServiceArr.size(); i++) {
            ((BleServerService) mServiceArr.get(i)).deleteService();
        }
        try
        {
            mService.unregisterServerProfileCallback(mAppHandle);
        } catch (Throwable t) {
            Log.e("BleServerProfile", "Unable to stop profile", t);
            return;
        }
    }
/*为连接设置最大传输单元*/
    public void setMtuSize(int connId, int mtuSize)
    {
        Log.i("BleServerProfile", "setMtuSize");
        mMtuMap.put(Integer.valueOf(connId), Integer.valueOf(mtuSize));
    }
/*为一个活跃的连接设置需要的加密等级*/
    public void setEncryption(String bdaddr, byte action)
    {
        try
        {
            mService.setEncryption(bdaddr, action);
        } catch (Throwable t) {
            Log.e("BleServerProfile", "Unable to set encryption for connection", t);
        }
    }
/*当已经请求一个后台连接时，定义本地设备扫描远程低功耗设备的强度*/
    public void setScanParameters(int scanInterval, int scanWindow)
    {
        try
        {
```

```
            mService.setScanParameters(scanInterval, scanWindow);
        } catch (Throwable t) {
            Log.e("BleServerProfile", "Unable to set scan parameters", t);
        }
    }
/*打开一个外设 GAP 客户端的连接*/
    public void open(String bdaddr, boolean isDirect)
    {
        try
        {
            mService.GATTServer_Open(mAppHandle, bdaddr, isDirect);
        } catch (Throwable t) {
            Log.e("BleServerProfile", "Unable to open Gatt connection", t);
        }
    }
/*取消一个正在进行中对外设 GATT 客户端的打开操作*/
    public void cancelOpen(String bdaddr, boolean isDirect)
    {
        try
        {
            mService.GATTServer_CancelOpen(mAppHandle, bdaddr, isDirect);
        } catch (Throwable t) {
            Log.e("BleServerProfile", "Unable to open Gatt connection", t);
            return;
        }
    }
/*关闭一个到远程低功耗规范客户端的连接*/
    public void close(String bdaddr)
    {
        try
        {
            mService.GATTServer_Close(((Integer) mConnMap.get(bdaddr))
                    .intValue());
        } catch (Throwable t) {
            Log.e("BleServerProfile", "Unable to open Gatt connection", t);
            return;
        }
    }
/*当一个客户端建立或断开连接时触发的回调函数*/
        public void onClientConnected(int connId, String bdaddr, boolean isConnected)
        {
            Log.i("BleServerProfile", "onClientConnCted addr is " + bdaddr + " connId is " + connId);

            mProfile.onClientConnected(bdaddr, isConnected);
            if (isConnected)
                mProfile.mConnMap.put(bdaddr, Integer.valueOf(connId));
            else
                mProfile.mConnMap.remove(bdaddr);
        }
    private class GattServiceConnection
            implements ServiceConnection
```

```
    {
        private Context context;

        private GattServiceConnection(Context c) {
            context = c;
        }

        public void onServiceConnected(ComponentName name, IBinder service)
        {
            Log.d("BleServerProfile", "Connected to GattService!");

            if (service != null)
                try {
                    BleServerProfile.this.mService = IBluetoothGatt.Stub.asInterface(service);
                    BleServerProfile.this.initProfile();
                } catch (Throwable t) {
                    Log.e("BleServerProfile", "Unable to get Binder to GattService", t);
                }
        }

        public void onServiceDisconnected(ComponentName name)
        {
            Log.d("BleServerProfile", "Disconnected from GattService!");
        }
    }
}
```

（6）创建低功耗服务

在 Broadcom 公司提供的源码中，文件 BleServerService.java 的功能是创建一个低功耗服务，这是服务器端角色上的低功耗规范的一部分。BleServerService 的派生类包含了一个或多个 BleCharacteristic 对象。在应用程序中，需要重写类 BleServerService 来实现一个服务。文件 BleServerService.java 的主要实现代码如下：

```
public abstract class BleServerService
{
    private final String TAG = "BleServerService";

    private HashMap<Integer, BleCharacteristic> mCharHdlMap = null;
    private HashMap<Integer, BleServerService> mServiceHdlMap = null;
    private HashMap<Integer, AttributeRequestInfo> mAttrReqMap = null;

    private ArrayList<BleCharacteristic> mCharQueue = null;
    private ArrayList<BleDescriptor> mDirtyDescQueue = null;
    private BleGattID mServiceId;
    private BleGattID mAppUuid;
    private BleServerProfile mProfileHandle;
    private IBluetoothGatt mService;
    private int mSvcHandle = -1;
    private byte mSupTransport;
    private BleServiceCallback mGattServiceCallback;
    private boolean isServiceAvailable = false;
```

```
    private int mSvcInstance = 0;

    private boolean isPrimary = false;
    private int mNumHandles;
    private final int CHAR_ADDED = 0;
    private final int CHAR_DESC_ADDED = 1;
    private final int ATTRIBUTE_WRITE = 2;
    private final int ATTRIBUTE_READ = 3;
    private final int HDL_VAL_INDICATION = 4;
    private final int HDL_VAL_NOTIFICATION = 5;
    private final int MTU_EXCHANGE = 6;
    private final int EXECUTE_WRITE = 7;

    private Handler mHandler = new Handler()
    {
        public void handleMessage(Message msg)
        {
        }

    };
/*构造函数，使用给定的 ID 构造一个低功耗服务*/
    public BleServerService(BleGattID serviceId, int numHandles)
    {
        /**
         * TODO: implement
         */
        this.mServiceId = serviceId;
        this.mNumHandles = numHandles;
        this.mSupTransport = 2;
        this.mGattServiceCallback = new BleServiceCallback(this);
        this.mCharHdlMap = new HashMap();
        this.mServiceHdlMap = new HashMap();
        this.mCharQueue = new ArrayList();
        this.mAttrReqMap = new HashMap();

        if (this.mServiceId.getServiceType() == -1)
            this.mServiceId.setServiceType(0);
        throw new RuntimeException("not implemented");


    }
/*构造函数，使用给定的 ID 构造一个新的低功耗服务*/
    public BleServerService(BleGattID serviceId, byte supTransport, int numHandles)
    {
        this.mServiceId = serviceId;
        this.mNumHandles = numHandles;
        this.mSupTransport = supTransport;
        this.mGattServiceCallback = new BleServiceCallback(this);

        this.mCharHdlMap = new HashMap();
        this.mCharQueue = new ArrayList();
```

```
        this.mServiceHdlMap = new HashMap();
        this.mAttrReqMap = new HashMap();

        if (this.mServiceId.getServiceType() == -1)
            this.mServiceId.setServiceType(0);
        throw new RuntimeException("not implemented");
    }
/*初始化服务*/
    protected void initService()
    {
        if (this.mService != null)
            try {
                this.mService.registerServerServiceCallback(this.mServiceId,
                        this.mAppUuid, this.mGattServiceCallback);
            } catch (Throwable t) {
                Log.e("BleServerService", "initService", t);
            }
    }
/*注册服务到蓝牙协议栈*/
    public void createService()
    {
        if (this.mService != null)
            try {
                this.mService.GATTServer_CreateService(this.mProfileHandle.getAppHandle(),
                        this.mServiceId, this.mNumHandles);
            } catch (Throwable t)
            {
                Log.e("BleServerService", "createService", t);
            }
    }
/*从蓝牙协议栈注销服务*/
    public void deleteService()
    {
        if (this.mService != null)
            try {
                this.mService.GATTServer_DeleteService(this.mSvcHandle);
            } catch (Throwable t) {
                Log.e("BleServerService", "deleteService ", t);
            }
    }
/*启用服务*/
    public void startService()
    {
        if (this.mService != null)
            try {
                this.mService.GATTServer_StartService(this.mSvcHandle, this.mSupTransport);
            } catch (Throwable t) {
                Log.e("BleServerService", "startService ", t);
            }
    }
/*停止服务*/
```

```
    public void stopService()
    {
        if (this.mService != null) {
            this.mProfileHandle.notifyAction(4);
            try {
                this.mService.unregisterServerServiceCallback(this.mSvcHandle);
                this.mService.GATTServer_StopService(this.mSvcHandle);
            } catch (Throwable t) {
                Log.e("BleServerService", "stopService ", t);
            }
        }
    }
/*为此服务添加一个包含的服务*/
    public void addIncludedService(BleServerService service)
    {
        if (this.mService != null)
            try {
                if (service.isRegistered()) {
                    this.mServiceHdlMap.put(Integer.valueOf(service.getServiceHandle()),
                            service);
                    this.mService.GATTServer_AddIncludedService(this.mSvcHandle,
                            service.getServiceHandle());
                }
                else {
                    Log.i("BleServerService",
                            "addIncludedService: Service to be included is not registered.");
                }
            } catch (Throwable t) {
                Log.e("BleServerService", "addIncludedService", t);
            }
    }
/*更新一个特性或描述符*/
    public void updateCharacteristic(BleCharacteristic charObj)
    {
        addCharacteristic(charObj);
    }
/*当客户端已经请求读或写一个本地特性属性后发送一个响应*/
    public void sendResponse(String address, int transId, byte[] data, int statusCode)
    {
        Log.d("BleServerService", "sendResponse() address = " + address + ", transId = "
                + transId + ",statusCode = " + statusCode);

        if (this.mService == null) {
            Log.e("BleServerService", "sendResponse(): error. GattService not available");
            return;
        }

        AttributeRequestInfo attrInfo = (AttributeRequestInfo) this.mAttrReqMap
                .remove(Integer.valueOf(transId));
        if (attrInfo == null)
        {
```

```
            Log.e("BleServerService",
                    "sendResponse() error. attrInfo not found with transId " + transId);
            return;
        }

        byte[] dataToSend = null;
        if (attrInfo.mOffset == 0) {
            dataToSend = data;
        } else {
            dataToSend = new byte[data.length - attrInfo.mOffset];
            System.arraycopy(data, attrInfo.mOffset, dataToSend, 0, dataToSend.length);
        }

        try
        {
            this.mService
                    .GATTServer_SendRsp(
                            attrInfo.mConnId,
                            attrInfo.mTransId,
                            (byte) statusCode,
                            attrInfo.mAttrHandle,
                            attrInfo.mOffset,
                            dataToSend,
                            (byte) 0,
                            false);
        } catch (Throwable t)
        {
            Log.e("BleServerService", "sendResponse(): error", t);
        }
    }
/*当客户端已经请求读或写一个本地特性属性后发送一个响应*/
    public void sendResponse(String address, int transId, int handle, int offset, byte[] data,
            int statusCode, boolean isWrite)
    {
        int connId = getConnId(address);
        if ((connId != -1) && (this.mService != null))
            try {
                this.mService.GATTServer_SendRsp(connId, transId, (byte) statusCode,
                        handle, offset, data, (byte) 0, isWrite);
            } catch (Throwable t)
            {
                Log.e("BleServerService", "sendResponse", t);
            }
    }
/*当本地属性改变时发送一个通知给客户端*/
    public void sendNotification(String address, int attrHandle, byte[] value)
    {
        int connId = getConnId(address);
        if ((connId != -1) && (this.mService != null))
            try {
                this.mService
```

```
                    .GATTServer_HandleValueNotification(connId, attrHandle, value);
        } catch (Throwable t) {
            Log.e("BleServerService", "sendNotification", t);
        }
    }
/*发送一个指示到客户端*/
    public void sendIndication(String address, int attrHandle, byte[] value)
    {
        int connId = getConnId(address);
        if ((connId != -1) && (this.mService != null))
            try {
                this.mService.GATTServer_HandleValueIndication(connId, attrHandle, value);
            } catch (Throwable t) {
                Log.e("BleServerService", "sendIndication", t);
            }
    }

/*添加新特性到此服务*/
    private void addCharacteristic(BleCharacteristic charObj, boolean addtoQueue) {
        Log.i("BleServerService", "GATTServer_AddCharacteristic");
        ArrayList dirtyDescQueue = charObj.getDirtyDescQueue();

        if ((this.mService == null) || (charObj == null) || (charObj.getID() == null)) {
            Log
                    .i("BleServerService",
                            "GattService/Characteristic object passed in is null.. Cannot add the chanaracteristic...");
            return;
        }
        try
        {
            if (charObj.isRegistered()) {
                Log.d("BleServerService", "Starting to add descriptors, dirtyDesc size ="
                        + dirtyDescQueue.size());
                boolean dirtyMask = charObj.isDirty();
                if (!dirtyDescQueue.isEmpty())
                {
                    BleDescriptor descObj = (BleDescriptor) dirtyDescQueue.get(0);
                    Log.i("BleServerService", "GATTServer_AddCharDescriptor");
                    this.mService.GATTServer_AddCharDescriptor(this.mSvcHandle,
                            descObj.getPermission(), descObj.mID);
                }
                else if (dirtyMask) {
                    Log.i("BleServerService", "GATTServer_AddCharValue");
                    HashMap<String, Integer> connMap = this.mProfileHandle.getConnMap();

                    int clientCfg;
                    for (Map.Entry<String, Integer> entry : connMap.entrySet()) {
                        String address = (String) entry.getKey();
                        int connId = ((Integer) entry.getValue()).intValue();
                        clientCfg = 0;
                    }
```

```
                    return;
                }
            } else {
                if (addtoQueue) {
                    synchronized (this) {
                        this.mCharQueue.add(charObj);
                        Log.e("BleServerService",
                                "Adding a new characteristic... SIZE IS " + this.mCharQueue.size()
                                        + "Uuid=" + charObj.getID());

                        if (this.mCharQueue.size() > 1)
                            return;
                    }
                }
                BleGattID uuid = charObj.getID();
                this.mService.GATTServer_AddCharacteristic(this.mSvcHandle, uuid,
                        charObj.getPermission(), charObj.getProperty(), charObj.isDirty(),
                        dirtyDescQueue.size());
            }
        } catch (Throwable t)
        {
            Log.e("BleServerService", "addCharacteristic", t);
        }
    }
/*这是一个回调函数，当添加一个包含的服务时触发*/
    public void onIncludedServiceAdded(byte status, BleServerService includedService)
    {
        Log.d("BleServerService", "OnIncludedServiceAdded : status=" + status
                + "Included service" + includedService.getUuid());
    }
/*当添加一个特性时调用*/
    public void onCharacteristicAdded(byte status, BleCharacteristic charObj)
    {
        Log.d("BleServerService", "OnCharacteristicAdded : Characteristic uuid = "
                + charObj.getID() + "status=" + status);
    }
/*当一个读或写操作的响应已经发送时调用*/
    public void onResponseSendCompleted(byte status, BleCharacteristic charObj)
    {
        Log.d("BleServerService", "onResponseSendCompleted : status=" + status);
    }
/*这是一个回调函数，当添加一个特性时调用*/
    public void onCharacteristicRead(String address, int transId, int attrHandle,
            BleCharacteristic charObj)
    {
        AttributeRequestInfo attrInfo = (AttributeRequestInfo) this.mAttrReqMap
                .remove(Integer.valueOf(transId));
        Log.d("BleServerService", "Inside onCharacteristicRead()");
        if (attrInfo == null)
```

```
{
    Log.e("BleServerService",
            "onCharacteristicRead() error. attrInfo not found with transId " + transId);
    return;
}
if (charObj == null)
{
    Log.e("BleServerService", "onCharacteristicRead() error. charObj is null");
    return;
}

byte[] data = charObj.getValueByHandle(attrHandle);
if (data == null) {
    Log.d("BleServerService", "Attribute not found with handle " + attrHandle);
    try
    {
        this.mService.GATTServer_SendRsp(attrInfo.mConnId,
                attrInfo.mTransId,
                (byte) 10,
                attrInfo.mAttrHandle,
                attrInfo.mOffset, null,
                (byte) 0,
                false);
    } catch (Throwable t)
    {
        Log.e("BleServerService", "onCharacteristicRead(): error", t);
    }
    return;
}

int dataLength = data == null ? 0 : data.length;

if (attrInfo.mOffset >= dataLength) {
    Log.e("BleServerService",
            "onCharacteristicRead() error. dataLength < attrInfo.mOffset");
    try
    {
        this.mService.GATTServer_SendRsp(
                attrInfo.mConnId,
                attrInfo.mTransId,
                (byte) 7,
                attrInfo.mAttrHandle,
                attrInfo.mOffset,
                null,
                (byte) 0,
                false);
    } catch (Throwable t)
    {
        Log.e("BleServerService", "onCharacteristicRead(): error", t);
    }
    return;
```

```
        }

        byte[] dataToSend = null;
        if (attrInfo.mOffset == 0) {
            dataToSend = data;
        } else {
            dataToSend = new byte[data.length - attrInfo.mOffset];
            System.arraycopy(data, attrInfo.mOffset, dataToSend, 0, dataToSend.length);
        }

        try
        {
            this.mService.GATTServer_SendRsp(attrInfo.mConnId, attrInfo.mTransId, (byte) 0,
                    attrInfo.mAttrHandle, attrInfo.mOffset, dataToSend, (byte) 0, false);
        } catch (Throwable t)
        {
            Log.e("BleServerService", "sendResponse(): error", t);
        }
    }
/*这是一个回调函数，当特性的写操作完成时被触发*/
    public void onCharacteristicWrite(String address, BleCharacteristic charObj)
    {
        Log.d("BleServerService", "onCharacteristicWrite : modified characteristic="
                + charObj.getID());
    }

    private class BleServiceCallback extends IBleServiceCallback.Stub {
        private BleServerService mGattService;

        public BleServiceCallback(BleServerService service) {
            this.mGattService = service;
        }
/*注册服务*/
        public void onServiceRegistered(byte status, BluetoothGattID svcId) {
            Log.i("BleServerService", "onServiceRegistered");
            if (status == 0)
            {
                this.mGattService.setServiceInstance(svcId.getInstanceID());
                this.mGattService.createService();
            } else {
                Log.e("BleServerService", "#######Service registration failed...");
                BleServerService.this.mProfileHandle.notifyAction(1);
            }
        }
/*创建服务*/
        public void onServiceCreated(byte status, int svcHandle) {
            Log.i("BleServerService", "onServiceCreated");
            if (status == 0) {
                this.mGattService.setServiceHandle(svcHandle);
                BleServerService.this.mProfileHandle.notifyAction(0);
```

```
            } else {
                BleServerService.this.mProfileHandle.notifyAction(1);
            }
        }
```

（7）描述低功耗蓝牙服务的特性

在 Broadcom 公司提供的源码中，文件 BleCharacteristic.java 的功能是描述低功耗蓝牙服务的特性。在特性中包含了描述符、实际值和元数据，提供了表现格式或便于阅读值的描述。文件 BleCharacteristic.java 的主要实现代码如下：

```
/*获取 GATT 的 ID 值*/
    private BleGattID getBleGattId(int handle)
    {
        for (Map.Entry<BleGattID, Integer> entry : mHandleMap.entrySet()) {
            if (handle == entry.getValue().intValue()) {
                return entry.getKey();
            }
        }
        return null;
    }

    /**
    *返回该特征的实例的 ID
    *实例 ID 的 BLE 配置文件和服务用于标识属于一个给定的实例的服务或轮廓的特征
    */
    public int getInstanceID()
    {
        return mID.getInstanceID();
    }

    /**
    *指定一个实例 ID 的这一特性
    *
    *@see {@link #getInstanceID()}
    */
    public void setInstanceID(int instanceID)
    {
        mID.setInstanceId(instanceID);
    }

    /**
    *根据特性向一个给定的偏移量设置原始值的字节
    */
    public byte setValue(byte[] value, int offset, int len, int handle, int totalsize,
            String address)
    {
        int uuid = -1;
        int uuidType = -1;
        Log.e("BleCharacteristic", "#### handle is " + handle + " total size is " + totalsize);

        BleGattID gattUuid = getBleGattId(handle);
```

```
        if (gattUuid == null) {
            Log.e("BleCharacteristic", "setValue: Invalid handle");
            return BleConstants.GATT_INVALID_HANDLE;
        }

        if (gattUuid.equals(mID)) {
            Log.i("BleCharacteristic", "##Writing a characteristic value..");
            Log.i("BleCharacteristic", "##offset=" + offset + " mMaxLength="
                    + mMaxLength + " totalsize=" + totalsize);
            return setValue(value, offset, len, gattUuid, totalsize, address);
        }
        BleDescriptor descObj = mDescriptorMap.get(gattUuid);
        if (descObj != null) {
            Log.i("BleCharacteristic", "##Writing descriptor value..");
            Log.i("BleCharacteristic",
                    "##offset=" + offset + " mMaxSize=" + descObj.getMaxLength() + " totalsize="
                            + totalsize + "desc uuid =" + descObj.getID());
            if (offset > descObj.getMaxLength())
                return BleConstants.GATT_INVALID_OFFSET;
            if (offset + totalsize > descObj.getMaxLength())
                return BleConstants.GATT_INVALID_ATTR_LEN;
            Log.i("BleCharacteristic", "find the user defined descriptor ");
            return descObj.setValue(value, offset, len, gattUuid, totalsize, address);
        }
        Log.e("BleCharacteristic", "Failed to write the value correctly!!!");
        return -127;
    }

    /**
    *一个给定的偏移量设置原始值的字节
    */
    @Override
    public byte setValue(byte[] value, int offset, int len, BleGattID gattUuid, int totalsize, String address)
    {
        return super.setValue(value, offset, len, gattUuid, totalsize, address);
    }

    /**
    *Gets the characteristic properties value (bit field).
    */
    public int getProperty()
    {
        return mProp;
    }

    /**
    *设置特性属性集
    */
    public void setProperty(int Prop)
    {
        mProp = Prop;
```

```
}

/**
*获取一个基于 UUID 的描述符
*/
public BleDescriptor getDescriptor(BleGattID descriptor)
{
    BleDescriptor descObj = mDescriptorMap.get(descriptor);
    if (descObj != null) {
        return descObj;
    }

    return null;
}

/**
*添加一个描述符对象
*/
public void addDescriptor(BleGattID descId, BleDescriptor descriptor)
{
    Log.d("BleCharacteristic", "Inside add descriptor");
    mDescriptorMap.put(descId, descriptor);
    mDirtyDescQueue.add(descriptor);
    descriptor.setCharRef(this);
}

/**
*添加一个描述符对象
*/
public void addDescriptor(BleDescriptor descriptor)
{
    mDescriptorMap.put(descriptor.mID, descriptor);
    mDirtyDescQueue.add(descriptor);
    descriptor.setCharRef(this);
}

/**
*返回所有用户定义的都包含在这一特性数组描述符
*/
public ArrayList<BleDescriptor> getAllDescriptors()
{
    ArrayList<BleDescriptor> descList = new ArrayList<BleDescriptor>();
    for (Map.Entry<BleGattID, BleDescriptor> entrySet : mDescriptorMap.entrySet()) {
        descList.add(entrySet.getValue());
    }
    return descList;
}

/**
*映射这个属性句柄值属性
*/
```

```
    public void addHandle(BleGattID uuid, int handle)
    {
        mHandleMap.put(uuid, Integer.valueOf(handle));
    }

    /**
    *返回一个对给定属性 ID 的处理结果
    */
    public int getHandle(BleGattID uuid)
    {
        Integer tmp;
        if ((tmp = mHandleMap.get(uuid)) != null) {
            return tmp.intValue();
        }
        return -1;
    }

    /**
    *设置所需的读/写这个属性的验证水平
    */
    @Override
    public void setAuthReq(byte AuthReq)
    {
        mAuthReq = AuthReq;
    }

    /**
    *返回是否允许签写这个特性
    */
    public boolean isAuthenticated()
    {
        return  (mProp  &  BleConstants.GATT_CHAR_PROP_BIT_AUTH)  ==  BleConstants.GATT_CHAR_
PROP_BIT_AUTH;
    }
    /**
     *返回一个基于先前分配的句柄值这一特性属性
     */
    @Override
    public byte[] getValueByHandle(int handle)
    {
        BleGattID gattUuid = getBleGattId(handle);
        if (gattUuid == null) {
            Log.w("BleCharacteristic", "Attribute UUID not found with handle " + handle);
            return null;
        }
        int uuidType = gattUuid.getUuidType();
        if (uuidType == BleConstants.GATT_UUID_TYPE_16)
            return getValueByUUID16(gattUuid);
        if (uuidType == BleConstants.GATT_UUID_TYPE_128) {
            return getValueByUUID128(gattUuid);
        }
```

```
        Log.w("BleCharacteristic", "Invalid UUID type.");
        return null;
    }

    /**
     *检索一个基于 16 位 UUID 这个特征的属性
     */
    public byte[] getValueByUUID16(BleGattID uuid)
    {
        int uuid16 = uuid.getUuid16();
        if (uuid16 == -1) {
            Log.w("BleCharacteristic", "Invalid UUID16.");
            return null;
        }
        int thisAttrUuid16 = mID == null ? -1 : mID.getUuid16();
        if (uuid16 == thisAttrUuid16)
        {
            return getValue();
        }
        BleDescriptor descObj = mDescriptorMap.get(uuid);
        if (descObj != null) {
            Log.d("BleCharacteristic", "Descriptor UUID = " + descObj.getID().getUuid16());
            return descObj.getValue();
        }
        Log.w("BleCharacteristic", "Attribute query not supported for uuid16 value "
                + uuid16);
        return null;
    }
    /**
     *检索一个基于 128 位 UUID 这个特征的属性*/
    public byte[] getValueByUUID128(BleGattID uuid)
    {
        UUID uuid128 = uuid.getUuid();
        if (uuid128 == null) {
            return null;
        }
        if ((mID != null) && (uuid128.equals(mID.getUuid()))) {
            return getValue();
        }
        BleDescriptor descObj = mDescriptorMap.get(uuid);
        if (descObj != null) {
            return descObj.getValue();
        }
        Log.w("BleCharacteristic", "Attribute query not supported for uuid128 value "
                + uuid128);
        return null;
    }
}
```

（8）低功耗描述符

在 Broadcom 公司提供的源码中，文件 BleDescriptor.java 是 BleCharacteristic 的一部分，功能是定

义了一个低功耗描述符。文件 BleDescriptor.java 的主要实现代码如下：

```
public class BleDescriptor extends BleAttribute
        implements Parcelable
{
    private static final String TAG = "BleDescriptor";
    private BleCharacteristic mCharObj;
    protected HashMap<String, Integer> mClientcfgMap = new HashMap();

    /** @hide */
    @SuppressWarnings({
            "unchecked", "rawtypes"
    })
    public static final Parcelable.Creator<BleDescriptor> CREATOR = new Parcelable.Creator()
    {
        public BleDescriptor createFromParcel(Parcel source) {
            return new BleDescriptor(source);
        }
        public BleDescriptor[] newArray(int size)
        {
            return new BleDescriptor[size];
        }
    };
    /**
     *从一个给定的偏移设置原始值字节的描述符
     * @return {@link BleConstants#GATT_SUCCESS if successful}
     */
    @Override
    public byte setValue(byte[] value, int offset, int length, BleGattID gattUuid,
            int totalSize, String address)
    {
        int uuidType = gattUuid.getUuidType();
        int uuid = -1;
        Log.e("BleDescriptor", "#### UUID type=" + gattUuid.getUuidType());
        if (uuidType == 2) {
            uuid = gattUuid.getUuid16();
            if (uuid == -1) {
                Log.e("BleDescriptor", "setValue: Invalid handle (UUID16 not found)");
                return 1;
            }
        }
        if (uuid == 10500) {
            Log.i("BleDescriptor", "##Writing a Presentation format..");
        } else if (uuid == 10498) {
            Log.i("BleDescriptor", "##Writing a characteristic client config");
            if (totalSize > this.mMaxLength)
                return 13;
            int valueInt = 0;
            for (int i = 0; i < length; i++) {
                int shift = (length - 1 - i) * 8;
                valueInt += ((value[i] & 0xFF) << shift);
            }
```

```
                this.mClientcfgMap.put(address, Integer.valueOf(valueInt));
            } else if (gattUuid.equals(this.mID)) {
                Log.i("BleDescriptor", "##Writing a descriptor value..");
                Log.i("BleDescriptor", "##offset=" + offset + " mMaxLength=" + this.mMaxLength
                        + " length=" + length);
                super.setValue(value, offset, length, gattUuid, totalSize, address);
            }
            this.mDirty = true;
            return 0;
        }
}
```

（9）标识低功耗蓝牙规范、服务和特性

在 Broadcom 公司提供的源码中，文件 BleGattID.java 的功能是定义了一个标识低功耗蓝牙规范、服务和特性的类，此类使用 16 位或 128 位的 UUIDs 来标识一个给定的低功耗蓝牙实体，这个实体包含规范、服务和特性。文件 BleGattID.java 的主要实现代码如下：

```
/**
 *标识一个蓝牙 GATT 特性或属性
 */
public final class BleGattID extends BluetoothGattID
        implements Parcelable
{
    private static final String BASE_UUID_TPL = "%08x-0000-1000-8000-00805f9b34fb";
    @SuppressWarnings({
            "rawtypes", "unchecked"
    })
    public static final Parcelable.Creator<BleGattID> CREATOR = new Parcelable.Creator() {
        public BleGattID createFromParcel(Parcel source) {
            int instId = source.readInt();
            int type = source.readInt();
            int serviceType = source.readInt();

            if (type == 16) {
                String sUuid = source.readString();
                return new BleGattID(instId, sUuid, serviceType);
            }
            int uuid = source.readInt();
            return new BleGattID(instId, uuid, serviceType);
        }

        public BleGattID[] newArray(int size)
        {
            return new BleGattID[size];
        }

    };
```

（10）为远程蓝牙设备提供额外信息

在 Broadcom 公司提供的源码中，文件 BleAdapter.java 的功能是为远程蓝牙设备提供额外的信息，能够判断远程设备是否是低功耗设备、BR/EDR 传统蓝牙设备或双模设备（同时支持低功耗和传统设备）。文件 BleAdapter.java 的主要实现代码如下：

```
/**
 *提供帮助的功能和相关的常数扩展蓝牙功能的低能耗信息
 */
public class BleAdapter
{
    private static final String TAG = "BleAdapter";
    private static final boolean D = true;

    private static final int API_LEVEL = 5;
    private static IBluetoothGatt mService;
    private GattServiceConnection mSvcConn;
    private Context mContext;

    /**
     * 设置远程 ACTION_FOUND 设备的额外信息
     *
     * @see {@link #DEVICE_TYPE_BREDR}, {@link #DEVICE_TYPE_BLE},
     *        {@link #DEVICE_TYPE_DUMO}
     */
    public static final String EXTRA_DEVICE_TYPE = "android.bluetooth.device.extra.DEVICE_TYPE";

    public static final byte DEVICE_TYPE_BREDR = 1;

    public static final byte DEVICE_TYPE_BLE = 2;
    public static final byte DEVICE_TYPE_DUMO = 3;
    public static final String ACTION_UUID = "android.bluetooth.le.device.action.UUID";
    public static final String EXTRA_UUID = "android.bluetooth.le.device.extra.UUID";
    public static final String EXTRA_DEVICE = "android.bluetooth.le.device.extra.DEVICE";
    private static boolean startService() {
        if (mService != null)
            return true;
        IBinder service = ServiceManager.getService(BleConstants.BLUETOOTH_LE_SERVICE);
        if (service != null)
            mService = IBluetoothGatt.Stub.asInterface(service);
        return mService != null;
    }
    /**
     * 构建一种新的 BleAdapter 对象
     */
    public BleAdapter(Context ctx) {
        this.mContext = ctx;
        if (startService()==false)
            throw new RuntimeException("failed connecting to service");
        this.init();
    }
    /**
     * 返回蓝牙设备的类型(LE、BR/EDR or dual-mode)
     *
     * @param device - The remote device who's type is to be determined
     * @return The type of the remote device
     * @see {@link #DEVICE_TYPE_BREDR}, {@link #DEVICE_TYPE_BLE},
     *        {@link #DEVICE_TYPE_DUMO}
```

```
 */
public static byte getDeviceType(BluetoothDevice device)
{
    if (!startService())
        throw new RuntimeException("service not available");
    if (device != null) {
        try {
            return mService.getDeviceType(device.getAddress());
        } catch (RemoteException e) {
            Log.e(TAG, "error", e);
        }
    }
    return 0;
}
/**
 * 启动远程设备中的蓝牙服务，发现使用{@link #ACTION_UUID}的意图
 *
 * @param deviceAddress - Bluetooth address of the remote device in
 *              00:11:22:33:44:55 format
 * @return true if the device discovery was started successfully
 */
public static boolean getRemoteServices(String deviceAddress)
{
    if (!startService())
        throw new RuntimeException("service not available");
    BluetoothAdapter adapter = BluetoothAdapter.getDefaultAdapter();
    if (adapter == null)
        return false;
    try {
        mService.getUUIDs(deviceAddress);
        return true;
    } catch (RemoteException e) {
        e.printStackTrace();
        if (D)
            Log.e(TAG, "error", e);
    }
    return false;
}
/**
 * 初始化 BleAdapter 对象，连接蓝牙 GATT 服务失效回调
 */
public void init()
{
    /**
     * TODO: implement
     */
    Log.d("BleAdapter", "init");
    Intent i = new Intent();
    i.setClassName("com.broadcom.bt.app.system", "com.broadcom.bt.app.system.GattService");
    mContext.bindService(i, mSvcConn, 1);
}
```

（11）保存和 GATT 相关的常量

在 Broadcom 公司提供的源码中，文件 BleConstants.java 的功能是定义保存各种和 GATT 相关的常量，这些常量用于表示各种和实现低功耗功能函数的属性和返回值。文件 BleConstants.java 的主要实现代码如下：

```
public abstract class BleConstants
{
    public static final int GATT_UNDEFINED = -1;
    public static final int GATT_SERVICE_CREATION_SUCCESS = 0;
    public static final int GATT_SERVICE_CREATION_FAILED = 1;
    public static final int GATT_SERVICE_START_SUCCESS = 2;
    public static final int GATT_SERVICE_START_FAILED = 3;
    public static final int GATT_SERVICE_STOPPED = 4;
    public static final int SERVICE_UNAVAILABLE = 1;
    public static final int GATT_SERVICE_PRIMARY = 0;
    public static final int GATT_SERVICE_SECONDARY = 1;
    public static final int GATT_SERVER_PROFILE_INITIALIZED = 0;
    public static final int GATT_SERVER_PROFILE_UP = 1;
    public static final int GATT_SERVER_PROFILE_DOWN = 2;
    public static final int GATT_SUCCESS = 0;
    public static final int GATT_INVALID_HANDLE = 1;
    public static final int GATT_READ_NOT_PERMIT = 2;
    public static final int GATT_WRITE_NOT_PERMIT = 3;
    public static final int GATT_INVALID_PDU = 4;
    public static final int GATT_INSUF_AUTHENTICATION = 5;
    public static final int GATT_REQ_NOT_SUPPORTED = 6;
    public static final int GATT_INVALID_OFFSET = 7;
    public static final int GATT_INSUF_AUTHORIZATION = 8;
    public static final int GATT_PREPARE_Q_FULL = 9;
    public static final int GATT_NOT_FOUND = 10;
    public static final int GATT_NOT_LONG = 11;
    public static final int GATT_INSUF_KEY_SIZE = 12;
    public static final int GATT_INVALID_ATTR_LEN = 13;
    public static final int GATT_ERR_UNLIKELY = 14;
    public static final int GATT_INSUF_ENCRYPTION = 15;
    public static final int GATT_UNSUPPORT_GRP_TYPE = 16;
    public static final int GATT_INSUF_RESOURCE = 17;
    public static final int GATT_ILLEGAL_PARAMETER = 135;
    public static final int GATT_NO_RESOURCES = 128;
    public static final int GATT_INTERNAL_ERROR = 129;
    public static final int GATT_WRONG_STATE = 130;
    public static final int GATT_DB_FULL = 131;
    public static final int GATT_BUSY = 132;
    public static final int GATT_ERROR = 133;
    public static final int GATT_CMD_STARTED = 134;
    public static final int GATT_PENDING = 136;
    public static final int GATT_AUTH_FAIL = 137;
    public static final int GATT_MORE = 138;
    public static final int GATT_INVALID_CFG = 139;
    public static final byte GATT_AUTH_REQ_NONE = 0;
    public static final byte GATT_AUTH_REQ_NO_MITM = 1;
```

```
public static final byte GATT_AUTH_REQ_MITM = 2;
public static final byte GATT_AUTH_REQ_SIGNED_NO_MITM = 3;
public static final byte GATT_AUTH_REQ_SIGNED_MITM = 4;
public static final int GATT_PERM_READ = 1;
public static final int GATT_PERM_READ_ENCRYPTED = 2;
public static final int GATT_PERM_READ_ENC_MITM = 4;
public static final int GATT_PERM_WRITE = 16;
public static final int GATT_PERM_WRITE_ENCRYPTED = 32;
public static final int GATT_PERM_WRITE_ENC_MITM = 64;
public static final int GATT_PERM_WRITE_SIGNED = 128;
public static final int GATT_PERM_WRITE_SIGNED_MITM = 256;
public static final byte GATT_CHAR_PROP_BIT_BROADCAST = 1;
public static final byte GATT_CHAR_PROP_BIT_READ = 2;
public static final byte GATT_CHAR_PROP_BIT_WRITE_NR = 4;
public static final byte GATT_CHAR_PROP_BIT_WRITE = 8;
public static final byte GATT_CHAR_PROP_BIT_NOTIFY = 16;
public static final byte GATT_CHAR_PROP_BIT_INDICATE = 32;
public static final byte GATT_CHAR_PROP_BIT_AUTH = 64;
public static final byte GATT_CHAR_PROP_BIT_EXT_PROP = -128;
public static final byte SVC_INF_INVALID = -1;
public static final int GATTC_TYPE_WRITE_NO_RSP = 1;
public static final int GATTC_TYPE_WRITE = 2;
public static final int GATT_FORMAT_RES = 0;
public static final int GATT_FORMAT_BOOL = 1;
public static final int GATT_FORMAT_2BITS = 2;
public static final int GATT_FORMAT_NIBBLE = 3;
public static final int GATT_FORMAT_UINT8 = 4;
public static final int GATT_FORMAT_UINT12 = 5;
public static final int GATT_FORMAT_UINT16 = 6;
public static final int GATT_FORMAT_UINT24 = 7;
public static final int GATT_FORMAT_UINT32 = 8;
public static final int GATT_FORMAT_UINT48 = 9;
public static final int GATT_FORMAT_UINT64 = 10;
public static final int GATT_FORMAT_UINT128 = 11;
public static final int GATT_FORMAT_SINT8 = 12;
public static final int GATT_FORMAT_SINT12 = 13;
public static final int GATT_FORMAT_SINT16 = 14;
public static final int GATT_FORMAT_SINT24 = 15;
public static final int GATT_FORMAT_SINT32 = 16;
public static final int GATT_FORMAT_SINT48 = 17;
public static final int GATT_FORMAT_SINT64 = 18;
public static final int GATT_FORMAT_SINT128 = 19;
public static final int GATT_FORMAT_FLOAT32 = 20;
public static final int GATT_FORMAT_FLOAT64 = 21;
public static final int GATT_FORMAT_SFLOAT = 22;
public static final int GATT_FORMAT_FLOAT = 23;
public static final int GATT_FORMAT_DUINT16 = 24;
public static final int GATT_FORMAT_UTF8S = 25;
public static final int GATT_FORMAT_UTF16S = 26;
public static final int GATT_FORMAT_STRUCT = 27;
public static final int GATT_FORMAT_MAX = 28;
```

```
    public static final String GATT_UUID_CHAR_EXT_PROP = "00002900-0000-1000-8000-00805f9b34fb";
    public static final String GATT_UUID_CHAR_DESCRIPTION = "00002901-0000-1000-8000-00805f9b34fb";
    public static final String GATT_UUID_CHAR_CLIENT_CONFIG = "00002902-0000-1000-8000-00805f9b34fb";
    public static final String GATT_UUID_CHAR_SRVR_CONFIG = "00002903-0000-1000-8000-00805f9b34fb";
    public static final String GATT_UUID_CHAR_PRESENT_FORMAT = "00002904-0000-1000-8000-00805f9b34fb";
    public static final String GATT_UUID_CHAR_AGG_FORMAT = "00002905-0000-1000-8000-00805f9b34fb";
    public static final int GATT_UUID_CHAR_EXT_PROP16 = 10496;
    public static final int GATT_UUID_CHAR_DESCRIPTION16 = 10497;
    public static final int GATT_UUID_CHAR_CLIENT_CONFIG16 = 10498;
    public static final int GATT_UUID_CHAR_SRVR_CONFIG16 = 10499;
    public static final int GATT_UUID_CHAR_PRESENT_FORMAT16 = 10500;
    public static final int GATT_UUID_CHAR_AGG_FORMAT16 = 10501;
    public static final int GATT_TRANSPORT_BREDR_LE = 2;
    public static final int GATT_TRANSPORT_BREDR = 1;
    public static final int GATT_TRANSPORT_LE = 0;
    public static final int GATT_UUID_TYPE_128 = 16;
    public static final int GATT_UUID_TYPE_32 = 4;
    public static final int GATT_UUID_TYPE_16 = 2;
    public static final int PREPARE_QUEUE_SIZE = 200;
    public static final int GATT_MAX_CHAR_VALUE_LENGTH = 100;
    public static final int GATT_CLIENT_CONFIG_NOTIFICATION_BIT = 1;
    public static final int GATT_CLIENT_CONFIG_INDICATION_BIT = 2;
    public static final int GATT_INVALID_CONN_ID = 65535;
    public static final int VALUE_DIRTY = 1;
    public static final int USER_DESCRIPTION_DIRTY = 2;
    public static final int EXT_PROP_DIRTY = 4;
    public static final int PRESENTATION_FORMAT_DIRTY = 8;
    public static final int CLIENT_CONFIG_DIRTY = 16;
    public static final int SERVER_CONFIG_DIRTY = 32;
    public static final int AGGREGATED_FORMAT_DIRTY = 64;
    public static final int USER_DESCRIPTOR_DIRTY = 128;
    public static final int ALL_DIRTY = 127;
    public static final byte GATT_ENCRYPT_NONE = 0;
    public static final byte GATT_ENCRYPT = 1;
    public static final byte GATT_ENCRYPT_NO_MITM = 2;
    public static final byte GATT_ENCRYPT_MITM = 3;
    static final String ACTION_OBSERVE_RESULT = "com.broadcom.bt.app.gatt.OBSERVE_RESULT";
    static final String ACTION_OBSERVE_COMPLETED = "com.broadcom.bt.app.gatt.OBSERVE_ COMPLETED";
    static final String EXTRA_ADDRESS = "ADDRESS";
    static final String EXTRA_ADDR_TYPE = "ADDR_TYPE";
    static final String EXTRA_RSSI = "RSSI";
    static final String EXTRA_ADV_DATA = "ADV_DATA";
    static final String EXTRA_NUM_RESULTS = "NUM_RESULTS";
    static final String GATT_SVC_PKG_NAME = "com.broadcom.bt.app.system";
    static final String GATT_SVC_NAME = "com.broadcom.bt.app.system.GattService";

    public static final String BLUETOOTH_LE_SERVICE = "com.manuelnaranjo.btle";
}
```

至此，Broadcom 公司推出的低功耗蓝牙协议栈 BlueDroid 的开发文档和 API 源码分析完毕。因为本书篇幅的限制，只是分析了主要的模块类，其他类的实现代码的功能和原理请读者参阅其源码中的注释说明。

# 第 3 篇

## 信息识别篇


- 第 6 章　语音识别技术详解
- 第 7 章　手势识别实战
- 第 8 章　在物联网设备中处理多媒体数据
- 第 9 章　GPS 地图定位

# 第 6 章　语音识别技术详解

语音识别是一门交叉学科，经过多年的发展，语音识别技术取得了显著进步，逐渐从实验室走向市场。据专家预计，在未来 10 年内，语音识别技术将进入工业、家电、通信、汽车电子、医疗、家庭服务、消费电子产品等各个领域。很多专家都认为语音识别技术是 2000—2010 年间信息技术领域十大重要的科技发展技术之一。语音识别技术是 Android SDK 中比较重要且比较新颖的一项技术，在 Android 穿戴设备应用中可以通过语音来控制设备。本章将详细讲解在 Android 物联网设备中使用语音识别技术的基本知识，为读者步入本书后面知识的学习打下基础。

## 6.1　语音识别技术基础

**知识点讲解：光盘:视频\知识点\第 6 章\语音识别技术基础.avi**

语音识别技术所涉及的领域包括信号处理、模式识别、概率论和信息论、发声机理和听觉机理、人工智能等。本节将详细讲解语音识别技术的基本知识。

### 6.1.1　语音识别的发展历史

1952 年，贝尔研究所 Davis 等研究员成功研制了世界上第一个能识别 10 个英文数字发音的实验系统。

1960 年，英国的 Denes 等人研究成功了世界上第一个计算机语音识别系统。

1970 年以后，大规模的语音识别得到了良好的发展契机，在小词汇量、孤立词的识别方面取得了实质性的进展。DARPA（Defense Advanced Research Projects Agency）是在 20 世纪 70 年代由美国国防部远景研究计划局资助的一项 10 年计划，其旨在支持语言理解系统的研究开发工作。

1980 年以后，研究的重点逐渐转向大词汇量、非特定人连续语音识别。在研究思路上也发生了重大变化，即由传统的基于标准模板匹配的技术思路开始转向基于统计模型（HMM）的技术思路。此外，再次提出了将神经网络技术引入语音识别问题的技术思路。美国国防部远景研究计划局又资助了一项为期 10 年的 DARPA 战略计划，其中包括噪声下的语音识别和会话（口语）识别系统，识别任务设定为“（1000 单词）连续语音数据库管理”。

1981 年，日本在第 5 代计算机计划中提出了有关语音识别“输入-输出”自然语言的宏伟目标，虽然没能实现预期目标，但是有关语音识别技术的研究有了大幅度的加强和进展。

1987 年起，日本又拟出新的国家项目：高级人机口语接口和自动电话翻译系统。

1990 年以后，DARPA 计划仍在持续进行中。其研究重点已转向识别装置中的自然语言处理部分，识别任务设定为“航空旅行信息检索”。并且在 20 世纪 90 年代，语音识别的系统框架方面并没有什么重大突破。但在语音识别技术的应用及产品化方面有很大的进展。

### 6.1.2　语音识别技术的发展历程

目前 IBM 语音研究小组在大词汇语音识别方面处于领先地位，在 20 世纪 70 年代开始了它的大词汇语音识别研究工作。AT&T 的贝尔研究所也开始了一系列有关非特定人语音识别的实验。这一研究历经 10 年，其成果是确立了如何制作用于非特定人语音识别的标准模板的方法。在这一时期所取得的重大进展有如下 3 点。

（1）隐式马尔科夫模型（HMM）技术的成熟和不断完善成为语音识别的主流方法。

（2）以知识为基础的语音识别的研究日益受到重视。在进行连续语音识别时，除了识别声学信息外，更多地利用各种语言知识，诸如构词、句法、语义、对话背景方面等知识来帮助进一步对语音作出识别和理解。同时在语音识别研究领域，还产生了基于统计概率的语言模型。

（3）人工神经网络在语音识别中的应用研究的兴起。在这些研究中，大部分采用基于反向传播算法（BP 算法）的多层感知网络。人工神经网络具有区分复杂的分类边界的能力，显然它十分有助于模式划分。特别是在电话语音识别方面，由于其有着广泛的应用前景，成了当前语音识别应用的一个热点。

另外，面向个人用途的连续语音听写机技术也日趋完善。这方面最具代表性的是 IBM 的 ViaVoice 和 Dragon 公司的 Dragon Dictate 系统。这些系统具有说话人自适应能力，新用户不需要对全部词汇进行训练，便可在使用中不断提高识别率。

## 6.2　Text-To-Speech 技术

**知识点讲解：光盘:视频\知识点\第 6 章\Text-To-Speech 技术.avi**

Text-To-Speech 简称 TTS，是 Android 1.6 版本中比较重要的新功能，能够将所指定的文本转成不同语言音频输出。TTS 可以方便地嵌入到游戏或者应用程序中，增强用户体验。本节将详细讲解在 Android 系统中使用 Text-To-Speech 技术的基本知识，为读者步入本书后面知识的学习打下基础。

### 6.2.1　Text-To-Speech 基础

TTS engine 依托于当前 Android Platform 所支持的如下 5 大语言：

☑ English

☑ French

☑ German

☑ Italian

☑ Spanish

TTS 可以将文本随意地转换成以上任意 5 种语言的语音输出。与此同时，对于个别的语言版本将取决于不同的时区，例如对于 English，在 TTS 中可以分别输出美式和英式两种不同的版本。既然能支持如此庞大的数据量，TTS 引擎对于资源的优化采取预加载的方法。根据一系列的参数信息从库中提取相应的资源，并加载到当前系统中。尽管当前大部分加载有 Android 操作系统的设备都通过这套引擎来提供 TTS 功能，但由于一些设备的存储空间非常有限，而影响到 TTS 最大限度地发挥功能，算是当前的一个瓶颈。为此开发小组引入了检测模块，让利用这项技术的应用程序或者游戏针对不同的设备可以有相应的优化调整，从而避免由于此项功能的限制，影响到整个应用程序的使用。比较稳妥的做法是让用户自行选择是否有足够的空间或者需求来加载此项资源，下面给出了一个标准的检测方法。

```
Intent checkIntent = new Intent();
checkIntent.setAction(TextToSpeech.Engine.ACTION_CHECK_TTS_DATA);
startActivityForResult(checkIntent, MY_DATA_CHECK_CODE);
```

如果当前系统允许创建一个 android.speech.tts.TextToSpeech 的 Object 对象，则说明已经提供 TTS 功能的支持，将检测返回结果中给出的 CHECK_VOICE_DATA_PASS 标记。如果系统不支持这项功能，那么用户可以选择是否加载这项功能，从而让设备支持输出多国语言语音的功能 Multi-lingual Talking。ACTION_INSTALL_TTS_DATA intent 将用户引入 Android market 中的 TTS 下载界面。下载完成后将自动完成安装，实现上述过程的完整代码（androidres.com）如下：

```
private TextToSpeech mTts;
protected void onActivityResult(int requestCode, int resultCode, Intent data) {
    if (requestCode == MY_DATA_CHECK_CODE) {
        if (resultCode == TextToSpeech.Engine.CHECK_VOICE_DATA_PASS) {
            mTts = new TextToSpeech(this, this);
        } else {
            Intent installIntent = new Intent();
            installIntent.setAction(
                TextToSpeech.Engine.ACTION_INSTALL_TTS_DATA);
            startActivity(installIntent);
        }
    }
}
```

TextToSpeech 实体和 OnInitListener 都需要引用当前 Activity 的 Context 作为构造参数。OnInitListener()的用处是通知系统当前 TTS Engine 已经加载完成，并处于可用状态。

### 6.2.2 Text-To-Speech 的实现流程

（1）检查 TTS 数据是否可用，例如下面的代码。

```
view plaincopy to clipboardprint?
//检查 TTS 数据是否已经安装并且可用
        Intent checkIntent = new Intent();
        checkIntent.setAction(TextToSpeech.Engine.ACTION_CHECK_TTS_DATA);
        startActivityForResult(checkIntent, REQ_TTS_STATUS_CHECK);
protected   void onActivityResult(int requestCode, int resultCode, Intent data) {
        if(requestCode == REQ_TTS_STATUS_CHECK)
        {
            switch (resultCode) {
            case TextToSpeech.Engine.CHECK_VOICE_DATA_PASS:
                //这个返回结果表明 TTS Engine 可以用
            {
                mTts = new TextToSpeech(this, this);
                Log.v(TAG, "TTS Engine is installed!");
            }
                break;
```

```
            case TextToSpeech.Engine.CHECK_VOICE_DATA_BAD_DATA:
                //需要的语音数据已损坏
            case TextToSpeech.Engine.CHECK_VOICE_DATA_MISSING_DATA:
                //缺少需要语言的语音数据
            case TextToSpeech.Engine.CHECK_VOICE_DATA_MISSING_VOLUME:
                //缺少需要语言的发音数据
            {
                //这 3 种情况都表明数据有错，重新下载安装需要的数据
                Log.v(TAG, "Need language stuff:"+resultCode);
                Intent dataIntent = new Intent();
                dataIntent.setAction(TextToSpeech.Engine.ACTION_INSTALL_TTS_DATA);
                startActivity(dataIntent);

            }
                break;
            case TextToSpeech.Engine.CHECK_VOICE_DATA_FAIL:
                //检查失败
            default:
                Log.v(TAG, "Got a failure. TTS apparently not available");
                break;
            }
        }
        else
        {
            //其他 Intent 返回的结果
        }
        }
```

（2）初始化 TTS，例如下面的代码。

```
view plaincopy to clipboardprint?
//实现 TTS 初始化接口
    @Override
    public void onInit(int status) {
        //TTS Engine 初始化完成
        if(status == TextToSpeech.SUCCESS)
        {
            int result = mTts.setLanguage(Locale.US);
            //设置发音语言
            if(result == TextToSpeech.LANG_MISSING_DATA || result == TextToSpeech.LANG_NOT_
SUPPORTED)
            //判断语言是否可用
            {
                Log.v(TAG, "Language is not available");
                speakBtn.setEnabled(false);
            }
            else
            {
    mTts.speak("This is an example of speech synthesis.", TextToSpeech.QUEUE_ADD, null);
                speakBtn.setEnabled(true);
            }
        }
    }
```

（3）设置发音语言，例如下面的代码。

```
view plaincopy to clipboardprint?
public void onItemSelected(AdapterView<?> parent, View view,
        int position, long id) {
    int pos = langSelect.getSelectedItemPosition();
    int result = -1;
    switch (pos) {
    case 0:
    {
        inputText.setText("I love you");
        result = mTts.setLanguage(Locale.US);
    }
        break;
    case 1:
    {
        inputText.setText("Je t'aime");
        result = mTts.setLanguage(Locale.FRENCH);
    }
        break;
    case 2:
    {
        inputText.setText("Ich liebe dich");
        result = mTts.setLanguage(Locale.GERMAN);
    }
        break;
    case 3:
    {
        inputText.setText("Ti amo");
        result = mTts.setLanguage(Locale.ITALIAN);
    }
        break;
    case 4:
    {
        inputText.setText("Te quiero");
        result = mTts.setLanguage(new Locale("spa", "ESP"));
    }
        break;
    default:
        break;
    }
    //设置发音语言
    if(result == TextToSpeech.LANG_MISSING_DATA || result == TextToSpeech.LANG_NOT_SUPPORTED)
    //判断语言是否可用
    {
        Log.v(TAG, "Language is not available");
        speakBtn.setEnabled(false);
    }
    else
    {
        speakBtn.setEnabled(true);
    }
}
```

（4）设置单击 Button 按钮发出声音，例如下面的代码。

```
view plaincopy to clipboardprint?
public void onClick(View v) {
    //朗读输入框中的内容
    mTts.speak(inputText.getText().toString(), TextToSpeech.QUEUE_ADD, null);
}
```

### 6.2.3　实战演练——使用 Text-To-Speech 实现语音识别

下面将通过一个具体实例的实现过程，详细讲解在 Android 物联网设备中使用 Text-To-Speech 技术实现语音识别的过程。本实例的功能是在屏幕中显示一个文本框供用户输入需要读的内容，单击按钮后开始读取文本框中输入的内容，本实例支持中文。

| 实　例 | 功　能 | 源 码 路 径 |
| --- | --- | --- |
| 实例 6-1 | 使用 Text-To-Speech 实现语音识别 | 光盘:\daima\6\TextSpeechEX |

（1）编写布局文件 main.xml，功能是在屏幕中显示一个可输入内容的文本框，具体实现代码如下：

```
<LinearLayout xmlns:android="http://schemas.android.com/apk/res/android"
    android:orientation="vertical" android:layout_width="fill_parent"
    android:layout_height="fill_parent">
    <EditText android:id="@+id/EditText01" android:text="hello guan"
        android:layout_width="fill_parent" android:layout_height="wrap_content">
    </EditText>
    <Button android:text="朗读" android:id="@+id/Button01"
        android:layout_width="wrap_content" android:layout_height="wrap_content">
    </Button>
</LinearLayout>
```

（2）编写程序文件 speechActivity.java，功能是根据用户在文本框中输入的内容实现语音识别，具体实现代码如下：

```
public class speechActivity extends Activity   implements TextToSpeech.OnInitListener{
    private TextToSpeech mSpeech;
    private Button btn;
    static final int TTS_CHECK_CODE = 0;
    private EditText mEditText;
    private TTS myTts;

    @Override
    public void onCreate(Bundle savedInstanceState) {
        super.onCreate(savedInstanceState);
        setContentView(R.layout.main);
        btn = (Button) findViewById(R.id.Button01);
        mEditText = (EditText) findViewById(R.id.EditText01);
        //btn.setEnabled(false);
        myTts = new TTS(this, ttsInitListener, true);
        mSpeech = new TextToSpeech(this, this);
        btn.setOnClickListener(new OnClickListener() {
            public void onClick(View v) {
                myTts.setLanguage("CHINESE");
                myTts.speak(mEditText.getText().toString(), 0, null);
```

```
                    mSpeech.speak(mEditText.getText().toString(),
                              TextToSpeech.QUEUE_FLUSH, null);
                }
            });
        }

        private TTS.InitListener ttsInitListener = new TTS.InitListener() {
                public void onInit(int version) {
                  myTts.setLanguage("CHINESE");
                  myTts.speak("爸爸", 0, null);
                }
        };


        public void onInit(int status) {
            if (status == TextToSpeech.SUCCESS) {
                int result = mSpeech.setLanguage(Locale.CHINA);
                if (result == TextToSpeech.LANG_MISSING_DATA
                        || result == TextToSpeech.LANG_NOT_SUPPORTED) {
                    Log.e("lanageTag", "not use");
                } else {
                    btn.setEnabled(true);
                    mSpeech.speak("我喜欢你", TextToSpeech.QUEUE_FLUSH,
                              null);
                }
            }
        }

        @Override
        protected void onDestroy() {
            if (mSpeech != null) {
                mSpeech.stop();
                mSpeech.shutdown();
            }
            super.onDestroy();
        }
}
```

执行后的效果如图 6-1 所示。

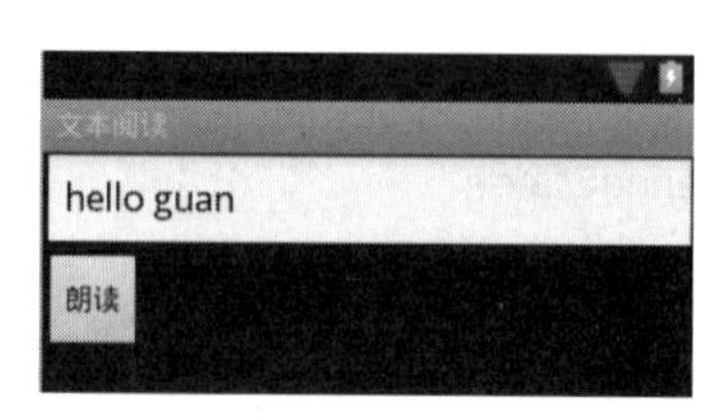


图 6-1　执行效果

### 6.2.4　实战演练——借助开源项目实现中文语音识别

Android 从 1.6 版本开始便支持 Text To Speech，使用的是 Pico 语音合成引擎，但是只支持 English、

French、German、Italian 和 Spanish 5 大语言，暂时没有对中文提供支持。因此使用 Android 默认的 TTS Engine 是无法朗读中文的。读者可以通过开源项目 eyes-free（http://code.google.com/p/eyes-free/）来实现。在安装开源项目 eyes-free 提供的 TTS Service Extended 的 APK 后，就可以在程序中使用开源项目 eyes-free 提供的 TTS library，并把 TTS Engine 设置为不是默认的 Pico，而是 eSpeak，这样就可以实现朗读中文功能。

下面将通过一个具体实例的实现过程，详细讲解在 Android 物联网设备中使用 Text-To-Speech 技术识别中文语音的过程。

| 实　例 | 功　能 | 源 码 路 径 |
|---|---|---|
| 实例 6-2 | 借助开源项目实现中文语音识别 | 光盘:\daima\6\TextSpeechEX |

（1）从网络中下载需要的开源 TTS library 的 jar 包，然后新建一个 Android 工程，工程名为 TTSCH，并且把下载的 jar 包放在 assets 文件夹下。右击工程，在弹出的快捷菜单中选择 properties | Java Build Path | Libraries | Add JARs 命令，然后向工程中添加进 assets 下的 jar 包。工程结构如图 6-2 所示。

图 6-2　工程结构

（2）编写布局文件 main.xml，功能是在屏幕中显示一个可输入内容的文本框，并在下方显示一个触发识别功能的单击按钮。文件 main.xml 的具体实现代码如下：

```
<?xml version="1.0" encoding="utf-8"?>
<LinearLayout xmlns:android="http://schemas.android.com/apk/res/android"
    android:orientation="vertical"
    android:layout_width="fill_parent"
    android:layout_height="fill_parent"
    >
    <EditText
     android:id="@+id/ttstext"
     android:layout_width="fill_parent"
     android:layout_height="wrap_content"
     android:text="可以说中文">
    </EditText>
     <Button
          android:id="@+id/ttsbtn"
          android:layout_width="fill_parent"
          android:layout_height="wrap_content"
          android:text="开始说"
          android:enabled="false">
     </Button>
</LinearLayout>
```

（3）编写 Java 程序文件，使用开源项目 eyes-free 的 TTS API 中的 TextToSpeechBeta 创建 TTS 对象。具体实现代码如下：

```
public class NiHaoTTS extends Activity implements OnInitListener{
    /** Called when the activity is first created. */
    private Button mBtn;
    private EditText mText;
    //使用 com.google.tts 包中的 TextToSpeechBeta
    private TextToSpeechBeta mTTS;

    private static final String TAG = "TTS Demo";
    private static final int REQ_TTS_STATUS_CHECK = 0;

    @Override
    public void onCreate(Bundle savedInstanceState) {
        super.onCreate(savedInstanceState);
        setContentView(R.layout.main);

        //检查 TTS 数据是否已经安装并且可用
        Intent checkIntent = new Intent();
        checkIntent.setAction(TextToSpeechBeta.Engine.ACTION_CHECK_TTS_DATA);
        startActivityForResult(checkIntent, REQ_TTS_STATUS_CHECK);

        mText = (EditText)findViewById(R.id.ttstext);
        mBtn = (Button) findViewById(R.id.ttsbtn);
        mBtn.setOnClickListener(new OnClickListener() {

                @Override
                public void onClick(View v) {
                    String ttsText = mText.getText().toString();
                    if(ttsText != "")
                    {
                        //读取文本框中的中文
                        mTTS.speak(ttsText, TextToSpeechBeta.QUEUE_ADD, null);
                    }
                }
        });
    }
  //实现 TTS 初始化接口
    @Override
    public void onInit(int status, int version) {
        Log.v(TAG, "version = " + String.valueOf(version));
        //判断 TTS 初始化的返回版本号，如果为-1，表示没有安装对应的 TTS 数据
        if(version == -1)
        {
            //提示安装所需的 TTS 数据
            alertInstallEyesFreeTTSData();
        }
        else
        {
            //TTS Engine 初始化完成
            if(status == TextToSpeechBeta.SUCCESS)
            {
```

```
                    Log.v(TAG, "success to init tts");
                    //设置 TTS 引擎，com.google.tts 即 eSpeak 支持的语言包含中文，使用 Android 系统默认
的 pico 可以设置为 com.svox.pico
                    mTTS.setEngineByPackageNameExtended("com.google.tts");
                    int result = mTTS.setLanguage(Locale.CHINA);
                    //设置发音语言
                    if(result == TextToSpeechBeta.LANG_MISSING_DATA || result == TextToSpeechBeta.
LANG_NOT_SUPPORTED)
                    //判断语言是否可用
                    {
                        Log.v(TAG, "Language is not available");
                        mBtn.setEnabled(false);
                    }
                    else
                    {
                        mTTS.speak("你好,朋友!", TextToSpeechBeta.QUEUE_ADD, null);
                        mBtn.setEnabled(true);
                    }
                }
                else
                {
                    Log.v(TAG, "failed to init tts");
                }
            }
        }

    protected   void onActivityResult(int requestCode, int resultCode, Intent data) {
        if(requestCode == REQ_TTS_STATUS_CHECK)
        {
            switch (resultCode) {
            case TextToSpeechBeta.Engine.CHECK_VOICE_DATA_PASS:
                //这个返回结果表明 TTS Engine 可以用
            {
                //使用的是 TextToSpeechBeta
               mTTS = new TextToSpeechBeta(this, this);
               Log.v(TAG, "TTS Engine is installed!");

            }

                break;
            case TextToSpeechBeta.Engine.CHECK_VOICE_DATA_BAD_DATA:
                //需要的语音数据已损坏
            case TextToSpeechBeta.Engine.CHECK_VOICE_DATA_MISSING_DATA:
                //缺少需要语言的语音数据
            case TextToSpeechBeta.Engine.CHECK_VOICE_DATA_MISSING_VOLUME:
                //缺少需要语言的发音数据
            {
                //这 3 种情况都表明数据有错，重新下载安装需要的数据
                Log.v(TAG, "Need language stuff:"+resultCode);
                Intent dataIntent = new Intent();
                dataIntent.setAction(TextToSpeechBeta.Engine.ACTION_INSTALL_TTS_DATA);
```

```
                startActivity(dataIntent);
            }
                break;
            case TextToSpeechBeta.Engine.CHECK_VOICE_DATA_FAIL:
                //检查失败
            default:
                Log.v(TAG, "Got a failure. TTS apparently not available");
                break;
            }
        }
        else
        {
            //其他 Intent 返回的结果
        }
    }
    //弹出对话框提示安装所需的 TTS 数据
    private void alertInstallEyesFreeTTSData()
    {
        Builder alertInstall = new AlertDialog.Builder(this)
            .setTitle("缺少需要的语音包")
            .setMessage("下载安装缺少的语音包")
            .setPositiveButton("确定", new DialogInterface.OnClickListener() {

                @Override
                public void onClick(DialogInterface dialog, int which) {
                    //下载 eyes-free 的语音数据包
                    String ttsDataUrl = "http://eyes-free.googlecode.com/files/tts_3.1_market.apk";
                    Uri ttsDataUri = Uri.parse(ttsDataUrl);
                    Intent ttsIntent = new Intent(Intent.ACTION_VIEW, ttsDataUri);
                    startActivity(ttsIntent);
                }
            })
            .setNegativeButton("取消", new DialogInterface.OnClickListener() {

                @Override
                public void onClick(DialogInterface dialog, int which) {
                    finish();
                }
            });
            alertInstall.create().show();

    }

    @Override
    protected void onDestroy() {
        super.onDestroy();
        if(mTTS!=null){
            mTTS.shutdown();
        }
    }
```

```
    @Override
    protected void onPause() {
        super.onPause();
        if(mTTS != null)
        {
            mTTS.stop();
        }
    }
}
```

执行后将先提示安装 APK 安装包，如图 6-3 所示。

安装完成后会在应用程序中看到对应的图标，并且在 TTS 的设置中也会增加 eSpeak TTS 项，如图 6-4 所示。

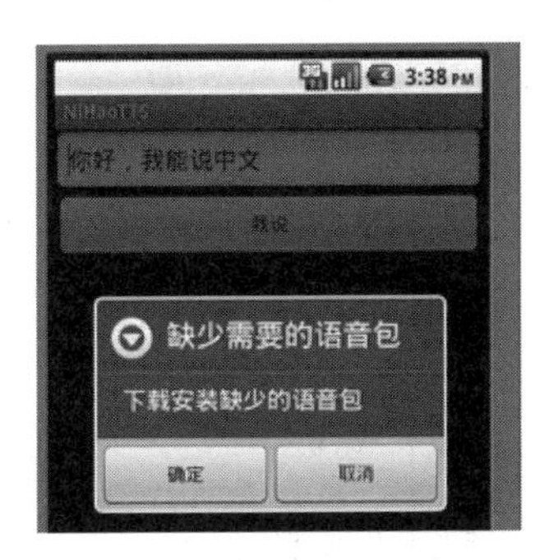


图 6-3　提示安装 APK 包

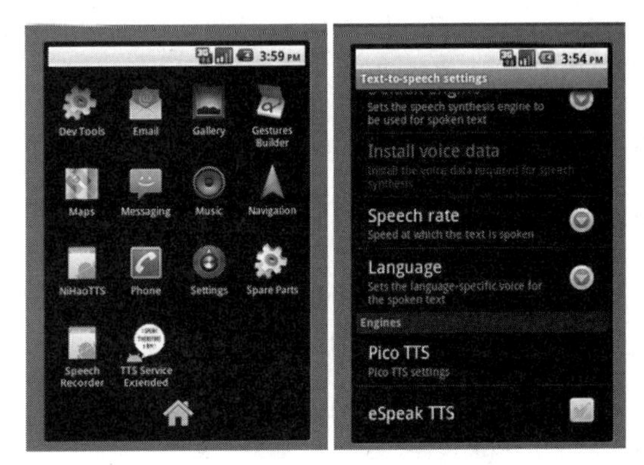


图 6-4　成功安装效果

安装后再次打开应用程序，这时会有个选择询问使用哪个 TTS，选择第二项，如图 6-5 所示。此时就可以在文本框中输入中文来识别了。

图 6-5　询问界面

# 6.3　Voice Recognition 技术详解

**知识点讲解：光盘:视频\知识点\第 6 章\Voice Recognition 技术详解.avi**

我们知道，苹果的 iPhone 语音识别用的是 Google 的技术，作为力推 Android 的 Google 自然会将其核心技术植入其中，并结合 Google 的云端技术将其发扬光大。本节将详细讲解 Voice Recognition 技术的基本知识。

## 6.3.1 Voice Recognition 技术基础

在 Android 中使用 Google 的 Voice Recognition 的方法极其简单，在 Android 自带的 API 例子中，是通过一个 Intent 的 Action 动作来完成的，主要有以下两种模式。

- ☑ ACTION_RECOGNIZE_SPEECH：一般语音识别，在这种模式下可以捕捉到语音处理后的文字列。
- ☑ ACTION_WEB_SEARCH：网络搜索。

在 API Demo 源码中提供的语音识别实例的具体实现代码如下：

```
package com.example.android.apis.app;

import com.example.android.apis.R;

import android.app.Activity;
import android.content.Intent;
import android.content.pm.PackageManager;
import android.content.pm.ResolveInfo;
import android.os.Bundle;
import android.speech.RecognizerIntent;
import android.view.View;
import android.view.View.OnClickListener;
import android.widget.ArrayAdapter;
import android.widget.Button;
import android.widget.ListView;

import java.util.ArrayList;
import java.util.List;

/**
 *用 API 开发的抽象语音识别代码
 */
public class VoiceRecognition extends Activity implements OnClickListener {

    private static final int VOICE_RECOGNITION_REQUEST_CODE = 1234;

    private ListView mList;

    /**
     *呼叫与活动首先被创建
     */
    @Override
    public void onCreate(Bundle savedInstanceState) {
        super.onCreate(savedInstanceState);
        //从它的 XML 布局描述的 UI
        setContentView(R.layout.voice_recognition);
```

```
        //得到最新互作用的显示项目
        Button speakButton = (Button) findViewById(R.id.btn_speak);
        mList = (ListView) findViewById(R.id.list);

        //查看公认活动是否存在
        PackageManager pm = getPackageManager();
        List<ResolveInfo> activities = pm.queryIntentActivities(
                new Intent(RecognizerIntent.ACTION_RECOGNIZE_SPEECH), 0);
        if (activities.size() != 0) {
            speakButton.setOnClickListener(this);
        } else {
            speakButton.setEnabled(false);
            speakButton.setText("Recognizer not present");
        }
    }

    /**
     *单击“开始识别”按钮后的处理事件
     */
    public void onClick(View v) {
        if (v.getId() == R.id.btn_speak) {
            startVoiceRecognitionActivity();
        }
    }
    /**
     *发送开始语音识别信号
     */
    private void startVoiceRecognitionActivity() {
        Intent intent = new Intent(RecognizerIntent.ACTION_RECOGNIZE_SPEECH);
        intent.putExtra(RecognizerIntent.EXTRA_LANGUAGE_MODEL,
                RecognizerIntent.LANGUAGE_MODEL_FREE_FORM);
        intent.putExtra(RecognizerIntent.EXTRA_PROMPT, "Speech recognition demo");
        startActivityForResult(intent, VOICE_RECOGNITION_REQUEST_CODE);
    }

    /**
     *处理识别结果
     */
    @Override
    protected void onActivityResult(int requestCode, int resultCode, Intent data) {
        if (requestCode == VOICE_RECOGNITION_REQUEST_CODE && resultCode == RESULT_OK) {

            ArrayList<String> matches = data.getStringArrayListExtra(
                    RecognizerIntent.EXTRA_RESULTS);
            mList.setAdapter(new ArrayAdapter<String>(this, android.R.layout.simple_list_item_1,
                    matches));
        }
        super.onActivityResult(requestCode, resultCode, data);
    }
}
```

上述代码保存在 Google 的 API 开源文件中，原理和实现代码十分简单，感兴趣的读者可以学习一下，上述源码执行后，用户通过单击 Speak!按钮，显示界面如图 6-6 所示；用户说完话后将提交到云端搜索，如图 6-7 所示；在云端搜索完成后将返回打印数据，如图 6-8 所示。

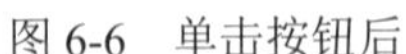
图 6-6　单击按钮后

图 6-7　说完后

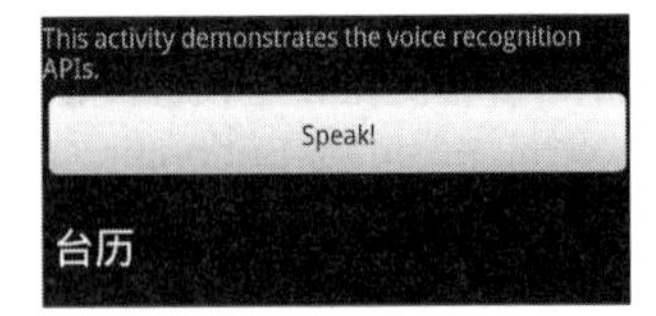


图 6-8　返回识别结果

## 6.3.2　实战演练——使用 Voice Recognition 实现语音识别

本节将通过一个具体实例的实现过程，详细讲解在 Android 物联网设备中使用 Voice Recognition 技术实现语音识别的过程。

| 实　例 | 功　能 | 源 码 路 径 |
| --- | --- | --- |
| 实例 6-3 | 使用 Voice Recognition 实现语音识别 | 光盘:\daima\6\yuyinEX |

本实例的功能是在屏幕中显示一个文本框供用户输入需要读的内容，单击按钮后开始读取文本框中输入的内容，本实例支持中文。本实例的具体实现流程如下：

（1）编写布局文件 activity_main.xml，功能是在屏幕中设置一个图标按钮，单击按钮后可以说话，并在下方列表中显示语音识别结果。文件 activity_main.xml 的具体实现代码如下：

```
<LinearLayout xmlns:android="http://schemas.android.com/apk/res/android"
    xmlns:tools="http://schemas.android.com/tools"
    android:layout_width="match_parent"
    android:layout_height="match_parent"
    android:orientation="vertical"
    android:paddingBottom="@dimen/activity_vertical_margin"
    android:paddingLeft="@dimen/activity_horizontal_margin"
    android:paddingRight="@dimen/activity_horizontal_margin"
    android:paddingTop="@dimen/activity_vertical_margin"
    tools:context=".MainActivity" >

    <TextView
        android:layout_width="match_parent"
        android:layout_height="wrap_content"
        android:gravity="center_horizontal"
        android:layout_margin="20dp"
        android:text="@string/start_voice"
        android:textSize="16sp" />

    <Button
        android:id="@+id/mic_button"
        android:layout_width="wrap_content"
```

```
        android:layout_height="wrap_content"
        android:layout_gravity="center_horizontal"
        android:background="@drawable/ic_mic" />

    <ListView
        android:id="@+id/result_list"
        android:layout_width="fill_parent"
        android:layout_height="0dip"
        android:layout_weight="1" />

</LinearLayout>
```

（2）编写主 Activity 文件 MainActivity.java，功能是监听用户单击屏幕的图标按钮，调用识别函数进行语音识别，并调用函数 onActivityResult 显示识别结果。文件 MainActivity.java 的具体实现代码如下：

```
public class MainActivity extends Activity {

    private static final int REQUEST_CODE = 1;
    private ListView mResultList;
    private Button mMicButton;

    @Override
    protected void onCreate(Bundle savedInstanceState) {
        super.onCreate(savedInstanceState);
        setContentView(R.layout.activity_main);

        mMicButton = (Button) findViewById(R.id.mic_button);
        mResultList = (ListView) findViewById(R.id.result_list);

        PackageManager pm = getPackageManager();
        List<ResolveInfo> activities = pm.queryIntentActivities(new Intent(RecognizerIntent.ACTION_RECOGNIZE_
SPEECH), 0);
        if (activities == null || activities.size() == 0) {
            mMicButton.setEnabled(false);
            Toast.makeText(getApplicationContext(), "Not Supported", Toast.LENGTH_LONG).show();
        }
        mMicButton.setOnClickListener(new OnClickListener() {
            @Override
            public void onClick(View v) {
                runVoiceRecognition();
            }
        });
    }

    private void runVoiceRecognition() {
        Intent intent = VoiceRecognitionIntentFactory.getFreeFormRecognizeIntent("Speak Now...");
        startActivityForResult(intent, REQUEST_CODE);
    }

    @Override
    protected void onActivityResult(int requestCode, int resultCode, Intent data) {
```

```
        if (requestCode == REQUEST_CODE && resultCode == RESULT_OK) {
            ArrayList<String> matches = data.getStringArrayListExtra(RecognizerIntent. EXTRA_RESULTS);
            mResultList.setAdapter(new ArrayAdapter<String>(this, android.R.layout.simple_list_item_1,
matches));
        }

        super.onActivityResult(requestCode, resultCode, data);
    }

    @Override
    public boolean onCreateOptionsMenu(Menu menu) {
        getMenuInflater().inflate(R.menu.voice, menu);
        return true;
    }

}
```

（3）编写文件 VoiceRecognitionIntentFactory.java，功能是调用 Android 内置的 Voice Recognition 技术实现语音识别功能，具体实现代码如下：

```
public class VoiceRecognitionIntentFactory {
    public static final int ACTION_GET_LANGUAGE_DETAILS_REQUEST_CODE = 88811;
    private static final int MAX_RESULTS = 100;

    private VoiceRecognitionIntentFactory() { }

    public static Intent getSimpleRecognizerIntent(String prompt)
    {
        Intent intent = getBlankRecognizeIntent();
        intent.putExtra(RecognizerIntent.EXTRA_LANGUAGE_MODEL,
RecognizerIntent.LANGUAGE_MODEL_WEB_SEARCH);
        intent.putExtra(RecognizerIntent.EXTRA_PROMPT, prompt);
        return intent;
    }

    public static Intent getBlankRecognizeIntent()
    {
        Intent intent = new Intent(RecognizerIntent.ACTION_RECOGNIZE_SPEECH);
        return intent;
    }

    public static Intent getFreeFormRecognizeIntent(String prompt){
         Intent intent = getBlankRecognizeIntent();
         intent.putExtra(RecognizerIntent.EXTRA_LANGUAGE_MODEL,
RecognizerIntent.LANGUAGE_MODEL_FREE_FORM);
         intent.putExtra(RecognizerIntent.EXTRA_PROMPT, prompt);
         return intent;
    }

    public static Intent getWebSearchRecognizeIntent()
    {
        Intent intent = new Intent(RecognizerIntent.ACTION_WEB_SEARCH);
        return intent;
```

```
    }

    public static Intent getHandsFreeRecognizeIntent()
    {
        Intent intent = new Intent(RecognizerIntent.ACTION_VOICE_SEARCH_HANDS_FREE);
        return intent;
    }

    public static Intent getPossilbeWebSearchRecognizeIntent(String prompt)
    {
        Intent intent = getWebSearchRecognizeIntent();
        intent.putExtra(RecognizerIntent.EXTRA_LANGUAGE_MODEL,
RecognizerIntent.LANGUAGE_MODEL_WEB_SEARCH);
        intent.putExtra(RecognizerIntent.EXTRA_PROMPT, prompt);
        intent.putExtra(RecognizerIntent.EXTRA_MAX_RESULTS, MAX_RESULTS);
        intent.putExtra(RecognizerIntent.EXTRA_WEB_SEARCH_ONLY, false);
        intent.putExtra(RecognizerIntent.EXTRA_PARTIAL_RESULTS, true);
        return intent;
    }

    public static Intent getLanguageDetailsIntent()
    {
        Intent intent = new Intent(RecognizerIntent.ACTION_GET_LANGUAGE_DETAILS);
        return intent;
    }
}
```

至此，整个实例介绍完毕，执行后的效果如图 6-9 所示。

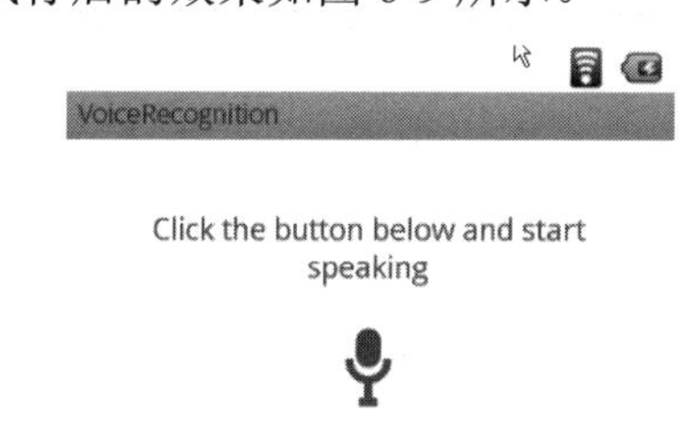


图 6-9　执行效果

## 6.4　实战演练——为设备中所有的 APP 实现语音提醒功能

**知识点讲解：光盘:视频\知识点\第 6 章\为设备中所有的 APP 实现语音提醒功能.avi**

在 Android 设备应用中，通常使用提醒功能来及时通知当前的状态信息，例如来短信时会在顶部状态栏中显示提示文字。本节将通过一个具体实例的实现过程，详细讲解将提醒文字转换为语音的方法，通过声音来实现状态提醒功能。本实例的功能十分强大，能够为当前设备中所有已经安装的 APP 程序实现语音提醒功能。

| 实　例 | 功　能 | 源 码 路 径 |
|---|---|---|
| 实例 6-4 | 将提醒文字转换为语音 | 光盘:\daima\6\voicenotifyCH |

本实例是一个开源的 Android 应用程序，使用文本在 Android 1.6 及以上设备的状态栏中发送文本通知消息时，实现辅助服务的语音通知功能，这样可以在不看屏幕的情况下也能及时获取通知信息的内容。在本实例的语音功能列表中，可以设置语音提醒功能的各个属性。

（1）编写布局文件 appwidget.xml，功能是在屏幕中显示一个指定的图片作为背景，具体实现代码如下：

```
<?xml version="1.0" encoding="utf-8"?>
<ImageButton xmlns:android="http://schemas.android.com/apk/res/android"
    android:id="@+id/button"
    android:layout_width="match_parent"
    android:layout_height="match_parent"
    android:background="@null"
    android:scaleType="fitCenter"
    android:src="@drawable/widget_disabled"/>
```

（2）编写文件 WidgetProvider.java，功能是通过布局文件 appwidget.xml 载入预制的部件，实现界面视图功能。文件 WidgetProvider.java 的具体实现代码如下：

```
public class WidgetProvider extends AppWidgetProvider {
    static final String ACTION_TOGGLE = "voicenotify.widget.TOGGLE",
                        ACTION_UPDATE = "voicenotify.widget.UPDATE";

    @Override
    public void onUpdate(Context context, AppWidgetManager appWidgetManager, int[] appWidgetIds) {
        for (int i = 0; i < appWidgetIds.length; i++) {
            RemoteViews views = new RemoteViews(context.getPackageName(), R.layout.appwidget);
            updateViews(context, views);
            appWidgetManager.updateAppWidget(appWidgetIds[i], views);
        }
    }

    @Override
    public void onReceive(Context context, Intent intent) {
        if (intent.getAction().equals(ACTION_TOGGLE)) {
            if (Service.isRunning()) {
                Toast.makeText(context,
                        Service.toggleSuspend() ? R.string.service_suspended : R.string.service_running,
                        Toast.LENGTH_SHORT).show();
            }
        } else if (intent.getAction().equals(ACTION_UPDATE)) {
            RemoteViews views = new RemoteViews(context.getPackageName(), R.layout.appwidget);
            updateViews(context, views);
            AppWidgetManager.getInstance(context).updateAppWidget(new ComponentName(context,
WidgetProvider.class), views);
        } else super.onReceive(context, intent);
    }

    private static void updateViews(Context context, RemoteViews views) {
        PendingIntent pendingIntent;
        if (Service.isRunning()) {
            pendingIntent = PendingIntent.getBroadcast(context, 0,
```

```
                    new Intent(context, WidgetProvider.class).setAction(ACTION_TOGGLE), 0);
            if (Service.isSuspended()) {
                views.setImageViewResource(R.id.button, R.drawable.widget_suspended);
            } else {
                views.setImageViewResource(R.id.button, R.drawable.widget_running);
            }
        } else {
            pendingIntent = PendingIntent.getActivity(context, 0, MainActivity.getAccessibilityIntent(), 0);
            views.setImageViewResource(R.id.button, R.drawable.widget_disabled);
        }
        views.setOnClickPendingIntent(R.id.button, pendingIntent);
    }
}
```

执行效果如图 6-10 所示。

图 6-10　执行效果

（3）编写文件 preferences.xml，功能是列表显示本语音识别系统的设置选项，例如 TTS 设置选项、TTS 流和 TTS 延时等选项。文件 preferences.xml 的具体实现代码如下：

```
<PreferenceScreen xmlns:android="http://schemas.android.com/apk/res/android">
    <Preference
        android:key="@string/key_status" />
    <Preference
        android:key="@string/key_appList"
        android:title="@string/app_list"
        android:summary="@string/app_list_summary" />
    <Preference
        android:key="@string/key_ttsSettings"
        android:title="@string/tts_settings"
        android:summary="@string/tts_settings_summary" />
    <EditTextPreference
        android:key="@string/key_ttsString"
        android:title="@string/tts_message"
        android:summary="@string/tts_message_summary"
        android:dialogMessage="@string/tts_message_dialog"
        android:defaultValue="%t: %m" />
    <ListPreference
        android:key="@string/key_ttsStream"
        android:title="@string/tts_stream"
```

```
        android:summary="@string/tts_stream_summary"
        android:entries="@array/stream_name"
        android:entryValues="@array/stream_value"
        android:defaultValue="3" />
    <EditTextPreference
        android:key="@string/key_ttsDelay"
        android:title="@string/tts_delay"
        android:summary="@string/tts_delay_summary"
        android:dialogMessage="@string/tts_delay_dialog_msg"
        android:digits="0123456789"
        android:inputType="number" />
    <EditTextPreference
        android:key="@string/key_tts_repeat"
        android:title="@string/tts_repeat"
        android:summary="@string/tts_repeat_summary"
        android:dialogMessage="@string/tts_repeat_dialog_msg"
        android:digits="0123456789"
        android:inputType="number" />
    <CheckBoxPreference
        android:key="@string/key_toasts"
        android:title="@string/speak_toasts"
        android:summary="@string/speak_toasts_summary" />
    <CheckBoxPreference
        android:key="@string/key_audio_focus"
        android:title="@string/audio_focus"
        android:summary="@string/audio_focus_summary"
        android:defaultValue="true" />
    <EditTextPreference
        android:key="@string/key_shake_threshold"
        android:title="@string/shake_to_silence"
        android:summary="@string/shake_to_silence_summary"
        android:dialogMessage="@string/shake_to_silence_dialog_msg"
        android:digits="0123456789"
        android:inputType="number"
        android:defaultValue="100" />
    <EditTextPreference
        android:key="@string/key_ignore_strings"
        android:title="@string/ignore_strings"
        android:summary="@string/ignore_strings_summary"
        android:dialogMessage="@string/ignore_strings_dialog_msg" />
    <CheckBoxPreference
        android:key="@string/key_ignore_empty"
        android:title="@string/ignore_empty"
        android:summaryOn="@string/ignore_empty_summary_on"
        android:summaryOff="@string/ignore_empty_summary_off"
        android:defaultValue="true" />
    <EditTextPreference
        android:key="@string/key_ignore_repeat"
        android:title="@string/ignore_repeat"
        android:summary="@string/ignore_repeat_summary"
        android:dialogMessage="@string/ignore_repeat_dialog_msg"
```

```
        android:digits="0123456789"
        android:inputType="number"
        android:defaultValue="10" />
    <Preference
        android:key="@string/key_device_state"
        android:title="@string/device_state"
        android:summary="@string/device_state_summary" />
    <Preference
        android:key="@string/key_quietStart"
        android:title="@string/quiet_start"
        android:summary="@string/quiet_start_summary" />
    <Preference
        android:key="@string/key_quietEnd"
        android:title="@string/quiet_end"
        android:summary="@string/quiet_end_summary" />
    <Preference
        android:key="@string/key_test"
        android:title="@string/test"
        android:summary="@string/test_summary" />
    <Preference
        android:key="@string/key_notify_log"
        android:title="@string/notify_log"
        android:summary="@string/notify_log_summary" />
    <Preference
        android:key="@string/key_support"
        android:title="@string/support"
        android:summary="@string/support_summary" />
</PreferenceScreen>
```

（4）编写文件 MainActivity.java，功能是载入文件 preferences.xml 中的控件元素实现界面布局，加载系统中各个设置选项的当前设置值。文件 MainActivity.java 的具体实现代码如下：

```
public class MainActivity extends PreferenceActivity implements OnPreferenceClickListener,
OnSharedPreferenceChangeListener {
    private static final int SDK_VERSION = Build.VERSION.SDK_INT;
    private Preference pStatus, pDeviceState, pQuietStart, pQuietEnd, pTest, pNotifyLog, pSupport;
    private static final int DLG_DEVICE_STATE = 0,
                             DLG_QUIET_START = 1,
                             DLG_QUIET_END = 2,
                             DLG_LOG = 3,
                             DLG_SUPPORT = 4,
                             DLG_DONATE = 5;
    private OnStatusChangeListener statusListener = new OnStatusChangeListener() {
        @Override
        public void onStatusChanged() {
            updateStatus();
        }
    };

    @Override
    protected void onCreate(Bundle savedInstanceState) {
        super.onCreate(savedInstanceState);
```

```
        Common.init(this);
        addPreferencesFromResource(R.xml.preferences);
        pStatus = findPreference(getString(R.string.key_status));
        pStatus.setOnPreferenceClickListener(this);
        pDeviceState = findPreference(getString(R.string.key_device_state));
        pDeviceState.setOnPreferenceClickListener(this);
        pQuietStart = findPreference(getString(R.string.key_quietStart));
        pQuietStart.setOnPreferenceClickListener(this);
        pQuietEnd = findPreference(getString(R.string.key_quietEnd));
        pQuietEnd.setOnPreferenceClickListener(this);
        pTest = findPreference(getString(R.string.key_test));
        pTest.setOnPreferenceClickListener(this);
        pNotifyLog = findPreference(getString(R.string.key_notify_log));
        pNotifyLog.setOnPreferenceClickListener(this);
        pSupport = findPreference(getString(R.string.key_support));
        pSupport.setOnPreferenceClickListener(this);
        findPreference(getString(R.string.key_appList)).setIntent(new Intent(this, AppList.class));
        Preference pTTS = findPreference(getString(R.string.key_ttsSettings));
        Intent ttsIntent = getTtsIntent();
        if (ttsIntent != null) {
            pTTS.setIntent(ttsIntent);
        } else {
            pTTS.setEnabled(false);
            pTTS.setSummary(R.string.tts_settings_summary_fail);
        }
        if (SDK_VERSION < 11) {
            getPreferenceScreen().removePreference(findPreference(getString(R.string.key_toasts)));
            if (SDK_VERSION < 8) {

    getPreferenceScreen().removePreference(findPreference(getString(R.string.key_audio_focus)));
            }
        }
    }

    static Intent getAccessibilityIntent() {
        Intent intent = new Intent();
        if (SDK_VERSION > 4) {
            intent.setAction(android.provider.Settings.ACTION_ACCESSIBILITY_SETTINGS);
        } else if (SDK_VERSION == 4) {
            intent.setAction(Intent.ACTION_MAIN);
            intent.setClassName("com.android.settings", "com.android.settings.AccessibilitySettings");
        }
        return intent;
    }

    private Intent getTtsIntent() {
        Intent intent = new Intent(Intent.ACTION_MAIN);
        if (isClassExist("com.android.settings.TextToSpeechSettings")) {
            if (SDK_VERSION >= 11 && SDK_VERSION <= 13) {
                intent.setAction(android.provider.Settings.ACTION_SETTINGS);
                intent.putExtra(EXTRA_SHOW_FRAGMENT,
```

```
"com.android.settings.TextToSpeechSettings");
                    intent.putExtra(EXTRA_SHOW_FRAGMENT_ARGUMENTS, intent.getExtras());
                } else intent.setClassName("com.android.settings", "com.android.settings.TextToSpeech Settings");
            } else if (isClassExist("com.android.settings.Settings$TextToSpeechSettingsActivity")) {
                if (SDK_VERSION == 14) {
                    intent.setAction(android.provider.Settings.ACTION_SETTINGS);
                    intent.putExtra(EXTRA_SHOW_FRAGMENT,
"com.android.settings.tts.TextToSpeechSettings");
                    intent.putExtra(EXTRA_SHOW_FRAGMENT_ARGUMENTS, intent.getExtras());
                } else intent.setClassName("com.android.settings", "com.android.settings.Settings$TextToSpeech
SettingsActivity");
            } else if (isClassExist("com.google.tv.settings.TextToSpeechSettingsTop")) {
                intent.setClassName("com.google.tv.settings",
"com.google.tv.settings.TextToSpeechSettingsTop");
            } else return null;
            return intent;
        }

        private boolean isClassExist(String name) {
            try {
                PackageInfo pkgInfo = getPackageManager().getPackageInfo(
                    name.substring(0, name.lastIndexOf(".")), PackageManager.GET_ACTIVITIES);
                if (pkgInfo.activities != null) {
                    for (int n = 0; n < pkgInfo.activities.length; n++) {
                        if (pkgInfo.activities[n].name.equals(name)) return true;
                    }
                }
            } catch (PackageManager.NameNotFoundException e) {}
            return false;
        }

        @Override
        public boolean onPreferenceClick(Preference preference) {
            if (preference == pStatus && Service.isRunning() && Service.isSuspended()) {
                Service.toggleSuspend();
                return true;
            } else if (preference == pDeviceState) {
                showDialog(DLG_DEVICE_STATE);
                return true;
            } else if (preference == pQuietStart) {
                showDialog(DLG_QUIET_START);
                return true;
            } else if (preference == pQuietEnd) {
                showDialog(DLG_QUIET_END);
                return true;
            } else if (preference == pTest) {
                if (!AppList.findOrAddApp(getPackageName(), this).getEnabled()) {
                    Toast.makeText(this, getString(R.string.test_ignored), Toast.LENGTH_LONG).show();
                }
                new Timer().schedule(new TimerTask() {
                    @Override
```

```
                public void run() {
                    Notification notification = new Notification(R.drawable.icon,
                        getString(R.string.test_notify_msg), System.currentTimeMillis());
                    notification.defaults |= Notification.DEFAULT_SOUND;
                    notification.flags |= Notification.FLAG_AUTO_CANCEL;
                    notification.setLatestEventInfo(MainActivity.this, getString(R.string.app_name), getString
(R.string.test),
                        PendingIntent.getActivity(MainActivity.this, 0, getIntent(), 0));

    ((NotificationManager)getSystemService(Context.NOTIFICATION_SERVICE)).notify(0, notification);
                }
            }, 5000);
            return true;
        } else if (preference == pNotifyLog) {
            showDialog(DLG_LOG);
            return true;
        } else if (preference == pSupport) {
            showDialog(DLG_SUPPORT);
            return true;
        }
        return false;
    }

    @Override
    protected Dialog onCreateDialog(int id) {
        int i;
        switch (id) {
        case DLG_DEVICE_STATE:
            final CharSequence[] items = getResources().getStringArray(R.array.device_states);
            return new AlertDialog.Builder(this)
            .setTitle(R.string.device_state_dialog_title)
            .setMultiChoiceItems(items,
                new boolean[] {
                    Common.getPrefs(this).getBoolean(Common.KEY_SPEAK_SCREEN_OFF, true),
                    Common.getPrefs(this).getBoolean(Common.KEY_SPEAK_SCREEN_ON, true),
                    Common.getPrefs(this).getBoolean(Common.KEY_SPEAK_HEADSET_OFF, true),
                    Common.getPrefs(this).getBoolean(Common.KEY_SPEAK_HEADSET_ON, true),
                    Common.getPrefs(this).getBoolean(Common.KEY_SPEAK_SILENT_ON, false)
                },
                new DialogInterface.OnMultiChoiceClickListener() {
                    @Override
                    public void onClick(DialogInterface dialog, int which, boolean isChecked) {
                        if (which == 0) {

    Common.getPrefs(MainActivity.this).edit().putBoolean(Common.KEY_SPEAK_SCREEN_OFF, isChecked).
commit();
                        } else if (which == 1) {

    Common.getPrefs(MainActivity.this).edit().putBoolean(Common.KEY_SPEAK_SCREEN_ON, isChecked).
commit();
                        } else if (which == 2) {
```

```
Common.getPrefs(MainActivity.this).edit().putBoolean(Common.KEY_SPEAK_HEADSET_OFF, isChecked).
commit();
                    } else if (which == 3) {

Common.getPrefs(MainActivity.this).edit().putBoolean(Common.KEY_SPEAK_HEADSET_ON,  isChecked).
commit();
                    } else if (which == 4) {

Common.getPrefs(MainActivity.this).edit().putBoolean(Common.KEY_SPEAK_SILENT_ON,  isChecked).
commit();
                    }
                }
            }
        ).create();
    case DLG_QUIET_START:
        i = Common.getPrefs(this).getInt(getString(R.string.key_quietStart), 0);
        return new TimePickerDialog(this, sTimeSetListener, i/60, i%60, false);
    case DLG_QUIET_END:
        i = Common.getPrefs(this).getInt(getString(R.string.key_quietEnd), 0);
        return new TimePickerDialog(this, eTimeSetListener, i/60, i%60, false);
    case DLG_LOG:
        return new AlertDialog.Builder(this)
        .setTitle(R.string.notify_log)
        .setView(new NotifyList(this))
        .setNeutralButton(android.R.string.ok, new DialogInterface.OnClickListener() {
            public void onClick(DialogInterface dialog, int id) {
                dialog.dismiss();
            }
        })
        .create();
    case DLG_SUPPORT:
        return new AlertDialog.Builder(this)
        .setTitle(R.string.support)
        .setItems(R.array.support_items, new DialogInterface.OnClickListener() {
            public void onClick(DialogInterface dialog, int item) {
                switch (item) {
                    case 0:
                        showDialog(DLG_DONATE);
                        break;
                    case 1:
                        Intent iMarket = new Intent(Intent.ACTION_VIEW, Uri.parse("market://details?
id=com. pilot51.voicenotify"));
                        iMarket.addFlags(Intent.FLAG_ACTIVITY_NEW_TASK);
                        try {
                            startActivity(iMarket);
                        } catch (ActivityNotFoundException e) {
                            e.printStackTrace();
                            Toast.makeText(getBaseContext(), R.string.error_market, Toast.LENGTH_
LONG).show();
                        }
```

```
                            break;
                        case 2: //Contact developer
                            Intent iEmail = new Intent(Intent.ACTION_SEND);
                            iEmail.setType("plain/text");
                            iEmail.putExtra(Intent.EXTRA_EMAIL, new String[] {getString(R.string.dev_
email)});
                            iEmail.putExtra(Intent.EXTRA_SUBJECT,
getString(R.string.email_subject));
                            String version = null;
                            try {
                                version = getPackageManager().getPackageInfo(getPackageName(),
0).versionName;
                            } catch (NameNotFoundException e) {
                                e.printStackTrace();
                            }
                            iEmail.putExtra(Intent.EXTRA_TEXT,
                                    getString(R.string.email_body,
                                            version,
                                            Build.VERSION.RELEASE,
                                            Build.ID,
                                            Build.MANUFACTURER + " " +
Build.BRAND + " " + Build.MODEL));
                            try {
                                startActivity(iEmail);
                            } catch (ActivityNotFoundException e) {
                                e.printStackTrace();
                                Toast.makeText(getBaseContext(), R.string.error_email, Toast.LENGTH_
LONG).show();
                            }
                            break;
                        case 3:
                            startActivity(new Intent(Intent.ACTION_VIEW, Uri.parse("http://getlocalization.
com/voicenotify")));
                            break;
                        case 4:
                            startActivity(new Intent(Intent.ACTION_VIEW, Uri.parse("https://github.com/
pilot51/voicenotify")));
                            break;
                    }
                }
            }).create();
        case DLG_DONATE:
            return new AlertDialog.Builder(this)
            .setTitle(R.string.donate)
            .setItems(R.array.donate_services, new DialogInterface.OnClickListener() {
                public void onClick(DialogInterface dialog, int item) {
                    switch (item) {
                        case 0:
                            showWalletDialog();
                            break;
                        case 1:
```

```
                        startActivity(new  Intent(Intent.ACTION_VIEW,  Uri.parse("https://paypal.com/
cgi-bin/webscr?"
                                    +"cmd=_donations&business=pilota51%40gmail%2ecom&lc=
US&item_name=Voice%20Notify&"
                                    + "no_note=0&no_shipping=1&currency_code=USD")));
                        break;
                    }
                }
            }).create();
        }
        return null;
    }

    private void showWalletDialog() {
        final Intent walletIntent = getWalletIntent();
        AlertDialog.Builder dlg = new AlertDialog.Builder(this)
        .setTitle(R.string.donate_wallet_title)
        .setMessage(R.string.donate_wallet_message)
        .setNegativeButton(android.R.string.cancel, null);
        if (walletIntent != null) {
            dlg.setPositiveButton(R.string.donate_wallet_launch_app, new DialogInterface.OnClickListener()
{
                @Override
                public void onClick(DialogInterface dialog, int which) {
                    startActivity(walletIntent);
                }
            });
        } else {
            dlg.setPositiveButton(R.string.donate_wallet_launch_web, new DialogInterface. OnClickListener()
{
                @Override
                public void onClick(DialogInterface dialog, int which) {
                    startActivity(new  Intent(Intent.ACTION_VIEW,  Uri.parse("https://wallet.google.com/
manage/# sendMoney:")));
                }
            });
        }
        dlg.show();
    }

    /**
     * @return The intent for Google Wallet, otherwise null if installation is not found
     */
    private Intent getWalletIntent() {
        try {
            getPackageManager().getPackageInfo("com.google.android.apps.walletnfcrel",PackageManager.
GET_ACTIVITIES);
            return new Intent(Intent.ACTION_MAIN)
                    .setComponent(new ComponentName("com.google.android.apps.walletnfcrel",
                                            "com.google.android.apps.wallet.WalletRootActivity"));
        } catch (PackageManager.NameNotFoundException e) {
            return null;
        }
```

```
    }

    private TimePickerDialog.OnTimeSetListener sTimeSetListener = new TimePickerDialog.OnTime
SetListener() {
        public void onTimeSet(TimePicker view, int hourOfDay, int minute) {
    Common.getPrefs(MainActivity.this).edit().putInt(getString(R.string.key_quietStart), hourOfDay * 60 +
minute).commit();
        }
    };
    private TimePickerDialog.OnTimeSetListener eTimeSetListener = new TimePickerDialog.OnTime
SetListener() {
        public void onTimeSet(TimePicker view, int hourOfDay, int minute) {
    Common.getPrefs(MainActivity.this).edit().putInt(getString(R.string.key_quietEnd), hourOfDay * 60 +
minute).commit();
        }
    };

    private void updateStatus() {
        if (Service.isSuspended() && Service.isRunning()) {
            pStatus.setTitle(R.string.service_suspended);
            pStatus.setSummary(R.string.status_summary_suspended);
            pStatus.setIntent(null);
        } else {
            pStatus.setTitle(Service.isRunning() ? R.string.service_running : R.string.service_disabled);
            pStatus.setSummary(R.string.status_summary_accessibility);
            pStatus.setIntent(getAccessibilityIntent());
        }
    }

    @Override
    protected void onResume() {
        super.onResume();
        Common.getPrefs(this).registerOnSharedPreferenceChangeListener(this);
        Service.registerOnStatusChangeListener(statusListener);
        updateStatus();
    }

    @Override
    protected void onPause() {
        Service.unregisterOnStatusChangeListener(statusListener);
        Common.getPrefs(this).unregisterOnSharedPreferenceChangeListener(this);
        super.onPause();
    }

    public void onSharedPreferenceChanged(SharedPreferences sp, String key) {
        if (key.equals(getString(R.string.key_ttsStream))) {
            Common.setVolumeStream(this);
        }
    }
}
```

在上述代码中，在设备屏幕界面中加载的是系统当前的设置选项值。通过上述代码会监听当前用

户对屏幕选项的操作，根据用户的操作可以调用处理事件重新设置自己想要的选项，执行效果如图 6-11 所示。

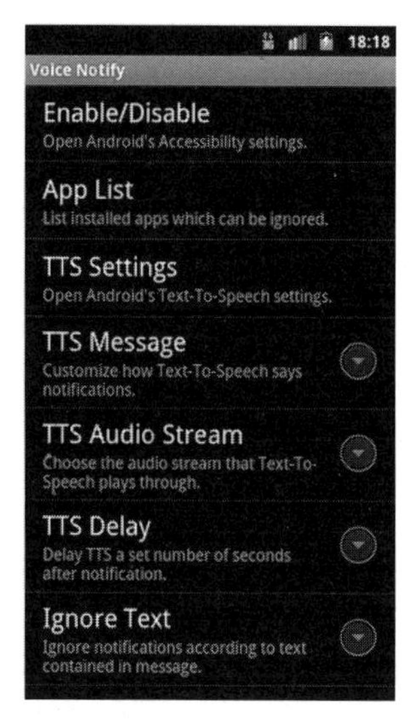


图 6-11　系统设置选项列表界面

（5）编写文件 app_list_item.xml，功能是通过列表显示当前系统中的 APP 程序。文件 app_list_item.xml 的具体实现代码如下：

```
<LinearLayout
    xmlns:android="http://schemas.android.com/apk/res/android"
    android:layout_width="match_parent"
    android:layout_height="wrap_content">
    <LinearLayout
        android:layout_width="0dp"
        android:layout_height="wrap_content"
        android:layout_weight="1"
        android:orientation="vertical">
        <TextView
            android:id="@+id/text1"
            android:layout_width="match_parent"
            android:layout_height="wrap_content"
            android:textSize="24sp" />
        <TextView
            android:id="@+id/text2"
            android:layout_width="match_parent"
            android:layout_height="wrap_content" />
    </LinearLayout>
    <CheckBox
        android:id="@+id/checkbox"
        android:layout_width="40dp"
        android:layout_height="40dp"
        android:layout_gravity="center"
        android:clickable="false"
        android:focusable="false"
        android:duplicateParentState="true" />
</LinearLayout>
```

（6）编写文件 AppList.java，功能是通过文件 app_list_item.xml 加载系统中安装的 APP，在列表显示的 APP 中可以设置是否忽略当前软件。如果忽略某一个软件，这个 APP 软件的通知信息将不会被语音识别提醒。文件 AppList.java 的具体实现代码如下：

```
public class AppList extends ListActivity {
    private ListView lv;
    private Adapter adapter;
    private static ArrayList<App> apps;
    private static boolean defEnable;
    private static final String KEY_DEFAULT_ENABLE = "defEnable";
    private static final int IGNORE_TOGGLE = 0, IGNORE_ALL = 1, IGNORE_NONE = 2;
    private static final Object SYNC_APPS = new Object();
    private static OnListUpdateListener listener;
    private static boolean isUpdating;

    @Override
    public void onCreate(Bundle savedInstanceState) {
        super.onCreate(savedInstanceState);
        Common.init(this);
        requestWindowFeature(Window.FEATURE_INDETERMINATE_PROGRESS);
        lv = getListView();
        lv.setTextFilterEnabled(true);
        adapter = new Adapter();
        listener = new OnListUpdateListener() {
            @Override
            public void onListUpdated() {
                runOnUiThread(new Runnable() {
                    public void run() {
                        adapter.setData(apps);
                    }
                });
            }
            @Override
            public void onUpdateCompleted() {
                runOnUiThread(new Runnable() {
                    public void run() {
                        setProgressBarIndeterminateVisibility(false);
                    }
                });
                listener = null;
            }
        };
        lv.setAdapter(adapter);
        lv.setOnItemClickListener(new OnItemClickListener() {
            @Override
            public void onItemClick(AdapterView<?> parent, View view, int position, long id) {
                setIgnore((App)adapter.getItem(position), IGNORE_TOGGLE);
                adapter.notifyDataSetChanged();
            }
        });
        defEnable = Common.getPrefs(this).getBoolean(KEY_DEFAULT_ENABLE, true);
        updateAppsList();
```

```
    }

    private interface OnListUpdateListener {
        void onListUpdated();
        void onUpdateCompleted();
    }
    private static void onListUpdated() {
        if (listener != null) listener.onListUpdated();
    }

    private void updateAppsList() {
        setProgressBarIndeterminateVisibility(true);
        if (isUpdating) {
            adapter.setData(apps);
            return;
        }
        isUpdating = true;
        new Thread(new Runnable() {
            public void run() {
                synchronized (SYNC_APPS) {
                    apps = Database.getApps();
                    onListUpdated();
                    final boolean isFirstLoad = apps.isEmpty();
                    PackageManager packMan = getPackageManager();

                    for (int a = apps.size() - 1; a >= 0; a--) {
                        App app = apps.get(a);
                        try {
                            packMan.getApplicationInfo(app.getPackage(), 0);
                        } catch (NameNotFoundException e) {
                            if (!isFirstLoad) app.remove();
                            apps.remove(a);
                            onListUpdated();
                        }
                    }

                    inst:for (ApplicationInfo appInfo : packMan.getInstalledApplications(0)) {
                        for (App app : apps) {
                            if (app.getPackage().equals(appInfo.packageName)) {
                                continue inst;
                            }
                        }
                        App app = new App(appInfo.packageName, String.valueOf(appInfo.loadLabel
(packMan)), defEnable);
                        apps.add(app);
                        onListUpdated();
                        if (!isFirstLoad) app.updateDb();
                    }

                    Collections.sort(apps, new Comparator<App>() {
                        @Override
                        public int compare(App app1, App app2) {
```

```
                    return app1.getLabel().compareToIgnoreCase(app2.getLabel());
                }
            });
            onListUpdated();
            if (isFirstLoad) Database.setApps(apps);
        }
        isUpdating = false;
        if (listener != null) listener.onUpdateCompleted();
    }
    }).start();
}

@Override
public boolean onCreateOptionsMenu(Menu menu) {
    super.onCreateOptionsMenu(menu);
    getMenuInflater().inflate(R.menu.app_list, menu);
    return true;
}

@Override
public boolean onOptionsItemSelected(MenuItem item) {
    switch (item.getItemId()) {
    case R.id.ignore_all:
        setDefaultEnable(false);
        massIgnore(IGNORE_ALL);
        return true;
    case R.id.ignore_none:
        setDefaultEnable(true);
        massIgnore(IGNORE_NONE);
        return true;
    case R.id.filter:

((InputMethodManager)getSystemService(Context.INPUT_METHOD_SERVICE)).toggleSoftInput(0, 0);
        return true;
    }
    return false;
}

/**
 * @param pkg Package name used to find {@link App} in current list or create a new one from system.
 * @param ctx Context required to get default enabled preference and to get package manager for
searching system.
 * @return Found or created {@link App}, otherwise null if app not found on system.
 */
static App findOrAddApp(String pkg, Context ctx) {
    synchronized (SYNC_APPS) {
        if (apps == null) {
            defEnable = Common.getPrefs(ctx).getBoolean(KEY_DEFAULT_ENABLE, true);
            apps = Database.getApps();
        }
        for (App app : apps) {
```

```
                if (app.getPackage().equals(pkg)) {
                    return app;
                }
            }
            try {
                PackageManager packMan = ctx.getPackageManager();
                App app = new App(pkg, packMan.getApplicationInfo(pkg, 0).loadLabel(packMan).
toString(), defEnable);
                apps.add(app.updateDb());
                return app;
            } catch (NameNotFoundException e) {
                e.printStackTrace();
                return null;
            }
        }
    }

    private void massIgnore(int ignoreType) {
        for (App app : apps) {
            setIgnore(app, ignoreType);
        }
        adapter.notifyDataSetChanged();
        new Thread(new Runnable() {
            public void run() {
                Database.setApps(apps);
            }
        }).start();
    }

    private void setIgnore(App app, int ignoreType) {
        if (!app.getEnabled() & (ignoreType == IGNORE_TOGGLE | ignoreType == IGNORE_NONE)) {
            app.setEnabled(true, ignoreType == IGNORE_TOGGLE);
            if (ignoreType == IGNORE_TOGGLE) {
                Toast.makeText(this, getString(R.string.app_is_not_ignored, app.getLabel()), Toast.LENGTH_
SHORT).show();
            }
        } else if (app.getEnabled() & (ignoreType == IGNORE_TOGGLE | ignoreType == IGNORE_ALL)) {
            app.setEnabled(false, ignoreType == IGNORE_TOGGLE);
            if (ignoreType == IGNORE_TOGGLE) {
                Toast.makeText(this, getString(R.string.app_is_ignored, app.getLabel()), Toast.LENGTH_
SHORT).show();
            }
        }
    }

    private void setDefaultEnable(boolean enable) {
        defEnable = enable;
        Common.getPrefs(this).edit().putBoolean(KEY_DEFAULT_ENABLE, defEnable).commit();
    }

    private class Adapter extends BaseAdapter implements Filterable {
```

```
private final ArrayList<App> baseData = new ArrayList<App>();
private final ArrayList<App> adapterData = new ArrayList<App>();
private LayoutInflater mInflater;
private SimpleFilter filter;

private Adapter() {
    mInflater = (LayoutInflater)getSystemService(Context.LAYOUT_INFLATER_SERVICE);
}

private void setData(ArrayList<App> list) {
    baseData.clear();
    baseData.addAll(list);
    refresh();
}

private void refresh() {
    adapterData.clear();
    adapterData.addAll(baseData);
    notifyDataSetChanged();
}

@Override
public int getCount() {
    return adapterData.size();
}

@Override
public Object getItem(int position) {
    return adapterData.get(position);
}

@Override
public long getItemId(int position) {
    return position;
}

@Override
public View getView(int position, View view, ViewGroup parent) {
    if (view == null) {
        view = mInflater.inflate(R.layout.app_list_item, parent, false);
    }
    ((TextView)view.findViewById(R.id.text1)).setText(adapterData.get(position).getLabel());
    ((TextView)view.findViewById(R.id.text2)).setText(adapterData.get(position).getPackage());
    ((CheckBox)view.findViewById(R.id.checkbox)).setChecked(adapterData.get(position).getEnabled());
    return view;
}

@Override
public Filter getFilter() {
    if (filter == null) filter = new SimpleFilter();
    return filter;
}
```

```
        private class SimpleFilter extends Filter {
            @Override
            protected FilterResults performFiltering(CharSequence prefix) {
                FilterResults results = new FilterResults();
                if (prefix == null || prefix.length() == 0) {
                    results.values = baseData;
                    results.count = baseData.size();
                } else {
                    String prefixString = prefix.toString().toLowerCase();
                    ArrayList<App> newValues = new ArrayList<App>();
                    for (App app : baseData) {
                        if (app.getLabel().toLowerCase().contains(prefixString)
                                || app.getPackage().toLowerCase().contains(prefixString)) {
                            newValues.add(app);
                        }
                    }
                    results.values = newValues;
                    results.count = newValues.size();
                }
                return results;
            }

            @SuppressWarnings("unchecked")
            @Override
            protected void publishResults(CharSequence constraint, FilterResults results) {
                adapterData.clear();
                adapterData.addAll((ArrayList<App>)results.values);
                if (results.count > 0) notifyDataSetChanged();
                else notifyDataSetInvalidated();
            }
        }
    }
}
```

在上述代码中，应用程序会监听用户对列表中某个应用程序复选框的操作，根据用户是否选中这个选项来设置是否忽略。执行效果如图 6-12 所示。

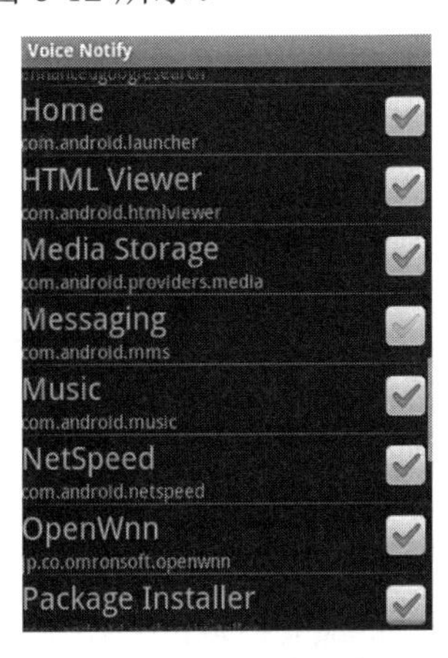


图 6-12　设置界面

（7）编写文件 notify_log_item.xml，功能是通过文本框控件显示当前选项的设置信息，具体实现代码如下：

```
<LinearLayout xmlns:android="http://schemas.android.com/apk/res/android"
    android:layout_width="match_parent"
    android:layout_height="wrap_content"
    android:orientation="vertical">
    <TextView
        android:id="@+id/time"
        android:layout_width="match_parent"
        android:layout_height="wrap_content"
        android:gravity="center" />
    <TextView
        android:id="@+id/title"
        android:layout_width="match_parent"
        android:layout_height="wrap_content"
        android:gravity="center"
        android:textSize="24sp" />
    <TextView
        android:id="@+id/message"
        android:layout_width="match_parent"
        android:layout_height="wrap_content"
        android:gravity="center"
        android:textSize="16sp" />
    <TextView
        android:id="@+id/ignore_reasons"
        android:layout_width="match_parent"
        android:layout_height="wrap_content"
        android:gravity="center" />
</LinearLayout>
```

（8）编写文件 NotifyList.java，功能是通过文件 notify_log_item.xml 载入布局控件来显示界面，在显示列表中根据选项的 ID 值显示对应的选项的辅助信息，包括时间、是否忽略、忽略原因、位置、提醒信息大小等。文件 NotifyList.java 的具体实现代码如下：

```
public class NotifyList extends ListView {
    private Resources res;
    private static ArrayList<NotifyItem> list = new ArrayList<NotifyItem>();
    private Adapter adapter;
    private static OnListChangeListener listener;
    private static final int HISTORY_LIMIT = 20;

    NotifyList(Context context) {
        super(context);
        res = getResources();
        setDivider(res.getDrawable(R.drawable.divider));
        adapter = new Adapter(context, list);
        setAdapter(adapter);
    }

    static void addNotification(App app, String message) {
        if (list.size() == HISTORY_LIMIT) {
```

```
            list.remove(list.size() - 1);
        }
        list.add(0, new NotifyItem(app, message));
        if (listener != null) {
            listener.onListChange();
        }
    }

    static void setLastIgnore(String ignoreReasons, boolean isNew) {
        if (list.isEmpty()) return;
        list.get(0).setIgnoreReasons(ignoreReasons, isNew);
        if (listener != null) {
            listener.onListChange();
        }
    }

    private static class NotifyItem {
        private App app;
        private String message, ignoreReasons, time;
        private SimpleDateFormat sdf = new SimpleDateFormat("HH:mm:ss");
        private boolean silenced;

        private NotifyItem(App app, String message) {
            this.app = app;
            this.message = message;
            time = sdf.format(Calendar.getInstance().getTime());
        }

        private App getApp() {
            return app;
        }

        private String getMessage() {
            return message;
        }

        private String getIgnoreReasons() {
            return ignoreReasons;
        }

        void setIgnoreReasons(String reasons, boolean isNew) {
            silenced = !isNew;
            ignoreReasons = reasons;
        }

        private String getTime() {
            return time;
        }
    }

    private static interface OnListChangeListener {
```

```
        void onListChange();
    }

    private class Adapter extends BaseAdapter {
        private ArrayList<NotifyItem> data;
        private LayoutInflater mInflater;

        private Adapter(final Context context, ArrayList<NotifyItem> list) {
            data = list;
            mInflater = (LayoutInflater)context.getSystemService(Context.LAYOUT_INFLATER_SERVICE);
            listener = new OnListChangeListener() {
                @Override
                public void onListChange() {
                    ((Activity)context).runOnUiThread(new Runnable() {
                        public void run() {
                            notifyDataSetChanged();
                        }
                    });
                }
            };
        }

        @Override
        public int getCount() {
            return data.size();
        }

        @Override
        public Object getItem(int position) {
            return data.get(position);
        }

        @Override
        public long getItemId(int position) {
            return position;
        }

        @Override
        public View getView(int position, View convertView, ViewGroup parent) {
            View view = convertView;
            if (view == null) {
                view = mInflater.inflate(R.layout.notify_log_item, parent, false);
            }
            final NotifyItem item = data.get(position);
            ((TextView)view.findViewById(R.id.time)).setText(item.getTime());
            ((TextView)view.findViewById(R.id.title)).setText(item.getApp().getLabel());
            TextView textView = (TextView)view.findViewById(R.id.message);
            if (item.getMessage().length() != 0) {
                textView.setText(item.getMessage());
                textView.setVisibility(TextView.VISIBLE);
            } else textView.setVisibility(TextView.GONE);
```

```
                textView = (TextView)view.findViewById(R.id.ignore_reasons);
                if (item.getIgnoreReasons() != null && item.getIgnoreReasons().length() != 0) {
                    textView.setText(item.getIgnoreReasons());
                    if (item.silenced) textView.setTextColor(Color.YELLOW);
                    else textView.setTextColor(Color.RED);
                    textView.setVisibility(TextView.VISIBLE);
                } else textView.setVisibility(TextView.GONE);
                view.setOnLongClickListener(new OnLongClickListener() {
                    @Override
                    public boolean onLongClick(View v) {
                        new AlertDialog.Builder(getContext())
                        .setTitle(res.getString(item.getApp().getEnabled()
                                                    ? R.string.ignore_app
                                                    : R.string.unignore_app,
                                                    item.getApp().getLabel()))
                        .setPositiveButton(R.string.yes, new DialogInterface.OnClickListener() {
                            @Override
                            public void onClick(DialogInterface dialog, int which) {
                                item.getApp().setEnabled(!item.getApp().getEnabled(), true);
                                Toast.makeText(getContext(), res.getString(item.getApp().getEnabled()
                                                    ? R.string.app_is_not_ignored
                                                    : R.string.app_is_ignored,
                                                    item.getApp().getLabel()),
                                                    Toast.LENGTH_SHORT).show();
                            }
                        })
                        .setNegativeButton(android.R.string.cancel, null)
                        .show();
                        return false;
                    }
                });
                return view;
            }
        }
}
```

（9）本实例中使用了 SQLite 数据库，在里面保存当前系统中各个选项的设置值。编写文件 Database.java，通过 SQL 语句实现对各个选项当前设置值的查询、修改和删除功能。文件 Database.java 的具体实现代码如下：

```
public class Database extends SQLiteOpenHelper {
    private static String TAG = Database.class.getSimpleName();
    private Context context;
    private static Database database;
    private static final int DB_VERSION = 1;
    private static final String
        OLD_FILE = "ignored_apps",
        DB_NAME = "apps.db",
        TABLE_NAME = "apps",
        COLUMN_PACKAGE = "package",
        COLUMN_LABEL = "name",
        COLUMN_ENABLED = "is_enabled",
```

```
        CREATE_TBL_APPS = "create table if not exists " + TABLE_NAME + "(" + BaseColumns._ID
                + " integer primary key autoincrement, " + COLUMN_PACKAGE + " text not null, "
                + COLUMN_LABEL + " text not null, " + COLUMN_ENABLED + " integer);";

    private Database(Context context) {
        super(context, DB_NAME, null, DB_VERSION);
        this.context = context.getApplicationContext();
        try {
            if (!context.getDatabasePath(DB_NAME).exists()
                        && new File(context.getFilesDir().toString() + File.separatorChar + OLD_FILE).exists()) {
                upgradeOldIgnores();
            }
        } catch (Exception e) {
            Log.w(TAG, "Error checking for old ignores to be transferred to database.");
            e.printStackTrace();
        }
    }

    /**
     *如果尚未初始化数据库<br />
     * Call {@link #getInstance()} to get the static instance.
     */
    static void init(Context context) {
        if (database != null) {
            Log.w(TAG, "Database already initialized!");
        } else database = new Database(context);
    }

    /** @返回先前初始化静态数据库实例，如果{@link #init(Context)}没有执行则返回 null*/
    static Database getInstance() {
        if (database == null) {
            Log.w(TAG, "Database not initialized!");
        }
        return database;
    }

    /**从旧的数据库副本忽略文件并删除旧的文件*/
    @SuppressWarnings("unchecked")
    void upgradeOldIgnores() {
        ArrayList<String> oldList = new ArrayList<String>();
        FileInputStream file = null;
        try {
            file = context.openFileInput(OLD_FILE);
        } catch (FileNotFoundException e) {}
        try {
            ObjectInputStream in = new ObjectInputStream(file);
            try {
                oldList = (ArrayList<String>)in.readObject();
            } catch (ClassNotFoundException e) {
                Log.e(TAG, "Error: Failed to read ignored_apps - Data appears corrupt");
                e.printStackTrace();
            }
```

```
            in.close();
        } catch (IOException e) {
            Log.e(TAG, "Error: Failed to read ignored_apps");
            e.printStackTrace();
        }
        ArrayList<App> newList = new ArrayList<App>();
        PackageManager packMan = context.getPackageManager();
        ApplicationInfo appInfo;
        for (String s : oldList) {
            try {
                appInfo = packMan.getApplicationInfo(s, PackageManager.GET_UNINSTALLED_ PACKAGES);
                newList.add(new  App(appInfo.packageName,  String.valueOf(appInfo.loadLabel  (packMan)),
false));
            } catch (NameNotFoundException e) {
                e.printStackTrace();
            }
        }
        setApps(newList);
        context.deleteFile(OLD_FILE);
    }

    /** @返回一个新的列表包含从数据库中所有的应用程序*/
    static synchronized ArrayList<App> getApps() {
        SQLiteDatabase db = database.getReadableDatabase();
        Cursor cursor = db.query(TABLE_NAME, null, null, null, null, null, COLUMN_LABEL + " COLLATE
NOCASE");
        ArrayList<App> list = new ArrayList<App>();
        while (cursor.moveToNext()) {
            list.add(new App(
                cursor.getString(cursor.getColumnIndex(COLUMN_PACKAGE)),
                cursor.getString(cursor.getColumnIndex(COLUMN_LABEL)),
                cursor.getInt(cursor.getColumnIndex(COLUMN_ENABLED)) == 1
            ));
        }
        cursor.close();
        db.close();
        return list;
    }

    /**
     *清除并设置数据库中的所有程序集
     * @param list The list of apps to add in the database.
     */
    static synchronized void setApps(ArrayList<App> list) {
        SQLiteDatabase db = database.getWritableDatabase();
        db.delete(TABLE_NAME, null, null);
        ContentValues values;
        for (App app : list) {
            values = new ContentValues();
            values.put(COLUMN_PACKAGE,   app.getPackage());
            values.put(COLUMN_LABEL,   app.getLabel());
            values.put(COLUMN_ENABLED,   app.getEnabled() ? 1 : 0);
```

```
                db.insert(TABLE_NAME, null, values);
            }
            db.close();
        }

        /**
         *如果没有找到匹配的，更新应用程序数据库中匹配的包的名称或补充
         * @param app The app to add or update in the database.
         */
        static synchronized void addOrUpdateApp(App app) {
            ContentValues values = new ContentValues();
            values.put(COLUMN_PACKAGE,   app.getPackage());
            values.put(COLUMN_LABEL,   app.getLabel());
            values.put(COLUMN_ENABLED,   app.getEnabled() ? 1 : 0);
            SQLiteDatabase db = database.getWritableDatabase();
            if (db.update(TABLE_NAME, values, COLUMN_PACKAGE + " = ?", new String[] {app.getPackage()})
== 0) {
                db.insert(TABLE_NAME, null, values);
            }
            db.close();
        }

        /**
         *更新应用程序启用的数据库匹配包名称值
         * @param app The app to update in the database
         */
        static synchronized void updateAppEnable(App app) {
            ContentValues values = new ContentValues();
            values.put(COLUMN_ENABLED,   app.getEnabled() ? 1 : 0);
            SQLiteDatabase db = database.getWritableDatabase();
            db.update(TABLE_NAME, values, COLUMN_PACKAGE + " = ?", new String[] {app.getPackage()});
            db.close();
        }

        /**
         *删除应用程序数据库匹配的包名
         * @param app The app to remove from the database
         */
        static synchronized void removeApp(App app) {
            SQLiteDatabase db = database.getWritableDatabase();
            db.delete(TABLE_NAME, COLUMN_PACKAGE + " = ?", new String[] {app.getPackage()});
            db.close();
        }

        @Override
        public void onCreate(SQLiteDatabase db) {
            db.execSQL(CREATE_TBL_APPS);
        }

        @Override
        public void onUpgrade(SQLiteDatabase db, int oldVersion, int newVersion) {}
}
```

（10）为了方便用户及时取消语音提醒功能，编写文件 Shake.java，功能是通过摇晃设备的方式取消语音提醒功能。文件 Shake.java 的具体实现代码如下：

```
public class Shake implements SensorEventListener {
    private final Context context;
    private final SensorManager manager;
    private final Sensor sensor;
    private OnShakeListener listener;
    private int threshold, overThresholdCount;
    private float accelCurrent, accelLast;

    Shake(Context c) {
        context = c;
        manager = (SensorManager)c.getSystemService(Context.SENSOR_SERVICE);
        sensor = manager.getDefaultSensor(Sensor.TYPE_ACCELEROMETER);
    }

    void enable() {
        if (listener == null) return;
        try {
            threshold = Integer.parseInt(Common.getPrefs(context).getStrin g(context.getString(R.string.
key_shake_threshold), null));
        } catch (NumberFormatException e) {
            return;
        }
        manager.registerListener(this, sensor, SensorManager.SENSOR_DELAY_NORMAL);
    }

    void disable() {
        manager.unregisterListener(this);
        accelCurrent = 0;
        accelLast = 0;
        overThresholdCount = 0;
    }

    void setOnShakeListener(OnShakeListener listener) {
        this.listener = listener;
    }

    @Override
    public void onAccuracyChanged(Sensor sensor, int accuracy) {}

    @Override
    public void onSensorChanged(SensorEvent event) {
        float x = event.values[0];
        float y = event.values[1];
        float z = event.values[2];
        accelCurrent = (float)Math.sqrt(x * x + y * y + z * z);
        float accel = accelCurrent - accelLast;
        if (accelLast != 0 && Math.abs(accel) > threshold / 10) {
            overThresholdCount++;
            if (overThresholdCount >= 2) {
```

```
                    listener.onShake();
                }
            } else {
                overThresholdCount = 0;
            }
            accelLast = accelCurrent;
        }

        interface OnShakeListener {
            void onShake();
        }
}
```

至此，整个实例介绍完毕。目前本实例只是支持英语、德语、法语、西班牙语、匈牙利语、俄语和意大利语，这一点从本项目的工程结构中可以看出，如图 6-13 所示。

另外，本项目已经在谷歌市场中发布，地址是 https://play.google.com/store/apps/details?id=com.pilot51.voicenotify，名称是 Voice Notify，读者可以下载安装并体会，如图 6-14 所示。

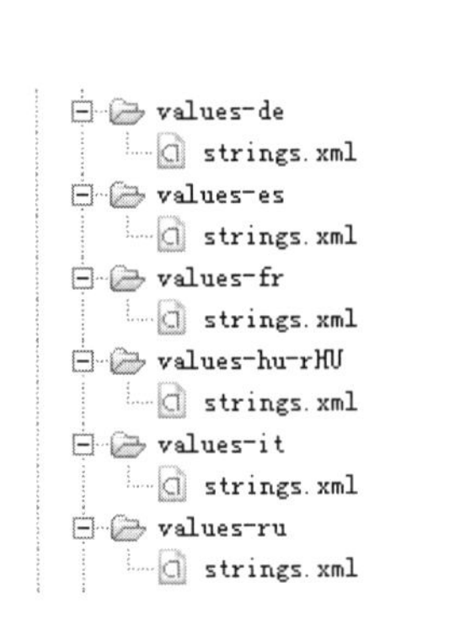


图 6-13　支持的语言

图 6-14　谷歌市场上的本项目

# 第 7 章　手势识别实战

手势识别技术是 Android SDK 中比较重要并且比较新颖的一项技术，在 Android 物联网设备应用中可以通过手势来灵活地操控设备的运行。本章将详细讲解在 Android 物联网设备中使用手势识别技术的基本知识和具体方法，为读者步入本书后面知识的学习打下基础。

## 7.1　Android 中的事件监听机制

**知识点讲解：光盘:视频\知识点\第 7 章\Android 中的事件监听机制.avi**

本书前面的内容中，已经多次用到了 Android 事件监听机制，例如监听是否单击屏幕、是否触摸单击了某个按钮等。本节将详细讲解 Android 系统中事件监听机制的基本知识。

### 7.1.1　Android 系统中的监听事件

在 Android 应用开发过程中，存在如下所示的常用监听事件。

（1）ListView 事件监听

☑ setOnItemSelectedListener：鼠标滚动时触发。

☑ setOnItemClickListener：点击时触发。

（2）EditText 事件监听

☑ setOnKeyListener：获取焦点时触发。

（3）RadioGroup 事件监听

☑ setOnCheckedChangeListener：点击时触发。

（4）CheckBox 事件监听

☑ setOnCheckedChangeListener：点击时触发。

（5）Spinner 事件监听

☑ setOnItemSelectedListener：点击时触发。

（6）DatePicker 事件监听

☑ onDateChangedListener：日期改变时触发。

（7）DatePickerDialog 事件监听

☑ onDateSetListener：设置日期时触发。

（8）TimePicker 事件监听

☑ onTimeChangedListener：时间改变时触发。

（9）TimePickerDialog 事件监听

☑ onTimeSetListener：设置时间时触发。

（10）Button、ImageButton 事件监听

☑ setOnClickListener：点击时触发。

（11）Menu 事件监听

☑ onOptionsItemSelected：点击时触发。

（12）Gallery 事件监听

☑ setOnItemClickListener：点击时触发。

（13）GridView 事件监听

☑ setOnItemClickListener：点击时触发。

### 7.1.2 Android 事件监听器的回调方法

在 Android 操作系统中，对于事件的处理是一个非常基础而且重要的操作，很多功能都需要对相关事件进行触发才能实现。例如 Android 事件监听器是视图 View 类的接口，包含一个单独的回调方法。这些方法将在视图中注册的监听器被用户界面操作触发时由 Android 框架调用。在现实应用中，如下所示的回调方法被包含在 Android 事件监听器接口中。

（1）onClick()

包含于 View.OnClickListener。当用户触摸这个 item（在触摸模式下），或者通过浏览键或跟踪球聚焦在这个 item 上，然后按下“确认”键或者按下跟踪球时被调用。

（2）onLongClick()

包含于 View.OnLongClickListener。当用户触摸并按住这个 item（在触摸模式下），或者通过浏览键或跟踪球聚焦在这个 item 上，然后保持按下“确认”键或者按下跟踪球（一秒钟）时被调用。

（3）onFocusChange()

包含于 View.OnFocusChangeListener。当用户使用浏览键或跟踪球浏览进入或离开这个 item 时被调用。

（4）onKey()

包含于 View.OnKeyListener。当用户聚焦在这个 item 上并按下或释放设备上的一个按键时被调用。

（5）onTouch()

包含于 View.OnTouchListener。当用户执行的动作被当作一个触摸事件时被调用，包括按下、释放，或者屏幕上任何的移动手势（在这个 item 的边界内）。

（6）onCreateContextMenu()

包含于 View.OnCreateContextMenuListener。当正在创建一个上下文菜单时被调用（作为持续的“长点击”动作的结果）。

上述方法是它们相应接口的唯一“住户”。要定义这些方法并处理事件，需要在活动中实现这个嵌套接口或定义它为一个匿名类。然后传递一个实例给各自的 View.set...Listener()方法。例如调用 setOnClickListener()并传递给它 OnClickListener 实现。

下面的代码演示了为一个按钮注册一个点击监听器的方法。

```
private OnClickListener mCorkyListener = new OnClickListener() {
public void onClick(View v) {
}
};
```

```
protected void onCreate(Bundle savedValues) {
...
Button button = (Button)findViewById(R.id.corky);
button.setOnClickListener(mCorkyListener);
...
}
```

此时可能会发现，把 OnClickListener 作为活动的一部分来实现会简便很多，这样可以避免额外的类加载和对象分配。例如下面的演示代码。

```
public class ExampleActivity extends Activity implements OnClickListener {
protected void onCreate(Bundle savedValues) {
...
Button button = (Button)findViewById(R.id.corky);
button.setOnClickListener(this);
}
public void onClick(View v) {
}
...
}
```

上述代码中的 onClick()回调没有返回值，但是一些其他 Android 事件监听器必须返回一个布尔值。原因和事件相关，具体原因如下所示。

☑ onLongClick()：返回一个布尔值来指示用户是否已经消费了这个事件而不应该再进一步处理它。也就是说，返回 true 表示用户已经处理了这个事件而且到此为止；返回 false 表示用户还没有处理和/或这个事件应该继续交给其他 on-click 监听器。

☑ onKey()：返回一个布尔值来指示用户是否已经消费了这个事件而不应该再进一步处理它。也就是说，返回 true 表示用户已经处理了这个事件而且到此为止；返回 false 表示用户还没有处理和/或这个事件应该继续交给其他 on-key 监听器。

☑ onTouch()：返回一个布尔值来指示用户的监听器是否已经消费了这个事件。重要的是这个事件可以有多个彼此跟随的动作。因此，如果当接收到向下动作事件时返回 false，那表明用户还没有消费这个事件而且对后续动作也不感兴趣。那么，用户将不会被该事件中的其他动作调用，例如手势或最后出现向上动作事件。

在 Android 应用中，按键事件总是递交给当前焦点所在的视图。它们从视图层次的顶层开始被分发，然后依次向下，直到到达恰当的目标。如果视图（或者一个子视图）当前拥有焦点，那么可以看到事件经由 dispatchKeyEvent()方法分发。除了视图截获按键事件外，还可以在活动中使用 onKeyDown()和 onKeyUp()来接收所有的事件。

**注意：** Android 将首先调用事件处理器，其次是类定义中合适的默认处理器。这样，当从这些事件监听器中返回 true 时会停止事件向其他 Android 事件监听器传播，并且也会阻塞视图中的此事件处理器的回调函数。所以，当返回 true 时需要确认是否希望终止这个事件。

### 7.1.3　Android 事件处理的两种模型

在 Android 系统中提供了两种方式的事件处理，分别是基于回调的事件处理和基于监听器的事件处理。对于基于监听器的事件处理而言，主要就是为 Android 界面组件绑定特定的事件监听器；对于基于回调的事件处理而言，主要的做法是重写 Android 组件特定的回调函数，Android 的大部分界面组件都提供了事件响应的回调函数，只要重写它们即可。

### 1. 基于监听器的事件处理

在基于监听器的事件处理的监听器模型中，主要涉及如下 3 类对象。

☑ 事件源 Event Source：产生事件的来源，通常是各种组件，如按钮、窗口等。

☑ 事件 Event：事件封装了界面组件上发生的特定事件的具体信息，如果监听器需要获取界面组件上所发生事件的相关信息，一般通过事件 Event 对象来传递。

☑ 事件监听器 Event Listener：负责监听事件源发生的事件，并对不同的事件做相应的处理。

基于监听器的事件处理机制是一种委派式 Delegation 的事件处理方式，事件源将整个事件委托给事件监听器，由监听器对事件进行响应处理。这种处理方式将事件源和事件监听器分离，有利于提供程序的可维护性。

例如，View 类中的 OnLongClickListener 监听器不需要传递事件，演示代码如下：

```
public interface OnLongClickListener {
    boolean onLongClick(View v);
}
```

例如，View 类中的 OnLongClickListener 监听器定义如下，需要传递事件 MotionEvent。

```
public interface OnTouchListener {
    boolean onTouch(View v, MotionEvent event);
}
```

### 2. 基于回调的事件处理

相比基于监听器的事件处理模型，基于回调的事件处理模型要简单一些，该模型中，事件源和事件监听器是合一的，也就是说没有独立的事件监听器存在。当用户在 GUI 组件上触发某事件时，由该组件自身特定的函数负责处理该事件。通常通过重写 Override 组件类的事件处理函数实现事件的处理。

例如，View 类实现了 KeyEvent.Callback 接口中的一系列回调函数，所以基于回调的事件处理机制通过自定义 View 来实现，自定义 View 时重写这些事件处理方法即可。具体演示代码如下：

```
public interface Callback {
    //几乎所有基于回调的事件处理函数都会返回一个 boolean 类型值，该返回值用于
    //标识该处理函数是否能完全处理该事件
//返回 true，表明该函数已完全处理该事件，该事件不会传播出去
//返回 false，表明该函数未完全处理该事件，该事件会传播出去
    boolean onKeyDown(int keyCode, KeyEvent event);
    boolean onKeyLongPress(int keyCode, KeyEvent event);
    boolean onKeyUp(int keyCode, KeyEvent event);
    boolean onKeyMultiple(int keyCode, int count, KeyEvent event);
}
```

由此可见，基于监听器的事件模型符合单一职责原则，事件源和事件监听器分开实现。Android 的事件处理机制保证基于监听器的事件处理会优先于基于回调的事件处理被触发。在某些特定情况下，基于回调的事件处理机制会更好地提高程序的内聚性。

## 7.1.4 基于自定义监听器的事件处理流程

在实际项目开发中，经常需要自定义监听器来实现自定义业务流程的处理，而且一般都不是基于

GUI 界面作为事件源的。这里以常见的 app 自动更新为例进行说明，在自动更新过程中，会存在两个状态：下载中和下载完成，而程序需要在这两个状态做不同的事情，“下载中”需要在 UI 界面上实时显示软件包下载的进度，“下载完成”后，取消进度条的显示。这里进行一个模拟，重点在于说明自定义监听器的事件处理流程。

（1）定义事件监听器，具体代码如下：

```
public interface DownloadListener {
    public void onDownloading(int progress);  //下载过程中的处理函数
    public void onDownloaded();               //下载完成的处理函数
}
```

（2）实现下载操作的工具类，具体代码如下：

```
public class DownloadUtils {
    private static DownloadUtils instance = null;
    private DownloadUtils() {
    }
    public static synchronized DownloadUtils instance() {
        if (instance == null) {
            instance = new DownloadUtils();
        }
        return instance;
    }

    private boolean isDownloading = true;

    private int progress = 0;

    //实际开发中这个函数需要传入 url 作为参数，以获取服务器端的安装包位置
    public void download(DownloadListener listener) throws InterruptedException {
        while (isDownloading) {
            listener.onDownloading(progress);
            //下载过程的简单模拟
            Thread.sleep(1000);
            progress += 10;
            if (progress >= 100) {
                isDownloading = false;
            }
        }
        //下载完成
        listener.onDownloaded();
    }
}
```

（3）在 main 函数中模拟事件源，具体代码如下：

```
public class DownloadUI {

    public static void main(String[] args) {
        try {
            DownloadUtils.instance().download(new MyDownloadListener());
        } catch (InterruptedException e) {
            e.printStackTrace();
        }
```

```
    }

    private static class MyDownloadListener implements DownloadListener {

        @Override
        public void onDownloading(int progress) {
            System.out.println("下载进度是：" + progress);
        }

        @Override
        public void onDownloaded() {
            System.out.println("下载完成");
        }

    }
}
```

运行上面的模拟程序，执行效果如图 7-1 所示。

```
下载进度是: 0
下载进度是: 10
下载进度是: 20
下载进度是: 30
下载进度是: 40
下载进度是: 50
下载进度是: 60
下载进度是: 70
下载进度是: 80
下载进度是: 90
下载完成
```

图 7-1 执行效果

# 7.2 手势识别技术介绍

**知识点讲解：光盘:视频\知识点\第 7 章\手势识别技术介绍.avi**

对于触摸屏设备来说，其消息传递机制包括按下、抬起和移动这几种，用户只需要简单地实现重载 onTouch 或者设置触摸监听器 setOnTouchListener 即可处理触摸事件。但是有时为了提高应用程序的用户体验，需要识别用户当前正在操作的手势。本节将详细讲解在 Android 设备中实现手势识别的基本知识。

## 7.2.1 手势识别类 GestureDetector

在 Android 系统中，专门提供了手势识别类 GestureDetector。在 Android 设备中，通过类 GestureDetector 可以识别很多的手势，通过其 nTouchEvent(event)方法可以完成不同手势的识别。类 GestureDetector 对外提供了两个接口：OnGestureListener 和 OnDoubleTapListener，另外还提供了一个内部类 SimpleOnGestureListener。

（1）GestureDetector.OnDoubleTapListener 接口：用来通知 DoubleTap 事件，类似于鼠标的双击事

件。此接口中各个成员的具体说明如下所示。

- ☑ onDoubleTap(MotionEvent e)：在二次双击 Touch down 时触发。
- ☑ onDoubleTapEvent(MotionEvent e)：通知 DoubleTap 手势中的事件，包含 down、up 和 move 事件（这里指的是在双击之间发生的事件，例如在同一个地方双击会产生 DoubleTap 手势，而在 DoubleTap 手势里面还会发生 down 和 up 事件，这两个事件由该函数通知）；双击的第二下 Touch down 和 up 都会触发，可用 e.getAction()区分。
- ☑ onSingleTapConfirmed(MotionEvent e)：用来判定该次点击是 SingleTap 而不是 DoubleTap，如果连续点击两次就是 DoubleTap 手势，如果只点击一次，系统等待一段时间后没有收到第二次点击则判定该次点击为 SingleTap 而不是 DoubleTap，然后触发 SingleTapConfirmed 事件。这个方法不同于 onSingleTapUp，它是在 GestureDetector 确信用户在第一次触摸屏幕后，没有紧跟着第二次触摸屏幕，也就是不是“双击”时触发。

（2）GestureDetector.OnGestureListener 接口：用来通知普通的手势事件，该接口有如下所示的 6 个回调函数。

- ☑ onDown(MotionEvent e)：down 事件。
- ☑ onSingleTapUp(MotionEvent e)：一次点击 up 事件，在 touch down 后没有滑动。
- ☑ onLongPress：用户长按触摸屏，由多个 MotionEvent ACTION_DOWN 触发。
- ☑ onShowPress(MotionEvent e)：down 事件发生而 move 或 up 还没发生前触发该事件。
- ☑ onFling(MotionEvent e1, MotionEvent e2, float velocityX, float velocityY)：滑动手势事件，Touch 了滑动一点距离后，在 ACTION_UP 时才会触发。各个参数的具体说明如下所示。
  - ➢ e1：第一个 ACTION_DOWN MotionEvent 并且只有一个。
  - ➢ e2：最后一个 ACTION_MOVE MotionEvent。
  - ➢ velocityX：X 轴上的移动速度，像素/秒。
  - ➢ velocityY：Y 轴上的移动速度，像素/秒，触发条件为 X 轴的坐标位移大于 FLING_MIN_DISTANCE，且移动速度大于 FLING_MIN_VELOCITY 个像素/秒。
- ☑ onScroll(MotionEvent e1, MotionEvent e2, float distanceX, float distanceY)：在屏幕上拖动事件。无论是用手拖动 view，或者是以抛的动作滚动，都会多次触发。这个方法在 ACTION_MOVE 动作发生时就会触发。

## 7.2.2　手势检测器类 GestureDetector

Android 系统的事件处理机制是基于 Listener（监听器）实现的，和触摸屏相关的事件是通过 onTouchListener 实现的。另外，在 Android 系统中，所有类 View 的子类都可以通过 setOnTouchListener()、setOnKeyListener()等方法来添加对某一类事件的监听器。并且 Listener 一般会以 Interface（接口）的方式来提供，其中包含一个或多个 abstract（抽象）方法，我们需要实现这些方法来完成 onTouch()、onKey()等操作。这样当给某个 View 设置了事件 Listener，并实现了其中的抽象方法以后，程序便可以在特定的事件被 Dispatch（调用）到该 View 时，通过 callback 函数给予对应的响应。

在 Android 开发应用中，有多种使用类 GestureDetector 的方法。

### 1. 第一种

（1）通过 GestureDetector 的构造方法将 SimpleOnGestureListener 对象传递进去，这样

GestureDetector 就能处理不同的手势。

```
public GestureDetector(Context context, GestureDetector.OnGestureListener listener)
```

（2）在 onTouch()方法中实现 OnTouchListener 监听，代码如下：

```
private OnTouchListener gestureTouchListener = new OnTouchListener() {
            public boolean onTouch(View v, MotionEvent event) {
        return gDetector.onTouchEvent(event);
    }
};
```

## 2．第二种

（1）使用如下所示的方法构建场景：

```
private GestureDetector mGestureDetector;
mGestureListener = new BookOnGestureListener();
```

（2）使用 new 新建构造出来的 GestureDetector 对象：

```
mGestureDetector = new GestureDetector(mGestureListener);
class BookOnGestureListener implements OnGestureListener {
```

（3）实现事件处理：

```
public boolean onTouchEvent(MotionEvent event) {
                mGestureListener.onTouchEvent(event);
}
```

## 3．第三种

（1）在当前类中创建一个 GestureDetector 实例。

```
private GestureDetector mGestureDetector;
```

（2）创建一个 Listener 来实时监听当前面板操作手势。

```
class LearnGestureListener extends GestureDetector.SimpleOnGestureListener
```

（3）在初始化时，将 Listener 实例关联当前的 GestureDetector 实例。

```
mGestureDetector = new GestureDetector(this, new LearnGestureListener());
```

（4）使用方法 onTouchEvent()作为入口检测，通过传递 MotionEvent 参数来监听操作手势。

```
mGestureDetector.onTouchEvent(event)
```

例如下面的演示代码：

```
private GestureDetector mGestureDetector;
@Override
public void onCreate(Bundle savedInstanceState) {
    super.onCreate(savedInstanceState);
    mGestureDetector = new GestureDetector(this, new LearnGestureListener());
}
@Override
public boolean onTouchEvent(MotionEvent event) {
    if (mGestureDetector.onTouchEvent(event))
        return true;
    else
        return false;
}
class LearnGestureListener extends GestureDetector.SimpleOnGestureListener{
    @Override
    public boolean onSingleTapUp(MotionEvent ev) {
```

```
        Log.d("onSingleTapUp",ev.toString());
        return true;
    }
    @Override
    public void onShowPress(MotionEvent ev) {
        Log.d("onShowPress",ev.toString());
    }
    @Override
    public void onLongPress(MotionEvent ev) {
        Log.d("onLongPress",ev.toString());
    }
    @Override
    public boolean onScroll(MotionEvent e1, MotionEvent e2, float distanceX, float distanceY) {
        Log.d("onScroll",e1.toString());
        return true;
    }
    @Override
    public boolean onDown(MotionEvent ev) {
        Log.d("onDownd",ev.toString());
        return true;
    }
    @Override
    public boolean onFling(MotionEvent e1, MotionEvent e2, float velocityX, float velocityY) {
        Log.d("d",e1.toString());
        Log.d("e2",e2.toString());
        return true;
    }
}
```

4．第四种

（1）创建一个 GestureDetector 的对象，传入 listener 对象，在接收到的 onTouchEvent 中将 event 传给 GestureDetector 进行分析，listener 会回调给相应的动作。

（2）通过 GestureDetector.SimpleOnGestureListener（Framework 帮我们简化了）实现了 OnGesture Listener 和 OnDoubleTapListener 两个接口类，只需要继承它并重写其中的回调即可。

（3）设置在第一次点击 down 时，给 Hanlder 发送一个延时的消息，例如延时 300ms。如果在 300ms 中发生了第二次点击的 down 事件，那么就认为是双击事件，并移除之前发送的延时消息。如果 300ms 后仍没有第二次的 down 消息，那么就判定为 SingleTapConfirmed 事件（当然，此时用户的手指应已完成第一次点击的 up 过程）。第三次点击的判定和双击的判定类似，只是多了一次发送延时消息的过程。

例如下面的演示代码：

```
private GestureDetector mGestureDetector;
@Override
public void onCreate(Bundle savedInstanceState) {
  super.onCreate(savedInstanceState);
  mGestureDetector = new GestureDetector(this, new MyGestureListener());
}
@Override
public boolean onTouchEvent(MotionEvent event) {
```

```
    return mGestureDetector.onTouchEvent(event);
}
class MyGestureListener extends GestureDetector.SimpleOnGestureListener{
    @Override
    public boolean onSingleTapUp(MotionEvent ev) {
        Log.d("onSingleTapUp",ev.toString());
        return true;
    }
    @Override
    public void onShowPress(MotionEvent ev) {
        Log.d("onShowPress",ev.toString());
    }
    @Override
    public void onLongPress(MotionEvent ev) {
        Log.d("onLongPress",ev.toString());
    }
...
}
```

### 7.2.3　手势识别处理事件和方法

在 Android 系统中实现手势识别功能时，通常通过如下所示 id 事件和方法实现。

（1）boolean onDoubleTap(MotionEvent e)：双击的第二下 Touch down 时触发。

（2）boolean onDoubleTapEvent(MotionEvent e)：双击的第二下 Touch down 和 up 都会触发，可用 e.getAction()区分。

（3）boolean onDown(MotionEvent e)：Touch down 时触发。

（4）boolean onFling(MotionEvent e1, MotionEvent e2, float velocityX, float velocityY)：Touch 了滑动一点距离后，up 时触发。

（5）void onLongPress(MotionEvent e)：Touch 了不移动一直 Touch down 时触发。

（6）boolean onScroll(MotionEvent e1, MotionEvent e2, float distanceX, float distanceY)：Touch 了滑动时触发。

（7）void onShowPress(MotionEvent e)：Touch 了还没有滑动时触发。

**注意：** onDown 和 onLongPress 的具体对比如下所示。

☑　onDown 只要 Touch down 一定立刻触发。

☑　而 Touch down 后过一会儿没有滑动先触发 onShowPress 再触发 onLongPress。

由此可见，Touch down 后一直不滑动，会按照 onDown→onShowPress→onLongPress 的顺序进行触发。

（8）boolean onSingleTapConfirmed(MotionEvent e)和 boolean onSingleTapUp(MotionEvent e)：这两个函数都是在 Touch down 后又没有滑动（onScroll），又没有长按（onLongPress），然后 Touch up 时触发。

（9）onDown→onSingleTapUp→onSingleTapConfirmed：点击一下非常快的（不滑动）Touch up。

（10）onDown→onShowPress→onSingleTapUp→onSingleTapConfirmed：点击一下稍微慢点的（不滑动）Touch up。

# 7.3　实战演练——通过点击的方式移动图片

**知识点讲解：光盘:视频\知识点\第 7 章\通过点击的方式移动图片.avi**

在触摸屏手机中，点击移动照片的功能十分常见。在本实例用 ImageView 控件来显示 Drawable 中的照片，在程序运行后将照片放在屏幕中央。通过 onTouchEvent 来处理点击、拖动、放开等事件来完成拖动图片的功能。并且设置了 ImageView 的点击监听事件，让用户在点击图片的同时恢复到图片的初始位置。本节将通过一个具体实例的实现过程，讲解在 Android 屏幕中通过点击的方式移动图片的方法和具体实现流程。

| 实　例 | 功　能 | 源 码 路 径 |
|---|---|---|
| 实例 7-1 | 在屏幕中通过点击的方式移动图片 | 光盘:\daima\7\moveEX |

编写主程序文件 example162.java，具体实现流程如下所示。

（1）通过 DisplayMetrics 获取屏幕对象，分别用 intScreenX 和 intScreenY 取得屏幕解析像素并分别设置图片的宽和高。具体代码如下：

```
public void onCreate(Bundle savedInstanceState)
{
  super.onCreate(savedInstanceState);
  setContentView(R.layout.main);

  /*取得屏幕对象*/
  DisplayMetrics dm = new DisplayMetrics();
  getWindowManager().getDefaultDisplay().getMetrics(dm);

  /*取得屏幕解析像素*/
  intScreenX = dm.widthPixels;
  intScreenY = dm.heightPixels;

  /*设置图片的宽和高*/
  intWidth = 100;
  intHeight = 100;
```

（2）将图片从 Drawable 中赋值给 ImageView 控件来呈现在屏幕中，并通过方法 RestoreButton() 初始化按钮使其位置居中。具体代码如下：

```
/*通过 findViewById 构造器创建 ImageView 对象*/
mImageView01 =(ImageView) findViewById(R.id.myImageView1);
/*将图片从 Drawable 赋值给 ImageView 来呈现*/
mImageView01.setImageResource(R.drawable.baby);

/*初始化按钮位置居中*/
RestoreButton();
```

（3）定义点击监听事件 setOnClickListener，当用户点击 ImageView 图片时将图片还原到初始位置显示。具体代码如下：

```
/*当点击 ImageView，还原初始位置*/
mImageView01.setOnClickListener(new Button.OnClickListener()
```

```
  {
    @Override
    public void onClick(View v)
    {
      RestoreButton();
    }
  });
}
```

（4）定义 onTouchEvent(MotionEvent event)覆盖触控事件。首先取得手指触控屏幕的位置，然后实现触控事件的处理，分别实现点击屏幕、移动位置和离开屏幕这 3 个动作处理。具体代码如下：

```
/*覆盖触控事件*/
public boolean onTouchEvent(MotionEvent event)
{
  /*取得手指触控屏幕的位置*/
  float x = event.getX();
  float y = event.getY();

  try
  {
    /*触控事件的处理*/
    switch (event.getAction())
    {
      /*点击屏幕*/
      case MotionEvent.ACTION_DOWN:
        picMove(x, y);
          break;
      /*移动位置*/
      case MotionEvent.ACTION_MOVE:
        picMove(x, y);
          break;
      /*离开屏幕*/
      case MotionEvent.ACTION_UP:
        picMove(x, y);
          break;
    }
  }catch(Exception e)
    {
      e.printStackTrace();
    }
  return true;
}
```

（5）定义方法 picMove(float x, float y)来移动屏幕中的图片，具体代码如下：

```
/*移动图片的方法*/
private void picMove(float x, float y)
{
  /*默认微调图片与指针的相对位置*/
  mX=x-(intWidth/2);
  mY=y-(intHeight/2);

  /*防止图片超过屏幕的相关处理*/
```

```
  /*防止屏幕向右超过屏幕*/
  if((mX+intWidth)>intScreenX)
  {
    mX = intScreenX-intWidth;
  }
  /*防止屏幕向左超过屏幕*/
  else if(mX<0)
  {
    mX = 0;
  }
  /*防止屏幕向下超过屏幕*/
  else if ((mY+intHeight)>intScreenY)
  {
    mY=intScreenY-intHeight;
  }
  /*防止屏幕向上超过屏幕*/
  else if (mY<0)
  {
    mY = 0;
  }
  /*通过 log  来查看图片位置*/
  Log.i("jay", Float.toString(mX)+","+Float.toString(mY));
  /*以 setLayoutParams()方法，重新安排 Layout 上的位置*/
  mImageView01.setLayoutParams
  (
    new AbsoluteLayout.LayoutParams
    (intWidth,intHeight,(int) mX,(int)mY)
  );
}
```

（6）定义方法 RestoreButton()来还原 ImageView 图片到初始位置，具体代码如下：

```
/*还原 ImageView 位置的事件处理*/
public void RestoreButton()
{
  intDefaultX = ((intScreenX-intWidth)/2);
  intDefaultY = ((intScreenY-intHeight)/2);
  /*Toast 还原位置坐标*/
  mMakeTextToast
  (
    "("+
    Integer.toString(intDefaultX)+
    ","+
    Integer.toString(intDefaultY)+")",true
  );

  /*以 setLayoutParams()方法，重新安排 Layout 上的位置*/
  mImageView01.setLayoutParams
  (
    new AbsoluteLayout.LayoutParams
    (intWidth,intHeight,intDefaultX,intDefaultY)
  );
}
```

执行后效果如图 7-2 所示，可以通过鼠标点击的方式移动图片的位置，如图 7-3 所示。

图 7-2　执行效果

图 7-3　移动图片

# 7.4　实战演练——实现各种手势识别

**知识点讲解：光盘:视频\知识点\第 7 章\实现各种手势识别.avi**

本节将通过一个具体实例的实现过程，讲解在 Android 系统中实现各种常见手势识别的方法和具体实现流程。

| 实　　例 | 功　　能 | 源 码 路 径 |
| --- | --- | --- |
| 实例 7-2 | 在屏幕中实现各种常见的手势识别 | 光盘:\daima\7\GestureEX |

## 7.4.1　布局文件 main.xml

布局文件 main.xml 非常简单，具体实现代码如下：

```
<LinearLayout xmlns:android="http://schemas.android.com/apk/res/android"
    android:orientation="vertical"
    android:layout_width="fill_parent"
    android:layout_height="fill_parent"
    >
<TextView
    android:layout_width="fill_parent"
    android:layout_height="wrap_content"
    android:text="@string/hello"
    />
</LinearLayout>
```

## 7.4.2　隐藏屏幕顶部的电池等图标和标题内容

主 Activity 的实现文件是 mainActivity.java，功能是为了更好地演示手势识别效果，将屏幕顶部的

电池等图标和标题内容隐藏。文件 mainActivity.java 的具体实现代码如下：

```
public class mainActivity extends Activity{
    private GestureDetector mGestureDetector;;
    @Override
    public void onCreate(Bundle savedInstanceState) {
        super.onCreate(savedInstanceState);
        mGestureDetector = new GestureDetector(this, new MyGestureListener());

        if(getRequestedOrientation()!=ActivityInfo.SCREEN_ORIENTATION_LANDSCAPE){
        setRequestedOrientation(ActivityInfo.SCREEN_ORIENTATION_LANDSCAPE);
        }

        this.getWindow().setFlags(WindowManager.LayoutParams.FLAG_FULLSCREEN,   WindowManager.
LayoutParams.FLAG_FULLSCREEN);
        //隐去电池等图标和一切修饰部分（状态栏部分）
        this.requestWindowFeature(Window.FEATURE_NO_TITLE);
        //隐去标题栏（程序的名字）
        setContentView(new MyView(this));
    }

    @Override
    public boolean onTouchEvent(MotionEvent event) {
      if (mGestureDetector.onTouchEvent(event))
            return true;
      else
            return false;
    }
}
```

## 7.4.3　监听触摸屏幕中各种常用的手势

编写文件 MyGestureListener.java，功能是监听触摸屏幕中各种常用的手势，具体实现代码如下：

```
public class MyGestureListener extends SimpleOnGestureListener implements
        OnGestureListener {
    @Override
    public boolean onDoubleTap(MotionEvent e) {
        MyView.x=e.getX();
        MyView.y=e.getY();
        return super.onDoubleTap(e);
    }
    @Override
    public boolean onDoubleTapEvent(MotionEvent e) {
        MyView.x=e.getX();
        MyView.y=e.getY();
        return super.onDoubleTapEvent(e);
    }
    @Override
    public boolean onDown(MotionEvent e) {
        MyView.x=e.getX();
        MyView.y=e.getY();
```

```
            return super.onDown(e);
        }
        @Override
        public boolean onFling(MotionEvent e1, MotionEvent e2, float velocityX,
                float velocityY) {
            MyView.x=e2.getX();
            MyView.y=e2.getY();
            return super.onFling(e1, e2, velocityX, velocityY);
        }
        @Override
        public void onLongPress(MotionEvent e) {
            MyView.x=e.getX();
            MyView.y=e.getY();
            super.onLongPress(e);
        }
        @Override
        public boolean onScroll(MotionEvent e1, MotionEvent e2, float distanceX,
                float distanceY) {
            MyView.x=e2.getX();
            MyView.y=e2.getY();
            return super.onScroll(e1, e2, distanceX, distanceY);
        }
        @Override
        public void onShowPress(MotionEvent e) {
            super.onShowPress(e);
        }
        @Override
        public boolean onSingleTapConfirmed(MotionEvent e) {
            MyView.x=e.getX();
            MyView.y=e.getY();
            return super.onSingleTapConfirmed(e);
        }
        @Override
        public boolean onSingleTapUp(MotionEvent e) {
            return super.onSingleTapUp(e);
        }
}
```

### 7.4.4 根据监听到的用户手势创建视图

编写文件 MyView.java，功能是根据监听到的用户手势创建不同的视图。文件 MyView.java 的具体实现代码如下：

```
public class MyView extends SurfaceView implements SurfaceHolder.Callback {
    SurfaceHolder holder;
    static float x;
    static float y;
    public MyView(Context context) {
        super(context);
        holder = this.getHolder();//获取 holder
        holder.addCallback(this);
```

```
    }

    @Override
    public void surfaceChanged(SurfaceHolder holder, int format, int width,
            int height) {

    }

    @Override
    public void surfaceCreated(SurfaceHolder holder) {
        new Thread(new MyThread()).start();
    }

    @Override
    public void surfaceDestroyed(SurfaceHolder holder) {

    }
    public class MyThread implements Runnable {
        Paint paint=new Paint();
        @Override
        public void run() {
            while(true){
                Canvas canvas = holder.lockCanvas(null);//获取画布
                paint.setColor(Color.BLACK);
                canvas.drawRect(0, 0, 320, 480, paint); //竖屏
                canvas.drawRect(0, 0, 480, 320, paint);
                paint.setColor(Color.GREEN);
                canvas.drawRect(x-5, y-5, x+5, y+5, paint);

                holder.unlockCanvasAndPost(canvas);//解锁画布，提交画好的图像
                try {
                    Thread.sleep(100);
                } catch (InterruptedException e) {
                    e.printStackTrace();
                }
            }
        }
    }
}
```

至此，整个实例介绍完毕，执行后的效果如图 7-4 所示。在真机中运行后，会实现手势识别功能。

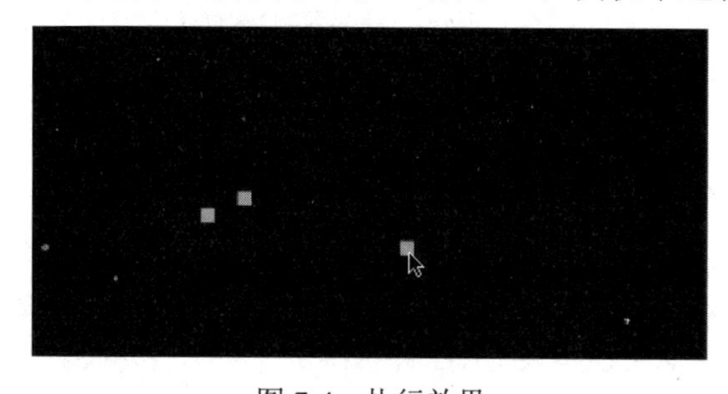

图 7-4 执行效果

# 7.5　实战演练——实现手势翻页效果

**知识点讲解：光盘:视频\知识点\第 7 章\实现手势翻页效果.avi**

本节将通过一个具体实例的实现过程，讲解在 Android 系统中实现手势翻页效果的方法和具体实现流程。

| 实　例 | 功　能 | 源 码 路 径 |
|---|---|---|
| 实例 7-3 | 在屏幕中实现手势翻页效果 | 光盘:\daima\7\MoveViewEX |

## 7.5.1　布局文件 main.xml

布局文件 main.xml 非常简单，具体实现代码如下：

```
<LinearLayout xmlns:android="http://schemas.android.com/apk/res/android"
    android:orientation="vertical"
    android:layout_width="fill_parent"
    android:layout_height="fill_parent"
    >
<TextView
    android:layout_width="fill_parent"
    android:layout_height="wrap_content"
    android:text="@string/hello"
    />
</LinearLayout>
```

## 7.5.2　监听手势

编写程序文件 MyViewGroup.java，根据用户触摸屏幕的操作来响应手势，通过 ViewFlipper 变化当前的显示内容，通过 GestureDetector 监听手势实现多页滑动展示效果。文件 MyViewGroup.java 的具体实现代码如下：

```
public class MyViewGroup extends ViewGroup implements OnGestureListener {

    private float mLastMotionY;//最后点击的点
    private GestureDetector detector;
    int move = 0;//移动距离
    int MAXMOVE = 850;//最大允许的移动距离
    private Scroller mScroller;
    int up_excess_move = 0;//往上多移的距离
    int down_excess_move = 0;//往下多移的距离
    private final static int TOUCH_STATE_REST = 0;
    private final static int TOUCH_STATE_SCROLLING = 1;
    private int mTouchSlop;
    private int mTouchState = TOUCH_STATE_REST;
    Context mContext;
```

```
public MyViewGroup(Context context) {
    super(context);
    mContext = context;
    setBackgroundResource(R.drawable.pic);
    mScroller = new Scroller(context);
    detector = new GestureDetector(this);

    final ViewConfiguration configuration = ViewConfiguration.get(context);
    //获得可以认为是滚动的距离
    mTouchSlop = configuration.getScaledTouchSlop();

    //添加子 View
    for (int i = 0; i < 48; i++) {
        final Button      MButton = new Button(context);
        MButton.setText("" + (i + 1));
        MButton.setOnClickListener(new OnClickListener() {

            public void onClick(View v) {
                Toast.makeText(mContext, MButton.getText(), Toast.LENGTH_SHORT).show();
            }
        });
        addView(MButton);
    }
}

@Override
public void computeScroll() {
    if (mScroller.computeScrollOffset()) {
        //返回当前滚动 X 方向的偏移
        scrollTo(0, mScroller.getCurrY());
        postInvalidate();
    }
}

@Override
public boolean onInterceptTouchEvent(MotionEvent ev) {
    final int action = ev.getAction();

    final float y = ev.getY();
    switch (ev.getAction())
    {
    case MotionEvent.ACTION_DOWN:

        mLastMotionY = y;
        mTouchState = mScroller.isFinished() ? TOUCH_STATE_REST
                : TOUCH_STATE_SCROLLING;
        break;
    case MotionEvent.ACTION_MOVE:
        final int yDiff = (int) Math.abs(y - mLastMotionY);
        boolean yMoved = yDiff > mTouchSlop;
        //判断是否是移动
```

```
                if (yMoved) {
                    mTouchState = TOUCH_STATE_SCROLLING;
                }
                break;
            case MotionEvent.ACTION_UP:
                mTouchState = TOUCH_STATE_REST;
                break;
            }
            return mTouchState != TOUCH_STATE_REST;
        }

        @Override
        public boolean onTouchEvent(MotionEvent ev) {

            // final int action = ev.getAction();

            final float y = ev.getY();
            switch (ev.getAction())
            {
            case MotionEvent.ACTION_DOWN:
                if (!mScroller.isFinished()) {
                    mScroller.forceFinished(true);
                    move = mScroller.getFinalY();
                }
                mLastMotionY = y;
                break;
            case MotionEvent.ACTION_MOVE:
                if (ev.getPointerCount() == 1) {

                    //随手指拖动的代码
                    int deltaY = 0;
                    deltaY = (int) (mLastMotionY - y);
                    mLastMotionY = y;
                    Log.d("move", "" + move);
                    if (deltaY < 0) {
                        //下移
                        //判断上移是否滑过头
                        if (up_excess_move == 0) {
                            if (move > 0) {
                                int move_this = Math.max(-move, deltaY);
                                move = move + move_this;
                                scrollBy(0, move_this);
                            } else if (move == 0) {//如果已经是最顶端，继续往下拉
                                Log.d("down_excess_move", "" + down_excess_move);
                                down_excess_move = down_excess_move - deltaY / 2;//记录下多往下拉
的值

                                scrollBy(0, deltaY / 2);
                            }
                        } else if (up_excess_move > 0)//之前有上移过头
                        {
                            if (up_excess_move >= (-deltaY)) {
```

```
                        up_excess_move = up_excess_move + deltaY;
                        scrollBy(0, deltaY);
                    } else {
                        up_excess_move = 0;
                        scrollBy(0, -up_excess_move);
                    }
                }
            } else if (deltaY > 0) {
                //上移
                if (down_excess_move == 0) {
                    if (MAXMOVE - move > 0) {
                        int move_this = Math.min(MAXMOVE - move, deltaY);
                        move = move + move_this;
                        scrollBy(0, move_this);
                    } else if (MAXMOVE - move == 0) {
                        if (up_excess_move <= 100) {
                            up_excess_move = up_excess_move + deltaY / 2;
                            scrollBy(0, deltaY / 2);
                        }
                    }
                } else if (down_excess_move > 0) {
                    if (down_excess_move >= deltaY) {
                        down_excess_move = down_excess_move - deltaY;
                        scrollBy(0, deltaY);
                    } else {
                        down_excess_move = 0;
                        scrollBy(0, down_excess_move);
                    }
                }
            }
        }
        break;
    case MotionEvent.ACTION_UP:
        //多滚是负数记录到 move 中
        if (up_excess_move > 0) {
            //多滚了要弹回去
            scrollBy(0, -up_excess_move);
            invalidate();
            up_excess_move = 0;
        }
        if (down_excess_move > 0) {
            //多滚了要弹回去
            scrollBy(0, down_excess_move);
            invalidate();
            down_excess_move = 0;
        }
        mTouchState = TOUCH_STATE_REST;
        break;
    }
    return this.detector.onTouchEvent(ev);
}
```

```
int Fling_move = 0;

public boolean onFling(MotionEvent e1, MotionEvent e2, float velocityX,
        float velocityY) {
    //随手指快速拨动的代码
    Log.d("onFling", "onFling");
    if (up_excess_move == 0 && down_excess_move == 0) {

        int slow = -(int) velocityY * 3 / 4;
        mScroller.fling(0, move, 0, slow, 0, 0, 0, MAXMOVE);
        move = mScroller.getFinalY();
        computeScroll();
    }
    return false;
}

public boolean onDown(MotionEvent e) {
    return true;
}

public boolean onScroll(MotionEvent e1, MotionEvent e2, float distanceX,
        float distanceY) {
    return false;
}

public void onShowPress(MotionEvent e) {
}

public boolean onSingleTapUp(MotionEvent e) {
    return false;
}

public void onLongPress(MotionEvent e) {
}

@Override
protected void onLayout(boolean changed, int l, int t, int r, int b) {
    int childTop = 0;
    int childLeft = 0;
    final int count = getChildCount();
    for (int i = 0; i < count; i++) {
        final View child = getChildAt(i);
        if (child.getVisibility() != View.GONE) {
            child.setVisibility(View.VISIBLE);
            child.measure(r - l, b - t);
            child
                    .layout(childLeft, childTop, childLeft + 80,
                            childTop + 80);
            if (childLeft < 160) {
                childLeft += 80;
```

```
                    } else {
                        childLeft = 0;
                        childTop += 80;
                    }
                }
            }
        }
}
```

执行之后将实现翻页效果，如图 7-5 所示。

图 7-5　执行效果

# 第8章　在物联网设备中处理多媒体数据

从 Android 2.2 版本以后，Android 对多媒体框架进行了很大的调整，抛弃了原来的 OpenCore 框架，改用 StageFright 框架。和 OpenCore 框架相比，StageFright 框架的最突出优点是封装简单，功能强大。在 Android 2.2 及以前，OpenCore 位于 external 目录下，在 Android 2.3 以后，多媒体的功能被放置到 frameworks/base/media 目录下。本章将详细讲解 OpenCore 框架和 StageFright 框架的基本知识，详细讲解在 Android 物联网设备中播放音乐、录制声音、实现振动、设置铃音的方法。

## 8.1　Android 多媒体系统架构基础

**知识点讲解：光盘:视频\知识点\第 8 章\Android 多媒体系统架构基础.avi**

Android 多媒体引擎和插件的基本层次结构如图 8-1 所示。

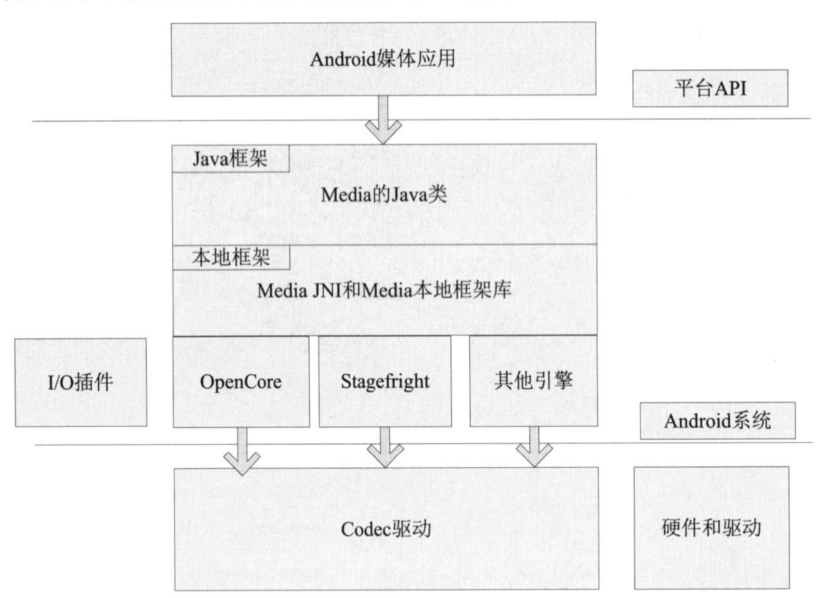


图 8-1　Android 多媒体引擎和插件的基本层次

Android 系统的多媒体框架结构如图 8-2 所示。

从多媒体应用的实现角度来看，多媒体系统主要包含如下两方面的内容。

（1）输入、输出环节：音频、视频纯数据流的输入、输出系统。

（2）中间处理环节：包括文件格式处理和编码/解码环节处理。

假如想要处理一个 MP3 文件，媒体播放器的处理流程是：将一个 MP3 格式的文件作为播放器的输入，将声音从播放器设备输出。在具体实现上，MP3 播放器经过了 MP3 格式文件解析、MP3 码流解码和 PCM 输出播放的过程。整个过程如图 8-3 所示。

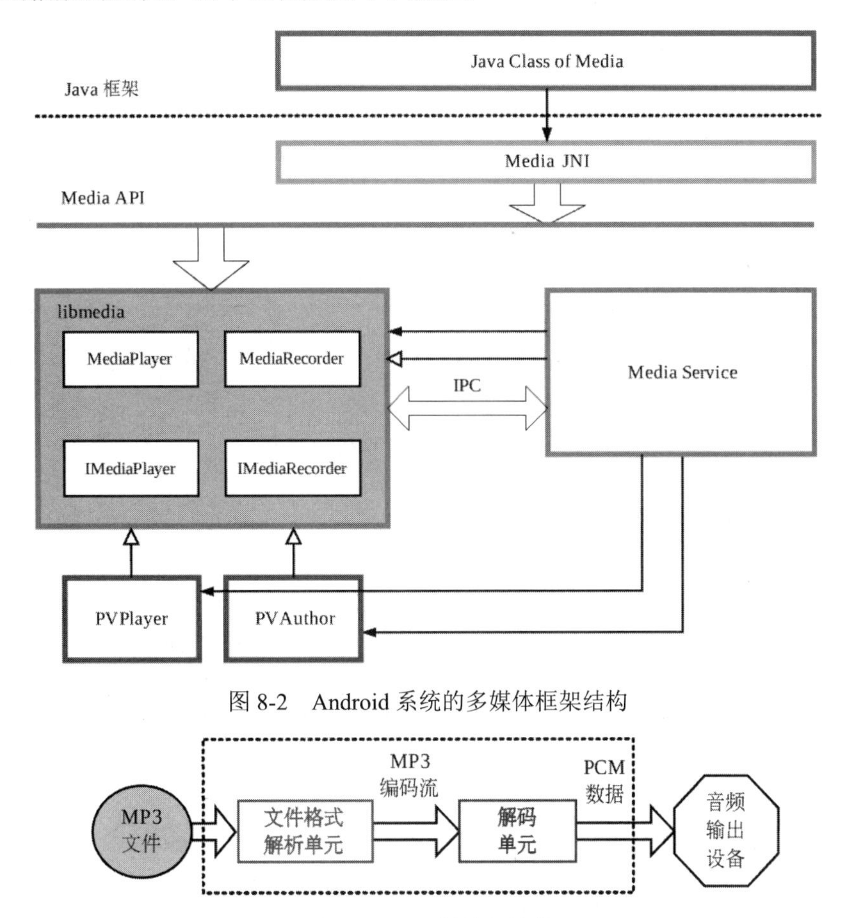


图 8-2　Android 系统的多媒体框架结构

图 8-3　MP3 播放器结构

## 8.1.1　OpenMax 框架介绍

OpenMax 是 NVIDIA 公司和 Khronos 在 2006 年联合推出的产品，是一款多媒体应用程序的框架标准。OpenMax 通过使用媒体加速组件，能够在开发、集成和编程环节中实现跨多操作系统和处理器硬件平台，提供了全面的流媒体编码/解码器功能和应用程序便携化处理。

OpenMax 的官方网站地址是 http://www.khronos.org/openmax/。

OpenMax 是一个多媒体应用程序的框架标准。其中，OpenMax IL（集成层）技术规格定义了媒体组件接口，以便在嵌入式器件的流媒体框架中快速集成加速编码/解码器。

在 Android 中，OpenMax IL 层通常被用在多媒体引擎插件中，Android 的多媒体引擎 OpenCore 和 StageFright 都可以使用 OpenMax 作为插件，主要用于编码和解码（Codec）处理。

在 Android 的框架层中定义了由 Android 封装的 OpenMax 接口，此接口和标准的接口类似，但是使用的是 C++类型接口，并且使用了 Android 的 Binder IPC 机制。StageFright 使用 Android 封装的 OpenMax 接口，OpenCore 没有使用此接口，而是使用其他形式对 OpenMax IL 层接口进行封装。

Android 中 OpenMax 的基本层次结构如图 8-4 所示。

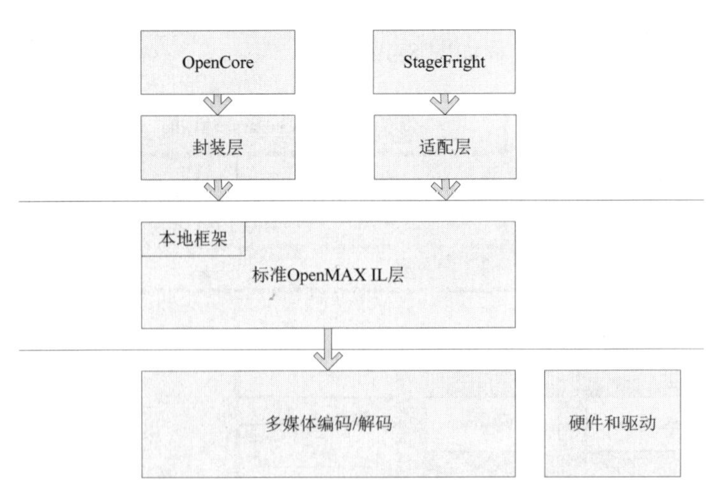


图 8-4　OpenMax 多媒体框架的层次结构

### 8.1.2　OpenCore 框架介绍

OpenCore 的另外一个常用的称呼是 PacketVideo，它是 Android 的多媒体核心。其实 PacketVideo 是一家公司的名称，而 OpenCore 是这套多媒体框架的软件层的名称。OpenCore 和其他 Android 程序库相比，OpenCore 的代码十分庞大，它是基于 C++实现的，在里面定义了全功能的操作系统移植层，各种基本的功能均被封装成类的形式，各层次之间的接口多使用继承等方式。OpenCore 的基本结构如图 8-5 所示。

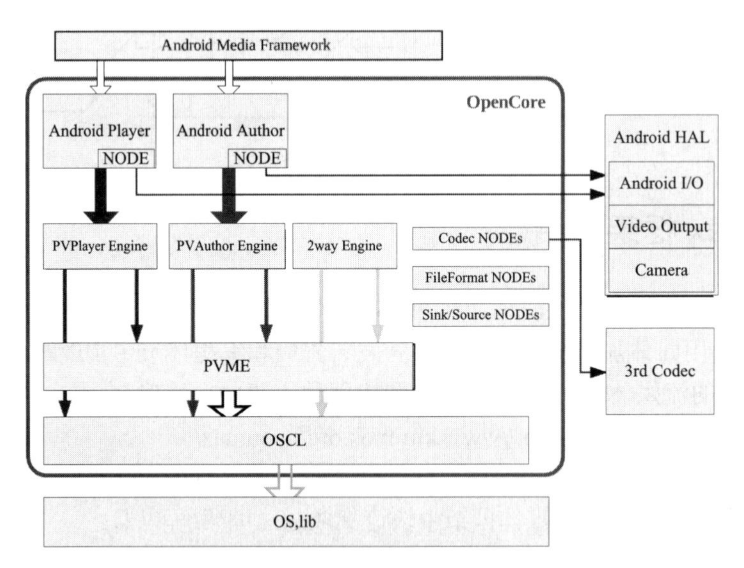


图 8-5　OpenCore 的层次结构

OpenCore 是一个多媒体的框架，主要包含如下所示的两方面内容。

☑　PVPlayer：提供了媒体播放器的功能，可以完成各种音频（Audio）、视频（Video）流的回放（Playback）功能。

☑　PVAuthor：提供媒体流记录的功能，可以完成各种音频（Audio）、视频（Video）流以及静态图像捕获功能。

PVPlayer 和 PVAuthor 以 SDK 的形式提供给开发者，可以在这个 SDK 之上构建多种应用程序和服务。在移动终端中常常使用的多媒体应用程序，例如媒体播放器、照相机、录像机、录音机等。

在图 8-5 所示的结构中，主要层次元素的介绍如下。

（1）OSCL

OSCL 是 Operating System Compatibility Library 的缩写，意为操作系统兼容库，在里面包含了一些操作系统底层的操作，为了更好地在不同操作系统之间移植。包含了基本数据类型、配置、字符串工具、I/O、错误处理、线程等内容，类似一个基础的 C++库。

（2）PVMF

PVMF 是 Packet Video Multimedia Framework 的缩写，意为 PV 多媒体框架，可以在框架内实现一个文件解析（parser）和组成（composer）、编解码的 NODE，也可以继承其通用的接口，在用户层实现一些 NODE。

（3）PVPlayer Engine

PVPlayer Engine 是 PVPlayer 引擎。

（4）PVAuthor Engine

PVAuthor Engine 是 PVAuthor 引擎。

除了上述 4 个元素外，其实在 OpenCore 中包含的内容还有很多。从播放的角度看，PVPlayer 输入的（Source）是文件或者网络媒体流，输出（Sink）的是音频/视频的输出设备，其基本功能包含了媒体流控制、文件解析、音频/视频流的解码（Decode）等方面的内容。除了从文件中播放媒体文件之外，还包含了与网络相关的 RTSP 流（Real Time Stream Protocol，实时流协议）。在媒体流记录方面，PVAuthor 的输入（Source）是照相机、麦克风等设备，输出（Sink）是各种文件，包含了流的同步、音频/视频流的编码（Encode）以及文件的写入等功能。

在使用 OpenCore SDK 时，有可能需要在应用程序层实现一个适配器（Adaptor），然后在适配器上实现具体的功能，对于 PVMF 的 NODE 也可以基于通用的接口，在上层实现，以插件的形式使用。

### 8.1.3　StageFright 框架介绍

从 Android 2.0 开始，Google 开始引入了架构稍微简单的 StageFright，并且从 Android 2.2 开始，几乎完全放弃了 OpenCore，转向主推 StageFright。

（1）StageFright 代码结构

StageFright 是一个轻量级的多媒体框架，其主要功能是基于 OpenMax 实现的。在 StageFright 中提供了媒体播放等接口，这些接口可以为 Android 框架层所使用。

在 Android 开源代码中，StageFright 的头文件路径是 frameworks/base/include/media/stagefright/。

实现 StageFright 功能的文件路径是 frameworks/base/media/libstagefright/。

实现 StageFright 播放器和录音器功能的文件路径是 frameworks/base/media/libmediaplayerservice/。

测试 StageFright 功能的代码路径是 frameworks/base/cmds/stagefright/。

（2）StageFright 实现 OpenMax 接口

StageFright 可以实现 Android 系统中的 OpenMax 接口，可以让 StageFright 引擎内的 OMXCode 调用实现的 OpenMax 接口，最终目的是使用 OpenMax IL 编码/解码功能。

# 8.2 Graphics 类详解

**知识点讲解：光盘:视频\知识点\第 8 章\Graphics 类详解.avi**

类 Graphics 是一个全能的绘图类，不但可以绘制 2D 图像，而且可以轻松地为这些图像填充不同的颜色。为什么 Graphics 这么强呢？原因是它有很多子类，通过这些子类可以实现不同的功能。本节将详细讲解在 Android 系统中使用 Graphics 类处理二维图像的知识，为读者步入本书后面知识的学习打下基础。

## 8.2.1 Graphics 类基础

在类 Graphics 中有如下 10 个非常重要的子类。

☑ Color 类
☑ Paint 类
☑ Canvas 画布
☑ Rect 矩形类
☑ NinePatch 类
☑ Matrix 类
☑ Bitmap 类
☑ BitmapFactory 类
☑ Typeface 类
☑ Shader 类

## 8.2.2 使用 Graphics 类

在 Android 系统中，类 Rect 的完整形式是 Android.Graphics.Rect，表示矩形区域。类 Rect 除了能够表示一个矩形区域位置描述外，还可以帮助计算图形之间是否发生碰撞关系，这一点对于 Android 游戏开发比较有用。在类 Rect 的方法成员中，主要通过如下 3 种重载方法来判断包含关系。

```
boolean contains(int left, int top, int right, int bottom)
boolean contains(int x, int y)
boolean contains(Rect r)
```

在上述构造方法中包含了 4 个参数，即 left、top、right 和 bottom，分别代表左、上、右、下 4 个方向，具体说明如下所示。

☑ left：矩形区域中左边的 X 坐标。
☑ top：矩形区域中顶部的 Y 坐标。
☑ right：矩形区域中右边的 X 坐标。
☑ bottom：矩形区域中底部的 Y 坐标。

例如下面代码的含义是，左上角的坐标是（150,75），右下角的坐标是（260,120）。

```
Rect(150, 75, 260, 120)
```

在 Android 系统中，另外一个矩形类是 RectF，此类和类 Rect 的用法几乎完全相同。两者的区别是精度不一样，Rect 是使用 int 类型作为数值，RectF 是使用 float 类型作为数值。在类 RectF 中包含了一个矩形的 4 个单精度浮点坐标，通过上、下、左、右 4 个边的坐标来表示一个矩形。这些坐标值属性可以被直接访问，使用 width 和 height 方法可以获取矩形的宽和高。

类 Rect 和类 RectF 提供的方法也不是完全一致，类 RectF 提供了如下所示的构造方法。

☑ RectF()：功能是构造一个无参的矩形。

☑ RectF(float left,float top,float right,float bottom)：功能是构造一个指定了 4 个参数的矩形。

☑ RectF(Rect F r)：功能是根据指定的 RectF 对象来构造一个 RectF 对象（对象的左边坐标不变）。

☑ RectF(Rect r)：功能是根据给定的 Rect 对象来构造一个 RectF 对象。

另外在类 RectF 中还提供了很多功能强大的方法，具体说明如下所示。

☑ Public Boolean contain(RectF r)：功能是判断一个矩形是否在此矩形内，如果在这个矩形内或者和这个矩形等价则返回 true，同样类似的方法还有 public Boolean contain(float left,float top,float right,float bottom)和 public Boolean contain(float x,float y)。

☑ Public void union(float x,float y)：功能是更新这个矩形，使它包含矩形自己和（x,y）这个点。

下面将通过一个具体的演示实例来讲解在 Android 中使用 Canvas 类的方法。

| 实　例 | 功　能 | 源码路径 |
|---|---|---|
| 实例 8-1 | 在 Android 设备中使用矩形类 Rect 和 RectF | 光盘:\daima\8\RectEX |

实例文件 RectL.java 的主要实现代码如下：

```
/*声明 Paint 对象*/
private Paint mPaint = null;
        private RectL_1 mGameView2 = null;
public RectL(Context context)
{
        super(context);
        /*构建对象*/
        mPaint = new Paint();

        mGameView2 = new RectL_1(context);

        /*开启线程*/
        new Thread(this).start();
}

public void onDraw(Canvas canvas)
{
        super.onDraw(canvas);

        /*设置画布为黑色背景*/
        canvas.drawColor(Color.BLACK);
        /*取消锯齿*/
        mPaint.setAntiAlias(true);

        mPaint.setStyle(Paint.Style.STROKE);
```

```
{
    /*定义矩形对象*/
    Rect rect1 = new Rect();
    /*设置矩形大小*/
    rect1.left = 5;
    rect1.top = 5;
    rect1.bottom = 25;
    rect1.right = 45;

    mPaint.setColor(Color.BLUE);
    /*绘制矩形*/
    canvas.drawRect(rect1, mPaint);

    mPaint.setColor(Color.RED);
    /*绘制矩形*/
    canvas.drawRect(50, 5, 90, 25, mPaint);

    mPaint.setColor(Color.YELLOW);
    /*绘制圆形(圆心 x,圆心 y,半径 r,p)*/
    canvas.drawCircle(40, 70, 30, mPaint);

    /*定义椭圆对象*/
    RectF rectf1 = new RectF();
    /*设置椭圆大小*/
    rectf1.left = 80;
    rectf1.top = 30;
    rectf1.right = 120;
    rectf1.bottom = 70;

    mPaint.setColor(Color.LTGRAY);
    /*绘制椭圆*/
    canvas.drawOval(rectf1, mPaint);

    /*绘制多边形*/
    Path path1 = new Path();

    /*设置多边形的点*/
    path1.moveTo(150+5, 80-50);
    path1.lineTo(150+45, 80-50);
    path1.lineTo(150+30, 120-50);
    path1.lineTo(150+20, 120-50);
    /*使这些点构成封闭的多边形*/
    path1.close();

    mPaint.setColor(Color.GRAY);
    /*绘制这个多边形*/
    canvas.drawPath(path1, mPaint);

    mPaint.setColor(Color.RED);
    mPaint.setStrokeWidth(3);
    /*绘制直线*/
    canvas.drawLine(5, 110, 315, 110, mPaint);
```

```
		}
		//绘制实心几何体
		mPaint.setStyle(Paint.Style.FILL);
		{
			/*定义矩形对象*/
			Rect rect1 = new Rect();
			/*设置矩形大小*/
			rect1.left = 5;
			rect1.top = 130+5;
			rect1.bottom = 130+25;
			rect1.right = 45;
			mPaint.setColor(Color.BLUE);
			/*绘制矩形*/
			canvas.drawRect(rect1, mPaint);

			mPaint.setColor(Color.RED);
			/*绘制矩形*/
			canvas.drawRect(50, 130+5, 90, 130+25, mPaint);
			mPaint.setColor(Color.YELLOW);
			/*绘制圆形(圆心 x,圆心 y,半径 r,p)*/
			canvas.drawCircle(40, 130+70, 30, mPaint);
			/*定义椭圆对象*/
			RectF rectf1 = new RectF();
			/*设置椭圆大小*/
			rectf1.left = 80;
			rectf1.top = 130+30;
			rectf1.right = 120;
			rectf1.bottom = 130+70;
			mPaint.setColor(Color.LTGRAY);
			/*绘制椭圆*/
			canvas.drawOval(rectf1, mPaint);
			/*绘制多边形*/
			Path path1 = new Path();
			/*设置多边形的点*/
			path1.moveTo(150+5, 130+80-50);
			path1.lineTo(150+45, 130+80-50);
			path1.lineTo(150+30, 130+120-50);
			path1.lineTo(150+20, 130+120-50);
			/*使这些点构成封闭的多边形*/
			path1.close();
			mPaint.setColor(Color.GRAY);
			/*绘制这个多边形*/
			canvas.drawPath(path1, mPaint);
			mPaint.setColor(Color.RED);
			mPaint.setStrokeWidth(3);
			/*绘制直线*/
			canvas.drawLine(5, 130+110, 315, 130+110, mPaint);
		}
		/*通过 ShapeDrawable 来绘制几何图形*/
		mGameView2.DrawShape(canvas);
	}
	//触笔事件
```

```
    public boolean onTouchEvent(MotionEvent event)
    {
        return true;
    }
    //按键按下事件
    public boolean onKeyDown(int keyCode, KeyEvent event)
    {
        return true;
    }
    //按键弹起事件
    public boolean onKeyUp(int keyCode, KeyEvent event)
    {
        return false;
    }
    public boolean onKeyMultiple(int keyCode, int repeatCount, KeyEvent event)
    {
        return true;
    }
    public void run()
    {
        while (!Thread.currentThread().isInterrupted())
        {
            try
            {
                Thread.sleep(100);
            }
            catch (InterruptedException e)
            {
                Thread.currentThread().interrupt();
            }
            //使用 postInvalidate 可以直接在线程中更新界面
            postInvalidate();
        }
    }
}
```

执行后的效果如图 8-6 所示。

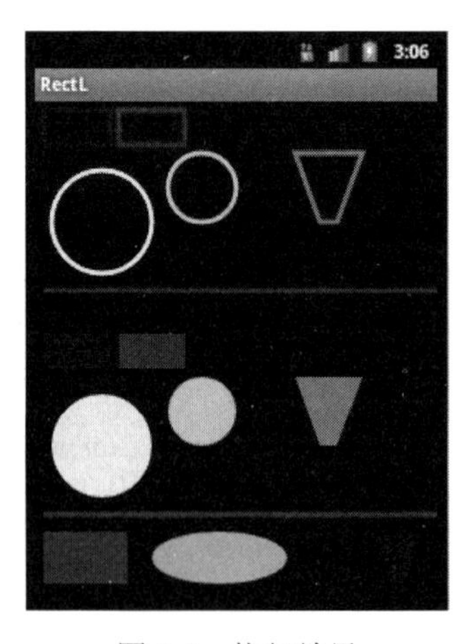


图 8-6　执行效果

# 8.3　二 维 动 画

**知识点讲解：光盘:视频\知识点\第 8 章\二维动画.avi**

在多媒体领域中，具有视觉冲击力的动画是程序员们永远谈论的话题之一。动画和简单的图像相比，更具有震撼性的效果。Android 提供了一套完整的动画框架，使得开发者可以用它来开发各种动画效果。

## 8.3.1　类 Drawable

Android SDK 提供了一个功能强大的类——Drawable，虽然 Drawable 是一个很抽象的概念，但是可以实现动画效果。市面中有很多知名度颇高的动画工具，例如 Flash，并且在手机中也可以使用这些工具实现动画效果。但是身为 Android 家族的产品，它去完成和动画有关的任务，并且取得的效果还不错。就这样久而久之，Drawable 深受开发者的青睐。

为了深入了解 Drawable 的基本知识，我们先通过一个简单的例子来认识它。在这个例子中，使用 Drawable 的子类 ShapeDrawable 绘制一幅图，具体实现流程如下：

（1）创建一个 OvalShape（椭圆）。

（2）使用刚创建的 OvalShape 构造一个 ShapeDrawable 对象 mDrawable。

（3）设置 mDrawable 的颜色。

（4）设置 mDrawable 的大小。

（5）将 mDrawable 画在 testViewCH 的画布上。

本例的具体代码如下：

```
public class testViewCH extends View {
private ShapeDrawable mDrawable;
public testViewCH(Context context) {
super(context);
int x = 10;
int y = 10;
int width = 300;
int height = 50;
mDrawable = new ShapeDrawable(new OvalShape());
mDrawable.getPaint().setColor(0xff74AC23);
mDrawable.setBounds(x, y, x + width, y + height);
}
protected void onDraw(Canvas canvas)
super.onDraw(canvas);
canvas.drawColor(Color.WHITE);//画白色背景
mDrawable.draw(canvas);
}
}
```

上述代码的执行效果如图 8-7 所示。

图 8-7　执行效果

例子虽然简单，但是却让我们明白了 Drawable 就是一个可画的对象，可能是一张位图（BitmapDrawable），也可能是一个图形（ShapeDrawable），还有可能是一个图层（LayerDrawable）。在项目中可以根据画图的需求，创建相应的可画对象，这样就可以将这个可画对象当作一块“画布（Canvas）”，在其上面操作可画对象，并最终将这种可画对象显示在画布上，有点类似于“内存画布”。

## 8.3.2　实现 Tween 动画效果

通过本章前面内容学习，了解了 Drawable 可以实现动画效果。其实 Drawable 的功能何止如此，它更加强大的功能是可以显示 Animation。在 Android SDK 中提供了如下两种 Animation。

（1）Tween Animation：通过对场景中的对象不断做平移、缩放、旋转等变换来产生动画效果。

（2）Frame Animation：和电影似的顺序播放事先做好的图像。

由此可见，Android 平台提供了如下两类动画。

☑ Tween 动画：用于对场景中的对象不断进行图像变换来产生动画效果，可以把对象进行缩小、放大、旋转和渐变等操作。

☑ Frame 动画：用于顺序播放事先做好的图像。

在使用 Animation 前需要学会定义 Animation 的方法，Animation 是以 XML 格式定义的，定义好的 XML 文件存放在 res\anim 目录中。Tween Animation 与 Frame Animation 的定义、使用都有很大的差异。

Tween 动画通过对 View 的内容完成一系列的图形变换，通过平移、缩放、旋转、改变透明度来实现动画效果。在 XML 文件中，Tween 动画主要包括以下 4 种动画效果。

☑ Alpha：渐变透明度动画效果。

☑ Scale：渐变尺寸伸缩动画效果。

☑ Translate：画面转移位置移动动画效果。

☑ Rotate：画面转移旋转动画效果。

在 Java 代码中，Tween 动画对应以下 4 种动画效果。

☑ AlphaAnimation：渐变透明度动画效果。

☑ ScaleAnimation：渐变尺寸伸缩动画效果。

☑ TranslateAnimation：画面转换位置移动动画效果。

☑ RotateAnimation：画面转移旋转动画效果。

Tween 动画是通过预先定义一组指令，这些指令指定了图形变换的类型、触发时间和持续时间。程序沿着时间线执行这些指令就可以实现动画效果。我们可以首先定义 Animation 动画对象，然后设置该动画的一些属性，最后通过 startAnimation()方法开始动画效果。

| 实　例 | 功　能 | 源码路径 |
| --- | --- | --- |
| 实例 8-2 | 在 Android 中实现 Tween 动画效果 | 光盘:\daima\8\TweenEX |

实例文件 TweenCH.java 的主要代码如下：

```
/*定义 Alpha 动画*/
```

```
private Animation mAnimationAlpha = null;

/*定义 Scale 动画*/
private Animation mAnimationScale = null;

/*定义 Translate 动画*/
private Animation mAnimationTranslate = null;

/*定义 Rotate 动画*/
private Animation mAnimationRotate= null;

/*定义 Bitmap 对象*/
Bitmap mBitQQ= null;

public example9(Context context)
{
	super(context);

	/*装载资源*/
	mBitQQ = ((BitmapDrawable) getResources().getDrawable(R.drawable.qq)).getBitmap();
}

public void onDraw(Canvas canvas)
{
	super.onDraw(canvas);

	/*绘制图片*/
	canvas.drawBitmap(mBitQQ, 0, 0, null);
}

public boolean onKeyUp(int keyCode, KeyEvent event)
{
	switch ( keyCode )
	{
	case KeyEvent.KEYCODE_DPAD_UP:
		/*创建 Alpha 动画*/
		mAnimationAlpha = new AlphaAnimation(0.1f, 1.0f);
		/*设置动画的时间*/
		mAnimationAlpha.setDuration(3000);
		/*开始播放动画*/
		this.startAnimation(mAnimationAlpha);
		break;
	case KeyEvent.KEYCODE_DPAD_DOWN:
		/*创建 Scale 动画*/
		mAnimationScale =new ScaleAnimation(0.0f, 1.0f, 0.0f, 1.0f,
									Animation.RELATIVE_TO_SELF, 0.5f,
									Animation.RELATIVE_TO_SELF, 0.5f);
		/*设置动画的时间*/
		mAnimationScale.setDuration(500);
		/*开始播放动画*/
		this.startAnimation(mAnimationScale);
```

```
                break;
            case KeyEvent.KEYCODE_DPAD_LEFT:
                /*创建 Translate 动画*/
                mAnimationTranslate = new TranslateAnimation(10, 100,10, 100);
                /*设置动画的时间*/
                mAnimationTranslate.setDuration(1000);
                /*开始播放动画*/
                this.startAnimation(mAnimationTranslate);
                break;
            case KeyEvent.KEYCODE_DPAD_RIGHT:
                /*创建 Rotate 动画*/
                mAnimationRotate=new RotateAnimation(0.0f, +360.0f,
                                                  Animation.RELATIVE_TO_SELF,0.5f,
                                                  Animation.RELATIVE_TO_SELF, 0.5f);
                /*设置动画的时间*/
                mAnimationRotate.setDuration(1000);
                /*开始播放动画*/
                this.startAnimation(mAnimationRotate);
                break;
        }
        return true;
    }
}
```

执行后可以通过键盘的上、下、左、右键实现动画效果，如图 8-8 所示。

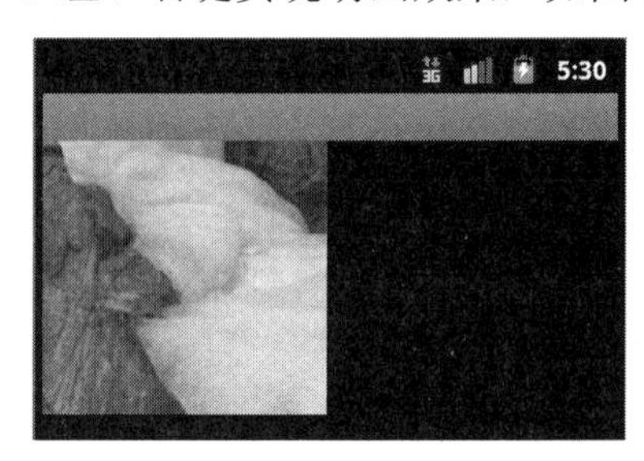


图 8-8　执行效果

## 8.3.3　实现 Frame 动画效果

在 Android SDK 开发应用中，可以使用类 AnimationDrawable 来定义并使用 Frame 动画。对应 Android SDK 的位置如下所示。

☑　Tween animation：android.view.animation 包。

☑　Frame animation：android.graphics.drawable.AnimationDrawable 类。

### 1．AnimationDrawable 介绍

AnimationDrawable 的功能是获取、设置动画的属性，里面最为常用的方法如下所示。

☑　int getDuration()：获取动画的时长。

☑　int getNumberOfFrames()：获取动画的帧数。

☑　boolean isOneShot()：获取 oneshot 属性。

- ☑ Void setOneShot(boolean oneshot)：设置 oneshot 属性。
- ☑ void inflate(Resurce r,XmlPullParser p,AttributeSet attrs)：增加、获取帧动画。
- ☑ Drawable getFrame(int index)：获取某帧的 Drawable 资源。
- ☑ void addFrame(Drawable frame,int duration) ：为当前动画增加帧（资源、持续时长）。
- ☑ void start()：开始动画。
- ☑ void run()：外界不能直接调用，使用 start()替代。
- ☑ boolean isRunning()：当前动画是否在运行。
- ☑ void stop()：停止当前动画。

### 2．Frame Animation 格式定义

我们既可以在 XML Resource 中定义 Frame Animation，也可以使用 AnimationDrawable 中的 API 来定义。由于 Tween Animation 和 Frame Animation 有着很大的不同，所以定义 XML 的格式也很不相同。

定义 Frame Animation 的格式是：首先是 animation-list 根节点，animation-list 根节点中包含多个 item 子节点，每个 item 节点定义一帧动画，定义当前帧的 drawable 资源和当前帧持续的时间。表 8-1 对节点中的元素进行了详细说明。

表 8-1　XML 属性元素说明

| XML 属性 | 说　明 |
| --- | --- |
| drawable | 当前帧引用的 drawable 资源 |
| duration | 当前帧显示的时间（毫秒为单位） |
| oneshot | 如果为 true，表示动画只播放一次停止在最后一帧上，如果设置为 false 表示动画循环播放 |
| variablePadding | 如果为真允许 drawable’s 根据被选择的现状而变动 |
| visible | 规定 drawable 的初始可见性，默认为 flase |

### 3．使用 Frame 动画

使用 Frame 动画的方法十分简单，只需要创建一个 AnimationDrawabledF 对象来表示 Frame 动画，然后通过 addFrame 方法把每一帧要显示的内容添加进去，最后通过 start()方法就可以播放这个动画，同时还可以通过 setOneShot()方法设置是否重复播放。

Frame 动画主要是通过 AnimationDrawable 类来实现的，用 start()和 stop()这两个重要的方法分别启动和停止动画。Frame 动画一般通过 XML 文件配置，在 Android 工程的 res/anim 目录下创建一个 XML 配置文件，该配置文件有一个<animation-list>根元素和若干个<item>子元素。

下面将通过一个具体的演示实例来讲解实现 Tween 动画的 4 种效果的方法。

| 题　目 | 目　的 | 源 码 路 径 |
| --- | --- | --- |
| 实例 8-3 | 演示 Tween 动画的 4 种动画效果 | 光盘:\daima\8\myActionAnimationEX |

本实例的具体实现流程如下：

（1）编写文件 my_alpha_action.xml，实现 Alpha 渐变透明度动画效果，主要实现代码如下：

```
<?xml version="1.0" encoding="utf-8"?>
<set xmlns:android="http://schemas.android.com/apk/res/android" >
<alpha
android:fromAlpha="0.1"
```

```
android:toAlpha="1.0"
android:duration="3000"
/>
<!-- 透明度控制动画效果 alpha
        浮点型值:
        fromAlpha 属性为动画起始时透明度
        toAlpha 属性为动画结束时透明度
        说明:
        0.0 表示完全透明
        1.0 表示完全不透明
        以上值取 0.0～1.0 之间的 float 数据类型的数字
        长整型值:
        duration 属性为动画持续时间
        说明:时间以毫秒为单位
-->
</set>
```

（2）编写文件 my_rotate_action.xml，实现 Rotate 画面转移旋转动画效果，主要实现代码如下：

```
<?xml version="1.0" encoding="utf-8"?>
<set xmlns:android="http://schemas.android.com/apk/res/android">
<rotate
        android:interpolator="@android:anim/accelerate_decelerate_interpolator"
        android:fromDegrees="0"
        android:toDegrees="+350"
        android:pivotX="50%"
        android:pivotY="50%"
        android:duration="3000" />
<!-- rotate 旋转动画效果
        属性:interpolator 指定一个动画的插入器
        在试验过程中，使用 android.res.anim 中的资源时发现有 3 种动画插入器
        accelerate_decelerate_interpolator   加速-减速动画插入器
        accelerate_interpolator              加速-动画插入器
        decelerate_interpolator              减速-动画插入器
        浮点型值:
        fromDegrees 属性为动画起始时物件的角度
        toDegrees 属性为动画结束时物件旋转的角度，可以大于 360 度
        当角度为负数，表示逆时针旋转
        当角度为正数，表示顺时针旋转
        (负数 from——to 正数:顺时针旋转)
        (负数 from——to 负数:逆时针旋转)
        (正数 from——to 正数:顺时针旋转)
        (正数 from——to 负数:逆时针旋转)
        pivotX 属性为动画相对于物件的 X 坐标的开始位置
        pivotY 属性为动画相对于物件的 Y 坐标的开始位置
        说明:以上两个属性值从 0%～100%中取值，50%为物件的 X 或 Y 方向坐标上的中点位置
        长整型值:duration 属性为动画持续时间，时间以毫秒为单位
-->
</set>
```

（3）编写文件 my_scale_action.xml，实现 Scale 渐变尺寸伸缩动画效果，主要实现代码如下：

```
<?xml version="1.0" encoding="utf-8"?>
<set xmlns:android="http://schemas.android.com/apk/res/android">
    <scale android:interpolator="@android:anim/accelerate_decelerate_interpolator"
```

```
            android:fromXScale="0.0"
            android:toXScale="1.4"
            android:fromYScale="0.0"
            android:toYScale="1.4"
            android:pivotX="50%"
            android:pivotY="50%"
            android:fillAfter="false"
            android:duration="700" />
</set>
<!-- 尺寸伸缩动画效果 scale
        属性：interpolator 指定一个动画的插入器
        有 3 种动画插入器
        accelerate_decelerate_interpolator   加速-减速动画插入器
        accelerate_interpolator              加速-动画插入器
        decelerate_interpolator              减速-动画插入器
        fromXScale 属性为动画起始时 X 坐标上的伸缩尺寸
        toXScale 属性为动画结束时 X 坐标上的伸缩尺寸
        fromYScale 属性为动画起始时 Y 坐标上的伸缩尺寸
        toYScale 属性为动画结束时 Y 坐标上的伸缩尺寸
        以上 4 种属性值

        0.0 表示收缩到没有
        1.0 表示正常无伸缩
        值小于 1.0 表示收缩
        值大于 1.0 表示放大
        pivotX 属性为动画相对于物件的 X 坐标的开始位置
        pivotY 属性为动画相对于物件的 Y 坐标的开始位置
        以上两个属性值从 0%～100%中取值，50%为物件的 X 或 Y 方向坐标上的中点位置
        duration 属性为动画持续时间，时间以毫秒为单位
        fillAfter 属性应当设置为 true，该动画转化在动画结束后被应用
-->
```

（4）编写文件 my_translate_action.xml，实现 Translate 画面转移位置移动动画效果，主要实现代码如下：

```
<?xml version="1.0" encoding="utf-8"?>
<set xmlns:android="http://schemas.android.com/apk/res/android">
<translate
android:fromXDelta="30"
android:toXDelta="-80"
android:fromYDelta="30"
android:toYDelta="300"
android:duration="2000"
/>
<!-- translate 位置转移动画效果
        fromXDelta 属性为动画起始时 X 坐标上的位置
        toXDelta 属性为动画结束时 X 坐标上的位置
        fromYDelta 属性为动画起始时 Y 坐标上的位置
        toYDelta 属性为动画结束时 Y 坐标上的位置
        没有指定 fromXType、toXType、fromYType 和 toYType 时，默认是以自己为相对参照物
        duration 属性为动画持续时间，时间以毫秒为单位
-->
</set>
```

（5）编写文件 myActionAnimation.java，使用 case 语句根据用户的选择来显示对应的动画效果，主要实现代码如下：

```
public void onCreate(Bundle savedInstanceState) {
    super.onCreate(savedInstanceState);
    setContentView(R.layout.main);
    button_alpha = (Button) findViewById(R.id.button_Alpha);
    button_alpha.setOnClickListener(this);
    button_scale = (Button) findViewById(R.id.button_Scale);
    button_scale.setOnClickListener(this);
    button_translate = (Button) findViewById(R.id.button_Translate);
    button_translate.setOnClickListener(this);
    button_rotate = (Button) findViewById(R.id.button_Rotate);
    button_rotate.setOnClickListener(this);
}
public void onClick(View button) {
    switch (button.getId()) {
    case R.id.button_Alpha: {
        myAnimation_Alpha = AnimationUtils.loadAnimation(this,R.anim.my_alpha_action);
        button_alpha.startAnimation(myAnimation_Alpha);
    }
        break;
    case R.id.button_Scale: {
        myAnimation_Scale= AnimationUtils.loadAnimation(this,R.anim.my_scale_action);
        button_scale.startAnimation(myAnimation_Scale);
    }
        break;
    case R.id.button_Translate: {
        myAnimation_Translate= AnimationUtils.loadAnimation(this,R.anim.my_translate_action);
        button_translate.startAnimation(myAnimation_Translate);
    }
        break;
    case R.id.button_Rotate: {
        myAnimation_Rotate= AnimationUtils.loadAnimation(this,R.anim.my_rotate_action);
        button_rotate.startAnimation(myAnimation_Rotate);
    }
        break;
    default:
        break;
    }
}
```

执行后的效果如图 8-9 所示。单击屏幕中的选项会显示对应的动画效果，例如单击“Translate 动画”选项后的效果如图 8-10 所示。

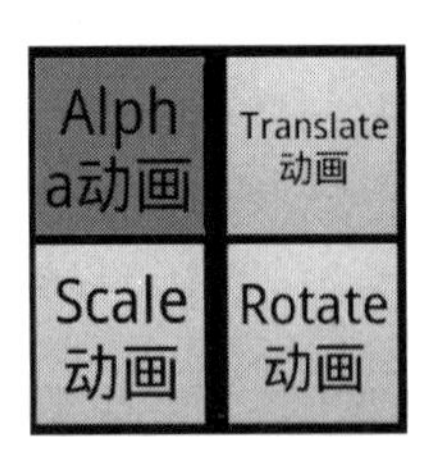


图 8-9　执行效果

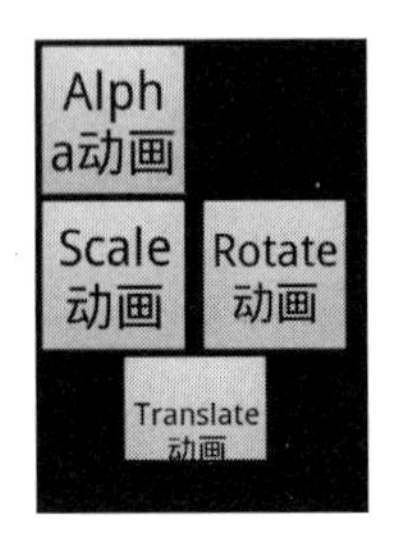


图 8-10　Translate 动画效果

# 8.4　OpenGL ES 详解

**知识点讲解：光盘:视频\知识点\第 8 章\OpenGL ES 详解.avi**

在 Android 系统中，通过 OpenGL 实现三维效果功能。本节将详细讲解 OpenGL 基础性知识，为读者步入后面知识的学习打下基础。

## 8.4.1　OpenGL ES 基础

OpenGL ES（OpenGL for Embedded Systems）是 OpenGL 三维图形 API 的子集，针对手机、PDA 和游戏主机等嵌入式设备而设计。该 API 由 Khronos 集团定义推广，Khronos 是一个图形软硬件行业协会，该协会主要关注图形和多媒体方面的开放标准。OpenGL ES 是从 OpenGL 裁剪定制而来的，去除了 glBegin/glEnd、四边形（GL_QUADS）、多边形（GL_POLYGONS）等复杂图元等许多非绝对必要的特性。经过多年发展，现在主要有两个版本，OpenGL ES 1.x 针对固定管线硬件，OpenGL ES 2.x 针对可编程管线硬件。OpenGL ES 1.0 是以 OpenGL 1.3 规范为基础的，OpenGL ES 1.1 是以 OpenGL 1.5 规范为基础的，它们分别又支持 common 和 common lite 两种 profile。lite profile 只支持定点实数，而 common profile 既支持定点数又支持浮点数。OpenGL ES 2.0 则是参照 OpenGL 2.0 规范定义的，common profile 发布于 2005 年 8 月，引入了对可编程管线的支持。

OpenGL ES 的最大意义是作为嵌入式 3D 图形算法标准。因为 OpenGL ES 是免授权费的、跨平台的、功能完善的 2D 和 3D 图形应用程序接口 API，它针对多种嵌入式系统实现了专门设计，这些设计包括控制台、移动电话、手持设备、家电设备和汽车。OpenGL ES 由精心定义的桌面 OpenGL 子集组成，创造了软件与图形加速器之间灵活强大的底层交互接口。OpenGL ES 包含浮点运算和定点运算系统描述以及 EGL 针对便携设备的本地视窗系统规范。OpenGL ES 1.x 面向功能固定的硬件所设计并提供加速支持、图形质量及性能标准。OpenGL ES 2.x 则提供包括遮盖器技术在内的全可编程 3D 图形算法。OpenGL ES-SC 专为有高安全性需求的特殊市场精心打造。

## 8.4.2　Android 用到 OpenGL ES

在 Android 系统中用到的是 OpenGL 图形函数库，它可以实现跨平台操作。Android 系统真正所用到的是它的一个子集 OpenGL ES（OpenGL for Embedded Systems（嵌入式系统）），这是 OpenGL 的嵌入式版本。

Android 系统使用 OpenGL 的标准接口来支持 3D 图形功能，Android 3D 图形系统也分为 Java 框架和本地代码两部分。本地代码主要实现 OpenGL 接口的库，在 Java 框架层，javax.microedition.khronos.opengles 是 Java 标准的 OpenGL 包，android.opengl 包提供了 OpenGL 系统和 Android GUI 系统之间的联系。

Android 的本地代码位于目录 frameworks/base/opengl 中。

JNI 代码位于目录 frameworks/base/core/com_google_android_gles_jni_GLImpl.cpp 和 frameworks/base/core/com_google_android_gles_jni_EGLImpl.cpp 中。

Java 类位于目录 opengl/java/javax/microedition/khronos 中。

### 8.4.3 OpenGL ES 的基本操作

#### 1．构造 OpenGL ES View

在 Andorid 系统中构造一个 OpenGL View 的方法非常简单，只需要完成如下两个方面的工作即可。

（1）GLSurfaceView

在 android.opengl、javax.microedition.khronos.egl、javax.microedition.khronos.opengles 和 java.nio 等包中，Android 系统提供了 OpenGL ES API 的主要定义。其中 GLSurfaceView 是这些包的核心类，主要功能如下：

☑ 串联 OpenGL ES 与 Android 的 View 层次结构之间的桥梁作用。

☑ 使 Open GL ES 库适应于 Anndroid 系统的 Activity 生命周期。

☑ 更加容易选择合适的 Frame Buffer 像素格式。

☑ 创建和管理单独绘图线程以达到平滑动画效果。

☑ 提供了方便使用的调试工具来跟踪 OpenGL ES 函数调用以帮助检查错误。

由此可见，编写 OpenGL ES 应用的第一步是从类 GLSurfaceView 开始的，在设置 GLSurfaceView 时，只需调用一个方法来设置 OpenGLView 用到的 GLSurfaceView.Renderer 即可。

```
public void setRenderer(GLSurfaceView.Renderer renderer)
```

（2）GLSurfaceView.Renderer

类 GLSurfaceView.Renderer 定义了一个统一图形绘制的接口，在里面定义了如下 3 个接口函数。

```
public void onSurfaceCreated(GL10 gl, EGLConfig config)
public void onDrawFrame(GL10 gl)
public void onSurfaceChanged(GL10 gl, int width, int height)
```

在上述接口函数中，各个参数的具体说明如下所示。

☑ onSurfaceCreated：在这个方法中主要用来设置一些绘制时不常变化的参数，如背景色、是否打开 z-buffer 等。

☑ onDrawFrame：定义实际的绘图操作。

☑ onSurfaceChanged：如果设备支持屏幕横向和纵向切换，这个方法将发生在横向<->纵向互换时。此时可以重新设置绘制的纵横比率。

#### 2．基本 3D 绘图处理

一个 3D 图形通常是由一些小的基本元素（顶点、边、面和多边形）构成，每个基本元素都可以单独来操作。下面将简要介绍 3D 绘图的一些基本构成要素，为步入本书后面知识的学习打下一个坚实的基础。

（1）顶点

顶点（Vertex）是 3D 建模时用到的最小构成元素，顶点定义为两条或是多条边交会的地方。在 3D 模型中一个顶点可以为多条边、面或是多边形所共享。一个顶点也可以代表一个点光源或是 Camera 的位置。图 8-11 中标识为黄色的点都是顶点（Vertex）。

在 Android 系统中，可以使用一个浮点数数组来定义一个顶点，通常将浮点数数组放在一个 Buffer

（java.nio）中以提高性能。图 8-12 中定义了 4 个顶点。

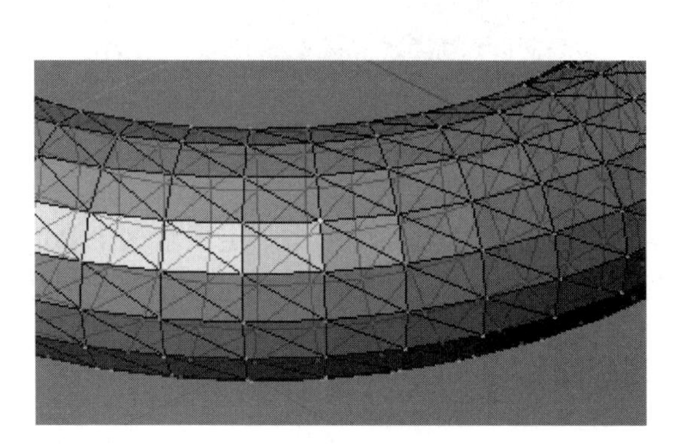

图 8-11　顶点

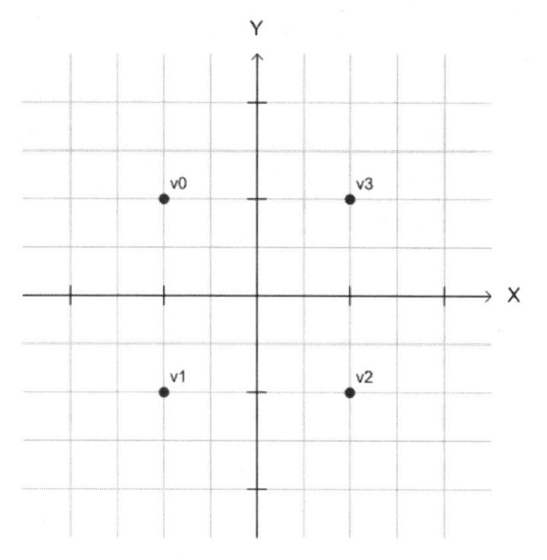


图 8-12　4 个顶点

在 Android 系统中，可以通过如下代码来定义图 8-13 中的 4 个顶点。

```
private float vertices[] = {
 -1.0f,   1.0f, 0.0f,   //0
 -1.0f, -1.0f, 0.0f,   //1
  1.0f, -1.0f, 0.0f,    //2
  1.0f,   1.0f, 0.0f,   //3
};
```

为了提高程序的性能，我们通常将这些数组存放到 java.io 定义的 Buffer 类中，代码如下：

```
ByteBuffer vbb = ByteBuffer.allocateDirect(vertices.length * 4);
vbb.order(ByteOrder.nativeOrder());
FloatBuffer vertexBuffer = vbb.asFloatBuffer();
vertexBuffer.put(vertices);
vertexBuffer.position(0);
```

在定义顶点之后，接下来需要将它们传给 OpenGL ES 库，此传递功能是通过 OpenGL ES 提供的“管道 Pipeline”机制实现的。此管道定义了一些“开关”来控制 OpenGL ES 支持的某些功能，在默认情况下这些功能是关闭的，如果需要使用 OpenGL ES 的这些功能，需要明确告知 OpenGL“管道”打开所需功能。

（2）边

边（Edge）定义为两个顶点之间的线段。边是面和多边形的边界线。在 3D 模型中，边可以被相邻的两个面或是多边形所共享。对一个边做变换将影响边相接的所有顶点、面或多边形。在 OpenGL 中，通常无须直接来定义一个边，而是通过顶点定义一个面，从而由面定义其所对应的 3 条边。可以通过修改边的两个顶点来更改一条边，图 8-13 中的黄色线段代表一条边。

（3）面

在 OpenGL ES 中的面（Face）特指一个三角形，由 3 个顶点和 3 条边构成，对一个面所做的变化影响到连接面的所有顶点和边以及面多边形。图 8-14 中的黄色区域代表一个面。

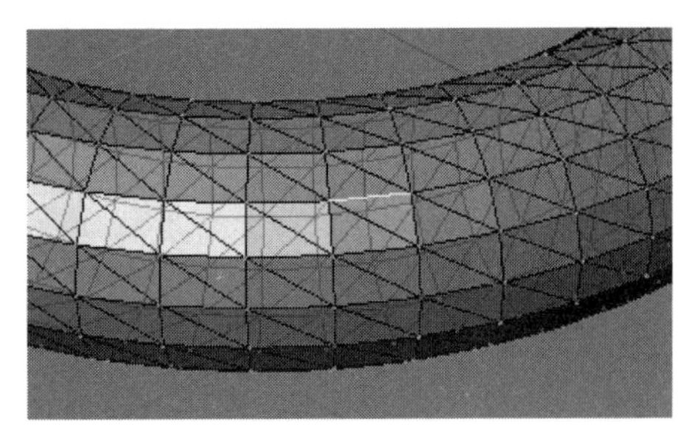

图 8-13　边

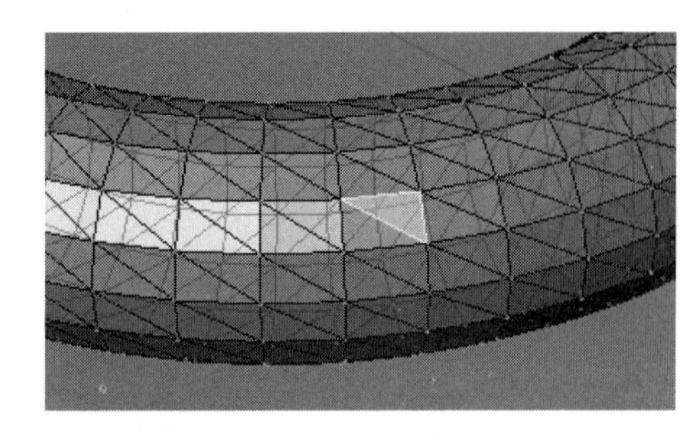

图 8-14　面

（4）多边形

多边形（Polygon）由多个面（三角形）拼接而成，在三维空间上，多边形不一定表示这个 Polygon 在同一平面上。这里使用默认的逆时针方向代表面前面（Front）。图 8-15 中的黄色区域是一个多边形。

在 Android 系统中使用顶点和 buffer 来定义多边形，图 8-16 中定义了一个正方形。

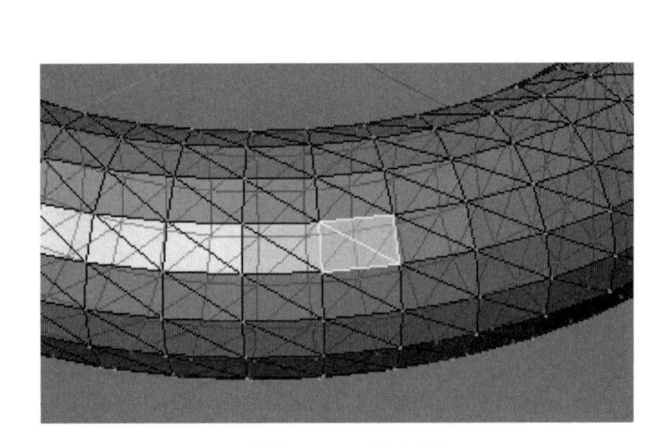

图 8-15　多边形

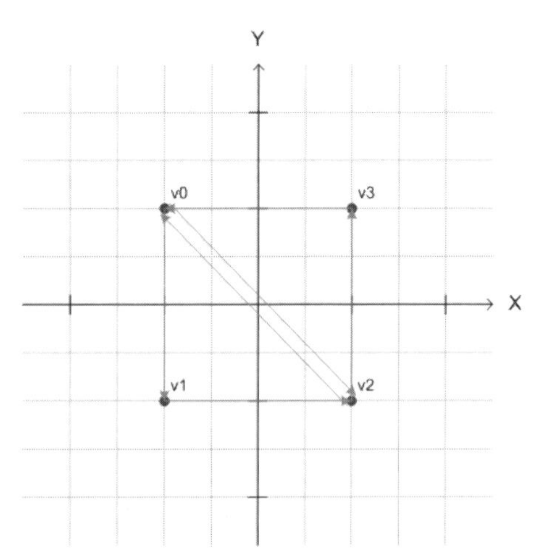


图 8-16　定义了一个正方形

图 8-16 中正方形对应的顶点和 buffer 的定义代码如下：

```
private short[] indices = { 0, 1, 2, 0, 2, 3 };
ByteBuffer ibb = ByteBuffer.allocateDirect(indices.length * 2);
ibb.order(ByteOrder.nativeOrder());
ShortBuffer indexBuffer = ibb.asShortBuffer();
indexBuffer.put(indices);
indexBuffer.position(0);
```

（5）Render 渲染

在定义好多边形后，需要了解和使用 OpenGL ES 的 API 来绘制并渲染（Render）这个多边形。在 OpenGL ES 中，提供了如下两种方法来绘制一个空间几何图形。

☑　public abstract void glDrawArrays(int mode, int first, int count)：使用 VetexBuffer 来绘制，顶点的顺序由 vertexBuffer 中的顺序指定。

☑　public abstract void glDrawElements(int mode, int count, int type, Buffer indices)：可以重新定义顶点的顺序，顶点的顺序由 indices Buffer 指定。

下面将通过一个具体实例来讲解使用 OpenGL ES 绘制一个正方形的方法。

| 实　例 | 功　能 | 源码路径 |
| --- | --- | --- |
| 实例 8-4 | 使用 OpenGL ES 绘制一个正方形 | 光盘:\daima\8\zhengEX |

本实例代码的具体实现流程如下：

（1）编写布局文件 main.xml，具体代码如下：

```
<?xml version="1.0" encoding="utf-8"?>
<LinearLayout xmlns:android="http://schemas.android.com/apk/res/android"
    android:orientation="vertical"
    android:layout_width="fill_parent"
    android:layout_height="fill_parent"
    >
<TextView
    android:layout_width="fill_parent"
    android:layout_height="wrap_content"
    android:text="@string/hello"
    />
</LinearLayout>
```

（2）编写文件 zheng.java，在此定义类 zheng，通过前面介绍的步骤在屏幕中绘制一个正方形。具体代码如下：

```
package ex.zheng;

import java.nio.ByteBuffer;
import java.nio.ByteOrder;
import java.nio.FloatBuffer;
import java.nio.ShortBuffer;

import javax.microedition.khronos.opengles.GL10;
public class zheng {
     private float vertices[] = {
                    -1.0f,   1.0f, 0.0f,
                    -1.0f, -1.0f, 0.0f,
                     1.0f, -1.0f, 0.0f,
                     1.0f,   1.0f, 0.0f,
          };
     private short[] indices = { 0, 1, 2, 0, 2, 3 };
     private FloatBuffer vertexBuffer;
     private ShortBuffer indexBuffer;

     public zheng() {
          ByteBuffer vbb = ByteBuffer.allocateDirect(vertices.length * 4);
          vbb.order(ByteOrder.nativeOrder());
          vertexBuffer = vbb.asFloatBuffer();
          vertexBuffer.put(vertices);
          vertexBuffer.position(0);
          ByteBuffer ibb = ByteBuffer.allocateDirect(indices.length * 2);
          ibb.order(ByteOrder.nativeOrder());
          indexBuffer = ibb.asShortBuffer();
          indexBuffer.put(indices);
          indexBuffer.position(0);
```

```
    }
    public void draw(GL10 gl) {
        gl.glFrontFace(GL8.GL_CCW);
        gl.glEnable(GL8.GL_CULL_FACE);
        gl.glCullFace(GL8.GL_BACK);
        gl.glEnableClientState(GL8.GL_VERTEX_ARRAY);
        gl.glVertexPointer(3, GL8.GL_FLOAT, 0,
                            vertexBuffer);
        gl.glDrawElements(GL8.GL_TRIANGLES, indices.length,
                    GL8.GL_UNSIGNED_SHORT, indexBuffer);
        gl.glDisableClientState(GL8.GL_VERTEX_ARRAY);
        gl.glDisable(GL8.GL_CULL_FACE);
    }
}
```

执行后的效果如图 8-17 所示。

图 8-17　执行效果

### 3. 变换 3D 坐标

OpenGL ES 使用了右手坐标系统，右手坐标系判断方法：在空间直角坐标系中，让右手拇指指向 X 轴的正方向，食指指向 Y 轴的正方向，如果中指能指向 Z 轴的正方向，则称这个坐标系为右手直角坐标系。在使用 OpenGL ES 绘制的 3D 坐标系统中，可以通过改变 transformations 的方式实现 3D 坐标的变换。

（1）使用 Translate 实现平移变换

在 OpenGL ES 中，使用方法 public abstract void glTranslatef(float x, float y, float z)实现坐标平移变换。例如通过坐标平移变换来移动 Android 屏幕中的三维图形，还可以进行多次平移变换，其结果为多个平移矩阵的累计结果，矩阵的顺序不重要，可以互换。

（2）使用 Rotate 实现旋转

在 OpenGL ES 中，使用方法 public abstract void glRotatef(float angle, float x, float y, float z)实现选择坐标旋转功能，单位为角度。（x,y,z）定义旋转的参照矢量方向，多次旋转的顺序非常重要。

假如选择一个骰子，首先按下列顺序选择 3 次。

```
gl.glRotatef(90f, 1.0f, 0.0f, 0.0f);
gl.glRotatef(90f, 0.0f, 1.0f, 0.0f);
gl.glRotatef(90f, 0.0f, 0.0f, 1.0f);
```

旋转过程如图 8-18 所示。

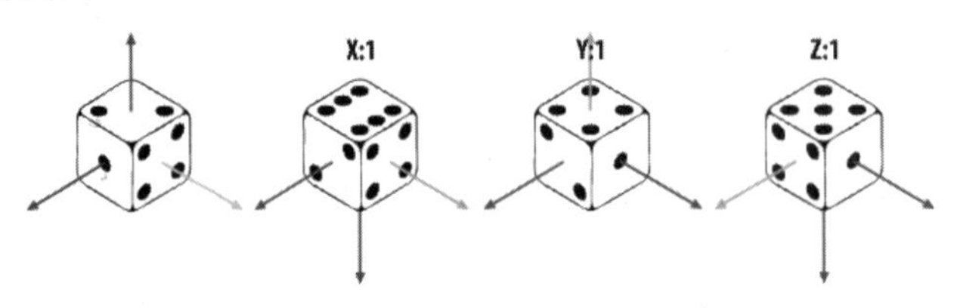


图 8-18　旋转骰子

如果想逆向旋转回原先的初始状态，需要进行如下旋转。

```
gl.glRotatef(90f, -1.0f, 0.0f, 0.0f);
gl.glRotatef(90f, 0.0f, -1.0f, 0.0f);
gl.glRotatef(90f, 0.0f, 0.0f, -1.0f);
```

逆向旋转过程如图 8-19 所示。

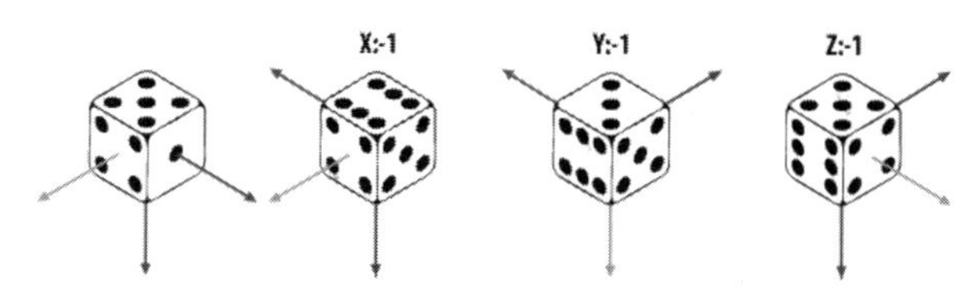


图 8-19　旋转骰子

（3）使用 Translate & Rotate 实现平移和旋转组合变换

在 OpenGL ES 中，在对 Mesh（网格，构成三维形体的基本单位）同时进行平移和选择变换时，坐标变换的顺序也直接影响最终的结果。在变换时，坐标变换都是相对于变换的 Mesh 本身的坐标系而进行的。

（4）Scale 缩放

在 OpenGL ES 中，使用方法 public abstract void glScalef (float x, float y, float z)实现缩放功能。

（5）矩阵操作，单位矩阵

在进行平移、旋转和缩放变换操作时，所有的变换都是针对当前的矩阵（与当前矩阵相乘）。在具体实现上，OpenGL ES 使用如下函数实现。

☑　public abstract void glLoadIdentity()：将当前矩阵恢复为最初的无变换的矩阵，用单位矩阵（无平移、缩放、旋转）实现。

☑　public abstract void glPushMatrix()和 public abstract void glPopMatrix()：在栈中保存当前矩阵和从栈中恢复所存矩阵。

进行坐标变换的一个好习惯是在变换前使用 glPushMatrix 保存当前矩阵，完成坐标变换操作后，再调用 glPopMatrix 恢复原先的矩阵设置。

### 4．添加颜色

使用 OpenGL ES 可以给绘制的图形填充颜色，OpenGL ES 支持的颜色模式是 RGBA 模式（红、绿、蓝、透明度）。在定义颜色时，OpenGL 使用 0…1 之间的浮点数表示。0 为 0，1 相当于 255（0xFF）。最简单的上色方法是顶点着色（Vertxt coloring）法，不但可以使用单色，而且也可以定义颜色渐变或者使用材质（类同于二维图形中各种 Brush 类型）。

（1）Flat coloring（单色）

Flat coloring 是通知 OpenGL 使用单一的颜色来渲染，OpenGL 将一直使用指定的颜色来渲染直到指定其他的颜色。指定颜色的方法为 glColor4f()，具体格式如下：

```
public abstract void glColor4f(float red, float green, float blue, float alpha)
```

默认的是 red、green 和 blue，值为 1，代表白色，这也是为什么在本章前面实例中绘制的正方形是白色的原因。可以创建一个新类 FlatColoredSquare 作为 Sequare 的子类，将它的 draw 重定义为如下格式。

```
public void draw(GL10 gl) {
 gl.glColor4f(0.5f, 0.5f, 1.0f, 1.0f);
 super.draw(gl);
}
```

然后将 OpenGLRenderer 的 square 类型改为 FlatColoredSquare。

```
private FlatColoredSquare square=new FlatColoredSquare();
```

此时如果编译运行，正方形颜色将变成蓝色。

（2）Smooth coloring（平滑颜色过渡）

当给每个顶点定义一个颜色值时，OpenGL ES 会自动在不同顶点颜色之间生成中间过渡颜色，即渐变色。

## 8.4.4 绘制图形

至此，在本章前面讲解的实例中，绘制的都不能算是三维效果图形。下面将详细讲解使用 OpenGL ES 绘制三维图形的方法，为读者步入本书后面知识的学习打下基础。

在现实世界中，Mesh（网格，三角面）是构成空间形体的基本元素，本章前面的正方形也是由两个 Mesh 构成的。下面将通过一个具体实例来讲解使用 Mesh 构成四面体和椎体等基本空间形体的方法。

| 实　例 | 功　能 | 源 码 路 径 |
| --- | --- | --- |
| 实例 8-5 | 使用 OpenGL ES 绘制三维效果图形 | 光盘:\daima\8\sanEX |

本实例的具体实现流程如下所示。

（1）编写文件 wang.java，在此文件中定义一个基类 wang，设置所有空间形体最基本的构成元素为 Mesh 类型（三角形网格）。文件 wang.java 的具体代码如下：

```
package ex.san.mesh;
import java.nio.ByteBuffer;
import java.nio.ByteOrder;
import java.nio.FloatBuffer;
import java.nio.ShortBuffer;
import javax.microedition.khronos.opengles.GL10;
import android.graphics.Bitmap;
import android.opengl.GLUtils;
public class wang {
	private FloatBuffer mVerticesBuffer = null;
	private ShortBuffer mIndicesBuffer = null;
	private FloatBuffer mTextureBuffer;
	private int mTextureId = -1;
	private Bitmap mBitmap;
	private boolean mShouldLoadTexture = false;
	private int mNumOfIndices = -1;
	private final float[] mRGBA = new float[] { 1.0f, 1.0f, 1.0f, 1.0f };
	private FloatBuffer mColorBuffer = null;
	public float x = 0;
	public float y = 0;
	public float z = 0;
	public float rx = 0;
	public float ry = 0;
	public float rz = 0;
	public void draw(GL10 gl) {
		gl.glFrontFace(GL8.GL_CCW);
		gl.glEnable(GL8.GL_CULL_FACE);
		gl.glCullFace(GL8.GL_BACK);
		gl.glEnableClientState(GL8.GL_VERTEX_ARRAY);
```

```
        gl.glVertexPointer(3, GL8.GL_FLOAT, 0, mVerticesBuffer);
        gl.glColor4f(mRGBA[0], mRGBA[1], mRGBA[2], mRGBA[3]);
        if (mColorBuffer != null) {
            gl.glEnableClientState(GL8.GL_COLOR_ARRAY);
            gl.glColorPointer(4, GL8.GL_FLOAT, 0, mColorBuffer);
        }
        if (mShouldLoadTexture) {
            loadGLTexture(gl);
            mShouldLoadTexture = false;
        }
        if (mTextureId != -1 && mTextureBuffer != null) {
            gl.glEnable(GL8.GL_TEXTURE_2D);
            gl.glEnableClientState(GL8.GL_TEXTURE_COORD_ARRAY);
            gl.glTexCoordPointer(2, GL8.GL_FLOAT, 0, mTextureBuffer);
            gl.glBindTexture(GL8.GL_TEXTURE_2D, mTextureId);
        }
        gl.glTranslatef(x, y, z);
        gl.glRotatef(rx, 1, 0, 0);
        gl.glRotatef(ry, 0, 1, 0);
        gl.glRotatef(rz, 0, 0, 1);
        gl.glDrawElements(GL8.GL_TRIANGLES, mNumOfIndices,
                GL8.GL_UNSIGNED_SHORT, mIndicesBuffer);
        gl.glDisableClientState(GL8.GL_VERTEX_ARRAY);
        if (mTextureId != -1 && mTextureBuffer != null) {
            gl.glDisableClientState(GL8.GL_TEXTURE_COORD_ARRAY);
        }
        gl.glDisable(GL8.GL_CULL_FACE);
    }
    protected void setVertices(float[] vertices) {
        ByteBuffer vbb = ByteBuffer.allocateDirect(vertices.length * 4);
        vbb.order(ByteOrder.nativeOrder());
        mVerticesBuffer = vbb.asFloatBuffer();
        mVerticesBuffer.put(vertices);
        mVerticesBuffer.position(0);
    }
    protected void setIndices(short[] indices) {
        ByteBuffer ibb = ByteBuffer.allocateDirect(indices.length * 2);
        ibb.order(ByteOrder.nativeOrder());
        mIndicesBuffer = ibb.asShortBuffer();
        mIndicesBuffer.put(indices);
        mIndicesBuffer.position(0);
        mNumOfIndices = indices.length;
    }
    protected void setTextureCoordinates(float[] textureCoords) { //New
        ByteBuffer byteBuf = ByteBuffer
                .allocateDirect(textureCoords.length * 4);
        byteBuf.order(ByteOrder.nativeOrder());
        mTextureBuffer = byteBuf.asFloatBuffer();
        mTextureBuffer.put(textureCoords);
        mTextureBuffer.position(0);
    }
```

```
    protected void setColor(float red, float green, float blue, float alpha) {
        mRGBA[0] = red;
        mRGBA[1] = green;
        mRGBA[2] = blue;
        mRGBA[3] = alpha;
    }
    protected void setColors(float[] colors) {
        ByteBuffer cbb = ByteBuffer.allocateDirect(colors.length * 4);
        cbb.order(ByteOrder.nativeOrder());
        mColorBuffer = cbb.asFloatBuffer();
        mColorBuffer.put(colors);
        mColorBuffer.position(0);
    }
    public void loadBitmap(Bitmap bitmap) {
        this.mBitmap = bitmap;
        mShouldLoadTexture = true;
    }
    private void loadGLTexture(GL10 gl) {
        int[] textures = new int[1];
        gl.glGenTextures(1, textures, 0);
        mTextureId = textures[0];
        gl.glBindTexture(GL8.GL_TEXTURE_2D, mTextureId);
        gl.glTexParameterf(GL8.GL_TEXTURE_2D, GL8.GL_TEXTURE_MIN_FILTER,
                GL8.GL_LINEAR);
        gl.glTexParameterf(GL8.GL_TEXTURE_2D, GL8.GL_TEXTURE_MAG_FILTER,
                GL8.GL_LINEAR);
        gl.glTexParameterf(GL8.GL_TEXTURE_2D, GL8.GL_TEXTURE_WRAP_S,
                GL8.GL_REPEAT);
        gl.glTexParameterf(GL8.GL_TEXTURE_2D, GL8.GL_TEXTURE_WRAP_T,
                GL8.GL_REPEAT);
        GLUtils.texImage2D(GL8.GL_TEXTURE_2D, 0, mBitmap, 0);
    }
}
```

在上述代码中，需要注意如下 5 点。

☑ setVertices：允许子类重新定义顶点坐标。

☑ setIndices：允许子类重新定义顶点的顺序。

☑ setColor/setColors：允许子类重新定义颜色。

☑ x、y、z：定义平移变换的参数。

☑ rx、ry、rz：定义旋转变换的参数。

（2）编写文件 fang.java，在此文件中定义类 fang，fang 可以有宽度、高度和深度，宽度定义为沿 X 轴方向的长度，深度定义为沿 Z 轴方向的长度，高度为 Y 轴方向。文件 fang.java 的具体代码如下：

```
package ex.san.mesh;
public class fang extends wang {
    public fang() {
        this(1, 1);
    }
    public fang(float width, float height) {
        float textureCoordinates[] = { 0.0f, 2.0f, 2.0f, 2.0f, 0.0f, 0.0f,
```

```
                    2.0f, 0.0f, };
        short[] indices = new short[] { 0, 1, 2, 1, 3, 2 };
        float[] vertices = new float[] { -0.5f, -0.5f, 0.0f, 0.5f, -0.5f, 0.0f,
                    -0.5f, 0.5f, 0.0f, 0.5f, 0.5f, 0.0f };
        setIndices(indices);
        setVertices(vertices);
        setTextureCoordinates(textureCoordinates);
    }
}
```

（3）编写文件 Group.java，在此文件中定义了类 Group，该类可以管理多个空间几何形体，如果把 Mesh 比作 Android 的 View，那么可以把 Group 看作 Android 的 ViewGroup。此处类 Group 的主要功能是把针对 Group 的操作（如 draw）分发到 Group 中的每个成员对应的操作（如 draw）。文件 Group.java 的具体代码如下：

```
package ex.san.mesh;
import java.util.Vector;
import javax.microedition.khronos.opengles.GL10;
public class Group extends wang {
    private final Vector<wang> mChildren = new Vector<wang>();
    @Override
    public void draw(GL10 gl) {
        int size = mChildren.size();
        for (int i = 0; i < size; i++)
            mChildren.get(i).draw(gl);
    }
    public void add(int location, wang object) {
        mChildren.add(location, object);
    }
    public boolean add(wang object) {
        return mChildren.add(object);
    }
    public void clear() {
        mChildren.clear();
    }
see java.util.Vector#get(int)
    public wang get(int location) {
        return mChildren.get(location);
    }
    public wang remove(int location) {
        return mChildren.remove(location);
    }
    public boolean remove(Object object) {
        return mChildren.remove(object);
    }
    public int size() {
        return mChildren.size();
    }
}
```

执行后将会在屏幕中形成一个三维效果图的图案，执行效果如图 8-20 所示。

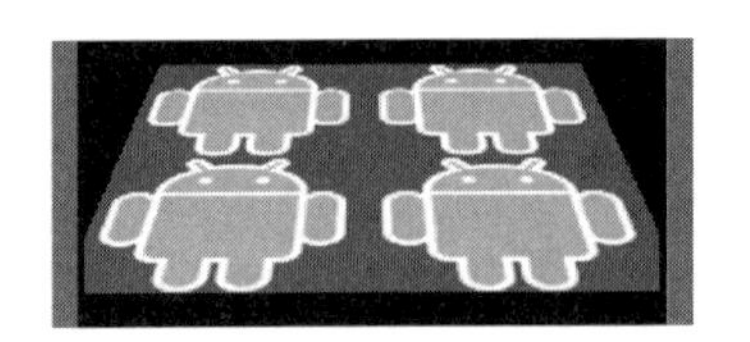

图 8-20　执行效果

# 8.5　音 频 开 发

**知识点讲解：光盘:视频\知识点\第 8 章\音频开发.avi**

在多媒体领域中，音频永远是主流应用之一。本节将详细讲解在 Android 平台中开发音频应用基本知识，为读者步入后面知识的学习打下基础。

## 8.5.1　音频接口类

在 Android 系统中，通过如下所示的接口类来实现音频功能。

☑ 音乐类型的音频资源：通过 MediaPlayer 来播放。
☑ 音调：通过 ToneGenerator 来播放。
☑ 提示音：通过 Ringtone 来播放。
☑ 游戏中的音频资源：通过 SoundPool 来播放。
☑ 录音功能：通过 MediaRecorder 和 AudioRecord 等来记录音频。

除了上述功能类之外，Android 还提供了实现音量调节和音频设备的管理等功能的类，这些类的具体说明如下所示。

☑ AudioManager：通过音频服务，为上层提供了音量和铃声模式控制的接口，铃声模式控制包括扬声器、耳机、蓝牙等是否打开，麦克风是否静音等。在开发多媒体应用时会经常用到 AudioManager。
☑ AudioSystem：提供了定义音频系统的基本类型和基本操作的接口，对应的 JNI 接口文件为 android_media_ AudioSystem.cpp。在 Android 音频系统中主要包括如下类型。
  ➢ STREAM_VOICE_CALL
  ➢ STREAM_SYSTEM
  ➢ STREAM_ RING
  ➢ STREAM_MUSIC
  ➢ STREAM_ALARM
  ➢ STREAM_NOTIFICATION
  ➢ STREAM_BLUETOOTH_SCO
  ➢ STREAM_SYSTEM_ENFORCED、
  ➢ STREAM_DTMF
  ➢ STREAM_TTS
☑ AudioTrack：直接为 PCM 数据提供支持，对应的 JNI 接口文件为 android_media_AudioTrack.cpp。

☑ AudioRecord：是音频系统的录音接口，默认的编码格式为 PCM_16_BIT，对应的 JNI 接口文件为 android media_AudioRecord.cpp。

☑ Ringtone 和 RingtoneManager：为铃声、提示音、闹钟等提供了快速播放以及管理的接口，实质是对媒体播放器提供了一个简单的封装。

☑ ToneGenerator：提供了对 DTMF 音（ITU-T Q.23），以及呼叫监督音（3GPP TS 22.001）、专用音（3GPP TS 31.111）中规定的音频的支持，根据呼叫状态和漫游状态，该文件产生的音频路径为下行音频或者传输给扬声器或耳机。对应的 JNI 接口文件为 android_media_ToneGenerator.cpp。其中 DTMF 音为 WAV 格式，相关的音频类型定义位于文件 ToneGenerator.h 中。

☑ SoundPool：能够播放音频流的组合音，主要被应用在游戏领域。对应的 JNI 接口为 android_media_SoundPool.cpp。

☑ SoundPool：可以从 APK 包中的资源文件或者文件系统中的文件将音频资源加载到内存中。在底层的实现上，SoundPool 通过媒体播放服务可以将音频资源解码为一个 16bit 的单声道或者立体声的 PCM 流，这使得应用避免了在回放过程中进行解码造成的延迟。

☑ 除了回放过程中延迟小的优点外，SoundPool 还能够对一定数量的音频流进行同时播放。当要播放的音频流数量超过 SoundPool 所设定的最大值时，SoundPool 将会停止已播放的一条低优先级的音频流。SoundPool 最大播放音频流数量的设置，可以避免 CPU 过载和影响 UI 体验。

☑ android.media.audiofx 包：这是从 Android 2.3 开始新增的包，提供了对单曲和全局的音效的支持，包括重低音、环绕音、均衡器、混响和可视化等声音特效。

## 8.5.2 AudioManager 控制铃声

在 Android 系统中，因为类 AudioManager 在包 android.Media 中定义，所以其地址是 android.Media.AudioManager，该类能够设置访问控制音量和铃声模式。

（1）方法

在类 AudioManager 中是通过方法实现音频功能的，其中最为常用的方法如下所示。

☑ adjustVolume(int direction, int flags)：用来控制手机音量大小，当传入的第一个参数为 AudioManager.ADJUST_LOWER 时，可将音量调小一个单位，传入 AudioManager.ADJUST_RAISE 时，则可以将音量调大一个单位。

☑ getMode()：返回当前音频模式。

☑ getRingerMode()：返回当前的铃声模式。

☑ getStreamVolume(int streamType)：取得当前手机的音量，最大值为 7，最小值为 0，当为 0 时，手机自动将模式调整为“振动模式”。

☑ setRingerMode(int ringerMode)：改变铃声模式。

（2）声音模式

手机都有声音模式，声音、静音还有振动，甚至振动加声音兼备，这些都是手机的基本功能。在 Android 手机中，我们同样可以通过 Android 的 SDK 提供的声音管理接口来管理手机声音模式以及调整声音大小，这就是 Android 中 AudioManager 的使用。

☑ 设置声音模式

```
//声音模式
AudioManager.setRingerMode(AudioManager.RINGER_MODE_NORMAL);
```

```
//静音模式
AudioManager.setRingerMode(AudioManager.RINGER_MODE_SILENT);
//振动模式
AudioManager.setRingerMode(AudioManager.RINGER_MODE_VIBRATE);
```

☑ 调整声音大小

```
//减小声音音量
AudioManager.adjustVolume(AudioManager.ADJUST_LOWER,  0);
//调大声音音量
AudioManager.adjustVolume(AudioManager.ADJUST_RAISE, 0);
```

（3）基本应用

AudioManager 类的常见应用如下所示。

☑ 实现音量控制，例如下面的代码。

```
//音量控制，初始化定义
AudioManager mAudioManager = (AudioManager) getSystemService(Context.AUDIO_SERVICE);
//最大音量
int maxVolume = mAudioManager.getStreamMaxVolume(AudioManager.STREAM_MUSIC);
//当前音量
int currentVolume = mAudioManager.getStreamVolume(AudioManager.STREAM_MUSIC);
```

☑ 控制音量大小，例如下面的代码。

```
if(isSilent){
        mAudioManager.setStreamVolume(AudioManager.STREAM_MUSIC, 0, 0);
    }else{
        mAudioManager.setStreamVolume(AudioManager.STREAM_MUSIC, tempVolume, 0); //tempVolume:
音量绝对值
    }
```

☑ 以一步步长控制音量的增减，并弹出系统默认音量控制条。例如下面的代码。

```
view sourceprint?
    //降低音量，调出系统音量控制
    if(flag == 0){
mAudioManager.adjustStreamVolume(AudioManager.STREAM_MUSIC,AudioManager.ADJUST_LOWER,
                                        AudioManager.FX_FOCUS_NAVIGATION_UP);
    }
    //增加音量，调出系统音量控制
    else if(flag == 1){
mAudioManager.adjustStreamVolume(AudioManager.STREAM_MUSIC,AudioManager.ADJUST_RAISE,
                                        AudioManager.FX_FOCUS_NAVIGATION_UP);
    }
```

（4）调节声音的基本步骤

在 Android 系统中使用 AudioManager 类调节声音的基本步骤如下。

① 通过系统服务获得声音管理器，例如下面的代码。

```
AudioManager audioManager =  (AudioManager)getSystemService(Service.AUDIO_SERVICE);
```

② 根据实际需要调用适当的方法，例如下面的代码。

```
audioManager.adjustStreamVolume(int streamType, int direction, int flags);
```

上述参数的具体说明如下所示。

☑ streamType：声音类型，可取下面的值。

- ➢ STREAM_VOICE_CALL：打电话时的声音。
- ➢ STREAM_SYSTEM：Android 系统声音。
- ➢ STREAM_RING：电话铃响。
- ➢ STREAM_MUSIC：音乐声音。
- ➢ STREAM_ALARM：警告声音。

☑ direction：调整音量的方向，可取下面的值。

- ➢ ADJUST_LOWER：调低音量。
- ➢ ADJUST_RAISE：调高音量。
- ➢ ADJUST_SAME：保持先前音量。

☑ flags：可选标志位。

③ 设置指定声音类型，例如下面的代码。

```
audioManager.setStreamMute(int streamType, boolean state)
```

通过上述方法设置指定声音类型（streamType）是否为静音。如果 state 为 true，则设置为静音；否则，不设置为静音。

④ 设置铃音模式，例如下面的代码。

```
audioManager.setRingerMode(int ringerMode);
```

通过上述方法设置铃音模式，可取的值如下所示。

☑ RINGER_MODE_NORMAL：铃音正常模式。

☑ RINGER_MODE_SILENT：铃音静音模式。

☑ RINGER_MODE_VIBRATE：铃音振动模式，即铃音为静音，启动振动。

⑤ 设置声音模式，例如下面的代码。

```
audioManager.setMode(int mode);
```

通过上述方法设置声音模式，可取的值如下所示。

☑ MODE_NORMAL：正常模式，即在没有铃音与电话的情况。

☑ MODE_RINGTONE：铃响模式。

☑ MODE_IN_CALL：接通电话模式。

☑ MODE_IN_COMMUNICATION：通话模式。

**注意**：*声音的调节是没有权限要求的。*

| 实　例 | 功　能 | 源 码 路 径 |
| --- | --- | --- |
| 实例 8-6 | 设置短信提示铃声 | 光盘:\daima\8\LingEX |

本实例的具体实现流程如下所示。

（1）在文件 main.xml 中设置 3 个按钮，分别实现启用、停止和设置间隔时间功能。主要代码如下：

```
<LinearLayout xmlns:android="http://schemas.android.com/apk/res/android"
android:orientation="vertical"
android:layout_width="fill_parent"
android:layout_height="fill_parent"
android:gravity="center"
>
    <Button
        android:id="@+id/startButton"
```

```
            android:text="@string/startButton"
            android:layout_width="fill_parent"
            android:layout_height="wrap_content" />
        <Button
            android:id="@+id/endButton"
            android:text="@string/endButton"
            android:layout_width="fill_parent"
            android:layout_height="wrap_content" />
        <Button
            android:id="@+id/configButton"
            android:text="@string/configButton"
            android:layout_width="fill_parent"
            android:layout_height="wrap_content" />
</LinearLayout>
```

（2）编写文件 lingCHService.java，开启一个 Service 监听短信的事件，在短信到达后进行声音播放的处理，牵涉到的主要是 Service、Broadcast、MediaPlayer 和 Preference。在此包含了存放铃声的 Map 和播放铃声等逻辑处理，通过 AudioManager 来暂时打开多媒体声音，播放完再关闭。文件 BellService.java 的主要代码如下：

```
public class lingCHService extends Service {
    //监听事件
    public static final String SMS_RECEIVED_ACTION = "android.provider.Telephony.SMS_RECEIVED";
    //铃声序列
    public static final int ONE_SMS = 1;
    public static final int TWO_SMS = 2;
    public static final int THREE_SMS = 3;
    public static final int FOUR_SMS = 4;
    public static final int FIVE_SMS = 5;

    private HashMap<Integer,Integer> bellMap;//铃声 Map
    private Date lastSMSTime;//上条短信时间
    private int currentBell;//当前应当播放铃声
    private boolean justStart=true;//是否是第一次启动，避免首次启动马上收到短信导致立即播放第二条铃声的
情况
    private AudioManager am;
    private int currentMediaStatus;
    private int currentMediaMax;

    public IBinder onBind(Intent intent) {
        return null;
    }

    @Override
    public void onCreate() {
        super.onCreate();
        IntentFilter filter = new IntentFilter();
        filter.addAction(SMS_RECEIVED_ACTION);
        Log.e("COOKIE", "Service start");
        //注册监听
        registerReceiver(messageReceiver, filter);
        //初始化 Map，根据之后改进可以替换其中的铃声
```

```
        bellMap = new HashMap<Integer,Integer>();
        bellMap.put(ONE_SMS, R.raw.holyshit);
        bellMap.put(TWO_SMS, R.raw.holydouble);
        bellMap.put(THREE_SMS, R.raw.holytriple);
        bellMap.put(FOUR_SMS, R.raw.holyultra);
        bellMap.put(FIVE_SMS, R.raw.holyrampage);
        //当前时间
        lastSMSTime=new Date(System.currentTimeMillis());
        //当前应当播放的铃声，初始为 1
        //之后根据间隔判断，若为 5 分钟之内则+1
        //若距离上一次超过 5 分钟，则重新置为 1
        currentBell=1;
    }

    @Override
    public void onStart(Intent intent, int startId) {
        super.onStart(intent, startId);
    }

    @Override
    public void onDestroy() {
        super.onDestroy();
        //取消监听
        unregisterReceiver(messageReceiver);
        Log.e("COOKIE", "Service end");
    }

    //设定广播
    private BroadcastReceiver messageReceiver = new BroadcastReceiver() {

        @Override
        public void onReceive(Context context, Intent intent) {
            String action = intent.getAction();
            if (action.equals(SMS_RECEIVED_ACTION)) {
                playBell(context, 0);
            }
        }

    };
    //播放音效
    private void playBell(Context context, int num) {
        //为防止用户当前模式关闭了 media 音效，先将 media 打开
        am=(AudioManager)getSystemService(Context.AUDIO_SERVICE);//获取音量控制
        currentMediaStatus=am.getStreamVolume(AudioManager.STREAM_MUSIC);
        currentMediaMax=am.getStreamMaxVolume(AudioManager.STREAM_MUSIC);
        am.setStreamVolume(AudioManager.STREAM_MUSIC, currentMediaMax, 0);
        //创建 MediaPlayer 进行播放
        MediaPlayer mp = MediaPlayer.create(context, getBellResource());
        mp.setOnCompletionListener(new musicCompletionListener());
        mp.start();
    }
```

```
    private class musicCompletionListener implements OnCompletionListener {
        @Override
        public void onCompletion(MediaPlayer mp) {
            //播放结束释放 mp 资源
            mp.release();
            //恢复用户之前的 media 模式
            am.setStreamVolume(AudioManager.STREAM_MUSIC, currentMediaStatus, 0);
        }
    }
    //获取当前应该播放的铃声
    private int getBellResource() {
        //判断时间间隔（毫秒）
        int preferenceInterval;
        long interval;
        Date curTime = new Date(System.currentTimeMillis());
        interval=curTime.getTime()-lastSMSTime.getTime();
        lastSMSTime=curTime;
        preferenceInterval=getPreferenceInterval();
        if(interval<preferenceInterval*60*1000&&!justStart){
            currentBell++;
            if(currentBell>5){
                currentBell=5;
            }
        }else{
            currentBell=1;
        }
        justStart=false;
        return bellMap.get(currentBell);
    }
    //获取 Preference 设置
    private int getPreferenceInterval(){
        SharedPreferences settings = PreferenceManager.getDefaultSharedPreferences(this);
        int interval=Integer.valueOf(settings.getString("interval_config", "5"));
//Log.v("COOKIE", "interval: "+interval);
        return interval;
    }
}
```

（3）编写文件 LingCHActivity.java，为屏幕中的 3 个 Button 设置相应的处理事件。主要代码如下：

```
public class lingCHService extends Service {
    //监听事件
    public static final String SMS_RECEIVED_ACTION = "android.provider.Telephony.SMS_RECEIVED";
    //铃声序列
    public static final int ONE_SMS = 1;
    public static final int TWO_SMS = 2;
    public static final int THREE_SMS = 3;
    public static final int FOUR_SMS = 4;
    public static final int FIVE_SMS = 5;

    private HashMap<Integer,Integer> bellMap;//铃声 Map
```

```
    private Date lastSMSTime;         //上条短信时间
    private int currentBell;          //当前应当播放铃声
    private boolean justStart=true;  //是否是第一次启动，避免首次启动马上收到短信导致立即播放第二条铃声
的情况
    private AudioManager am;
    private int currentMediaStatus;
    private int currentMediaMax;

    public IBinder onBind(Intent intent) {
        return null;
    }

    @Override
    public void onCreate() {
        super.onCreate();
        IntentFilter filter = new IntentFilter();
        filter.addAction(SMS_RECEIVED_ACTION);
        Log.e("COOKIE", "Service start");
        //注册监听
        registerReceiver(messageReceiver, filter);
        //初始化 Map，根据之后改进可以替换其中的铃声
        bellMap = new HashMap<Integer,Integer>();
        bellMap.put(ONE_SMS, R.raw.holyshit);
        bellMap.put(TWO_SMS, R.raw.holydouble);
        bellMap.put(THREE_SMS, R.raw.holytriple);
        bellMap.put(FOUR_SMS, R.raw.holyultra);
        bellMap.put(FIVE_SMS, R.raw.holyrampage);
        //当前时间
        lastSMSTime=new Date(System.currentTimeMillis());
        //当前应当播放的铃声，初始为 1
        //之后根据间隔判断，若为 5 分钟之内则+1
        //若距离上一次超过 5 分钟，则重新置为 1
        currentBell=1;
    }

    @Override
    public void onStart(Intent intent, int startId) {
        super.onStart(intent, startId);
    }

    @Override
    public void onDestroy() {
        super.onDestroy();
        //取消监听
        unregisterReceiver(messageReceiver);
        Log.e("COOKIE", "Service end");
    }

    //设定广播
    private BroadcastReceiver messageReceiver = new BroadcastReceiver() {
```

```
        @Override
        public void onReceive(Context context, Intent intent) {
            String action = intent.getAction();
            if (action.equals(SMS_RECEIVED_ACTION)) {
                playBell(context, 0);
            }
        }

    };
    //播放音效
    private void playBell(Context context, int num) {
        //为防止用户当前模式关闭了 media 音效，先将 media 打开
        am=(AudioManager)getSystemService(Context.AUDIO_SERVICE);//获取音量控制
        currentMediaStatus=am.getStreamVolume(AudioManager.STREAM_MUSIC);
        currentMediaMax=am.getStreamMaxVolume(AudioManager.STREAM_MUSIC);
        am.setStreamVolume(AudioManager.STREAM_MUSIC, currentMediaMax, 0);
        //创建 MediaPlayer 进行播放
        MediaPlayer mp = MediaPlayer.create(context, getBellResource());
        mp.setOnCompletionListener(new musicCompletionListener());
        mp.start();
    }

    private class musicCompletionListener implements OnCompletionListener {
        @Override
        public void onCompletion(MediaPlayer mp) {
            //播放结束释放 mp 资源
            mp.release();
            //恢复用户之前的 media 模式
            am.setStreamVolume(AudioManager.STREAM_MUSIC, currentMediaStatus, 0);
        }
    }
    //获取当前应该播放的铃声
    private int getBellResource() {
        //判断时间间隔（毫秒）
        int preferenceInterval;
        long interval;
        Date curTime = new Date(System.currentTimeMillis());
        interval=curTime.getTime()-lastSMSTime.getTime();
        lastSMSTime=curTime;
        preferenceInterval=getPreferenceInterval();
        if(interval<preferenceInterval*60*1000&&!justStart){
            currentBell++;
            if(currentBell>5){
                currentBell=5;
            }
        }else{
            currentBell=1;
        }
        justStart=false;
        return bellMap.get(currentBell);
    }
```

```
    //获取 Preference 设置
    private int getPreferenceInterval(){
        SharedPreferences settings = PreferenceManager.getDefaultSharedPreferences(this);
        int interval=Integer.valueOf(settings.getString("interval_config", "5"));
//Log.v("COOKIE", "interval: "+interval);
        return interval;
    }
}
```

执行之后在屏幕中单击按钮可以设置对应的铃声，效果如图 8-21 所示。

图 8-21　执行效果

# 8.6　录音详解

**知识点讲解：光盘:视频\知识点\第 8 章\录音详解.avi**

在 Android 系统中，录制音频和视频最常采用的是 MediaRecorder 接口。为了录制音频文件，需要设置音频源、输出格式、录制时间、编码格式等。类 AudioRecord 在 Java 应用程序中管理音频资源，用来记录从平台音频输入设备产生的数据。通过 AudioRecord 对象来完成“pulling”（读取）数据。通过以下几个方法负责立即从 AudioRecord 对象读取：read(byte[],int,int)、read(short[],int,int)或 read(ByteBuffer, int)。无论使用哪种音频格式，使用 AudioRecord 是最方便的。本节将详细讲解在 Android 系统中实现录音功能的方法。

## 8.6.1　使用 MediaRecorder 接口录制音频

在创建 AudioRecord 对象时，AudioRecord 会初始化，并和音频缓冲区连接，用来缓冲新的音频数据。根据构造时指定的缓冲区大小，来决定 AudioRecord 能够记录多长的数据。从硬件设备读取的数据应小于整个记录缓冲区。

MediaRecorder 的内部类是 AudioRecord.OnRecordPositionUpdateListener，当 AudioRecord 收到一个由 setNotificationMarkerPosition(int)设置的通知标志，或由 setPositionNotificationPeriod(int)设置的周期更新记录的进度状态时，回调此接口。

（1）常量

MediaRecorder 中常用常量如下所示。

☑ public static final int ERROR：表示操作失败，常量值：-1(0xffffffff)。
☑ public static final int ERROR_BAD_VALUE：表示使用了一个不合理的值导致的失败，常量值为-2(0xfffffffe)。
☑ public static final int ERROR_INVALID_OPERATION：表示不恰当的方法导致的失败，常量值为-3(0xfffffffd)。
☑ public static final int RECORDSTATE_RECORDING：指示 AudioRecord 录制状态为“正在录制”，常量值为 3(0x00000003)。
☑ public static final int RECORDSTATE_STOPPED：指示 AudioRecord 录制状态为“不在录制”，常量值为 1(0x00000001)。
☑ public static final int STATE_INITIALIZED：指示 AudioRecord 准备就绪，常量值为 1(0x00000001)。
☑ public static final int STATE_UNINITIALIZED：指示 AudioRecord 状态没有初始化成功，常量值为 0(0x00000000)。
☑ public static final int SUCCESS：表示操作成功，常量值为 0(0x00000000)。

（2）构造函数

MediaRecorder 中的构造函数是 AudioRecord，具体格式如下：

```
public AudioRecord(int audioSource, int sampleRateInHz, int channelConfig, int audioFormat, int buffer SizeInBytes)
```

各个参数的具体说明如下。

☑ audioSource：录制源。
☑ sampleRateInHz：默认采样率，单位为 Hz。44100Hz 是当前唯一能保证在所有设备上工作的采样率，在一些设备上还有 22050、16000 或 11025。
☑ channelConfig：描述音频通道设置。
☑ audioFormat：音频数据保证支持此格式。
☑ bufferSizeInBytes：在录制过程中，音频数据写入缓冲区的总数（字节）。从缓冲区读取的新音频数据总会小于此值。getMinBufferSize(int,int,int)会返回 AudioRecord 实例创建成功后的最小缓冲区。设置的值比 getMinBufferSize()还小则会导致初始化失败。

（3）公共方法

☑ public int getAudioFormat()：返回设置的音频数据格式。
☑ public int getAudioSource()：返回音频录制源。
☑ public int getChannelConfiguration()：返回设置的频道设置。请参见 CHANNEL_IN_MONO 和 CHANNEL_IN_STEREO。
☑ public int getChannelCount()：返回设置的频道数目。
☑ public static int getMinBufferSize(int sampleRateInHz, int channelConfig, int audioFormat)：返回成功创建 AudioRecord 对象所需要的最小缓冲区大小。其参数介绍如下。
   ➢ sampleRateInHz：默认采样率，单位为 Hz。
   ➢ channelConfig：描述音频通道设置。
   ➢ audioFormat：音频数据保证支持此格式。

如果硬件不支持录制参数，或输入了一个无效的参数，则返回 ERROR_BAD_VALUE，如果硬件查询到输出属性没有实现，或最小缓冲区用 byte 表示，则返回 ERROR。

**注意**：这个大小并不保证在负荷下的流畅录制，应根据预期的频率来选择更高的值，AudioRecord 实例在推送新数据时使用此值。

- ☑ public int getNotificationMarkerPosition()：返回通知，标记框架中的位置。
- ☑ public int getPositionNotificationPeriod()：返回通知，更新框架中的时间位置。
- ☑ public int getRecordingState()：返回 AudioRecord 实例的录制状态。
- ☑ public int getSampleRate()：返回设置的音频数据样本采样率，单位为 Hz。
- ☑ public int getState()：返回 AudioRecord 实例的状态。 这点非常有用，用在 AudioRecord 实例创建成功后，检查初始化属性。
- ☑ public int read(short[] audioData, int offsetInShorts, int sizeInShorts)：从音频硬件录制缓冲区读取数据。上述参数的具体说明如下所示。
  - ➢ audioData：写入的音频录制数据。
  - ➢ offsetInShorts：目标数组 audioData 的起始偏移量。
  - ➢ sizeInShorts ：请求读取的数据大小。

返回值是一个 short 型数据，表示读取到的数据，如果对象属性没有初始化，则返回 ERROR_INVALID_OPERATION；如果参数不能解析成有效的数据或索引，则返回 ERROR_BAD_VALUE。 返回数值不会超过 sizeInShorts。

- ☑ public int read(byte[] audioData, int offsetInBytes, int sizeInBytes)：从音频硬件录制缓冲区读取数据。
  - ➢ audioData：写入的音频录制数据。
  - ➢ offsetInBytes：audioData 的起始偏移值，单位为 byte。
  - ➢ sizeInBytes：读取的最大字节数。

返回值是读入缓冲区的总 byte 数，如果对象属性没有初始化，则返回 ERROR_INVALID_OPERATION；如果参数不能解析成有效的数据或索引，则返回 ERROR_BAD_VALUE。读取的总 byte 数不会超过 sizeInBytes。

- ☑ public int read(ByteBuffer audioBuffer, int sizeInBytes)：从音频硬件录制缓冲区读取数据，直接复制到指定缓冲区。如果 audioBuffer 不是直接的缓冲区，此方法总是返回 0。上述参数的具体说明如下所示。
  - ➢ audioBuffer：存储写入音频录制数据的缓冲区。
  - ➢ sizeInBytes：请求的最大字节数。

返回值是读入缓冲区的总 byte 数，如果对象属性没有初始化，则返回 ERROR_INVALID_OPERATION；如果参数不能解析成有效的数据或索引，则返回 ERROR_BAD_VALUE。读取的总 byte 数不会超过 sizeInBytes。

- ☑ public void release()：释放本地 AudioRecord 资源。一般对象不能经常使用此方法，而且在调用 release()后，必须设置引用为 null。
- ☑ public int setNotificationMarkerPosition(int markerInFrames)：如果设置了 setRecordPosition Update Listener(OnRecordPositionUpdateListener)或 setRecordPositionUpdateListener(OnRecord Position UpdateListener, Handler)，则通知监听者设置位置标记。参数 markerInFrames 表示在框架中快速标记位置，返回值是返回错误或成功代码。

☑ public int setPositionNotificationPeriod(int periodInFrames)：如果设置了 setRecordPosition UpdateListener(OnRecordPositionUpdateListener)或 setRecordPositionUpdateListener(OnRecord PositionUpdateListener, Handler)，则通知监听者设置时间标记。参数 markerInFrames 表示在框架中快速更新时间标记，返回值是返回错误或成功代码，请参见 SUCCESS 和 ERROR_INVALID_OPERATION。

☑ public void setRecordPositionUpdateListener(AudioRecord.OnRecordPositionUpdateListener listener, Handler handler)：当之前设置的标志已经成立，或者周期录制位置更新时，设置处理监听者。使用此方法将 Handler 和其他的线程联系起来来接收 AudioRecord 事件，比创建 AudioTrack 实例更好一些。参数 handler 用来接收事件通知消息。

☑ public void setRecordPositionUpdateListener(AudioRecord.OnRecordPositionUpdateListener listener)：当之前设置的标志已经成立，或者周期录制位置更新时，设置处理监听者。

☑ public void startRecording()：AudioRecord 实例开始进行录制。

4．受保护方法

AudioRecord 中的受保护方法是 protected void finalize()，用于通知 VM 回收此对象内存。方法 finalize()只能用在运行的应用程序没有任何线程再使用此对象，来告诉垃圾回收器回收此对象。此方法用于释放系统资源，由垃圾回收器清除此对象。默认没有实现，由 VM 来决定，但子类根据需要可重写 finalize()。在执行期间，调用此方法可能会立即抛出未定义异常，但是可以忽略。

**注意：** VM 保证对象可以一次或多次调用 finalize()，但并不保证 finalize()会马上执行。例如，对象 B 的 finalize()可能延迟执行，等待对象 A 的 finalize()延迟回收 A 的内存。为了安全起见，请查看 ReferenceQueue，在它里面提供了更多的控制 VM 的垃圾回收。

## 8.6.2 使用 AudioRecord 接口录音

类 AudioRecord 在 Java 应用程序中管理音频资源，用来记录从平台音频输入设备产生的数据。通过 AudioRecord 对象来完成“pulling”（读取）数据。通过以下几个方法负责立即从 AudioRecord 对象读取：read(byte[], int, int)、read(short[], int, int)或 read(ByteBuffer, int)。无论使用哪种音频格式，使用 AudioRecord 是最方便的。

在创建 AudioRecord 对象时，AudioRecord 会初始化，并和音频缓冲区连接，用来缓冲新的音频数据。根据构造时指定的缓冲区大小，来决定 AudioRecord 能够记录多长的数据。从硬件设备读取的数据应小于整个记录缓冲区。

（1）常量

AudioRecord 中包含的常量如下所示。

☑ public static final int ERROR：表示操作失败，常量值为-1(0xffffffff)。

☑ public static final int ERROR_BAD_VALUE：表示使用了一个不合理的值导致的失败，常量值为-2(0xfffffffe)。

☑ public static final int ERROR_INVALID_OPERATION：表示不恰当的方法导致的失败，常量值为-3(0xfffffffd)。

☑ public static final int RECORDSTATE_RECORDING：指示 AudioRecord 录制状态为“正在录

制”，常量值为 3(0x00000003)。

☑ public static final int RECORDSTATE_STOPPED：指示 AudioRecord 录制状态为“不在录制”，常量值为 1(0x00000001)。

☑ public static final int STATE_INITIALIZED：指示 AudioRecord 准备就绪，常量值为 1(0x00000001)。

☑ public static final int STATE_UNINITIALIZED：指示 AudioRecord 状态没有初始化成功，常量值为 0(0x00000000)。

☑ public static final int SUCCESS：表示操作成功，常量值为 0(0x00000000)。

（2）构造函数

AudioRecord 中的构造函数是 AudioRecord，格式如下：

```
public AudioRecord (int audioSource, int sampleRateInHz, int channelConfig, int audioFormat, int buffer SizeInBytes)
```

各个参数的具体说明如下所示。

☑ audioSource：录制源。

☑ sampleRateInHz：默认采样率，单位为 Hz。44100Hz 是当前唯一能保证在所有设备上工作的采样率，在一些设备上还有 22050、16000 或 11025。

☑ channelConfig：描述音频通道设置。

☑ audioFormat：音频数据保证支持此格式。

☑ bufferSizeInBytes：在录制过程中，音频数据写入缓冲区的总数（字节）。从缓冲区读取的新音频数据总会小于此值。用 getMinBufferSize(int, int, int)返回 AudioRecord 实例创建成功后的最小缓冲区，如果其设置的值比 getMinBufferSize()还小则会导致初始化失败。

（3）公共方法

AudioRecord 中的公共方法如下所示。

☑ public int getAudioFormat()：返回设置的音频数据格式。参见 ENCODING_PCM_16BIT 和 ENCODING_PCM_8BIT。

☑ public int getAudioSource()：返回音频录制源。

☑ public int getChannelConfiguration()：返回设置的频道设置。参见 CHANNEL_IN_MONO 和 CHANNEL_IN_STEREO。

☑ public int getChannelCount()：返回设置的频道数目。

☑ public static int getMinBufferSize(int sampleRateInHz, int channelConfig, int audioFormat)：返回成功创建 AudioRecord 对象所需要的最小缓冲区大小。需要注意的是，这个大小并不能保证在负荷下的流畅录制，应根据预期的频率来选择更高的值，AudioRecord 实例在推送新数据时使用此值。

上述参数的具体说明如下所示。

☑ sampleRateInHz：默认采样率，单位为 Hz。

☑ channelConfig：描述音频通道设置。

☑ audioFormat：音频数据保证支持此格式。参见 ENCODING_PCM_16BIT。

如果硬件不支持录制参数，或输入了一个无效的参数，则返回 ERROR_BAD_VALUE，如果硬件查询到输出属性没有实现，或最小缓冲区用 byte 表示，则返回 ERROR。

☑ public int getNotificationMarkerPosition()：返回通知，标记框架中的位置。

☑ public int getPositionNotificationPeriod()：返回通知，更新框架中的时间位置。

☑ public int getRecordingState()：返回 AudioRecord 实例的录制状态。

☑ public int getSampleRate()：返回设置的音频数据样本采样率，单位为 Hz。

☑ public int getState()：返回 AudioRecord 实例的状态。用在 AudioRecord 实例创建成功后，检查初始化属性。它能肯定请求到了合适的硬件资源。

☑ public int read(short[] audioData, int offsetInShorts, int sizeInShorts)：从音频硬件录制缓冲区读取数据。上述参数的具体说明如下所示。

  ➢ audioData：写入的音频录制数据。
  ➢ offsetInShorts：目标数组 audioData 的起始偏移量。
  ➢ sizeInShorts：请求读取的数据大小。

返回值是 short 型数据，表示读取到的数据，如果对象属性没有初始化，则返回 ERROR_INVALID_OPERATION；如果参数不能解析成有效的数据或索引，则返回 ERROR_BAD_VALUE。返回数值不会超过 sizeInShorts。

☑ public int read(byte[] audioData, int offsetInBytes, int sizeInBytes)：从音频硬件录制缓冲区读取数据。上述参数的具体说明如下所示。

  ➢ audioData：写入的音频录制数据。
  ➢ offsetInBytes：audioData 的起始偏移值，单位为 byte。
  ➢ sizeInBytes：读取的最大字节数。

读入缓冲区的总 byte 数，如果对象属性没有初始化，则返回 ERROR_INVALID_OPERATION；如果参数不能解析成有效的数据或索引，则返回 ERROR_BAD_VALUE。读取的总 byte 数不会超过 sizeInBytes。

☑ public int read(ByteBuffer audioBuffer, int sizeInBytes)：从音频硬件录制缓冲区读取数据，直接复制到指定缓冲区。如果 audioBuffer 不是直接的缓冲区，此方法总是返回 0。上述参数的具体说明如下所示。

  ➢ audioBuffer：存储写入音频录制数据的缓冲区。
  ➢ sizeInBytes：请求的最大字节数。

读入缓冲区的总 byte 数，如果对象属性没有初始化，则返回 ERROR_INVALID_OPERATION；如果参数不能解析成有效的数据或索引，则返回 ERROR_BAD_VALUE。读取的总 byte 数不会超过 sizeInBytes。

☑ public void release()：释放本地 AudioRecord 资源。对象不能经常使用此方法，而且在调用 release()后，必须设置引用为 null。

☑ public int setNotificationMarkerPosition(int markerInFrames)：如果设置了 setRecordPosition UpdateListener(OnRecordPositionUpdateListener)或 setRecordPositionUpdate Listener(OnRecordPosition UpdateListener, Handler)，则通知监听者设置位置标记。参数 markerInFrames 表示在框架中快速标记位置。

☑ public int setPositionNotificationPeriod(int periodInFrames)：如果设置了 setRecordPositionUpdate Listener(OnRecordPositionUpdateListener)或 setRecordPositionUpdateListener(OnRecordPosition UpdateListener, Handler)，则通知监听者设置时间标记。参数 markerInFrames 表示在框架中快速更新时间标记。

☑ public void setRecordPositionUpdateListener(AudioRecord.OnRecordPositionUpdateListener listener, Handler handler)：当之前设置的标志已经成立，或者周期录制位置更新时，设置处理监听者。使用此方法将 Handler 和其他的线程联系起来来接收 AudioRecord 事件，比创建 AudioTrack 实例更好一些。参数 handler 用来接收事件通知消息。

☑ public void setRecordPositionUpdateListener(AudioRecord.OnRecordPositionUpdateListener listener)：当之前设置的标志已经成立，或者周期录制位置更新时，设置处理监听者。

☑ public void startRecording()：表示 AudioRecord 实例开始进行录制。

（4）受保护方法

在 AudioRecord 中受保护方法是 protected void finalize()，此方法用于通知 VM 回收此对象内存。只能被用在运行的应用程序没有任何线程再使用此对象，来告诉垃圾回收器回收此对象。

方法 finalize()用于释放系统资源，由垃圾回收器清除此对象。默认没有实现，由 VM 来决定，但子类根据需要可重写 finalize()。在执行期间，调用此方法可能会立即抛出未定义异常，但是可以忽略。

**注意**：VM 保证对象可以一次或多次调用 finalize()，但并不保证 finalize()会马上执行。例如，对象 B 的 finalize()可能延迟执行，等待对象 A 的 finalize()延迟回收 A 的内存。为了安全起见，请查看 ReferenceQueue，它提供了更多的控制 VM 的垃圾回收。

另外，需要在 Activity 的线程中创建 AudioRecord 对象，可以在独立的线程中读取数据，否则像华为 U8800 之类手机录音时会出错。

# 8.7　在物联网设备中播放音乐

**知识点讲解：光盘:视频\知识点\第 8 章\在物联网设备中播放音乐.avi**

在整个手机体系中，播放音频功能才是真正的核心，例如 MP3 播放。本节将详细讲解在 Android 物联网设备中播放音乐的知识，为读者步入本书后面知识的学习打下基础。

## 8.7.1　使用 AudioTrack 播放音频

要想学好 AudioTrack API，读者可以从分析 Android 源码中的 Java 源码做起。其具体代码如下。

```
//根据采样率、采样精度和单双声道来取得 frame 的大小
int bufsize = AudioTrack.getMinBufferSize(8000,                    //每秒 8K 个点
  AudioFormat.CHANNEL_CONFIGURATION_STEREO,                        //双声道
AudioFormat.ENCODING_PCM_16BIT);                                   //一个采样点 16 比特-2 个字节

//创建 AudioTrack
AudioTrack trackplayer = new AudioTrack(AudioManager.STREAM_MUSIC, 8000,
  AudioFormat.CHANNEL_CONFIGURATION_ STEREO,
  AudioFormat.ENCODING_PCM_16BIT,
  bufsize,
AudioTrack.MODE_STREAM);
 trackplayer.play() ;                                              //开始
trackplayer.write(bytes_pkg, 0, bytes_pkg.length) ;                //向 track 中写数据
...
trackplayer.stop();                                                //停止播放
trackplayer.release();                                             //释放底层资源
```

在上述 AudioTrack 代码中，有 MODE_STATIC 和 MODE_STREAM 两种分类。STREAM 的意思

是由用户在应用程序通过 write（写）方式把数据一次一次地写到 AudioTrack 中。这个和在 Socket 中发送数据一样，应用层从某个地方获取数据，例如通过编/解码得到 PCM 数据，然后写入到 AudioTrack。这种方式的坏处就是总是在 Java 层和 Native 层交互，效率损失较大。

而 STATIC 的意思是一开始创建时就把音频数据放到一个固定的 buffer，然后直接传给 AudioTrack，后续就不用一次次地 write 了。AudioTrack 会自己播放这个 buffer 中的数据。这种方法对于铃声等内存占用较小、延时要求较高的声音来说很适用。

## 8.7.2 使用 MediaPlayer 播放音频

MediaPlayer 的功能比较强大，不但可以播放音频，而且可以播放视频，并且还可以通过 VideoView 播放视频。和 VideoView 相比，MediaPlayer 的优点非常多，例如简单易用。但是 MediaPlayer 也不是万能的，也有缺点——播放视频时需要 SurfaceView 帮忙。但是总体来说，SurfaceView 比普通的自定义 View 更有绘图上的优势，它支持完全的 OpenGL ES 库。

MediaPlayer 能被用来控制音频/视频文件或流媒体的回放，可以在 VideoView 中找到关于如何使用这个类中的方法的例子。使用 MediaPlayer 实现音频/视频播放的基本步骤如下：

（1）生成 MediaPlayer 对象，根据播放文件从不同的地方使用不同的生成方式（参考 MediaPlayer API 即可）。

（2）得到 MediaPlayer 对象后，根据实际需要调用不同的方法，如 start()、stop()、pause()和 release()等。

读者需要注意的是，在不需要播放时要及时释放掉与 MediaPlayer 对象相连接的播放文件，因为直接使用 MediaPlayer 对象一般都是进行音频播放。

## 8.7.3 使用 SoundPool 播放音频

SoundPool 在 Android 系统中的地位一般，但是 Android 还偏偏离不开这种类型。原因是 MediaPlayer 适合播放长点的音频。而 SoundPool 能够播放一些短的反应速度要求高的声音，如游戏中的爆破声。正是这一功能，所以 SoundPool 被保留了下来。

### 1. 主要特点

（1）SoundPool 使用了独立的线程来载入音乐文件，不会阻塞 UI 主线程的操作。但是这里如果音效文件过大没有载入完成，调用 play()方法时可能产生严重的后果，这里 Android SDK 提供了一个 SoundPool.OnLoadCompleteListener 类来帮助我们了解媒体文件是否载入完成，重载 onLoadComplete (SoundPool soundPool, int sampleId, int status)方法即可获得。

（2）从上面的 onLoadComplete()方法可以看出该类有很多参数，例如类似 id，SoundPool 在 load 时可以处理多个媒体一次初始化并放入内存中，这里效率比 MediaPlayer 高了很多。

（3）SoundPool 类支持同时播放多个音效，这对于游戏来说是十分必要的，而 MediaPlayer 类是同步执行的，只能一个文件一个文件地播放。

### 2. 载入音效的方法

☑ int load(Context context, int resId, int priority)：从 APK 资源载入。

☑ int load(FileDescriptor fd, long offset, long length, int priority)：从 FileDescriptor 对象载入。

☑ int load(AssetFileDescriptor afd, int priority)：从 Asset 对象载入。

☑ int load(String path, int priority)：从完整文件路径名载入。

## 8.8　为物联网设备实现振动功能

**知识点讲解：光盘:视频\知识点\第 8 章\为物联网设备实现振动功能.avi**

无论是智能手机还是普通手机，几乎每一款手机都具备振动功能。在 Android 系统中，振动功能是通过类 Vibrator 实现的，读者可以在 SDK 中的 android.os.Vibrator 找到相关的描述。振动方法的语法格式如下：

```
vibrate(long[] pattern, int repeat)
```

☑ long[] pattern：是一个节奏数组，如{1, 200}。

☑ repeat：是重复次数，-1 为不重复，而数字直接表示的是具体的数字，和一般-1 表示无限不同。

在使用振动功能之前，需要先在 manifest 中加入下面的权限。

```
<uses-permission android:name="android.permission.VIBRATE"/>
```

在设置振动（Vibration）事件时，必须要知道命令其振动的时间长短、振动事件的周期等。因为在 Android 中设置的数值都是以毫秒（1000 毫秒=1 秒）来做计算的，所以在做设置时，必须要注意设置时间的长短，如果设置的时间值太小，会感觉不出来。

要让手机振动，需创建 Vibrator 对象，通过调用 vibrate()方法来达到振动的目的，在 Vibrator 的构造器中有 4 个参数，前 3 个的值是设置振动的大小，可以把数值改成一大一小，这样就可以明显感觉出振动的差异，而最后一个值是设置振动的时间。

在 Android 系统中，开发振动应用程序的基本流程如下所示。

（1）在 manifest 文件中声明振动权限。

（2）通过系统服务获得手机振动服务，例如下面的代码。

```
Vibrator vibrator = (Vibrator)getSystemService(VIBRATOR_SERVICE);
```

（3）得到振动服务后检测 vibrator 是否存在，例如下面的代码。

```
vibrator.hasVibrator();
```

通过上述代码可以检测当前硬件是否有 vibrator，如果有，返回 true，如果没有，返回 false。

（4）根据实际需要进行适当的调用，例如下面的代码。

```
vibrator.vibrate(long milliseconds);
```

通过上述代码开始启动 vibrator 持续 milliseconds 毫秒。

（5）编写下面的代码。

```
vibrator.vibrate(long[] pattern, int repeat);
```

这样以 pattern 方式重复 repeat 次启动 vibrator。pattern 的形式如下：

```
new   long[]{arg1,arg2,arg3,arg4......}
```

在上述格式中，其中以两个一组的如 arg1 和 arg2 为一组、arg3 和 arg4 为一组，每一组的前一个代表等待多少毫秒启动 vibrator，后一个代表 vibrator 持续多少毫秒停止，之后往复即可。repeat 表示重复次数，当其为-1 时，表示不重复只以 pattern 的方式运行一次。

（6）停止振动，代码如下。

```
vibrator.cancel();
```

## 8.9 实战闹钟功能

**知识点讲解：光盘:视频\知识点\第 8 章\实战闹钟功能.avi**

在 Android 系统中是通过 AlarmManage 实现闹钟功能的，对应 AlarmManage 有一个 AlarmManagerServie 服务程序，该服务程序才是真正提供闹铃服务的，它主要维护应用程序注册下来的各类闹铃并适时设置即将触发的闹铃给闹铃设备。在 Android 系统中，Linux 实现的设备名为“/dev/alarm”，并且一直监听闹铃设备，一旦有闹铃触发或者是闹铃事件发生，AlarmManagerServie 服务程序就会遍历闹铃列表找到相应的注册闹铃并发出广播。该服务程序在系统启动时被系统服务程序 System_service 启动并初始化闹铃设备（/dev/alarm）。当然，在 Java 层的 AlarmManagerService 与 Linux Alarm 驱动程序接口之间还有一层封装，那就是 JNI。

AlarmManager 将应用与服务分割开来后，使得应用程序开发者不用关心具体的服务，而是直接通过 AlarmManager 来使用这种服务。AlarmManager 与 AlarmManagerServie 之间是通过 Binder 来通信的，它们之间是多对一的关系。

在 Android 系统中，AlarmManage 提供了 3 个接口 5 种类型的闹铃服务。其中 3 个接口如下所示。

```
//取消已经注册的与参数匹配的闹铃
void cancel(PendingIntent operation)
//注册一个新的闹铃
void set( int type, long triggerAtTime, PendingIntent operation)
//注册一个重复类型的闹铃
void setRepeating( int type, long triggerAtTime, long interval, PendingIntent operation)
//设置时区
void setTimeZone(String timeZone)
```

5 个闹铃类型如下所示。

```
public static final int ELAPSED_REALTIME
//当系统进入睡眠状态时，这种类型的闹铃不会唤醒系统。直到系统下次被唤醒才传递它，该闹铃所用的时间是相对时间，是从系统启动后开始计时的，包括睡眠时间，可以通过调用 SystemClock.elapsedRealtime()获得。系统值是 3(0x00000003)
        public static final int ELAPSED_REALTIME_WAKEUP
        //能唤醒系统，用法同 ELAPSED_REALTIME，系统值是 2(0x00000002)
        public static final int RTC

//当系统进入睡眠状态时，这种类型的闹铃不会唤醒系统。直到系统下次被唤醒才传递它，该闹铃所用的时间是绝对时间，所用时间是 UTC 时间，可以通过调用 System.currentTimeMillis()获得。系统值是 1(0x00000001)
        public static final int RTC_WAKEUP
        //能唤醒系统，用法同 RTC 类型，系统值为 0(0x00000000)
        Public static final int POWER_OFF_WAKEUP
//能唤醒系统，它是一种关机闹铃，就是说设备在关机状态下也可以唤醒系统，所以把它称之为关机闹铃。使用方法同 RTC 类型，系统值为 4(0x00000004)
```

# 第 9 章　GPS 地图定位

Map 地图对大家来说应该不算陌生，谷歌地图更是名扬大地，被广泛用于商业、民用和军用项目中。作为谷歌官方旗下产品之一的 Android 系统，可以非常方便地使用 Google 地图实现位置定位功能。本章将详细讲解在 Android 设备中使用位置服务和地图 API 的基本流程，为读者步入本书后面知识的学习打下基础。

## 9.1　位 置 服 务

**知识点讲解：光盘:视频\知识点\第 9 章\位置服务.avi**

在 Android 系统中，可以使用谷歌地图获取当前的位置信息，Android 系统可以无缝地支持 GPS 和谷歌网络地图。在现实应用中，通常将各种不同的定位技术称之为 LBS（意为基于位置的服务，是 Location Based Service 的缩写），它是通过电信移动运营商的无线电通信网络（如 GSM 网、CDMA 网）或外部定位方式（如 GPS）获取移动终端用户的位置信息（地理坐标或大地坐标），在 GIS（Geographic Information System，地理信息系统）平台的支持下，为用户提供相应服务的一种增值业务。本节将详细讲解在 Android 物联网设备中实现位置服务的基本知识。

### 9.1.1　类 location 详解

在 Android 设备中，可以使用类 android.location 来实现定位功能。

（1）Google Map API

Android 系统提供了一组访问 Google MAP 的 API，借助 Google MAP 及定位 API，就可以在地图上显示用户当前的地理位置。在 Android 中定义了一个名为 com.google.android.maps 的包，其中包含了一系列用于在 Google Map 上显示、控制和层叠信息的功能类，下面是该包中最重要的几个类。

☑ MapActivity：用于显示 Google MAP 的 Activity 类，它需要连接底层网络。
☑ MapView：用于显示地图的 View 组件，它必须和 MapActivity 配合使用。
☑ MapController：用于控制地图的移动。
☑ Overlay：是一个可显示于地图之上的可绘制的对象。
☑ GeoPoint：是一个包含经纬度位置的对象。

（2）Android Location API

在 Android 设备中，实现定位功能的相关类如下所示。

☑ LocationManager：提供访问定位服务的功能，也提供了获取最佳定位提供者的功能。另外，临近警报功能（前面所说的那种功能）也可以借助该类来实现。
☑ LocationProvider：是定位提供者的抽象类。定位提供者具备周期性报告设备地理位置的功能。

☑ LocationListener：提供定位信息发生改变时的回调功能。必须事先在定位管理器中注册监听器对象。

☑ Criteria：使得应用能够通过在 LocationProvider 中设置的属性来选择合适的定位提供者。

## 9.1.2 实现定位服务功能

在 Android 设备中，实现定位处理的基本流程如下所示。

（1）编写 Activity 类

这一步的目的是使用 Google Map API 来显示地图，然后使用定位 API 来获取设备的当前定位信息，以在 Google Map 上设置设备的当前位置，用户定位会随着用户的位置移动而发生改变。

首先需要一个继承 MapActivity 的 Activity 类，例如下面的代码。

```
class MyGPSActivity extends MapActivity {
        ......
        }
```

要成功引用 Google MAP API，还必须先在 AndroidManifest.xml 中定义如下信息。

```
<uses-library android:name="com.google.android.maps"/>
```

（2）使用 MapView

要先在设备屏幕中显示地图，需要将 MapView 加入到应用中来。例如在布局文件（main.xml）中加入如下所示的代码。

```
<com.google.android.maps.MapView
                android:id="@+id/myGMap"
                android:layout_width="fill_parent"
                android:layout_height="fill_parent"
                android:enabled="true"
                android:clickable="true"
                android:apiKey="API_Key_String"
/>
```

另外，要使用 Google Map 服务，还需要一个 API key。可以通过如下方式获取 API key：

☑ 找到 USER_HOME\Local Settings\Application Data\Android 目录下的 debug.keystore 文件。

☑ 使用 keytool 工具来生成认证信息（MD5），使用如下命令行：

```
keytool -list -alias androiddebugkey -keystore <path_to_debug_keystore>.keystore -storepass
android -keypass android
```

☑ 打开 Sign Up for the Android Maps API 页面，输入之前生成的认证信息（MD5）将获取到用户的 API key。

☑ 替换上面 AndroidManifest.xml 配置文件中 API_Key_String 为刚才获取的 API key。

**注意：**上面获取 API key 的介绍比较简单，后面将会通过一个具体实例来演示获取 API key 的方法。

继续补全 MyGPSActivity 类的代码，在此使用 MapView，例如下面的代码。

```
class MyGPSActivity extends MapActivity {
    @Override
    public void onCreate(Bundle savedInstanceState) {
    //创建并初始化地图
        gMapView = (MapView) findViewById(R.id.myGMap);
            GeoPoint p = new GeoPoint((int) (lat * 1000000), (int) (long * 1000000));
```

```
            gMapView.setSatellite(true);
            mc = gMapView.getController();
            mc.setCenter(p);
            mc.setZoom(14);
    }
}
```

另外，必须先设置一些权限后才能使用定位信息，在文件 AndroidManifest.xml 中的配置方式如下：

```
<uses-permission android:name="android.permission.INTERNET"></uses-permission>
<uses-permission android:name="android.permission.ACCESS_COARSE_LOCATION"></uses-permission>
<uses-permission android:name="android.permission.ACCESS_FINE_LOCATION"></uses-permission>
```

（3）实现定位管理器

可以使用 Context.getSystemService()方法实现定位管理器功能，并传入 Context.LOCATION_SERVICE 参数来获取定位管理器的实例。例如下面的代码。

```
LocationManager lm = (LocationManager) getSystemService(Context.LOCATION_SERVICE);
```

将原先的 MyGPSActivity 作一些修改，让它实现一个 LocationListener 接口，使其能够监听定位信息的改变：

```
class MyGPSActivity extends MapActivity implements LocationListener {
…………
public void onLocationChanged(Location location) {}
public void onProviderDisabled(String provider) {}
public void onProviderEnabled(String provider) {}
public void onStatusChanged(String provider, int status, Bundle extras) {}
protected boolean isRouteDisplayed() {
return false;
}
}
```

初始化 LocationManager，并在它的 onCreate()方法中注册定位监听器。例如下面的代码。

```
@Override
public void onCreate(Bundle savedInstanceState) {
LocationManager lm = (LocationManager)getSystemService(Context.LOCATION_SERVICE);
lm.requestLocationUpdates(LocationManager.GPS_PROVIDER, 1000L, 500.0f, this);
        }
```

这样代码中的方法 onLocationChanged()会在用户的位置发生 500 米距离的改变之后进行调用。这里默认使用的 LocationProvider 是“gps”（GSP_PROVIDER），但是可以根据需要，使用特定的 Criteria 对象调用 LocationManger 类的 getBestProvider()方法获取其他的 LocationProvider。以下代码是 onLocationChanged()方法的参考实现。

```
  public void onLocationChanged(Location location) {
    if (location != null) {
    double lat = location.getLatitude();
    double lng = location.getLongitude();
      p = new GeoPoint((int) lat * 1000000, (int) lng * 1000000);
      mc.animateTo(p);
    }
}
```

通过上面的代码，获取了当前的新位置并在地图上更新位置显示。还可以为应用程序添加一些诸如缩放效果、地图标注和文本等功能。

（4）添加缩放控件

```
//将缩放控件添加到地图上
ZoomControls zoomControls =   (ZoomControls) gMapView.getZoomControls();
zoomControls.setLayoutParams(new ViewGroup.LayoutParams(LayoutParams.WRAP_CONTENT,
LayoutParams.WRAP_CONTENT));
gMapView.addView(zoomControls);
gMapView.displayZoomControls(true);
```

（5）添加 Map Overlay

最后一步是添加 Map Overlay，例如通过下面的代码可以定义一个 overlay。

```
  class MyLocationOverlay extends com.google.android.maps.Overlay {
    public boolean draw(Canvas canvas, MapView mapView, boolean shadow, long when) {
  super.draw(canvas, mapView, shadow);
  Paint paint = new Paint();
  //将经纬度转换成实际屏幕坐标
  Point myScreenCoords = new Point();
  mapView.getProjection().toPixels(p, myScreenCoords);
  paint.setStrokeWidth(1);
  paint.setARGB(255, 255, 255, 255);
  paint.setStyle(Paint.Style.STROKE);
  Bitmap bmp = BitmapFactory.decodeResource(getResources(), R.drawable.marker);
  canvas.drawBitmap(bmp, myScreenCoords.x, myScreenCoords.y, paint);
  canvas.drawText("how are you…", myScreenCoords.x, myScreenCoords.y, paint);
  return true;
}
}
```

通过上面的 Overlay 会在地图上显示一段文本，接下来可以把这个 Overlay 添加到地图上去。

```
MyLocationOverlay myLocationOverlay = new MyLocationOverlay();
List<Overlay> list = gMapView.getOverlays();
list.add(myLocationOverlay);
```

## 9.1.3　实战演练——在 Android 设备中实现 GPS 定位

下面将通过具体实例来演示在 Android 设备中实现 GPS 定位功能的基本流程。

| 题　　目 | 目　　的 | 源 码 路 径 |
|---|---|---|
| 实例 9-1 | 用 GPS 定位技术获取当前的位置信息 | 光盘:\daima\9\GPSLocationEX |

本实例的具体实现流程如下所示。

（1）在文件 AndroidManifest.xml 中添加 ACCESS_FINE_LOCATION 权限，具体代码如下：

```
<uses-permission android:name="android.permission.ACCESS_FINE_LOCATION"/>
```

（2）在 onCreate(Bundle savedInstanceState)中获取当前位置信息，通过 LocationManager 周期性获得当前设备的一个类。要想获取 LocationManager 实例，必须调用 Context.getSystemService()方法并传入服务名 LOCATION_SERVICE("location")。创建 LocationManager 实例后可以通过调用 getLastKnown Location()方法将上一次 LocationManager 获得有效位置信息以 Location 对象的形式返回。getLastKnownLocation()方法需要传入一个字符串参数来确定使用定位服务类型，本实例传入的是静态常量 LocationManager.GPS_PROVIDER，这表示使用 GPS 技术定位。最后还需要使用 Location 对象将

位置信息以文本方式显示到用户界面。具体实现代码如下：

```
public void onCreate(Bundle savedInstanceState) {
    super.onCreate(savedInstanceState);
    setContentView(R.layout.main);
    LocationManager locationManager;
    String serviceName = Context.LOCATION_SERVICE;
    locationManager = (LocationManager)getSystemService(serviceName);
    Criteria criteria = new Criteria();
    criteria.setAccuracy(Criteria.ACCURACY_FINE);
    criteria.setAltitudeRequired(false);
    criteria.setBearingRequired(false);
    criteria.setCostAllowed(true);
    criteria.setPowerRequirement(Criteria.POWER_LOW);
    String provider = locationManager.getBestProvider(criteria, true);

    Location location = locationManager.getLastKnownLocation(provider);
    updateWithNewLocation(location);
    /*每隔 1000ms 更新一次*/
    locationManager.requestLocationUpdates(provider, 2000, 10,
          locationListener);
}
```

（3）定义方法 updateWithNewLocation(Location location)更新显示用户界面，具体代码如下：

```
    private void updateWithNewLocation(Location location) {
     String latLongString;
     TextView myLocationText;
     myLocationText = (TextView)findViewById(R.id.myLocationText);
     if (location != null) {
     double lat = location.getLatitude();
     double lng = location.getLongitude();
     latLongString = "纬度是:" + lat + "\n 经度是:" + lng;
     } else {
     latLongString = "失败";
     }
     myLocationText.setText("获取的当前位置是:\n" +
     latLongString);
    }
}
```

（4）定义 LocationListener 对象 locationListener，当坐标改变时触发此函数。如果 Provider 传进相同的坐标，它就不会被触发。具体代码如下：

```
private final LocationListener locationListener = new LocationListener() {
 public void onLocationChanged(Location location) {
 updateWithNewLocation(location);
 }
 public void onProviderDisabled(String provider){
 updateWithNewLocation(null);
 }
 public void onProviderEnabled(String provider){ }
 public void onStatusChanged(String provider, int status,
 Bundle extras){ }
 };
```

下面开始测试，因为模拟器上没有 GPS 设备，所以需要在 Eclipse 的 DDMS 工具中提供模拟的 GPS 数据。即依次选择 DDMS | Emulator Control 命令，在弹出的对话框中找到 Location Control 选项，在此输入坐标，完成后单击 Send 按钮，如图 9-1 所示。

图 9-1　设置坐标

因为用到了 Google API，所以要在项目中引入 Google API，右击项目，在弹出的快捷菜单中选择 Properties 命令，在弹出的对话框中选择 Google APIs 版本，如图 9-2 所示。

Project Build Target

| Target Name | Vendor | Platform | API ... |
| --- | --- | --- | --- |
| ☐ Android 1.1 | Android Open Source Project | 1.1 | 2 |
| ☐ Android 1.5 | Android Open Source Project | 1.5 | 3 |
| ☐ Android 1.6 | Android Open Source Project | 1.6 | 4 |
| ☐ Android 2.0 | Android Open Source Project | 2.0 | 5 |
| ☐ Android 2.0.1 | Android Open Source Project | 2.0.1 | 6 |
| ☐ Android 2.1-update1 | Android Open Source Project | 2.1-update1 | 7 |
| ☐ Android 2.2 | Android Open Source Project | 2.2 | 8 |
| ☐ Android 2.2 | Android Open Source Project | 2.2 | 8 |
| ☐ Google APIs | Google Inc. | 2.2 | 8 |
| ☐ GALAXY Tab Addon | Samsung Electronics Co., Ltd. | 2.2 | 8 |
| ☐ Android 2.3 | Android Open Source Project | 2.3 | 9 |
| ☑ Google APIs | Google Inc. | 2.3 | 9 |

图 9-2　引用 Google APIs

这样模拟器运行后会显示当前的坐标，如图 9-3 所示。

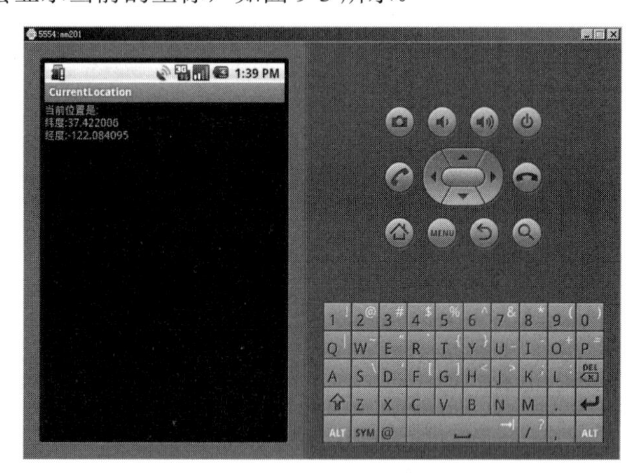


图 9-3　执行效果

## 9.2　随时更新位置信息

**知识点讲解：光盘:视频\知识点\第 9 章\随时更新位置信息.avi**

随着物联网设备的移动，GPS 的位置信息也会发生变化，此时可以通过编程的方式来及时获取并

更新当前的位置信息。本节将详细讲解随时更新位置信息的基本知识。

## 9.2.1　库 Maps 中的类

在库 Maps 中提供了十几个类，通过这些类可以实现位置更新功能。在这些库类中，最为常用的类包括 MapController、MapView 和 MapActivity 等。

（1）MapController

控制地图移动、伸缩，以某个 GPS 坐标为中心，控制 MapView 中的 View 组件，管理 Overlay，提供 View 的基本功能。使用多种地图模式如地图模式（某些城市可实时对交通状况进行更新）、卫星模式、街景模式）来查看 Google Map。常用方法有 animateTo(GeoPoint point)、setCenter(GeoPoint point) 和 setZoom(int zoomLevel)等。

（2）MapView

MapView 是用来显示地图的 view，它派生自 android.view.ViewGroup。当 MapView 获得焦点时，可以控制地图的移动和缩放。Android 中的地图可以以不同的形式显示出来，如街景模式、卫星模式等。

MapView 只能被 MapActivity 来创建，这是因为 MapView 需要通过后台的线程来连接网络或者文件系统，而这些线程要由 MapActivity 来管理。常用方法有 getController()、getOverlays()、setSatellite(boolean)、setTraffic(boolean)、setStreetView(boolean)和 setBuiltInZoomControls(boolean)等。

（3）MapActivity

MapActivity 是一个抽象类，任何想要显示 MapView 的 Activity 都需要派生自 MapActivity。并且在其派生类的 onCreate()中，都要创建一个 MapView 实例，可以通过 MapViewconstructor（然后添加到 View 中 ViewGroup.addView(View)）或者 layout XML 来创建。

（4）Overlay

Overlay 是覆盖到 MapView 的最上层，可以扩展其 ondraw 接口，自定义在 MapView 中显示一些自己的东西。MapView 通过 MapView.getOverlays()对 Overlay 进行管理。

除了 Overlay 这个基类，Google 还扩展了如下两个比较有用的 Overlay。

☑ MylocationOverlay：集成了 Android.location 中接收当前坐标的接口，集成 SersorManager 中 CompassSensor 的接口。只需要 enableMyLocation()，enableCompass 即可让程序拥有实时的 MyLocation 以及 Compass 功能（Activity.onResume()中）。

☑ ItemlizedOverlay：管理一个 OverlayItem 链表，用图片等资源在地图上作风格相同的标记。

（5）Projection

MapView 中 GPS 坐标与设备坐标的转换（GeoPoint 和 Point）。

## 9.2.2　使用 LocationManager 监听位置

LocationManager 支持监听器模式，通过调用 requestLocationUpdates()方法能够为其设置一个位置监听器 LocationListener。同时方法 requestLocationUpdates()还需要指定要使用的位置服务类型、位置更新时间和最新位移，这样可以确保在满足用户需求的前提下最低的电量消耗。

例如在下面的代码中，设置了更新位置信息的最小间隔为 2 秒，位移变化在 10 米以上。如果 GPS 位置超过 10 米，且时间间隔超过 2 秒时，LocationListener 的回调方法 onLocationChanged()就会被调用，应用程序可以通过 onLocationChanged 来反映位置信息的变化。

```
public class CurrentLocationWithMap extends MapActivity {
    public void onCreate(Bundle savedInstanceState) {
        super.onCreate(savedInstanceState);
        setContentView(R.layout.main);
        LocationManager locationManager;
        String context = Context.LOCATION_SERVICE;
        locationManager = (LocationManager)getSystemService(context);
        //String provider = LocationManager.GPS_PROVIDER;
        /*Location Provider 查询条件*/
        Criteria criteria = new Criteria();
        criteria.setAccuracy(Criteria.ACCURACY_FINE);
        criteria.setAltitudeRequired(false);
        criteria.setBearingRequired(false);
        criteria.setCostAllowed(true);
        criteria.setPowerRequirement(Criteria.POWER_LOW);
        String provider = locationManager.getBestProvider(criteria, true);
        Location location = locationManager.getLastKnownLocation(provider);
        updateWithNewLocation(location);
        /*设置更新位置信息的最小间隔为 2 秒，位移变化在 10 米以上*/
        locationManager.requestLocationUpdates(provider, 2000, 10,
                locationListener);
    }
   /*Location  发生变化时被调用*/
   private final LocationListener locationListener = new LocationListener() {
    public void onLocationChanged(Location location) {
    updateWithNewLocation(location);
    }
    public void onProviderDisabled(String provider){
    updateWithNewLocation(null);
    }
    public void onProviderEnabled(String provider){ }
    public void onStatusChanged(String provider, int status,
    Bundle extras){ }
   };
    private void updateWithNewLocation(Location location) {
     String latLongString;
     TextView myLocationText;
     myLocationText = (TextView)findViewById(R.id.myLocationText);
     if (location != null) {
          double lat = location.getLatitude();
          double lng = location.getLongitude();
          latLongString = "纬度是:" + lat + "\n 经度是:" + lng;
          ctrlMap.animateTo(new GeoPoint((int)(lat*1E6),(int)(lng*1E6)));
     } else {
          latLongString = "失败";
     }
     myLocationText.setText("获取的当前位置是:\n" +latLongString);

    }
```

在使用类 LocationManager 时，通常需要用到如下所示的方法。

☑ getLatitude()：获取经度值。

☑ getLongitude()：获取纬度值。

☑ getAltitude()：获取海拔值。

### 9.2.3　实战演练——监听当前设备的坐标和海拔

本节将通过具体实例来演示在 Android 设备中显示当前位置的坐标和海拔的基本方法。

| 题　　目 | 目　　的 | 源 码 路 径 |
| --- | --- | --- |
| 实例 9-2 | 显示当前位置的坐标和海拔 | 光盘:\daima\9\GPSEX |

本实例的具体实现流程如下所示。

（1）在文件 AndroidManifest.xml 中添加 ACCESS_FINE_LOCATION 权限和 ACCESS_LOCATION_EXTRA_COMMANDS 权限，具体代码如下：

```
<uses-permission android:name="android.permission.ACCESS_FINE_LOCATION" />
<uses-permission android:name="android.permission.ACCESS_LOCATION_EXTRA_COMMANDS"/>
```

（2）编写布局文件 main.xml，设置在屏幕中分别显示当前位置的经度、纬度、速度和海拔等信息。文件 main.xml 的具体实现代码如下：

```
<LinearLayout xmlns:android="http://schemas.android.com/apk/res/android"
    android:layout_width="fill_parent"
    android:layout_height="fill_parent"
    android:background="#008080"
    android:id="@+id/mainlayout" android:orientation="vertical">

    <gps.mygps.paintview android:id="@+id/iddraw"
     android:layout_width="fill_parent"
     android:layout_height="300dip"
    />

    <TableLayout android:layout_width="fill_parent"
          android:layout_height="wrap_content">

     <TableRow>
          <TextView       android:id="@+id/speed"
                    android:layout_width="wrap_content"
                    android:layout_height="wrap_content"
                    android:text="速度"
                    style="@style/smalltext"
                    android:gravity="center"
                    android:layout_weight="33"/>
          <TextView       android:id="@+id/altitude"
                    android:layout_width="wrap_content"
                    android:layout_height="wrap_content"
                    android:text="海拔"
                    style="@style/smalltext"
                    android:gravity="center"
```

```
            android:layout_weight="33"/>
        <TextView      android:id="@+id/bearing"
            android:layout_width="wrap_content"
            android:layout_height="wrap_content"
            android:text="航向"
            style="@style/smalltext"
            android:gravity="center"
            android:layout_weight="34"/>
    </TableRow>
    <TableRow>
        <TextView      android:id="@+id/speedvalue"
            android:layout_width="wrap_content"
            android:layout_height="wrap_content"
            style="@style/normaltext"
            android:gravity="center"
            android:layout_weight="33"/>
        <TextView      android:id="@+id/altitudevalue"
            android:layout_width="wrap_content"
            android:layout_height="wrap_content"
            style="@style/normaltext"
            android:layout_weight="33"
            android:gravity="center"/>
        <TextView      android:id="@+id/bearvalue"
            android:layout_width="wrap_content"
            android:layout_height="wrap_content"
            style="@style/normaltext"
            android:gravity="center"
            android:layout_weight="34"/>
    </TableRow>
</TableLayout>

<TableLayout android:layout_width="fill_parent"
        android:layout_height="wrap_content">
    <TableRow>
        <TextView      android:layout_width="wrap_content"
            android:layout_height="wrap_content"
            android:text="维度"
            android:gravity="center"
            android:layout_weight="50"
            style="@style/smalltext"/>
        <TextView      android:layout_width="wrap_content"
            android:layout_height="wrap_content"
            android:text="卫星"
            android:gravity="center"
            android:layout_weight="50"
            style="@style/smalltext"/>
        <TextView      android:layout_width="wrap_content"
            android:layout_height="wrap_content"
            android:text="经度"
            style="@style/smalltext"
            android:gravity="center"
```

```
                    android:layout_weight="50"/>
    </TableRow>
    <TableRow>
        <TextView      android:id="@+id/latitudevalue"
                    android:layout_width="wrap_content"
                    android:layout_height="wrap_content"
                    style="@style/normaltext"
                    android:gravity="center"
                    android:layout_weight="33"/>
        <TextView      android:id="@+id/satellitevalue"
                    android:layout_width="wrap_content"
                    android:layout_height="wrap_content"
                    style="@style/normaltext"
                    android:gravity="center"
                    android:layout_weight="33"/>
        <TextView      android:id="@+id/longitudevalue"
                    android:layout_width="wrap_content"
                    android:layout_height="wrap_content"
                    style="@style/normaltext"
                    android:gravity="center"
                    android:layout_weight="34"/>
    </TableRow>
    </TableLayout>
    <TableLayout android:layout_width="fill_parent"
        android:layout_height="wrap_content">
        <TableRow>
            <TextView      android:id="@+id/time"
                        android:layout_width="wrap_content"
                        android:layout_height="wrap_content"
                        android:text="时间:"
                        style="@style/normaltext"
                        />
            <TextView      android:id="@+id/timevalue"
                        android:layout_width="wrap_content"
                        android:layout_height="wrap_content"
                        style="@style/normaltext"
                        />
        </TableRow>
    </TableLayout>

    <RelativeLayout android:layout_width="fill_parent"
            android:layout_height="wrap_content">
    <Button android:id="@+id/close"
        android:layout_width="wrap_content"
        android:layout_height="wrap_content"
        android:text="关闭"
        android:textSize="20sp"
        android:layout_alignParentRight="true"></Button>
    <Button    android:id="@+id/open"
        android:layout_height="wrap_content"
        android:layout_width="wrap_content"
```

```
            android:text="打开"
            android:textSize="20sp"
                android:layout_toLeftOf="@id/close"></Button>
    </RelativeLayout>

    <TextView android:id="@+id/error"
            android:layout_width="fill_parent"
            android:layout_height="wrap_content"
            style="@style/smalltext"
            />
</LinearLayout>
```

（3）编写程序文件 Mygps.java，功能是监听用户单击屏幕按钮的事件，获取当前位置的定位信息。文件 Mygps.java 的具体实现代码如下：

```
public class Mygps extends Activity {

    protected static final String TAG = null;
    //位置类
    private Location location;
    //定位管理类
    private LocationManager locationManager;
    private String provider;
    //监听卫星变量
    private GpsStatus gpsStatus;
    Iterable<GpsSatellite> allSatellites;
    float satellitedegree[][] = new float[24][3];

    float alimuth[] = new   float[24];
    float elevation[] = new float[24];
    float snr[] = new float[24];

    private boolean status=false;
    protected Iterator<GpsSatellite> Iteratorsate;
    private float bear;

    //获取手机屏幕分辨率的类
    private DisplayMetrics dm;


    paintview layout;
    Button openbutton;
    Button closebutton;
    TextView latitudeview;
    TextView longitudeview;
    TextView altitudeview;
    TextView speedview;
    TextView timeview;
    TextView errorview;
    TextView bearingview;
    TextView satcountview;
```

```
    @Override
    public void onCreate(Bundle savedInstanceState) {
        super.onCreate(savedInstanceState);

      requestWindowFeature(Window.FEATURE_NO_TITLE);
      getWindow().setFlags(WindowManager.LayoutParams.FLAG_FULLSCREEN,
WindowManager.LayoutParams.FLAG_FULLSCREEN);

      setContentView(R.layout.main);

        findview();

        openbutton.setOnClickListener(new View.OnClickListener() {

          @Override
          public void onClick(View v) {
                if(!status)
                {
                     openGPSSettings();
                     getLocation();
                     status = true;
                }
          }
    });

        closebutton.setOnClickListener(new View.OnClickListener() {

                @Override
                public void onClick(View v) {
                     closeGps();
                }
          });
    }

    private void findview() {
     openbutton = (Button)findViewById(R.id.open);
     closebutton = (Button)findViewById(R.id.close);
     latitudeview = (TextView)findViewById(R.id.latitudevalue);
     longitudeview = (TextView)findViewById(R.id.longitudevalue);
     altitudeview = (TextView)findViewById(R.id.altitudevalue);
     speedview = (TextView)findViewById(R.id.speedvalue);
     timeview = (TextView)findViewById(R.id.timevalue);
     errorview = (TextView)findViewById(R.id.error);
     bearingview = (TextView)findViewById(R.id.bearvalue);
     layout=(gps.mygps.paintview)findViewById(R.id.iddraw);
     satcountview = (TextView)findViewById(R.id.satellitevalue);
     }


     protected void closeGps() {
```

```
// dm = new DisplayMetrics();
// getWindowManager().getDefaultDisplay().getMetrics(dm);
// heightp = dm.heightPixels;
// widthp = dm.widthPixels;
// alimuth[0] = 60;
// elevation[0] = 20;
// snr[0] = 60;
// alimuth[1] = 260;
// elevation[1] = 10;
// snr[1] = 50;
// layout.redraw(240,alimuth,elevation,snr, widthp,heightp, 2);

        if(status == true)
        {
            locationManager.removeUpdates(locationListener);
            locationManager.removeGpsStatusListener(statusListener);
            errorview.setText("");
            latitudeview.setText("");
            longitudeview.setText("");
            speedview.setText("");
            timeview.setText("");
            altitudeview.setText("");
            bearingview.setText("");
            satcountview.setText("");
            status = false;
        }
    }



    //定位监听类负责监听位置信息的变化情况
    private final LocationListener locationListener = new LocationListener()
    {

        @Override
        public void onLocationChanged(Location location)
        {
        //获取 GPS 信息和位置提供者 provider 中的位置信息
// location = locationManager.getLastKnownLocation(provider);
      //通过 GPS 获取位置
        updateToNewLocation(location);
            //showInfo(getLastPosition(), 2);
        }

        @Override
        public void onProviderDisabled(String arg0)
        {

        }
```

```
        @Override
        public void onProviderEnabled(String arg0)
        {

        }

        @Override
        public void onStatusChanged(String arg0, int arg1, Bundle arg2)
        {
        updateToNewLocation(null);
        }
    };

    //添加监听卫星
    private final GpsStatus.Listener statusListener= new GpsStatus.Listener(){

        @Override
        public void onGpsStatusChanged(int event) {
            //TODO Auto-generated method stub
            //获取 GPS 卫星信息
            gpsStatus = locationManager.getGpsStatus(null);

            switch(event)
            {
            case GpsStatus.GPS_EVENT_STARTED:

            break;
                //第一次定位时间
            case GpsStatus.GPS_EVENT_FIRST_FIX:

            break;
                //收到的卫星信息
            case GpsStatus.GPS_EVENT_SATELLITE_STATUS:
                DrawMap();

            break;

            case GpsStatus.GPS_EVENT_STOPPED:
            break;
            }
        }
    };
    private int heightp;
    private int widthp;

    private void openGPSSettings()
    {

// dm = new DisplayMetrics();
// getWindowManager().getDefaultDisplay().getMetrics(dm);
```

```
// heightp = dm.heightPixels;
// widthp = dm.widthPixels;
// alimuth[0] = 60;
// elevation[0] = 70;
// snr[0] = 60;
// alimuth[1] = 260;
// elevation[1] = 10;
// snr[1] = 50;
// layout.redraw(50,alimuth,elevation,snr, widthp,heightp, 2);

    //获取位置管理服务
    locationManager = (LocationManager)this.getSystemService(Context.LOCATION_SERVICE);
    if (locationManager.isProviderEnabled(android.location.LocationManager.GPS_PROVIDER))
    {
        Toast.makeText(this, "GPS 模块正常", Toast.LENGTH_SHORT).show();
        return;
    }
    status = false;
    Toast.makeText(this, "请开启 GPS！", Toast.LENGTH_SHORT).show();
    Intent intent = new Intent(Settings.ACTION_SECURITY_SETTINGS);
    startActivityForResult(intent,0); //此为设置完成后返回的获取界面     }
   }


    protected void DrawMap() {
        //TODO Auto-generated method stub

        int i = 0;
        //获取屏幕信息
        dm = new DisplayMetrics();
        getWindowManager().getDefaultDisplay().getMetrics(dm);
        heightp = dm.heightPixels;
        widthp = dm.widthPixels;

        //获取卫星信息
        allSatellites = gpsStatus.getSatellites();
        Iteratorsate = allSatellites.iterator();

        while(Iteratorsate.hasNext())
        {
            GpsSatellite satellite = Iteratorsate.next();
            alimuth[i] = satellite.getAzimuth();
            elevation[i] = satellite.getElevation();
            snr[i] = satellite.getSnr();
            i++;
        }
        satcountview.setText(""+i);
    layout.redraw(bear,alimuth,elevation,snr, widthp,heightp, i);
        layout.invalidate();
    }
```

```
private void getLocation()
{

//查找到服务信息，位置数据标准类
Criteria criteria = new Criteria();
//查询精度:高
criteria.setAccuracy(Criteria.ACCURACY_FINE);
//是否查询海拔:是
criteria.setAltitudeRequired(true);
//是否查询方位角:是
criteria.setBearingRequired(true);
//是否允许付费
criteria.setCostAllowed(true);
//电量要求:底
criteria.setPowerRequirement(Criteria.POWER_LOW);
//是否查询速度:是
criteria.setSpeedRequired(true);

provider = locationManager.getBestProvider(criteria, true);


//获取 GPS 信息，获取位置提供者 provider 中的位置信息
location = locationManager.getLastKnownLocation(provider);
 //通过 GPS 获取位置
updateToNewLocation(location);
//设置监听器，自动更新的最小时间为间隔 N 秒（1 秒为 1×1000，这样写主要为了方便）或最小位移变化超
过 N 米
//实时获取位置提供者 provider 中的数据，一旦发生位置变化，立即通知应用程序
locationManager.requestLocationUpdates(provider, 1000, 0,locationListener);
//监听卫星
locationManager.addGpsStatusListener(statusListener);
}


private void updateToNewLocation(Location location)
{

if (location != null)
{
    bear = location.getBearing();
    double   latitude = location.getLatitude();         //纬度
    double longitude= location.getLongitude();          //经度
    float GpsSpeed = location.getSpeed();               //速度
    long GpsTime = location.getTime();                  //时间
    Date date = new Date(GpsTime);

    DateFormat df = new SimpleDateFormat("yyyy-MM-dd HH:mm:ss");

    double GpsAlt = location.getAltitude();             //海拔
    latitudeview.setText("" + latitude);
    longitudeview.setText("" + longitude);
```

```
            speedview.setText(""+GpsSpeed);
            timeview.setText(""+df.format(date));
            altitudeview.setText(""+GpsAlt);
            bearingview.setText(""+bear);

            }
        else
        {
            errorview.setText("无法获取地理信息");
            }
        }
}
```

本实例在模拟器中的执行效果如图 9-4 所示。

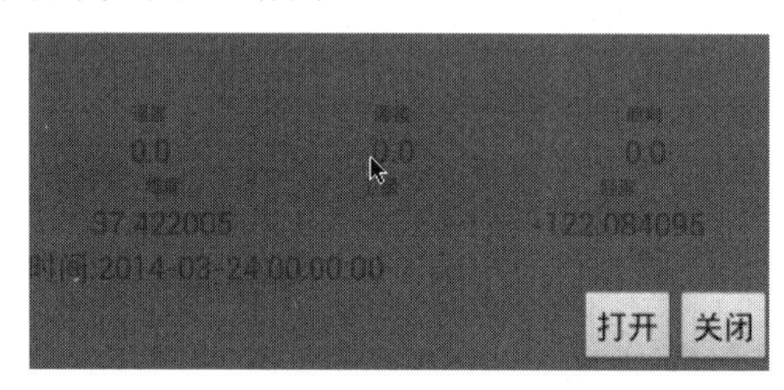


图 9-4　在模拟器中的执行效果

# 9.3　在设备中使用地图

**知识点讲解：光盘:视频\知识点\第 9 章\在设备中使用地图.avi**

在 Android 设备中可以直接使用 Google 地图，可以用地图的形式显示位置信息。下面将详细讲解在 Android 设备中使用 Google 地图的方法。

## 9.3.1　添加 Google Map 密钥

Android 系统中提供了一个 map 包（com.google.android.maps），通过其中的 MapView 可以方便地利用 Google 地图资源来进行编程，可以在 Android 设备中调用 Google 地图。在使用 Google 地图之前需要进行如下所示的配置工作。

（1）添加 maps.jar

在 Android SDK 中，以 JAR 库的形式提供了和 Google Map 有关的 API，此 JAR 库位于 android-sdk-windows\add-ons\google_apis-4 目录下。要把 maps.jar 添加到项目中，可以在项目属性的 Android 栏中指定使用包含 Google API 的 Target 作为项目的构建目标，如图 9-5 所示。

（2）将地图嵌入到应用

通过使用 MapActivity 和 MapView 控件，可以轻松地将地图嵌入到应用程序中。在此步骤中，需要将 Google API 添加到构建路径中。方法是在图 9-6 所示界面中选择 Java Build Path 节点，然后在 Target 中选中 Google APIs 复选框，设置项目中包含 Google API。

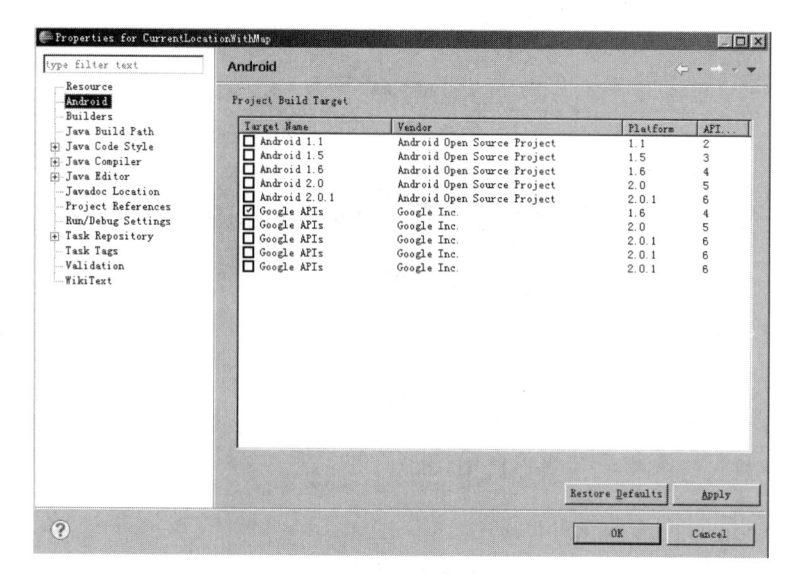


图 9-5　在项目中包含 Google API

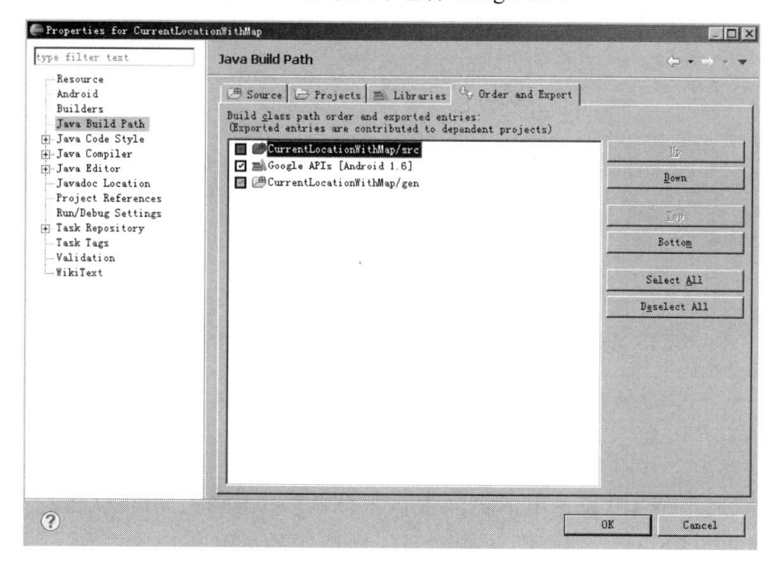


图 9-6　将 Google API 添加到构建路径

（3）获取 Map API 密钥

在利用 MapView 之前，必须要先申请一个 Android Map API Key。具体步骤如下。

第 1 步：找到 debug.keystore 文件，通常位于如下目录中。

C:\Documents and Settings\你的当前用户\Local Settings\Application Data\Android

第 2 步：获取 MD5 指纹。运行 cmd.exe，执行如下命令获取 MD5 指纹。

```
>keytool -list -alias androiddebugkey -keystore "debug.keystore 的路径" -storepass android -keypass android
```

例如笔者机器输入如下命令：

```
keytool -list -alias androiddebugkey -keystore "C:\Documents and Settings\Administrator\.android\debug.keystore"
-storepass android -keypass android
```

此时系统会提示输入 keystore 密码，这时输入“android”，系统就会输出我们申请到的 MD5 认证指纹，如图 9-7 所示。

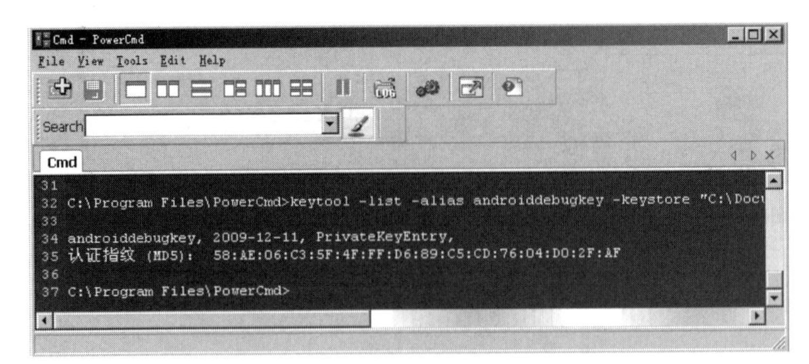


图 9-7　获取的认证指纹

**注意：** 因为在 CMD 中不能直接复制、粘贴，这样很影响我们的编程效率，所以笔者使用了第三方软件 PowerCmd 来代替机器中自带的 CMD 工具。

第 3 步：申请 Android map 的 API Key。

打开浏览器，输入网址 http://code.google.com/intl/zh-CN/android/maps-api-signup.html，如图 9-8 所示。

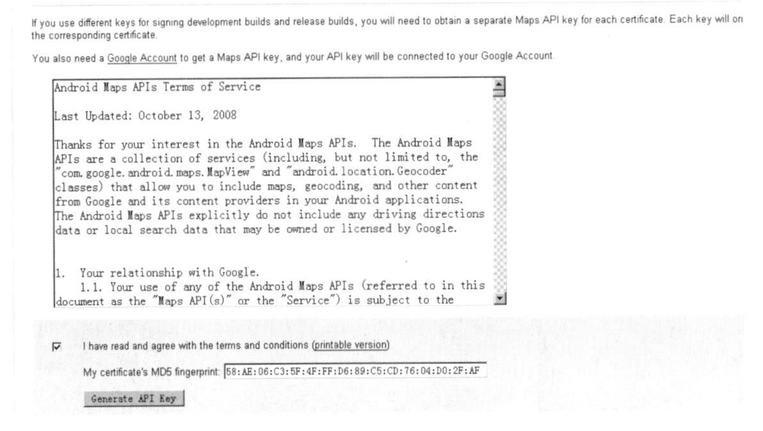


图 9-8　申请主页

在 Google 的 android map API Key 申请页面上输入图 9-8 中得到的 MD5 认证指纹，单击 Generate API Key 按钮后即可转到如图 9-9 所示的画面，得到我们申请到的 API Key。

图 9-9　得到的 API Key

至此，就成功地获取了一个 API Key。

## 9.3.2　使用 Map API 密钥

当申请到一个 Android Map API Key 后，接下来可以使用 Map API 密钥实现编程，具体实现流程如下。

（1）在 AndroidManifest.xml 中声明权限

在 Android 系统中，如果程序执行需要读取到安全敏感的项目，那么必须在 AndroidManifest.xml 中声明相关权限请求，例如这个地图程序需要从网络读取相关数据。所以必须声明 android.permission.INTERNET 权限。具体方法是在文件 AndroidManifest.xml 中添加如下代码。

```
<uses-permission android:name="android.permission.INTERNET" />
```

另外，因为 maps 类不是 Android 启动的默认类，所以还需要在文件 AndroidManifest.xml 的 application 标签中声明要用 maps 类：

```
<uses-library android:name="com.google.android.maps" />
```

下面是基本的 AndroidManifest.xml 文件代码：

```
<manifest xmlns:android="http://schemas.android.com/apk/res/android"
    <application android:icon="@drawable/icon" android:label="@string/app_name">
     <uses-library android:name="com.google.android.maps" />
    </application>
 <uses-permission android:name="android.permission.INTERNET" />
</manifest>
```

（2）在布局文件 main.xml 中规划 UI 界面

假设要显示杭州的卫星地图，并在地图上方有 5 个按钮，分别可以放大地图、缩小地图或者切换显示模式（卫星、交通和街景）。即整个界面主要由两部分组成，上面是一排 5 个按钮，下面是 Map View。

Android 中的 LinearLayout 可以互相嵌套，在此可以把上面 5 个按钮放在一个子 LinearLayout 里边（子 LinearLayout 的指定可以由 android:addStatesFromChildren="true" 实现），然后再把这个子 LinearLayout 加到外面的父 LinearLayout 中。具体实现代码如下：

```
*为了简化篇幅，去掉一些不是重点说明的属性
<LinearLayout xmlns:android="http://schemas.android.com/apk/res/android"
 android:orientation="vertical" android:layout_width="fill_parent"
 android:layout_height="fill_parent">

 <LinearLayout android:layout_width="fill_parent"
  android:addStatesFromChildren="true"                /*说明是子 Layout
  android:gravity="center_vertical"                   /*这个子 Layout 中的按钮是横向排列
  >

  <Button android:id="@+id/ZoomOut"
   android:text="放大"
   android:layout_width="wrap_content"
   android:layout_height="wrap_content"
   android:layout_marginTop="5dip"                    /*下面的 4 个属性，指定了按钮的相对位置
   android:layout_marginLeft="30dip"
   android:layout_marginRight="5dip"
   android:layout_marginBottom="5dip"
```

```
    android:padding="5dip" />

  /*其余 4 个按钮省略
 </LinearLayout>
 <com.google.android.maps.MapView
  android:id="@+id/map"
  android:layout_width="fill_parent"
  android:layout_height="fill_parent"
  android:enabled="true"
  android:clickable="true"
  android:apiKey="在此输入 9.3.1 节申请的 API KEY"                /*必须加上 9.3.1 节申请的 API KEY
 />

</LinearLayout>
```

（3）设置主文件的这个类必须继承于 MapActivity

```
public class Mapapp extends MapActivity {
```

onCreate()函数的核心代码如下：

```
 public void onCreate(Bundle icicle) {
 //取得地图 View
   myMapView = (MapView) findViewById(R.id.map);
   //设置为卫星模式
   myMapView.setSatellite(true);
   //地图初始化的点:杭州
   GeoPoint p = new GeoPoint((int) (30.27 * 1000000),
      (int) (120.16 * 1000000));
   //取得地图 View 的控制
   MapController mc = myMapView.getController();
   //定位到杭州
   mc.animateTo(p);
   //设置初始化倍数
   mc.setZoom(DEFAULT_ZOOM_LEVEL);
}
```

然后编写缩放按钮的处理代码，具体如下：

```
btnZoomIn.setOnClickListener(new View.OnClickListener() {
 public void onClick(View view) {
    myMapView.getController().setZoom(myMapView.getZoomLevel() - 1);
});
```

地图模式的切换由下面的代码实现：

```
 btnSatellite.setOnClickListener(new View.OnClickListener() {
 public void onClick(View view) {
   myMapView.setSatellite(true);                //卫星模式为 True
   myMapView.setTraffic(false);                 //交通模式为 False
   myMapView.setStreetView(false);              //街景模式为 False
 }
});
```

至此，就完成了第一个使用 Map API 的应用程序。

### 9.3.3　实战演练——在 Android 设备中使用谷歌地图实现定位

本实例的功能是在 Android 设备中使用谷歌地图实现定位功能。

| 题　　目 | 目　　的 | 源 码 路 径 |
| --- | --- | --- |
| 实例 9-3 | 在 Android 设备中使用谷歌地图实现定位 | 光盘:\daima\9\LocationMapEX |

其具体实现流程如下所示。

（1）在布局文件 main.xml 中插入两个 Button，分别实现对地图的“放大”和“缩小”，然后通过 ToggleButton 控制是否显示卫星地图，最后设置申请的 API Key。具体代码如下：

```
<?xml version="1.0" encoding="utf-8"?>
<LinearLayout xmlns:android="http://schemas.android.com/apk/res/android"
    android:orientation="vertical"
    android:layout_width="fill_parent"
    android:layout_height="fill_parent"
    >
<TextView
     android:id="@+id/myLocationText"
    android:layout_width="fill_parent"
    android:layout_height="wrap_content"
    />
<LinearLayout
    android:orientation="horizontal"
    android:layout_width="fill_parent"
    android:layout_height="wrap_content" >
    <Button
        android:id="@+id/in"
        android:layout_width="fill_parent"
        android:layout_height="wrap_content"
        android:layout_weight="1"
        android:text="放大地图" />
    <Button
        android:id="@+id/out"
        android:layout_width="fill_parent"
        android:layout_height="wrap_content"
        android:layout_weight="1"
        android:text="缩小地图" />
</LinearLayout>
<ToggleButton
    android:id="@+id/switchMap"
    android:layout_width="wrap_content"
    android:layout_height="wrap_content"
    android:textOff="卫星开关"
    android:textOn="卫星开关"/>
<com.google.android.maps.MapView
    android:id="@+id/myMapView"
    android:layout_width="fill_parent"
    android:layout_height="fill_parent"
```

```
        android:clickable="true"
        android:apiKey="0by7ffx8jX0A_LWXeKCMTWAh8CqHAlqvzetFqjQ"
        />
</LinearLayout>
```

（2）在文件 AndroidManifest.xml 中分别声明 android.permission.INTERNET 和 INTERNET 权限。具体代码如下：

```
<?xml version="1.0" encoding="utf-8"?>
<manifest xmlns:android="http://schemas.android.com/apk/res/android"
      package="com.UserCurrentLocationMap"
      android:versionCode="1"
      android:versionName="1.0.0">
    <application android:icon="@drawable/icon" android:label="@string/app_name">
        <activity android:name=".UserCurrentLocationMap"
                  android:label="@string/app_name">
            <intent-filter>
                <action android:name="android.intent.action.MAIN" />
                <category android:name="android.intent.category.LAUNCHER" />
            </intent-filter>
        </activity>
        <uses-library android:name="com.google.android.maps"/>
    </application>
    <uses-permission android:name="android.permission.INTERNET"/>
     <uses-permission android:name="android.permission.ACCESS_FINE_LOCATION"/>
</manifest>
```

（3）编写主程序文件 CurrentLocationWithMap.java，具体实现流程如下所示。

☑ 通过方法 onCreate()将 MapView 绘制到屏幕上。因为 MapView 只能继承自 MapActivity 中的活动，所以必须用方法 onCreate()将 MapView 绘制到屏幕上，并同时覆盖方法 isRouteDisplayed()，它表示是否需要在地图上绘制导航线路。主要代码如下：

```
package com.UserCurrentLocationMap;
…………………………………………………
public class CurrentLocationWithMap extends MapActivity {
    MapView map;

    MapController ctrlMap;
    Button inBtn;
    Button outBtn;
    ToggleButton switchMap;
        @Override
     protected boolean isRouteDisplayed() {
          return false;
     }
```

☑ 定义方法 onCreate()，首先引入主布局 main.xml，并通过方法 findViewById()获得 MapView 对象的引用，接着调用 getOverlays()方法获取其 Overylay 链表，并将构建好的 MyLocationOverlay 对象添加到链表中去。其中 MyLocationOverlay 对象调用的 enableMyLocation()方法表示尝试通过位置服务来获取当前的位置。具体代码如下：

```
@Override
public void onCreate(Bundle savedInstanceState) {
     super.onCreate(savedInstanceState);
```

```
    setContentView(R.layout.main);

    map = (MapView)findViewById(R.id.myMapView);
    List<Overlay> overlays = map.getOverlays();
    MyLocationOverlay myLocation = new MyLocationOverlay(this,map);
    myLocation.enableMyLocation();
    overlays.add(myLocation);
```

☑ 为“放大”和“缩小”这两个按钮设置处理程序，首先通过方法 getController()获取 MapView 的 MapController 对象，然后在“放大”和“缩小”两个按钮单击事件监听器的回放方法中，根据按钮的不同实现对 MapView 的缩放。具体代码如下：

```
ctrlMap = map.getController();
inBtn = (Button)findViewById(R.id.in);
outBtn = (Button)findViewById(R.id.out);
OnClickListener listener = new OnClickListener() {
    @Override
    public void onClick(View v) {
        switch (v.getId()) {
        case R.id.in:                                    /*如果是缩小*/
            ctrlMap.zoomIn();
            break;
        case R.id.out:                                   /*如果是放大*/
            ctrlMap.zoomOut();
            break;
        default:
            break;
        }
    }
};
inBtn.setOnClickListener(listener);
outBtn.setOnClickListener(listener);
```

☑ 通过方法 onCheckedChanged()获取是否选择了 switchMap，如果选择了则显示卫星地图。首先通过方法 findViewById()获取对应 id 的 ToggleButton 对象的引用，然后调用 setOnChecked ChangeListener()方法，设置对事件监听器选中的事件进行处理。根据 ToggleButton 是否被选中，进而通过 setSatellite()方法启用或禁用卫星地图功能。具体代码如下：

```
switchMap = (ToggleButton)findViewById(R.id.switchMap);
switchMap.setOnCheckedChangeListener(new OnCheckedChangeListener() {
    @Override
    public void onCheckedChanged(CompoundButton cBtn, boolean isChecked) {
        if (isChecked == true) {
            map.setSatellite(true);
        } else {
            map.setSatellite(false);
        }
    }
});
```

☑ 通过 LocationManager 获取当前的位置，然后通过 getBestProvider()方法来获取和查询条件，最后设置更新位置信息的最小间隔为 2 秒，位移变化在 10 米以上。具体代码如下：

```
    LocationManager locationManager;
```

```
    String context = Context.LOCATION_SERVICE;
    locationManager = (LocationManager)getSystemService(context);
    //String provider = LocationManager.GPS_PROVIDER;

    Criteria criteria = new Criteria();
    criteria.setAccuracy(Criteria.ACCURACY_FINE);
    criteria.setAltitudeRequired(false);
    criteria.setBearingRequired(false);
    criteria.setCostAllowed(true);
    criteria.setPowerRequirement(Criteria.POWER_LOW);
    String provider = locationManager.getBestProvider(criteria, true);

    Location location = locationManager.getLastKnownLocation(provider);
    updateWithNewLocation(location);
    locationManager.requestLocationUpdates(provider, 2000, 10,
            locationListener);
}
```

☑ 设置回调方法何时被调用，具体代码如下：

```
private final LocationListener locationListener = new LocationListener() {
 public void onLocationChanged(Location location) {
 updateWithNewLocation(location);
 }
 public void onProviderDisabled(String provider){
 updateWithNewLocation(null);
 }
 public void onProviderEnabled(String provider){ }
 public void onStatusChanged(String provider, int status,
 Bundle extras){ }
 };
```

☑ 定义方法 updateWithNewLocation(Location location)来显示地理信息和地图信息，具体代码如下：

```
    private void updateWithNewLocation(Location location) {
          String latLongString;
          TextView myLocationText;
          myLocationText = (TextView)findViewById(R.id.myLocationText);
          if (location != null) {
                double lat = location.getLatitude();
                double lng = location.getLongitude();
                latLongString = "纬度是:" + lat + "\n 经度是:" + lng;

                ctrlMap.animateTo(new GeoPoint((int)(lat*1E6),(int)(lng*1E6)));
          } else {
                latLongString = "获取失败";
          }
          myLocationText.setText("当前的位置是:\n" +
          latLongString);

    }
}
```

至此，整个实例全部介绍完毕，在图 9-10 中选定一个经度和纬度位置后，可以显示此位置的定位信息，并且定位信息分别以文字和地图的形式显示出来，如图 9-11 所示。

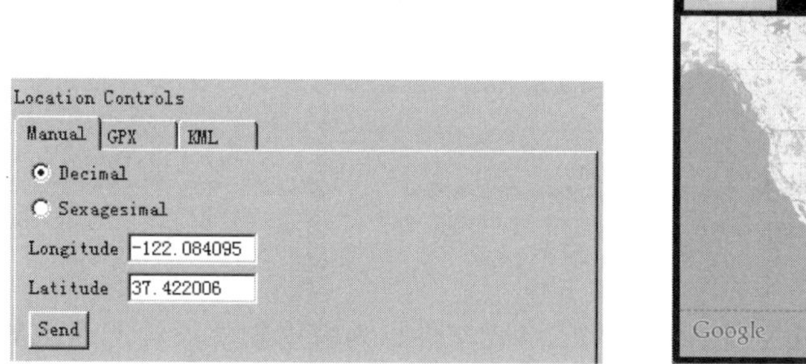


图 9-10　指定位置

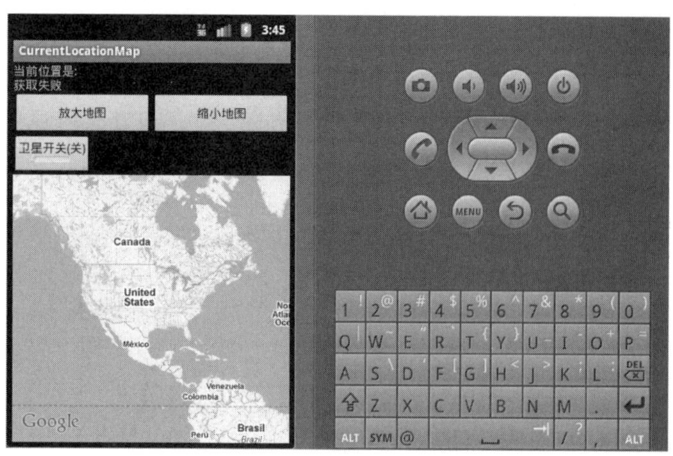


图 9-11　显示对应信息

单击“放大地图”和“缩小地图”按钮后，能控制地图的大小显示，如图 9-12 所示。打开卫星视图后，可以显示此位置范围对应的卫星地图，如图 9-13 所示。

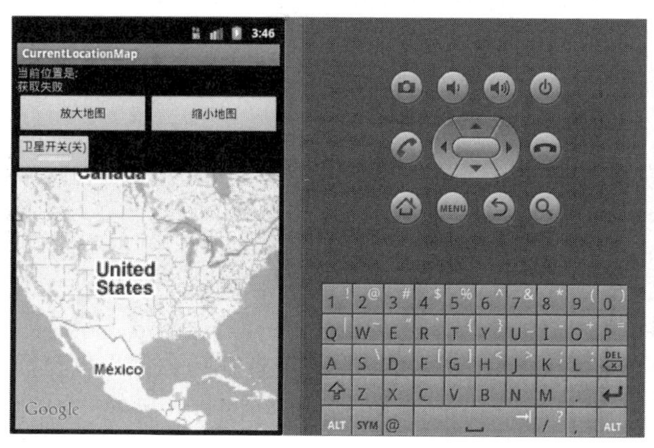

图 9-12　放大后效果

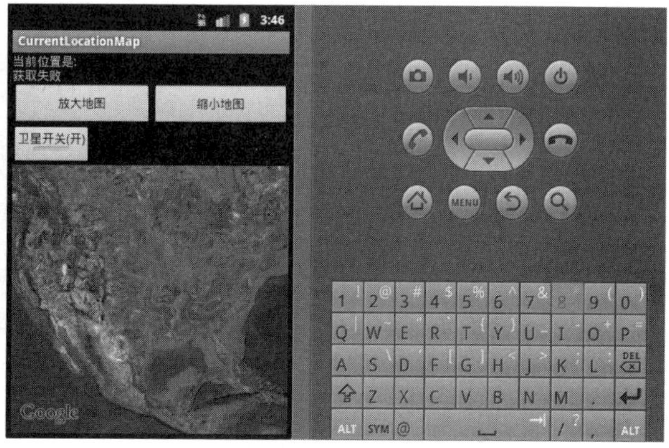


图 9-13　卫星地图

# 9.4　接近警报

**知识点讲解：光盘:视频\知识点\第 9 章\接近警报.avi**

在 Android 系统中，可以使用 LocationManager 来设置接近警报功能。此功能和本章前面讲解的地图定位功能类似，但是可以在物联网设备进入或离开某一个指定区域时发送通知应用，而并不是在新位置时才发送通知程序。本节将详细讲解在 Android 系统中实现接近警报应用的方法。

## 9.4.1　类 Geocoder 基础

在现实世界中，地图和定位服务通常使用经纬度来精确地指出地理位置。在 Android 系统中，提供了地理编码类 Geocoder 来转换经纬度和现实世界的地址。地理编码是一个街道、地址或者其他位置（经度、纬度）转化为坐标的过程。反向地理编码是将坐标转换为地址（经度、纬度）的过程。一组反向地理编码结果间可能会有所差异。例如在一个结果中可能包含最临近建筑的完整街道地址，而另一个可能只包含城市名称和邮政编码。Geocoder 要求的后端服务并没有包含在基本的 Android 框架中。如果没有此后端服务，执行 Geocoder 的查询方法将返回一个空列表。使用 isPresent()方法，以确定 Geocoder 是否能够正常执行。

在 Android 系统中，类 Geocoder 的继承关系如下：

```
public final class Geocoder extends Object
java.lang.Object
android.location.Geocoder
```

在 Android 系统中，类 Geocoder 的主要功能如下所示。

（1）设置模拟器以支持定位服务

GPS 数据格式有 GPX 和 KML 两种，其中 GPX 是一个 XML 格式文件，为应用软件设计的通用的 GPS 数据格式，可以用来描述路点、轨迹和路程。而 KML 是基于 XML（eXtensible Markup Language，可扩展标记语言）语法标准的一种标记语言（Markup Language），采用标记结构，含有嵌套的元素和属性。由 Google 旗下的 Keyhole 公司发展并维护，用来表达地理标记。

LBS 是 Location Based Service 的缩写，是一个总称，用来描述用于找到设备当前位置的不同技术。主要包含如下所示的两个元素。

- ☑ locationManager：用于提供 LBS 的钩子 hook，获得当前位置，跟踪移动，设置移入和移出指定区域的接近警报。
- ☑ LocationProviders：提供多种定位方式供开发者选择，常用的定位方式有 GPS（GPS_PROVIDER）定位和 NETWORK(NETWORK_PROVIDER)定位。

```
String providerName =LocationManager.GPS_PROVIDER;
LocationProvidergpsProvider;
gpsProvider =locationManager.getProvider(providerName);
```

在 Eclipse 开发环境中，依次选择 DDMS | Location Controls 命令可以设置位置变化数据，以在模拟器中测试应用程序，如图 9-14 所示。使用 ManualTab，可以指定特定的纬度/经度对。另外，KML 和 GPX 可以载入 KML 和 GPX 文件。一旦加载，可以跳转到特定的航点（位置）或顺序播放每个位置。

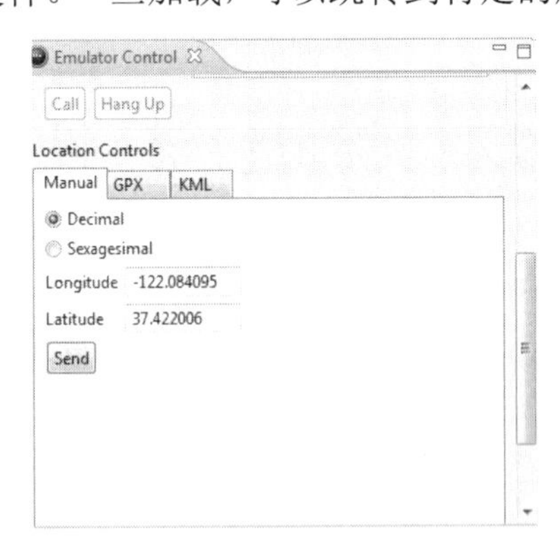


图 9-14　选择 DDMS | Location Controls 命令设置位置变化数据

也可以用类 Criteria 设置符合要求的 Provider 的条件查询（精度=精确/粗略，能耗=高/中/低，成本，返回海拔，速度，方位的能力），例如下面的代码。

```
Criteria criteria = newCriteria();
criteria.setAccuracy(Criteria.ACCURACY_COARSE);
```

```
criteria.setPowerRequirement(Criteria.POWER_LOW);
criteria.setAltitudeRequired(false);
criteria.setBearingRequired(false);
criteria.setSpeedRequired(false);
criteria.setCostAllowed(true);
String bestProvider = locationManager.getBestProvider(criteria, true);//或者用 getProviders 返回所有可能匹配的
Provider
List<String>matchingProviders = locationManager.getProviders(criteria,false);
```

在使用 LocationManager 前，需要将 uses-permission 加到 manifest 文件中以支持对 LBS 硬件的访问。GPS 需要 finepermission 权限，Network 需要 coarsepermission 权限。

```
<uses-permissionandroid:name="android.permission.ACCESS_FINE_LOCATION"/>
<uses-permissionandroid:name="android.permission.ACCESS_COARSE_LOCATION"/>
```

使用 getLastKnownLocation()方法可以获得最新的位置。

```
String provider =LocationManager.GPS_PROVIDER;
Location location = locationManager.getLastKnownLocation(provider);
```

（2）跟踪运动（TrackingMovement）

☑ 可以使用 requestLocationUpdates()方法取得最新的位置变化，为优化性能可指定位置变化的最小时间（毫秒）和最小距离（米）。当超出最小时间和距离值时，Location Listener 将触发 onLocationChanged 事件。

```
locationManager.requestLocationUpdates(provider,t, distance,myLocationListener);
```

☑ 用 RomoveUpdates()方法停止位置更新。

☑ 大多数 GPS 硬件都明显地消耗电能。

（3）邻近警告（Proximity Alerts）

通过邻近警告功能让应用程序设置触发器，当用户在地理位置上移动或超出设定距离时触发。

☑ 可用 PendingIntent 定义 Proximity Alert 触发时广播的 Intent。

☑ 为了处理 proximityalert，需要创建 BroadcastReceiver，并重写 onReceive()方法。例如下面的代码。

```
public classProximityIntentReceiver extends BroadcastReceiver {
    @Override
    public void onReceive(Context context, Intent intent) {
    String key =LocationManager.KEY_PROXIMITY_ENTERING;
    Booleanentering = intent.getBooleanExtra(key, false);
    [ ...perform proximity alert actions ... ]
    }
    }
```

☑ 要想启动监听，需要注册这个 Receiver。

```
IntentFilter filter =new IntentFilter(TREASURE_PROXIMITY_ALERT);
registerReceiver(newProximityIntentReceiver(), filter);
```

## 9.4.2　Geocoder 的公共构造器和公共方法

在 Android 系统中，类 Geocoder 包含了如下所示的公共构造器。

（1）public Geocoder(Context context, Local local)：功能是根据给定的语言环境构造一个 Geocoder 对象。各个参数的具体说明如下。

☑ context：当前的上下文对象。

☑ local：当前语言环境。

（2）public Geocoder(Context context)：功能是根据给定的系统默认语言环境构造一个 Geocoder 对象。参数 context 表示当前的上下文对象。

在 Android 系统中，类 Geocoder 包含了如下所示的公共方法。

（1）public List<Address>getFromLocation(double latitude, double longitude, int maxResults)：功能是根据给定的经纬度返回一个描述此区域的地址数组。返回的地址将根据构造器提供的语言环境进行本地化。

返回值：一组地址对象，如果没有找到匹配项，或者后台服务无效的话则返回 null 或者空序列。也可能通过网络获取，返回结果是一个最好的估计值，但不能保证其完全正确。

各个参数的具体说明如下。

☑ latitude：纬度。

☑ longitude：经度。

☑ maxResults：要返回的最大结果数，推荐 1～5。

包含的异常如下。

☑ IllegalArgumentException：纬度小于-90 或者大于 90。

☑ IllegalArgumentException：经度小于-180 或者大于 180。

☑ IOException：如果没有网络或者 I/O 错误。

（2）public List<Address>getFromLocationName(String locationName, int maxResults, double lowerLeftLatitude, double lowerLeftLongitude, double upperRightLatitude, double upperRightLongitude)：功能是返回一个由给定的位置名称参数所描述的地址数组。名称参数可以是一个位置名称，例如“Dalvik, Iceland”，也可以是一个地址，例如“1600 Amphitheatre Parkway, Mountain View, CA”，也可以是一个机场代号，例如“SFO”……返回的地址将根据构造器提供的语言环境进行本地化。

也可以指定一个搜索边界框，该边界框由左下方坐标经纬度和右上方坐标经纬度确定。

返回值：一组地址对象，如果没有找到匹配项，或者后台服务无效的话则返回 null 或者空序列。也有可能是通过网络获取。返回结果是一个最好的估计值，但不能保证其完全正确。通过 UI 主线程的后台线程来调用这个方法可能更加有用。

各个参数的具体说明如下。

☑ locationName：用户提供的位置描述。

☑ maxResults：要返回的最大结果数，推荐 1～5。

☑ lowerLeftLatitude：左下角纬度，用来设定矩形范围。

☑ lowerLeftLongitude：左下角经度，用来设定矩形范围。

☑ upperRightLatitude：右上角纬度，用来设定矩形范围。

☑ upperRightLongitude：右上角经度，用来设定矩形范围。

包含的异常如下。

☑ IllegalArgumentException：如果位置描述为空。

☑ IllegalArgumentException：如果纬度小于-90 或者大于 90。

☑ IllegalArgumentException：如果经度小于-180 或者大于 180。

☑ IOException：如果没有网络或者 I/O 错误。

（3）public List<Address>getFromLocationName(String locationName, int maxResults)：功能是返回一个由给定的位置名称参数所描述的地址数组。名称参数可以是一个位置名称，例如“Dalvik, Iceland”，也可以是一个地址，例如“1600 Amphitheatre Parkway, Mountain View, CA”，也可以是一个机场代号，例如“SFO”，返回的地址将根据构造器提供的语言环境进行本地化。在现实应用中，通过 UI 主线程的后台线程来调用这个方法可能会更加有用。

返回值：一组地址对象，如果没有找到匹配项，或者后台服务无效的话则返回 null 或者空序列。也有可能是通过网络获取。返回结果是一个最好的估计值，但不能保证其完全正确。

各个参数的具体说明如下。

☑　locationName：用户提供的位置描述。

☑　maxResults：要返回的最大结果数，推荐 1～5。

包含的异常如下。

☑　IllegalArgumentException：如果位置描述为空。

☑　IOException：如果没有网络或者 I/O 错误。

（4）public static boolean isPresent()：如果 Geocoder 的方法 getFromLocation()和方法 getFromLcationName()都实现，则返回 true。当没有网络连接时，这些方法仍然可能返回空值或者空序列。

## 9.4.3　实战演练——接近某个位置时实现自动提醒

本实例的功能是当穿戴设备接近某个位置时实现自动提醒。本实例源码是开源代码，来源于地址 https://github.com/gast-lib/gast-lib，读者可以自行登录并下载。

| 实　　例 | 功　　能 | 源 码 路 径 |
| --- | --- | --- |
| 实例 9-4 | 接近某个位置时实现自动提醒 | https://github.com/gast-lib/gast-lib |

本实例的具体实现流程如下。

（1）编写主界面 GeocodeActivity 用于从用户处获取目标位置的信息，并使用 LocationManager 对象对位置进行编码处理。主界面 GeocodeActivity 的布局文件是 geocode.xml，功能是供用户选择一个目标位置信息，具体实现代码如下：

```
<RelativeLayout xmlns:android="http://schemas.android.com/apk/res/android"
    android:layout_width="match_parent"
    android:layout_height="match_parent"
    android:orientation="vertical" >

    <TextView android:id="@+id/enterLocationLabel"
        android:layout_height="wrap_content"
        android:layout_width="wrap_content"
        android:text="@string/enterLocationLabel"
        android:layout_alignParentTop="true" />

    <EditText android:id="@+id/enterLocationValue"
        android:layout_height="wrap_content"
        android:layout_width="match_parent"
        android:layout_below="@id/enterLocationLabel"
        android:text="springfield" />
```

```
    <Button android:id="@+id/lookupLocationButton"
        android:layout_width="match_parent"
        android:layout_height="wrap_content"
        android:text="@string/lookupLocationButton"
        android:layout_below="@id/enterLocationValue"
        android:onClick="onLookupLocationClick" />

    <Button android:id="@+id/okButton"
        android:layout_width="match_parent"
        android:layout_height="wrap_content"
        android:layout_alignParentBottom="true"
        android:text="@android:string/ok"
        android:onClick="onOkClick" />
    <ListView android:id="@android:id/list"
        android:layout_width="match_parent"
        android:layout_height="match_parent"
        android:drawSelectorOnTop="false"
        android:layout_above="@id/okButton"
        android:layout_below="@id/lookupLocationButton"
        android:choiceMode="singleChoice" />
</RelativeLayout>
```

主界面 GeocodeActivity 的的程序文件是 GeocodeActivity.java，功能是对设备中提供的地址实现地理编码和反向地理编码处理。当单击 Lookup Location 按钮时会运行 onLookupLocationClick()方法，在此方法中调用了类 Geocode。文件 GeocodeActivity.java 的具体实现代码如下：

```
public class GeocodeActivity extends ListActivity
{
    private static final String TAG = "GeocodeActivity";
    private static final int MAX_ADDRESSES = 30;

    @Override
    protected void onCreate(Bundle savedInstanceState)
    {
        super.onCreate(savedInstanceState);

        setContentView(R.layout.geocode);
    }

    public void onLookupLocationClick(View view)
    {
        if (Geocoder.isPresent())
        {
            EditText addressText = (EditText) findViewById(R.id.enterLocationValue);

            try
            {
                List<Address> addressList = new Geocoder(this).getFromLocationName(addressText. getText().
toString(), MAX_ADDRESSES);

                List<AddressWrapper> addressWrapperList = new ArrayList<AddressWrapper>();
```

```
                for (Address address : addressList)
                {
                    addressWrapperList.add(new AddressWrapper(address));
                }

                setListAdapter(new ArrayAdapter<AddressWrapper>(this, android.R.layout.simple_list_item_
single_choice, addressWrapperList));
            }
            catch (IOException e)
            {
                Log.e(TAG, "Could not geocode address", e);

                new AlertDialog.Builder(this)
                    .setMessage(R.string.geocodeErrorMessage)
                    .setTitle(R.string.geocodeErrorTitle)
                    .setPositiveButton(android.R.string.ok, new DialogInterface.OnClickListener()
                    {
                        @Override
                        public void onClick(DialogInterface dialog, int which)
                        {
                            dialog.dismiss();
                        }
                    }).show();
            }
        }
    }

    public void onOkClick(View view)
    {
        ListView listView = getListView();

        Intent intent = getIntent();
        if (listView.getCheckedItemPosition() != ListView.INVALID_POSITION)
        {
            AddressWrapper addressWrapper = (AddressWrapper)listView.getItemAtPosition (listView.
getCheckedItemPosition());

            intent.putExtra("name", addressWrapper.toString());
            intent.putExtra("latitude", addressWrapper.getAddress().getLatitude());
            intent.putExtra("longitude", addressWrapper.getAddress().getLongitude());
        }

        this.setResult(RESULT_OK, intent);
        finish();
    }

    private static class AddressWrapper
    {
        private Address address;
```

```
        public AddressWrapper(Address address)
        {
            this.address = address;
        }

        @Override
        public String toString()
        {
            StringBuilder stringBuilder = new StringBuilder();

            for (int i = 0; i < address.getMaxAddressLineIndex(); i++)
            {
                stringBuilder.append(address.getAddressLine(i));

                if ((i + 1) < address.getMaxAddressLineIndex())
                {
                    stringBuilder.append(", ");
                }
            }

            return stringBuilder.toString();
        }

        public Address getAddress()
        {
            return address;
        }
    }
}
```

在上述代码中，调用方法 isPresent()检验当前设备是否支持地理编码和反向地理编码功能。方法 fromLocationName()能够解析输入的位置字符串和标志性建筑的坐标，这一功能是通过网络查询来实现的。得到查询结果后，会在下方使用 ListView 列表进行展示。

（2）开始实现接近警报设置界面，此功能通过 LocationManager 实现。接近警报设置界面的布局文件是 proximity_alert.xml，提供了两个单选按钮供用户选择设置类型，并且可以设置允许的接近读取半径值。文件 proximity_alert.xml 的具体实现代码如下：

```
<RelativeLayout
  xmlns:android="http://schemas.android.com/apk/res/android"
  android:orientation="vertical"
  android:layout_width="match_parent"
  android:layout_height="match_parent">

    <TextView android:id="@+id/locationLabel"
        android:layout_width="wrap_content"
        android:layout_height="wrap_content"
        android:text="@string/locationLabel"
        android:layout_alignParentTop="true"
        android:layout_alignParentLeft="true"
        style="@style/apptext" />
```

```
<TextView android:id="@+id/locationValue"
    android:layout_width="wrap_content"
    android:layout_height="wrap_content"
    android:layout_alignTop="@id/locationLabel"
    android:layout_toRightOf="@id/locationLabel"
    android:text="@string/none"
    android:paddingLeft="5dip"
    style="@style/apptext" />

<TextView android:id="@+id/latitudeLabel"
    android:layout_width="wrap_content"
    android:layout_height="wrap_content"
    android:text="@string/latitudeLabel"
    android:layout_below="@id/locationValue"
    android:layout_alignParentLeft="true"
    style="@style/apptext" />

<TextView android:id="@+id/latitudeValue"
    android:layout_width="wrap_content"
    android:layout_height="wrap_content"
    android:layout_alignTop="@id/latitudeLabel"
    android:layout_toRightOf="@id/latitudeLabel"
    android:text="@string/none"
    android:paddingLeft="5dip"
    style="@style/apptext" />

<TextView android:id="@+id/longitudeLabel"
    android:layout_width="wrap_content"
    android:layout_height="wrap_content"
    android:text="@string/longitudeLabel"
    android:layout_below="@id/latitudeLabel"
    android:layout_alignParentLeft="true"
    style="@style/apptext" />

<TextView android:id="@+id/longitudeValue"
    android:layout_width="wrap_content"
    android:layout_height="wrap_content"
    android:layout_alignTop="@id/longitudeLabel"
    android:layout_toRightOf="@id/longitudeLabel"
    android:text="@string/none"
    android:paddingLeft="5dip"
    style="@style/apptext" />

<TextView android:id="@+id/radiusLabel"
    android:layout_width="wrap_content"
    android:layout_height="wrap_content"
    android:text="@string/radiusLabel"
    android:layout_below="@id/longitudeLabel"
    android:layout_alignParentLeft="true"
    style="@style/apptext" />
```

```
<EditText android:id="@+id/radiusValue"
    android:layout_width="wrap_content"
    android:layout_height="wrap_content"
    android:layout_alignTop="@id/radiusLabel"
    android:layout_toRightOf="@id/radiusLabel"
    android:text="10"
    android:paddingLeft="5dip"
    style="@style/apptext"
    android:inputType="number" />

<Button
    android:layout_height="wrap_content"
    android:layout_width="match_parent"
    android:text="@string/setLocation"
    android:onClick="onSetLocationClick"
    android:layout_below="@id/radiusValue" />

<LinearLayout android:id="@+id/buttons"
    android:layout_width="match_parent"
    android:layout_height="wrap_content"
    android:orientation="horizontal"
    android:layout_alignParentBottom="true" >

     <Button android:id="@+id/setProximityAlert"
          android:layout_height="wrap_content"
          android:layout_width="match_parent"
          android:text="@string/setProximityAlert"
          android:onClick="onSetProximityAlertClick"
          android:enabled="false"
          android:layout_weight="1" />

     <Button android:id="@+id/clearProximityAlert"
          android:layout_height="wrap_content"
          android:layout_width="match_parent"
          android:text="@string/clearProximityAlert"
          android:onClick="onClearProximityAlertClick"
          android:enabled="false"
          android:layout_weight="1" />
</LinearLayout>

<RadioGroup android:id="@+id/proximityTypeRadioGroup"
    android:layout_height="wrap_content"
    android:layout_width="match_parent"
    android:orientation="horizontal"
    android:layout_above="@id/buttons" >

    <RadioButton android:id="@+id/androidProximityAlert"
        android:layout_height="wrap_content"
        android:layout_width="wrap_content"
        android:text="@string/androidProximityAlertTypeLabel"
        android:layout_weight="1"
```

```
            style="@style/apptext" />
        <RadioButton android:id="@+id/customProximityAlert"
            android:layout_height="wrap_content"
            android:layout_width="wrap_content"
            android:text="@string/customProximityAlertTypeLabel"
            android:layout_weight="1"
            style="@style/apptext" />
    </RadioGroup>

    <TextView
        android:layout_above="@id/proximityTypeRadioGroup"
        style="@style/apptext"
        android:text="Select Proximity Alert Type" />
</RelativeLayout>
```

接近警报设置界面的 Activity 是 ProximityAlertActivity.java，具体实现流程如下：

☑ 首先通过 onSetProximityAlertClick 读取输入的半径值，当来到预定目标的这一半径范围之内时会发出警报。

☑ 读取半径值后调用 locationManager.addProximityAlert 来传递坐标、半径、失效和广播 Intent 等参数。

☑ 编写方法 onClearProximityAlertClick()用于处理单击 Clear Proximity Alert 按钮的动作，可以在不需要时关闭接近警报功能，这样做的好处是可以避免这个接近警报过期。

```
public class ProximityAlertActivity extends Activity
{
    private static final String USE_ANDROID_PROXIMITY_TYPE_KEY = "useAndroidProximityTypeKey";

    private LocationManager locationManager;
    private PendingIntent pendingIntent;
    private SharedPreferences preferences;
    private RadioButton androidProximityTypeRadioButton;
    private Button setProximityAlert;
    private Button clearProximityAlert;
    private double latitude = Double.MAX_VALUE;
    private double longitude = Double.MAX_VALUE;

    @Override
    protected void onCreate(Bundle savedInstanceState)
    {
        super.onCreate(savedInstanceState);
        setContentView(R.layout.proximity_alert);

        locationManager = (LocationManager) getSystemService(LOCATION_SERVICE);

        pendingIntent = ProximityPendingIntentFactory.createPendingIntent(this);

        preferences = getPreferences(MODE_PRIVATE);
        androidProximityTypeRadioButton =
                (RadioButton)findViewById(R.id.androidProximityAlert);

        setProximityAlert = (Button) findViewById(R.id.setProximityAlert);
```

```
        clearProximityAlert = (Button) findViewById(R.id.clearProximityAlert);
    }

    @Override
    protected void onResume()
    {
        super.onResume();

        if (preferences.getBoolean(USE_ANDROID_PROXIMITY_TYPE_KEY, true))
        {
            androidProximityTypeRadioButton.setChecked(true);
        }
        else
        {
            ((RadioButton)findViewById(R.id.customProximityAlert)).setChecked(true);
        }
    }

    @Override
    protected void onPause()
    {
        super.onPause();

        locationManager.removeProximityAlert(pendingIntent);
        preferences.edit().putBoolean(USE_ANDROID_PROXIMITY_TYPE_KEY, androidProximityType Radio
Button.isChecked()).commit();
    }

    public void onSetProximityAlertClick(View view)
    {
        EditText radiusView = (EditText)findViewById(R.id.radiusValue);
        int radius =
                Integer.parseInt(radiusView.getText().toString());

        if (androidProximityTypeRadioButton.isChecked())
        {
            locationManager.addProximityAlert(latitude,
                                              longitude,
                                              radius,
                                              -1,
                                              pendingIntent);
        }
        else
        {
            Criteria criteria = new Criteria();
            criteria.setAccuracy(Criteria.ACCURACY_COARSE);
            Intent intent = new Intent(this, ProximityAlertService.class);
            intent.putExtra(ProximityAlertService.LATITUDE_INTENT_KEY, latitude);
            intent.putExtra(ProximityAlertService.LONGITUDE_INTENT_KEY, longitude);
            intent.putExtra(ProximityAlertService.RADIUS_INTENT_KEY, (float)radius);
            startService(intent);
```

```
        }

        setProximityAlert.setEnabled(false);
        clearProximityAlert.setEnabled(true);
    }

    public void onClearProximityAlertClick(View view)
    {
        if (androidProximityTypeRadioButton.isChecked())
        {
            locationManager.removeProximityAlert(pendingIntent);
        }

        setProximityAlert.setEnabled(true);
        clearProximityAlert.setEnabled(false);
    }

    public void onSetLocationClick(View view)
    {
        startActivityForResult(new Intent(this, GeocodeActivity.class), 1);
    }

    @Override
    protected void onActivityResult(int requestCode, int resultCode, Intent data)
    {
        super.onActivityResult(requestCode, resultCode, data);

        if (resultCode == RESULT_OK
                && data != null
                && data.hasExtra("name")
                && data.hasExtra("latitude")
                && data.hasExtra("longitude"))
        {
            latitude = data.getDoubleExtra("latitude", Double.MAX_VALUE);
            longitude = data.getDoubleExtra("longitude", Double.MAX_VALUE);

            ((TextView)findViewById(R.id.locationValue)).setText(data.getStringExtra("name"));
            ((TextView)findViewById(R.id.latitudeValue)).setText(String.valueOf(latitude));
            ((TextView)findViewById(R.id.longitudeValue)).setText(String.valueOf(longitude));

            setProximityAlert.setEnabled(true);
            clearProximityAlert.setEnabled(false);
        }
    }
}
```

（3）开始实现发送接近警报响应信息模块。当穿戴设备进入或离开预定坐标的指定半径区域时，接近警报会广播发送一个 Intent。上述功能是通过文件 ProximityAlertBroadcastReceiver.java 实现的，主要代码如下：

```
public class ProximityAlertBroadcastReceiver extends LocationBroadcastReceiver
{
```

```
    private static final int NOTIFICATION_ID = 9999;

    @Override
    public void onEnteringProximity(Context context)
    {
        displayNotification(context, "Entering Proximity");
    }

    @Override
    public void onExitingProximity(Context context)
    {
     displayNotification(context, "Exiting Proximity");
    }

    private void displayNotification(Context context, String message)
    {
     NotificationManager notificationManager = (NotificationManager)context.getSystemService(Context.
NOTIFICATION_ SERVICE);

        PendingIntent pi = PendingIntent.getActivity(context, 0, new Intent(), 0);

        Notification notification = new Notification(R.drawable.icon, message, System.currentTimeMillis());
        notification.setLatestEventInfo(context, "GAST", "Proximity Alert", pi);

        notificationManager.notify(NOTIFICATION_ID, notification);
    }
}
```

这样当设备进入和离开预定义区域时，会发送一个通知信息。

（4）因为穿戴设备的电池通常是有限的，而接近警报功能需要长时间开启，所以很费电。为了解决上述局限性，在接近警报设置界面中提供了两个单选按钮供用户选择设置类型，其中 Custom 类型可供用户自定义设置。编写文件 ProximityAlertActivity.java，实现对默认接近警报的优化，不但限制了对 GPS 的使用，而且降低了请求位置更新功能的频率。设置在最后才会将服务注册到网络提供者，以获取位置更新功能，这样可以长时间使用网络提供者，并且只有在网络提供者无法对设备和目标区域之间的距离做出准确计算的情况下才会启用 GPS。当计算完两者的距离之后，需要通过服务来确定是否要广播一个 Intent 来发出接近警报。如果不需要，则取消当前位置的更新请求，并重新计算最短距离，并用新的距离来注册位置更新请求。文件 ProximityAlertActivity.java 的具体实现代码如下：

```
public class ProximityAlertActivity extends Activity
{
    private static final String USE_ANDROID_PROXIMITY_TYPE_KEY = "useAndroidProximityTypeKey";

    private LocationManager locationManager;
    private PendingIntent pendingIntent;
    private SharedPreferences preferences;
    private RadioButton androidProximityTypeRadioButton;
    private Button setProximityAlert;
    private Button clearProximityAlert;
    private double latitude = Double.MAX_VALUE;
    private double longitude = Double.MAX_VALUE;
```

```
    @Override
    protected void onCreate(Bundle savedInstanceState)
    {
        super.onCreate(savedInstanceState);
        setContentView(R.layout.proximity_alert);

        locationManager = (LocationManager) getSystemService(LOCATION_SERVICE);

        pendingIntent = ProximityPendingIntentFactory.createPendingIntent(this);

        preferences = getPreferences(MODE_PRIVATE);
        androidProximityTypeRadioButton =
                (RadioButton)findViewById(R.id.androidProximityAlert);

        setProximityAlert = (Button) findViewById(R.id.setProximityAlert);
        clearProximityAlert = (Button) findViewById(R.id.clearProximityAlert);
    }

    @Override
    protected void onResume()
    {
        super.onResume();

        if (preferences.getBoolean(USE_ANDROID_PROXIMITY_TYPE_KEY, true))
        {
            androidProximityTypeRadioButton.setChecked(true);
        }
        else
        {
            ((RadioButton)findViewById(R.id.customProximityAlert)).setChecked(true);
        }
    }

    @Override
    protected void onPause()
    {
        super.onPause();

        locationManager.removeProximityAlert(pendingIntent);
        preferences.edit().putBoolean(USE_ANDROID_PROXIMITY_TYPE_KEY,
androidProximityTypeRadioButton.isChecked()).commit();
    }

    public void onSetProximityAlertClick(View view)
    {
        EditText radiusView = (EditText)findViewById(R.id.radiusValue);
        int radius =
                Integer.parseInt(radiusView.getText().toString());

        if (androidProximityTypeRadioButton.isChecked())
```

```
        {
            locationManager.addProximityAlert(latitude,
                                              longitude,
                                              radius,
                                              -1,
                                              pendingIntent);
        }
        else
        {
            Criteria criteria = new Criteria();
            criteria.setAccuracy(Criteria.ACCURACY_COARSE);
            Intent intent = new Intent(this, ProximityAlertService.class);
            intent.putExtra(ProximityAlertService.LATITUDE_INTENT_KEY, latitude);
            intent.putExtra(ProximityAlertService.LONGITUDE_INTENT_KEY, longitude);
            intent.putExtra(ProximityAlertService.RADIUS_INTENT_KEY, (float)radius);
            startService(intent);
        }

        setProximityAlert.setEnabled(false);
        clearProximityAlert.setEnabled(true);
    }

    public void onClearProximityAlertClick(View view)
    {
        if (androidProximityTypeRadioButton.isChecked())
        {
            locationManager.removeProximityAlert(pendingIntent);
        }

        setProximityAlert.setEnabled(true);
        clearProximityAlert.setEnabled(false);
    }

    public void onSetLocationClick(View view)
    {
        startActivityForResult(new Intent(this, GeocodeActivity.class), 1);
    }

    @Override
    protected void onActivityResult(int requestCode, int resultCode, Intent data)
    {
        super.onActivityResult(requestCode, resultCode, data);

        if (resultCode == RESULT_OK
                && data != null
                && data.hasExtra("name")
                && data.hasExtra("latitude")
                && data.hasExtra("longitude"))
        {
            latitude = data.getDoubleExtra("latitude", Double.MAX_VALUE);
            longitude = data.getDoubleExtra("longitude", Double.MAX_VALUE);
```

```
            ((TextView)findViewById(R.id.locationValue)).setText(data.getStringExtra("name"));
            ((TextView)findViewById(R.id.latitudeValue)).setText(String.valueOf(latitude));
            ((TextView)findViewById(R.id.longitudeValue)).setText(String.valueOf(longitude));

            setProximityAlert.setEnabled(true);
            clearProximityAlert.setEnabled(false);
        }
    }
}
```

至此，整个实例介绍完毕。对于本实例的具体实现源码，读者可以参阅开源站点的源码。

# 第 4 篇

## 传感器应用篇

- 第 10 章　Android 传感器系统架构详解
- 第 11 章　光线传感器和磁场传感器
- 第 12 章　加速度传感器、方向传感器和陀螺仪传感器
- 第 13 章　旋转向量传感器、距离传感器和气压传感器
- 第 14 章　温度传感器和湿度传感器

# 第 10 章 Android 传感器系统架构详解

传感器是近年来随着物联网这一概念的流行而推出的，现在人们已经逐渐地认识了传感器这一概念。其实传感器在大家日常的生活中经常见到甚至是用到，例如楼宇的声控楼梯灯和马路上的路灯等。本章将详细讲解 Android 系统中传感器系统的基本知识，为读者步入本书后面知识的学习打下基础。

## 10.1 Android 传感器系统概述

**知识点讲解：光盘:视频\知识点\第 10 章\Android 传感器系统概述.avi**

在 Android 系统中提供的主要传感器有加速度传感器、磁场、方向、陀螺仪、光线、压力、温度和接近等。传感器系统会主动对上层报告传感器精度和数据的变化，并且提供了设置传感器精度的接口，这些接口可以在 Java 应用和 Java 框架中使用。

Android 传感器系统的基本层次结构如图 10-1 所示。

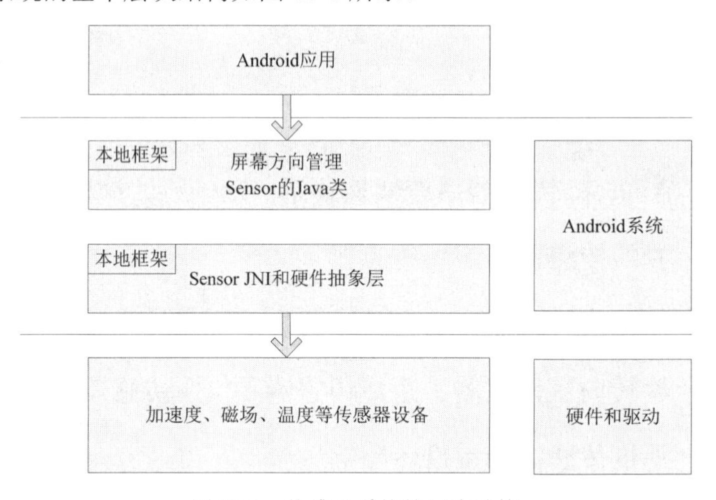


图 10-1 传感器系统的层次结构

根据图 10-1 所示的结构，Android 传感器系统从上到下分别是 Java 应用层、Java 框架对传感器的应用、传感器类、传感器硬件抽象层和传感器驱动。图 10-1 中各个层的具体说明如下所示。

（1）传感器系统的 Java 部分，代码路径是 frameworks/base/include/core/java/android/hardware。

此部分对应的实现文件是 Sensor*.java。

（2）传感器系统的 JNI 部分，代码路径是 frameworks/base/core/jni/android_hardware_SensorManager.cpp。

在此部分中提供了对类 android.hardware.Sensor.Manage 的本地支持。

（3）传感器系统 HAL 层，头文件路径是 hardware/libhardware/include/hardware/sensors.h。

在 Android 系统中，传感器系统的硬件抽象层需要特意编码实现。

（4）驱动层，代码路径是 kernel/driver/hwmon/$(PROJECT)/sensor。

在库 sensor.so 中提供了如下所示的 8 个 API 函数。

☑ 控制方面：在结构体 ensors_control_device_t 中定义，包括如下所示的函数。
  - int (*open_data_source)(struct sensors_control_device_t *dev)
  - int (*activate)(struct sensors_control_device_t *dev, int handle, int enabled)
  - int (*set_delay)(struct sensors_control_device_t *dev, int32_t ms)
  - int (*wake)(struct sensors_control_device_t *dev)

☑ 数据方面：在结构体 sensors_data_device_t 中定义，包括如下所示的函数。
  - int (*data_open)(struct sensors_data_device_t *dev, int fd)
  - int (*data_close)(struct sensors_data_device_t *dev)
  - int (*poll)(struct sensors_data_device_t *dev, sensors_data_t* data)

☑ 模块方面：在结构体 sensors_module_t 中定义，包括如下一个函数。
  - int (*get_sensors_list)(struct sensors_module_t* module, struct sensor_t const** list)

在 Android 系统的 Java 层中，Sensor 的状态是由 SensorService 来负责控制的，其 Java 代码和 JNI 代码分别位于如下文件中。

☑ frameworks/base/services/java/com/android/server/SensorService.java

☑ frameworks/base/services/jni/com_android_server_SensorService.cpp

SensorManager 负责在 Java 层 Sensor 的数据控制，它的 Java 代码和 JNI 代码分别位于如下文件中。

☑ frameworks/base/core/java/android/hardware/SensorManager.java

☑ frameworks/base/core/jni/android_hardware_SensorManager.cpp

在 Android 的 Framework 中，是通过文件 sensorService.java 和 sensorManager.java 实现与 Sensor 传感器通信的。文件 sensorService.java 的通信功能是通过 JNI 调用 sensorService.cpp 中的方法实现的。

文件 sensorManager.java 的具体通信功能是通过 JNI 调用 sensorManager.cpp 中的方法实现的。文件 sensorService.cpp 和 sensorManager.cpp 通过文件 hardware.c 与 sensor.so 通信。其中文件 sensorService.cpp 实现对 sensor 的状态控制，文件 sensorManager.cpp 实现对 sensor 的数据控制。

库 sensor.so 通过 ioctl 控制 sensor driver 的状态，通过打开 sensor driver 对应的设备文件读取 G-sensor 采集的数据。

## 10.2　Java 层详解

**知识点讲解：光盘:视频\知识点\第 10 章\Java 层详解.avi**

在 Android 系统中，传感器系统的 Java 部分的实现文件是\sdk\apps\SdkController\src\com\android\tools\sdkcontroller\activities\SensorActivity.java。

通过阅读文件 SensorActivity.java 的源码可知，在应用程序中使用传感器需要用到 hardware 包中的 SensorManager、SensorListener 等相关的类，具体实现代码如下：

```
public class SensorActivity extends BaseBindingActivity
        implements android.os.Handler.Callback {
```

```
@SuppressWarnings("hiding")
public static String TAG = SensorActivity.class.getSimpleName();
private static boolean DEBUG = true;

private static final int MSG_UPDATE_ACTUAL_HZ = 0x31415;

private TableLayout mTableLayout;
private TextView mTextError;
private TextView mTextStatus;
private TextView mTextTargetHz;
private TextView mTextActualHz;
private SensorChannel mSensorHandler;

private final Map<MonitoredSensor, DisplayInfo> mDisplayedSensors =
    new HashMap<SensorChannel.MonitoredSensor, SensorActivity.DisplayInfo>();
private final android.os.Handler mUiHandler = new android.os.Handler(this);
private int mTargetSampleRate;
private long mLastActualUpdateMs;

/**第一次创建 Activity 时调用*/
@Override
public void onCreate(Bundle savedInstanceState) {
    super.onCreate(savedInstanceState);
    setContentView(R.layout.sensors);
    mTableLayout = (TableLayout) findViewById(R.id.tableLayout);
    mTextError   = (TextView) findViewById(R.id.textError);
    mTextStatus = (TextView) findViewById(R.id.textStatus);
    mTextTargetHz = (TextView) findViewById(R.id.textSampleRate);
    mTextActualHz = (TextView) findViewById(R.id.textActualRate);
    updateStatus("Waiting for connection");

    mTextTargetHz.setOnKeyListener(new OnKeyListener() {
        @Override
        public boolean onKey(View v, int keyCode, KeyEvent event) {
            updateSampleRate();
            return false;
        }
    });
    mTextTargetHz.setOnFocusChangeListener(new OnFocusChangeListener() {
        @Override
        public void onFocusChange(View v, boolean hasFocus) {
            updateSampleRate();
        }
    });
}

@Override
protected void onResume() {
    if (DEBUG) Log.d(TAG, "onResume");
    //BaseBindingActivity 绑定后台服务
    super.onResume();
    updateError();
```

```
}

@Override
protected void onPause() {
    if (DEBUG) Log.d(TAG, "onPause");
    super.onPause();
}

@Override
protected void onDestroy() {
    if (DEBUG) Log.d(TAG, "onDestroy");
    super.onDestroy();
    removeSensorUi();
}

// ----------

@Override
protected void onServiceConnected() {
    if (DEBUG) Log.d(TAG, "onServiceConnected");
    createSensorUi();
}

@Override
protected void onServiceDisconnected() {
    if (DEBUG) Log.d(TAG, "onServiceDisconnected");
    removeSensorUi();
}

@Override
protected ControllerListener createControllerListener() {
    return new SensorsControllerListener();
}

// ----------

private class SensorsControllerListener implements ControllerListener {
    @Override
    public void onErrorChanged() {
        runOnUiThread(new Runnable() {
            @Override
            public void run() {
                updateError();
            }
        });
    }

    @Override
    public void onStatusChanged() {
        runOnUiThread(new Runnable() {
            @Override
            public void run() {
```

```
                ControllerBinder binder = getServiceBinder();
                if (binder != null) {
                    boolean connected = binder.isEmuConnected();
                    mTableLayout.setEnabled(connected);
                    updateStatus(connected ? "Emulated connected" : "Emulator disconnected");
                }
            }
        });
    }
}

private void createSensorUi() {
    final LayoutInflater inflater = getLayoutInflater();

    if (!mDisplayedSensors.isEmpty()) {
        removeSensorUi();
    }

    mSensorHandler = (SensorChannel) getServiceBinder().getChannel(Channel.SENSOR_CHANNEL);
    if (mSensorHandler != null) {
        mSensorHandler.addUiHandler(mUiHandler);
        mUiHandler.sendEmptyMessage(MSG_UPDATE_ACTUAL_HZ);

        assert mDisplayedSensors.isEmpty();
        List<MonitoredSensor> sensors = mSensorHandler.getSensors();
        for (MonitoredSensor sensor : sensors) {
            final TableRow row = (TableRow) inflater.inflate(R.layout.sensor_row,
                                                             mTableLayout,
                                                             false);
            mTableLayout.addView(row);
            mDisplayedSensors.put(sensor, new DisplayInfo(sensor, row));
        }
    }
}

private void removeSensorUi() {
    if (mSensorHandler != null) {
        mSensorHandler.removeUiHandler(mUiHandler);
        mSensorHandler = null;
    }
    mTableLayout.removeAllViews();
    for (DisplayInfo info : mDisplayedSensors.values()) {
        info.release();
    }
    mDisplayedSensors.clear();
}

private class DisplayInfo implements CompoundButton.OnCheckedChangeListener {
    private MonitoredSensor mSensor;
    private CheckBox mChk;
    private TextView mVal;
```

```
        public DisplayInfo(MonitoredSensor sensor, TableRow row) {
            mSensor = sensor;

            mChk = (CheckBox) row.findViewById(R.id.row_checkbox);
            mChk.setText(sensor.getUiName());
            mChk.setEnabled(sensor.isEnabledByEmulator());
            mChk.setChecked(sensor.isEnabledByUser());
            mChk.setOnCheckedChangeListener(this);

            //初始化显示该传感器的文本框
            mVal = (TextView) row.findViewById(R.id.row_textview);
            mVal.setText(sensor.getValue());
        }

        /**
         *为相关的复选框选中状态进行变化处理。当复选框被选中时会监听传感器变化
         *如果不加以控制会取消传感器的变化
         */
        @Override
        public void onCheckedChanged(CompoundButton buttonView, boolean isChecked) {
            if (mSensor != null) {
                mSensor.onCheckedChanged(isChecked);
            }
        }

        public void release() {
            mChk = null;
            mVal = null;
            mSensor = null;

        }

        public void updateState() {
            if (mChk != null && mSensor != null) {
                mChk.setEnabled(mSensor.isEnabledByEmulator());
                mChk.setChecked(mSensor.isEnabledByUser());
            }
        }

        public void updateValue() {
            if (mVal != null && mSensor != null) {
                mVal.setText(mSensor.getValue());
            }
        }
    }

    /**实现回调处理程序*/
    @Override
    public boolean handleMessage(Message msg) {
        DisplayInfo info = null;
        switch (msg.what) {
        case SensorChannel.SENSOR_STATE_CHANGED:
```

```
                info = mDisplayedSensors.get(msg.obj);
                if (info != null) {
                    info.updateState();
                }
                break;
            case SensorChannel.SENSOR_DISPLAY_MODIFIED:
                info = mDisplayedSensors.get(msg.obj);
                if (info != null) {
                    info.updateValue();
                }
                if (mSensorHandler != null) {
                    updateStatus(Integer.toString(mSensorHandler.getMsgSentCount()) + " events sent");

                    //如果值已经修改则更新 "actual rate"
                    long ms = mSensorHandler.getActualUpdateMs();
                    if (ms != mLastActualUpdateMs) {
                        mLastActualUpdateMs = ms;
                        String hz = mLastActualUpdateMs <= 0 ? "--" :
                                            Integer.toString((int) Math.ceil(1000. / ms));
                        mTextActualHz.setText(hz);
                    }
                }
                break;
            case MSG_UPDATE_ACTUAL_HZ:
                if (mSensorHandler != null) {
                    //如果值已经修改则更新 "actual rate"
                    long ms = mSensorHandler.getActualUpdateMs();
                    if (ms != mLastActualUpdateMs) {
                        mLastActualUpdateMs = ms;
                        String hz = mLastActualUpdateMs <= 0 ? "--" :
                                            Integer.toString((int) Math.ceil(1000. / ms));
                        mTextActualHz.setText(hz);
                    }
                    mUiHandler.sendEmptyMessageDelayed(MSG_UPDATE_ACTUAL_HZ, 1000 /*1s*/);
                }
            }
            return true;
        }

        private void updateStatus(String status) {
            mTextStatus.setVisibility(status == null ? View.GONE : View.VISIBLE);
            if (status != null) mTextStatus.setText(status);
        }

        private void updateError() {
            ControllerBinder binder = getServiceBinder();
            String error = binder == null ? "" : binder.getServiceError();
            if (error == null) {
                error = "";
            }
```

```
        mTextError.setVisibility(error.length() == 0 ? View.GONE : View.VISIBLE);
        mTextError.setText(error);
    }

    private void updateSampleRate() {
        String str = mTextTargetHz.getText().toString();
        try {
            int hz = Integer.parseInt(str.trim());

            if (hz <= 0 || hz > 50) {
                hz = 50;
            }

            if (hz != mTargetSampleRate) {
                mTargetSampleRate = hz;
                if (mSensorHandler != null) {
                    mSensorHandler.setUpdateTargetMs(hz <= 0 ? 0 : (int)(1000.0f / hz));
                }
            }
        } catch (Exception ignore) {}
    }
}
```

通过上述代码可知，整个 Java 层利用大家熟悉的观察者模式对传感器的数据进行了监听处理。

# 10.3　Frameworks 层详解

**知识点讲解：光盘:视频\知识点\第 10 章\Frameworks 层详解.avi**

在 Android 系统中，Frameworks 层是 Android 系统提供的应用程序开发接口和应用程序框架，与应用程序的调用是通过类实例化或类继承进行的。对应用程序来说，最重要的就是把 SensorListener 注册到 SensorManager 上，从而才能以观察者身份接收到数据的变化，因此，我们把目光落在 SensorManager 的构造函数、RegisterListener 函数和通知机制相关的代码上。在 Android 传感器系统中，Frameworks 层的代码路径是 frameworks/base/include/core/java/android/hardware。本节将详细讲解传感器系统的 Frameworks 层的具体实现流程。

## 10.3.1　监听传感器的变化

在 Android 传感器系统的 Frameworks 层中，文件 SensorListener.java 用于监听从 Java 应用层中传递过来的变化。文件 SensorListener.java 比较简单，具体代码如下：

```
package android.hardware;
@Deprecated
public interface SensorListener {
    public void onSensorChanged(int sensor, float[] values);
    public void onAccuracyChanged(int sensor, int accuracy);
}
```

## 10.3.2 注册监听

当文件 SensorListener.java 监听到变化之后，会通过文件 SensorManager.java 来向服务注册监听变化，并调度 Sensor 的具体任务。例如在开发 Android 传感器应用程序时，在上层的通用开发流程如下所示。

（1）通过 getSystemService(SENSOR_SERVICE);语句得到传感器服务。这样得到一个用来管理分配调度处理 Sensor 工作的 SensorManager。SensorManager 并不服务运行于后台，真正属于 Sensor 的系统服务是 SensorService，在终端下的#service list 中可以看到 sensorservice: [android.gui.Sensor Server]。

（2）通过 getDefaultSensor(Sensor.TYPE_GRAVITY);语句得到传感器类型，当然还有各种千奇百怪的传感器，具体可以查阅 Android 官网的 API 或者源码中的文件 Sensor.java。

（3）注册监听器 SensorEventListener。在应用程序中打开一个监听接口，专门用于处理传感器的数据。

（4）通过回调函数 onSensorChanged 和 onAccuracyChanged 实现实时监听。例如对重力感应器的 x、y、z 值经算法变换得到左右、上下、前后方向等，就由这个回调函数实现。

综上所述，传感器顶层的处理流程如图 10-2 所示。

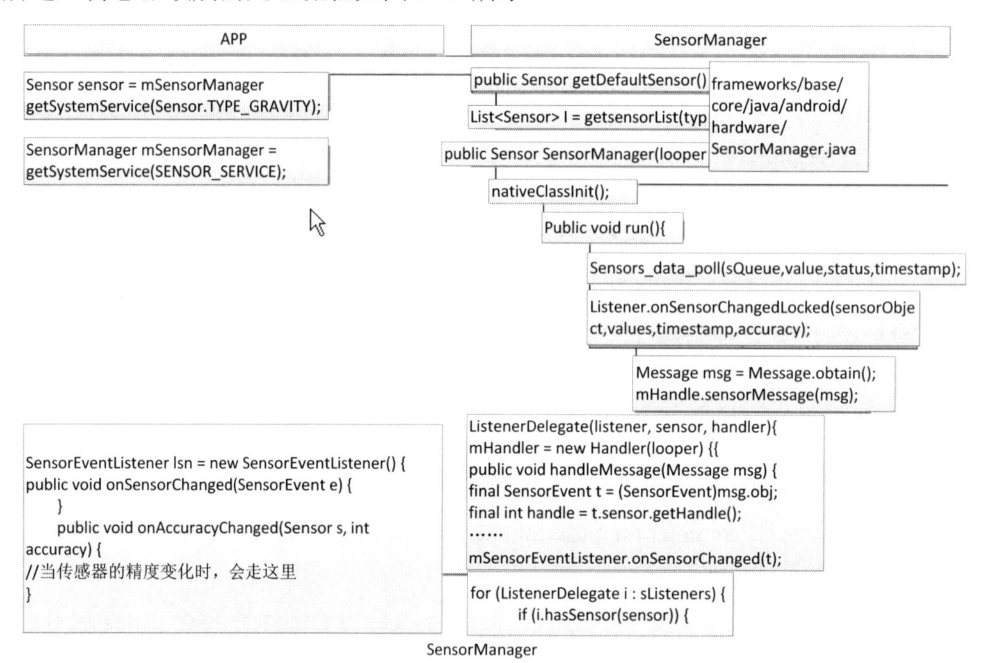


图 10-2　传感器顶层的处理流程

文件 SensorManager.java 的具体实现流程如下：

（1）定义类 SensorManager，然后设置各种传感器的初始变量值，具体代码如下：

```
public abstract class SensorManager {
    protected static final String TAG = "SensorManager";
    private static final float[] mTempMatrix = new float[16];

    private final SparseArray<List<Sensor>> mSensorListByType =
            new SparseArray<List<Sensor>>();
```

```
private LegacySensorManager mLegacySensorManager;
@Deprecated
public static final int SENSOR_ORIENTATION = 1 << 0;
@Deprecated
public static final int SENSOR_ACCELEROMETER = 1 << 1;
@Deprecated
public static final int SENSOR_TEMPERATURE = 1 << 2;
@Deprecated
public static final int SENSOR_MAGNETIC_FIELD = 1 << 3;
@Deprecated
public static final int SENSOR_LIGHT = 1 << 4;
@Deprecated
public static final int SENSOR_PROXIMITY = 1 << 5;
@Deprecated
public static final int SENSOR_TRICORDER = 1 << 6;
@Deprecated
public static final int SENSOR_ORIENTATION_RAW = 1 << 7;
@Deprecated
public static final int SENSOR_ALL = 0x7F;
@Deprecated
public static final int SENSOR_MIN = SENSOR_ORIENTATION;
@Deprecated
public static final int SENSOR_MAX = ((SENSOR_ALL + 1)>>1);

@Deprecated
public static final int DATA_X = 0;
@Deprecated
public static final int DATA_Y = 1;
@Deprecated
public static final int DATA_Z = 2;
@Deprecated
public static final int RAW_DATA_INDEX = 3;
@Deprecated
public static final int RAW_DATA_X = 3;
@Deprecated
public static final int RAW_DATA_Y = 4;
@Deprecated
public static final int RAW_DATA_Z = 5;

public static final float STANDARD_GRAVITY = 9.80665f;

public static final float GRAVITY_SUN = 275.0f;
public static final float GRAVITY_MERCURY = 3.70f;

public static final float GRAVITY_VENUS = 8.87f;
public static final float GRAVITY_EARTH = 9.80665f;
public static final float GRAVITY_MOON = 1.6f;
public static final float GRAVITY_MARS = 3.71f;
public static final float GRAVITY_JUPITER = 23.12f;
public static final float GRAVITY_SATURN = 8.96f;
public static final float GRAVITY_URANUS = 8.69f;
```

```
public static final float GRAVITY_NEPTUNE = 11.0f;
public static final float GRAVITY_PLUTO = 0.6f;
public static final float GRAVITY_DEATH_STAR_I = 0.000000353036145f;
public static final float GRAVITY_THE_ISLAND = 4.815162342f;


/**对地球表面的最大磁场*/
public static final float MAGNETIC_FIELD_EARTH_MAX = 60.0f;
/**对地球表面的最小磁场*/
public static final float MAGNETIC_FIELD_EARTH_MIN = 30.0f;
/**标准大气压*/
public static final float PRESSURE_STANDARD_ATMOSPHERE = 1013.25f;
public static final float LIGHT_SUNLIGHT_MAX = 120000.0f;
public static final float LIGHT_SUNLIGHT = 110000.0f;
public static final float LIGHT_SHADE = 20000.0f;
public static final float LIGHT_OVERCAST = 10000.0f;
public static final float LIGHT_SUNRISE = 400.0f;
public static final float LIGHT_CLOUDY = 100.0f;
public static final float LIGHT_FULLMOON = 0.25f;
public static final float LIGHT_NO_MOON = 0.001f;


/**尽可能快地获得传感器数据*/
public static final int SENSOR_DELAY_FASTEST = 0;
/**适合游戏速度*/
public static final int SENSOR_DELAY_GAME = 1;
/**适合于用户接口速率*/
public static final int SENSOR_DELAY_UI = 2;
/**（默认值）适合屏幕方向的变化*/
public static final int SENSOR_DELAY_NORMAL = 3;
/**
 *返回的值，该传感器是不可信的，需要进行校准或环境不允许读数
 */
public static final int SENSOR_STATUS_UNRELIABLE = 0;
/**
 *该传感器是报告的低精度的数据，与环境的校准是必要的
 */
public static final int SENSOR_STATUS_ACCURACY_LOW = 1;
/**
 *这种传感器是精确的平均频率的报告数据，与环境的校准可以提高精确度
 */
public static final int SENSOR_STATUS_ACCURACY_MEDIUM = 2;

/**该传感报告准确性最大的数据*/
public static final int SENSOR_STATUS_ACCURACY_HIGH = 3;
public static final int AXIS_X = 1;
public static final int AXIS_Y = 2;
public static final int AXIS_Z = 3;
public static final int AXIS_MINUS_X = AXIS_X | 0x80;
public static final int AXIS_MINUS_Y = AXIS_Y | 0x80;
public static final int AXIS_MINUS_Z = AXIS_Z | 0x80;
```

（2）定义各种设备类型方法和设备数据的方法，这些方法非常重要，在编写的应用程序中，可以通过 AIDL 接口远程调用（RPC）的方式得到 SensorManager。这样通过在类 SensorManager 中的方法，可以得到底层的各种传感器数据。上述方法的具体实现代码如下：

```
public int getSensors() {
    return getLegacySensorManager().getSensors();
}
public List<Sensor> getSensorList(int type) {
    //第一次缓存返回列表
    List<Sensor> list;
    final List<Sensor> fullList = getFullSensorList();
    synchronized (mSensorListByType) {
        list = mSensorListByType.get(type);
        if (list == null) {
            if (type == Sensor.TYPE_ALL) {
                list = fullList;
            } else {
                list = new ArrayList<Sensor>();
                for (Sensor i : fullList) {
                    if (i.getType() == type)
                        list.add(i);
                }
            }
            list = Collections.unmodifiableList(list);
            mSensorListByType.append(type, list);
        }
    }
    return list;
}
public Sensor getDefaultSensor(int type) {
    List<Sensor> l = getSensorList(type);
    return l.isEmpty() ? null : l.get(0);
}
@Deprecated
public boolean registerListener(SensorListener listener, int sensors) {
    return registerListener(listener, sensors, SENSOR_DELAY_NORMAL);
}
@Deprecated
public boolean registerListener(SensorListener listener, int sensors, int rate) {
    return getLegacySensorManager().registerListener(listener, sensors, rate);
}
@Deprecated
public void unregisterListener(SensorListener listener) {
    unregisterListener(listener, SENSOR_ALL | SENSOR_ORIENTATION_RAW);
}
@Deprecated
public void unregisterListener(SensorListener listener, int sensors) {
    getLegacySensorManager().unregisterListener(listener, sensors);
}
public void unregisterListener(SensorEventListener listener, Sensor sensor) {
```

```
        if (listener == null || sensor == null) {
            return;
        }
        unregisterListenerImpl(listener, sensor);
    }
    public void unregisterListener(SensorEventListener listener) {
        if (listener == null) {
            return;
        }
        unregisterListenerImpl(listener, null);
    }
    protected abstract void unregisterListenerImpl(SensorEventListener listener, Sensor sensor);
    public boolean registerListener(SensorEventListener listener, Sensor sensor, int rate) {
        return registerListener(listener, sensor, rate, null);
    }
    public boolean registerListener(SensorEventListener listener, Sensor sensor, int rate,
            Handler handler) {
        if (listener == null || sensor == null) {
            return false;
        }

        int delay = -1;
        switch (rate) {
            case SENSOR_DELAY_FASTEST:
                delay = 0;
                break;
            case SENSOR_DELAY_GAME:
                delay = 20000;
                break;
            case SENSOR_DELAY_UI:
                delay = 66667;
                break;
            case SENSOR_DELAY_NORMAL:
                delay = 200000;
                break;
            default:
                delay = rate;
                break;
        }
        return registerListenerImpl(listener, sensor, delay, handler);
    }
    protected abstract boolean registerListenerImpl(SensorEventListener listener, Sensor sensor,
            int delay, Handler handler);

    public static boolean getRotationMatrix(float[] R, float[] I,
            float[] gravity, float[] geomagnetic) {
        float Ax = gravity[0];
        float Ay = gravity[1];
        float Az = gravity[2];
        final float Ex = geomagnetic[0];
```

```
final float Ey = geomagnetic[1];
final float Ez = geomagnetic[2];
float Hx = Ey*Az - Ez*Ay;
float Hy = Ez*Ax - Ex*Az;
float Hz = Ex*Ay - Ey*Ax;
final float normH = (float)Math.sqrt(Hx*Hx + Hy*Hy + Hz*Hz);
if (normH < 0.1f) {
    //设备靠近自由下落，或接近北磁极，典型值是> 100
    return false;
}
final float invH = 1.0f / normH;
Hx *= invH;
Hy *= invH;
Hz *= invH;
final float invA = 1.0f / (float)Math.sqrt(Ax*Ax + Ay*Ay + Az*Az);
Ax *= invA;
Ay *= invA;
Az *= invA;
final float Mx = Ay*Hz - Az*Hy;
final float My = Az*Hx - Ax*Hz;
final float Mz = Ax*Hy - Ay*Hx;
if (R != null) {
    if (R.length == 9) {
        R[0] = Hx;    R[1] = Hy;     R[2] = Hz;
        R[3] = Mx;    R[4] = My;     R[5] = Mz;
        R[6] = Ax;    R[7] = Ay;     R[8] = Az;
    } else if (R.length == 16) {
        R[0] = Hx;    R[1] = Hy;     R[2] = Hz;     R[3]  = 0;
        R[4] = Mx;    R[5] = My;     R[6] = Mz;     R[7]  = 0;
        R[8] = Ax;    R[9] = Ay;     R[10] = Az;    R[11] = 0;
        R[12] = 0;    R[13] = 0;     R[14] = 0;     R[15] = 1;
    }
}
if (I != null) {
    //通过投射地磁向量到 Z（重力）和 X（地磁矢量的水平分量）计算轴的倾斜基质
    final float invE = 1.0f / (float)Math.sqrt(Ex*Ex + Ey*Ey + Ez*Ez);
    final float c = (Ex*Mx + Ey*My + Ez*Mz) * invE;
    final float s = (Ex*Ax + Ey*Ay + Ez*Az) * invE;
    if (I.length == 9) {
        I[0] = 1;     I[1] = 0;     I[2] = 0;
        I[3] = 0;     I[4] = c;     I[5] = s;
        I[6] = 0;     I[7] =-s;     I[8] = c;
    } else if (I.length == 16) {
        I[0] = 1;     I[1] = 0;     I[2] = 0;
        I[4] = 0;     I[5] = c;     I[6] = s;
        I[8] = 0;     I[9] =-s;     I[10]= c;
        I[3] = I[7] = I[11] = I[12] = I[13] = I[14] = 0;
        I[15] = 1;
    }
}
```

```
        return true;
    }
    public static float getInclination(float[] I) {
        if (I.length == 9) {
            return (float)Math.atan2(I[5], I[4]);
        } else {
            return (float)Math.atan2(I[6], I[5]);
        }
    }

    public static boolean remapCoordinateSystem(float[] inR, int X, int Y,
            float[] outR)
    {
        if (inR == outR) {
            final float[] temp = mTempMatrix;
            synchronized(temp) {
                if (remapCoordinateSystemImpl(inR, X, Y, temp)) {
                    final int size = outR.length;
                    for (int i=0 ; i<size ; i++)
                        outR[i] = temp[i];
                    return true;
                }
            }
        }
        return remapCoordinateSystemImpl(inR, X, Y, outR);
    }

    private static boolean remapCoordinateSystemImpl(float[] inR, int X, int Y,
            float[] outR)
    {
        /*
         * X 和 Y 定义一个旋转矩阵'R':
         *
         *  (X==1)?((X&0x80)?-1:1):0      (X==2)?((X&0x80)?-1:1):0      (X==3)?((X&0x80)?-1:1):0
         *  (Y==1)?((Y&0x80)?-1:1):0      (Y==2)?((Y&0x80)?-1:1):0      (Y==3)?((X&0x80)?-1:1):0
         *                                  r[0] ^ r[1]
         * 其中第三行是前两行的向量积
         *
         */

        final int length = outR.length;
        if (inR.length != length)
            return false;   //invalid parameter
        if ((X & 0x7C)!=0 || (Y & 0x7C)!=0)
            return false;   //invalid parameter
        if (((X & 0x3)==0) || ((Y & 0x3)==0))
            return false;   //no axis specified
        if ((X & 0x3) == (Y & 0x3))
            return false;   //same axis specified
```

```
        int Z = X ^ Y;

        final int x = (X & 0x3)-1;
        final int y = (Y & 0x3)-1;
        final int z = (Z & 0x3)-1;

        final int axis_y = (z+1)%3;
        final int axis_z = (z+2)%3;
        if (((x^axis_y)|(y^axis_z)) != 0)
            Z ^= 0x80;

        final boolean sx = (X>=0x80);
        final boolean sy = (Y>=0x80);
        final boolean sz = (Z>=0x80);

        final int rowLength = ((length==16)?4:3);
        for (int j=0 ; j<3 ; j++) {
            final int offset = j*rowLength;
            for (int i=0 ; i<3 ; i++) {
                if (x==i)    outR[offset+i] = sx ? -inR[offset+0] : inR[offset+0];
                if (y==i)    outR[offset+i] = sy ? -inR[offset+1] : inR[offset+1];
                if (z==i)    outR[offset+i] = sz ? -inR[offset+2] : inR[offset+2];
            }
        }
        if (length == 16) {
            outR[3] = outR[7] = outR[11] = outR[12] = outR[13] = outR[14] = 0;
            outR[15] = 1;
        }
        return true;
    }
    public static float[] getOrientation(float[] R, float values[]) {

        if (R.length == 9) {
            values[0] = (float)Math.atan2(R[1], R[4]);
            values[1] = (float)Math.asin(-R[7]);
            values[2] = (float)Math.atan2(-R[6], R[8]);
        } else {
            values[0] = (float)Math.atan2(R[1], R[5]);
            values[1] = (float)Math.asin(-R[9]);
            values[2] = (float)Math.atan2(-R[8], R[10]);
        }
        return values;
    }
    public static float getAltitude(float p0, float p) {
        final float coef = 1.0f / 5.255f;
        return 44330.0f * (1.0f - (float)Math.pow(p/p0, coef));
    }

    public static void getAngleChange( float[] angleChange, float[] R, float[] prevR) {
        float rd1=0,rd4=0, rd6=0,rd7=0, rd8=0;
```

```
float ri0=0,ri1=0,ri2=0,ri3=0,ri4=0,ri5=0,ri6=0,ri7=0,ri8=0;
float pri0=0, pri1=0, pri2=0, pri3=0, pri4=0, pri5=0, pri6=0, pri7=0, pri8=0;

if(R.length == 9) {
    ri0 = R[0];
    ri1 = R[1];
    ri2 = R[2];
    ri3 = R[3];
    ri4 = R[4];
    ri5 = R[5];
    ri6 = R[6];
    ri7 = R[7];
    ri8 = R[8];
} else if(R.length == 16) {
    ri0 = R[0];
    ri1 = R[1];
    ri2 = R[2];
    ri3 = R[4];
    ri4 = R[5];
    ri5 = R[6];
    ri6 = R[8];
    ri7 = R[9];
    ri8 = R[10];
}

if(prevR.length == 9) {
    pri0 = prevR[0];
    pri1 = prevR[1];
    pri2 = prevR[2];
    pri3 = prevR[3];
    pri4 = prevR[4];
    pri5 = prevR[5];
    pri6 = prevR[6];
    pri7 = prevR[7];
    pri8 = prevR[8];
} else if(prevR.length == 16) {
    pri0 = prevR[0];
    pri1 = prevR[1];
    pri2 = prevR[2];
    pri3 = prevR[4];
    pri4 = prevR[5];
    pri5 = prevR[6];
    pri6 = prevR[8];
    pri7 = prevR[9];
    pri8 = prevR[10];
}

//计算出我们需要的旋转差矩阵的部分
//rd[i][j] = pri[0][i] * ri[0][j] + pri[1][i] * ri[1][j] + pri[2][i] * ri[2][j];
```

```
        rd1 = pri0 * ri1 + pri3 * ri4 + pri6 * ri7; //rd[0][1]
        rd4 = pri1 * ri1 + pri4 * ri4 + pri7 * ri7; //rd[1][1]
        rd6 = pri2 * ri0 + pri5 * ri3 + pri8 * ri6; //rd[2][0]
        rd7 = pri2 * ri1 + pri5 * ri4 + pri8 * ri7; //rd[2][1]
        rd8 = pri2 * ri2 + pri5 * ri5 + pri8 * ri8; //rd[2][2]

        angleChange[0] = (float)Math.atan2(rd1, rd4);
        angleChange[1] = (float)Math.asin(-rd7);
        angleChange[2] = (float)Math.atan2(-rd6, rd8);
    }
    public static void getRotationMatrixFromVector(float[] R, float[] rotationVector) {

        float q0;
        float q1 = rotationVector[0];
        float q2 = rotationVector[1];
        float q3 = rotationVector[2];

        if (rotationVector.length == 4) {
            q0 = rotationVector[3];
        } else {
            q0 = 1 - q1*q1 - q2*q2 - q3*q3;
            q0 = (q0 > 0) ? (float)Math.sqrt(q0) : 0;
        }

        float sq_q1 = 2 * q1 * q1;
        float sq_q2 = 2 * q2 * q2;
        float sq_q3 = 2 * q3 * q3;
        float q1_q2 = 2 * q1 * q2;
        float q3_q0 = 2 * q3 * q0;
        float q1_q3 = 2 * q1 * q3;
        float q2_q0 = 2 * q2 * q0;
        float q2_q3 = 2 * q2 * q3;
        float q1_q0 = 2 * q1 * q0;

        if(R.length == 9) {
            R[0] = 1 - sq_q2 - sq_q3;
            R[1] = q1_q2 - q3_q0;
            R[2] = q1_q3 + q2_q0;

            R[3] = q1_q2 + q3_q0;
            R[4] = 1 - sq_q1 - sq_q3;
            R[5] = q2_q3 - q1_q0;

            R[6] = q1_q3 - q2_q0;
            R[7] = q2_q3 + q1_q0;
            R[8] = 1 - sq_q1 - sq_q2;
        } else if (R.length == 16) {
            R[0] = 1 - sq_q2 - sq_q3;
            R[1] = q1_q2 - q3_q0;
            R[2] = q1_q3 + q2_q0;
```

```
            R[3] = 0.0f;

            R[4] = q1_q2 + q3_q0;
            R[5] = 1 - sq_q1 - sq_q3;
            R[6] = q2_q3 - q1_q0;
            R[7] = 0.0f;

            R[8] = q1_q3 - q2_q0;
            R[9] = q2_q3 + q1_q0;
            R[10] = 1 - sq_q1 - sq_q2;
            R[11] = 0.0f;

            R[12] = R[13] = R[14] = 0.0f;
            R[15] = 1.0f;
        }
    }
    public static void getQuaternionFromVector(float[] Q, float[] rv) {
        if (rv.length == 4) {
            Q[0] = rv[3];
        } else {
            Q[0] = 1 - rv[0]*rv[0] - rv[1]*rv[1] - rv[2]*rv[2];
            Q[0] = (Q[0] > 0) ? (float)Math.sqrt(Q[0]) : 0;
        }
        Q[1] = rv[0];
        Q[2] = rv[1];
        Q[3] = rv[2];
    }
    public boolean requestTriggerSensor(TriggerEventListener listener, Sensor sensor) {
        return requestTriggerSensorImpl(listener, sensor);
    }
    protected abstract boolean requestTriggerSensorImpl(TriggerEventListener listener,
            Sensor sensor);
    public boolean cancelTriggerSensor(TriggerEventListener listener, Sensor sensor) {
        return cancelTriggerSensorImpl(listener, sensor, true);
    protected abstract boolean cancelTriggerSensorImpl(TriggerEventListener listener,
            Sensor sensor, boolean disable);
    private LegacySensorManager getLegacySensorManager() {
        synchronized (mSensorListByType) {
            if (mLegacySensorManager == null) {
                Log.i(TAG, "This application is using deprecated SensorManager API which will "
                        + "be removed someday.  Please consider switching to the new API.");
                mLegacySensorManager = new LegacySensorManager(this);
            }
            return mLegacySensorManager;
        }
    }
}
```

上述方法的功能非常重要，其实就是我们在开发传感器应用程序时用到的 API 接口。有关上述方法的具体说明，读者可以查阅官网 SDK API 中对于类 android.hardware.SensorManager 的具体说明，如图 10-3 所示。

Introduction
App Components
App Resources
App Manifest
User Interface
Animation and Graphics
Computation
Media and Camera
Location and Sensors
Connectivity
Text and Input
Data Storage
Administration

| Sensor | Type | Description | Common Uses |
|---|---|---|---|
| TYPE_ACCELEROMETER | Hardware | Measures the acceleration force in m/s$^2$ that is applied to a device on all three physical axes (x, y, and z), including the force of gravity. | Motion detection (shake, tilt, etc.). |
| TYPE_AMBIENT_TEMPERATURE | Hardware | Measures the ambient room temperature in degrees Celsius (°C). See note below. | Monitoring air temperatures. |
| TYPE_GRAVITY | Software or Hardware | Measures the force of gravity in m/s$^2$ that is applied to a device on all three physical axes (x, y, z). | Motion detection (shake, tilt, etc.). |
| TYPE_GYROSCOPE | Hardware | Measures a device's rate of rotation in rad/s around each of the three physical axes (x, y, and z). | Rotation detection (spin, turn, etc.). |
| TYPE_LIGHT | Hardware | Measures the ambient light level (illumination) in lx. | Controlling screen brightness. |
| TYPE_LINEAR_ACCELERATION | Software or Hardware | Measures the acceleration force in m/s$^2$ that is applied to a device on all three physical axes (x, y, and z), excluding the force of | Monitoring acceleration along a single |

图 10-3 Android SDK API 中对 android.hardware.SensorManager 的具体说明

# 10.4 JNI 层详解

**知识点讲解：光盘:视频\知识点\第 10 章\JNI 层详解.avi**

在 Android 系统中，传感器系统的 JNI 部分的代码路径是 frameworks/base/core/jni/android_hardware_SensorManager.cpp。

在此文件中提供了对类 android.hardware.Sensor.Manage 的本地支持。上层和 JNI 层的调用关系如图 10-4 所示。

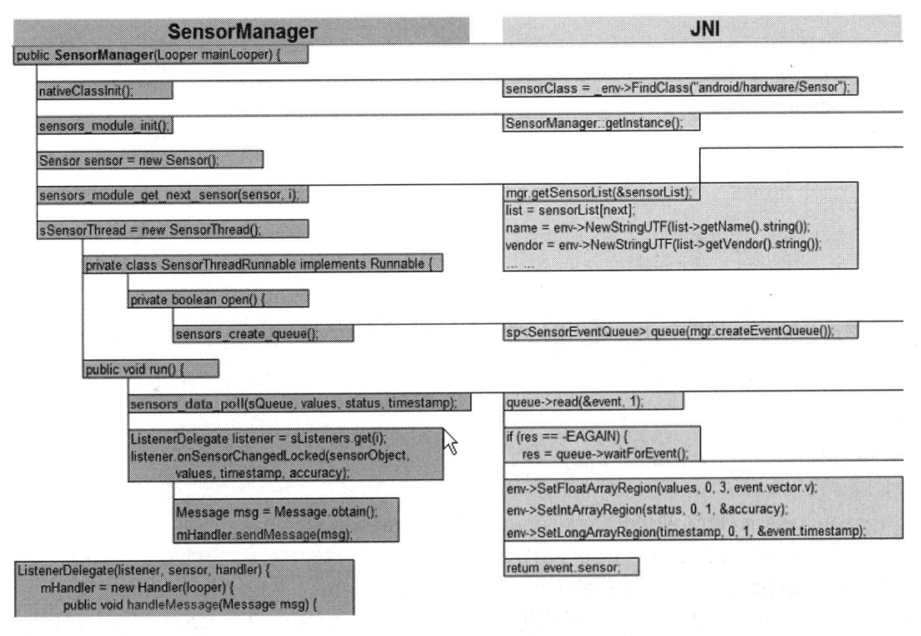


图 10-4 上层和 JNI 层的调用关系

在图 10-4 所示的调用关系中涉及了如下所示的 API 接口方法。

☑ nativeClassInit()：在 JNI 层得到 android.hardware.Sensor 的 JNI 环境指针。
☑ sensors_module_init()：通过 JNI 调用本地框架，得到 SensorService，SensorService 初始化控制流各功能。
☑ new Sensor()：建立一个 Sensor 对象，具体可查阅官网 API android.hardware.Sensor。
☑ sensors_module_get_next_sensor()：上层得到设备支持的所有 Sensor，并放入 SensorList 链表。
☑ new SensorThread()：创建 Sensor 线程，当应用程序 registerListener()注册监听器时开启线程 run()，注意当没有数据变化时线程会阻塞。

## 10.4.1 实现本地函数

文件 android_hardware_SensorManager.cpp 的功能是实现文件 SensorManager.java 中的 native（本地）函数，主要是通过调用文件 SenrsorManager.cpp 和 SensorEventQueue.cpp 中的相关类来完成相关工作的。文件 android_hardware_SensorManager.cpp 的具体实现代码如下：

```
static struct {
    jclass clazz;
    jmethodID dispatchSensorEvent;
} gBaseEventQueueClassInfo;

namespace android {

struct SensorOffsets
{
    jfieldID    name;
    jfieldID    vendor;
    jfieldID    version;
    jfieldID    handle;
    jfieldID    type;
    jfieldID    range;
    jfieldID    resolution;
    jfieldID    power;
    jfieldID    minDelay;
} gSensorOffsets;


/*
 *下面的方法是不是线程安全的和不打算用的
 */

static void
nativeClassInit (JNIEnv *_env, jclass _this)
{
    jclass sensorClass = _env->FindClass("android/hardware/Sensor");
    SensorOffsets& sensorOffsets = gSensorOffsets;
    sensorOffsets.name = _env->GetFieldID(sensorClass,"mName", "Ljava/lang/String;");
    sensorOffsets.vendor = _env->GetFieldID(sensorClass,"mVendor", "Ljava/lang/String;");
    sensorOffsets.version = _env->GetFieldID(sensorClass,"mVersion", "I");
```

```
    sensorOffsets.handle = _env->GetFieldID(sensorClass,"mHandle", "I");
    sensorOffsets.type = _env->GetFieldID(sensorClass,"mType", "I");
    sensorOffsets.range = _env->GetFieldID(sensorClass,"mMaxRange", "F");
    sensorOffsets.resolution = _env->GetFieldID(sensorClass,"mResolution","F");
    sensorOffsets.power = _env->GetFieldID(sensorClass,"mPower", "F");
    sensorOffsets.minDelay = _env->GetFieldID(sensorClass,"mMinDelay",    "I");
}

static jint
nativeGetNextSensor(JNIEnv *env, jclass clazz, jobject sensor, jint next)
{
    SensorManager& mgr(SensorManager::getInstance());

    Sensor const* const* sensorList;
    size_t count = mgr.getSensorList(&sensorList);
    if (size_t(next) >= count)
        return -1;

    Sensor const* const list = sensorList[next];
    const SensorOffsets& sensorOffsets(gSensorOffsets);
    jstring name = env->NewStringUTF(list->getName().string());
    jstring vendor = env->NewStringUTF(list->getVendor().string());
    env->SetObjectField(sensor, sensorOffsets.name, name);
    env->SetObjectField(sensor, sensorOffsets.vendor, vendor);
    env->SetIntField(sensor, sensorOffsets.version, list->getVersion());
    env->SetIntField(sensor, sensorOffsets.handle, list->getHandle());
    env->SetIntField(sensor, sensorOffsets.type, list->getType());
    env->SetFloatField(sensor, sensorOffsets.range, list->getMaxValue());
    env->SetFloatField(sensor, sensorOffsets.resolution, list->getResolution());
    env->SetFloatField(sensor, sensorOffsets.power, list->getPowerUsage());
    env->SetIntField(sensor, sensorOffsets.minDelay, list->getMinDelay());

    next++;
    return size_t(next) < count ? next : 0;
}

//----------------------------------------------------------------------------

class Receiver : public LooperCallback {
    sp<SensorEventQueue> mSensorQueue;
    sp<MessageQueue> mMessageQueue;
    jobject mReceiverObject;
    jfloatArray mScratch;
public:
    Receiver(const sp<SensorEventQueue>& sensorQueue,
            const sp<MessageQueue>& messageQueue,
            jobject receiverObject, jfloatArray scratch) {
        JNIEnv* env = AndroidRuntime::getJNIEnv();
        mSensorQueue = sensorQueue;
        mMessageQueue = messageQueue;
        mReceiverObject = env->NewGlobalRef(receiverObject);
```

```
        mScratch = (jfloatArray)env->NewGlobalRef(scratch);
    }
    ~Receiver() {
        JNIEnv* env = AndroidRuntime::getJNIEnv();
        env->DeleteGlobalRef(mReceiverObject);
        env->DeleteGlobalRef(mScratch);
    }
    sp<SensorEventQueue> getSensorEventQueue() const {
        return mSensorQueue;
    }

    void destroy() {
        mMessageQueue->getLooper()->removeFd( mSensorQueue->getFd() );
    }

private:
    virtual void onFirstRef() {
        LooperCallback::onFirstRef();
        mMessageQueue->getLooper()->addFd(mSensorQueue->getFd(), 0,
                ALOOPER_EVENT_INPUT, this, mSensorQueue.get());
    }

    virtual int handleEvent(int fd, int events, void* data) {
        JNIEnv* env = AndroidRuntime::getJNIEnv();
        sp<SensorEventQueue> q = reinterpret_cast<SensorEventQueue *>(data);
        ssize_t n;
        ASensorEvent buffer[16];
        while ((n = q->read(buffer, 16)) > 0) {
            for (int i=0 ; i<n ; i++) {

                env->SetFloatArrayRegion(mScratch, 0, 16, buffer[i].data);

                env->CallVoidMethod(mReceiverObject,
                        gBaseEventQueueClassInfo.dispatchSensorEvent,
                        buffer[i].sensor,
                        mScratch,
                        buffer[i].vector.status,
                        buffer[i].timestamp);

                if (env->ExceptionCheck()) {
                    ALOGE("Exception dispatching input event.");
                    return 1;
                }
            }
        }
        if (n<0 && n != -EAGAIN) {
        }

        return 1;
    }
};
```

```
static jint nativeInitSensorEventQueue(JNIEnv *env, jclass clazz, jobject eventQ, jobject msgQ, jfloatArray
scratch) {
    SensorManager& mgr(SensorManager::getInstance());
    sp<SensorEventQueue> queue(mgr.createEventQueue());

    sp<MessageQueue> messageQueue = android_os_MessageQueue_getMessageQueue(env, msgQ);
    if (messageQueue == NULL) {
        jniThrowRuntimeException(env, "MessageQueue is not initialized.");
        return 0;
    }

    sp<Receiver> receiver = new Receiver(queue, messageQueue, eventQ, scratch);
    receiver->incStrong((void*)nativeInitSensorEventQueue);
    return jint(receiver.get());
}

static jint nativeEnableSensor(JNIEnv *env, jclass clazz, jint eventQ, jint handle, jint us) {
    sp<Receiver> receiver(reinterpret_cast<Receiver *>(eventQ));
    return receiver->getSensorEventQueue()->enableSensor(handle, us);
}

static jint nativeDisableSensor(JNIEnv *env, jclass clazz, jint eventQ, jint handle) {
    sp<Receiver> receiver(reinterpret_cast<Receiver *>(eventQ));
    return receiver->getSensorEventQueue()->disableSensor(handle);
}

static void nativeDestroySensorEventQueue(JNIEnv *env, jclass clazz, jint eventQ, jint handle) {
    sp<Receiver> receiver(reinterpret_cast<Receiver *>(eventQ));
    receiver->destroy();
    receiver->decStrong((void*)nativeInitSensorEventQueue);
}


//----------------------------------------------------------------------------

static JNINativeMethod gSystemSensorManagerMethods[] = {
    {"nativeClassInit",
            "()V",
            (void*)nativeClassInit },

    {"nativeGetNextSensor",
            "(Landroid/hardware/Sensor;I)I",
            (void*)nativeGetNextSensor },
};

static JNINativeMethod gBaseEventQueueMethods[] = {
    {"nativeInitBaseEventQueue",

"(Landroid/hardware/SystemSensorManager$BaseEventQueue;Landroid/os/MessageQueue;[F)I",
            (void*)nativeInitSensorEventQueue },
```

```
    {"nativeEnableSensor",
            "(III)I",
            (void*)nativeEnableSensor },

    {"nativeDisableSensor",
            "(II)I",
            (void*)nativeDisableSensor },

    {"nativeDestroySensorEventQueue",
            "(I)V",
            (void*)nativeDestroySensorEventQueue },
};

};

using namespace android;

#define FIND_CLASS(var, className) \
        var = env->FindClass(className); \
        LOG_FATAL_IF(! var, "Unable to find class " className); \
        var = jclass(env->NewGlobalRef(var));

#define GET_METHOD_ID(var, clazz, methodName, methodDescriptor) \
        var = env->GetMethodID(clazz, methodName, methodDescriptor); \
        LOG_FATAL_IF(! var, "Unable to find method " methodName);

int register_android_hardware_SensorManager(JNIEnv *env)
{
    jniRegisterNativeMethods(env, "android/hardware/SystemSensorManager",
            gSystemSensorManagerMethods, NELEM(gSystemSensorManagerMethods));

    jniRegisterNativeMethods(env, "android/hardware/SystemSensorManager$BaseEventQueue",
            gBaseEventQueueMethods, NELEM(gBaseEventQueueMethods));

    FIND_CLASS(gBaseEventQueueClassInfo.clazz,
"android/hardware/SystemSensorManager$BaseEventQueue");

    GET_METHOD_ID(gBaseEventQueueClassInfo.dispatchSensorEvent,
            gBaseEventQueueClassInfo.clazz,
            "dispatchSensorEvent", "(I[FIJ)V");

    return 0;
}
```

### 10.4.2　处理客户端数据

文件 frameworks/native/libs/gui/SensorManager.cpp 提供了对传感器数据部分的操作，实现了 sensor_data_XXX()格式的函数。另外在 Native 层的客户端，文件 SensorManager.cpp 还负责与服务端 SensorService.cpp

之间的通信工作。文件 SensorManager.cpp 的具体实现代码如下：

```
// ----------------------------------------------------------------------------
namespace android {
// ----------------------------------------------------------------------------

ANDROID_SINGLETON_STATIC_INSTANCE(SensorManager)

SensorManager::SensorManager()
    : mSensorList(0)
{
    assertStateLocked();
}

SensorManager::~SensorManager()
{
    free(mSensorList);
}

void SensorManager::sensorManagerDied()
{
    Mutex::Autolock _l(mLock);
    mSensorServer.clear();
    free(mSensorList);
    mSensorList = NULL;
    mSensors.clear();
}

status_t SensorManager::assertStateLocked() const {
    if (mSensorServer == NULL) {
        const String16 name("sensorservice");
        for (int i=0 ; i<4 ; i++) {
            status_t err = getService(name, &mSensorServer);
            if (err == NAME_NOT_FOUND) {
                usleep(250000);
                continue;
            }
            if (err != NO_ERROR) {
                return err;
            }
            break;
        }

        class DeathObserver : public IBinder::DeathRecipient {
            SensorManager& mSensorManger;
            virtual void binderDied(const wp<IBinder>& who) {
                ALOGW("sensorservice died [%p]", who.unsafe_get());
                mSensorManger.sensorManagerDied();
            }
        public:
            DeathObserver(SensorManager& mgr) : mSensorManger(mgr) { }
```

```
        };

        mDeathObserver = new DeathObserver(*const_cast<SensorManager *>(this));
        mSensorServer->asBinder()->linkToDeath(mDeathObserver);

        mSensors = mSensorServer->getSensorList();
        size_t count = mSensors.size();
        mSensorList = (Sensor const**)malloc(count * sizeof(Sensor*));
        for (size_t i=0 ; i<count ; i++) {
            mSensorList[i] = mSensors.array() + i;
        }
    }

    return NO_ERROR;
}
ssize_t SensorManager::getSensorList(Sensor const* const** list) const
{
    Mutex::Autolock _l(mLock);
    status_t err = assertStateLocked();
    if (err < 0) {
        return ssize_t(err);
    }
    *list = mSensorList;
    return mSensors.size();
}

Sensor const* SensorManager::getDefaultSensor(int type)
{
    Mutex::Autolock _l(mLock);
    if (assertStateLocked() == NO_ERROR) {
        for (size_t i=0 ; i<mSensors.size() ; i++) {
            if (mSensorList[i]->getType() == type)
                return mSensorList[i];
        }
    }
    return NULL;
}

sp<SensorEventQueue> SensorManager::createEventQueue()
{
    sp<SensorEventQueue> queue;

    Mutex::Autolock _l(mLock);
    while (assertStateLocked() == NO_ERROR) {
        sp<ISensorEventConnection> connection =
                mSensorServer->createSensorEventConnection();
        if (connection == NULL) {
            ALOGE("createEventQueue: connection is NULL. SensorService died.");
            continue;
        }
        queue = new SensorEventQueue(connection);
```

```
            break;
        }
        return queue;
}

// -----------------------------------------------------------------------
};
```

## 10.4.3　处理服务端数据

文件 frameworks/native/services/sensorservice/SensorService.cpp 能够实现 Sensor 真正的后台服务，是服务端的数据处理中心。在Android的传感器系统中，SensorService作为一个轻量级的System Service，在 SystemServer 内运行，在 system_init<system_init.cpp>中调用了 SensorService::instantiate()。具体来说，SensorService 的主要功能如下：

（1）通过 SensorService::instantiate 创建实例对象，并增加到 ServiceManager 中，然后创建并启动线程，并执行 threadLoop。

（2）threadLoop 从 sensor 驱动获取原始数据，然后通过 SensorEventConnection 把事件发送给客户端。

（3）BnSensorServer 的成员函数负责让客户端获取 sensor 列表和创建 SensorEventConnection。

文件 SensorService.cpp 的具体实现代码如下：

```
namespace android {

const char* SensorService::WAKE_LOCK_NAME = "SensorService";

SensorService::SensorService()
    : mInitCheck(NO_INIT)
{
}

void SensorService::onFirstRef()
{
    ALOGD("nuSensorService starting...");

    SensorDevice& dev(SensorDevice::getInstance());

    if (dev.initCheck() == NO_ERROR) {
        sensor_t const* list;
        ssize_t count = dev.getSensorList(&list);
        if (count > 0) {
            ssize_t orientationIndex = -1;
            bool hasGyro = false;
            uint32_t virtualSensorsNeeds =
                    (1<<SENSOR_TYPE_GRAVITY) |
                    (1<<SENSOR_TYPE_LINEAR_ACCELERATION) |
                    (1<<SENSOR_TYPE_ROTATION_VECTOR);

            mLastEventSeen.setCapacity(count);
            for (ssize_t i=0 ; i<count ; i++) {
```

```
            registerSensor( new HardwareSensor(list[i]) );
            switch (list[i].type) {
                case SENSOR_TYPE_ORIENTATION:
                    orientationIndex = i;
                    break;
                case SENSOR_TYPE_GYROSCOPE:
                    hasGyro = true;
                    break;
                case SENSOR_TYPE_GRAVITY:
                case SENSOR_TYPE_LINEAR_ACCELERATION:
                case SENSOR_TYPE_ROTATION_VECTOR:
                    virtualSensorsNeeds &= ~(1<<list[i].type);
                    break;
            }
        }

        //它是安全的，在这里实例化 SensorFusion 对象
        //如果被实例化后，H/W 传感器已注册
        const SensorFusion& fusion(SensorFusion::getInstance());

        if (hasGyro) {
            //总是实例化 Android 的虚拟传感器。因为它们是实例化落后于 HAL 的传感器，它们不会干扰
应用程序，除非它们看起来特别像它们的名字

            registerVirtualSensor( new RotationVectorSensor() );
            registerVirtualSensor( new GravitySensor(list, count) );
            registerVirtualSensor( new LinearAccelerationSensor(list, count) );

            //这是选项
            registerVirtualSensor( new OrientationSensor() );
            registerVirtualSensor( new CorrectedGyroSensor(list, count) );
        }

        mUserSensorList = mSensorList;

        if (hasGyro) {
            registerVirtualSensor( new GyroDriftSensor() );
        }

        if (hasGyro &&
                (virtualSensorsNeeds & (1<<SENSOR_TYPE_ROTATION_VECTOR))) {
            if (orientationIndex >= 0) {
                mUserSensorList.removeItemsAt(orientationIndex);
            }
        }

        //调试传感器列表
        for (size_t i=0 ; i<mSensorList.size() ; i++) {
            switch (mSensorList[i].getType()) {
                case SENSOR_TYPE_GRAVITY:
                case SENSOR_TYPE_LINEAR_ACCELERATION:
                case SENSOR_TYPE_ROTATION_VECTOR:
                    if (strstr(mSensorList[i].getVendor().string(), "Google")) {
```

```
                        mUserSensorListDebug.add(mSensorList[i]);
                    }
                    break;
                default:
                    mUserSensorListDebug.add(mSensorList[i]);
                    break;
            }
        }

        run("SensorService", PRIORITY_URGENT_DISPLAY);
        mInitCheck = NO_ERROR;
    }
  }
}

void SensorService::registerSensor(SensorInterface* s)
{
    sensors_event_t event;
    memset(&event, 0, sizeof(event));

    const Sensor sensor(s->getSensor());
    //添加到传感器列表（返回给客户端）
    mSensorList.add(sensor);
    //加入到我们的手柄 - > SensorInterface 映射
    mSensorMap.add(sensor.getHandle(), s);
    //创建 mLastEventSeen 数组中的一个条目
    mLastEventSeen.add(sensor.getHandle(), event);
}

void SensorService::registerVirtualSensor(SensorInterface* s)
{
    registerSensor(s);
    mVirtualSensorList.add( s );
}

SensorService::~SensorService()
{
    for (size_t i=0 ; i<mSensorMap.size() ; i++)
        delete mSensorMap.valueAt(i);
}

static const String16 sDump("android.permission.DUMP");

status_t SensorService::dump(int fd, const Vector<String16>& args)
{
    const size_t SIZE = 1024;
    char buffer[SIZE];
    String8 result;
    if (!PermissionCache::checkCallingPermission(sDump)) {
        snprintf(buffer, SIZE, "Permission Denial: "
                "can't dump SurfaceFlinger from pid=%d, uid=%d\n",
                IPCThreadState::self()->getCallingPid(),
```

```
                IPCThreadState::self()->getCallingUid());
        result.append(buffer);
    } else {
        Mutex::Autolock _l(mLock);
        snprintf(buffer, SIZE, "Sensor List:\n");
        result.append(buffer);
        for (size_t i=0 ; i<mSensorList.size() ; i++) {
            const Sensor& s(mSensorList[i]);
            const sensors_event_t& e(mLastEventSeen.valueFor(s.getHandle()));
            snprintf(buffer, SIZE,
                    "%-48s| %-32s | 0x%08x | maxRate=%7.2fHz | "
                    "last=<%5.1f,%5.1f,%5.1f>\n",
                    s.getName().string(),
                    s.getVendor().string(),
                    s.getHandle(),
                    s.getMinDelay() ? (1000000.0f / s.getMinDelay()) : 0.0f,
                    e.data[0], e.data[1], e.data[2]);
            result.append(buffer);
        }
        SensorFusion::getInstance().dump(result, buffer, SIZE);
        SensorDevice::getInstance().dump(result, buffer, SIZE);

        snprintf(buffer, SIZE, "%d active connections\n",
                mActiveConnections.size());
        result.append(buffer);
        snprintf(buffer, SIZE, "Active sensors:\n");
        result.append(buffer);
        for (size_t i=0 ; i<mActiveSensors.size() ; i++) {
            int handle = mActiveSensors.keyAt(i);
            snprintf(buffer, SIZE, "%s (handle=0x%08x, connections=%d)\n",
                    getSensorName(handle).string(),
                    handle,
                    mActiveSensors.valueAt(i)->getNumConnections());
            result.append(buffer);
        }
    }
    write(fd, result.string(), result.size());
    return NO_ERROR;
}

void SensorService::cleanupAutoDisabledSensor(const sp<SensorEventConnection>& connection,
        sensors_event_t const* buffer, const int count) {
    SensorInterface* sensor;
    status_t err = NO_ERROR;
    for (int i=0 ; i<count ; i++) {
        int handle = buffer[i].sensor;
        if (getSensorType(handle) == SENSOR_TYPE_SIGNIFICANT_MOTION) {
            if (connection->hasSensor(handle)) {
                sensor = mSensorMap.valueFor(handle);
                err = sensor ?sensor->resetStateWithoutActuatingHardware(connection.get(), handle)
                        : status_t(BAD_VALUE);
                if (err != NO_ERROR) {
```

```
                    ALOGE("Sensor Inteface: Resetting state failed with err: %d", err);
                }
                cleanupWithoutDisable(connection, handle);
            }
        }
    }
}

bool SensorService::threadLoop()
{
    ALOGD("nuSensorService thread starting...");

    const size_t numEventMax = 16;
    const size_t minBufferSize = numEventMax + numEventMax * mVirtualSensorList.size();
    sensors_event_t buffer[minBufferSize];
    sensors_event_t scratch[minBufferSize];
    SensorDevice& device(SensorDevice::getInstance());
    const size_t vcount = mVirtualSensorList.size();

    ssize_t count;
    bool wakeLockAcquired = false;
    const int halVersion = device.getHalDeviceVersion();
    do {
        count = device.poll(buffer, numEventMax);
        if (count<0) {
            ALOGE("sensor poll failed (%s)", strerror(-count));
            break;
        }

        //TODO()：添加一个标志，用该传感器的定义来表示
        //在这里可以唤醒 AP 传感器
        for (int i = 0; i < count; i++) {
            if (getSensorType(buffer[i].sensor) == SENSOR_TYPE_SIGNIFICANT_MOTION) {
                acquire_wake_lock(PARTIAL_WAKE_LOCK, WAKE_LOCK_NAME);
                wakeLockAcquired = true;
                break;
            }
        }

        recordLastValue(buffer, count);

        //处理虚拟传感器
        if (count && vcount) {
            sensors_event_t const * const event = buffer;
            const DefaultKeyedVector<int, SensorInterface*> virtualSensors(
                    getActiveVirtualSensors());
            const size_t activeVirtualSensorCount = virtualSensors.size();
            if (activeVirtualSensorCount) {
                size_t k = 0;
                SensorFusion& fusion(SensorFusion::getInstance());
                if (fusion.isEnabled()) {
                    for (size_t i=0 ; i<size_t(count) ; i++) {
```

```
                    fusion.process(event[i]);
                }
            }
            for (size_t i=0 ; i<size_t(count) && k<minBufferSize ; i++) {
                for (size_t j=0 ; j<activeVirtualSensorCount ; j++) {
                    if (count + k >= minBufferSize) {
                        ALOGE("buffer too small to hold all events: "
                                "count=%u, k=%u, size=%u",
                                count, k, minBufferSize);
                        break;
                    }
                    sensors_event_t out;
                    SensorInterface* si = virtualSensors.valueAt(j);
                    if (si->process(&out, event[i])) {
                        buffer[count + k] = out;
                        k++;
                    }
                }
            }
            if (k) {
                //记录最近合成值
                recordLastValue(&buffer[count], k);
                count += k;
                //缓冲器由时间戳排序
                sortEventBuffer(buffer, count);
            }
        }
    }

    //处理的 RotationVector 传感器向后兼容
    if (halVersion < SENSORS_DEVICE_API_VERSION_1_0) {
        for (int i = 0; i < count; i++) {
            if (getSensorType(buffer[i].sensor) == SENSOR_TYPE_ROTATION_VECTOR) {
                buffer[i].data[4] = -1;
            }
        }
    }

    //发送我们的活动给客户
    const SortedVector< wp<SensorEventConnection> > activeConnections(
            getActiveConnections());
    size_t numConnections = activeConnections.size();
    for (size_t i=0 ; i<numConnections ; i++) {
        sp<SensorEventConnection> connection(
                activeConnections[i].promote());
        if (connection != 0) {
            connection->sendEvents(buffer, count, scratch);
            cleanupAutoDisabledSensor(connection, buffer, count);
        }
    }

    //我们已经读取的数据，上层需要控制 wakelock
```

```
            if (wakeLockAcquired) release_wake_lock(WAKE_LOCK_NAME);

        } while (count >= 0 || Thread::exitPending());

        ALOGW("Exiting SensorService::threadLoop => aborting...");
        abort();
        return false;
}

void SensorService::recordLastValue(
            sensors_event_t const * buffer, size_t count)
{
        Mutex::Autolock _l(mLock);

        //为每个传感器记录最近的事件
        int32_t prev = buffer[0].sensor;
        for (size_t i=1 ; i<count ; i++) {
            //这个缓冲区的最后一个事件记录每个传感器类型
            int32_t curr = buffer[i].sensor;
            if (curr != prev) {
                mLastEventSeen.editValueFor(prev) = buffer[i-1];
                prev = curr;
            }
        }
        mLastEventSeen.editValueFor(prev) = buffer[count-1];
}

void SensorService::sortEventBuffer(sensors_event_t* buffer, size_t count)
{
        struct compar {
            static int cmp(void const* lhs, void const* rhs) {
                sensors_event_t const* l = static_cast<sensors_event_t const*>(lhs);
                sensors_event_t const* r = static_cast<sensors_event_t const*>(rhs);
                return l->timestamp - r->timestamp;
            }
        };
        qsort(buffer, count, sizeof(sensors_event_t), compar::cmp);
}

SortedVector< wp<SensorService::SensorEventConnection> >
SensorService::getActiveConnections() const
{
        Mutex::Autolock _l(mLock);
        return mActiveConnections;
}

DefaultKeyedVector<int, SensorInterface*>
SensorService::getActiveVirtualSensors() const
{
        Mutex::Autolock _l(mLock);
        return mActiveVirtualSensors;
}
```

```
String8 SensorService::getSensorName(int handle) const {
    size_t count = mUserSensorList.size();
    for (size_t i=0 ; i<count ; i++) {
        const Sensor& sensor(mUserSensorList[i]);
        if (sensor.getHandle() == handle) {
            return sensor.getName();
        }
    }
    String8 result("unknown");
    return result;
}

int SensorService::getSensorType(int handle) const {
    size_t count = mUserSensorList.size();
    for (size_t i=0 ; i<count ; i++) {
        const Sensor& sensor(mUserSensorList[i]);
        if (sensor.getHandle() == handle) {
            return sensor.getType();
        }
    }
    return -1;
}


Vector<Sensor> SensorService::getSensorList()
{
    char value[PROPERTY_VALUE_MAX];
    property_get("debug.sensors", value, "0");
    if (atoi(value)) {
        return mUserSensorListDebug;
    }
    return mUserSensorList;
}

sp<ISensorEventConnection> SensorService::createSensorEventConnection()
{
    uid_t uid = IPCThreadState::self()->getCallingUid();
    sp<SensorEventConnection> result(new SensorEventConnection(this, uid));
    return result;
}

void SensorService::cleanupConnection(SensorEventConnection* c)
{
    Mutex::Autolock _l(mLock);
    const wp<SensorEventConnection> connection(c);
    size_t size = mActiveSensors.size();
    ALOGD_IF(DEBUG_CONNECTIONS, "%d active sensors", size);
    for (size_t i=0 ; i<size ; ) {
        int handle = mActiveSensors.keyAt(i);
        if (c->hasSensor(handle)) {
            ALOGD_IF(DEBUG_CONNECTIONS, "%i: disabling handle=0x%08x", i, handle);
            SensorInterface* sensor = mSensorMap.valueFor( handle );
```

```
            ALOGE_IF(!sensor, "mSensorMap[handle=0x%08x] is null!", handle);
            if (sensor) {
                sensor->activate(c, false);
            }
        }
        SensorRecord* rec = mActiveSensors.valueAt(i);
        ALOGE_IF(!rec, "mActiveSensors[%d] is null (handle=0x%08x)!", i, handle);
        ALOGD_IF(DEBUG_CONNECTIONS,
                "removing connection %p for sensor[%d].handle=0x%08x",
                c, i, handle);

        if (rec && rec->removeConnection(connection)) {
            ALOGD_IF(DEBUG_CONNECTIONS, "... and it was the last connection");
            mActiveSensors.removeItemsAt(i, 1);
            mActiveVirtualSensors.removeItem(handle);
            delete rec;
            size--;
        } else {
            i++;
        }
    }
    mActiveConnections.remove(connection);
    BatteryService::cleanup(c->getUid());
}

status_t SensorService::enable(const sp<SensorEventConnection>& connection,
        int handle)
{
    if (mInitCheck != NO_ERROR)
        return mInitCheck;

    Mutex::Autolock _l(mLock);
    SensorInterface* sensor = mSensorMap.valueFor(handle);
    SensorRecord* rec = mActiveSensors.valueFor(handle);
    if (rec == 0) {
        rec = new SensorRecord(connection);
        mActiveSensors.add(handle, rec);
        if (sensor->isVirtual()) {
            mActiveVirtualSensors.add(handle, sensor);
        }
    } else {
        if (rec->addConnection(connection)) {
            //该传感器已经启动，如果新应用需要连接使用
            //则立即发送使用请求以获取传感器的值
            if (sensor->getSensor().getMinDelay() == 0) {
                sensors_event_t scratch;
                sensors_event_t& event(mLastEventSeen.editValueFor(handle));
                if (event.version == sizeof(sensors_event_t)) {
                    connection->sendEvents(&event, 1);
                }
            }
        }
```

```
    }

    if (connection->addSensor(handle)) {
        BatteryService::enableSensor(connection->getUid(), handle);
        if (mActiveConnections.indexOf(connection) < 0) {
            mActiveConnections.add(connection);
        }
    } else {
        ALOGW("sensor %08x already enabled in connection %p (ignoring)",
            handle, connection.get());
    }

    //设置使用传感器
    status_t err = sensor ? sensor->activate(connection.get(), true) : status_t(BAD_VALUE);

    if (err != NO_ERROR) {
        //启用失败，在 SensorDevice 复位状态
        status_t resetErr = sensor ? sensor->resetStateWithoutActuatingHardware(connection.get(),
                handle) : status_t(BAD_VALUE);
        //启用失败，处于复位状态
        cleanupWithoutDisable(connection, handle);
    }
    return err;
}

status_t SensorService::disable(const sp<SensorEventConnection>& connection,
        int handle)
{
    if (mInitCheck != NO_ERROR)
        return mInitCheck;

    status_t err = cleanupWithoutDisable(connection, handle);
    if (err == NO_ERROR) {
        SensorInterface* sensor = mSensorMap.valueFor(handle);
        err = sensor ? sensor->activate(connection.get(), false) : status_t(BAD_VALUE);
    }
    return err;
}

status_t SensorService::cleanupWithoutDisable(const sp<SensorEventConnection>& connection,
        int handle) {
    Mutex::Autolock _l(mLock);
    SensorRecord* rec = mActiveSensors.valueFor(handle);
    if (rec) {
        //看是否变为无效
        if (connection->removeSensor(handle)) {
            BatteryService::disableSensor(connection->getUid(), handle);
        }
        if (connection->hasAnySensor() == false) {
            mActiveConnections.remove(connection);
        }
```

```
            //传感器是否变为无效
            if (rec->removeConnection(connection)) {
                mActiveSensors.removeItem(handle);
                mActiveVirtualSensors.removeItem(handle);
                delete rec;
            }
            return NO_ERROR;
        }
        return BAD_VALUE;
}

status_t SensorService::setEventRate(const sp<SensorEventConnection>& connection,
            int handle, nsecs_t ns)
{
        if (mInitCheck != NO_ERROR)
            return mInitCheck;

        SensorInterface* sensor = mSensorMap.valueFor(handle);
        if (!sensor)
            return BAD_VALUE;

        if (ns < 0)
            return BAD_VALUE;

        nsecs_t minDelayNs = sensor->getSensor().getMinDelayNs();
        if (ns < minDelayNs) {
            ns = minDelayNs;
        }

        if (ns < MINIMUM_EVENTS_PERIOD)
            ns = MINIMUM_EVENTS_PERIOD;

        return sensor->setDelay(connection.get(), handle, ns);
}

// ---------------------------------------------------------------------------

SensorService::SensorRecord::SensorRecord(
            const sp<SensorEventConnection>& connection)
{
        mConnections.add(connection);
}

bool SensorService::SensorRecord::addConnection(
            const sp<SensorEventConnection>& connection)
{
        if (mConnections.indexOf(connection) < 0) {
            mConnections.add(connection);
            return true;
        }
        return false;
}
```

```
bool SensorService::SensorRecord::removeConnection(
        const wp<SensorEventConnection>& connection)
{
    ssize_t index = mConnections.indexOf(connection);
    if (index >= 0) {
        mConnections.removeItemsAt(index, 1);
    }
    return mConnections.size() ? false : true;
}

// ---------------------------------------------------------------------------

SensorService::SensorEventConnection::SensorEventConnection(
        const sp<SensorService>& service, uid_t uid)
    : mService(service), mChannel(new BitTube()), mUid(uid)
{
}

SensorService::SensorEventConnection::~SensorEventConnection()
{
    ALOGD_IF(DEBUG_CONNECTIONS, "~SensorEventConnection(%p)", this);
    mService->cleanupConnection(this);
}

void SensorService::SensorEventConnection::onFirstRef()
{
}

bool SensorService::SensorEventConnection::addSensor(int32_t handle) {
    Mutex::Autolock _l(mConnectionLock);
    if (mSensorInfo.indexOf(handle) < 0) {
        mSensorInfo.add(handle);
        return true;
    }
    return false;
}

bool SensorService::SensorEventConnection::removeSensor(int32_t handle) {
    Mutex::Autolock _l(mConnectionLock);
    if (mSensorInfo.remove(handle) >= 0) {
        return true;
    }
    return false;
}

bool SensorService::SensorEventConnection::hasSensor(int32_t handle) const {
    Mutex::Autolock _l(mConnectionLock);
    return mSensorInfo.indexOf(handle) >= 0;
}

bool SensorService::SensorEventConnection::hasAnySensor() const {
    Mutex::Autolock _l(mConnectionLock);
    return mSensorInfo.size() ? true : false;
```

```
}

status_t SensorService::SensorEventConnection::sendEvents(
        sensors_event_t const* buffer, size_t numEvents,
        sensors_event_t* scratch)
{
    //筛选出的事件不用于此连接
    size_t count = 0;
    if (scratch) {
        Mutex::Autolock _l(mConnectionLock);
        size_t i=0;
        while (i<numEvents) {
            const int32_t curr = buffer[i].sensor;
            if (mSensorInfo.indexOf(curr) >= 0) {
                do {
                    scratch[count++] = buffer[i++];
                } while ((i<numEvents) && (buffer[i].sensor == curr));
            } else {
                i++;
            }
        }
    } else {
        scratch = const_cast<sensors_event_t *>(buffer);
        count = numEvents;
    }

    // ASensorEvent 和 sensors_event_t 是同一类型
    ssize_t size = SensorEventQueue::write(mChannel,
            reinterpret_cast<ASensorEvent const*>(scratch), count);
    if (size == -EAGAIN) {
        return size;
    }

    return size < 0 ? status_t(size) : status_t(NO_ERROR);
}

sp<BitTube> SensorService::SensorEventConnection::getSensorChannel() const
{
    return mChannel;
}

status_t SensorService::SensorEventConnection::enableDisable(
        int handle, bool enabled)
{
    status_t err;
    if (enabled) {
        err = mService->enable(this, handle);
    } else {
        err = mService->disable(this, handle);
    }
    return err;
}
```

```
status_t SensorService::SensorEventConnection::setEventRate(
        int handle, nsecs_t ns)
{
    return mService->setEventRate(this, handle, ns);
}

// ---------------------------------------------------------------------------
};
```

通过上述实现代码，可以了解 SensorService 服务的创建、启动过程，整个过程的 C/S 通信架构如图 10-5 所示。

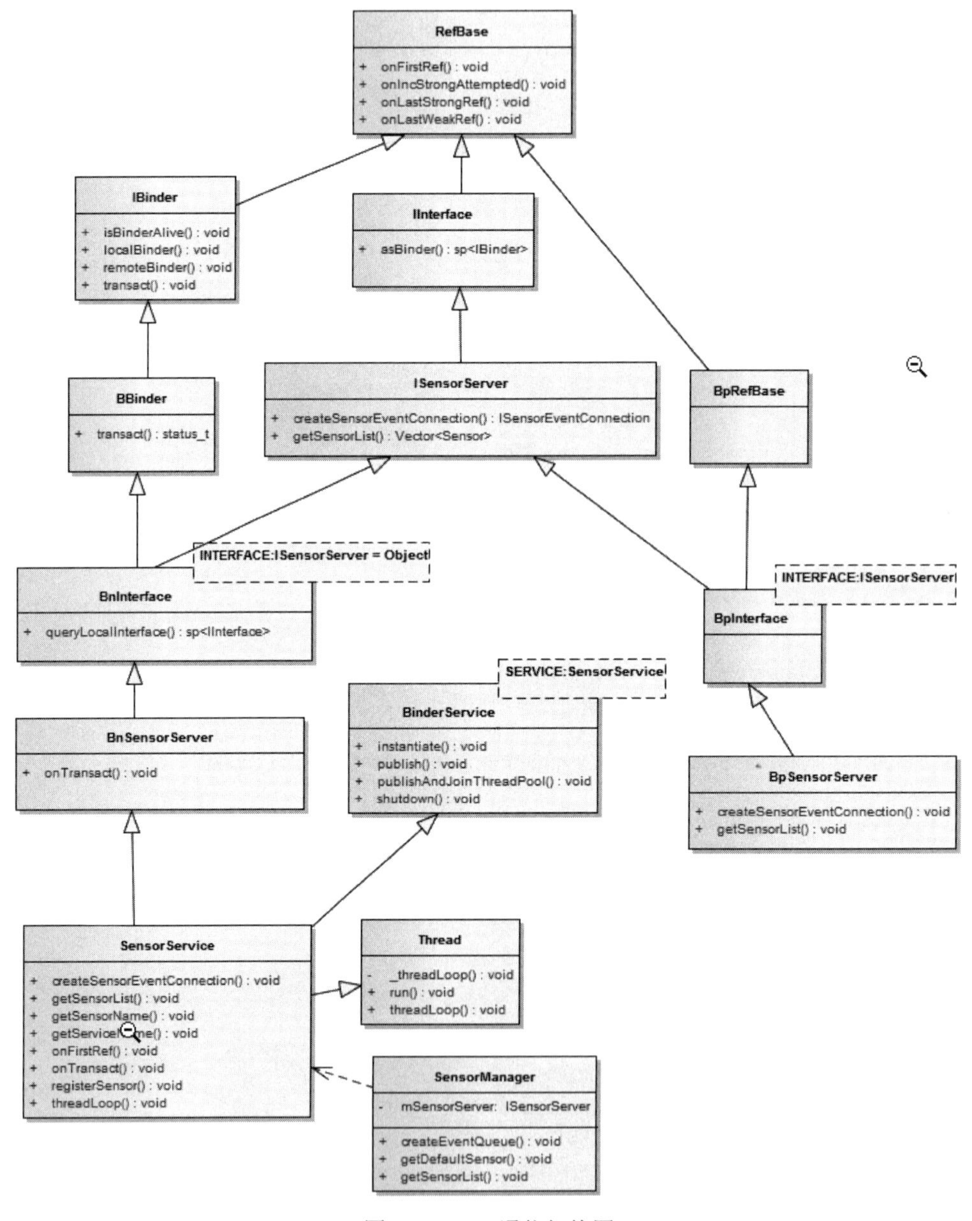


图 10-5　C/S 通信架构图

需要注意的是，并没有在系统中用 BpSensorServer，即使从 ISensorServer.cpp 中把它删除也不会对 Sensor 的工作有任何影响。这是因为它的工作已经被文件 SensorManager.cpp 所取代，ServiceManager 会直接获取上面文件 System_init 中添加的 SensorService 对象。

## 10.4.4　封装 HAL 层的代码

在 Android 系统中，通过文件 frameworks/native/services/sensorservice/SensorDevice.cpp 封装了 HAL 层的代码，此文件的主要功能如下：

- ☑ 获取 sensor 列表（getSensorList）。
- ☑ 获取 sensor 事件（poll）。
- ☑ Enable 或 Disable sensor（activate）。
- ☑ 设置 delay 时间。

文件 SensorDevice.cpp 的具体实现代码如下：

```
namespace android {
// ---------------------------------------------------------------------------

ANDROID_SINGLETON_STATIC_INSTANCE(SensorDevice)

SensorDevice::SensorDevice()
    :   mSensorDevice(0),
        mSensorModule(0)
{
    status_t err = hw_get_module(SENSORS_HARDWARE_MODULE_ID,
            (hw_module_t const**)&mSensorModule);

    ALOGE_IF(err, "couldn't load %s module (%s)",
            SENSORS_HARDWARE_MODULE_ID, strerror(-err));

    if (mSensorModule) {
        err = sensors_open(&mSensorModule->common, &mSensorDevice);

        ALOGE_IF(err, "couldn't open device for module %s (%s)",
                SENSORS_HARDWARE_MODULE_ID, strerror(-err));

        if (mSensorDevice) {
            sensor_t const* list;
            ssize_t count = mSensorModule->get_sensors_list(mSensorModule, &list);
            mActivationCount.setCapacity(count);
            Info model;
            for (size_t i=0 ; i<size_t(count) ; i++) {
                mActivationCount.add(list[i].handle, model);
                mSensorDevice->activate(mSensorDevice, list[i].handle, 0);
            }
        }
    }
}
```

```
void SensorDevice::dump(String8& result, char* buffer, size_t SIZE)
{
    if (!mSensorModule) return;
    sensor_t const* list;
    ssize_t count = mSensorModule->get_sensors_list(mSensorModule, &list);

    snprintf(buffer, SIZE, "%d h/w sensors:\n", int(count));
    result.append(buffer);

    Mutex::Autolock _l(mLock);
    for (size_t i=0 ; i<size_t(count) ; i++) {
        const Info& info = mActivationCount.valueFor(list[i].handle);
        snprintf(buffer, SIZE, "handle=0x%08x, active-count=%d, rates(ms)={ ",
                list[i].handle,
                info.rates.size());
        result.append(buffer);
        for (size_t j=0 ; j<info.rates.size() ; j++) {
            snprintf(buffer, SIZE, "%4.1f%s",
                    info.rates.valueAt(j) / 1e6f,
                    j<info.rates.size()-1 ? ", " : "");
            result.append(buffer);
        }
        snprintf(buffer, SIZE, " }, selected=%4.1f ms\n",    info.delay / 1e6f);
        result.append(buffer);
    }
}

ssize_t SensorDevice::getSensorList(sensor_t const** list) {
    if (!mSensorModule) return NO_INIT;
    ssize_t count = mSensorModule->get_sensors_list(mSensorModule, list);
    return count;
}

status_t SensorDevice::initCheck() const {
    return mSensorDevice && mSensorModule ? NO_ERROR : NO_INIT;
}

ssize_t SensorDevice::poll(sensors_event_t* buffer, size_t count) {
    if (!mSensorDevice) return NO_INIT;
    ssize_t c;
    do {
        c = mSensorDevice->poll(mSensorDevice, buffer, count);
    } while (c == -EINTR);
    return c;
}

status_t SensorDevice::resetStateWithoutActuatingHardware(void *ident, int handle)
{
    if (!mSensorDevice) return NO_INIT;
    Info& info( mActivationCount.editValueFor(handle));
    Mutex::Autolock _l(mLock);
```

```
    info.rates.removeItem(ident);
    return NO_ERROR;
}

status_t SensorDevice::activate(void* ident, int handle, int enabled)
{
    if (!mSensorDevice) return NO_INIT;
    status_t err(NO_ERROR);
    bool actuateHardware = false;

    Info& info( mActivationCount.editValueFor(handle) );


    ALOGD_IF(DEBUG_CONNECTIONS,
            "SensorDevice::activate: ident=%p, handle=0x%08x, enabled=%d, count=%d",
            ident, handle, enabled, info.rates.size());

    if (enabled) {
        Mutex::Autolock _l(mLock);
        ALOGD_IF(DEBUG_CONNECTIONS, "... index=%ld",
                info.rates.indexOfKey(ident));

        if (info.rates.indexOfKey(ident) < 0) {
            info.rates.add(ident, DEFAULT_EVENTS_PERIOD);
            if (info.rates.size() == 1) {
                actuateHardware = true;
            }
        } else {
            //传感器已经激活此 IDENT
        }
    } else {
        Mutex::Autolock _l(mLock);
        ALOGD_IF(DEBUG_CONNECTIONS, "... index=%ld",
                info.rates.indexOfKey(ident));

        ssize_t idx = info.rates.removeItem(ident);
        if (idx >= 0) {
            if (info.rates.size() == 0) {
                actuateHardware = true;
            }
        } else {
            //没有启用这个传感器的 IDENT
        }
    }

    if (actuateHardware) {
        ALOGD_IF(DEBUG_CONNECTIONS, "\t>>> actuating h/w");

        err = mSensorDevice->activate(mSensorDevice, handle, enabled);
        ALOGE_IF(err, "Error %s sensor %d (%s)",
                enabled ? "activating" : "disabling",
```

```
                handle, strerror(-err));
    }

    { //范围为锁
        Mutex::Autolock _l(mLock);
        nsecs_t ns = info.selectDelay();
        mSensorDevice->setDelay(mSensorDevice, handle, ns);
    }

    return err;
}

status_t SensorDevice::setDelay(void* ident, int handle, int64_t ns)
{
    if (!mSensorDevice) return NO_INIT;
    Mutex::Autolock _l(mLock);
    Info& info( mActivationCount.editValueFor(handle) );
    status_t err = info.setDelayForIdent(ident, ns);
    if (err < 0) return err;
    ns = info.selectDelay();
    return mSensorDevice->setDelay(mSensorDevice, handle, ns);
}

int SensorDevice::getHalDeviceVersion() const {
    if (!mSensorDevice) return -1;

    return mSensorDevice->common.version;
}

// ---------------------------------------------------------------------------

status_t SensorDevice::Info::setDelayForIdent(void* ident, int64_t ns)
{
    ssize_t index = rates.indexOfKey(ident);
    if (index < 0) {
        ALOGE("Info::setDelayForIdent(ident=%p, ns=%lld) failed (%s)",
                ident, ns, strerror(-index));
        return BAD_INDEX;
    }
    rates.editValueAt(index) = ns;
    return NO_ERROR;
}

nsecs_t SensorDevice::Info::selectDelay()
{
    nsecs_t ns = rates.valueAt(0);
    for (size_t i=1 ; i<rates.size() ; i++) {
        nsecs_t cur = rates.valueAt(i);
        if (cur < ns) {
            ns = cur;
        }
```

```
    }
    delay = ns;
    return ns;
}

// ----------------------------------------------------------------------------
};
```

这样 SensorSevice 会把任务交给 SensorDevice，而 SensorDevice 会调用标准的抽象层接口。由此可见，Sensor 架构的抽象层接口是最标准的一种，它很好地实现了抽象层与本地框架的分离。

## 10.4.5　消息队列处理

在 Android 传感器系统中，文件 frameworks/native/libs/gui/SensorEventQueue.cpp 的功能是处理消息。文件 SensorEventQueue.cpp 能够在创建其实例时传入 SensorEventConnection 实例，SensorEventConnection 继承于 ISensorEventConnection。SensorEventConnection 其实是客户端调用 SensorService 的 createSensorEventConnection()方法创建的，是客户端与服务端沟通的桥梁，通过这个桥梁可以完成如下所示的功能。

☑ 获取管道的句柄。

☑ 往管道读写数据。

☑ 通知服务端对 Sensor 使能。

文件 frameworks/native/libs/gui/SensorEventQueue.cpp 的具体实现代码如下：

```
// ----------------------------------------------------------------------------
namespace android {
// ----------------------------------------------------------------------------

SensorEventQueue::SensorEventQueue(const sp<ISensorEventConnection>& connection)
    : mSensorEventConnection(connection)
{
}

SensorEventQueue::~SensorEventQueue()
{
}

void SensorEventQueue::onFirstRef()
{
    mSensorChannel = mSensorEventConnection->getSensorChannel();
}

int SensorEventQueue::getFd() const
{
    return mSensorChannel->getFd();
}


ssize_t SensorEventQueue::write(const sp<BitTube>& tube,
        ASensorEvent const* events, size_t numEvents) {
```

```
    return BitTube::sendObjects(tube, events, numEvents);
}

ssize_t SensorEventQueue::read(ASensorEvent* events, size_t numEvents)
{
    return BitTube::recvObjects(mSensorChannel, events, numEvents);
}

sp<Looper> SensorEventQueue::getLooper() const
{
    Mutex::Autolock _l(mLock);
    if (mLooper == 0) {
        mLooper = new Looper(true);
        mLooper->addFd(getFd(), getFd(), ALOOPER_EVENT_INPUT, NULL, NULL);
    }
    return mLooper;
}

status_t SensorEventQueue::waitForEvent() const
{
    const int fd = getFd();
    sp<Looper> looper(getLooper());

    int events;
    int32_t result;
    do {
        result = looper->pollOnce(-1, NULL, &events, NULL);
        if (result == ALOOPER_POLL_ERROR) {
            ALOGE("SensorEventQueue::waitForEvent error (errno=%d)", errno);
            result = -EPIPE; //unknown error, so we make up one
            break;
        }
        if (events & ALOOPER_EVENT_HANGUP) {
            ALOGE("SensorEventQueue::waitForEvent error HANGUP");
            result = -EPIPE; //unknown error, so we make up one
            break;
        }
    } while (result != fd);

    return  (result == fd) ? status_t(NO_ERROR) : result;
}

status_t SensorEventQueue::wake() const
{
    sp<Looper> looper(getLooper());
    looper->wake();
    return NO_ERROR;
}

status_t SensorEventQueue::enableSensor(Sensor const* sensor) const {
    return mSensorEventConnection->enableDisable(sensor->getHandle(), true);
```

```
}

status_t SensorEventQueue::disableSensor(Sensor const* sensor) const {
    return mSensorEventConnection->enableDisable(sensor->getHandle(), false);
}

status_t SensorEventQueue::enableSensor(int32_t handle, int32_t us) const {
    status_t err = mSensorEventConnection->enableDisable(handle, true);
    if (err == NO_ERROR) {
        mSensorEventConnection->setEventRate(handle, us2ns(us));
    }
    return err;
}

status_t SensorEventQueue::disableSensor(int32_t handle) const {
    return mSensorEventConnection->enableDisable(handle, false);
}

status_t SensorEventQueue::setEventRate(Sensor const* sensor, nsecs_t ns) const {
    return mSensorEventConnection->setEventRate(sensor->getHandle(), ns);
}

// ---------------------------------------------------------------------------
};
```

由此可见，SensorManager 负责控制流，通过 C/S 的 Binder 机制与 SensorService 实现通信。具体过程如图 10-6 所示。

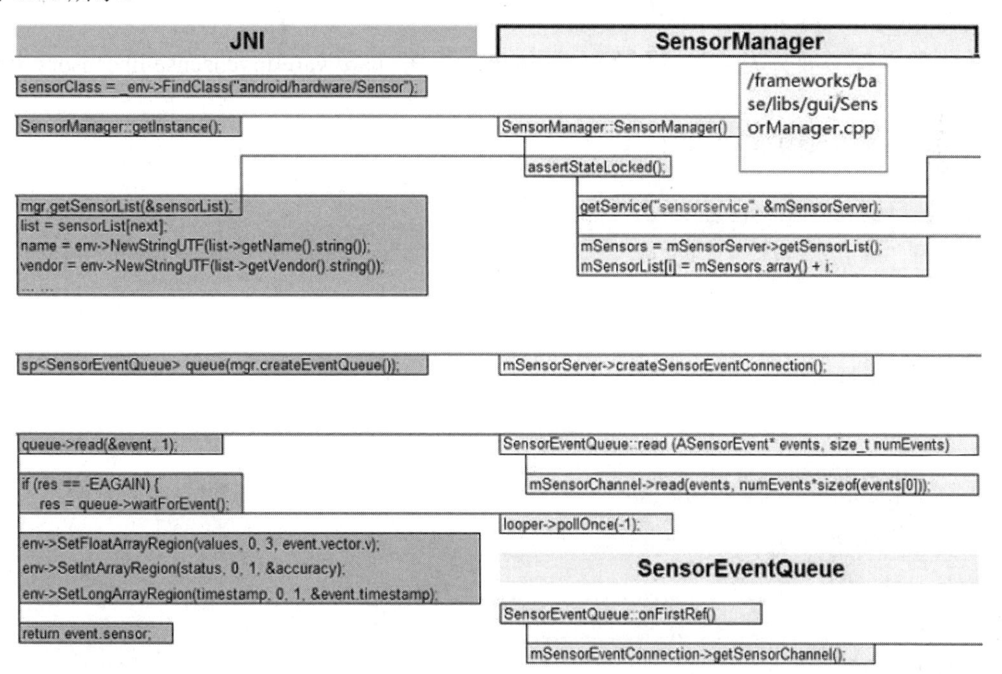


图 10-6　SensorManager 控制流的处理流程

而 SensorEventQueue 负责数据流，功能是通过管道机制来读写底层的数据。具体过程如图 10-7 所示。

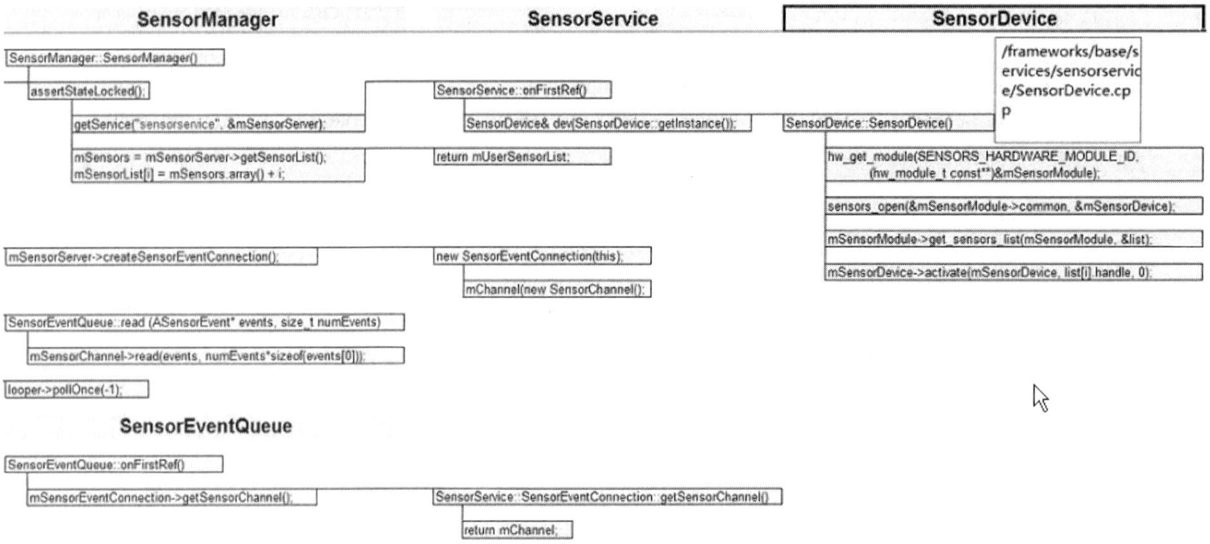

SensorManager
SensorService
SensorDevice
/frameworks/base/services/sensorservice/SensorDevice.cpp
SensorManager::SensorManager()
assertStateLocked();
getService("sensorservice", &mSensorServer);
mSensors = mSensorServer->getSensorList();
mSensorList[i] = mSensors.array() + i;
SensorService::onFirstRef()
SensorDevice& dev(SensorDevice::getInstance());
return mUserSensorList;
SensorDevice::SensorDevice()
hw_get_module(SENSORS_HARDWARE_MODULE_ID, (hw_module_t const**)&mSensorModule);
sensors_open(&mSensorModule->common, &mSensorDevice);
mSensorModule->get_sensors_list(mSensorModule, &list);
mSensorDevice->activate(mSensorDevice, list[i].handle, 0);
mSensorServer->createSensorEventConnection();
new SensorEventConnection(this);
mChannel(new SensorChannel();
SensorEventQueue::read (ASensorEvent* events, size_t numEvents)
mSensorChannel->read(events, numEvents*sizeof(events[0]));
looper->pollOnce(-1);
SensorEventQueue
SensorEventQueue::onFirstRef()
mSensorEventConnection->getSensorChannel();
SensorService::SensorEventConnection::getSensorChannel()
return mChannel;

图 10-7　SensorEventQueue 数据流的处理流程

# 10.5　HAL 层详解

**知识点讲解：光盘:视频\知识点\第 10 章\HAL 层详解.avi**

在 Android 系统中，HAL 层提供了 Android 独立于具体硬件的抽象接口。其中 HAL 层的头文件路径是 hardware/libhardware/include/hardware/sensors.h。

而具体实现文件需要开发者个人编写，具体可以参考 hardware\invensense\libsensors_iio\sensors_mpl.cpp。

文件 sensors.h 的主要实现代码如下：

```
typedef struct {
    union {
        float v[3];
        struct {
            float x;
            float y;
            float z;
        };
        struct {
            float azimuth;
            float pitch;
            float roll;
        };
    };
    int8_t status;
    uint8_t reserved[3];
} sensors_vec_t;

/**
```

```
 *未校准陀螺仪和磁场数据事件
 */
typedef struct {
  union {
    float uncalib[3];
    struct {
      float x_uncalib;
      float y_uncalib;
      float z_uncalib;
    };
  };
  union {
    float bias[3];
    struct {
      float x_bias;
      float y_bias;
      float z_bias;
    };
  };
} uncalibrated_event_t;

/**
 *各种类型的传感器数据中的联合
 *可以返回
 */
typedef struct sensors_event_t {
    int32_t version;
    int32_t sensor;

    /* 传感器类型 */
    int32_t type;
    int32_t reserved0;

    /* 时间，单位微秒 */
    int64_t timestamp;
    union {
        float data[16];

        sensors_vec_t acceleration;
        sensors_vec_t magnetic;
        sensors_vec_t orientation;
        sensors_vec_t gyro;
        float temperature;
        float distance;

        float light;

        float pressure;

        float relative_humidity;
```

```
            uint64_t step_counter;

            uncalibrated_event_t uncalibrated_gyro;

            uncalibrated_event_t uncalibrated_magnetic;
        };
        uint32_t reserved1[4];
} sensors_event_t;

struct sensor_t;
struct sensors_module_t {
        struct hw_module_t common;
        int (*get_sensors_list)(struct sensors_module_t* module,
                    struct sensor_t const** list);
};

struct sensor_t {
        const char* name;

        const char* vendor;
        int version;
        int handle;

        int type;

        float maxRange;

        float resolution;

        float power;
        int32_t minDelay;

        void* reserved[8];
};

struct sensors_poll_device_t {
        struct hw_device_t common;
        int (*activate)(struct sensors_poll_device_t *dev,
                    int handle, int enabled);
        int (*setDelay)(struct sensors_poll_device_t *dev,
                    int handle, int64_t ns);
        int (*poll)(struct sensors_poll_device_t *dev,
                    sensors_event_t* data, int count);
};
typedef struct sensors_poll_device_1 {
        union {
            struct sensors_poll_device_t v0;

            struct {
                    struct hw_device_t common;
                    int (*activate)(struct sensors_poll_device_t *dev,
```

```
                    int handle, int enabled);

            int (*setDelay)(struct sensors_poll_device_t *dev,
                    int handle, int64_t period_ns);

            int (*poll)(struct sensors_poll_device_t *dev,
                    sensors_event_t* data, int count);
        };
    };
    int (*batch)(struct sensors_poll_device_1* dev,
            int handle, int flags, int64_t period_ns, int64_t timeout);

    void (*reserved_procs[8])(void);

} sensors_poll_device_1_t;


/**用于打开和关闭传感器装置*/

static inline int sensors_open(const struct hw_module_t* module,
        struct sensors_poll_device_t** device) {
    return module->methods->open(module,
            SENSORS_HARDWARE_POLL, (struct hw_device_t**)device);
}

static inline int sensors_close(struct sensors_poll_device_t* device) {
    return device->common.close(&device->common);
}

static inline int sensors_open_1(const struct hw_module_t* module,
        sensors_poll_device_1_t** device) {
    return module->methods->open(module,
            SENSORS_HARDWARE_POLL, (struct hw_device_t**)device);
}

static inline int sensors_close_1(sensors_poll_device_1_t* device) {
    return device->common.close(&device->common);
}

__END_DECLS

#endif  //ANDROID_SENSORS_INTERFACE_H
```

而具体的实现文件是 Linux Kernel 层，也就是具体的硬件设备驱动程序，例如可以将其命名为 sensors.c，然后编写如下定义 struct sensors_poll_device_t 的代码。

```
struct sensors_poll_device_t {
    struct hw_device_t common;

    //激活/停用一个传感器
    int (*activate)(struct sensors_poll_device_t *dev,
```

```
                int handle, int enabled);

    //对于一个给定的传感器，设置在传感器事件之间的时间延迟，单位微秒
    int (*setDelay)(struct sensors_poll_device_t *dev,
                int handle, int64_t ns);

    //返回传感器数据的数组
    int (*poll)(struct sensors_poll_device_t *dev,
                sensors_event_t* data, int count);
};
```

也可以编写如下定义 struct sensors_module_t 的代码。

```
struct sensors_module_t {
    struct hw_module_t common;

    /**
     *枚举所有可用的传感器。这份名单是在“名单”返回
     *@传感器在列表中返回数
     */
    int (*get_sensors_list)(struct sensors_module_t* module,
                struct sensor_t const** list);
};
```

也可以编写如下定义 struct sensor_t 的代码。

```
struct sensor_t {
    const char* name;
    int version;
    int handle;
    int type;
    float maxRange;
    float resolution;
    float power;
    int32_t minDelay;
    void* reserved[8];
};
```

也可以编写如下定义 struct sensors_event_t 的代码。

```
typedef struct {
    union {
        float v[3];
        struct {
            float x;
            float y;
            float z;
        };
        struct {
            float azimuth;
            float pitch;
            float roll;
        };
    };
    int8_t status;
    uint8_t reserved[3];
```

```
} sensors_vec_t;
typedef struct sensors_event_t {
    int32_t version;

    int32_t sensor;

    int32_t type;
    int32_t reserved0;

    int64_t timestamp;

    union {
        float data[16];
        sensors_vec_t acceleration;

        sensors_vec_t magnetic;

        sensors_vec_t orientation;

        sensors_vec_t gyro;

        float temperature;

        float distance;

        float light;

        float pressure;

        float relative_humidity;
    };
    uint32_t reserved1[4];
} sensors_event_t;
```

也可以编写如下定义 struct sensors_module_t 的代码。

```
static const struct sensor_t sSensorList[] = {
        { "MMA8452Q 3-axis Accelerometer",
                "Freescale Semiconductor",
                1, SENSORS_HANDLE_BASE+ID_A,
                SENSOR_TYPE_ACCELEROMETER, 4.0f*9.81f, (4.0f*9.81f)/256.0f, 0.2f, 0, { } },
        { "AK8975 3-axis Magnetic field sensor",
                "Asahi Kasei",
                1, SENSORS_HANDLE_BASE+ID_M,
                SENSOR_TYPE_MAGNETIC_FIELD, 2000.0f, 1.0f/16.0f, 6.8f, 0, { } },
        { "AK8975 Orientation sensor",
                "Asahi Kasei",
                1, SENSORS_HANDLE_BASE+ID_O,
                SENSOR_TYPE_ORIENTATION, 360.0f, 1.0f, 7.0f, 0, { } },

    { "ST 3-axis Gyroscope sensor",
            "STMicroelectronics",
```

```
            1, SENSORS_HANDLE_BASE+ID_GY,
            SENSOR_TYPE_GYROSCOPE, RANGE_GYRO, CONVERT_GYRO, 6.1f, 1190, { } },

    { "AL3006Proximity sensor",
          "Dyna Image Corporation",
          1, SENSORS_HANDLE_BASE+ID_P,
          SENSOR_TYPE_PROXIMITY,
          PROXIMITY_THRESHOLD_CM, PROXIMITY_THRESHOLD_CM,
          0.5f, 0, { } },

          { "AL3006 light sensor",
                    "Dyna Image Corporation",
                    1, SENSORS_HANDLE_BASE+ID_L,
                    SENSOR_TYPE_LIGHT, 10240.0f, 1.0f, 0.5f, 0, { } },

};

static int open_sensors(const struct hw_module_t* module, const char* name,
          struct hw_device_t** device);

static int sensors__get_sensors_list(struct sensors_module_t* module,
          struct sensor_t const** list)
{
     *list = sSensorList;
     return ARRAY_SIZE(sSensorList);
}

static struct hw_module_methods_t sensors_module_methods = {
     .open = open_sensors
};

const struct sensors_module_t HAL_MODULE_INFO_SYM = {
     .common = {
          .tag = HARDWARE_MODULE_TAG,
          .version_major = 1,
          .version_minor = 0,
          .id = SENSORS_HARDWARE_MODULE_ID,
          .name = "MMA8451Q & AK8973A & gyro Sensors Module",
          .author = "The Android Project",
          .methods = &sensors_module_methods,
     },
     .get_sensors_list = sensors__get_sensors_list
};

static int open_sensors(const struct hw_module_t* module, const char* name,
          struct hw_device_t** device)
{
     return init_nusensors(module, device); //待后面讲解
}
```

至此，整个 Android 系统中传感器模块的源码分析完毕。由此可见，整个传感器系统的总体调用关系如图 10-8 所示。

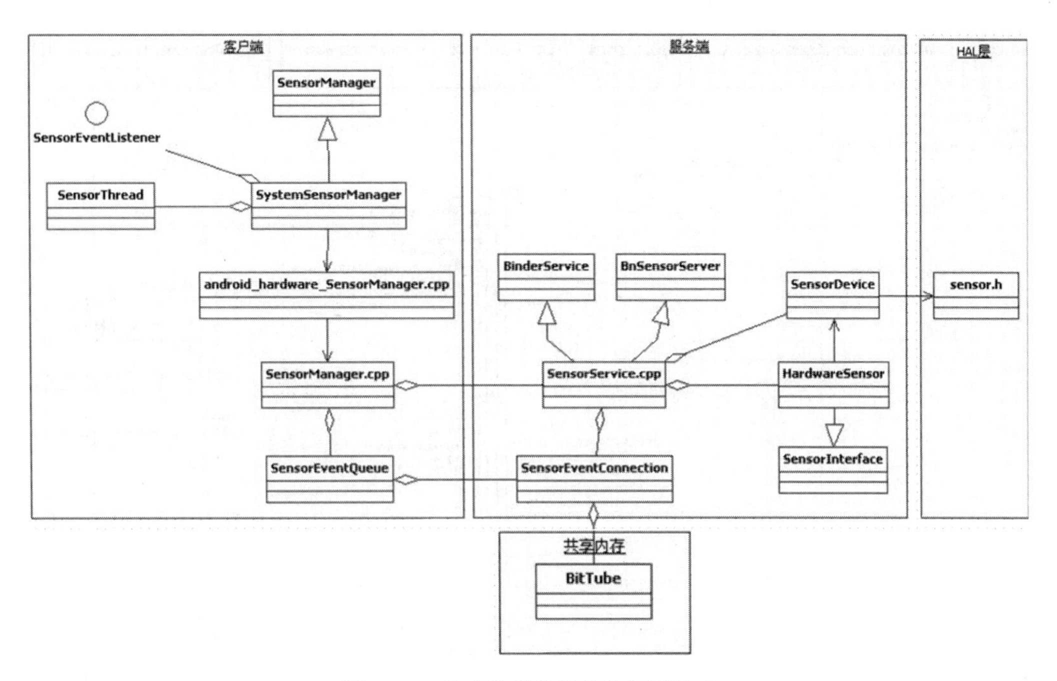


图 10-8　传感器系统的总体调用关系

客户端读取数据时的调用时序如图 10-9 所示。

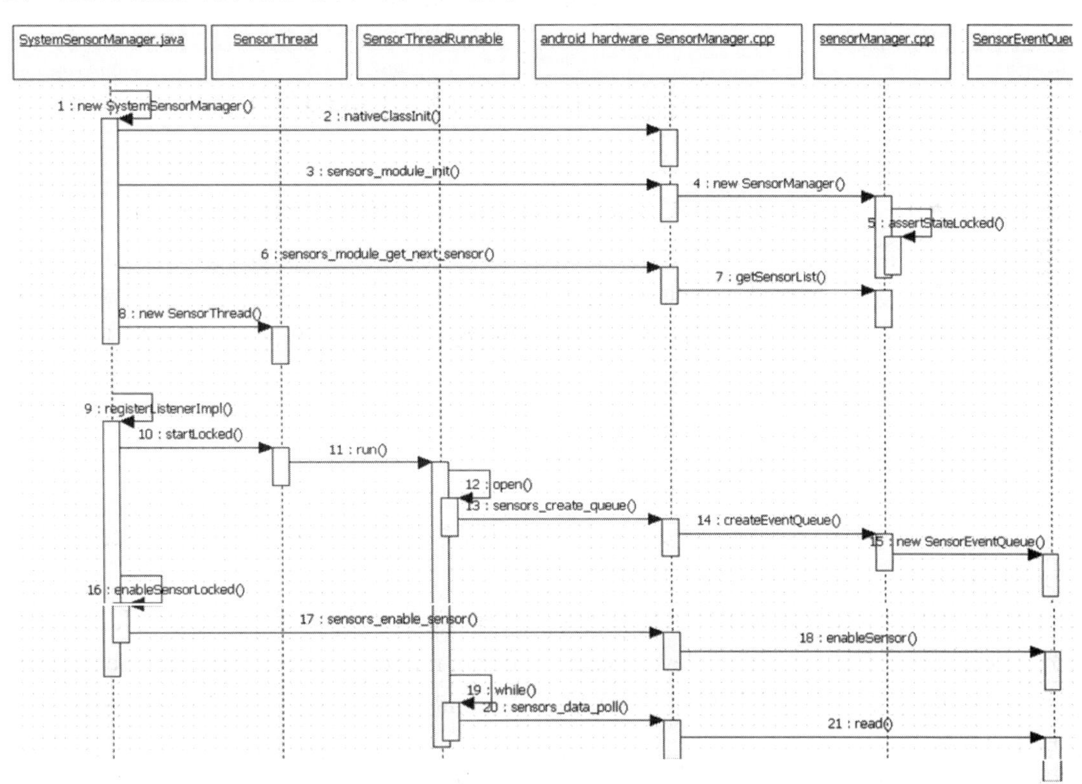


图 10-9　客户端读取数据时的调用时序图

服务器端的调用时序如图 10-10 所示。

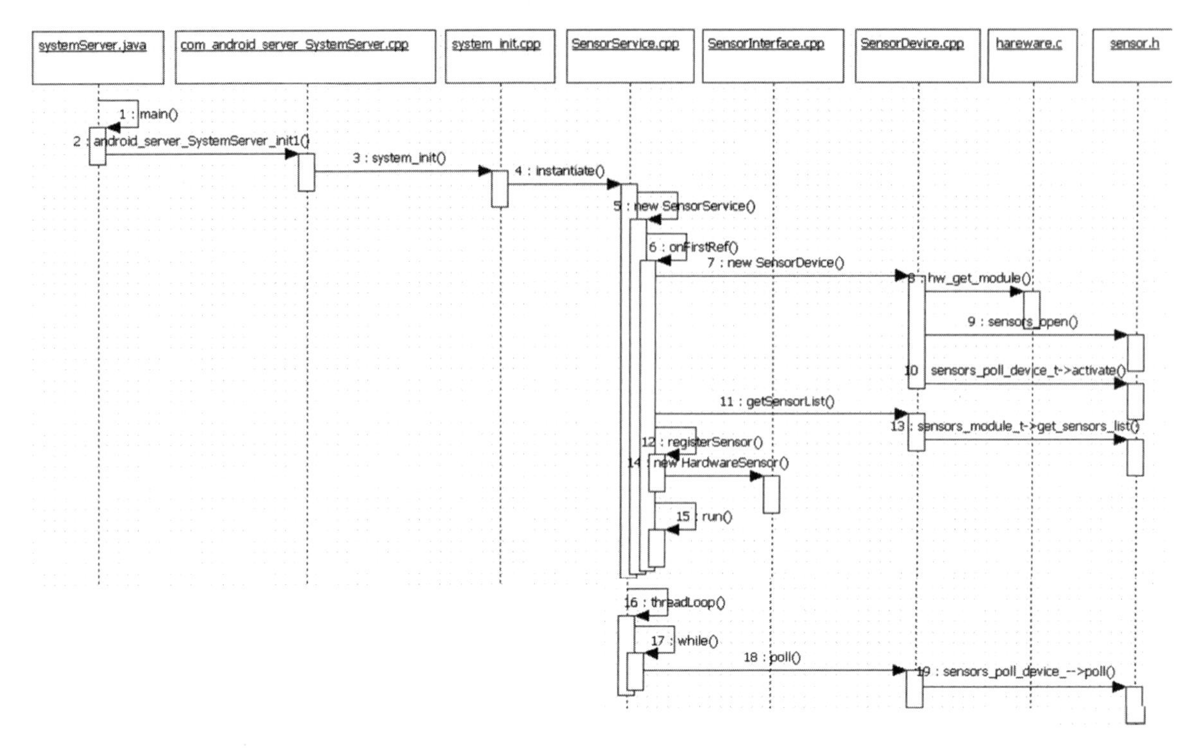


图 10-10　服务器端的调用时序图

# 10.6　Android 传感器应用开发基础

**知识点讲解：光盘:视频\知识点\第 10 章\Android 传感器应用开发基础.avi**

在本章前面的内容中，已经详细讲解了 Android 系统中传感器系统的架构知识。在现实应用中，传感器系统在物联网设备、可穿戴设备和家具设备中得到了广泛的应用。本节将详细讲解开发 Android 传感器应用程序的基础知识，介绍使用传感器技术开发物联网设备应用程序的基本流程，为读者步入本书后面知识的学习打下坚实的基础。

## 10.6.1　查看包含的传感器

在安装 Android SDK 后，依次打开安装目录中的帮助文件 android SDK/sdk/docs/reference/android/hardware/Sensor.html。

在此文件中列出了 Android 传感器系统所包含的所有传感器类型，如图 10-11 所示。

另外，也可以直接在线登录 http://developer.android.com/reference/android/hardware/Sensor.html 来查看。由此可见，在当前最新（笔者写稿时最新）版本 Android 4.4 中一共提供了 18 种传感器 API。各个类型的具体说明如下。

（1）TYPE_ACCELEROMETER：加速度传感器，单位是 $m/s^2$，测量应用于设备 X、Y、Z 轴上的加速度，又叫做 G-sensor。

（2）TYPE_AMBIENT_TEMPERATURE：温度传感器，单位是℃，能够测量并返回当前的温度。

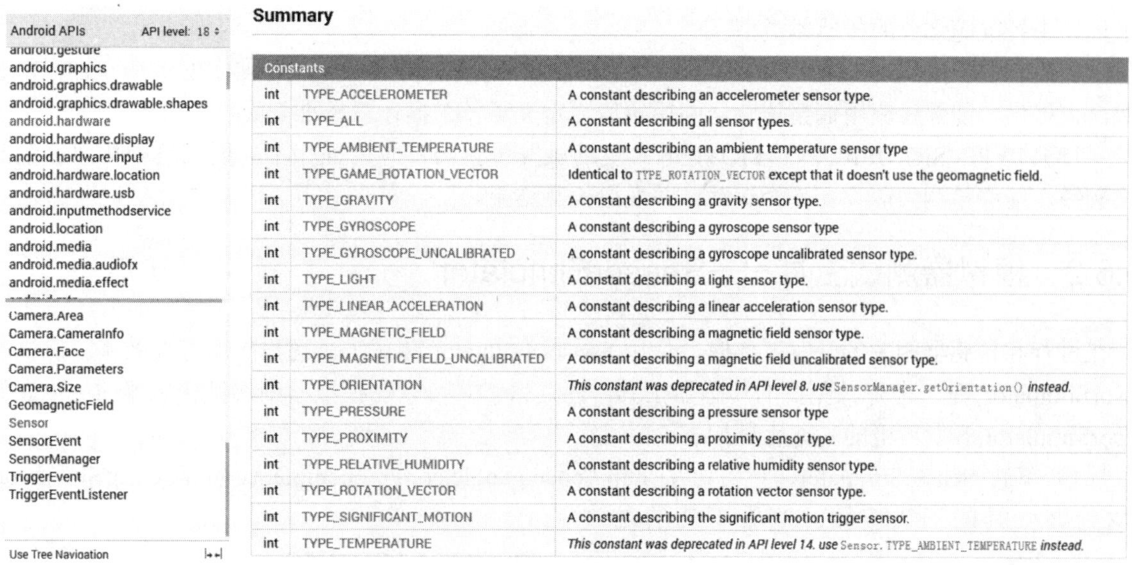

Android APIs　API level: 18

android.gesture
android.graphics
android.graphics.drawable
android.graphics.drawable.shapes
android.hardware
android.hardware.display
android.hardware.input
android.hardware.location
android.hardware.usb
android.inputmethodservice
android.location
android.media
android.media.audiofx
android.media.effect

Camera.Area
Camera.CameraInfo
Camera.Face
Camera.Parameters
Camera.Size
GeomagneticField
Sensor
SensorEvent
SensorManager
TriggerEvent
TriggerEventListener

Use Tree Navigation

**Summary**

Constants

| | | |
|---|---|---|
| int | TYPE_ACCELEROMETER | A constant describing an accelerometer sensor type. |
| int | TYPE_ALL | A constant describing all sensor types. |
| int | TYPE_AMBIENT_TEMPERATURE | A constant describing an ambient temperature sensor type |
| int | TYPE_GAME_ROTATION_VECTOR | Identical to TYPE_ROTATION_VECTOR except that it doesn't use the geomagnetic field. |
| int | TYPE_GRAVITY | A constant describing a gravity sensor type. |
| int | TYPE_GYROSCOPE | A constant describing a gyroscope sensor type |
| int | TYPE_GYROSCOPE_UNCALIBRATED | A constant describing a gyroscope uncalibrated sensor type. |
| int | TYPE_LIGHT | A constant describing a light sensor type. |
| int | TYPE_LINEAR_ACCELERATION | A constant describing a linear acceleration sensor type. |
| int | TYPE_MAGNETIC_FIELD | A constant describing a magnetic field sensor type. |
| int | TYPE_MAGNETIC_FIELD_UNCALIBRATED | A constant describing a magnetic field uncalibrated sensor type. |
| int | TYPE_ORIENTATION | *This constant was deprecated in API level 8. use* SensorManager.getOrientation() *instead.* |
| int | TYPE_PRESSURE | A constant describing a pressure sensor type |
| int | TYPE_PROXIMITY | A constant describing a proximity sensor type. |
| int | TYPE_RELATIVE_HUMIDITY | A constant describing a relative humidity sensor type. |
| int | TYPE_ROTATION_VECTOR | A constant describing a rotation vector sensor type. |
| int | TYPE_SIGNIFICANT_MOTION | A constant describing the significant motion trigger sensor. |
| int | TYPE_TEMPERATURE | *This constant was deprecated in API level 14. use* Sensor.TYPE_AMBIENT_TEMPERATURE *instead.* |

图 10-11　Android 传感器系统的类型

（3）TYPE_GRAVITY：重力传感器，单位是 m/s²，用于测量设备 X、Y、Z 轴上的重力，也叫 GV-sensor，地球上的数值是 9.8m/s²，也可以设置其他星球。

（4）TYPE_GYROSCOPE：陀螺仪传感器，单位是 rad/s，能够测量设备 X、Y、Z 3 个轴的角加速度数据。

（5）TYPE_LIGHT：光线感应传感器，单位是 lx，能够检测周围的光线强度，在手机系统中主要用于调节 LCD 亮度。

（6）TYPE_LINEAR_ACCELERATION：线性加速度传感器，单位是 m/s²，能够获取加速度传感器去除重力的影响得到的数据。

（7）TYPE_MAGNETIC_FIELD：磁场传感器，单位是 μT（微特斯拉），能够测量设备周围 3 个物理轴（x,y,z）的磁场。

（8）TYPE_ORIENTATION：方向传感器，用于测量设备围绕 3 个物理轴（x,y,z）的旋转角度，在新版本中已经使用 SensorManager.getOrientation()替代。

（9）TYPE_PRESSURE：气压传感器，单位是 hPa（百帕斯卡），能够返回当前环境下的压强。

（10）TYPE_PROXIMITY：距离传感器，单位是 cm，能够测量某个对象到屏幕的距离。可以在打电话时判断人耳到电话屏幕的距离，以关闭屏幕而达到省电功能。

（11）TYPE_RELATIVE_HUMIDITY：湿度传感器，单位是%，能够测量周围环境的相对湿度。

（12）TYPE_ROTATION_VECTOR：旋转向量传感器，旋转矢量代表设备的方向，是一个将坐标轴和角度混合计算得到的数据。

（13）TYPE_TEMPERATURE：温度传感器，在新版本中被 TYPE_AMBIENT_TEMPERATURE 替换。

（14）TYPE_ALL：返回所有的传感器类型。

（15）TYPE_GAME_ROTATION_VECTOR：除了不能使用地磁场之外，和 TYPE_ROTATION_VECTOR 的功能完全相同。

（16）TYPE_GYROSCOPE_UNCALIBRATED：提供了能够让应用调整传感器的原始值，定义了一个描述未校准陀螺仪的传感器类型。

（17）TYPE_MAGNETIC_FIELD_UNCALIBRATED：和 TYPE_GYROSCOPE_UNCALIBRATED 相似，也提供了能够让应用调整传感器的原始值，定义了一个描述未校准陀螺仪的传感器类型。

（18）TYPE_SIGNIFICANT_MOTION：运动触发传感器，应用程序不需要为这种传感器触发任何唤醒锁。能够检测当前设备是否运动，并发送检测结果。

## 10.6.2 模拟器测试工具——SensorSimulator

在进行和传感器相关的开发工作时，使用 SensorSimulator 测试工具可以提高开发效率。测试工具 SensorSimulator 是一个开源免费的传感器工具，通过该工具可以在模拟器中调试传感器的应用。搭建 SensorSimulator 开发环境的基本流程如下：

（1）下载 SensorSimulator，读者可从 http://code.google.com/p/openintents/wiki/SensorSimulator 网站找到该工具的下载链接。笔者下载的是 sensorsimulator-1.1.1.zip 版本，如图 10-12 所示。

图 10-12　下载 sensorsimulator-1.1.1.zip

（2）将下载好的 SensorSimulator 解压到本地根目录，例如 C 盘的根目录。

（3）向模拟器安装 SensorSimulatorSettings-1.1.1.apk。首先在操作系统中依次选择“开始”|“运行”命令，进入“运行”对话框。

（4）在“运行”对话框中输入“cmd”进入 cmd 命令行，之后通过 cd 命令将当前目录导航到 SensorSimulatorSettings-1.1.1.apk 目录下，然后输入下列命令向模拟器安装该 apk。

```
adb install SensorSimulatorSettings-1.1.1.apk
```

需要注意的是，安装 apk 时，一定要保证模拟器正在运行才可以，安装成功后会输出 Success 提示，如图 10-13 所示。

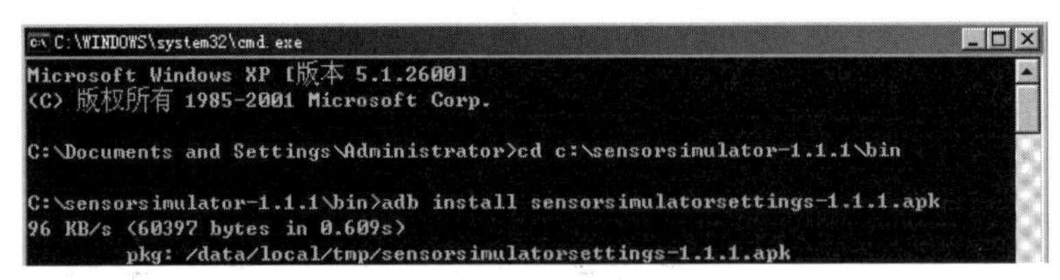


图 10-13　安装 apk

下面开始配置应用程序，假设要在项目 jiaSCH 中使用 SensorSimulator，则配置流程如下：

（1）在 Eclipse 中打开项目 jiaSCH，然后为该项目添加 JAR 包，使其能够使用 SensorSimulator 工具的类和方法。添加方法非常简单，在 Eclipse 的 Package Explorer 中找到该项目的文件夹 jiaSCH，然后右击该文件夹并在弹出的快捷菜单中选择 Properties 命令，弹出如图 10-14 所示的 Properties for jiaS 窗口。

（2）选择 Java Build Path 选项，然后选择 Libraries 选项卡，如图 10-15 所示。

（3）单击 Add External JARs 按钮，在弹出的 JAR Selection 对话框中找到 Sensorsimulator 安装目

录下的 sensorsimulator-lib-1.1.1.jar，并将其添加到该项目中，如图 10-16 所示。

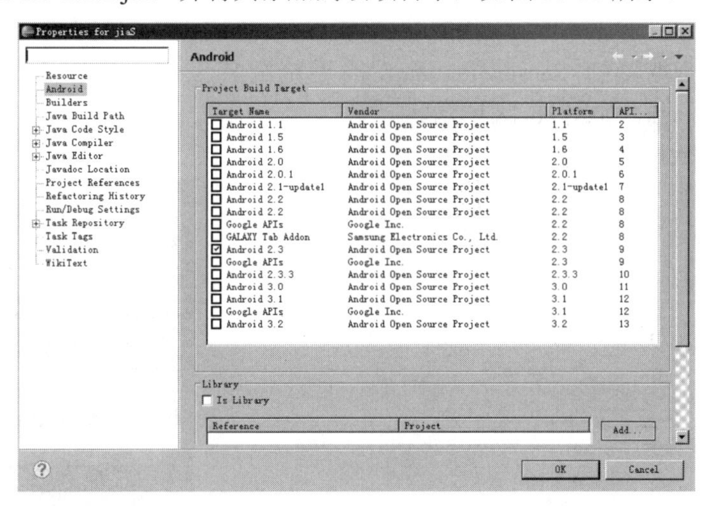


图 10-14　Properties for jiaS 窗口

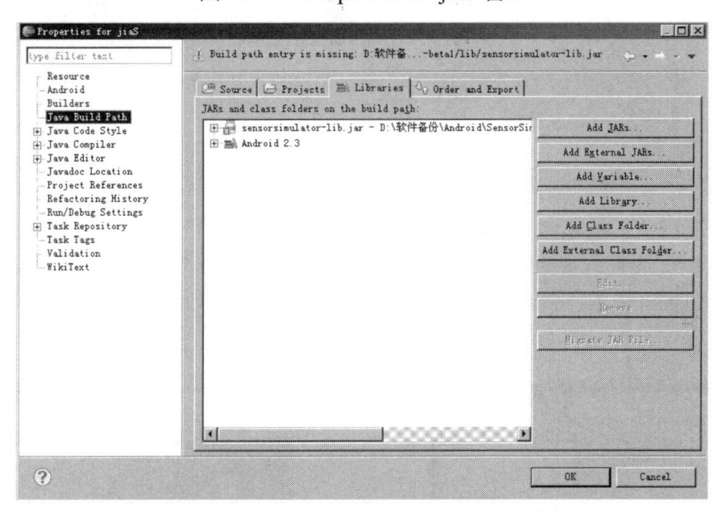


图 10-15　Libraries 选项卡

图 10-16　添加需要的 JAR 包

（4）开始启动 sensorsimulator.jar，并对手机模拟器上的 SensorSimulatorSettings 进行必要的配置。首先在 C:\sensorsimulator-1.1.1\bin 目录下找到 sensorsimulator.jar 并启动，运行后的界面如图 10-17 所示。

（5）进行手机模拟器和 SensorSimulator 的连接配置工作，运行手机模拟器上安装好的 SensorSimulatorSettings.apk，如图 10-18 所示。

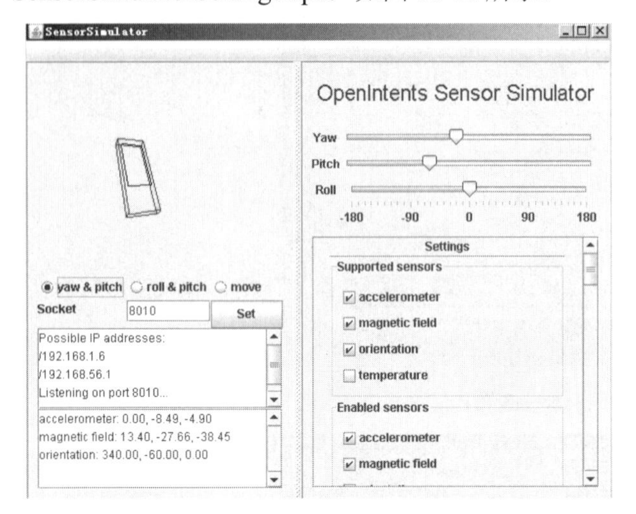


图 10-17　传感器的模拟器

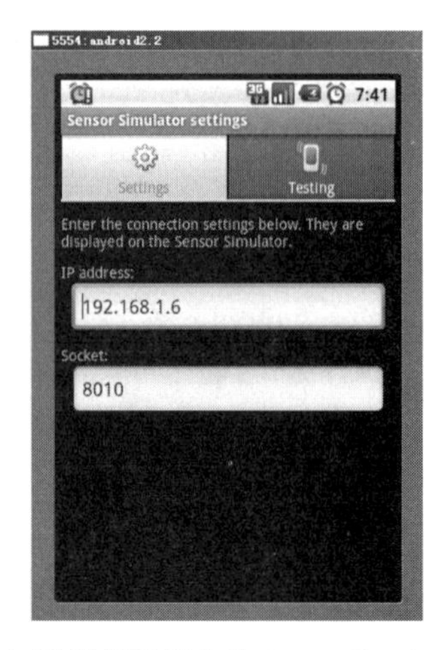


图 10-18　运行手机模拟器上的 SensorSimulatorSettings.apk

（6）在图 10-18 中输入 SensorSimulator 启动时显示的 IP 地址和端口号，单击 Testing 按钮后会进入测试连接界面，如图 10-19 所示。

（7）单击 Connect 按钮进入下一界面，如图 10-20 所示。在此界面中可以选择需要监听的传感器，如果能够从传感器中读取到数据，说明 SensorSimulator 与手机模拟器连接成功，可以测试自己开发的应用程序了。

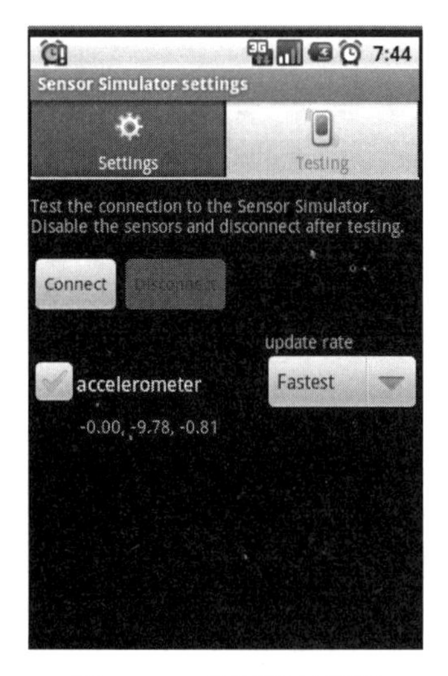


图 10-19　测试连接界面

图 10-20　连接界面

至此，使用 Eclipse 结合 SensorSimulator 配置传感器应用程序的基本流程介绍完毕。

## 10.6.3　实战演练——检测当前设备支持的传感器

本实例将演示在 Android 设备中检测当前设备支持传感器类型的方法。

| 实　例 | 功　能 | 源 码 路 径 |
| --- | --- | --- |
| 实例 10-1 | 检测当前设备支持的传感器 | 光盘:\daima\10\SensorEX |

本实例的功能是检测当前设备支持的传感器类型，具体实现流程如下。

（1）布局文件 main.xml 的具体实现代码如下：

```
<linearlayout android:layout_height="fill_parent" android:layout_width="fill_parent" android:orientation="vertical"
xmlns:android="http://schemas.android.com/apk/res/android">
<textview android:layout_height="wrap_content"
 android:layout_width="fill_parent" android:text=""
 android:id="@+id/TextView01"
>
</textview>
</linearlayout>
```

（2）主程序文件 MainActivity.java 的具体实现代码如下：

```
public class MainActivity extends Activity {

	@SuppressWarnings("deprecation")
	 @Override
	public void onCreate(Bundle savedInstanceState) {
		super.onCreate(savedInstanceState);
		setContentView(R.layout.main);

		//准备显示信息的 UI 组件
		final TextView tx1 = (TextView) findViewById(R.id.TextView01);

		//从系统服务中获得传感器管理器
		SensorManager sm = (SensorManager) getSystemService(Context.SENSOR_SERVICE);

		//从传感器管理器中获得全部的传感器列表
		List<Sensor> allSensors = sm.getSensorList(Sensor.TYPE_ALL);

		//显示有多少个传感器
		tx1.setText("经检测该手机有" + allSensors.size() + "个传感器，它们分别是：\n");

		//显示每个传感器的具体信息
		for (Sensor s : allSensors) {

			String tempString = "\n" + "  设备名称：" + s.getName() + "\n" + "  设备版本：" +
s.getVersion() + "\n" + " 供应商：" + s.getVendor() + "\n";

			switch (s.getType()) {
			case Sensor.TYPE_ACCELEROMETER:
				tx1.setText(tx1.getText().toString() + s.getType() + " 加速度传感器 accelerometer"
+ tempString);
				break;
```

```
                        case Sensor.TYPE_GYROSCOPE:
                                tx1.setText(tx1.getText().toString() + s.getType() + " 陀螺仪传感器 gyroscope"
+ tempString);
                                break;
                        case Sensor.TYPE_LIGHT:
                                tx1.setText(tx1.getText().toString()+s.getType()+"环境光线传感器 light" + tempString);
                                break;
                        case Sensor.TYPE_MAGNETIC_FIELD:
                                tx1.setText(tx1.getText().toString() + s.getType() + " 电磁场传感器 magnetic
field" + tempString);
                                break;
                        case Sensor.TYPE_ORIENTATION:
                                tx1.setText(tx1.getText().toString() + s.getType() + " 方向传感器 orientation" +
tempString);
                                break;
                        case Sensor.TYPE_PRESSURE:
                                tx1.setText(tx1.getText().toString()+s.getType()+" 压力传感器 pressure"+tempString);
                                break;
                        case Sensor.TYPE_PROXIMITY:
                                tx1.setText(tx1.getText().toString() + s.getType() + " 距离传感器 proximity"
+ tempString);
                                break;
                        case Sensor.TYPE_AMBIENT_TEMPERATURE :
                                tx1.setText(tx1.getText().toString() + s.getType() + " 温度传感器 temperature" +
tempString);
                                break;
                        default:
                                tx1.setText(tx1.getText().toString() + s.getType()+" 未知传感器" + tempString);
                                break;
                        }
                }

        }
}
```

上述实例代码需要在真机中运行，执行后将会列表显示当前设备所支持的传感器类型，如图 10-21 所示。

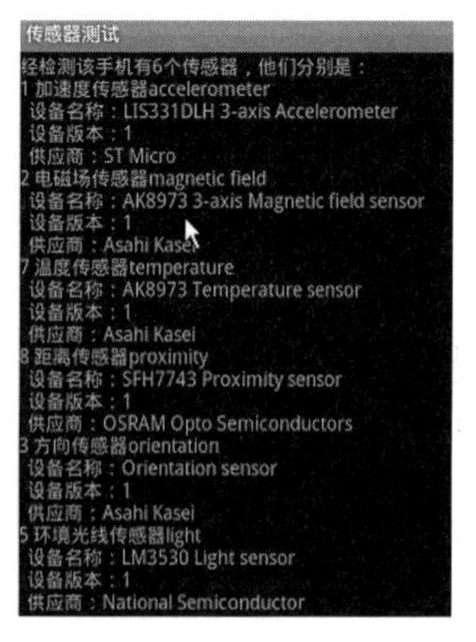


图 10-21　执行效果

# 第 11 章　光线传感器和磁场传感器

在 Android 系统中支持多种传感器（Sensor），传感器系统可以让智能手机的功能更加丰富多彩。Android 系统支持多种传感器，有的传感器已经在 Android 的框架中使用，大多数传感器由应用程序来使用，使用传感器可以开发出包括游戏在内的很多新奇的应用。在 Android 系统中支持的传感器包括加速度传感器（Accelerometer）、姿态传感器（Orientation）、磁场传感器（Magnetic Field）和光传感器（Light）等。本章将详细讲解在 Android 物联网设备中使用光线传感器和磁场传感器的基本知识。

## 11.1　光线传感器详解

**知识点讲解：光盘:视频\知识点\第 11 章\光线传感器详解.avi**

在现实应用中，光线传感器能够根据手机所处环境的光线来调节手机屏幕的亮度和键盘灯。例如在光线充足的地方屏幕会很亮，键盘灯就会关闭。相反如果在暗处，键盘灯就会亮，屏幕较暗（与屏幕亮度的设置也有关系），这样既保护了眼睛又节省了能量。光线传感器在进入睡眠模式时会发出蓝色周期性闪动的光，非常美观。本节将详细讲解 Android 系统光线传感器的基本知识。

### 11.1.1　光线传感器介绍

在物联网设备中，光线传感器通常位于前摄像头旁边的一个小点，如果在光线充足的情况下（室外或者是灯光充足的室内），大约在 2～3 秒之后键盘灯会自动熄灭，即使再操作机器键盘灯也不会亮，除非到了光线比较暗的地方才会自动亮起来。如果在光线充足的情况下用手将光线感应器遮上，在 2～3 秒后键盘灯会自动亮起来，在此过程中光线感应器起到了一个节电的功能。

要想在 Android 物联网设备中监听光线传感器，需要掌握如下所示的监听方法。

（1）registerListenr(SensorListenerlistenr,int sensors,int rate)：已过时。

（2）registerListenr(SensorListenerlistenr,int sensors)：已过时。

（3）registerListenr(SensorEventListenerlistenr,Sensor sensors,int rate)：注册一个需要监听的传感器。

（4）registerListenr(SensorEventListenerlistenr,Sensor sensors,int rate,Handlerhandler)：因为 SensorListener 已经过时，所以相应的注册方法也过时了。

在上述方法中，各个参数的具体说明如下。

☑　Listener：相应监听器的引用。

☑　Sensor：相应感应器的引用。

☑　Rate：感应器的反应速度，这个必须是系统提供的 4 个常量之一。

- SENSOR_DELAY_NORMAL：匹配屏幕方向的变化。
- SENSOR_DELAY_UI：匹配用户接口。

➢ SENSOR_DELAY_GAME：匹配游戏。

➢ SENSOR_DELAY_FASTEST：匹配所能达到的最快。

开发光传感器应用时需要监测 SENSOR_LIGHT，例如下面的代码。

```
private SensorListener mySensorListener = new SensorListener(){
@Override
 public void onAccuracyChanged(int sensor, int accuracy) {}     //重写 onAccuracyChanged()方法
 @Override
 public void onSensorChanged(int sensor, float[] values) {       //重写 onSensorChanged()方法
     if(sensor == SensorManager.SENSOR_LIGHT){                   //只检查光强度的变化
         myTextView1.setText("光的强度为："+values[0]);          //将光的强度显示到 TextView
             }
         }
};
@Override
protected void onResume() {                                      //重写 onResume 方法
         mySensorManager.registerListener(                       //注册监听
                 mySensorListener,                               //监听器 SensorListener 对象
                 SensorManager.SENSOR_LIGHT,                     //传感器的类型为光的强度
                 SensorManager.SENSOR_DELAY_UI                   //频率
                 );
         super.onResume();
}
```

在上述代码中，通过 if 语句判断是否为光的强度改变事件。在代码中只对光强度改变事件进行处理，将得到的光强度显示在屏幕中。光传感器只得到一个数据，而并不像其他传感器那样得到的是 X、Y、Z 3 个方向上的分量。

在注册监听时，通过传入 SensorManager.SENSOR_LIGHT 来通知系统只注册光传感器。

## 11.1.2 使用光线传感器的方法

在 Android 物联网设备中，使用光线传感器的基本流程如下：

（1）通过一个 SensorManager 来管理各种感应器，要想获得这个管理器的引用，必须通过如下所示的代码来实现。

```
(SensorManager)getSystemService(Context.SENSOR_SERVICE);
```

（2）在 Android 系统中，所有的感应器都属于 Sensor 类的一个实例，并没有继续细分下去，所以 Android 对于感应器的处理几乎是一模一样的。既然都是 Sensor 类，那么怎么获得相应的感应器呢？这时就需要通过 SensorManager 来获得，可以通过如下所示的代码来确定我们要获得感应器的类型。

```
sensorManager.getDefaultSensor(Sensor.TYPE_LIGHT);
```

通过上述代码获得光线感应器的引用。

（3）在获得相应的传感器的引用后可以来感应光线强度的变化，此时需要通过监听传感器的方式来获得变化，监听功能通过前面介绍的监听方法实现。Android 提供了两个监听方式，一个是 SensorEventListener，另一个是 SensorListener，后者已经在 Android API 上显示过时。

（4）在 Android 中注册传感器后，就说明启用了传感器。使用感应器是相当耗电的，这也是为什么传感器的应用没有那么广泛的主要原因，所以必须在不需要它时及时将其关掉。在 Android 中通过如下所示的注销方法来关闭。

☑ unregisterListener(SensorEventListenerlistener)

☑ unregisterListener(SensorEventListenerlistener,Sensor sensor)

（5）使用 SensorEventListener 来具体实现，在 Android 物联网设备中有如下两种实现该监听器的方法。

☑ onAccuracyChanged(Sensor sensor, int accuracy)：是反应速度变化的方法，也就是是 rate 变化时的方法。

☑ onSensorChanged(SensorEvent event)：是传感器的值变化的相应方法。

需要注意的是，上述两个方法会同时响应。也就是说，当感应器发生变化时，这两个方法会一起被调用。上述方法中 accuracy 的值是 4 个常量，对应的整数如下所示。

☑ SENSOR_DELAY_NORMAL：3。

☑ SENSOR_DELAY_UI：2。

☑ SENSOR_DELAY_GAME：1。

☑ SENSOR_DELAY_FASTEST：0。

而类 SensorEvent 有 4 个成员变量，具体说明如下所示。

☑ Accuracy：精确值。

☑ Sensor：发生变化的感应器。

☑ Timestamp：发生的时间，单位是纳秒。

☑ Values：发生变化后的值，这是一个长度为 3 的数组。

光线传感器只需要 values[0]的值，其他两个都为 0。而 values[0]就是开发光线传感器所需要的，单位是 klux，表示光线强度。

### 11.1.3　实战演练——获取设备中光线传感器的值

本实例将演示在 Android 物联网设备中使用光线传感器的方法。

| 实　例 | 功　能 | 源码路径 |
| --- | --- | --- |
| 实例 11-1 | 获取设备中光线传感器的值 | 光盘:\daima\11\SensorEX |

本实例的功能是获取设备中光线传感器的值，具体实现流程如下所示。

（1）编写布局文件 activity_main.xml，具体实现代码如下：

```
<RelativeLayout xmlns:android="http://schemas.android.com/apk/res/android"
    xmlns:tools="http://schemas.android.com/tools"
    android:layout_width="match_parent"
    android:layout_height="match_parent"
    android:paddingBottom="@dimen/activity_vertical_margin"
    android:paddingLeft="@dimen/activity_horizontal_margin"
    android:paddingRight="@dimen/activity_horizontal_margin"
    android:paddingTop="@dimen/activity_vertical_margin"
    tools:context=".MainActivity" >
    <TextView
        android:layout_width="wrap_content"
        android:layout_height="wrap_content"
        android:text="@string/hello_world" />
</RelativeLayout>
```

（2）编写程序文件 MainActivity.java，具体实现代码如下：

```
package com.example.sensor;

import android.hardware.Sensor;
import android.hardware.SensorEvent;
import android.hardware.SensorEventListener;
import android.hardware.SensorListener;
import android.hardware.SensorManager;
import android.os.Bundle;
import android.renderscript.Sampler.Value;
import android.app.Activity;
import android.view.Menu;
import android.widget.TextView;

public class MainActivity extends Activity implements SensorEventListener   {



    private SensorManager sensor;
    private TextView text;
    @Override
    protected void onCreate(Bundle savedInstanceState) {
        super.onCreate(savedInstanceState);
        setContentView(R.layout.activity_main);
        sensor = (SensorManager)getSystemService(SENSOR_SERVICE);
        text = (TextView)findViewById(R.id.textView1);
    }

    @Override
    public boolean onCreateOptionsMenu(Menu menu) {
        getMenuInflater().inflate(R.menu.activity_main, menu);
        return true;
    }

    @Override
    protected void onPause() {
        sensor.unregisterListener(this);
        super.onPause();
    }

    @Override
    protected void onResume() {

    sensor.registerListener(this,sensor.getDefaultSensor(Sensor.TYPE_LIGHT),SensorManager.SENSOR_D
ELAY_GAME);
        super.onResume();
    }

    @Override
    protected void onStop() {
```

```
        sensor.unregisterListener(this);
        super.onStop();
    }


    @Override
    public void onAccuracyChanged(Sensor sensor, int accuracy) {

    }

    @Override
    public void onSensorChanged(SensorEvent event) {
        float[] values = event.values;
        int sensorType = event.sensor.TYPE_LIGHT;
        if(sensorType==Sensor.TYPE_LIGHT)
        {
            text.setText(String.valueOf(values[0]));
        }
    }
}
```

在真机中执行后，将会显示设备中光线传感器的值。

### 11.1.4　实战演练——显示设备中光线传感器的强度

本实例将演示在 Android 物联网设备中使用光线传感器的方法。

| 实　例 | 功　能 | 源 码 路 径 |
|---|---|---|
| 实例 11-2 | 显示设备中光线传感器的强度 | 光盘:\daima\11\qiangEX |

本实例的功能是显示设备中光线传感器的强度值，具体实现流程如下所示。

（1）在 Eclipse 工程中引入两个开源开发包，如图 11-1 所示。

图 11-1　引入开发包

（2）编写布局文件 main.xml，具体实现代码如下：

```
<LinearLayout xmlns:android="http://schemas.android.com/apk/res/android"
    xmlns:tools="http://schemas.android.com/tools"
    android:layout_width="match_parent"
    android:layout_height="match_parent"
    android:orientation="vertical" >

    <TextView
        android:id="@+id/myTextView1"
        android:layout_width="wrap_content"
        android:layout_height="wrap_content"
     />
```

```
</LinearLayout>
```

（3）编写值文件 string.xml，具体实现代码如下：

```
<resources>
    <string name="app_name">Sample</string>
    <string name="title">光传感器</string>
    <string name="hello_world">Hello world!</string>
    <string name="menu_settings">Settings</string>
</resources>
```

（4）编写主程序文件 MainActivity.java，具体实现代码如下：

```
package com.example.qiang;
import org.openintents.sensorsimulator.hardware.Sensor;
import android.app.Activity;
import android.hardware.SensorManager;
import android.os.Bundle;
import android.widget.TextView;

public class MainActivity extends Activity implements android.hardware.SensorEventListener {

    private TextView myTextView1;

    private SensorManager mySensorManager;

    @Override
    public void onCreate(Bundle savedInstanceState) {
        super.onCreate(savedInstanceState);
        setContentView(R.layout.main);
        myTextView1 = (TextView) findViewById(R.id.myTextView1);
        mySensorManager = (SensorManager) getSystemService(SENSOR_SERVICE);

    }

    @Override
    protected void onResume() {
        mySensorManager.registerListener(
                this,
                mySensorManager.getDefaultSensor(Sensor.TYPE_LIGHT),
                SensorManager.SENSOR_DELAY_GAME
                );
        super.onResume();
    }

    @Override
    protected void onStop() {
        mySensorManager.unregisterListener(this);
        super.onStop();
    }
    @Override
    protected void onPause() {
        mySensorManager.unregisterListener(this);
```

```
            super.onPause();
        }

        @Override
        public void onAccuracyChanged(android.hardware.Sensor sensor, int accuracy) {

        }

        @Override
        public void onSensorChanged(android.hardware.SensorEvent event) {
            float[] values = event.values;
            int sensorType = event.sensor.TYPE_LIGHT;
            if (sensorType == Sensor.TYPE_LIGHT) {
                myTextView1.setText("当前光的强度为："+values[0]);
            }
        }
    }
```

# 11.2　磁场传感器详解

**知识点讲解：光盘:视频\知识点\第 11 章\磁场传感器详解.avi**

在现实应用中，经常需要检测 Android 设备的方向，例如设备的朝向和移动方向。在 Android 系统中，通常使用重力传感器、加速度传感器、磁场传感器和旋转矢量传感器来检测设备的方向。本节将详细讲解在 Android 设备中使用磁场传感器检测设备方向的基本知识，为读者步入本书后面知识的学习打下基础。

## 11.2.1　什么是磁场传感器

磁场传感器是可以将各种磁场及其变化的量转变成电信号输出的装置。自然界和人类社会生活的许多地方都存在磁场或与磁场相关的信息。磁场传感器是利用人工设置的永久磁体产生的磁场，可以作为许多种信息的载体，被广泛用于探测、采集、存储、转换、复现和监控各种磁场和磁场中承载的各种信息的任务。在当今的信息社会中，磁场传感器已成为信息技术和信息产业中不可缺少的基础元件。目前，人们已研制出利用各种物理、化学和生物效应的磁场传感器，并已在科研、生产和社会生活的各个方面得到广泛应用，承担起探究种种信息的任务。

在现实市面中，最早磁场传感器是伴随测磁仪器的进步而逐步发展的。在众多的测试磁场方法中，大多都是将磁场信息变成电信号进行测量。在测磁仪器中“探头”或“取样装置”就是磁场传感器。随着信息产业、工业自动化、交通运输、电力电子技术、办公自动化、家用电器、医疗仪器等的飞速发展和电子计算机应用的普及，需用大量的传感器将需进行测量和控制的非电参量，转换成可与计算机兼容的信号，作为它们的输入信号，这就给磁场传感器的快速发展提供了机会，形成了相当可观的磁场传感器产业。

## 11.2.2 磁场传感器的分类

在现实应用中，磁场传感器的主要分类如下所示。

（1）薄膜磁致电阻传感器

铁磁性物质在磁化过程中，它的电阻值沿磁化方向增加，并达到饱和的现象称为磁阻效应。薄膜磁阻元件是利用薄膜工艺和微细加工技术，将 NiPe、NiCo 合金用真空蒸镇或溯射工艺沉积到硅片或铁氧体基片上，通过微细加工技术制成一定形状的磁咀图形，形成三端式、四端式以及多端式器件。BMber 结构桥式电路磁阻元件具有灵敏度高、工作频率特性好、温度稳定性好、结构简单、体积小等特点。可制成高密度磁咀磁头、磁性编码器、磁阻位移传感器、磁阻电流传感器等。

（2）磁阻敏感器

物质在磁场中电阻发生变化的现象称为磁阻效应。对于铁、钴、镍及其合金等强磁性金属，当外加磁场平行于磁体内部磁化方向时，电阻几乎不随外加磁场变化；当外加磁场偏离金属的内磁化方向时，此类金属的电阻值将减小，这就是强磁金属的各向异性磁阻效应。

（3）电涡流式传感器

近年来，国内外正发展建立在电涡流效应原理上的传感器，即电涡流式传感器。这种传感器不但具有测量线性范围大、灵敏度高、结构简单、抗干扰能力强、不受油污等介质的影响等优点，而且又具有无损、非接触测量的特点，目前正广泛地应用于工业各部门中的位移、尺寸、厚度、振动、转速、压力、电导率、温度、波面等测量，以及探测金属材料和加工件表面裂纹及缺陷。

（4）磁性液体加速度传感器

磁性液体作为一种新型的纳米功能材料，一经问世便走到科学技术发展的前沿，目前科学家们已经将这种新型功能材料应用到广阔的领域中。以此为基础的磁性液体传感器技术也引起了国际技术领域广泛的关注。

（5）磁性液体水平传感器

磁性液体传感器的研究与应用起源于美国。早在 1983 年美国新墨西哥州阿尔帕克基应用技术公司就与美国空军签订了合同，美国新墨西哥州阿尔帕克基应用技术公司便开始研制基于磁性液体动力学原理（MHD）的主动式和被动式传感器。磁性液体传感器应用领域很广，既可以应用到民用上，也可以应用到军工上，有其他传感器所代替不了的功能，正因为如此，国外比较早地意识到开发磁性液体传感器 的意义，并且已开始了研究和生产，并将该种传感器应用到航空、航天、宇航站等尖端军事领域。美国、法国、德国、俄罗斯、日本和罗马尼亚等国家已开始利用磁性液体来制作各种传感器。磁性液体水平传感器对控制机器人工作状态，使太阳能栅板保持朝向太阳，使抛物天线持续朝向通信系统中的人造卫星等方面有着重要的应用。目前国外已研制出单轴、双轴、3 轴等类型的磁性液体水平传感器，而我国有关磁性液体传感器的研究尚处于实验和探索阶段。

## 11.2.3 Android 系统中的磁场传感器

在 Android 系统中，磁场传感器 TYPE_MAGNETIC_FIELD，单位是 μT（微特斯拉），能够测量设备周围 3 个物理轴（x,y,z）的磁场。在 Android 设备中，磁场传感器主要用于感应周围的磁感应强度，在注册监听器后主要用于捕获如下 3 个参数：

☑ values[0]

☑ values[1]

☑ values[2]

上述 3 个参数分别代表磁感应强度在空间坐标系中 3 个方向轴上的分量。所有数据的单位为 μT，即微特斯拉。

在 Android 系统中，磁场传感器主要包含如下所示的公共方法。

☑ int getFifoMaxEventCount()：返回该传感器可以处理事件的最大值。如果该值为 0，表示当前模式不支持此传感器。

☑ int getFifoReservedEventCount()：保留传感器在批处理模式中 FIFO 的事件数，给出了一个保证可以分批事件的最小值。

☑ float getMaximumRange()：传感器单元的最大范围。

☑ int getMinDelay()：最小延迟。

☑ String getName()：获取传感器的名称。

☑ floa getPower()：获取传感器电量。

☑ float getResolution()：获得传感器的分辨率。

☑ int getType()：获取传感器的类型。

☑ String getVendor()：获取传感器的供应商字符串。

☑ Int etVersion()：获取该传感器模块版本。

☑ String toString()：返回一个对当前传感器的字符串描述。

### 11.2.4　实战演练——获取磁场传感器的 3 个分量

本实例将演示在 Android 设备中使用磁场传感器的方法。

| 实　　例 | 功　　能 | 源 码 路 径 |
| --- | --- | --- |
| 实例 11-3 | 使用磁场传感器 | 光盘:\daima\11\cichangEX |

本实例的实现文件是 cichangLI.java，在此文件中定义了监听器类对象和注册监听的方法。主要代码如下所示。

```
public class cichangLI extends Activity {
    TextView myTextView1;//x 方向磁场分量
    TextView myTextView2;//y 方向磁场分量
    TextView myTextView3;//z 方向磁场分量
    //SensorManager mySensorManager;//引用 SensorManager 对象
    SensorManagerSimulator mySensorManager;//声明 SensorManagerSimulator 对象，调试时用
    @Override
    public void onCreate(Bundle savedInstanceState) {//重写 onCreate()方法
        super.onCreate(savedInstanceState);
        setContentView(R.layout.main);//当前的用户界面
        myTextView1 = (TextView) findViewById(R.id.myTextView1);//得到 myTextView1 引用
        myTextView2 = (TextView) findViewById(R.id.myTextView2);//得到 myTextView2 引用
        myTextView3 = (TextView) findViewById(R.id.myTextView3);//得到 myTextView3 引用
        //调试时用
        mySensorManager = SensorManagerSimulator.getSystemService(this, SENSOR_SERVICE);
```

```
        mySensorManager.connectSimulator();                         //连接 Simulator 服务器
    }
    @SuppressWarnings("deprecation")
     private SensorListener mySensorListener = new SensorListener(){
     @Override
     public void onAccuracyChanged(int sensor, int accuracy) {}     //重写 onAccuracyChanged()方法
     @Override
     public void onSensorChanged(int sensor, float[] values) {      //重写 onSensorChanged()方法
         if(sensor == SensorManager.SENSOR_MAGNETIC_FIELD){  //检查磁场的变化
             myTextView1.setText("x 方向的磁场分量为："+values[0]);//数据显示在 TextView
             myTextView2.setText("y 方向的磁场分量为："+values[1]);//数据显示在 TextView
             myTextView3.setText("z 方向的磁场分量为："+values[2]);//数据显示在 TextView
         }
     }
    };
    @Override
    protected void onResume() {                                     //重写 onResume()方法
     mySensorManager.registerListener(                              //注册监听
             mySensorListener,                                      //监听器 SensorListener 对象
             SensorManager.SENSOR_MAGNETIC_FIELD,                   //传感器的类型为加速度
             SensorManager.SENSOR_DELAY_UI                          //传感器事件传递的频度
             );
     super.onResume();
    }
    @Override
    protected void onPause() {                                      //重写 onPause()方法
//取消注册监听器
     mySensorManager.unregisterListener((SensorEventListener) mySensorListener);
     super.onPause();
    }
}
```

因为本实例比较简单，是根据 SensorSimulator 中附带的开源代码改编的，所以在此不再进行详细介绍，读者只需阅读本书附带光盘中的源码即可。

### 11.2.5 实战演练——演示常用传感器的基本用法

本实例将演示在 Android 设备中使用常用传感器的基本方法。

| 实　例 | 功　能 | 源 码 路 径 |
| --- | --- | --- |
| 实例 11-4 | 使用常用传感器 | 光盘:\daima\11\HelloEX |

（1）实现布局文件

布局文件 main.xml 的功能是使用 ListView 控件列表显示常用的传感器类型，具体实现代码如下：

```
<LinearLayout xmlns:android="http://schemas.android.com/apk/res/android"
android:orientation="vertical"
android:layout_width="fill_parent"
android:layout_height="fill_parent">
    <ListView android:id="@+id/ListView01"
              android:layout_width="wrap_content"
```

```
            android:layout_height="wrap_content"/>
</LinearLayout>
```

（2）实现程序文件

主 Activity 的实现文件是 HelloSensor.java，功能是响应用户选择的传感器来执行对应的处理程序，具体实现代码如下：

```
public class HelloSensor extends Activity {
    //5 个范例的菜单名称和应用程序 Class
    private Object[] activities = {
        "Compass", CompassDemo.class,
        "Orientation",OrientationDemo.class,
        "Accelerometer",AccelerometerDemo.class,
        "Magnetic Field",MagneticFieldDemo.class,
        "Temperature",TemperatureDemo.class,
    };
    //HelloSensor 主程式
    @Override
    public void onCreate(Bundle savedInstanceState) {
        super.onCreate(savedInstanceState);
        setContentView(R.layout.main);
        //建立 5 个范例菜单名称的数组 list
        CharSequence[] list = new CharSequence[activities.length / 2];
        for (int i = 0; i < list.length; i++) {
            list[i] = (String)activities[i * 2];
        }
        //将 5 个范例菜单名称放置在 listView
        ArrayAdapter<CharSequence> adapter = new ArrayAdapter<CharSequence>(this, android.R.layout.
simple_list_item_1, list);
        ListView listView = (ListView)findViewById(R.id.ListView01);
        listView.setAdapter(adapter);
        //按下菜单名称指向相关的应用程序 Class
        listView.setOnItemClickListener(new OnItemClickListener() {
            public void onItemClick(AdapterView<?> parent, View view, int position, long id) {
                Intent intent = new Intent(HelloSensor.this, (Class<?>)activities[position * 2 + 1]);
                startActivity(intent);
            }
        });
    }
}
```

当在主 Activity 界面选择 CompassDemo 选项后，会执行文件 CompassDemo.java 启动罗盘传感器，具体实现代码如下：

```
public class CompassDemo extends Activity implements SensorEventListener {
    private SensorManager sensorManager;
    private MySurfaceView view;
    private Object[] orientation = {
            "Rotate Z-axis Orientation", "Rotate X-axis Orientation","Rotate Y-axis Orientation",
    };
    //CompassDemo 主程序
    @Override
    public void onCreate(Bundle savedInstanceState) {
```

```
        super.onCreate(savedInstanceState);
        sensorManager = (SensorManager)getSystemService(SENSOR_SERVICE);
        view = new MySurfaceView(this);
        setContentView(view);
    }
    @Override
    protected void onResume() {
        super.onResume();
        List<Sensor> sensors = sensorManager.getSensorList(Sensor.TYPE_ORIENTATION);
        if (sensors.size() > 0) {
            sensorManager.registerListener(this, sensors.get(0), SensorManager.SENSOR_DELAY_ NORMAL);
        }
    }
    @Override
    protected void onPause() {
        super.onPause();
        sensorManager.unregisterListener(this);
    }
    public void onAccuracyChanged(Sensor sensor, int accuracy) {
    }
    public void onSensorChanged(SensorEvent event) {
        view.onValueChanged(event.values);
    }
    class MySurfaceView extends SurfaceView implements SurfaceHolder.Callback {
        private Bitmap bitmap,curBitmap;
        private float x, y, z, delta=-8;
        private int curWidth, curHeight;
        public MySurfaceView(Context context) {
            super(context);
            getHolder().addCallback(this);
            bitmap = BitmapFactory.decodeResource(getResources(), R.drawable.compass);
        }
        public void surfaceChanged(SurfaceHolder holder, int format, int width, int height) {
            x = getWidth()/2;
            y = getHeight()/2;
            onValueChanged(new float[3]);
        }
        public void surfaceCreated(SurfaceHolder holder) {
        }
        public void surfaceDestroyed(SurfaceHolder holder) {
        }
        @SuppressWarnings("static-access")
        void onValueChanged(float[] values) {
            Canvas canvas = getHolder().lockCanvas();
            if (canvas != null) {
                Paint paint = new Paint();
                paint.setAntiAlias(true);
                paint.setColor(Color.BLUE);
                paint.setTextSize(24);
                canvas.drawColor(Color.WHITE);
                canvas.save();
```

```
                    Matrix matrix = new Matrix();
                    curWidth = (int) (bitmap.getWidth()* 1);
                    curHeight =   (int) (bitmap.getHeight()* 1);
                    curBitmap = bitmap.createScaledBitmap(bitmap, curWidth, curHeight, false);
                    matrix.setRotate(-values[0]+delta, x , y );
                    canvas.setMatrix(matrix);
                    canvas.drawBitmap(curBitmap, x-curWidth/2, y-curHeight/2, null);
                    canvas.restore();
                    for (int i = 0; i < values.length; i++) {
                        canvas.drawText(orientation[i] + ": "   + values[i], 0, paint.getTextSize() * (i + 1), paint);
                    }
                    getHolder().unlockCanvasAndPost(canvas);
                }
            }
        }
}
```

当在主 Activity 界面选择 AccelerometerDemo 选项后，会执行文件 AccelerometerDemo.java 启动加速计传感器，具体实现代码如下：

```
public class AccelerometerDemo extends Activity implements SensorEventListener {
    private SensorManager sensorManager;
    private MySurfaceView view;
    private Object[] accelerometer = {
            "X-axis Accelerometer", "Y-axis Accelerometer","Z-axis Accelerometer",
    };
    //AccelerometerDemo 主程序
    @Override
    public void onCreate(Bundle savedInstanceState) {
        super.onCreate(savedInstanceState);
        sensorManager = (SensorManager)getSystemService(SENSOR_SERVICE);
        view = new MySurfaceView(this);
        setContentView(view);
    }
    @Override
    protected void onResume() {
        super.onResume();
        List<Sensor> sensors = sensorManager.getSensorList(Sensor.TYPE_ACCELEROMETER);
        if (sensors.size() > 0) {
            sensorManager.registerListener(this, sensors.get(0), SensorManager.SENSOR_DELAY_FASTEST);
        }
    }
    @Override
    protected void onPause() {
        super.onPause();
        sensorManager.unregisterListener(this);
    }
    public void onAccuracyChanged(Sensor sensor, int accuracy) {
    }
    public void onSensorChanged(SensorEvent event) {
        view.onValueChanged(event.values);
    }
```

```
    class MySurfaceView extends SurfaceView implements SurfaceHolder.Callback {
        private Bitmap bitmap, curBitmap;
        private float x, y, z;
        private int curWidth, curHeight;
        public MySurfaceView(Context context) {
            super(context);
            getHolder().addCallback(this);
            bitmap = BitmapFactory.decodeResource(getResources(), R.drawable.android);
        }
        public void surfaceChanged(SurfaceHolder holder, int format, int width, int height) {
            x = (getWidth() - bitmap.getWidth()) / 2;
            y = (getHeight() - bitmap.getHeight()) / 2;
            onValueChanged(new float[3]);
        }
        public void surfaceCreated(SurfaceHolder holder) {
        }
        public void surfaceDestroyed(SurfaceHolder holder) {
        }
        @SuppressWarnings("static-access")
        void onValueChanged(float[] values) {
            z = 2 + values[2]/5;
            curWidth = (int) (bitmap.getWidth()* z);
            curHeight =   (int) (bitmap.getHeight()* z);
            curBitmap = bitmap.createScaledBitmap(bitmap, curWidth, curHeight, false);
            x = (getWidth() - curWidth) / 2;
            y = (getHeight() - curHeight) / 2;
            x -= values[0]*10;
            y += values[1]*10;
            Canvas canvas = getHolder().lockCanvas();
            if (canvas != null) {
                Paint paint = new Paint();
                paint.setAntiAlias(true);
                paint.setColor(Color.BLUE);
                paint.setTextSize(24);
                canvas.drawColor(Color.WHITE);
                canvas.drawBitmap(curBitmap, x, y, null);
                for (int i = 0; i < values.length; i++) {
                    canvas.drawText(accelerometer[i] + ": " + values[i], 0, paint.getTextSize() * (i + 1),
paint);
                }
                getHolder().unlockCanvasAndPost(canvas);
            }
        }
    }
}
```

当在主 Activity 界面选择 Magnetic Field 选项后，会执行文件 MagneticFieldDemo.java 启动磁场传感器，具体实现代码如下：

```
public class MagneticFieldDemo extends Activity implements SensorEventListener {
    private SensorManager sensorManager;
    private MySurfaceView view;
```

```
private float max;
private Object[] magnetic = {
        "X-axis Maganetic Field", "Y-axis Maganetic Field","Z-axis Maganetic Field",
};
//MagneticFieldDemo 主程序
@Override
public void onCreate(Bundle savedInstanceState) {
    super.onCreate(savedInstanceState);
    sensorManager = (SensorManager)getSystemService(SENSOR_SERVICE);
    view = new MySurfaceView(this);
    setContentView(view);
}
@Override
protected void onResume() {
    super.onResume();
    List<Sensor> sensors = sensorManager.getSensorList(Sensor.TYPE_MAGNETIC_FIELD);
    if (sensors.size() > 0) {
        Sensor sensor = sensors.get(0);
        max = sensor.getMaximumRange();
        sensorManager.registerListener(this, sensor, SensorManager.SENSOR_DELAY_NORMAL);
    }
}
@Override
protected void onPause() {
    super.onPause();
    sensorManager.unregisterListener(this);
}
public void onAccuracyChanged(Sensor sensor, int accuracy) {
}
public void onSensorChanged(SensorEvent event) {
    view.onValueChanged(event.values);
}
class MySurfaceView extends SurfaceView implements SurfaceHolder.Callback {
    public MySurfaceView(Context context) {
        super(context);
        getHolder().addCallback(this);
    }
    public void surfaceChanged(SurfaceHolder holder, int format, int width, int height) {
        onValueChanged(new float[3]);
    }
    public void surfaceCreated(SurfaceHolder holder) {
    }
    public void surfaceDestroyed(SurfaceHolder holder) {
    }
    void onValueChanged(float[] values) {
        Canvas canvas = getHolder().lockCanvas();
        if (canvas != null) {
            Paint paint = new Paint();
            paint.setAntiAlias(true);
            paint.setColor(Color.BLUE);
            paint.setTextSize(24);
```

```
                canvas.drawColor(Color.WHITE);
                for (int i = 0; i < values.length; i++) {
                    canvas.drawText(magnetic[i] + ": " + values[i], 0, paint.getTextSize() * (i + 1), paint);
                }
                canvas.drawText("max: " + max, 0, paint.getTextSize() * 5, paint);
                getHolder().unlockCanvasAndPost(canvas);
            }
        }
    }
}
```

当在主 Activity 界面选择 Orientation 选项后，会执行文件 OrientationDemo.java 启动方向传感器，具体实现代码如下：

```
public class OrientationDemo extends Activity implements SensorEventListener {
    private SensorManager sensorManager;
    private MySurfaceView view;
    private Object[] orientation = {
            "Rotate Z-axis Orientation", "Rotate X-axis Orientation","Rotate Y-axis Orientation",
    };
    //OrientationDemo 主程序
    @Override
    public void onCreate(Bundle savedInstanceState) {
        super.onCreate(savedInstanceState);
        sensorManager = (SensorManager)getSystemService(SENSOR_SERVICE);
        view = new MySurfaceView(this);
        setContentView(view);
    }
    @Override
    protected void onResume() {
        super.onResume();
        List<Sensor> sensors = sensorManager.getSensorList(Sensor.TYPE_ORIENTATION);
        if (sensors.size() > 0) {
            sensorManager.registerListener(this, sensors.get(0), SensorManager.SENSOR_DELAY_NORMAL);
        }
    }
    @Override
    protected void onPause() {
        super.onPause();
        sensorManager.unregisterListener(this);
    }
    public void onAccuracyChanged(Sensor sensor, int accuracy) {
    }
    public void onSensorChanged(SensorEvent event) {
        view.onValueChanged(event.values);
    }
    class MySurfaceView extends SurfaceView implements SurfaceHolder.Callback {
        private Bitmap bitmap, bitmap1, bitmap2,curBitmap;
        private float x, y, z;
        private float x1=130, y1=160;
        private int curWidth, curHeight;
        public MySurfaceView(Context context) {
```

```
            super(context);
            getHolder().addCallback(this);
            bitmap = BitmapFactory.decodeResource(getResources(), R.drawable.androidplate);
            bitmap1 = BitmapFactory.decodeResource(getResources(), R.drawable.androidheight);
            bitmap2 = BitmapFactory.decodeResource(getResources(), R.drawable.androidwidth);
        }
        public void surfaceChanged(SurfaceHolder holder, int format, int width, int height) {
            x = getWidth()/2;
            y = getHeight()/2;
            onValueChanged(new float[3]);
        }
        public void surfaceCreated(SurfaceHolder holder) {
        }
        public void surfaceDestroyed(SurfaceHolder holder) {
        }
        @SuppressWarnings("static-access")
        void onValueChanged(float[] values) {
            Canvas canvas = getHolder().lockCanvas();
            if (canvas != null) {
                Paint paint = new Paint();
                paint.setAntiAlias(true);
                paint.setColor(Color.BLUE);
                paint.setTextSize(24);
                canvas.drawColor(Color.WHITE);
                canvas.save();
                Matrix matrix = new Matrix();
                curWidth = (int) (bitmap.getWidth()* 1);
                curHeight =   (int) (bitmap.getHeight()* 1);
                curBitmap = bitmap.createScaledBitmap(bitmap, curWidth, curHeight, false);
                matrix.setRotate(-values[0], x , y );
                canvas.setMatrix(matrix);
                canvas.drawBitmap(curBitmap, x-curWidth/2, y-curHeight/2, null);
                matrix.setRotate(values[1], x-x1 , y+y1 );
                canvas.setMatrix(matrix);
                canvas.drawBitmap(bitmap1, x-x1-bitmap1.getWidth()/2, y+y1-bitmap1.getHeight()/2, null);
                matrix.setRotate(-values[2], x+x1 , y+y1 );
                canvas.setMatrix(matrix);
                canvas.drawBitmap(bitmap2, x+x1-bitmap2.getWidth()/2, y+y1-bitmap2.getHeight()/2, null);
                canvas.restore();
                for (int i = 0; i < values.length; i++) {
                    canvas.drawText(orientation[i] + ": "   + values[i], 0, paint.getTextSize() * (i + 1), paint);
                }
                getHolder().unlockCanvasAndPost(canvas);
            }
        }
    }
}
```

当在主 Activity 界面选择 Temperature 选项后，会执行文件 TemperatureDemo.java 启动温度传感器，具体实现代码如下：

```
public class TemperatureDemo extends Activity implements SensorEventListener {
    private SensorManager sensorManager;
```

```
private MySurfaceView view;
private String vendor = "UNKNOWN";
private String name = "UNKNOWN";
private int version = 0;
private Object[] temperature = {
        "Temperature", "No Data","No Data",
};
//TemperatureDemo 主程序
@Override
public void onCreate(Bundle savedInstanceState) {
    super.onCreate(savedInstanceState);
    sensorManager = (SensorManager)getSystemService(SENSOR_SERVICE);
    view = new MySurfaceView(this);
    setContentView(view);
}
@Override
protected void onResume() {
    super.onResume();
    List<Sensor> sensors = sensorManager.getSensorList(Sensor.TYPE_TEMPERATURE);
    if (sensors.size() > 0) {
        Sensor sensor = sensors.get(0);
        vendor = sensor.getVendor();
        name = sensor.getName();
        version = sensor.getVersion();
        sensorManager.registerListener(this, sensor, SensorManager.SENSOR_DELAY_NORMAL);
    }
}
@Override
protected void onPause() {
    super.onPause();
    sensorManager.unregisterListener(this);
}
public void onAccuracyChanged(Sensor sensor, int accuracy) {
}
public void onSensorChanged(SensorEvent event) {
    view.onValueChanged(event.values);
}
class MySurfaceView extends SurfaceView implements SurfaceHolder.Callback {
    public MySurfaceView(Context context) {
        super(context);
        getHolder().addCallback(this);
    }
    public void surfaceChanged(SurfaceHolder holder, int format, int width, int height) {
        onValueChanged(new float[3]);
    }
    public void surfaceCreated(SurfaceHolder holder) {
    }
    public void surfaceDestroyed(SurfaceHolder holder) {
    }
    void onValueChanged(float[] values) {
        Canvas canvas = getHolder().lockCanvas();
```

```
            if (canvas != null) {
                Paint paint = new Paint();
                paint.setAntiAlias(true);
                paint.setColor(Color.BLUE);
                paint.setTextSize(24);
                canvas.drawColor(Color.WHITE);
                for (int i = 0; i < values.length; i++) {
                    canvas.drawText(temperature[i] + ": " + values[i], 0, paint.getTextSize() * (i + 1), paint);
                }
                canvas.drawText("Vender: "+vendor, 0, paint.getTextSize() * 5, paint);
                canvas.drawText("Name: "+name, 0, paint.getTextSize() * 6, paint);
                canvas.drawText("Version: "+ version, 0, paint.getTextSize() * 7, paint);
                getHolder().unlockCanvasAndPost(canvas);
            }
        }
    }
}
```

至此，整个实例介绍完毕，执行后的效果如图 11-2 所示。

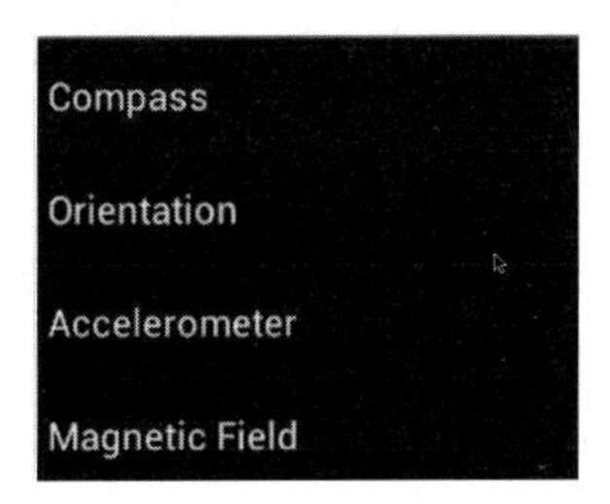


图 11-2　执行效果

# 第 12 章　加速度传感器、方向传感器和陀螺仪传感器

在 Android 物联网设备应用程序开发过程中，经常需要检测设备的运动数据，例如设备的运动速率和运动距离等。这些数据对于健身类物联网设备来说，都是十分重要的数据，例如健身手表可以及时测试晨练的运动距离和速率。在 Android 系统中，通常使用加速度传感器、线性加速度传感器和距离传感器来检测设备的运动数据。本章将详细讲解在 Android 物联网设备中使用加速度传感器、方向传感器和陀螺仪传感器的方法，为读者步入本书后面知识的学习打下基础。

## 12.1　加速度传感器详解

**知识点讲解：光盘:视频\知识点\第 12 章\加速度传感器详解.avi**

在现实应用中，加速度传感器可以帮助机器人了解它现在身处的环境，能够分辨出是在登山，还是在下山，是否摔倒等。一个好的程序员能够使用加速度传感器来分辨出上述情形，加速度传感器甚至可以用来分析发动机的振动。本节将简要讲解加速度传感器的基础性知识。

### 12.1.1　加速度传感器的分类

在实际应用过程中，可以将加速度传感器分为如下所示的 4 类。

（1）压电式

压电式加速度传感器又称压电加速度计。它也属于惯性式传感器。压电式加速度传感器的原理是利用压电陶瓷或石英晶体的压电效应，在加速度计受振时，质量块加在压电元件上的力也随之变化。当被测振动频率远低于加速度计的固有频率时，则力的变化与被测加速度成正比。

（2）压阻式

基于世界领先的 MEMS 硅微加工技术，压阻式加速度传感器具有体积小、低功耗等特点，易于集成在各种模拟和数字电路中，广泛应用于汽车碰撞实验、测试仪器、设备振动监测等领域。加速度传感器网为客户提供压阻式加速度传感器/压阻加速度计各品牌的型号、参数、原理、价格、接线图等信息。

（3）电容式

电容式加速度传感器是基于电容原理的极距变化型的电容传感器。电容式加速度传感器/电容式加速度计是对比较通用的加速度传感器。在某些领域无可替代，如安全气囊、手机移动设备等。电容式加速度传感器/电容式加速度计采用了微机电系统（MEMS）工艺，在大量生产时变得经济，从而保证了较低的成本。

（4）伺服式

伺服式加速度传感器是一种闭环测试系统，具有动态性能好、动态范围大和线性度好等特点。其工作原理，传感器的振动系统由“m-k”系统组成，与一般加速度计相同，但质量 m 上还接着一个电磁线圈，当基座上有加速度输入时，质量块偏离平衡位置，该位移大小由位移传感器检测出来，经伺服放大器放大后转换为电流输出，该电流流过电磁线圈，在永久磁铁的磁场中产生电磁恢复力，力图使质量块保持在仪表壳体中原来的平衡位置上，所以伺服加速度传感器在闭环状态下工作。由于有反馈作用，增强了抗干扰的能力，提高了测量精度，扩大了测量范围，伺服加速度测量技术广泛地应用于惯性导航和惯性制导系统中，在高精度的振动测量和标定中也有应用。

## 12.1.2 加速度传感器的主要应用领域

在计算机领域中，加速度传感器可以测量牵引力产生的加速度。例如在 IBM Thinkpad 笔记本电脑中就内置了加速度传感器，能够动态地监测出笔记本在使用中的振动。根据这些振动数据，系统会智能地选择关闭硬盘还是让其继续运行，这样可以最大程度地保护里面的数据。所以，加速度传感器主要应用在手柄振动/摇晃、仪器仪表、汽车制动启动、地震、报警系统、玩具、结构物、环境监视、工程测振、地质勘探、铁路、桥梁、大坝的振动测试与分析，还有鼠标，高层建筑结构动态特性和安全保卫振动侦察上。下面将详细讲解加速度传感器的主要应用领域。

（1）汽车安全

加速度传感器主要用于汽车安全气囊、防抱死系统、牵引控制系统等安全性能方面。在汽车安全应用中，加速度计的快速反应非常重要。安全气囊应在什么时候弹出要迅速确定，所以加速度计必须在瞬间做出反应。通过采用可迅速达到稳定状态而不是振动不止的传感器设计可以缩短器件的反应时间。其中，压阻式加速度传感器由于在汽车工业中的广泛应用而发展最快。

（2）游戏控制

加速度传感器可以检测上下左右的倾角的变化，因此通过前后倾斜手持设备来实现对游戏中物体的前后左右的方向控制，就变得很简单。

（3）图像自动翻转

用加速度传感器检测手持设备的旋转动作及方向，实现所要显示图像的转正。

（4）电子指南针倾斜校正

磁传感器是通过测量磁通量的大小来确定方向的。当磁传感器发生倾斜时，通过磁传感器的地磁通量将发生变化，从而使方向指向产生误差。因此，如果不带倾斜校正的电子指南针，需要用户水平放置。而利用加速度传感器可以测量倾角的这一原理，可以对电子指南针的倾斜进行补偿。

（5）GPS 导航系统死角的补偿

GPS 系统是通过接收三颗呈 120 度分布的卫星信号来最终确定物体的方位的。在一些特殊的场合和地貌，如隧道、高楼林立、丛林地带，GPS 信号会变弱甚至完全失去，这也就是所谓的死角。而通过加装加速度传感器及以前我们所通用的惯性导航，便可以进行系统死区的测量。对加速度传感器进行一次积分，就变成了单位时间里的速度变化量，从而测出在死区内物体的移动。

（6）计步器功能

加速度传感器可以检测交流信号以及物体的振动，人在走动时会产生一定规律性的振动，而加速度传感器可以检测振动的过零点，从而计算出人所走的步或跑步所走的步数，从而计算出人所移动的位移。并且利用一定的公式可以计算出卡路里的消耗。

（7）防手抖功能

用加速度传感器检测手持设备的振动/晃动幅度，当振动/晃动幅度过大时锁住照相快门，使所拍摄的图像永远是清晰的。

（8）闪信功能

通过挥动手持设备实现在空中显示文字，用户可以自己编写显示的文字。这个闪信功能是利用人们的视觉残留现象，用加速度传感器检测挥动的周期，实现所显示文字的准确定位。

（9）硬盘保护

利用加速度传感器检测自由落体状态，从而对迷你硬盘实施必要的保护。大家知道，硬盘在读取数据时，磁头与碟片之间的间距很小，因此，外界的轻微振动就会对硬盘产生很坏的后果，使数据丢失。而利用加速度传感器可以检测自由落体状态。当检测到自由落体状态时，让磁头复位，以减少硬盘的受损程度。

（10）设备或终端姿态检测

加速度传感器和陀螺仪通常称为惯性传感器，常用于各种设备或终端中实现姿态检测、运动检测等，也就很适合玩体感游戏的人群。加速度传感器利用重力加速度，可以用于检测设备的倾斜角度，但是它会受到运动加速度的影响，使倾角测量不够准确，所以通常需利用陀螺仪和磁传感器补偿。同时磁传感器测量方位角时，也是利用地磁场，当系统中电流变化或周围有导磁材料时，以及当设备倾斜时，测量出的方位角也不准确，这时需要用加速度传感器（倾角传感器）和陀螺仪进行补偿。

（11）智能产品

加速度传感器在微信功能中的创新功能突破了电子产品的千遍一律，这个功能的实现来源于传感器的方向、加速表、光线、磁场、临近性、温度等参数的特性。这个原理是手机里面集成的加速度传感器，它能够分别测量 X、Y、Z 3 个方面的加速度值，X 方向值的大小代表手机水平移动，Y 方向值的大小代表手机垂直移动，Z 方向值的大小代表手机的空间垂直方向，天空的方向为正，地球的方向为负，然后把相关的加速度值传输给操作系统，通过判断其大小变化，就能知道同时玩微信的朋友。

## 12.1.3 线性加速度传感器的原理

线性加速度传感器是加速度传感器的一种，单独分开讲的原因是为了和螺旋仪传感器分开。陀螺仪是测角速度的，加速度是测线性加速度的。其中前者利用了惯性原理，而后者利用了力平衡原理。线性加速度传感器利用了惯性原理：

```
A(加速度)=F(惯性力)/M(质量)
```

我们只需要测量 F 即可。怎么测量 F？只要用电磁力去平衡这个力就可以得到 F 对应于电流的关系，只需要用实验去标定这个比例系数就行了。多数加速度传感器是根据压电效应的原理来工作的。所谓的压电效应就是“对于不存在对称中心的异极晶体加在晶体上的外力除了使晶体发生形变以外，还将改变晶体的极化状态，在晶体内部建立电场，这种由于机械力作用使介质发生极化的现象称为正压电效应”。

一般加速度传感器就是利用了其内部的由于加速度造成的晶体变形这个特性。由于这个变形会产生电压，只要计算出产生电压和所施加的加速度之间的关系，就可以将加速度转化成电压输出。当然，还有很多其他方法来制作加速度传感器，如压阻技术、电容效应、热气泡效应和光效应，但是其最基本的原理都是由于加速度产生某个介质产生变形，通过测量其变形量并用相关电路转化成电压输出。每种技术都有各自的机会和问题。

压阻式加速度传感器由于在汽车工业中的广泛应用而发展最快。由于安全性越来越成为汽车制造商的卖点，这种附加系统也越来越多。压阻式加速度传感器 2000 年的市场规模约为 4.2 亿美元，根据有关调查，预计其市值将按年平均 4.1%的速度增长，至 2007 年达到 5.6 亿美元。这其中，欧洲市场的速度最快，因为欧洲是许多安全气囊和汽车生产企业的所在地。

压电技术主要在工业上用来防止机器故障，使用这种传感器可以检测机器潜在的故障以达到自保护，以及避免对工人产生意外伤害，这种传感器具有用户，尤其是质量行业的用户所追求的可重复性、稳定性和自生性。但是在许多新的应用领域，很多用户尚无使用这类传感器的意识，销售商冒险进入这种尚待开发的市场会麻烦很多，因为终端用户对由于使用这种传感器而带来的问题和解决方法都认识不多。如果这些问题能够得到解决，将会促进压电传感器得到更快的发展。2002 年压电传感器市值为 3 亿美元，预计其年增长率将达到 4.9%，到 2007 年达到 4.2 亿美元。

使用加速度传感器有时会碰到低频场合测量时输出信号出现失真的情况，用多种测量判断方法一时找不出故障出现的原因，经过分析总结，导致测量结果失真的因素主要是：系统低频响应差，系统低频信噪比差，外界环境对测量信号的影响。所以，只要出现加速度传感器低频测量信号失真情况，对比以上 3 点查看是哪个因素造成的，有针对性地进行解决。

在 Android 系统中，线性加速度传感器的类型是 TYPE_LINEAR_ACCELERATION，单位是 $m/s^2$，能够获取加速度传感器去除重力的影响得到的数据。

## 12.1.4　Android 系统中的加速度传感器

在 Android 系统中，加速度传感器是 TYPE_ACCELEROMETER，单位是 $m/s^2$，能够测量应用于设备 X、Y、Z 轴上的加速度，又叫做 G-sensor。在开发过程中，通过 Android 的加速度传感器可以取得 X、Y、Z 3 个轴的加速度。在 Android 系统中，在类 SensorManager 中定义了很多星体的重力加速度值，如表 12-1 所示。

表 12-1　类 SensorManager 被定义的各新星体的重力加速度值

| 常　量　名 | 说　明 | 实 际 的 值 |
|---|---|---|
| GRAVITY_DEATH_STAR_1 | 死亡星 | 3.5303614E-7 |
| GRAVITY_EARTH | 地球 | 9.80665 |
| GRAVITY_JUPITER | 木星 | 23.12 |
| GRAVITY_MARS | 火星 | 3.71 |
| GRAVITY_MERCURY | 水星 | 3.7 |
| GRAVITY_MOON | 月亮 | 1.6 |
| GRAVITY_NEPTUNE | 海王星 | 12.0 |
| GRAVITY_PLUTO | 冥王星 | 0.6 |
| GRAVITY_SATURN | 土星 | 8.96 |
| GRAVITY_SUN | 太阳 | 275.0 |
| GRAVITY_THE_ISLAND | 岛屿星 | 4.815162 |
| GRAVITY_URANUS | 天王星 | 8.69 |
| GRAVITY_VENUS | 金星 | 8.87 |

通常来说，从加速度传感器获取的值，拿手机等智能设备的人的手振动或放在摇晃的场所时，受振动影响设备的值增幅变化是存在的。手的摇动、轻微振动的影响属于长波形式，去掉这种长波干扰的影响，可以取得高精度的值。去掉这种长波，这种过滤机叫 Low-Pass Filter。Low-Pass Filter 机制有如下所示的 3 种封装方法。

☑ 从抽样数据中取得中间的值的方法。

☑ 最近取得的加速度的值每个很少变化的方法。

☑ 从抽样数据中取得中间的值的方法。

在 Android 应用中，有时需要获取瞬间加速度值，例如类似计步器、作用力测定的应用开发时，如果想检测出加速度急剧的变化。此时的处理和 Low-Pass Filter 处理相反，需要去掉短周波的影响，这样可以取得数据。像这种去掉短周波的影响的过滤器叫做 High-Pass Filter。

### 12.1.5 实战演练——获取 X、Y、Z 轴的加速度值

本实例将演示在 Android 中使用加速度传感器的方法。

| 实　　例 | 功　　能 | 源 码 路 径 |
| --- | --- | --- |
| 实例 12-1 | 获取 X、Y、Z 轴的加速度值 | 光盘:\daima\12\jiaEX |

本实例的具体实现流程如下所示。

（1）编写布局文件 main.xml，主要代码如下：

```
<?xml version="1.0" encoding="utf-8"?>                    <!--声明 xml 的版本以及编码格式-->
<LinearLayout xmlns:android="http://schemas.android.com/apk/res/android"
    android:orientation="vertical"
    android:layout_width="fill_parent"
    android:layout_height="fill_parent">                  <!--添加一个垂直的线性布局-->
    <TextView
        android:id="@+id/title"
        android:gravity="center_horizontal"
        android:textSize="20px"
        android:layout_width="fill_parent"
        android:layout_height="wrap_content"
        android:text="@string/title"/>                    <!--添加一个 TextView 控件-->
    <TextView
        android:id="@+id/myTextView1"
        android:textSize="18px"
        android:layout_width="fill_parent"
        android:layout_height="wrap_content"
        android:text="@string/myTextView1"/>              <!--添加一个 TextView 控件-->
    <TextView
        android:id="@+id/myTextView2"
        android:textSize="18px"
        android:layout_width="fill_parent"
        android:layout_height="wrap_content"
        android:text="@string/myTextView2"/>              <!--添加一个 TextView 控件-->
    <TextView
        android:id="@+id/myTextView3"
```

```
        android:textSize="18px"
        android:layout_width="fill_parent"
        android:layout_height="wrap_content"
        android:text="@string/myTextView3"/>            <!-- 添加一个 TextView 控件 -->
</LinearLayout>
```

（2）编写主程序文件 jiaSLI.java，此文件的具体实现流程如下：

☑ 声明 3 个 TextView 的引用，分别用来显示 3 个方向上的加速度。
☑ 声明对 SensorManager 对象的引用，此处因使用的是 SensorSimulator 工具来模拟传感器。
☑ 设置当前的用户界面，然后得到 XML 文件中配置的各个控件的引用。
☑ 初始化 SensorManager 对象，同样因为调试的原因用专用代码替代。
☑ 初始化监听器类，并重写了该类中的两个方法。
☑ 在 onSensorChanged 方法中只处理加速度的变化，并将得到的数值显示到 TextView 中。
☑ 重写类 Activity 的 onResume 方法，在该方法中为 SensorManager 添加监听，还需要重写 onPause 方法，在该方法中取消注册的监听器。

文件 jiaSLI.java 的主要实现代码如下：

```
public class jiaSCH extends Activity {
    TextView myTextView1;//x 方向加速度
    TextView myTextView2;//y 方向加速度
    TextView myTextView3;//z 方向加速度
    //SensorManager mySensorManager;//SensorManager 对象引用
    SensorManagerSimulator mySensorManager;                //声明 SensorManagerSimulator 对象，调试时用
    @Override
    public void onCreate(Bundle savedInstanceState) {      //重写 onCreate()方法
        super.onCreate(savedInstanceState);
        setContentView(R.layout.main);//设置当前的用户界面
        myTextView1 = (TextView) findViewById(R.id.myTextView1);      //得到 myTextView1 引用
        myTextView2 = (TextView) findViewById(R.id.myTextView2);      //得到 myTextView2 引用
        myTextView3 = (TextView) findViewById(R.id.myTextView3);      //得到 myTextView3 引用
        //调试时用
        mySensorManager = SensorManagerSimulator.getSystemService(this, SENSOR_SERVICE);
        mySensorManager.connectSimulator();                           //连接 Simulator 服务器
    }
    private SensorListener mySensorListener = new SensorListener(){
     @Override
     public void onAccuracyChanged(int sensor, int accuracy) {}       //重写 onAccuracyChanged()方法
     @Override
     public void onSensorChanged(int sensor, float[] values) {        //重写 onSensorChanged()方法
          if(sensor == SensorManager.SENSOR_ACCELEROMETER){//只检查加速度的变化
                myTextView1.setText("x 方向上的加速度为："+values[0]);//将提取的 x 数据显示到 TextView
                myTextView2.setText("y 方向上的加速度为："+values[1]);//将提取的 y 数据显示到 TextView
                myTextView3.setText("z 方向上的加速度为："+values[2]);//将提取的 z 数据显示到 TextView
          }
     }
    };
    @Override
    protected void onResume() {                                       //重写 onResume()方法
     mySensorManager.registerListener(                                //注册监听
                mySensorListener,                                     //监听 SensorListener 对象
```

```
                SensorManager.SENSOR_ACCELEROMETER,            //设置传感器的类型为加速度
                SensorManager.SENSOR_DELAY_UI                  //用传感器事件传递频度
                );
     super.onResume();
    }
    @Override
    protected void onPause() {                                 //重写 onPause()方法
     mySensorManager.unregisterListener(mySensorListener);     //取消注册监听器
     super.onPause();
    }
}
```

（3）为了调试本实例代码，需要为该程序添加网络权限，因为 SensorSimulator 安装在 Android 模拟器中的客户端，需要和桌面端的服务器进行通信。

## 12.1.6 实战演练——实现仿微信“摇一摇”效果

本实例的功能是实现仿微信的“摇一摇”效果。

| 实　例 | 功　能 | 源码路径 |
|---|---|---|
| 实例 12-2 | 在设备中实现仿微信“摇一摇”效果 | 光盘:\daima\12\ShakeEX |

本实例的布局文件是 shake.xml，在界面上方设置了一个图片，在下方设置了按钮控件和文本控件，具体实现代码如下：

```
<LinearLayout xmlns:android="http://schemas.android.com/apk/res/android"
    android:layout_width="fill_parent"
    android:layout_height="fill_parent"
    android:orientation="vertical" >

    <RelativeLayout
        android:layout_width="fill_parent"
        android:layout_height="fill_parent"
        android:layout_centerInParent="true" >

        <ImageView
            android:id="@+id/shakeBg"
            android:layout_width="wrap_content"
            android:layout_height="wrap_content"
            android:layout_centerInParent="true"
            android:src="@drawable/shake_all" />

        <LinearLayout
          android:layout_width="fill_parent"
          android:layout_height="wrap_content"
          android:layout_centerInParent="true"
          android:orientation="vertical" >

            <RelativeLayout
              android:id="@+id/shakeImgUp"
              android:layout_width="fill_parent"
```

```
            android:layout_height="190dp"
            android:background="#111111">
              <ImageView
                android:layout_width="wrap_content"
                android:layout_height="wrap_content"
                android:layout_alignParentBottom="true"
                android:layout_centerHorizontal="true"
                android:src="@drawable/shake_up"
                />
          </RelativeLayout>
          <RelativeLayout
            android:id="@+id/shakeImgDown"
            android:layout_width="fill_parent"
            android:layout_height="190dp"
            android:background="#111111">
              <ImageView
                android:layout_width="wrap_content"
                android:layout_height="wrap_content"
                android:layout_centerHorizontal="true"
                android:src="@drawable/shake_down"
                />
          </RelativeLayout>
      </LinearLayout>
</RelativeLayout>

<RelativeLayout
        android:id="@+id/shake_title_bar"
        android:layout_width="fill_parent"
        android:layout_height="45dp"
        android:background="@drawable/title_bar"
        android:gravity="center_vertical"   >
            <Button
            android:layout_width="70dp"
            android:layout_height="wrap_content"
            android:layout_centerVertical="true"
            android:text="返回"
            android:textSize="14sp"
            android:textColor="#fff"
            android:onClick="shake_activity_back"
            android:background="@drawable/title_btn_back"/>
            <TextView
            android:layout_width="wrap_content"
            android:layout_height="wrap_content"
            android:text="摇一摇"
            android:layout_centerInParent="true"
            android:textSize="20sp"
            android:textColor="#ffffff" />
            <ImageButton
            android:layout_width="67dp"
            android:layout_height="wrap_content"
            android:layout_alignParentRight="true"
```

```
                    android:layout_centerVertical="true"
                    android:layout_marginRight="5dp"
                    android:src="@drawable/mm_title_btn_menu"
                        android:background="@drawable/title_btn_right"
                        android:onClick="linshi"
                        />
        </RelativeLayout>


    <SlidingDrawer
        android:id="@+id/slidingDrawer1"
        android:layout_width="match_parent"
        android:layout_height="match_parent"
        android:content="@+id/content"
        android:handle="@+id/handle" >
        <Button
            android:id="@+id/handle"
            android:layout_width="wrap_content"
            android:layout_height="wrap_content"

            android:background="@drawable/shake_report_dragger_up" />
        <LinearLayout
            android:id="@+id/content"
            android:layout_width="match_parent"
            android:layout_height="match_parent"
            android:background="#f9f9f9" >
            <ImageView
                android:layout_width="match_parent"
                android:layout_height="wrap_content"
                android:scaleType="fitXY"
                android:src="@drawable/shake_line_up" />
        </LinearLayout>
    </SlidingDrawer>
</LinearLayout>
```

程序文件 shakeActivity.java 实现了主 Activity，核心功能是监听设备的摇动方向，定义了摇一摇动画，并设置摇动过程中的振动效果。文件 shakeActivity.java 的主要实现代码如下：

```
public class shakeActivity extends Activity{

    ShakeListener mShakeListener = null;
    Vibrator mVibrator;
    private RelativeLayout mImgUp;
    private RelativeLayout mImgDn;
    private RelativeLayout mTitle;

    private SlidingDrawer mDrawer;
    private Button mDrawerBtn;

    @Override
    public void onCreate(Bundle savedInstanceState) {
        super.onCreate(savedInstanceState);
```

```
        setContentView(R.layout.shake);
        //drawerSet ();//设置 drawer 监听切换按钮的方向

        mVibrator = (Vibrator)getApplication().getSystemService(VIBRATOR_SERVICE);

        mImgUp = (RelativeLayout) findViewById(R.id.shakeImgUp);
        mImgDn = (RelativeLayout) findViewById(R.id.shakeImgDown);
        mTitle = (RelativeLayout) findViewById(R.id.shake_title_bar);

        mDrawer = (SlidingDrawer) findViewById(R.id.slidingDrawer1);
        mDrawerBtn = (Button) findViewById(R.id.handle);
        mDrawer.setOnDrawerOpenListener(new OnDrawerOpenListener()
        {    public void onDrawerOpened()
             {
    mDrawerBtn.setBackgroundDrawable(getResources().getDrawable(R.drawable.shake_down));
                  TranslateAnimation titleup = new TranslateAnimation(Animation. RELATIVE_TO_SELF,
0f,Animation.RELATIVE_TO_SELF,0f,Animation.RELATIVE_TO_SELF,0f,Animation.RELATIVE_TO_SELF,-1.0f);
                  titleup.setDuration(200);
                  titleup.setFillAfter(true);
                  mTitle.startAnimation(titleup);
             }
        });
         /*设定 SlidingDrawer 被关闭的事件处理*/
        mDrawer.setOnDrawerCloseListener(new OnDrawerCloseListener()
        {    public void onDrawerClosed()
             {
    mDrawerBtn.setBackgroundDrawable(getResources().getDrawable(R.drawable.shake_report_dragger_up));
                  TranslateAnimation titledn = new TranslateAnimation(Animation.RELATIVE_TO_SELF,0f,
Animation.RELATIVE_TO_SELF,0f,Animation.RELATIVE_TO_SELF,-1.0f,Animation.RELATIVE_TO_SELF,0f);
                  titledn.setDuration(200);
                  titledn.setFillAfter(false);
                  mTitle.startAnimation(titledn);
             }
        });

        mShakeListener = new ShakeListener(this);
       mShakeListener.setOnShakeListener(new OnShakeListener() {
             public void onShake() {
                  //Toast.makeText(getApplicationContext(), "抱歉，暂时没有找到在同一时刻摇一摇的人。\n
再试一次吧！ ", Toast.LENGTH_SHORT).show();
                  startAnim();   //开始摇一摇手掌动画
                  mShakeListener.stop();
                  startVibrato(); //开始振动
                  new Handler().postDelayed(new Runnable(){
                       @Override
                       public void run(){
                            //Toast.makeText(getApplicationContext(), "抱歉，暂时没有找到\n 在同一时刻摇
一摇的人。\n 再试一次吧！ ", 500).setGravity(Gravity.CENTER,0,0).show();
                            Toast mtoast;
```

```
                                    mtoast = Toast.makeText(getApplicationContext(),
                                            "抱歉，暂时没有找到\n 在同一时刻摇一摇的人。\n 再试一次吧！", 10);
                                    //mtoast.setGravity(Gravity.CENTER, 0, 0);
                                    mtoast.show();
                                    mVibrator.cancel();
                                    mShakeListener.start();
                            }
                    }, 2000);
            }
        });
    }
    public void startAnim () {    //定义摇一摇动画
        AnimationSet animup = new AnimationSet(true);
        TranslateAnimation mytranslateanimup0 = new TranslateAnimation(Animation.RELATIVE_TO_
SELF,0f,Animation.RELATIVE_TO_SELF,0f,Animation.RELATIVE_TO_SELF,0f,Animation.RELATIVE_TO_SE
LF,-0.5f);
        mytranslateanimup0.setDuration(1000);
        TranslateAnimation mytranslateanimup1 = new TranslateAnimation(Animation.RELATIVE_TO_SELF,
0f,Animation.RELATIVE_TO_SELF,0f,Animation.RELATIVE_TO_SELF,0f,Animation.RELATIVE_TO_SELF,+0.5f);
        mytranslateanimup1.setDuration(1000);
        mytranslateanimup1.setStartOffset(1000);
        animup.addAnimation(mytranslateanimup0);
        animup.addAnimation(mytranslateanimup1);
        mImgUp.startAnimation(animup);

        AnimationSet animdn = new AnimationSet(true);
        TranslateAnimation mytranslateanimdn0 = new TranslateAnimation(Animation.RELATIVE_TO_SELF,
0f,Animation.RELATIVE_TO_SELF,0f,Animation.RELATIVE_TO_SELF,0f,Animation.RELATIVE_TO_SELF,+0.5f);
        mytranslateanimdn0.setDuration(1000);
        TranslateAnimation mytranslateanimdn1 = new TranslateAnimation(Animation.RELATIVE_TO_SELF,
0f,Animation.RELATIVE_TO_SELF,0f,Animation.RELATIVE_TO_SELF,0f,Animation.RELATIVE_TO_SELF,-0.5f);
        mytranslateanimdn1.setDuration(1000);
        mytranslateanimdn1.setStartOffset(1000);
        animdn.addAnimation(mytranslateanimdn0);
        animdn.addAnimation(mytranslateanimdn1);
        mImgDn.startAnimation(animdn);
    }
    public void startVibrato(){                                   //定义振动
        mVibrator.vibrate( new long[]{500,200,500,200}, -1); //第一个{ }里面是节奏数组， 第二个参数是重复
次数，-1 为不重复，非-1 为从 pattern 的指定下标开始重复
    }

    public void shake_activity_back(View v) {                     //标题栏返回按钮
        this.finish();
    }
    public void linshi(View v) {                                  //标题栏
        startAnim();
    }
    @Override
    protected void onDestroy() {
        super.onDestroy();
```

```
        if (mShakeListener != null) {
            mShakeListener.stop();
        }
    }
}
```

上述代码实现了与设备的交互控制——振动，振动是一种提醒或替换铃声的事件，通过上述代码可以了解到如何触发手机振动事件，虽然振动是手机默认的模式，但通过程序的辅助，可以做更精密的控制，诸如振动周期、持续时间等。在设置振动（Vibration）事件中，必须要知道命令其振动的时间长短、振动事件的周期等。而在 Android 中设置的数值都是以毫秒（1000 毫秒=1 秒）来做计算的，所以在做设置时需要注意，如果设置的时间值太小，会感觉不出来。要想让设备振动，需创建 Vibrator 对象，通过调用 vibrate 方法来达到振动的目的，在 Vibrator 的构造器中有 4 个参数，前 3 个的值是设置振动的大小，在这里可以把数值改成一大一小，这样就可以明显地感觉出振动的差异，而最后一个值是设置振动的时间。

程序文件 ShakeListener.java 的功能是为设备实现了一个设备摇晃的监听器，这一功能是通过传感器实现的。文件 ShakeListener.java 的主要实现代码如下：

```
public class ShakeListener implements SensorEventListener {
    //速度阈值，当摇晃速度达到这个值后产生作用
    private static final int SPEED_SHRESHOLD = 3000;
    //两次检测的时间间隔
    private static final int UPTATE_INTERVAL_TIME = 70;
    //传感器管理器
    private SensorManager sensorManager;
    //传感器
    private Sensor sensor;
    //重力感应监听器
    private OnShakeListener onShakeListener;
    //上下文
    private Context mContext;
    //手机上一个位置时重力感应坐标
    private float lastX;
    private float lastY;
    private float lastZ;
    //上次检测时间
    private long lastUpdateTime;

    //构造器
    public ShakeListener(Context c) {
        //获得监听对象
        mContext = c;
        start();
    }

    //开始
    public void start() {
        //获得传感器管理器
        sensorManager = (SensorManager) mContext
                .getSystemService(Context.SENSOR_SERVICE);
        if (sensorManager != null) {
```

```
            //获得重力传感器
            sensor = sensorManager.getDefaultSensor(Sensor.TYPE_ACCELEROMETER);
        }
        //注册
        if (sensor != null) {
            sensorManager.registerListener(this, sensor,
                    SensorManager.SENSOR_DELAY_GAME);
        }

    }

    //停止检测
    public void stop() {
        sensorManager.unregisterListener(this);
    }

    //设置重力感应监听器
    public void setOnShakeListener(OnShakeListener listener) {
        onShakeListener = listener;
    }

    //重力感应器感应获得变化数据
    public void onSensorChanged(SensorEvent event) {
        //现在检测时间
        long currentUpdateTime = System.currentTimeMillis();
        //两次检测的时间间隔
        long timeInterval = currentUpdateTime - lastUpdateTime;
        //判断是否达到了检测时间间隔
        if (timeInterval < UPTATE_INTERVAL_TIME)
            return;
        //现在的时间变成 last 时间
        lastUpdateTime = currentUpdateTime;

        //获得 x,y,z 坐标
        float x = event.values[0];
        float y = event.values[1];
        float z = event.values[2];

        //获得 x,y,z 的变化值
        float deltaX = x - lastX;
        float deltaY = y - lastY;
        float deltaZ = z - lastZ;

        //将现在的坐标变成 last 坐标
        lastX = x;
        lastY = y;
        lastZ = z;

        double speed = Math.sqrt(deltaX * deltaX + deltaY * deltaY + deltaZ
                * deltaZ)
                / timeInterval * 10000;
```

```
            Log.v("thelog", "===========log===================");
            //达到速度阈值，发出提示
            if (speed >= SPEED_SHRESHOLD) {
                onShakeListener.onShake();
            }
        }

        public void onAccuracyChanged(Sensor sensor, int accuracy) {

        }

        //摇晃监听接口
        public interface OnShakeListener {
            public void onShake();
        }

}
```

在真机中执行后将会实现和微信“摇一摇”类似的效果，如图 12-1 所示。

图 12-1　“摇一摇”效果

# 12.2　方向传感器详解

**知识点讲解：光盘:视频\知识点\第 12 章\方向传感器详解.avi**

在 Android 设备中，经常需要检测设备的方向，例如设备的朝向和移动方向。在 Android 系统中，通常使用重力传感器、加速度传感器、磁场传感器和旋转矢量传感器来检测设备的方向。本节将详细讲解在 Android 设备中使用方向传感器检测设备方向的基本知识，为读者步入本书后面知识的学习打下基础。

## 12.2.1　方向传感器基础

在现实世界中，方向传感器通过对力敏感的传感器，感受手机等设备在变换姿势时的重心变化，使手机等设备光标变化位置，从而实现选择的功能。方向传感器运用了欧拉角的知识，欧拉角的基本思想是将角位移分解为绕 3 个互相垂直轴的 3 个旋转组成的序列。其实，任意 3 个轴和任意顺序都可以，但最有意义的是使用笛卡儿坐标系并按一定的顺序所组成的旋转序列。

在学习欧拉角知识之前先介绍几种不同概念的坐标系，以便于读者理解欧拉角知识。

（1）世界坐标系

世界坐标系是一个特殊的坐标系，建立了描述其他坐标系所需要的参考框架。能够用世界坐标系描述其他坐标系的位置，而不能用更大的、外部的坐标系来描述世界坐标系。例如，“向西”“向东”等词汇就是世界坐标系中的描述词汇。

（2）物体坐标系

物体坐标系是和特定物体相关联的坐标系，每个物体都有它们独立的坐标系。当物体移动或改变方向时，和该物体相关联的坐标系将随之移动或改变方向。例如，“向左”“向右”等词汇就是物体坐标系中的描述词汇。

（3）摄像机坐标系

摄像机坐标系是和观察者密切相关的坐标系。在摄像机坐标系中，摄像机在原点，x 轴向右，z 轴向前（朝向屏幕内或摄像机方向），y 轴向上（不是世界的上方而是摄像机本身的上方）。

（4）惯性坐标系

惯性坐标系是为了简化世界坐标系到物体坐标系的转换而引入的一种新的坐标系。惯性坐标系的原点和物体坐标系的原点重合，但惯性坐标系的轴平行于世界坐标系的轴。

在欧拉角中，表示一个物体的方位用“Yaw-Pitch-Roll”约定。在这个系统中，一个方位被定义为一个 Yaw 角、一个 Pitch 角和一个 Ron 角。欧拉角的基本思想是让物体开始于“标准”方位，目的是使物体坐标轴和惯性坐标轴对齐。在标准方位上，让物体做 Yaw、Pitch 和 Roll 旋转，最后物体到达我们想要描述的方位。

（5）Yaw 轴

Yaw 轴是 3 个方向轴中唯一不变的轴，其方向总是竖直向上，和世界坐标系中的 Z 轴是等同的，也就是重力加速度 g 的反方向。

（6）Pitch 轴

Pitch 轴方向依赖于手机沿 Yaw 轴的转动情况，即当手机沿 Yaw 转过一定的角度后，Pitch 轴也相应围绕 Yaw 轴转动相同的角度。Pitch 轴的位置依赖于手机沿 Yaw 轴转过的角度，好比 Yaw 轴和 Pitch 轴是两根焊死在一起成 90。

## 12.2.2　Android 中的方向传感器

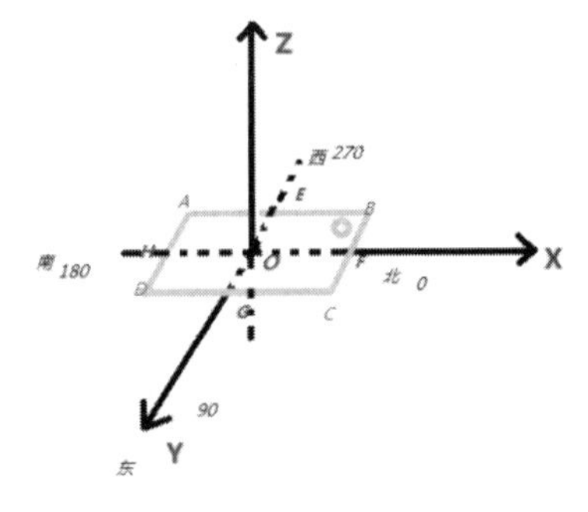


图 12-2　参数 X、Y、Z

在 Android 系统中，方向传感器的类型是 TYPE_ORIENTATION，用于测量设备围绕 3 个物理轴（x,y,z）的旋转角度，在新版本中已经使用 SensorManager.getOrientation()替代。Android 系统中的方向传感器在生活中的典型应用例子是指南针，接下来先来简单介绍一下传感器中 3 个参数 X、Y、Z 的含义，如图 12-2 所示。

如图 12-2 所示，绿色部分表示一个手机，带有小圈那一头是手机头部，各个部分的具体说明如下所示。

☑　传感器中的 X：如图 12-2 所示，规定 X 正半轴为北，手机

头部指向 OF 方向，此时 X 的值为 0。如果手机头部指向 OG 方向，此时 X 值为 90，指向 OH 方向，X 值为 180，指向 OE，X 值为 270。

☑ 传感器中的 Y：现在将手机沿着 BC 轴慢慢向上抬起，即手机头部不动，尾部慢慢向上翘起来，直到 AD 跑到 BC 右边并落在 XOY 平面上，Y 的值将从 0～180 之间变动，如果手机沿着 AD 轴慢慢向上抬起，即手机尾部不动，直到 BC 跑到 AD 左边并且落在 XOY 平面上，Y 的值将从 0～-180 之间变动，这就是方向传感器中 Y 的含义。

☑ 传感器中的 Z：现在将手机沿着 AB 轴慢慢向上抬起，即手机左边框不动，右边框慢慢向上翘起来，直到 CD 跑到 AB 右边并落在 XOY 平面上，Z 的值将从 0～180 之间变动，如果手机沿着 CD 轴慢慢向上抬起，即手机右边框不动，直到 AB 跑到 CD 左边并且落在 XOY 平面上，Z 的值将从 0～-180 之间变动，这就是方向传感器中 Z 的含义。

### 12.2.3　实战演练——测试当前设备的 3 个方向值

本实例将演示在 Android 中使用方向传感器的方法。

| 实　例 | 功　能 | 源 码 路 径 |
|---|---|---|
| 实例 12-3 | 使用方向传感器测试当前设备的 3 个方向值 | 光盘:\daima\12\fangxiangEX |

（1）实现布局文件

编写布局文件 main.xml，主要代码如下：

```
<TextView
    android:id="@+id/title"
    android:gravity="center_horizontal"
    android:textSize="20px"
    android:layout_width="fill_parent"
    android:layout_height="wrap_content"
    android:text="@string/title"/><!-- 添加一个 TextView 控件 -->
<TextView
    android:id="@+id/myTextView1"
    android:textSize="18px"
    android:layout_width="fill_parent"
    android:layout_height="wrap_content"
    android:text="@string/myTextView1"/><!-- 添加一个 TextView 控件 -->
<TextView
    android:id="@+id/myTextView2"
    android:textSize="18px"
    android:layout_width="fill_parent"
    android:layout_height="wrap_content"
    android:text="@string/myTextView2"/><!-- 添加一个 TextView 控件 -->
<TextView
    android:id="@+id/myTextView3"
    android:textSize="18px"
    android:layout_width="fill_parent"
    android:layout_height="wrap_content"
    android:text="@string/myTextView3"/><!-- 添加一个 TextView 控件 -->
```

（2）实现主程序文件

编写主程序文件 zitaiLI.java，此文件的具体实现流程如下所示。

☑ 声明 3 个分别用来显示 Yaw、Pitch、Roll 的 TextView 的引用。

☑ 声明 SensorManager 引用，而因调试原因用 SensorManagerSimulator 代替。

☑ 重写 Activity 的 onCreate()方法，在该方法中设置当前的用户界面，然后得到各个控件的引用，并初始化 SensorManager 或 SensorManagerSimulator。

☑ 初始化监听器类的对象，在重写的 onSensorChanged()方法中只对姿态 SENSOR_ORIENTATION 变化进行处理，将 3 个姿态值显示到 TextView 中。

☑ 重写 Activity 的 onResume()方法，在该方法中为 SensorManager 注册监听，此处传入的传感器类型为 SENSOR_ORIENTATION，表示只读取姿态数据。

文件 zitaiCH.java 的主要代码如下：

```
public class zitaiLI extends Activity {
    TextView myTextView1;
    TextView myTextView2;
    TextView myTextView3;
    //声明 SensorManagerSimulator 对象，调试时用
    SensorManagerSimulator mySensorManager;
    @Override
    public void onCreate(Bundle savedInstanceState) {                      //重写 onCreate()方法
        super.onCreate(savedInstanceState);
        setContentView(R.layout.main);                                     //设置用户界面
        myTextView1 = (TextView) findViewById(R.id.myTextView1);           //myTextView1 的引用
        myTextView2 = (TextView) findViewById(R.id.myTextView2);           //myTextView2 的引用
        myTextView3 = (TextView) findViewById(R.id.myTextView3);           //myTextView3 的引用
        //mySensorManager =
        //(SensorManager)getSystemService(SENSOR_SERVICE);                 //获得 SensorManager
        //调试时用
        mySensorManager = SensorManagerSimulator.getSystemService(this, SENSOR_SERVICE);
        mySensorManager.connectSimulator();                                //与 Simulator 服务器连接
    }
    private SensorListener mySensorListener = new SensorListener(){
        @Override
        public void onAccuracyChanged(int sensor, int accuracy) {}         //重写 onAccuracyChanged()方法
        @Override
        public void onSensorChanged(int sensor, float[] values) {          //重写 onSensorChanged()方法
            if(sensor == SensorManager.SENSOR_ORIENTATION){                //检查姿态变化
                myTextView1.setText("Yaw 为："+values[0]);                 //TextView 数据显示
                myTextView2.setText("Pitch 为："+values[1]);               //TextView 数据显示
                myTextView3.setText("Roll 为："+values[2]);                //TextView 数据显示
            }
        }
    };
    @Override
    protected void onResume() {                                            //重写 onResume()方法
        mySensorManager.registerListener(                                  //注册监听
                mySensorListener,                                          //监听器 SensorListener 对象
                SensorManager.SENSOR_ORIENTATION,                          //姿态传感器的类型
```

```
                SensorManager.SENSOR_DELAY_UI                //传感器事件传递频度
                );
        super.onResume();
    }
    @Override
    protected void onPause() {                               //重写 onPause()方法
        mySensorManager.unregisterListener(mySensorListener);    //取消注册监听器
        super.onPause();
    }
}
```

因为此实例比较简单，是根据 SensorSimulator 中附带的开源代码改编的，所以在此不再进行详细介绍，读者只需阅读本书附带光盘中的源码即可。

### 12.2.4　实战演练——开发一个指南针程序

在接下来的实例中，将演示在 Android 中使用方向传感器开发指南针应用程序的方法。在本实例中首先准备一张指南针素材图片，该图片上方向指南针指向北方。接下来开发一个检测方向的传感器，传感器程序可以检测到手机顶部绕 Z 转过多少度。在实例中需添加一张图片，并让图片总是反转方向传感器返回的第一个角度值。

| 实　例 | 功　能 | 源 码 路 径 |
|---|---|---|
| 实例 12-4 | 在设备中开发指南针程序 | 光盘:\daima\12\zhiEX |

（1）实现布局文件

编写布局文件 main.xml，功能是插入准备好的素材图片，主要实现代码如下：

```
<LinearLayout xmlns:android="http://schemas.android.com/apk/res/android"
    android:orientation="vertical"
    android:layout_width="fill_parent"
    android:layout_height="fill_parent"
    android:background="#fff"
    >
<ImageView
    android:id="@+id/znzImage"
    android:layout_width="fill_parent"
    android:layout_height="fill_parent"
    android:scaleType="fitCenter"
    android:src="@drawable/compass" />
</LinearLayout>
```

（2）实现程序文件

编写程序文件 Zhinanzheng.java，使用传感器获取设备的旋转角度值，并根据这个值返回指南针的角度。文件 Zhinanzheng.java 的具体实现代码如下：

```
public class Zhinanzheng extends Activity implements SensorEventListener{

    ImageView image;                    //指南针图片
    float currentDegree = 0f;           //指南针图片转过的角度

    SensorManager mSensorManager;  //管理器
```

```
    /** Called when the activity is first created. */
    @Override
    public void onCreate(Bundle savedInstanceState) {
        super.onCreate(savedInstanceState);
        setContentView(R.layout.main);

        image = (ImageView)findViewById(R.id.znzImage);
        mSensorManager = (SensorManager)getSystemService(SENSOR_SERVICE); //获取管理服务
    }

    @Override
    protected void onResume(){
        super.onResume();
        //注册监听器
        mSensorManager.registerListener(this,
                mSensorManager.getDefaultSensor(Sensor.TYPE_ORIENTATION), SensorManager. SENSOR_
DELAY_GAME);
    }

    //取消注册
    @Override
    protected void onPause(){
        mSensorManager.unregisterListener(this);
        super.onPause();

    }

    @Override
    protected void onStop(){
        mSensorManager.unregisterListener(this);
        super.onStop();

    }

    //传感器值改变
    @Override
    public void onAccuracyChanged(Sensor sensor, int accuracy) {

    }
    //精度改变
    @Override
    public void onSensorChanged(SensorEvent event) {
        //获取触发 event 的传感器类型
        int sensorType = event.sensor.getType();

        switch(sensorType){
        case Sensor.TYPE_ORIENTATION:
            float degree = event.values[0]; //获取 z 转过的角度
            //定义旋转动画对象 ra
            RotateAnimation ra = new RotateAnimation(currentDegree,-degree,Animation.RELATIVE_TO_
SELF,0.5f
```

```
                    ,Animation.RELATIVE_TO_SELF,0.5f);
            ra.setDuration(100);//动画持续时间
            image.startAnimation(ra);
            currentDegree = -degree;
            break;

        }
    }
}
```

执行后的效果如图 12-3 所示。

图 12-3　执行效果

# 12.3　陀螺仪传感器详解

**知识点讲解：光盘:视频\知识点\第 12 章\陀螺仪传感器详解.avi**

陀螺仪传感器是一个基于自由空间移动和手势的定位和控制系统。例如我们可以在假想的平面上移动鼠标，屏幕上的光标就会随之跟着移动，并且可以绕着链接画圈和点击按键。又如当我们正在演讲或离开桌子时，这些操作都能够很方便地实现。陀螺仪传感器已经被广泛运用于手机、平板等移动便携设备上，在将来设备中也会陆续使用陀螺仪传感器。本节将详细讲解在 Android 设备中使用陀螺仪传感器的基本知识，为读者步入本书后面知识的学习打下基础。

## 12.3.1　陀螺仪传感器基础

陀螺仪的原理是，当一个旋转物体的旋转轴所指的方向在不受外力影响时是不会改变的。根据这个道理，可以用陀螺仪来保持方向。然后用多种方法读取轴所指示的方向，并自动将数据信号传给控制系统。在现实生活中，骑自行车便是利用了这个原理。轮子转得越快越不容易倒，因为车轴有一股保持水平的力量。现代陀螺仪是一个能够精确地确定运动物体的方位的仪器，也是在现代航空、航海、航天和国防工业中广泛使用的一种惯性导航仪器。传统的惯性陀螺仪主要部分有机械式的陀螺仪，而机械式的陀螺仪对工艺结构的要求很高。20 世纪 70 年代提出了现代光纤陀螺仪的基本设想，到 20 世纪 80 年代以后，光纤陀螺仪就得到了非常迅速的发展，激光谐振陀螺仪也有了很大的发展。光纤陀螺仪具有结构紧凑、灵敏度高和工作可靠等特点。光纤陀螺仪在很多的领域已经完全取代了机械式的传统的陀螺仪，成为现代导航仪器中的关键部件。

根据框架的数目和支承的形式以及附件的性质进行划分，陀螺仪传感器的主要类型如下所示。

（1）二自由度陀螺仪：只有一个框架，使转子自转轴具有一个转动自由度。根据二自由度陀螺仪中所使用的反作用力矩的性质，可以把这种陀螺仪分为如下所示的 3 种类型。

☑ 积分陀螺仪（它使用的反作用力矩是阻尼力矩）。
☑ 速率陀螺仪（它使用的反作用力矩是弹性力矩）。
☑ 无约束陀螺仪（它仅有惯性反作用力矩）。

另外，除了机、电框架式陀螺仪外还出现了某些新型陀螺仪，例如静电式自由转子陀螺仪、挠性陀螺仪和激光陀螺仪等。

（2）三自由度陀螺仪：具有内、外两个框架，使转子自转轴具有两个转动自由度。在没有任何力矩装置时，它就是一个自由陀螺仪。

在当前技术水平条件下，陀螺仪传感器主要被用于如下两个领域。

（1）国防工业

陀螺仪传感器原本是运用到直升机模型上的，而它已经被广泛运用于手机这类移动便携设备上，不仅如此，现代陀螺仪是一种能够精确地确定运动物体的方位的仪器，所以陀螺仪传感器是现代航空、航海、航天和国防工业应用中必不可少的控制装置。陀螺仪传感器是法国的物理学家莱昂·傅科在研究地球自转时命名的，到如今一直是航空和航海上航行姿态及速率等最方便实用的参考仪表。

（2）开门报警器

陀螺仪传感器可以测量开门的角度，当门被打开一个角度后回发出报警声，或者结合 GPRS 模块发送短信以提醒门被打开了。另外，陀螺仪传感器集成了加速度传感器的功能，当门被打开的瞬间，将产生一定的加速度值，陀螺仪传感器将会测量到这个加速度值，达到预设的门槛值后，将发出报警声，或者结合 GPRS 模块发送短信以提醒门被打开了。报警器内还可以集成雷达感应测量功能，主要有人进入房间内移动时就会被雷达测量到。双重保险提醒防盗，可靠性高，误报率低，非常适合重要场合的防盗报警。

## 12.3.2　Android 中的陀螺仪传感器

在 Android 系统中，陀螺仪传感器的类型是 TYPE_GYROSCOPE，单位是 rad/s，能够测量设备 X、Y、Z 3 个轴的角加速度数据。Android 中的陀螺仪传感器又名为 Gyro-sensor 角速度器，它利用内部振动机械结构侦测物体转动所产生的角速度，进而计算出物体移动的角度。侦测水平改变的状态，但无法计算移动的激烈程度。下面将详细讲解 Android 中陀螺仪传感器的基本知识。

（1）陀螺仪传感器和加速度传感器的对比

在 Android 的传感器系统中，陀螺仪传感器和加速度传感器非常类似，两者的区别如下所示。

☑ 加速度传感器：用于测量加速度，借助一个 3 轴加速度计可以测得一个固定平台相对地球表面的运动方向，但是一旦平台运动起来，情况就会变得复杂很多。如果平台做自由落体，加速度计测得的加速度值为零。如果平台朝某个方向做加速度运动，各个轴向加速度值会含有重力产生的加速度值，使得无法获得真正的加速度值。例如，安装在 60° 横滚角飞机上的 3 轴加速度计会测得 2G 的垂直加速度值，而事实上飞机相对地区表面是 60° 的倾角。因此，

单独使用加速度计无法使飞机保持一个固定的航向。

☑ 陀螺仪传感器：用于测量机体围绕某个轴向的旋转角速率值。当使用陀螺仪测量飞机机体轴向的旋转角速率时，如果飞机在旋转，测得的值为非零值，飞机不旋转时，测量的值为零。因此，在 60° 横滚角的飞机上的陀螺仪测得的横滚角速率值为零，同样在飞机做水平直线飞行时的角速率值为零。可以通过角速率值的时间积分来估计当前的横滚角度，前提是没有误差的累积。陀螺仪测量的值会随时间漂移，经过几分钟甚至几秒钟定会累积出额外的误差来，而最终会导致对飞机当前相对水平面横滚角度完全错误的认知。因此，单独使用陀螺仪也无法保持飞机的特定航向。

综上所述，加速度传感器在较长时间的测量值（确定飞机航向）是正确的，而在较短时间内由于信号噪声的存在而有误差。陀螺仪传感器在较短时间内会比较准确，而在较长时间内则会因为漂移的存在而产生误差。因此，需要两者（相互调整）来确保航向的正确。

（2）物联网设备中的陀螺仪传感器

在物联网设备中，三自由度陀螺仪是一个可以识别设备，能够相对于地面绕 X、Y、Z 轴转动角度的感应器。无论是可穿戴设备，还是智能手机、平板电脑，通过使用陀螺仪传感器可以实现很多好玩的应用，例如指南针。

在实际开发过程中，可以用一个磁场感应器（Magnetic Sensor）来实现陀螺仪。磁场感应器是用来测量磁场感应强度的。一个 3 轴的磁 Sensor IC 可以得到当前环境下 X、Y 和 Z 方向上的磁场感应强度，对于 Android 中间层来说就是读取该感应器测量到的这 3 个值。当需要时，上报给上层应用程序。磁感应强度的单位是 T（特斯拉）或者是 Gs（高斯），1T 等于 10000Gs。

在了解陀螺仪之前，需要先了解 Android 系统定义坐标系的方法，在文件/hardware/libhardware/include/hardware/sensors.h 中进行了定义。

在上述文件 sensors.h 中，有一个如图 12-4 所示的效果图。

图 12-4 中表示设备的正上方是 Y 轴方向，右边是 X 轴方向，垂直设备屏幕平面向上的是 Z 轴方向，这个很重要。因为应用程序就是根据这样的定义来写的，所以我们报给应用的数据要和这个定义符合。另外，还需要清楚磁 Sensor 芯片贴在板上的坐标系。我们从芯片读出数据后要把芯片的坐标系转换为设备的实际坐标系。除非芯片贴在板上刚好和设备的 X、Y、Z 轴方向一致。

陀螺仪的实现是根据磁场感应强度的 3 个值计算出另外 3 个值。当需要时可以计算出这 3 个值上报给应用程序，这样就实现了陀螺仪的功能。下面将详细讲解这 3 个值的具体含义和计算方法。

（1）Azimuth 方位角：即绕 Z 轴转动的角度，0 度=正北。假设 Y 轴指向地磁正北方，直升机正前方的方向如图 12-5 所示。

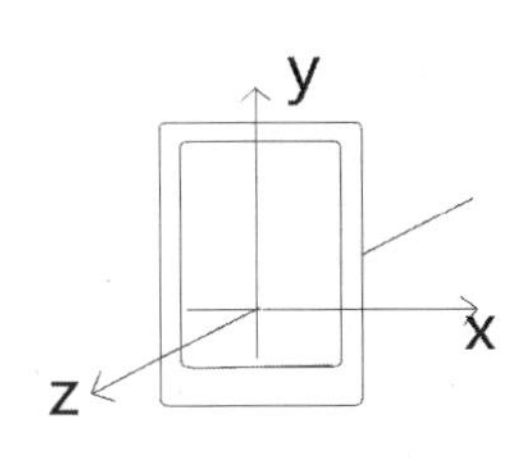


图 12-4　Android 系统定义的坐标系

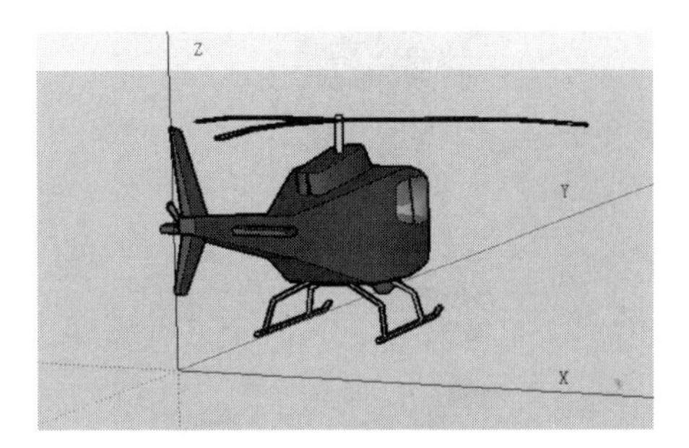


图 12-5　Azimuth 方位角

90 度=正东时如图 12-6 所示。180 度=正南时如图 12-7 所示。

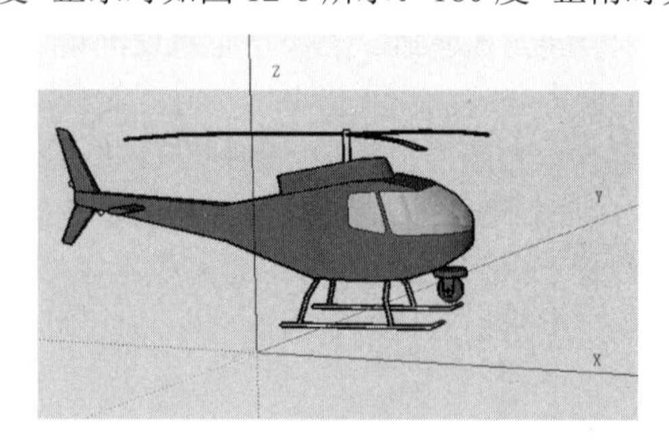


图 12-6　90 度=正东

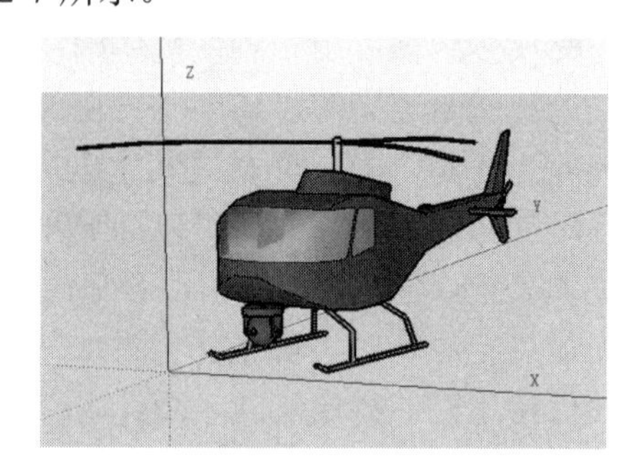


图 12-7　180 度=正南

270 度=正西时如图 12-8 所示。

在这种情况下，通过计算 X 和 Y 方向的磁感应强度的反正切的方式就可以得到方位角。要想实现指南针，只需要用这个值即可（不考虑设备非水平的情况）。

（2）Pitch 仰俯：绕 X 轴转动的角度（-180<=pitch<=180），如果设备水平放置，前方向下俯就是正值，如图 12-9 所示。

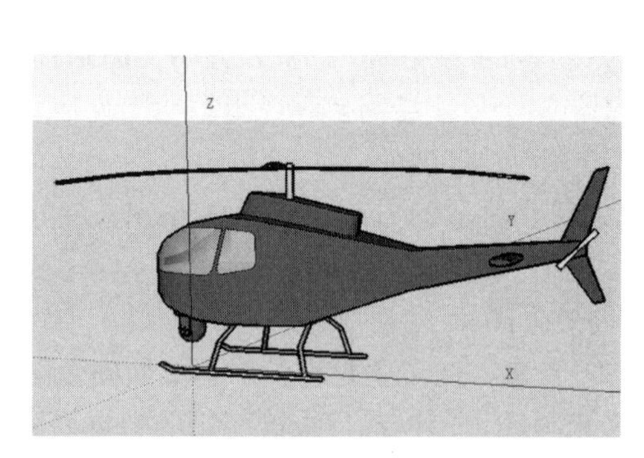


图 12-8　270 度=正西

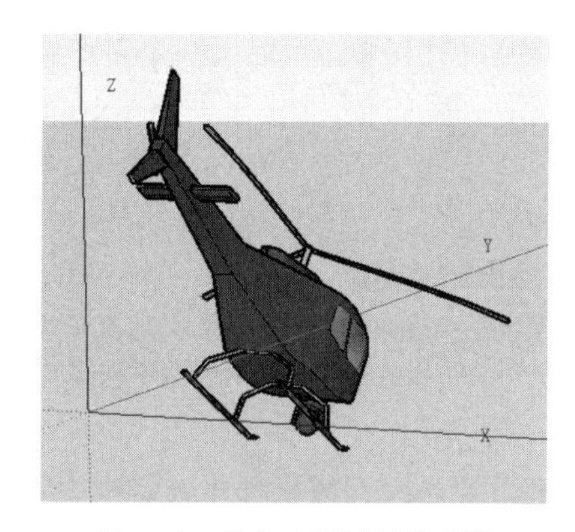


图 12-9　前方向下俯就是正值

前方向上仰就是负值，如图 12-10 所示。

在这种情况下，计算磁 Sensor 的 Y 和 Z 反正切就可以得到此角度值，如图 12-11 所示。

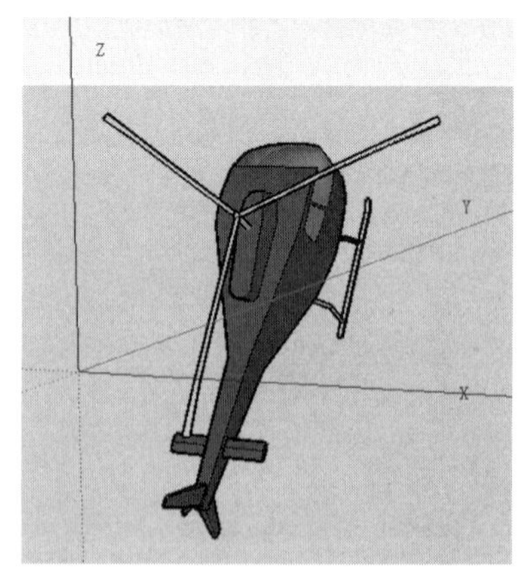


图 12-10　前方向上仰就是负值

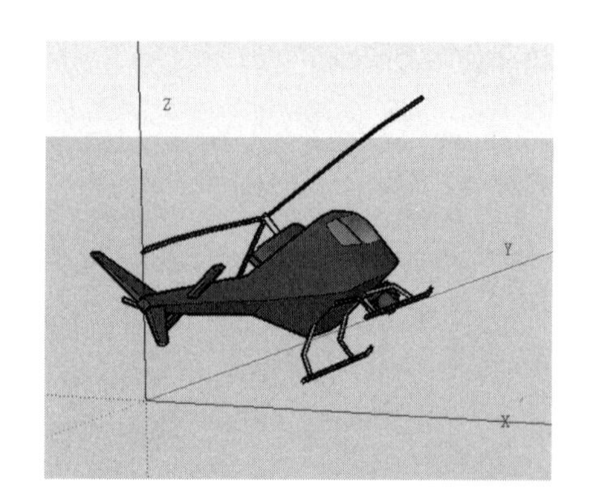


图 12-11　计算磁 Sensor 的 Y 和 Z 反正切

# 12.4　实战演练——联合使用加速度传感器和陀螺仪传感器

**知识点讲解：光盘:视频\知识点\第 12 章\联合使用加速度传感器和陀螺仪传感器.avi**

本节将详细讲解在 Android 设备中联合使用加速度传感器和陀螺仪传感器的方法。本节的陀螺仪程序演示了如何使用陀螺仪传感器测量 Android 设备的旋转角度的方法。本实例是一个基于 Android 的 Java 应用程序，实例算法的 JIST 可以应用于几乎任何硬件/语言组合来确定线性加速度。本陀螺仪实例包含 Android 的类演示了如何使用 Sensor.TYPE_GYROSCOPE 和 Sensor.TYPE_GYROSCOPE_UNCALIBRATED 两种传感器的方法。这包括集成的传感器的输出随着时间的推移来描述角度变化的设备、初始化旋转矩阵、新的旋转矩阵的 Concatination 与初始旋转矩阵和用于级联的旋转矩阵提供角度。

**注意：** 陀螺仪类型 Sensor.TYPE_GYROSCOPE 会因设备的移动而提供误差补偿。本陀螺仪项目还提供了一个陀螺仪传感器功能，提供了更强大和可靠的设备的 Rotation 测量值。本传感器实例融合使用了加速度传感器、磁传感器和陀螺仪传感器，联合使用上述传感器来实现设备的快速旋转或外部振动旋转的测量工作。

| 实　例 | 功　能 | 源码路径 |
| --- | --- | --- |
| 实例 12-5 | 联合使用加速度传感器和陀螺仪传感器 | 光盘:\daima\12\GyroscopeEX |

## 12.4.1　系统介绍界面

系统介绍界面的布局文件是 introduction_layout.xml，功能是在屏幕中显示系统介绍信息，介绍了这个系统的主要功能。文件 introduction_layout.xml 的主要实现代码如下：

```
<?xml version="1.0" encoding="utf-8"?>
<RelativeLayout xmlns:android="http://schemas.android.com/apk/res/android"
    android:layout_width="match_parent"
    android:layout_height="match_parent"
    android:background="#88666666" >

    <ImageView
        android:id="@+id/introduction_image"
        android:layout_width="match_parent"
        android:layout_height="match_parent"
        android:layout_centerInParent="true"
        android:src="@drawable/gyroscope_explorer_introduction_0" />

    <Button
        android:id="@+id/button_confirm"
        android:layout_width="250dp"
        android:layout_height="wrap_content"
        android:layout_alignParentBottom="true"
        android:layout_centerInParent="true"
        android:layout_marginBottom="10dp"
        android:layout_marginTop="10dp"
```

```
        android:background="@drawable/confirm_button_background"
        android:text="@string/confirm_label"
        android:textAppearance="?android:attr/textAppearanceLarge"
        android:textColor="@color/white"
        android:textStyle="bold" />
</RelativeLayout>
```

文件 IntroductionActivity.java 的功能是调用系统介绍界面的布局控件，监听用户单击屏幕底部的按钮，根据单击按钮在屏幕中显示不同的介绍信息界面。文件 IntroductionActivity.java 的具体实现代码如下：

```
public class IntroductionActivity extends Activity implements OnClickListener
{
    private final static int INTRODUCTION_1 = 1;
    private final static int INTRODUCTION_2 = 2;
    private final static int INTRODUCTION_3 = 3;
    private final static int INTRODUCTION_FINISHED = 4;
    private boolean hasRun;
    private int introductionCount = 0;
    @Override
    public void onCreate(Bundle savedInstanceState)
    {
        super.onCreate(savedInstanceState);
        readHintsPrefs();
        if (!hasRun)
        {
            setContentView(R.layout.introduction_layout);

            Button button = (Button) this.findViewById(R.id.button_confirm);
            button.setOnClickListener(this);
        }
        else
        {
            startGyroscopeActivity();
        }
    }

    private void startGyroscopeActivity()
    {
        hasRun = true;

        writeHintsPrefs();

        Intent intent = new Intent();
        intent.setClass(this, GyroscopeActivity.class);
        startActivity(intent);
    }

    private void readHintsPrefs()
    {
        SharedPreferences prefs = getSharedPreferences(PreferenceNames.HINTS,
                Activity.MODE_PRIVATE);

        hasRun = prefs.getBoolean(HintsPreferences.FIRST_RUN_HINTS_ENABLED,
```

```
                    false);
    }

    private void writeHintsPrefs()
    {
        SharedPreferences.Editor editor = getSharedPreferences(
                    PreferenceNames.HINTS, Activity.MODE_PRIVATE).edit();
        editor.putBoolean(HintsPreferences.FIRST_RUN_HINTS_ENABLED, hasRun);
        editor.commit();
    }

    @Override
    public void onClick(View arg0)
    {
        introductionCount++;
        ImageView introView = (ImageView) this
                    .findViewById(R.id.introduction_image);

        switch (introductionCount)
        {
        case INTRODUCTION_1:
            introView
                        .setImageResource(R.drawable.gyroscope_explorer_introduction_1);
            break;
        case INTRODUCTION_2:
            introView
                        .setImageResource(R.drawable.gyroscope_explorer_introduction_2);
            break;
        case INTRODUCTION_3:
            introView
                        .setImageResource(R.drawable.gyroscope_explorer_introduction_3);
            break;
        case INTRODUCTION_FINISHED:
            startGyroscopeActivity();
            break;
        }
    }
}
```

系统介绍界面执行后的效果如图 12-12 所示，单击 I got this 按钮会显示不同的信息介绍界面。

图 12-12　系统主界面

## 12.4.2 系统主界面

离开系统介绍界面会进入到系统主界面，在系统主界面中将以图形化的方式显示当前设备的陀螺仪的 X、Y 和 Z 值。其中在屏幕下方显示的是 TYPE_GYROSCOPE_UNCALIBRATED 的值，这表示未经过校准的陀螺仪传感器类型，没有经过陀螺偏移补偿。然而，这样的陀螺漂移偏差值分别退还给结果值，可以把它们用于自定义校准，在屏幕上方显示经过校验的陀螺仪值。系统主界面的布局文件是 activity_gyroscope.xml，具体实现代码如下：

```
<LinearLayout xmlns:android="http://schemas.android.com/apk/res/android"
    xmlns:tools="http://schemas.android.com/tools"
    android:layout_width="match_parent"
    android:layout_height="match_parent"
    android:orientation="vertical"
    tools:context=".GyroscopeActivity" >

    <RelativeLayout
        android:id="@+id/layout_calibrated_header"
        android:layout_width="match_parent"
        android:layout_height="wrap_content"
        android:layout_marginTop="5dp" >

        <RelativeLayout
            android:layout_width="wrap_content"
            android:layout_height="wrap_content"
            android:layout_centerHorizontal="true" >

            <TextView
                android:id="@+id/label_calibrated_filter_name"
                android:layout_width="wrap_content"
                android:layout_height="wrap_content"
                android:fontFamily="sans-serif-condensed"
                android:text="@string/sensor_calibrated_name"
                android:textAppearance="?android:attr/textAppearanceSmall"
                android:textColor="@color/dark_orange" />

            <TextView
                android:id="@+id/label_calibrated_filter_description"
                android:layout_width="wrap_content"
                android:layout_height="wrap_content"
                android:layout_toRightOf="@+id/label_calibrated_filter_name"
                android:fontFamily="sans-serif-condensed"
                android:text="@string/sensor_label"
                android:textAppearance="?android:attr/textAppearanceSmall" />
        </RelativeLayout>
    </RelativeLayout>

    <RelativeLayout
        android:id="@+id/layout_calibrated_status"
        android:layout_width="match_parent"
```

```
    android:layout_height="wrap_content"
    android:layout_marginTop="5dp" >

    <RelativeLayout
        android:layout_width="wrap_content"
        android:layout_height="wrap_content"
        android:layout_centerHorizontal="true" >

        <TextView
            android:id="@+id/label_calibrated_status"
            android:layout_width="wrap_content"
            android:layout_height="wrap_content"
            android:fontFamily="sans-serif-condensed"
            android:text="@string/sensor_unavailable"
            android:textAppearance="?android:attr/textAppearanceMedium"
            android:textColor="@color/dark_red" />
    </RelativeLayout>
</RelativeLayout>

<RelativeLayout
    android:id="@+id/layout_calibrated_statistics"
    android:layout_width="match_parent"
    android:layout_height="wrap_content" >

    <RelativeLayout
        android:layout_width="wrap_content"
        android:layout_height="wrap_content"
        android:layout_centerHorizontal="true"
        android:layout_centerVertical="true" >

        <TableLayout
            android:id="@+id/table_calibrated_statistics_left"
            android:layout_width="wrap_content"
            android:layout_height="wrap_content" >

            <TableRow
                android:id="@+id/table_calibrated_statistics_left_row_0"
                android:layout_width="fill_parent"
                android:layout_height="wrap_content"
                android:padding="2dip" >

                <RelativeLayout
                    android:layout_width="wrap_content"
                    android:layout_height="wrap_content"
                    android:layout_marginRight="20dp"
                    android:layout_weight="1" >

                    <TextView
                        android:id="@+id/label_x_axis_calibrated"
                        android:layout_width="wrap_content"
                        android:layout_height="wrap_content"
```

```
                android:layout_alignParentLeft="true"
                android:fontFamily="sans-serif-condensed"
                android:text="@string/label_x_axis"
                android:textAppearance="?android:attr/textAppearanceMedium" />

            <TextView
                android:id="@+id/value_x_axis_calibrated"
                android:layout_width="wrap_content"
                android:layout_height="wrap_content"
                android:layout_toRightOf="@+id/label_x_axis_calibrated"
                android:fontFamily="sans-serif-condensed"
                android:text="@string/value_default"
                android:textAppearance="?android:attr/textAppearanceMedium" />
        </RelativeLayout>

        <RelativeLayout
            android:layout_width="fill_parent"
            android:layout_height="wrap_content"
            android:layout_weight="1" >

            <TextView
                android:id="@+id/label_y_axis_calibrated"
                android:layout_width="wrap_content"
                android:layout_height="wrap_content"
                android:layout_alignParentLeft="true"
                android:fontFamily="sans-serif-condensed"
                android:text="@string/label_y_axis"
                android:textAppearance="?android:attr/textAppearanceMedium" />

            <TextView
                android:id="@+id/value_y_axis_calibrated"
                android:layout_width="wrap_content"
                android:layout_height="wrap_content"
                android:layout_toRightOf="@+id/label_y_axis_calibrated"
                android:fontFamily="sans-serif-condensed"
                android:text="@string/value_default"
                android:textAppearance="?android:attr/textAppearanceMedium" />
        </RelativeLayout>

        <RelativeLayout
            android:layout_width="fill_parent"
            android:layout_height="wrap_content"
            android:layout_marginLeft="20dp"
            android:layout_weight="1" >

            <TextView
                android:id="@+id/label_z_axis_calibrated"
                android:layout_width="wrap_content"
                android:layout_height="wrap_content"
                android:layout_alignParentLeft="true"
                android:fontFamily="sans-serif-condensed"
```

```
                        android:text="@string/label_z_axis"
                        android:textAppearance="?android:attr/textAppearanceMedium" />

                    <TextView
                        android:id="@+id/value_z_axis_calibrated"
                        android:layout_width="wrap_content"
                        android:layout_height="wrap_content"
                        android:layout_toRightOf="@+id/label_z_axis_calibrated"
                        android:fontFamily="sans-serif-condensed"
                        android:text="@string/value_default"
                        android:textAppearance="?android:attr/textAppearanceMedium" />
                </RelativeLayout>
            </TableRow>
        </TableLayout>
    </RelativeLayout>
</RelativeLayout>

<View
    android:layout_width="fill_parent"
    android:layout_height="1dp"
    android:background="@android:color/darker_gray" />

<RelativeLayout
    android:layout_width="match_parent"
    android:layout_height="wrap_content"
    android:layout_marginTop="5dp" >

    <com.kircherelectronics.gyroscopeexplorer.activity.gauge.flat.GaugeBearingFlat
        android:id="@+id/gauge_bearing_calibrated"
        android:layout_width="140dp"
        android:layout_height="140dp"
        android:layout_alignParentLeft="true"
        android:layout_marginLeft="20dp" />

    <com.kircherelectronics.gyroscopeexplorer.activity.gauge.flat.GaugeRotationFlat
        android:id="@+id/gauge_tilt_calibrated"
        android:layout_width="140dp"
        android:layout_height="140dp"
        android:layout_alignParentRight="true"
        android:layout_marginRight="20dp" />
</RelativeLayout>

<RelativeLayout
    android:id="@+id/layout_raw_header"
    android:layout_width="match_parent"
    android:layout_height="wrap_content"
    android:layout_marginTop="20dp" >

    <RelativeLayout
        android:layout_width="wrap_content"
        android:layout_height="wrap_content"
```

```
            android:layout_centerHorizontal="true" >

            <TextView
                android:id="@+id/label_raw_filter_name"
                android:layout_width="wrap_content"
                android:layout_height="wrap_content"
                android:fontFamily="sans-serif-condensed"
                android:text="@string/sensor_uncalibrated_name"
                android:textAppearance="?android:attr/textAppearanceSmall"
                android:textColor="@color/dark_orange" />

            <TextView
                android:id="@+id/label_raw_filter_description"
                android:layout_width="wrap_content"
                android:layout_height="wrap_content"
                android:layout_toRightOf="@+id/label_raw_filter_name"
                android:fontFamily="sans-serif-condensed"
                android:text="@string/sensor_label"
                android:textAppearance="?android:attr/textAppearanceSmall" />
        </RelativeLayout>
    </RelativeLayout>

        <RelativeLayout
        android:id="@+id/layout_uncalibrated_status"
        android:layout_width="match_parent"
        android:layout_height="wrap_content"
        android:layout_marginTop="5dp" >

        <RelativeLayout
            android:layout_width="wrap_content"
            android:layout_height="wrap_content"
            android:layout_centerHorizontal="true" >

            <TextView
                android:id="@+id/label_uncalibrated_status"
                android:layout_width="wrap_content"
                android:layout_height="wrap_content"
                android:fontFamily="sans-serif-condensed"
                android:text="@string/sensor_unavailable"
                android:textAppearance="?android:attr/textAppearanceMedium"
                android:textColor="@color/dark_red" />
        </RelativeLayout>
    </RelativeLayout>

    <RelativeLayout
        android:id="@+id/layout_raw_statistics"
        android:layout_width="match_parent"
        android:layout_height="wrap_content" >

        <RelativeLayout
            android:layout_width="wrap_content"
```

```
android:layout_height="wrap_content"
android:layout_centerHorizontal="true"
android:layout_centerVertical="true" >

<TableLayout
    android:id="@+id/table_raw_statistics_left"
    android:layout_width="wrap_content"
    android:layout_height="wrap_content" >

    <TableRow
        android:id="@+id/table_raw_statistics_left_row_0"
        android:layout_width="fill_parent"
        android:layout_height="wrap_content"
        android:padding="2dip" >

        <RelativeLayout
            android:layout_width="wrap_content"
            android:layout_height="wrap_content"
            android:layout_marginRight="20dp"
            android:layout_weight="1" >

            <TextView
                android:id="@+id/label_x_axis_raw"
                android:layout_width="wrap_content"
                android:layout_height="wrap_content"
                android:layout_alignParentLeft="true"
                android:fontFamily="sans-serif-condensed"
                android:text="@string/label_x_axis"
                android:textAppearance="?android:attr/textAppearanceMedium" />

            <TextView
                android:id="@+id/value_x_axis_raw"
                android:layout_width="wrap_content"
                android:layout_height="wrap_content"
                android:layout_toRightOf="@+id/label_x_axis_raw"
                android:fontFamily="sans-serif-condensed"
                android:text="@string/value_default"
                android:textAppearance="?android:attr/textAppearanceMedium" />
        </RelativeLayout>

        <RelativeLayout
            android:layout_width="fill_parent"
            android:layout_height="wrap_content"
            android:layout_weight="1" >

            <TextView
                android:id="@+id/label_y_axis_raw"
                android:layout_width="wrap_content"
                android:layout_height="wrap_content"
                android:layout_alignParentLeft="true"
                android:fontFamily="sans-serif-condensed"
```

```
                        android:text="@string/label_y_axis"
                        android:textAppearance="?android:attr/textAppearanceMedium" />

                    <TextView
                        android:id="@+id/value_y_axis_raw"
                        android:layout_width="wrap_content"
                        android:layout_height="wrap_content"
                        android:layout_toRightOf="@+id/label_y_axis_raw"
                        android:fontFamily="sans-serif-condensed"
                        android:text="@string/value_default"
                        android:textAppearance="?android:attr/textAppearanceMedium" />
                </RelativeLayout>

                <RelativeLayout
                    android:layout_width="fill_parent"
                    android:layout_height="wrap_content"
                    android:layout_marginLeft="20dp"
                    android:layout_weight="1" >

                    <TextView
                        android:id="@+id/label_z_axis_raw"
                        android:layout_width="wrap_content"
                        android:layout_height="wrap_content"
                        android:layout_alignParentLeft="true"
                        android:fontFamily="sans-serif-condensed"
                        android:text="@string/label_z_axis"
                        android:textAppearance="?android:attr/textAppearanceMedium" />

                    <TextView
                        android:id="@+id/value_z_axis_raw"
                        android:layout_width="wrap_content"
                        android:layout_height="wrap_content"
                        android:layout_toRightOf="@+id/label_z_axis_raw"
                        android:fontFamily="sans-serif-condensed"
                        android:text="@string/value_default"
                        android:textAppearance="?android:attr/textAppearanceMedium" />
                </RelativeLayout>
            </TableRow>
        </TableLayout>
    </RelativeLayout>
</RelativeLayout>

<View
    android:layout_width="fill_parent"
    android:layout_height="1dp"
    android:background="@android:color/darker_gray" />
<RelativeLayout
    android:layout_width="match_parent"
    android:layout_height="wrap_content"
    android:layout_marginTop="5dp" >
    <com.kircherelectronics.gyroscopeexplorer.activity.gauge.flat.GaugeBearingFlat
```

```
            android:id="@+id/gauge_bearing_raw"
            android:layout_width="140dp"
            android:layout_height="140dp"
            android:layout_alignParentLeft="true"
            android:layout_marginLeft="20dp" />
        <com.kircherelectronics.gyroscopeexplorer.activity.gauge.flat.GaugeRotationFlat
            android:id="@+id/gauge_tilt_raw"
            android:layout_width="140dp"
            android:layout_height="140dp"
            android:layout_alignParentRight="true"
            android:layout_marginRight="20dp" />
    </RelativeLayout>
    <RelativeLayout
        android:layout_width="match_parent"
        android:layout_height="fill_parent" >
        <RelativeLayout
            android:layout_width="match_parent"
            android:layout_height="wrap_content"
            android:layout_above="@+id/color_bar" >
            <RelativeLayout
                android:layout_width="wrap_content"
                android:layout_height="wrap_content"
                android:layout_centerHorizontal="true" >
                <ImageView
                    android:id="@+id/image_developer_icon"
                    android:layout_width="wrap_content"
                    android:layout_height="wrap_content"
                    android:layout_marginRight="4dp"
                    android:src="@drawable/ke_icon" />
                <TextView
                    android:id="@+id/label_developer_description"
                    android:layout_width="wrap_content"
                    android:layout_height="wrap_content"
                    android:layout_toRightOf="@+id/image_developer_icon"
                    android:fontFamily="sans-serif-condensed"
                    android:text="@string/developer_url"
                    android:textAppearance="?android:attr/textAppearanceSmall" />
            </RelativeLayout>
        </RelativeLayout>
        <RelativeLayout
            android:id="@+id/color_bar"
            android:layout_width="match_parent"
            android:layout_height="wrap_content"
            android:layout_alignParentBottom="true"
            android:layout_marginBottom="5dp"
            android:layout_marginTop="5dp" >
            <RelativeLayout
                android:layout_width="wrap_content"
                android:layout_height="wrap_content"
                android:layout_centerHorizontal="true" >
                <ImageView
```

```
                    android:id="@+id/image_color_bar"
                    android:layout_width="wrap_content"
                    android:layout_height="wrap_content"
                    android:src="@drawable/color_bar" />
            </RelativeLayout>
        </RelativeLayout>
    </RelativeLayout>
</LinearLayout>
```

文件 GyroscopeActivity.java 调用了系统主界面布局文件中的控件，显示当前设备的陀螺仪传感器的值。在显示传感器数据值时，是通过表盘样式的图形化界面显示的。并且还能够监听用户在系统主界面中的单击动作，并根据单击动作来设置执行对应的事件处理程序。文件 GyroscopeActivity.java 的具体实现代码如下：

```
public class GyroscopeActivity extends Activity implements SensorEventListener,
        FusedGyroscopeSensorListener
{
    public static final float EPSILON = 0.000000001f;
    private static final String tag = GyroscopeActivity.class.getSimpleName();
    private static final float NS2S = 1.0f / 1000000000.0f;
    private static final int MEAN_FILTER_WINDOW = 10;
    private static final int MIN_SAMPLE_COUNT = 30;
    private boolean hasInitialOrientation = false;
    private boolean stateInitializedCalibrated = false;
    private boolean stateInitializedRaw = false;
    private boolean useFusedEstimation = false;
    private boolean useRadianUnits = false;
    //表盘形式视图
    private GaugeBearingFlat gaugeBearingCalibrated;
    private GaugeBearingFlat gaugeBearingRaw;
    private GaugeRotationFlat gaugeTiltCalibrated;
    private GaugeRotationFlat gaugeTiltRaw;
    private DecimalFormat df;
    //校准值
    private float[] currentRotationMatrixCalibrated;
    private float[] deltaRotationMatrixCalibrated;
    private float[] deltaRotationVectorCalibrated;
    private float[] gyroscopeOrientationCalibrated;
    //未校准值
    private float[] currentRotationMatrixRaw;
    private float[] deltaRotationMatrixRaw;
    private float[] deltaRotationVectorRaw;
    private float[] gyroscopeOrientationRaw;
    //加速度计和磁力计的基础旋转矩阵
    private float[] initialRotationMatrix;
    //加速度矢量
    private float[] acceleration;
    //磁场矢量
    private float[] magnetic;
    private FusedGyroscopeSensor fusedGyroscopeSensor;
    private int accelerationSampleCount = 0;
    private int magneticSampleCount = 0;
```

```
private long timestampOldCalibrated = 0;
private long timestampOldRaw = 0;
private MeanFilter accelerationFilter;
private MeanFilter magneticFilter;
//需要 SensorManager 来注册传感器事件
private SensorManager sensorManager;
private TextView xAxisRaw;
private TextView yAxisRaw;
private TextView zAxisRaw;
private TextView xAxisCalibrated;
private TextView yAxisCalibrated;
private TextView zAxisCalibrated;
@Override
protected void onCreate(Bundle savedInstanceState)
{
	super.onCreate(savedInstanceState);
	setContentView(R.layout.activity_gyroscope);
	initUI();
	initMaths();
	initSensors();
	initFilters();
};
@Override
public boolean onCreateOptionsMenu(Menu menu)
{
	getMenuInflater().inflate(R.menu.gyroscope, menu);
	return true;
}
/**
 * 监听用户选择的处理菜单的 id
 * */
@Override
public boolean onOptionsItemSelected(MenuItem item)
{
	switch (item.getItemId())
	{
	//重置所有信息
	case R.id.action_reset:
		reset();
		restart();
		return true;
		//复位所有信息
	case R.id.action_config:
		Intent intent = new Intent();
		intent.setClass(this, ConfigActivity.class);
		startActivity(intent);
		return true;
	default:
		return super.onOptionsItemSelected(item);
	}
}
```

```
public void onResume()
{
    super.onResume();
    readPrefs();
    restart();
}
public void onPause()
{
    super.onPause();
    reset();
}
@Override
public void onSensorChanged(SensorEvent event)
{
    if (event.sensor.getType() == Sensor.TYPE_ACCELEROMETER)
    {
        onAccelerationSensorChanged(event.values, event.timestamp);
    }
    if (event.sensor.getType() == Sensor.TYPE_MAGNETIC_FIELD)
    {
        onMagneticSensorChanged(event.values, event.timestamp);
    }
    if (event.sensor.getType() == Sensor.TYPE_GYROSCOPE)
    {
        onGyroscopeSensorChanged(event.values, event.timestamp);
    }
    if (event.sensor.getType() == Sensor.TYPE_GYROSCOPE_UNCALIBRATED)
    {
        onGyroscopeSensorUncalibratedChanged(event.values, event.timestamp);
    }
}
@Override
public void onAngularVelocitySensorChanged(float[] angularVelocity,
        long timeStamp)
{
    gaugeBearingCalibrated.updateBearing(angularVelocity[0]);
    gaugeTiltCalibrated.updateRotation(angularVelocity);
    TextView status = (TextView) this
            .findViewById(R.id.label_calibrated_status);
    status.setText(R.string.sensor_active);
    int color = getResources().getColor(R.color.light_green);
    status.setTextColor(color);
    if (useRadianUnits)
    {
        xAxisCalibrated.setText(df.format(angularVelocity[0]));
        yAxisCalibrated.setText(df.format(angularVelocity[1]));
        zAxisCalibrated.setText(df.format(angularVelocity[2]));
    }
    else
    {
        xAxisCalibrated.setText(df.format(Math
```

```
                    .toDegrees(angularVelocity[0])));
        yAxisCalibrated.setText(df.format(Math
                    .toDegrees(angularVelocity[1])));
        zAxisCalibrated.setText(df.format(Math
                    .toDegrees(angularVelocity[2])));
    }
}
public void onAccelerationSensorChanged(float[] acceleration, long timeStamp)
{
    //从设备传感器获取原始磁场值的本地副本
    System.arraycopy(acceleration, 0, this.acceleration, 0,
                acceleration.length);
    //使用均值滤波来平滑输入传感器
    this.acceleration = accelerationFilter.filterFloat(this.acceleration);
    //统计接收到的样本数
    accelerationSampleCount++;
    //唯一确定的初始方向的加速度传感器和磁传感器
    //有足够的时间由平均滤波器进行平滑处理
    //如果方向尚未确定，只需要一次
    if (accelerationSampleCount > MIN_SAMPLE_COUNT
                && magneticSampleCount > MIN_SAMPLE_COUNT
                && !hasInitialOrientation)
    {
        calculateOrientation();
    }
}
public void onGyroscopeSensorChanged(float[] gyroscope, long timestamp)
{
    //不启动，直到第一个加速度计/磁方位已被取得
    if (!hasInitialOrientation)
    {
        return;
    }
    //基于陀螺仪的旋转矩阵的初始化
    if (!stateInitializedCalibrated)
    {
        currentRotationMatrixCalibrated = matrixMultiplication(
                    currentRotationMatrixCalibrated, initialRotationMatrix);
        stateInitializedCalibrated = true;
        TextView status = (TextView) this
                    .findViewById(R.id.label_calibrated_status);
        status.setText(R.string.sensor_active);
        int color = getResources().getColor(R.color.light_green);
        status.setTextColor(color);
    }
    //这个时间步长的三角形旋转值，将和目前的从旋转陀螺采样的数据计算后相乘
    if (timestampOldCalibrated != 0 && stateInitializedCalibrated)
    {
        final float dT = (timestamp - timestampOldCalibrated) * NS2S;
        float axisX = gyroscope[0];
        float axisY = gyroscope[1];
```

```
            float axisZ = gyroscope[2];
            //计算出这个实例的角速度
            float omegaMagnitude = (float) Math.sqrt(axisX * axisX + axisY
                    * axisY + axisZ * axisZ);
            if (omegaMagnitude > EPSILON)
            {
                axisX /= omegaMagnitude;
                axisY /= omegaMagnitude;
                axisZ /= omegaMagnitude;
            }
            //根据时间步长获得一个增量旋转，即计算围绕此轴旋转的角速度值
            //将获取的旋转矩阵值转换成一个四元数
            float thetaOverTwo = omegaMagnitude * dT / 2.0f;
            float sinThetaOverTwo = (float) Math.sin(thetaOverTwo);
            float cosThetaOverTwo = (float) Math.cos(thetaOverTwo);
            deltaRotationVectorCalibrated[0] = sinThetaOverTwo * axisX;
            deltaRotationVectorCalibrated[1] = sinThetaOverTwo * axisY;
            deltaRotationVectorCalibrated[2] = sinThetaOverTwo * axisZ;
            deltaRotationVectorCalibrated[3] = cosThetaOverTwo;
            SensorManager.getRotationMatrixFromVector(
                    deltaRotationMatrixCalibrated,
                    deltaRotationVectorCalibrated);
            currentRotationMatrixCalibrated = matrixMultiplication(
                    currentRotationMatrixCalibrated,
                    deltaRotationMatrixCalibrated);
            SensorManager.getOrientation(currentRotationMatrixCalibrated,
                    gyroscopeOrientationCalibrated);
        }
        timestampOldCalibrated = timestamp;
        gaugeBearingCalibrated.updateBearing(gyroscopeOrientationCalibrated[0]);
        gaugeTiltCalibrated.updateRotation(gyroscopeOrientationCalibrated);
        if (useRadianUnits)
        {
            xAxisCalibrated.setText(df
                    .format(gyroscopeOrientationCalibrated[0]));
            yAxisCalibrated.setText(df
                    .format(gyroscopeOrientationCalibrated[1]));
            zAxisCalibrated.setText(df
                    .format(gyroscopeOrientationCalibrated[2]));
        }
        else
        {
            xAxisCalibrated.setText(df.format(Math
                    .toDegrees(gyroscopeOrientationCalibrated[0])));
            yAxisCalibrated.setText(df.format(Math
                    .toDegrees(gyroscopeOrientationCalibrated[1])));
            zAxisCalibrated.setText(df.format(Math
                    .toDegrees(gyroscopeOrientationCalibrated[2])));
        }
    }
    public void onGyroscopeSensorUncalibratedChanged(float[] gyroscope,
```

```
        long timestamp)
{
    //不启动，直到第一个加速度计/磁方位已被获取
    if (!hasInitialOrientation)
    {
        return;
    }
    if (!stateInitializedRaw)
    {
        currentRotationMatrixRaw = matrixMultiplication(
                currentRotationMatrixRaw, initialRotationMatrix);
        stateInitializedRaw = true;
        TextView status = (TextView) this
                .findViewById(R.id.label_uncalibrated_status);
        status.setText(R.string.sensor_active);
        int color = getResources().getColor(R.color.light_green);
        status.setTextColor(color);
    }
    if (timestampOldRaw != 0 && stateInitializedRaw)
    {
        final float dT = (timestamp - timestampOldRaw) * NS2S;
        float axisX = gyroscope[0];
        float axisY = gyroscope[1];
        float axisZ = gyroscope[2];
        float omegaMagnitude = (float) Math.sqrt(axisX * axisX + axisY
                * axisY + axisZ * axisZ);
        //规范化旋转向量，如果它足够大，以获得轴
        if (omegaMagnitude > EPSILON)
        {
            axisX /= omegaMagnitude;
            axisY /= omegaMagnitude;
            axisZ /= omegaMagnitude;
        }
        float thetaOverTwo = omegaMagnitude * dT / 2.0f;
        float sinThetaOverTwo = (float) Math.sin(thetaOverTwo);
        float cosThetaOverTwo = (float) Math.cos(thetaOverTwo);
        deltaRotationVectorRaw[0] = sinThetaOverTwo * axisX;
        deltaRotationVectorRaw[1] = sinThetaOverTwo * axisY;
        deltaRotationVectorRaw[2] = sinThetaOverTwo * axisZ;
        deltaRotationVectorRaw[3] = cosThetaOverTwo;
        SensorManager.getRotationMatrixFromVector(deltaRotationMatrixRaw,
                deltaRotationVectorRaw);
        currentRotationMatrixRaw = matrixMultiplication(
                currentRotationMatrixRaw, deltaRotationMatrixRaw);
        SensorManager.getOrientation(currentRotationMatrixRaw,
                gyroscopeOrientationRaw);
    }
    timestampOldRaw = timestamp;
    gaugeBearingRaw.updateBearing(gyroscopeOrientationRaw[0]);
    gaugeTiltRaw.updateRotation(gyroscopeOrientationRaw);
    if (useRadianUnits)
```

```
        {
            xAxisRaw.setText(df.format(gyroscopeOrientationRaw[0]));
            yAxisRaw.setText(df.format(gyroscopeOrientationRaw[1]));
            zAxisRaw.setText(df.format(gyroscopeOrientationRaw[2]));
        }
        else
        {
            xAxisRaw.setText(df.format(Math
                    .toDegrees(gyroscopeOrientationRaw[0])));
            yAxisRaw.setText(df.format(Math
                    .toDegrees(gyroscopeOrientationRaw[1])));
            zAxisRaw.setText(df.format(Math
                    .toDegrees(gyroscopeOrientationRaw[2])));
        }
    }
    public void onMagneticSensorChanged(float[] magnetic, long timeStamp)
    {
        //从设备传感器获取原始磁场值的本地副本
        System.arraycopy(magnetic, 0, this.magnetic, 0, magnetic.length);
        //使用均值滤波来平滑输入传感器
        this.magnetic = magneticFilter.filterFloat(this.magnetic);
        magneticSampleCount++;
    }
    /**
     *计算当前方向角度的加速度计和磁力计的输出值
     */
    private void calculateOrientation()
    {
        hasInitialOrientation = SensorManager.getRotationMatrix(
                initialRotationMatrix, null, acceleration, magnetic);
        if (hasInitialOrientation)
        {
            sensorManager.unregisterListener(this,
                    sensorManager.getDefaultSensor(Sensor.TYPE_ACCELEROMETER));
            sensorManager.unregisterListener(this,
                    sensorManager.getDefaultSensor(Sensor.TYPE_MAGNETIC_FIELD));
        }
    }
    /**
     *初始化平均过滤器
     */
    private void initFilters()
    {
        accelerationFilter = new MeanFilter();
        accelerationFilter.setWindowSize(MEAN_FILTER_WINDOW);
        magneticFilter = new MeanFilter();
        magneticFilter.setWindowSize(MEAN_FILTER_WINDOW);
    }
    /**
     *初始化所需的数字数据结构
     */
```

```
private void initMaths()
{
        acceleration = new float[3];
        magnetic = new float[3];
        initialRotationMatrix = new float[9];
        deltaRotationVectorCalibrated = new float[4];
        deltaRotationMatrixCalibrated = new float[9];
        currentRotationMatrixCalibrated = new float[9];
        gyroscopeOrientationCalibrated = new float[3];
        //初始化当前旋转矩阵为单位矩阵
        currentRotationMatrixCalibrated[0] = 1.0f;
        currentRotationMatrixCalibrated[4] = 1.0f;
        currentRotationMatrixCalibrated[8] = 1.0f;
        deltaRotationVectorRaw = new float[4];
        deltaRotationMatrixRaw = new float[9];
        currentRotationMatrixRaw = new float[9];
        gyroscopeOrientationRaw = new float[3];
        //初始化当前旋转矩阵为单位矩阵
        currentRotationMatrixRaw[0] = 1.0f;
        currentRotationMatrixRaw[4] = 1.0f;
        currentRotationMatrixRaw[8] = 1.0f;
}
/**
 *初始化传感器
 */
private void initSensors()
{
        sensorManager = (SensorManager) this
                .getSystemService(Context.SENSOR_SERVICE);
        fusedGyroscopeSensor = new FusedGyroscopeSensor();
}
/**
 * 初始化 UI 界面
 */
private void initUI()
{
        //获取一个十进制格式的文本意见
        df = new DecimalFormat("#.##");
        //初始化原始（未校准）文本视图
        xAxisRaw = (TextView) this.findViewById(R.id.value_x_axis_raw);
        yAxisRaw = (TextView) this.findViewById(R.id.value_y_axis_raw);
        zAxisRaw = (TextView) this.findViewById(R.id.value_z_axis_raw);
        //初始化校准的文本视图
        xAxisCalibrated = (TextView) this
                .findViewById(R.id.value_x_axis_calibrated);
        yAxisCalibrated = (TextView) this
                .findViewById(R.id.value_y_axis_calibrated);
        zAxisCalibrated = (TextView) this
                .findViewById(R.id.value_z_axis_calibrated);
        //初始化原始（未校准）的文本视图
        gaugeBearingRaw = (GaugeBearingFlat) findViewById(R.id.gauge_bearing_raw);
```

```
        gaugeTiltRaw = (GaugeRotationFlat) findViewById(R.id.gauge_tilt_raw);
        //初始化校准的仪表视图
        gaugeBearingCalibrated = (GaugeBearingFlat) findViewById(R.id.gauge_bearing_calibrated);
        gaugeTiltCalibrated = (GaugeRotationFlat) findViewById(R.id.gauge_tilt_calibrated);
    }
    /**
     *矩阵相乘 a*b
     * @param a
     * @param b
     * @return a*b
     */
    private float[] matrixMultiplication(float[] a, float[] b)
    {
        float[] result = new float[9];
        result[0] = a[0] * b[0] + a[1] * b[3] + a[2] * b[6];
        result[1] = a[0] * b[1] + a[1] * b[4] + a[2] * b[7];
        result[2] = a[0] * b[2] + a[1] * b[5] + a[2] * b[8];
        result[3] = a[3] * b[0] + a[4] * b[3] + a[5] * b[6];
        result[4] = a[3] * b[1] + a[4] * b[4] + a[5] * b[7];
        result[5] = a[3] * b[2] + a[4] * b[5] + a[5] * b[8];
        result[6] = a[6] * b[0] + a[7] * b[3] + a[8] * b[6];
        result[7] = a[6] * b[1] + a[7] * b[4] + a[8] * b[7];
        result[8] = a[6] * b[2] + a[7] * b[5] + a[8] * b[8];
        return result;
    }
    /**
     * 重启传感器
     */
    @TargetApi(Build.VERSION_CODES.JELLY_BEAN_MR2)
    private void restart()
    {
        sensorManager.registerListener(this,
                sensorManager.getDefaultSensor(Sensor.TYPE_ACCELEROMETER),
                SensorManager.SENSOR_DELAY_FASTEST);
        sensorManager.registerListener(this,
                sensorManager.getDefaultSensor(Sensor.TYPE_MAGNETIC_FIELD),
                SensorManager.SENSOR_DELAY_FASTEST);
        if (!useFusedEstimation)
        {
            boolean enabled = sensorManager.registerListener(this,
                    sensorManager.getDefaultSensor(Sensor.TYPE_GYROSCOPE),
                    SensorManager.SENSOR_DELAY_FASTEST);
            if (!enabled)
            {
                showGyroscopeNotAvailableAlert();
            }
        }
        if (Utils.hasKitKat())
        {
            sensorManager.registerListener(this, sensorManager
                    .getDefaultSensor(Sensor.TYPE_GYROSCOPE_UNCALIBRATED),
```

```
                    SensorManager.SENSOR_DELAY_FASTEST);
        }
        if (useFusedEstimation)
        {
            boolean hasGravity = sensorManager.registerListener(
                    fusedGyroscopeSensor,
                    sensorManager.getDefaultSensor(Sensor.TYPE_GRAVITY),
                    SensorManager.SENSOR_DELAY_FASTEST);
            if (!hasGravity)
            {
                sensorManager.registerListener(fusedGyroscopeSensor,
                        sensorManager
                                .getDefaultSensor(Sensor.TYPE_ACCELEROMETER),
                        SensorManager.SENSOR_DELAY_FASTEST);
            }
            sensorManager.registerListener(fusedGyroscopeSensor,
                    sensorManager.getDefaultSensor(Sensor.TYPE_MAGNETIC_FIELD),
                    SensorManager.SENSOR_DELAY_FASTEST);
            boolean enabled = sensorManager.registerListener(
                    fusedGyroscopeSensor,
                    sensorManager.getDefaultSensor(Sensor.TYPE_GYROSCOPE),
                    SensorManager.SENSOR_DELAY_FASTEST);
            if (!enabled)
            {
                showGyroscopeNotAvailableAlert();
            }
            TextView label = (TextView) this
                    .findViewById(R.id.label_calibrated_filter_name);
            label.setText("Fused"
                    + getResources().getString(R.string.sensor_calibrated_name));
            fusedGyroscopeSensor.registerObserver(this);
        }
        else
        {
            TextView label = (TextView) this
                    .findViewById(R.id.label_calibrated_filter_name);
            label.setText(getResources().getString(
                    R.string.sensor_calibrated_name));
        }
    }
    /**
     *删除所有的传感器数据信息
     */
    @TargetApi(Build.VERSION_CODES.JELLY_BEAN_MR2)
    private void reset()
    {
        sensorManager.unregisterListener(this,
                sensorManager.getDefaultSensor(Sensor.TYPE_ACCELEROMETER));
        sensorManager.unregisterListener(this,
                sensorManager.getDefaultSensor(Sensor.TYPE_MAGNETIC_FIELD));
        if (!useFusedEstimation)
```

```
        {
            sensorManager.unregisterListener(this,
                    sensorManager.getDefaultSensor(Sensor.TYPE_GYROSCOPE));
        }
        if (Utils.hasKitKat())
        {
            sensorManager.unregisterListener(this, sensorManager
                    .getDefaultSensor(Sensor.TYPE_GYROSCOPE_UNCALIBRATED));
        }
        if (useFusedEstimation)
        {
            sensorManager.unregisterListener(fusedGyroscopeSensor,
                    sensorManager.getDefaultSensor(Sensor.TYPE_GRAVITY));
            sensorManager.unregisterListener(fusedGyroscopeSensor,
                    sensorManager.getDefaultSensor(Sensor.TYPE_ACCELEROMETER));
            sensorManager.unregisterListener(fusedGyroscopeSensor,
                    sensorManager.getDefaultSensor(Sensor.TYPE_MAGNETIC_FIELD));
            sensorManager.unregisterListener(fusedGyroscopeSensor,
                    sensorManager.getDefaultSensor(Sensor.TYPE_GYROSCOPE));
            fusedGyroscopeSensor.removeObserver(this);
        }
        initMaths();
        accelerationSampleCount = 0;
        magneticSampleCount = 0;
        hasInitialOrientation = false;
        stateInitializedCalibrated = false;
        stateInitializedRaw = false;
    }
    private void readPrefs()
    {
        SharedPreferences prefs = PreferenceManager
                .getDefaultSharedPreferences(this);
        useFusedEstimation = prefs.getBoolean(ConfigActivity.FUSION_PREFERENCE,
                false);
        useRadianUnits = prefs
                .getBoolean(ConfigActivity.UNITS_PREFERENCE, true);
        Log.d(tag, "Fusion: " + String.valueOf(useFusedEstimation));
        Log.d(tag, "Units Radians: " + String.valueOf(useRadianUnits));
    }
    private void showGyroscopeNotAvailableAlert()
    {
        AlertDialog.Builder alertDialogBuilder = new AlertDialog.Builder(this);
        //设置标题
        alertDialogBuilder.setTitle("Gyroscope Not Available");
        //设置对话框消息
        alertDialogBuilder
                .setMessage(
                        "Your device is not equipped with a gyroscope or it is not responding...")
                .setCancelable(false)
                .setNegativeButton("I'll look around...",
                        new DialogInterface.OnClickListener()
```

```
                        {
                            public void onClick(DialogInterface dialog, int id)
                            {
                                dialog.cancel();
                            }
                        });
        //创建警报对话框
        AlertDialog alertDialog = alertDialogBuilder.create();
        alertDialog.show();
    }

    @Override
    public void onAccuracyChanged(Sensor sensor, int accuracy)
    {

    }
}
```

至此，整个实例介绍完毕，因为是在模拟器中运行，所以系统主界面的执行效果如图 12-13 所示。在有陀螺仪传感器的物联网真机设备中运行本程序，会显示正确的预期效果。

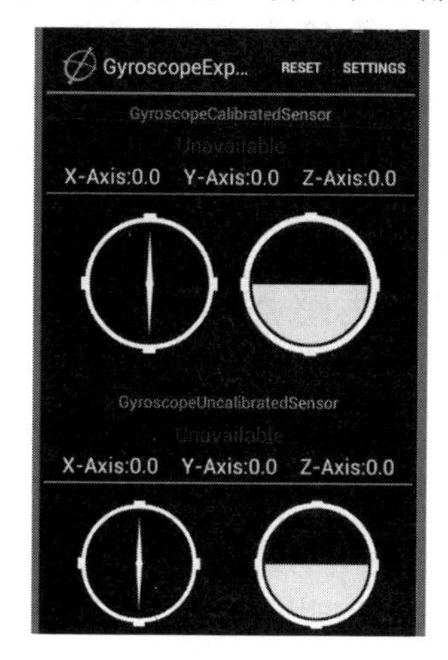


图 12-13　执行效果

# 第 13 章　旋转向量传感器、距离传感器和气压传感器

在 Android 设备应用程序开发过程中，经常使用旋转向量传感器、距离传感器和气压传感器来检测当前设备所处的环境参数。本章将详细讲解在 Android 设备中使用旋转向量传感器、距离传感器和气压传感器的方法。

## 13.1　旋转向量传感器详解

**知识点讲解：光盘:视频\知识点\第 13 章\旋转向量传感器详解.avi**

在 Android 系统中，旋转向量传感器的值是 TYPE_ROTATION_VECTOR，旋转矢量代表设备的方向，是一个将坐标轴和角度混合计算得到的数据。本节将详细讲解 Android 系统中旋转向量传感器的基本知识。

### 13.1.1　Android 中的旋转向量传感器

Android 旋转向量传感器的具体说明如表 13-1 所示。

表 13-1　Android 旋转向量传感器的具体说明

| 传　感　器 | 传感器事件数据 | 说　　明 | 测 量 单 位 |
|---|---|---|---|
| TYPE_ROTATION_VECTOR | SensorEvent.values[0] | 旋转向量沿 X 轴的部分（x×sin(θ/2)） | 无 |
| | SensorEvent.values[1] | 旋转向量沿 Y 轴的部分（y×sin(θ/2)） | |
| | SensorEvent.values[2] | 旋转向量沿 Z 轴的部分（z×sin(θ/2)） | |
| | SensorEvent.values[3] | 旋转向量的数值部分（(cos(θ/2)） | |

由表 13-1 可知，RV-sensor 能够输出如下所示的 3 个数据：

- ☑ x×sin(θ/2)
- ☑ y×sin(θ/2)
- ☑ z×sin(θ/2)

则 sin(theta/2)表示 RV 的数量级，RV 的方向与轴旋转的方向相同，这样 RV 的 3 个数值与 cos(theta/2)组成一个四元组。而 RV 的数据没有单位，使用的坐标系与加速度相同。例如下面的演示代码。

```
sensors_event_t.data[0] = x*sin(theta/2)
sensors_event_t.data[1] = y*sin(theta/2)
sensors_event_t.data[2] = z*sin(theta/2)
sensors_event_t.data[3] = cos(theta/2)
```

GV、LA 和 RV 的数值物理传感器没有直接给出，需要 G-sensor、O-sensor 和 Gyro-sensor 经过算法计算后得出。

由此可见，旋转向量代表了设备的方位，这个方位结果由角度和坐标轴信息组成，在里面包含了设备围绕坐标轴（X、Y、Z）旋转的角度 θ。例如下面的代码演示了获取默认的旋转向量传感器的方法。

```
private SensorManager mSensorManager;
private Sensor mSensor;
...
mSensorManager = (SensorManager) getSystemService(Context.SENSOR_SERVICE);
mSensor = mSensorManager.getDefaultSensor(Sensor.TYPE_ROTATION_VECTOR);
```

在 Android 系统中，旋转向量的 3 个元素等于四元组的后 3 个部分（$\cos(\theta/2)$、$x\times\sin(\theta/2)$、$y\times\sin(\theta/2)$、$z\times\sin(\theta/2)$），没有单位。X、Y、Z 轴的具体定义与加速度传感器的相同。旋转向量传感器的坐标系如图 13-1 所示。

上述坐标系具有如下所示的特点。

☑ X：定义为向量积 Y×Z。它是以设备当前位置为切点的地球切线，方向向东。

☑ Y：以设备当前位置为切点的地球切线，指向地磁北极。

☑ Z：与地平面垂直，指向天空。

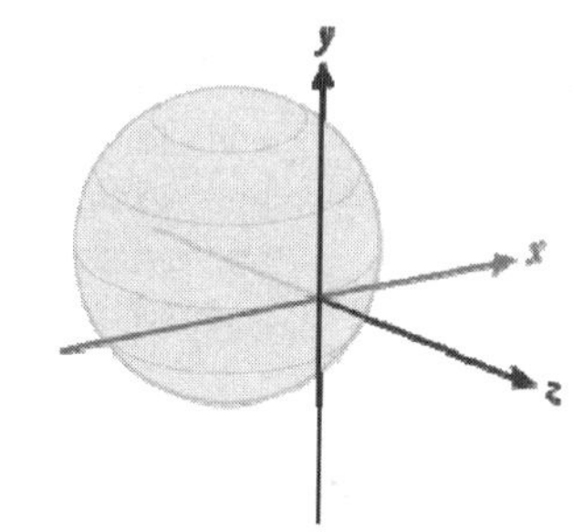


图 13-1　旋转向量传感器的坐标系

## 13.1.2　实战演练——确定设备当前的方向

本实例联合使用了旋转向量传感器、磁场传感器、重力传感器和加速度传感器，功能是获取当前设备的方向。

| 实　　例 | 功　　能 | 源 码 路 径 |
| --- | --- | --- |
| 实例 13-1 | 确定设备当前的具体方向 | https://github.com/gast-lib/gast-lib |

本实例的功能是当设备接近某个位置时实现自动提醒。本实例源码是开源代码，来源于地址 https://github.com/gast-lib/gast-lib/，读者可以自行登录并下载。

（1）实现主 Activity

本实例的主 Activity 是 DetermineOrientationActivity，通过布局文件 determine_orientation.xml 实现布局，在屏幕中提供一组单选按钮供用户选择所需要的传感器，并在屏幕下方显示传感器返回的数据。布局文件 determine_orientation.xml 的具体实现代码如下：

```
<RelativeLayout xmlns:android="http://schemas.android.com/apk/res/android"
    android:layout_width="match_parent"
    android:layout_height="match_parent"
    android:orientation="vertical" >

    <RadioGroup android:id="@+id/sensorSelector"
        android:layout_width="match_parent"
        android:layout_height="wrap_content"
        android:layout_alignParentTop="true" >

        <RadioButton android:id="@+id/gravitySensor"
            android:layout_width="match_parent"
```

```
            android:layout_height="wrap_content"
            android:text="@string/gravitySensorLabel"
            android:checked="true"
            android:onClick="onSensorSelectorClick" />

        <RadioButton android:id="@+id/accelerometerMagnetometer"
            android:layout_width="match_parent"
            android:layout_height="wrap_content"
            android:text="@string/accelerometerMagnetometerLabel"
            android:checked="false"
            android:onClick="onSensorSelectorClick" />

        <RadioButton android:id="@+id/gravityMagnetometer"
            android:layout_width="match_parent"
            android:layout_height="wrap_content"
            android:text="@string/gravityMagnetometerLabel"
            android:checked="false"
            android:onClick="onSensorSelectorClick" />

        <RadioButton android:id="@+id/rotationVector"
            android:layout_width="match_parent"
            android:layout_height="wrap_content"
            android:text="@string/rotationVectorLabel"
            android:checked="false"
            android:onClick="onSensorSelectorClick" />
    </RadioGroup>

    <ToggleButton android:id="@+id/ttsNotificationsToggleButton"
        android:layout_width="wrap_content"
        android:layout_height="wrap_content"
        android:text="@string/speakOrientationLabel"
        android:checked="true"
        android:layout_below="@id/sensorSelector"
        android:textOn="@string/ttsNotificationsOn"
        android:textOff="@string/ttsNotificationsOff"
        android:onClick="onTtsNotificationsToggleButtonClicked" />

    <TextView android:id="@+id/selectedSensorLabel"
        android:layout_width="wrap_content"
        android:layout_height="wrap_content"
        android:text="@string/selectedSensorLabel"
        android:layout_below="@id/ttsNotificationsToggleButton"
        android:layout_marginRight="5dip" />

    <TextView android:id="@+id/selectedSensorValue"
        android:layout_width="wrap_content"
        android:layout_height="wrap_content"
        android:layout_toRightOf="@id/selectedSensorLabel"
        android:layout_alignTop="@id/selectedSensorLabel"
        android:layout_alignBottom="@id/selectedSensorLabel" />
```

```
<TextView android:id="@+id/orientationLabel"
    android:layout_width="wrap_content"
    android:layout_height="wrap_content"
    android:text="@string/orientationLabel"
    android:layout_below="@id/selectedSensorValue"
    android:layout_marginRight="5dip" />

<TextView android:id="@+id/orientationValue"
    android:layout_width="wrap_content"
    android:layout_height="wrap_content"
    android:layout_toRightOf="@id/orientationLabel"
    android:layout_alignTop="@id/orientationLabel"
    android:layout_alignBottom="@id/orientationLabel" />

<TextView android:id="@+id/sensorXLabel"
    android:layout_width="wrap_content"
    android:layout_height="wrap_content"
    android:layout_below="@id/orientationValue"
    android:layout_marginRight="5dip" />

<TextView android:id="@+id/sensorXValue"
    android:layout_width="wrap_content"
    android:layout_height="wrap_content"
    android:layout_toRightOf="@id/sensorXLabel"
    android:layout_alignTop="@id/sensorXLabel"
    android:layout_alignBottom="@id/sensorXLabel" />

<TextView android:id="@+id/sensorYLabel"
    android:layout_width="wrap_content"
    android:layout_height="wrap_content"
    android:layout_below="@id/sensorXLabel"
    android:layout_marginRight="5dip" />

<TextView android:id="@+id/sensorYValue"
    android:layout_width="wrap_content"
    android:layout_height="wrap_content"
    android:layout_toRightOf="@id/sensorYLabel"
    android:layout_alignTop="@id/sensorYLabel"
    android:layout_alignBottom="@id/sensorYLabel" />

<TextView android:id="@+id/sensorZLabel"
    android:layout_width="wrap_content"
    android:layout_height="wrap_content"
    android:layout_below="@id/sensorYLabel"
    android:layout_marginRight="5dip" />

<TextView android:id="@+id/sensorZValue"
    android:layout_width="wrap_content"
    android:layout_height="wrap_content"
    android:layout_toRightOf="@id/sensorZLabel"
    android:layout_alignTop="@id/sensorZLabel"
```

```
        android:layout_alignBottom="@id/sensorZLabel" />
</RelativeLayout>
```

主 Activity 程序文件 DetermineOrientationActivity.java 的功能是获取 SensorManager 的引用，根据用户单选按钮的选择来注册这个传感器，然后调用这个传感器来获取数据。假如选择的是重力传感器，则会注册重力传感器，然后获取数组 SensorEvent.Values 中的 X、Y 和 Z 轴上的重力大小。当选择使用旋转向量传感器时，会调用方法 determineOrientation(rotationMatrix)根据给出的旋转矩阵计算出具体的方向。当使用者确定了设备的具体朝向时，会使用文本转语音的功能来提醒用户。本实例的语音提醒功能是通过 TTS 机制实现的，有关语音提醒方面的知识将在本书后面的章节中进行详细讲解。文件 DetermineOrientationActivity.java 的具体实现代码如下：

```
protected void onCreate(Bundle savedInstanceState)
{
    super.onCreate(savedInstanceState);
    super.setContentView(R.layout.determine_orientation);

    getWindow().addFlags(WindowManager.LayoutParams.FLAG_KEEP_SCREEN_ON);

    ttsParams = new HashMap<String, String>();
    ttsParams.put(Engine.KEY_PARAM_STREAM, String.valueOf(TTS_STREAM));

    this.setVolumeControlStream(TTS_STREAM);

    sensorManager = (SensorManager) getSystemService(SENSOR_SERVICE);

    sensorSelector = (RadioGroup) findViewById(R.id.sensorSelector);
    selectedSensorValue = (TextView) findViewById(R.id.selectedSensorValue);
    orientationValue = (TextView) findViewById(R.id.orientationValue);
    sensorXLabel = (TextView) findViewById(R.id.sensorXLabel);
    sensorXValue = (TextView) findViewById(R.id.sensorXValue);
    sensorYLabel = (TextView) findViewById(R.id.sensorYLabel);
    sensorYValue = (TextView) findViewById(R.id.sensorYValue);
    sensorZLabel = (TextView) findViewById(R.id.sensorZLabel);
    sensorZValue = (TextView) findViewById(R.id.sensorZValue);
    ttsNotificationsToggleButton =
            (ToggleButton) findViewById(R.id.ttsNotificationsToggleButton);

    preferences = getPreferences(MODE_PRIVATE);
    ttsNotifications =
            preferences.getBoolean(TTS_NOTIFICATION_PREFERENCES_KEY, true);
}

@Override
protected void onResume()
{
    super.onResume();

    ttsNotificationsToggleButton.setChecked(ttsNotifications);
    updateSelectedSensor();
}
```

```
@Override
protected void onPause()
{
    super.onPause();

    sensorManager.unregisterListener(this);

    if (tts != null)
    {
        tts.shutdown();
    }
}

@Override
public void onSensorChanged(SensorEvent event)
{
    float[] rotationMatrix;

    switch (event.sensor.getType())
    {
        case Sensor.TYPE_GRAVITY:
            sensorXLabel.setText(R.string.xAxisLabel);
            sensorXValue.setText(String.valueOf(event.values[0]));

            sensorYLabel.setText(R.string.yAxisLabel);
            sensorYValue.setText(String.valueOf(event.values[1]));

            sensorZLabel.setText(R.string.zAxisLabel);
            sensorZValue.setText(String.valueOf(event.values[2]));

            sensorYLabel.setVisibility(View.VISIBLE);
            sensorYValue.setVisibility(View.VISIBLE);
            sensorZLabel.setVisibility(View.VISIBLE);
            sensorZValue.setVisibility(View.VISIBLE);

            if (selectedSensorId == R.id.gravitySensor)
            {
                if (event.values[2] >= GRAVITY_THRESHOLD)
                {
                    onFaceUp();
                }
                else if (event.values[2] <= (GRAVITY_THRESHOLD * -1))
                {
                    onFaceDown();
                }
            }
            else
            {
                accelerationValues = event.values.clone();
                rotationMatrix = generateRotationMatrix();
```

```
                if (rotationMatrix != null)
                {
                    determineOrientation(rotationMatrix);
                }
            }

            break;
        case Sensor.TYPE_ACCELEROMETER:
            accelerationValues = event.values.clone();
            rotationMatrix = generateRotationMatrix();

            if (rotationMatrix != null)
            {
                determineOrientation(rotationMatrix);
            }
            break;
        case Sensor.TYPE_MAGNETIC_FIELD:
            magneticValues = event.values.clone();
            rotationMatrix = generateRotationMatrix();

            if (rotationMatrix != null)
            {
                determineOrientation(rotationMatrix);
            }
            break;
        case Sensor.TYPE_ROTATION_VECTOR:

            rotationMatrix = new float[16];
            SensorManager.getRotationMatrixFromVector(rotationMatrix,
                    event.values);
            determineOrientation(rotationMatrix);
            break;
    }
}

@Override
public void onAccuracyChanged(Sensor sensor, int accuracy)
{
    Log.d(TAG,
            String.format("Accuracy for sensor %s = %d",
            sensor.getName(), accuracy));
}
private float[] generateRotationMatrix()
{
    float[] rotationMatrix = null;

    if (accelerationValues != null && magneticValues != null)
    {
        rotationMatrix = new float[16];
        boolean rotationMatrixGenerated;
        rotationMatrixGenerated =
```

```
                SensorManager.getRotationMatrix(rotationMatrix,
                null,
                accelerationValues,
                magneticValues);

        if (!rotationMatrixGenerated)
        {
            Log.w(TAG, getString(R.string.rotationMatrixGenFailureMessage));

            rotationMatrix = null;
        }
    }

    return rotationMatrix;
}

/**
 *使用最后一个读取加速度和重力的值，以确定
 *设备是正面朝上还是朝下
 *
 * @param rotationMatrix The rotation matrix to use if the orientation
 * calculation
 */
private void determineOrientation(float[] rotationMatrix)
{
    float[] orientationValues = new float[3];
    SensorManager.getOrientation(rotationMatrix, orientationValues);

    double azimuth = Math.toDegrees(orientationValues[0]);
    double pitch = Math.toDegrees(orientationValues[1]);
    double roll = Math.toDegrees(orientationValues[2]);

    sensorXLabel.setText(R.string.azimuthLabel);
    sensorXValue.setText(String.valueOf(azimuth));

    sensorYLabel.setText(R.string.pitchLabel);
    sensorYValue.setText(String.valueOf(pitch));

    sensorZLabel.setText(R.string.rollLabel);
    sensorZValue.setText(String.valueOf(roll));

    sensorYLabel.setVisibility(View.VISIBLE);
    sensorYValue.setVisibility(View.VISIBLE);
    sensorZLabel.setVisibility(View.VISIBLE);
    sensorZValue.setVisibility(View.VISIBLE);

    if (pitch <= 10)
    {
        if (Math.abs(roll) >= 170)
        {
            onFaceDown();
```

```
            }
            else if (Math.abs(roll) <= 10)
            {
                onFaceUp();
            }
        }
    }

    /**
     *处理程序设备是正面朝上
     */
    private void onFaceUp()
    {
        if (!isFaceUp)
        {
            if (tts != null && ttsNotificationsToggleButton.isChecked())
            {
                tts.speak(getString(R.string.faceUpText),
                        TextToSpeech.QUEUE_FLUSH,
                        ttsParams);
            }

            orientationValue.setText(R.string.faceUpText);
            isFaceUp = true;
        }
    }

    /**
     * Handler for device being face down
     */
    private void onFaceDown()
    {
        if (isFaceUp)
        {
            if (tts != null && ttsNotificationsToggleButton.isChecked())
            {
                tts.speak(getString(R.string.faceDownText),
                        TextToSpeech.QUEUE_FLUSH,
                        ttsParams);
            }

            orientationValue.setText(R.string.faceDownText);
            isFaceUp = false;
        }
    }

    /**
     * Updates the views for when the selected sensor is changed
     */
    private void updateSelectedSensor()
    {
```

```
    sensorManager.unregisterListener(this);

    selectedSensorId = sensorSelector.getCheckedRadioButtonId();
    if (selectedSensorId == R.id.accelerometerMagnetometer)
    {
        sensorManager.registerListener(this,
                sensorManager.getDefaultSensor(Sensor.TYPE_ACCELEROMETER),
                RATE);

        sensorManager.registerListener(this,
                sensorManager.getDefaultSensor(Sensor.TYPE_MAGNETIC_FIELD),
                RATE);
    }
    else if (selectedSensorId == R.id.gravityMagnetometer)
    {
        sensorManager.registerListener(this,
                sensorManager.getDefaultSensor(Sensor.TYPE_GRAVITY),
                RATE);

        sensorManager.registerListener(this,
                sensorManager.getDefaultSensor(Sensor.TYPE_MAGNETIC_FIELD),
                RATE);
    }
    else if ((selectedSensorId == R.id.gravitySensor))
    {
        sensorManager.registerListener(this,
                sensorManager.getDefaultSensor(Sensor.TYPE_GRAVITY),
                RATE);
    }
    else
    {
        sensorManager.registerListener(this,
                sensorManager.getDefaultSensor(Sensor.TYPE_ROTATION_VECTOR),
                RATE);
    }

    RadioButton selectedSensorRadioButton =
            (RadioButton) findViewById(selectedSensorId);
    selectedSensorValue.setText(selectedSensorRadioButton.getText());
}

/**
 * Handles click event for the sensor selector
 *
 * @param view The view that was clicked
 */
public void onSensorSelectorClick(View view)
{
    updateSelectedSensor();
}
```

```
    /**
     * Handles click event for the TTS toggle button
     *
     * @param view The view for the toggle button
     */
    public void onTtsNotificationsToggleButtonClicked(View view)
    {
        ttsNotifications = ((ToggleButton) view).isChecked();
        preferences.edit()
            .putBoolean(TTS_NOTIFICATION_PREFERENCES_KEY, ttsNotifications)
            .commit();
    }

    @Override
    public void onSuccessfulInit(TextToSpeech tts)
    {
        super.onSuccessfulInit(tts);
        this.tts = tts;
    }

    @Override
    protected void receiveWhatWasHeard(List<String> heard, float[] confidenceScores)
    {
    }
}
```

（2）获取设备的旋转向量

编写文件 NorthFinder.java，首先获取设备的旋转向量，并将旋转向量的坐标映射到摄像头的轴上。如果取消了对方法 remapCoordinateSystem 的调用，则将当前设备指向北方，而并不是将后置摄像头指向北方。除此之外，在此文件中还使用 OpenGL 改变了屏幕的颜色，当后置摄像头指向北方时（允许误差 20° 以内），将屏幕颜色从红色变为绿色。文件 NorthFinder.java 的具体实现代码如下：

```
public class NorthFinder extends Activity implements SensorEventListener
{
    private static final int ANGLE = 20;

    private TextView tv;
    private GLSurfaceView mGLSurfaceView;
    private MyRenderer mRenderer;
    private SensorManager mSensorManager;
    private Sensor mRotVectSensor;
    private float[] orientationVals = new float[3];

    private final float[] mRotationMatrix = new float[16];

    @Override
    protected void onCreate(Bundle savedInstanceState)
    {
        super.onCreate(savedInstanceState);

        setContentView(R.layout.sensors_north_main);
```

```
        mRenderer = new MyRenderer();
        mGLSurfaceView = (GLSurfaceView) findViewById(R.id.glsurfaceview);
        mGLSurfaceView.setRenderer(mRenderer);

        tv = (TextView) findViewById(R.id.tv);

        mSensorManager = (SensorManager) getSystemService(SENSOR_SERVICE);
        mRotVectSensor =
                mSensorManager.getDefaultSensor(Sensor.TYPE_ROTATION_VECTOR);

    }

    @Override
    protected void onResume()
    {
        super.onResume();
        mSensorManager.registerListener(this, mRotVectSensor, 10000);
    }

    @Override
    protected void onPause()
    {
        super.onPause();
        mSensorManager.unregisterListener(this);
    }

    @Override
    public void onSensorChanged(SensorEvent event)
    {
        if (event.sensor.getType() == Sensor.TYPE_ROTATION_VECTOR)
        {
            SensorManager.getRotationMatrixFromVector(mRotationMatrix,
                    event.values);
            SensorManager
                    .remapCoordinateSystem(mRotationMatrix,
                            SensorManager.AXIS_X, SensorManager.AXIS_Z,
                            mRotationMatrix);
            SensorManager.getOrientation(mRotationMatrix, orientationVals);

            orientationVals[0] = (float) Math.toDegrees(orientationVals[0]);
            orientationVals[1] = (float) Math.toDegrees(orientationVals[1]);
            orientationVals[2] = (float) Math.toDegrees(orientationVals[2]);

            tv.setText(" Yaw: " + orientationVals[0] + "\n Pitch: "
                    + orientationVals[1] + "\n Roll (not used): "
                    + orientationVals[2]);

        }
    }
```

```
    @Override
    public void onAccuracyChanged(Sensor sensor, int accuracy)
    {
    }

    class MyRenderer implements GLSurfaceView.Renderer
    {
        public void onDrawFrame(GL10 gl)
        {
            gl.glClear(GL10.GL_COLOR_BUFFER_BIT);

            if (orientationVals[0] < ANGLE && orientationVals[0] > -ANGLE
                    && orientationVals[1] < ANGLE
                    && orientationVals[1] > -ANGLE)
            {
                gl.glClearColor(0, 1, 0, 1);
            }
            else
            {
                gl.glClearColor(1, 0, 0, 1);
            }
        }

        @Override
        public void onSurfaceChanged(GL10 gl, int width, int height)
        {
        }

        @Override
        public void onSurfaceCreated(GL10 gl, EGLConfig config)
        {
        }
    }
}
```

至此，整个实例的核心代码介绍完毕。为节省本书篇幅，其余的代码将不再进行详细讲解。

# 13.2　距离传感器详解

**知识点讲解：光盘:视频\知识点\第 13 章\距离传感器详解.avi**

在 Android 设备应用程序开发过程中，通过使用距离传感器可以测试设备的移动距离。本节将详细讲解在 Android 设备中使用距离传感器检测运动数据的基本知识。

## 13.2.1　距离传感器介绍

在 Android 系统中，需要使用加速度传感器、线性加速度传感器和距离传感器来检测设备的运动数据。在当前的技术条件下，距离传感器是指利用“飞行时间法”（Flying Time）的原理来实现测量距

离，以实现检测物体距离的一种传感器。“飞行时间法”（Flying Time）是通过发射特别短的光脉冲，并测量此光脉冲从发射到被物体反射回来的时间，通过测量时间间隔来计算与物体之间的距离。

在现实世界中，距离传感器在智能手机中的应用比较常见。一般触屏智能手机在默认设置下都会有一个延时锁屏的设置，就是在一段时间内，手机检测不到任何操作，就会进入锁屏状态。这样是有一定好处的。手机作为移动终端的一种，追求低功耗是设计的目标之一。延时锁屏既可以避免不必要的能量消耗，又能保证不丢失重要信息。另外，在使用触屏手机设备时，当接电话时距离传感器会起作用，当脸靠近屏幕时屏幕灯会熄灭，并自动锁屏，这样可以防止脸误操作。当脸离开屏幕时屏幕灯会自动开启，并且自动解锁。

除了被广泛应用于手机设备之外，距离传感器还被用于野外环境（山体情况、峡谷深度等）、飞机高度检测、矿井深度、物料高度测量等领域。并且在野外应用领域中，主要用于检测山体情况和峡谷深度等。而对飞机高度测量功能是通过检测飞机在起飞后和降落至地面之前时距离地面的高度，并将结果实时显示在控制面板上。也可以使用距离传感器测量物料各点高度，用于计算物料的体积。在显示应用中，用于飞机高度和物料高度的距离传感器有 LDM301 系列，用于野外应用的距离传感器有 LDM4x 系列。

在当前的物联网设备应用中，距离传感器被应用于智能皮带中。在皮带扣中嵌入了距离传感器，当把皮带调整至合适宽度、卡好皮带扣后，如果皮带在 10 秒钟内没有重新解开，传感器就会自动生成本次的腰围数据。皮带与皮带扣连接处的其中一枚铆钉将被数据传输装置所替代。当将智能手机放在铆钉处保持两秒钟静止，手机中的自我健康管理 App 会被自动激活，并获取本次腰围数据。

## 13.2.2　Android 系统中的距离传感器

在 Android 系统中，距离传感器也被称为 P-Sensor，值是 TYPE_PROXIMITY，单位是 cm，能够测量某个对象到屏幕的距离。可以在打电话时判断人耳到电话屏幕的距离，以关闭屏幕而达到省电功能。

P-Sensor 主要用于在通话过程中防止用户误操作屏幕，接下来以通话过程为例来讲解电话程序对 P-Sensor 的操作流程。

（1）在启动电话程序时，在.java 文件中新建了一个 P-Sensor 的 wakeLock 对象。例如下面的代码。

```
mProximityWakeLock = pm.newWakeLock(
PowerManager.PROXIMITY_SCREEN_OFF_WAKE_LOCK, LOG_TAG
);
```

对象 wakeLock 的功能是请求控制屏幕的点亮或熄灭。

（2）在电话状态发生改变时，例如接通了电话，调用.java 文件中的方法根据当前电话的状态来决定是否打开 P-Sensor。如果在通话过程中，电话是 OFF-HOOK 状态时打开 P-Sensor。例如下面的演示代码。

```
if (!mProximityWakeLock.isHeld()) {
                    if (DBG) Log.d(LOG_TAG, "updateProximitySensorMode: acquiring...");
                    mProximityWakeLock.acquire();
                }
```

在上述代码中，mProximityWakeLock.acquire()会调用到另外的方法打开 P-Sensor，这个另外的方法会判断当前手机有没有 P-Sensor。如果有，就会向 SensorManager 注册一个 P-Sensor 监听器。这样当 P-Sensor 检测到手机和人体距离发生改变时，就会调用服务监听器进行处理。同样，当电话挂断时，电话模块会去调用方法取消 P-Sensor 监听器。

在 Android 系统中，PowerManagerService 中的 P-Sensor 监听器会进行实时监听工作，当 P-Sensor 检测到距离有变化时就会进行监听。具体监听过程的代码如下：

```
SensorEventListener mProximityListener = new SensorEventListener() {
        public void onSensorChanged(SensorEvent event) {
            long milliseconds = SystemClock.elapsedRealtime();
            synchronized (mLocks) {
                float distance = event.values[0];                    //检测到手机和人体的距离
                long timeSinceLastEvent = milliseconds - mLastProximityEventTime;   //这次检测和上次检测
的时间差
                mLastProximityEventTime = milliseconds;     //更新上一次检测的时间
                mHandler.removeCallbacks(mProximityTask);
                boolean proximityTaskQueued = false;

                //compare against getMaximumRange to support sensors that only return 0 or 1
                boolean active = (distance >= 0.0 && distance < PROXIMITY_THRESHOLD &&
                        distance < mProximitySensor.getMaximumRange());   //如果距离小于某一个距离阈
值，默认是 5.0f，说明手机和脸部距离贴近，应该熄灭屏幕

                if (mDebugProximitySensor) {
                    Slog.d(TAG, "mProximityListener.onSensorChanged active: " + active);
                }
                if (timeSinceLastEvent < PROXIMITY_SENSOR_DELAY) {
                    mProximityPendingValue = (active ? 1 : 0);
                    mHandler.postDelayed(mProximityTask, PROXIMITY_SENSOR_DELAY – timeSinceLast
Event);
                    proximityTaskQueued = true;
                } else {
                    mProximityPendingValue = -1;
                    proximityChangedLocked(active);       //熄灭屏幕操作
                }

                boolean held = mProximityPartialLock.isHeld();
                if (!held && proximityTaskQueued) {
                    mProximityPartialLock.acquire();
                } else if (held && !proximityTaskQueued) {
                    mProximityPartialLock.release();
                }
            }
        }

        public void onAccuracyChanged(Sensor sensor, int accuracy) {
        }
    };
```

由上述代码可知，在监听时会首先通过 float distance = event.values[0];获取变化的距离。如果发现检测这次距离变化和上次距离变化时间差，例如小于系统设置的阈值则不会去熄灭屏幕。过于频繁的操作系统会忽略掉。如果感觉 P-Sensor 不够灵敏，就可以修改如下所示的系统默认值。

```
private static final int PROXIMITY_SENSOR_DELAY = 1000;
```

将上述值改小后就会发现 P-Sensor 会变得灵敏很多。

如果 P-Sensor 检测到这次距离变化小于系统默认值，并且这次是一次正常的变化，那么需要通过

如下代码熄灭屏幕。

```
proximityChangedLocked(active);
```

此处会判断 P-Sensor 是否可以用，如果不可用则返回，并忽略这次距离变化。

```
if (!mProximitySensorEnabled) {
                Slog.d(TAG, "Ignoring proximity change after sensor is disabled");
                return;
        }
```

如果一切都满足，则调用如下代码灭灯。

```
goToSleepLocked(SystemClock.uptimeMillis(),
                WindowManagerPolicy.OFF_BECAUSE_OF_PROX_SENSOR);
```

## 13.2.3　实战演练——实现自动锁屏功能

在当前的 Android 智能设备中，特别是触屏智能手机设备，在默认设置下都会有一个延时锁屏的功能，通过此功能设置后，如果在一段时间内设置手机检测不到任何操作，手机会自动进入锁屏状态。通过延时锁屏功能，不但可以避免不必要的能量消耗，而且又能保证不丢失重要的信息。

| 实　例 | 功　能 | 源码路径 |
|---|---|---|
| 实例 13-2 | 使用距离传感器实现自动锁屏功能 | 光盘:\daima\13\AutoLockEX |

本实例的功能是使用距离传感器实现自动锁屏功能，具体实现流程如下所示。

（1）编写布局文件 activity_main.xml，功能是在屏幕中分别设置“启动服务”、“停止服务”和“退出”3 个按钮，具体实现代码如下：

```
<RelativeLayout xmlns:android="http://schemas.android.com/apk/res/android"
    xmlns:tools="http://schemas.android.com/tools"
    android:layout_width="match_parent"
    android:layout_height="match_parent"
    android:paddingBottom="@dimen/activity_vertical_margin"
    android:paddingLeft="@dimen/activity_horizontal_margin"
    android:paddingRight="@dimen/activity_horizontal_margin"
    android:paddingTop="@dimen/activity_vertical_margin"
    tools:context=".MainActivity" >

    <TextView
        android:id="@+id/title_tv"
        android:layout_centerHorizontal="true"
        android:layout_width="wrap_content"
        android:layout_height="wrap_content"
        android:textSize="20sp"
        android:text="@string/title" />

     <Button
        android:id="@+id/start"
        android:layout_below="@id/title_tv"
        android:layout_centerHorizontal="true"
        android:layout_width="fill_parent"
        android:layout_height="wrap_content"
        android:textSize="20sp"
```

```
        android:text="@string/start" />

    <Button
        android:id="@+id/stop"
        android:layout_below="@id/start"
        android:layout_centerHorizontal="true"
        android:layout_width="fill_parent"
        android:layout_height="wrap_content"
        android:textSize="20sp"
        android:text="@string/stop" />

    <Button
        android:id="@+id/exit"
        android:layout_below="@id/stop"
        android:layout_centerHorizontal="true"
        android:layout_width="fill_parent"
        android:layout_height="wrap_content"
        android:textSize="20sp"
        android:text="@string/exit" />

    <TextView
        android:id="@+id/sensortitle_tv"
        android:layout_below="@id/exit"
        android:layout_centerHorizontal="true"
        android:layout_width="wrap_content"
        android:layout_height="wrap_content"
        android:textSize="20sp"
        android:text="@string/sensorinfo" />

     <TextView
        android:id="@+id/sensorinfo_tv"
        android:layout_below="@id/sensortitle_tv"
        android:layout_centerHorizontal="true"
        android:layout_width="fill_parent"
        android:layout_height="wrap_content"
        android:textSize="20sp"
        android:text="@string/sensorinfo" />
</RelativeLayout>
```

（2）编写文件 MainActivity.java，在启动时显示传感器名和版本号，并根据用户的按钮操作执行对应的事件处理程序。文件 MainActivity.java 的具体实现代码如下：

```
public class MainActivity extends Activity {

    private Button start;
    private Button stop;
    private Button exit;
    private TextView sensorinfo_tv;
    private Intent intent;
    private SensorManager sm = null;
    private Sensor promixty = null;
```

```
@Override
protected void onCreate(Bundle savedInstanceState) {
    super.onCreate(savedInstanceState);
    setContentView(R.layout.activity_main);
    if (null == sm) {
        sm = (SensorManager) getSystemService(SENSOR_SERVICE);      //获取传感器管理类
        promixty = sm.getDefaultSensor(Sensor.TYPE_PROXIMITY);      //获取距离传感器
    }
    String sensorInfo;
    if (null != promixty) {
        sensorInfo = "传感器名称：" + promixty.getName() + "\n"
                + "设备版本：" + promixty.getVersion() + "\n" + "供应商："
                + promixty.getVendor() + "\n";
    }
    else {
        sensorInfo = "无法获取距离传感器信息，可能是您的手机不支持该传感器。";
    }
    initUI();
    intent = new Intent("org.hq.autoLockService");
    start.setOnClickListener( new OnClickListener() {
        @Override
        public void onClick(View v) {
            //开启服务
            start();
        }
    });
    stop.setOnClickListener(new OnClickListener() {
        @Override
        public void onClick(View v) {
            //停止服务
            stop();
        }
    });
    exit.setOnClickListener( new OnClickListener() {

        @Override
        public void onClick(View v) {
            //结束本次 Activity
            finish();
        }
    });
    sensorinfo_tv.setText(sensorInfo);
}

private void initUI(){
    start = (Button) super.findViewById(R.id.start);
    stop = (Button) super.findViewById(R.id.stop);
    exit = (Button) super.findViewById(R.id.exit);
    sensorinfo_tv = (TextView) super.findViewById(R.id.sensorinfo_tv);
}
```

```
    private void start(){
        Bundle bundle = new Bundle();
        bundle.putInt("distance", 3);
        bundle.putBoolean("activited", true);
        intent.putExtras(bundle);
        startService(intent);
        //结束本次 Activity
        //finish();
    }
    //终止服务
    private void stop(){
        stopService(intent);
        Toast.makeText(this, "已停止后台服务。", Toast.LENGTH_SHORT).show();
    }

    @Override
    public boolean onCreateOptionsMenu(Menu menu) {
        getMenuInflater().inflate(R.menu.main, menu);
        return true;
    }

}
```

（3）编写文件 AutoLockService.java 实现自动锁屏服务，通过距离传感器来监听距离，自动进入锁屏状态。文件 AutoLockService.java 的具体实现代码如下：

```
public class AutoLockService extends Service implements SensorEventListener {

    private SensorManager sm = null;
    private Sensor promixty = null;
    //默认启用锁屏
    private static boolean ACTIVITED = true;
    //锁屏距离（单位：厘米）
    private static int LOCK_DIST = 3;

    @Override
    public IBinder onBind(Intent intent) {
        return null;
    }

    @Override
    public void onCreate() {
        super.onCreate();
        if (null == sm) {
            sm = (SensorManager) getSystemService(SENSOR_SERVICE);     //获取传感器管理类
            promixty = sm.getDefaultSensor(Sensor.TYPE_PROXIMITY);     //获取距离传感器
        }
        //显示距离传感器信息
        if (null != promixty) {
            Toast.makeText(this, "已创建后台服务。", Toast.LENGTH_SHORT).show();
        } else {
            Toast.makeText(this, "无法找到距离传感器", Toast.LENGTH_SHORT).show();
```

```
        }
    }

    @Override
    public void onDestroy() {
        super.onDestroy();
        if( null != sm ){
            //撤销监听器
            sm.unregisterListener(this);
        }
    }

    @Override
    public void onStart(Intent intent, int startId) {
        if (intent != null) {
            Bundle bundle = intent.getExtras();
            Toast.makeText(this, "后台服务已启动。", Toast.LENGTH_SHORT).show();
            //从 Intent 中获取设置参数
            if (bundle != null) {
                int dist = bundle.getInt("distance");
                ACTIVITED = bundle.getBoolean("activited");
                if (dist > 0 && dist < 9) {
                    LOCK_DIST = dist;
                }
            }
            //注册监听器
            sm.registerListener(this, promixty,SensorManager.SENSOR_DELAY_NORMAL);
        }
    }

    //监听精度变化
    @Override
    public void onAccuracyChanged(Sensor sensor, int accuracy) {
        Toast.makeText(this, "距离传感器 promixty 精度变为" + accuracy,
                Toast.LENGTH_SHORT).show();
    }

    @Override
    public void onSensorChanged(SensorEvent event) {
        if (event.values[0] < LOCK_DIST) //距离小于 5，锁屏
        {
            if (ACTIVITED) {
                lockScreen();
            }
        }
    }

    //跳至锁屏页面
    private void lockScreen() {
        Intent intent = new Intent();
        //在 Activity 之外启动，要加上 FLAG_ACTIVITY_NEW_TASK flag
```

```
            intent.setFlags( Intent.FLAG_ACTIVITY_NEW_TASK );
            intent.setClass(this, LockScreen.Controller.class);
            startActivity(intent);
        }
}
```

（4）编写文件 LockScreen.java，功能是实现锁屏功能，在锁屏之前需要先获取锁屏权限。文件 LockScreen.java 的具体实现代码如下：

```
public class LockScreen extends DeviceAdminReceiver {
    static final int RESULT_ENABLE = 1;

    public static class Controller extends Activity {

        DevicePolicyManager mDPM;
        ComponentName mDeviceAdminSample;

        @Override
        protected void onCreate(Bundle savedInstanceState) {
            super.onCreate(savedInstanceState);

            //首先要获得 Android 设备管理代理
            mDPM = (DevicePolicyManager) getSystemService(Context.DEVICE_POLICY_SERVICE);

            //LockScreen 继承自 DeviceAdminReceiver
            mDeviceAdminSample = new ComponentName(Controller.this,
                    LockScreen.class);
            //得到当前设备管理器有没有激活
            boolean active = mDPM.isAdminActive(mDeviceAdminSample);
            if (!active) {
                //如果没有激活，就去提示用户激活（第一次运行程序时）
                getAdmin();
            } else {
                //如果已经激活，就执行立即锁屏
                mDPM.lockNow();
            }
            // killMyself，锁屏之后就立即 kill 掉 Activity，避免资源的浪费
            //android.os.Process.killProcess(android.os.Process.myPid());
            finish();
        }

        //获取锁屏权限
        public void getAdmin() {
            Intent intent = new Intent(
                    DevicePolicyManager.ACTION_ADD_DEVICE_ADMIN);
            intent.putExtra(DevicePolicyManager.EXTRA_DEVICE_ADMIN,
                    mDeviceAdminSample);
            intent.putExtra(DevicePolicyManager.EXTRA_ADD_EXPLANATION,
                    "欢迎您的使用！在第一次使用时，请授予该程序锁屏权限。");
            startActivityForResult(intent, RESULT_ENABLE);
        }
    }
}
```

（5）在文件 AndroidManifest.xml 中声明权限，特别是需要注册一个广播接收者，具体实现代码如下：

```
<!-- 注册锁屏 Activity -->
<activity android:name="org.lock.LockScreen$Controller" >
</activity>

<service
    android:name="org.lock.AutoLockService"
    android:permission="android.permission.BIND_ACCESSIBILITY_SERVICE"
    android:enabled="true" >
    <intent-filter>
        <action android:name="org.lock.autoLockService" />
    </intent-filter>
</service>
<receiver
    android:name="org.lock.LockScreen"
    android:permission="android.permission.BIND_DEVICE_ADMIN" >
    <meta-data
        android:name="android.app.device_admin"
        android:resource="@xml/device_admin_sample" />

    <intent-filter>
        <action android:name="android.app.action.DEVICE_ADMIN_ENABLED" />
    </intent-filter>
</receiver>
```

至此，整个实例介绍完毕，执行后的效果如图 13-2 所示。

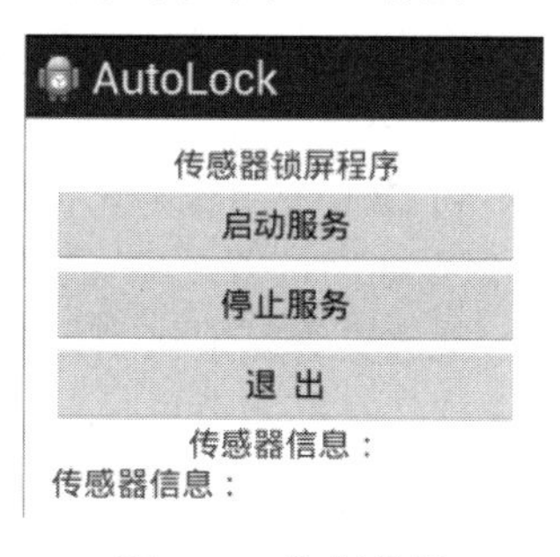


图 13-2　执行效果

# 13.3　气压传感器详解

**知识点讲解：光盘:视频\知识点\第 13 章\气压传感器详解.avi**

在 Android 设备开发应用过程中，通常需要使用设备来感知当前所处环境的信息，例如气压、GPS、海拔、湿度和温度。在 Android 系统中，专门提供了气压传感器、海拔传感器、湿度传感器和温度传感器来支持上述功能。本节将详细讲解在 Android 设备中使用气压传感器的基本知识，为读者步入本书后面知识的学习打下基础。

### 13.3.1 气压传感器基础

在现实应用中，气压传感器主要用于测量气体的绝对压强，主要适用于与气体压强相关的物理实验，如气体定律等，也可以在生物和化学实验中测量干燥、无腐蚀性的气体压强。气压传感器的原理比较简单，其主要的传感元件是一个对气压传感器内的强弱敏感的薄膜和一个顶针开控制，电路方面它连接了一个柔性电阻器。当被测气体的压力降低或升高时，这个薄膜变形带动顶针，同时该电阻器的阻值将会改变。从传感元件取得 0～5V 的信号电压，经过 A/D 转换由数据采集器接受，然后数据采集器以适当的形式把结果传送给计算机。

在现实应用中，很多气压传感器的主要部件为变容式硅膜盒。当该变容硅膜盒外界大气压力发生变化时发生顶针动作，而单晶硅膜盒会随着发生弹性变形，从而引起硅膜盒平行板电容器电容量的变化来控制气压传感器。

国标 GB7665-87 对传感器的定义是："能感受规定的被测量并按照一定的规律转换成可用信号的器件或装置，通常由敏感元件和转换元件组成"。而气压传感器是一种检测装置，能感受到被测量的信息，并能将检测感受到的信息按一定规律变换成为电信号或其他所需形式的信息输出，以满足信息的传输、处理、存储、显示、记录和控制等要求，是实现自动化检测和控制的首要环节。

### 13.3.2 气压传感器在智能手机中的应用

随着智能手机设备的发展，气压传感器得到了大力的普及。气压传感器首次在智能手机上使用是在 Galaxy Nexus 上，而之后推出的一些 Android 旗舰手机中也包含了这一传感器，像 Galaxy SIII、Galaxy Note2 也都有。对于喜欢登山的人来说，都会非常关心自己所处的高度。海拔高度的测量方法，一般常用的有两种方式，一是通过 GPS 全球定位系统，二是通过测出大气压，然后根据气压值计算出海拔高度。由于受到技术和其他方面原因的限制，GPS 计算海拔高度一般误差都会有 10 米左右，而如果在树林里或者是在悬崖下面，有时甚至接收不到 GPS 卫星信号。同时当用户处于楼宇内时，内置感应器可能会无法接收到 GPS 信号，从而不能够识别地理位置。配合气压传感器、加速计、陀螺仪等就能够实现精确定位。这样当用户在商场购物时，能够更容易地找到目标商品。

另外在汽车导航领域中，经常会有人抱怨在高架桥里导航会出错。例如在高架桥上时，GPS 说右转，而实际上右边根本没有右转出口，这主要是 GPS 无法判断用户是在桥上还是桥下而造成的错误导航。一般高架桥上下两层的高度都会有几米到十几米的距离，而 GPS 的误差可能会有几十米，所以发生上面的事情也就可以理解了。此时如果在手机中增加一个气压传感器就不一样了，它的精度可以做到 1 米的误差，这样就可以很好地辅助 GPS 来测量出所处的高度，错误导航的问题也就容易解决了。

而气压的方式可选择的范围会广些，而且可以把成本控制在比较低的水平。另外像 Galaxy Nexus 等手机的气压传感器还包括温度传感器，它可以捕捉到温度来对结果进行修正，以增加测量结果的精度。所以在手机原有 GPS 的基础上再增加气压传感器的功能，可以让三维定位更加精准。

在 Android 系统中，气压传感器的类型是 TYPE_PRESSURE，单位是 hPa（百帕斯卡），能够返回当前环境下的压强。

### 13.3.3 实战演练——开发一个 Android 气压计

本实例将详细讲解开发一个 Web 版气压测试系统的方法。

| 实　例 | 功　能 | 源码路径 |
| --- | --- | --- |
| 实例 13-3 | 开发一个 Android 气压计 | 光盘:\daima\13\barometerEX |

（1）编写插件调用文件

编写插件调用文件 plugin.xml，具体代码如下：

```
<plugin xmlns="http://apache.org/cordova/ns/plugins/1.0"
        id="org.dartlang.phonegap.barometer"
    version="0.0.2">

    <name>Device Barometer</name>
    <description>Cordova Device Barometer Plugin</description>
    <license>LICENSE</license>
    <keywords>cordova,device,barometer</keywords>
    <repo>https://github.com/zanderso/cordova-plugin-barometer</repo>
    <issue></issue>

    <js-module src="www/Pressure.js" name="Pressure">
        <clobbers target="Pressure" />
    </js-module>

    <js-module src="www/barometer.js" name="barometer">
        <clobbers target="navigator.barometer" />
    </js-module>

    <!-- android -->
    <platform name="android">

        <config-file target="res/xml/config.xml" parent="/*">
            <feature name="Barometer">
                <param name="android-package" value="org.dartlang.phonegap.barometer. BarometerListener"/>
            </feature>
        </config-file>

        <source-file src="src/android/BarometerListener.java" target-dir="src/org/dartlang/phonegap/barometer" />

    </platform>
</plugin>
```

（2）编写 Cordova 插件文件

编写 Cordova 插件文件 barometer.js 设置可以访问气压计的数据，具体实现代码如下：

```
var argscheck = require('cordova/argscheck'),
    utils = require("cordova/utils"),
    exec = require("cordova/exec"),
    Acceleration = require('./Pressure');

var running = false;

var timers = {};

var listeners = [];
```

```
var pressure = null;

//告知本地开始测试获取数据
function start() {
    exec(function(a) {
        var tempListeners = listeners.slice(0);
        pressure = new Pressure(a.val, a.timestamp);
        for (var i = 0, l = tempListeners.length; i < l; i++) {
            tempListeners[i].win(pressure);
        }
    }, function(e) {
        var tempListeners = listeners.slice(0);
        for (var i = 0, l = tempListeners.length; i < l; i++) {
            tempListeners[i].fail(e);
        }
    }, "Barometer", "start", []);
    running = true;
}

//告知本地停止获取数据
function stop() {
    exec(null, null, "Barometer", "stop", []);
    running = false;
}

//在监听数组中添加回调
function createCallbackPair(win, fail) {
    return {win:win, fail:fail};
}

function removeListeners(l) {
    var idx = listeners.indexOf(l);
    if (idx > -1) {
        listeners.splice(idx, 1);
        if (listeners.length === 0) {
            stop();
        }
    }
}

var barometer = {
    /**
     * 异步获取当前的气压
     *
     * @param {Function} successCallback       The function to call when the pressure data is available
     * @param {Function} errorCallback         The function to call when there is an error getting the pressure
data. (OPTIONAL)
     * @param {BarometerOptions} options       The options for getting the barometer data such as frequency.
(OPTIONAL)
     */
```

```
    getCurrentPressure: function(successCallback, errorCallback, options) {
        argscheck.checkArgs('fFO', 'barometer.getCurrentPressure', arguments);

        var p;
        var win = function(a) {
            removeListeners(p);
            successCallback(a);
        };
        var fail = function(e) {
            removeListeners(p);
            errorCallback && errorCallback(e);
        };

        p = createCallbackPair(win, fail);
        listeners.push(p);

        if (!running) {
            start();
        }
    },

    /**
     *在一个指定的事件间隔内不间断地异步获取气压值
     *
     * @param {Function} successCallback
     * @param {Function} errorCallback          The function to call when there is an error getting the pressure
data. (OPTIONAL)
     * @param {BarometerOptions} options        The options for getting the barometer data such as frequency.
(OPTIONAL)
     * @return String                           The watch id that must be passed to #clearWatch to stop
watching.
     */
    watchPressure: function(successCallback, errorCallback, options) {
        argscheck.checkArgs('fFO', 'barometer.watchPressure', arguments);
        var frequency = (options && options.frequency && typeof options.frequency == 'number') ? options.
frequency : 10000;

        //读取压力数值
        var id = utils.createUUID();

        var p = createCallbackPair(function(){}, function(e) {
            removeListeners(p);
            errorCallback && errorCallback(e);
        });
        listeners.push(p);

        timers[id] = {
            timer:window.setInterval(function() {
                if (pressure) {
                    successCallback(pressure);
                }
```

```
                }, frequency),
                listeners:p
            };

            if (running) {
                if (pressure) {
                    successCallback(pressure);
                }
            } else {
                start();
            }

            return id;
        },

        /**
         *清除指定的气压表
         *
         * @param {String} id          The id of the watch returned from #watchPressure.
         */
        clearWatch: function(id) {
            if (id && timers[id]) {
                window.clearInterval(timers[id].timer);
                removeListeners(timers[id].listeners);
                delete timers[id];
            }
        }
};
module.exports = barometer;
```

（3）定义每个时间点的压力值

编写文件 Pressure.js 定义每个时间点的压力值，具体代码如下：

```
var Pressure = function(val, timestamp) {
    this.val = val;
    this.timestamp = timestamp || (new Date()).getTime();
};

module.exports = Pressure;
```

（4）监听传感器传来的和存储的新压力值

在 Android 平台下编写 Java 程序文件 BarometerListener.java，功能是监听传感器传来的和存储的新压力值，具体实现代码如下：

```
/**
 * 监听传感器传来的和存储的新压力值
 */
public class BarometerListener extends CordovaPlugin implements SensorEventListener {

    public static int STOPPED = 0;
    public static int STARTING = 1;
    public static int RUNNING = 2;
    public static int ERROR_FAILED_TO_START = 3;
```

```
    private float pressure;                              //最近的压力值
    private long timestamp;                              //最近的时间值
    private int status;                                  //监听级别
    private int accuracy = SensorManager.SENSOR_STATUS_UNRELIABLE;

    private SensorManager sensorManager;                 //SensorManager 对象
    private Sensor mSensor;                              //传递气压传感器

    private CallbackContext callbackContext;             //跟踪 JS 回调的上下文
.

    private Handler mainHandler=null;
    private Runnable mainRunnable = new Runnable() {
        public void run() {
            BarometerListener.this.timeout();
        }
    };

    /**
     * 创建一个气压监听
     */
    public BarometerListener() {
        this.pressure = 0;
        this.timestamp = 0;
        this.setStatus(BarometerListener.STOPPED);
    }

    /**
     *初始化处理，得到和活动相关的路径
.
     *
     * @param cordova The context of the main Activity.
     * @param webView The associated CordovaWebView.
     */
    @Override
    public void initialize(CordovaInterface cordova, CordovaWebView webView) {
        super.initialize(cordova, webView);
        this.sensorManager = (SensorManager) cordova.getActivity().getSystemService(Context.SENSOR_
SERVICE);
    }

    /**
     *执行获取请求
     *
     * @param action          The action to execute.
     * @param args                The exec() arguments.
     * @param callbackId          The callback id used when calling back into JavaScript.
     * @return                    Whether the action was valid.
     */
    public boolean execute(String action, JSONArray args, CallbackContext callbackContext) {
```

```
        if (action.equals("start")) {
            this.callbackContext = callbackContext;
            if (this.status != BarometerListener.RUNNING) {
                this.start();
            }
        }
        else if (action.equals("stop")) {
            if (this.status == BarometerListener.RUNNING) {
                this.stop();
            }
        } else {
            return false;
        }

        PluginResult result = new PluginResult(PluginResult.Status.NO_RESULT, "");
        result.setKeepCallback(true);
        callbackContext.sendPluginResult(result);
        return true;
    }

    /**
     *当气压传感器被关闭时则停止监听
     */
    public void onDestroy() {
        this.stop();
    }

    //--------------------------------------------------------------------------
    // 下面是本地方法
    //--------------------------------------------------------------------------
    //
    /**
     *开始监听传感器
     *
     * @return              status of listener
    */
    private int start() {
        //如果已经开始或运行则返回
        if ((this.status == BarometerListener.RUNNING) || (this.status == BarometerListener.STARTING)) {
            return this.status;
        }

        this.setStatus(BarometerListener.STARTING);

        //从传感器获取气压
        List<Sensor> list = this.sensorManager.getSensorList(Sensor.TYPE_PRESSURE);

        if ((list != null) && (list.size() > 0)) {
          this.mSensor = list.get(0);
          this.sensorManager.registerListener(this, this.mSensor, SensorManager.SENSOR_DELAY_UI);
          this.setStatus(BarometerListener.STARTING);
```

```
        } else {
            this.setStatus(BarometerListener.ERROR_FAILED_TO_START);
            this.fail(BarometerListener.ERROR_FAILED_TO_START, "No sensors found to register barometer
listening to.");
            return this.status;
        }

        //设置一个超时回调的主线程
        stopTimeout();
        mainHandler = new Handler(Looper.getMainLooper());
        mainHandler.postDelayed(mainRunnable, 2000);

        return this.status;
    }
    private void stopTimeout() {
        if(mainHandler!=null){
            mainHandler.removeCallbacks(mainRunnable);
        }
    }
    /**
     *停止监听
     */
    private void stop() {
        stopTimeout();
        if (this.status != BarometerListener.STOPPED) {
            this.sensorManager.unregisterListener(this);
        }
        this.setStatus(BarometerListener.STOPPED);
        this.accuracy = SensorManager.SENSOR_STATUS_UNRELIABLE;
    }

    /**
     *如果传感器还没有开始则返回一个错误
     *
     * Called two seconds after starting the listener.
     */
    private void timeout() {
        if (this.status == BarometerListener.STARTING) {
            this.setStatus(BarometerListener.ERROR_FAILED_TO_START);
            this.fail(BarometerListener.ERROR_FAILED_TO_START, "Barometer could not be started.");
        }
    }

    /**
     * Called when the accuracy of the sensor has changed.
     *
     * @param sensor
     * @param accuracy
     */
    public void onAccuracyChanged(Sensor sensor, int accuracy) {
        if (sensor.getType() != Sensor.TYPE_PRESSURE) {
```

```
            return;
        }

        if (this.status == BarometerListener.STOPPED) {
            return;
        }
        this.accuracy = accuracy;
    }

    /**
     *传感器监听事件
     *
     * @param SensorEvent event
     */
    public void onSensorChanged(SensorEvent event) {
        if (event.sensor.getType() != Sensor.TYPE_PRESSURE) {
            return;
        }

        if (this.status == BarometerListener.STOPPED) {
            return;
        }
        this.setStatus(BarometerListener.RUNNING);

        if (this.accuracy >= SensorManager.SENSOR_STATUS_ACCURACY_MEDIUM) {

            this.timestamp = System.currentTimeMillis();
            this.pressure = event.values[0];

            this.win();
        }
    }

    @Override
    public void onReset() {
        if (this.status == BarometerListener.RUNNING) {
            this.stop();
        }
    }

    //发送一个错误到 JS 文件
    private void fail(int code, String message) {
        JSONObject errorObj = new JSONObject();
        try {
            errorObj.put("code", code);
            errorObj.put("message", message);
        } catch (JSONException e) {
            e.printStackTrace();
        }
        PluginResult err = new PluginResult(PluginResult.Status.ERROR, errorObj);
        err.setKeepCallback(true);
```

```
        callbackContext.sendPluginResult(err);
    }

    private void win() {
        PluginResult result = new PluginResult(PluginResult.Status.OK, this.getPressureJSON());
        result.setKeepCallback(true);
        callbackContext.sendPluginResult(result);
    }

    private void setStatus(int status) {
        this.status = status;
    }
    private JSONObject getPressureJSON() {
        JSONObject r = new JSONObject();
        try {
            r.put("val", this.pressure);
            r.put("timestamp", this.timestamp);
        } catch (JSONException e) {
            e.printStackTrace();
        }
        return r;
    }
}
```

至此，整个实例介绍完毕，这样便成功构建了一个简单的气压计模型。读者可以以此为基础进行拓展，设计一个自己喜欢的气压计程序。

# 第 14 章　温度传感器和湿度传感器

温度传感器（Temperature Transducer）是指能感受温度并转换成可用输出信号的传感器。和测量重量、温度一样，选择湿度传感器首先要确定测量范围。除了气象、科研部门外，从事温、湿度测控的一般不需要全湿程（0-100%RH）测量。在当今的信息时代，传感器技术与计算机技术、自动控制技术紧密结合。测量的目的在于控制，测量范围与控制范围合称使用范围。当然，对不需要从事测控系统的应用者来说，直接选择通用型湿度仪即可。本章将详细讲解 Android 设备中温度传感器和湿度传感器的基本知识，为读者步入本书后面知识的学习打下基础。

## 14.1　温度传感器详解

**知识点讲解：光盘:视频\知识点\第 14 章\温度传感器详解.avi**

从 17 世纪初人们开始利用温度进行测量。在半导体技术的支持下，20 世纪相继开发了半导体热电偶传感器、PN 结温度传感器和集成温度传感器。与之相应，根据波与物质的相互作用规律，相继开发了声学温度传感器、红外传感器和微波传感器。温度传感器是五花八门的各种传感器中最为常用的一种，现代的温度传感器外形非常小，这样更加让它广泛应用在生产实践的各个领域中，也为人们的生活提供了无数的便利和功能。

### 14.1.1　温度传感器介绍

温度传感器有 4 种主要类型：热电偶、热敏电阻、电阻温度检测器（RTD）和 IC 温度传感器。IC 温度传感器又包括模拟输出和数字输出两种类型。在现实世界中，温度传感器是温度测量仪表的核心部分，品种繁多。按测量方式可以分为接触式和非接触式两大类，按照传感器材料及电子元件特性分为热电阻和热电偶两类。

在当前的技术水平条件下，温度传感器的主要原理如下：

（1）金属膨胀原理设计的传感器

金属在环境温度变化后会产生一个相应的延伸，因此传感器可以以不同方式对这种反应进行信号转换。

（2）双金属片式传感器

双金属片由两片不同膨胀系数的金属贴在一起而组成，随着温度变化，材料 A 比另外一种金属膨胀程度要高，引起金属片弯曲。弯曲的曲率可以转换成一个输出信号。

（3）双金属杆和金属管传感器

随着温度的升高，金属管（材料 A）长度增加，而不膨胀钢杆（金属 B）的长度并不增加，这样由于位置的改变，金属管的线性膨胀就可以进行传递。反过来，这种线性膨胀可以转换成一个输出信号。

（4）液体和气体的变形曲线设计的传感器

在温度变化时，液体和气体同样会相应产生体积的变化。

综上所述，多种类型的结构可以把这种膨胀的变化转换成位置的变化，这样产生位置的变化可以输出为电位计、感应偏差、挡流板等形式的结果。

## 14.1.2　Android 系统中的温度传感器

在 Android 系统中，早期版本的温度传感器值是 TYPE_TEMPERATURE，在新版本中被 TYPE_AMBIENT_TEMPERATURE 替换。Android 温度传感器的单位是℃，能够测量并返回当前的温度。

在 Android 内核平台中自带了大量的传感器源码，读者可以在 Rexsee 的开源社区 http://www.rexsee.com/找到相关的原生代码。其中使用温度传感器相关的原生代码如下：

```
package rexsee.sensor;

import rexsee.core.browser.JavascriptInterface;
import rexsee.core.browser.RexseeBrowser;
import android.content.Context;
import android.hardware.Sensor;
import android.hardware.SensorEvent;
import android.hardware.SensorEventListener;
import android.hardware.SensorManager;

public class RexseeSensorTemperature implements JavascriptInterface {

	private static final String INTERFACE_NAME = "Temperature";
	@Override
	public String getInterfaceName() {
		return mBrowser.application.resources.prefix + INTERFACE_NAME;
	}
	@Override
	public JavascriptInterface getInheritInterface(RexseeBrowser childBrowser) {
		return this;
	}
	@Override
	public JavascriptInterface getNewInterface(RexseeBrowser childBrowser) {
		return new RexseeSensorTemperature(childBrowser);
	}

	public static final String EVENT_ONTEMPERATURECHANGED = "onTemperatureChanged";

	private final Context mContext;
	private final RexseeBrowser mBrowser;
	private final SensorManager mSensorManager;
	private final SensorEventListener mSensorListener;
	private final Sensor mSensor;

	private int mRate = SensorManager.SENSOR_DELAY_NORMAL;
```

```
    private int mCycle = 100; //milliseconds
    private int mEventCycle = 100; //milliseconds
    private float mAccuracy = 0;

    private long lastUpdate = -1;
    private long lastEvent = -1;

    private float value = -999f;

    public RexseeSensorTemperature(RexseeBrowser browser) {
            mContext = browser.getContext();
            mBrowser = browser;
            browser.eventList.add(EVENT_ONTEMPERATURECHANGED);

            mSensorManager = (SensorManager)
mContext.getSystemService(Context.SENSOR_SERVICE);

            mSensor = mSensorManager.getDefaultSensor(Sensor.TYPE_TEMPERATURE);

            mSensorListener = new SensorEventListener() {
                    @Override
                    public void onAccuracyChanged(Sensor sensor, int accuracy) {
                    }
                    @Override
                    public void onSensorChanged(SensorEvent event) {
                            if (event.sensor.getType() != Sensor.TYPE_TEMPERATURE) return;
                            long curTime = System.currentTimeMillis();
                            if (lastUpdate == -1 || (curTime - lastUpdate) > mCycle) {
                                    lastUpdate = curTime;
                                    float lastValue = value;
                                    value = event.values[SensorManager.DATA_X];
                                    if (lastEvent == -1 || (curTime - lastEvent) > mEventCycle) {
                                            if (Math.abs(value - lastValue) > mAccuracy) {
                                                    lastEvent = curTime;
                                                    mBrowser.eventList.run(EVENT_
ONTEMPERATURECHANGED);
                                            }
                                    }
                            }
                    }
            };

    }

    public String getLastKnownValue() {
            return (value == -999) ? "null" : String.valueOf(value);
    }

    public void setRate(String rate) {
            mRate = SensorRate.getInt(rate);
    }
```

```
    public String getRate() {
        return SensorRate.getString(mRate);
    }
    public void setCycle(int milliseconds) {
        mCycle = milliseconds;
    }
    public int getCycle() {
        return mCycle;
    }
    public void setEventCycle(int milliseconds) {
        mEventCycle = milliseconds;
    }
    public int getEventCycle() {
        return mEventCycle;
    }
    public void setAccuracy(float value) {
        mAccuracy = Math.abs(value);
    }
    public float getAccuracy() {
        return mAccuracy;
    }

    public boolean isReady() {
        return (mSensor == null) ? false : true;
    }
    public void start() {
        if (isReady()) {
            mSensorManager.registerListener(mSensorListener, mSensor, mRate);
        } else {
            mBrowser.exception(getInterfaceName(), "Temperature sensor is not found.");
        }
    }
    public void stop() {
        if (isReady()) {
            mSensorManager.unregisterListener(mSensorListener);
        }
    }
}
```

### 14.1.3 实战演练——开发一个 Android 温度计

本实例将详细讲解在 Android 设备中实现温度计功能的方法。

| 实　例 | 功　能 | 源 码 路 径 |
|---|---|---|
| 实例 14-1 | 开发一个 Android 温度计系统 | 光盘:\daima\14\wenduEX |

（1）实现布局文件

编写布局文件 main.xml，通过 TextView 控件显示获取的温度值，具体实现代码如下：

```
<LinearLayout xmlns:android="http://schemas.android.com/apk/res/android"
```

```
    android:orientation="vertical"
    android:layout_width="fill_parent"
    android:layout_height="fill_parent"><!--添加一个垂直的线性布局-->
    <TextView
        android:id="@+id/title"
        android:gravity="center_horizontal"
        android:textSize="20px"
        android:layout_width="fill_parent"
        android:layout_height="wrap_content"
        android:text="@string/title"/><!--添加一个 TextView 控件-->
    <TextView
        android:id="@+id/myTextView1"
        android:textSize="18px"
        android:layout_width="fill_parent"
        android:layout_height="wrap_content"
        android:text="@string/myTextView1"/><!--添加一个 TextView 控件-->
</LinearLayout>
```

（2）检测温度传感器的温度变化

为该项目添加 SensorSimulator 工具的 JAR 包，编写文件 activity.java 来检测温度传感器的温度变化，并设置在 onSensorChanged()方法中只对 SENSOR_TEMPERATURE 即温度变化进行检测。文件 activity.java 的具体实现代码如下：

```
import org.openintents.sensorsimulator.hardware.SensorManagerSimulator;

import wendu.R;
import android.app.Activity;
import android.hardware.SensorListener;
import android.hardware.SensorManager;
import android.os.Bundle;
import android.widget.TextView;
public class activity extends Activity {
    TextView myTextView1;//当前温度
    //SensorManager mySensorManager;//SensorManager 对象引用
    SensorManagerSimulator mySensorManager;//声明 SensorManagerSimulator 对象，调试时用
    @Override
    public void onCreate(Bundle savedInstanceState) {//重写 onCreate()方法
        super.onCreate(savedInstanceState);
        setContentView(R.layout.main);//设置当前的用户界面
        myTextView1 = (TextView) findViewById(R.id.myTextView1);//得到 myTextView1 的引用
        //mySensorManager = (SensorManager)getSystemService(SENSOR_SERVICE);
        //获得 SensorManager
        //调试时用
        mySensorManager = SensorManagerSimulator.getSystemService(this, SENSOR_SERVICE);
        mySensorManager.connectSimulator();                    //与 Simulator 连接
    }
    private SensorListener mySensorListener = new SensorListener(){
     @Override
     public void onAccuracyChanged(int sensor, int accuracy) {}      //重写 onAccuracyChanged()方法
     @Override
     public void onSensorChanged(int sensor, float[] values) {       //重写 onSensorChanged()方法
         if(sensor == SensorManager.SENSOR_TEMPERATURE){//只检查温度的变化
```

```
                myTextView1.setText("当前的温度为："+values[0]); //将当前温度显示到 TextView
            }
        }
    };
    @Override
    protected void onResume() {//重写 onResume()方法
     mySensorManager.registerListener(//注册监听
                mySensorListener, //监听器 SensorListener 对象
                SensorManager.SENSOR_TEMPERATURE,//传感器的类型为温度
                SensorManager.SENSOR_DELAY_UI//传感器事件传递的频度
                );
     super.onResume();
    }
    @Override
    protected void onPause() {//重写 onPause()方法
     mySensorManager.unregisterListener(mySensorListener);//取消注册监听器
     super.onPause();
    }
}
```

至此，整个实例介绍完毕。运行电脑端的 sensorsimulator 工具，调整工具中的参数使其模拟温度传感器。在 Environment Sensor 中选择 Temperature，在 Quick Setting 面板中对温度进行快速设置，在传感器模拟器中会显示当前所有的模拟参数，运行后会显示获取的温度值。

### 14.1.4　实战演练——测试电池的温度

本实例将详细讲解测试 Android 设备的电源温度的方法。本实例在现实中可能不会有市场效益，但是演示了测试电源温度的方法，对初学者来说具有指导意义。

| 实　　例 | 功　　能 | 源 码 路 径 |
| --- | --- | --- |
| 实例 14-2 | 测试电池的温度 | 光盘:\daima\14\ThermoEX |

（1）实现布局文件

编写布局文件 main.xml，功能是通过按钮控件来控制测温功能，具体实现代码如下：

```
<?xml version="1.0" encoding="utf-8"?>
<LinearLayout xmlns:android="http://schemas.android.com/apk/res/android"
      xmlns:mako = "http://schemas.android.com/apk/res/com.mako"
      android:orientation="horizontal"
      android:layout_width="match_parent"
      android:layout_height="match_parent"
      >
      <LinearLayout
            android:orientation="vertical"
            android:layout_height="match_parent"
            android:layout_width="62dp"
            android:gravity="bottom|center_horizontal"
            >
            <ZoomControls android:id = "@+id/zoomControl"
                  android:rotation="90"
```

```
                android:layout_width="wrap_content"
                android:layout_height="wrap_content"
                android:gravity="center_horizontal"
                android:orientation="vertical" />
            <com.mako.RecCheckButton android:id="@+id/recCheckButton"
                android:layout_width="wrap_content"
                android:layout_height="wrap_content" />
        </LinearLayout>
        <com.mako.TemperatureDataView
            android:layout_width="match_parent"
            android:layout_height="match_parent" />
        <!-- <com.mako.FrogListView
            android:layout_width="match_parent"
            android:layout_height="match_parent" /> -->
</LinearLayout>
```

（2）监听用户在主屏幕的操作

编写文件 mainact.java，功能是监听用户在主屏幕的操作，根据操作来执行对应的响应事件处理函数。文件 mainact.java 的具体实现代码如下：

```
public class mainact extends Activity{
    protected ArrayList<ArrayList<Datum>> temperaturesBaseRecordings;
    protected LinkedList<SummaryLayer> summaries;
    public void beginRecording(){
    }

    public void haltRecording(){
    }

    @Override
    public void onCreate(Bundle savedInstanceState){
        super.onCreate(savedInstanceState);
        setContentView(R.layout.main);
    }

    @Override
    public void onRestoreInstanceState(Bundle state){
        super.onRestoreInstanceState(state);
    }
    @Override
    public void onSaveInstanceState(Bundle state){
        super.onSaveInstanceState(state);
    }
    @Override
    public void onResume(){
        super.onResume();
    }
    @Override
    public void onRestart(){
        super.onRestart();
    }
    @Override
```

```
    public void onDestroy(){
        super.onDestroy();
    }
}
```

（3）实现基准数据处理

编写文件 Datum.java 实现基准数据处理，具体实现代码如下：

```
public class Datum implements Parcelable {
    public float temperature;
    public static final int
     CONTAMINATED_BY_ENGAGEMENT=1, /*that is, if it's being used, Likely to draw more power and
raise battery temp*/
     CONTAMINATED_BY_CHARGING=2,
     CONTAMINATED_BY_EMPTIES=4,
     NO_DATA = 5;
    public int flags;
    public Datum(float temp, int flag){temperature = temp; flags = flag;}

    public int describeContents(){return 0;}
    public void writeToParcel(Parcel out, int flaggings){
        out.writeInt(flags);
        out.writeFloat(temperature);
    }
    public static final Parcelable.Creator<Datum> CREATOR = new Parcelable.Creator<Datum>() {
        public Datum createFromParcel(Parcel in){
            float tempor = in.readFloat();
            int flagings = in.readInt();
            return new Datum(tempor, flagings);
        }
        public Datum [] newArray(int size){
            return new Datum[size];
        }
    };
    @Override
    public Datum clone(){
        return new Datum(temperature, flags);
    }
}
```

（4）绘制基准视图界面

编写文件 DatumView.java，功能是调用基准数据类 Datum 实现绘制基准视图界面功能，具体实现代码如下：

```
class DatumView extends LinearLayout{
    protected Datum v;
    protected TextView tv;
    protected FlagImage fi;
    public static final float defaultHeightInches = (float)0.4;
    protected static BitmapDrawable[] flagBitmaps = new BitmapDrawable[8];
    protected static BitmapDrawable nodatBitmap;
    public static Paint backGroundPaint;
    public static Paint haskPaint;
    public static Paint contaminatedEmpty;
```

```
    public static Paint contaminatedUsage;
    public static Paint contaminatedCharging;
    public static Path laef;
    protected class FlagImage extends ImageView{
        public FlagImage(Context cont, AttributeSet attrs, int defStyle){
            super(cont, attrs, defStyle); init(cont); }
        public FlagImage(Context cont, AttributeSet attrs){
            super(cont, attrs); init(cont); }
        public FlagImage(Context cont){
            super(cont); init(cont); }
        protected int flags;
        protected void init(Context cont){
            takeFlag(Datum.NO_DATA);
        }
        protected Bitmap paintBlank(int w, int h){
            Bitmap surf = Bitmap.createBitmap(w, h, Bitmap.Config.ARGB_8888);
            Canvas c = new Canvas(surf);
            c.drawPaint(backGroundPaint);
            float haskrad = (h*8)/10;
            float xo = w/2 - haskrad, yo = h/2 - haskrad;
            Path haskPath = new Path();
            RectF arcRect = new RectF(xo,yo,xo + haskrad,yo + haskrad);
            float arcRad = 64;
            haskPath.addArc(arcRect, -45 - arcRad/2, arcRad);
            haskPath.addArc(arcRect, -(45+180) + arcRad/2, -arcRad);
            haskPath.close();
            c.drawPath(haskPath, haskPaint);
            return surf;
        }
        protected Bitmap paintFlags(int w, int h, int flaggings){
            Bitmap surf = Bitmap.createBitmap(w, h, Bitmap.Config.ARGB_8888);
            Canvas c = new Canvas(surf);
            c.drawPaint(backGroundPaint);
            float separation = 2;
            float oneFlagSpan =
                (w <= 5*separation)?
                    0:
                    (h < 2*separation + (w - 5*separation)/3)?
                        h - 2*separation:
                        (w - 5*separation)/3;
            float fullWidth = oneFlagSpan*3 + 5*separation;
            float fullHeight = oneFlagSpan + 2*separation;
            Path mlaef = new Path(laef);
            Matrix transf = new Matrix(); transf.setRectToRect(
                new RectF(0,0,1,1),
                new RectF((h - fullHeight)/2 + separation, (w - fullWidth)/2 + separation,
separation+oneFlagSpan, separation+oneFlagSpan),
                Matrix.ScaleToFit.CENTER);
            mlaef.transform(transf);
            if((flaggings & Datum.CONTAMINATED_BY_ENGAGEMENT) != 0) c.drawPath(mlaef,
contaminatedUsage);
```

```
                mlaef.offset(oneFlagSpan + separation, 0);
                if((flaggings & Datum.CONTAMINATED_BY_CHARGING) != 0) c.drawPath(mlaef,
contaminatedCharging);
                mlaef.offset(oneFlagSpan + separation, 0);
                if((flaggings & Datum.CONTAMINATED_BY_EMPTIES) != 0) c.drawPath(mlaef,
contaminatedEmpty);
                return surf;
            }
            protected void installFlag(){
                if(flags == 0) return;
                if((flags & Datum.NO_DATA) != 0){
                    if(nodatBitmap != null) setImageDrawable(nodatBitmap);
                    else
                        setImageDrawable(( nodatBitmap = new BitmapDrawable(
                            getContext().getResources(),
                            paintBlank(getWidth(), getHeight())) ));
                }else{
                    int flagIndex = flags&(7);
                    if(flagBitmaps[flagIndex] != null) setImageDrawable(flagBitmaps[flagIndex]);
                    else
                        setImageDrawable(( flagBitmaps[flagIndex] = new BitmapDrawable(
                            getContext().getResources(),
                            paintFlags(getWidth(), getHeight(), flags)) ));
                }
            }
            public void takeFlag(int inFlag){
                flags = inFlag
                if(getHeight() != 0 && getWidth() != 0){
                    installFlag();
                }
            }
            @Override
            public void onSizeChanged(int oldw, int oldh, int w, int h){
                for(int i = 0; i < flagBitmaps.length; ++i) flagBitmaps[i] = null;
                nodatBitmap = null;
                takeFlag(flags);
            }
        }
        public DatumView(Context cont, AttributeSet attrs, int defStyle){
            super(cont, attrs, defStyle);
            init(cont);
        }
        public DatumView(Context cont, AttributeSet attrs){
            super(cont, attrs);
            init(cont);
        }
        public DatumView(Context cont){
            super(cont);
            init(cont);
        }
        protected void init(Context cont){
```

```
        Log.v("frog", "DatumView created");
        if(backGroundPaint == null){
            backGroundPaint = new Paint();
            backGroundPaint.setColor(0xffb0b0b0);
        }
        if(haskPaint == null){
            haskPaint = new Paint(Paint.ANTI_ALIAS_FLAG);
            haskPaint.setColor(0xffb0b0b0);
        }
        if(contaminatedCharging == null){
            contaminatedCharging = new Paint(Paint.ANTI_ALIAS_FLAG);
            contaminatedCharging.setColor(0xffffff00);
        }
        if(contaminatedUsage == null){
            contaminatedUsage = new Paint(Paint.ANTI_ALIAS_FLAG);
            contaminatedUsage.setColor(0xff00ff00);
        }
        if(contaminatedEmpty == null){
            contaminatedEmpty = new Paint(Paint.ANTI_ALIAS_FLAG);
            contaminatedEmpty.setColor(0xff00ffff);
        }
        if(laef == null){
            float cornerDeg = (float)0.1;
            laef = new Path();
            laef.moveTo(0,cornerDeg);
            laef.lineTo(cornerDeg,0);
            laef.lineTo(1,0);
            laef.lineTo(1,1 - cornerDeg);
            laef.lineTo(1 - cornerDeg,1);
            laef.lineTo(0,1);
            laef.close();
        }
        fi = new FlagImage(cont);
        fi.takeFlag(Datum.NO_DATA);
        tv = new TextView(cont);
        LinearLayout.LayoutParams tvLParams = new LinearLayout.LayoutParams(0,
ViewGroup.LayoutParams.WRAP_CONTENT);  tvLParams.gravity  =  android.view.Gravity.LEFT;tvLParams.
weight = 1;
        LinearLayout.LayoutParams fiLParams = new LinearLayout.LayoutParams(0, ViewGroup. LayoutParams.
MATCH_PARENT, 0); tvLParams.gravity = android.view.Gravity.RIGHT; tvLParams.weight = 0;
        addView( tv,   tvLParams );
        addView( fi,   fiLParams );
    }
    protected void installDatum(Datum in){
        v=in;
        if(v == null || (v.flags & Datum.NO_DATA) != 0){
        }else{
            tv.setText(""+v.temperature);
            fi.takeFlag(v.flags);
        }
    }
}
```

（5）记录测试的数据

编写文件 recording.java，功能是记录测试的数据，并将各个记录的值保存到 DB 数据库中。文件 recording.java 的具体实现代码如下：

```
public class recording extends Service{
    public static final long MEASURE_INTERVAL_IN_MILLISECONDS = 1000* 60* 5;
    protected int currentTemperature;
    protected boolean isCharging;
    protected boolean isBeingUsed;
    protected ScheduledThreadPoolExecutor timer;
    protected ScheduledFuture<?> future;
    protected Runnable recordingTask = new Runnable(){
        @Override public void run(){

        }
    };
    protected BroadcastReceiver batteryInfoReceiver = new BroadcastReceiver(){
        @Override
        public void onReceive(Context context, Intent intent){
            currentTemperature = intent.getIntExtra(BatteryManager.EXTRA_TEMPERATURE,0);
            isCharging = ((intent.getIntExtra(BatteryManager.EXTRA_STATUS,0) & BatteryManager.
BATTERY_STATUS_CHARGING) != 0);
        }};
    protected BroadcastReceiver screenIsOnReciever = new BroadcastReceiver(){
        @Override
        public void onReceive(Context context, Intent intent){    isBeingUsed = true;    }};
    protected BroadcastReceiver screenIsOffReciever = new BroadcastReceiver(){
        @Override
        public void onReceive(Context context, Intent intent){    isBeingUsed = false;    }};
    @Override public int onStartCommand(Intent intent, int flags, int startId){
        this.registerReceiver(batteryInfoReceiver, new IntentFilter(Intent.ACTION_BATTERY_CHANGED));
        this.registerReceiver(screenIsOnReciever,   new IntentFilter(Intent.ACTION_SCREEN_ON)); //using
SCREEN_ON instead of USER_PRESENT as screen is likely to tax the battery even without active use. I
suppose the variable this binds to is a misnomer
        this.registerReceiver(screenIsOffReciever, new IntentFilter(Intent.ACTION_SCREEN_OFF));
        timer = new ScheduledThreadPoolExecutor(1);
        return START_STICKY;
    }
    public void startRecording(){
        //open db
        //create db?
        //int initialDelay =
        future = timer.scheduleAtFixedRate(
            recordingTask,
            0,
            MEASURE_INTERVAL_IN_MILLISECONDS,
            java.util.concurrent.TimeUnit.MILLISECONDS );
    }
    public void haltRecording(){
        future.cancel(false);
    }
```

```
    @Override
    public IBinder onBind(Intent intent){return null;}
}
```

（6）生成图形化温度数据视图

编写文件 TemperatureDataView.java，功能是将存储记录的数字生成一个图形化的温度数据视图。文件 TemperatureDataView.java 的具体实现代码如下：

```
class TemperatureDataView extends ScrollView{
    protected RelativeLayout layout;
    protected View upperView;
    protected static final int upperViewID=0;
    protected View lowerView;
    protected static final int lowerViewID=1;
    protected int upperChild=-1, lowerChild=-1;
    protected int highestID=2;
    protected int dbLength;
    protected boolean seesEnd;
    protected int[] viewsPageCycle;
    protected int viewsPageEye;
    protected int basalRecord;
    protected int pageSize;
    protected int nPages;
    protected boolean sizeGiven = false;
    protected ExecutorService pool;
    protected Future<ArrayList<Datum>> lowerAccess;
    protected Future<ArrayList<Datum>> upperAccess;
    public static final float defaultItemHeightInches = (float)0.4;
    protected int itemHeight;
    protected RelativeLayout.LayoutParams defItemParams(){
        return new RelativeLayout.LayoutParams(ViewGroup.LayoutParams.MATCH_PARENT,
itemHeight);
    }

    protected class AccessTask implements Callable<ArrayList<Datum>>{
        int first, n;
        Datum defdat = new Datum(0, Datum.CONTAMINATED_BY_ENGAGEMENT |
Datum.CONTAMINATED_BY_CHARGING | Datum.CONTAMINATED_BY_EMPTIES);
        public AccessTask(int firstID, int nItems){ //, DB db){
            first = firstID; n = nItems;
        }
        public ArrayList<Datum> call(){
            ArrayList<Datum> ret = new ArrayList<Datum>(n);

            for(Datum dat : ret) dat = defdat.clone();
            return ret;
        }
    }

    protected void increaseUpperPadSize(){
        RelativeLayout.LayoutParams newUppersLayoutParams = new
RelativeLayout.LayoutParams(ViewGroup.LayoutParams.MATCH_PARENT,
```

```
upperView.getHeight()+pageSize*itemHeight);
            newUppersLayoutParams.addRule(RelativeLayout.ALIGN_PARENT_TOP);
            layout.updateViewLayout(upperView, newUppersLayoutParams);
      }
      protected void accessLower(){
            if(lowerAccess == null)
                  lowerAccess = pool.submit(new AccessTask(basalRecord+nPages*pageSize, pageSize));
      }
      protected void finalizeShiftViewDown(){
            ArrayList<Datum> resultList;
            if(lowerAccess == null){
                  accessLower();
            }
            try{
                  resultList = lowerAccess.get();
            }catch(InterruptedException ex){throw new RuntimeException("db access thread interrupted?");}
            catch(ExecutionException ex){throw new RuntimeException("db access thread interrupted?");}

            increaseUpperPadSize();
            int viewsPageCycleIter = viewsPageEye*pageSize;
            int end = viewsPageCycleIter+pageSize;
            int previousLayoutID =
                  (viewsPageEye == 0)?
                        viewsPageCycle[nPages*pageSize - 1]:
                        viewsPageCycle[viewsPageEye*pageSize -1];
            Iterator<Datum> iter = resultList.iterator();
            while(viewsPageCycleIter != end){
                  layout.removeViewAt(viewsPageCycle[viewsPageCycleIter]);
                  int newLayoutID = highestID++;
                  viewsPageCycle[viewsPageCycleIter] = newLayoutID;
                  RelativeLayout.LayoutParams params = defItemParams();
                  params.addRule(RelativeLayout.BELOW, previousLayoutID);
                  DatumView dv = new DatumView(getContext());
                  dv.installDatum(iter.next());
                  layout.addView(dv, newLayoutID, params);
                  ++viewsPageCycleIter;
                  previousLayoutID = newLayoutID;
            }
            /*//decrease the lower pad size and reattach it to previousLayoutID; //no. size only changes when
dbLength changes
            RelativeLayout.LayoutParams params = new
RelativeLayout.LayoutParams(ViewGroup.LayoutParams.MATCH_PARENT,
lowerView.getHeight()-pageSize*itemHeight);
            params.addRule(RelativeLayout.BELOW, previousLayoutID);
            layout.updateViewLayout(lowerView, params);*/
            ++viewsPageEye; if(viewsPageEye == nPages) viewsPageEye = 0;
            basalRecord+=pageSize;
            lowerAccess = null;
      }
      protected void acceptFirstSizing(){
            pageSize = this.getHeight()/itemHeight;
            nPages = 3;
```

```
        assert pageSize > 0 : "it looks like size isn't assigned before initialization :(. just make pagesize big or
something";
        viewsPageCycle = new int[nPages*pageSize];
        for(int i = 0; i < viewsPageCycle.length ; ++i) viewsPageCycle[i] = -1;
        viewsPageEye = 0;
        layout = new RelativeLayout(getContext());
        seesEnd = false;
        lowerView = new View(getContext());
        upperView = new View(getContext());
        RelativeLayout.LayoutParams params = new
RelativeLayout.LayoutParams(ViewGroup.LayoutParams.MATCH_PARENT, 0);
        params.addRule(RelativeLayout.ALIGN_PARENT_TOP);
        layout.addView(upperView, upperViewID, params);
        dbLength = 50;
        params = new RelativeLayout.LayoutParams(ViewGroup.LayoutParams.MATCH_PARENT,
dbLength*itemHeight);
        params.addRule(RelativeLayout.ALIGN_PARENT_TOP);
        layout.addView(lowerView, lowerViewID, params);
        addView(
            layout,
            new ViewGroup.LayoutParams(
                ViewGroup.LayoutParams.MATCH_PARENT,
                ViewGroup.LayoutParams.WRAP_CONTENT));
        if(dbLength*itemHeight < getHeight()) seesEnd=true;
        relocate(0);
        scrollTo(0,0);
    }
    protected void init(Context cont){
        pool = Executors.newFixedThreadPool(2);
        {
        android.util.DisplayMetrics metrics = new android.util.DisplayMetrics();

    ((WindowManager)cont.getSystemService(Context.WINDOW_SERVICE)).getDefaultDisplay().getMetrics(
metrics);
        itemHeight = (int)(defaultItemHeightInches * metrics.densityDpi);
        }
    }
    public TemperatureDataView(Context cont, AttributeSet attrs, int defStyle){
        super(cont, attrs, defStyle);
        init(cont);
    }
    public TemperatureDataView(Context cont, AttributeSet attrs){
        super(cont, attrs);
        init(cont);
    }
    public TemperatureDataView(Context cont){
        super(cont);
        init(cont);
    }
    protected void addDatumBegin(Datum v){
        DatumView view = new DatumView(getContext());
        view.installDatum(v);
```

```
        RelativeLayout.LayoutParams params = defItemParams();
        if(upperChild != -1) params.addRule(RelativeLayout.ABOVE, upperChild);
        else params.addRule(RelativeLayout.ALIGN_PARENT_TOP, upperChild);
        upperChild = highestID++;
        layout.addView(view, upperChild, params);
    }
    protected void addDatumEnd(Datum v){
        DatumView view = new DatumView(getContext());
        view.installDatum(v);
        RelativeLayout.LayoutParams params = defItemParams();
        if(lowerChild != -1) params.addRule(RelativeLayout.BELOW, lowerChild);
        else params.addRule(RelativeLayout.ALIGN_PARENT_BOTTOM, lowerChild);
        lowerChild = highestID++;
        layout.addView(view, lowerChild, params);
    }

    public void notifyFreshDatum(){
        ++dbLength;
        if(seesEnd){

        }else{
            RelativeLayout.LayoutParams params = new
RelativeLayout.LayoutParams(ViewGroup.LayoutParams.MATCH_PARENT, dbLength*itemHeight);
            params.addRule(RelativeLayout.ALIGN_PARENT_BOTTOM);
            layout.updateViewLayout(lowerView, params);
        }
    }

    @Override
    public void onSizeChanged(int oldw, int oldh, int w, int h){
        if(!sizeGiven){
            acceptFirstSizing();
            sizeGiven=true;
        }

    }

    protected void relocate(int y){ //initializes cached Datums for the new location
        if(y < 0) y=0;
        for(int i = 0; i < nPages*pageSize; ++i){
            if(viewsPageCycle[i] != -1) layout.removeViewAt(viewsPageCycle[i]);
        }
        viewsPageEye = 0;
        basalRecord = (y/(itemHeight*pageSize) - itemHeight*(pageSize/2))*pageSize;
        highestID = 2;
        ArrayList<Datum> data = (new AccessTask(basalRecord, nPages*pageSize)).call();
        int validStart = (basalRecord < 0)? -basalRecord : 0;
        RelativeLayout.LayoutParams upperLayoutParams;
        if(validStart == 0){
            upperLayoutParams = new
RelativeLayout.LayoutParams(ViewGroup.LayoutParams.MATCH_PARENT, basalRecord*itemHeight);
        }else{
```

```
                upperLayoutParams = new
RelativeLayout.LayoutParams(ViewGroup.LayoutParams.MATCH_PARENT, 0);
            }
            upperLayoutParams.addRule(RelativeLayout.ALIGN_PARENT_TOP);
            layout.updateViewLayout(upperView, upperLayoutParams);
            int prevID = upperViewID;
            for(; validStart < data.size(); ++validStart){
                Datum cur = data.get(validStart);
                DatumView curv = new DatumView(getContext());
                if(cur != null) curv.installDatum(cur);
                int id = highestID++;
                RelativeLayout.LayoutParams params = defItemParams();
                params.addRule(RelativeLayout.BELOW, prevID);
                layout.addView(curv, id, params);
                viewsPageCycle[validStart] = id;
                prevID = id;
            }
            upperChild = 2;
            lowerChild = prevID;
            if(lowerAccess != null) lowerAccess.cancel(true);
            lowerAccess = null;
            if(upperAccess != null) upperAccess.cancel(true);
            upperAccess = null;
        }

        @Override
        public void scrollTo(int x, int y){
            int loc = layout.getScrollY();
            if(loc < y){
                int height = getHeight();
                if(y+height > (basalRecord+(int)(nPages*0.75)*pageSize)*itemHeight){
                    if(y+height > (basalRecord+nPages*pageSize)*itemHeight){
                        finalizeShiftViewDown();
                    }else{
                        accessLower();
                    }
                }
            }
            super.scrollTo(x,y);
        }
        //public void selectDatabase()
        //public void pushFreshDatum(Datum in)
}
```

至此，整个实例介绍完毕，在模拟器中的执行效果如图 14-1 所示，在真机中会显示获取的温度值。

图 14-1　执行效果

# 14.2　湿度传感器详解

**知识点讲解：光盘:视频\知识点\第 14 章\湿度传感器详解.avi**

人类的生存和社会活动与湿度密切相关。随着现代化的实现，很难找出一个与湿度无关的领域来。由于应用领域不同，对湿度传感器的技术要求也不同。本节将详细讲解使用 Android 湿度传感器的基本知识。

## 14.2.1　Android 系统中的湿度传感器

在 Android 系统中，湿度传感器的值是 TYPE_RELATIVE_HUMIDITY，单位是%，能够测量周围环境的相对湿度。Android 系统中的湿度与光线、气压、温度传感器的使用方式相同，可以从湿度传感器读取到相对湿度的原始数据。而且，如果设备同时提供了湿度传感器（TYPE_RELATIVE_HUMIDITY）和温度传感器（TYPE_AMBIENT_TEMPERATURE），那么就可以用这两个数据流来计算出结露点和绝对湿度。

（1）结露点

结露点是在固定的气压下，空气中所含的气态水达到饱和而凝结成液态水所需要降至的温度。以下给出了计算结露点温度的公式：

$$t_d(t,RH) = T_n \cdot \frac{\ln(RH/100\%) + m \cdot t/(T_n+t)}{m - [\ln(RH/100\%) + m \cdot t/(T_n+t)]}$$

在上述公式中，各个参数的具体说明如下：

- ☑ $t_d$ = 结露点温度，单位是摄氏度℃。
- ☑ t = 当前温度，单位是摄氏度℃。
- ☑ RH = 当前相对湿度，单位是百分比（%）。
- ☑ m = 18.62。
- ☑ $T_n$ = 243.12。

（2）绝对湿度

绝对湿度是在一定体积的干燥空气中含有的水蒸气的质量。绝对湿度的计量单位是克/立方米。以下给出了计算绝对湿度的公式：

$$d_v(t,RH) = 218.7 \cdot \frac{(RH/100\%) \cdot A \cdot \exp(m \cdot t/(T_n+t))}{273.15 + t}$$

在上述公式中，各个参数的具体说明如下：

- ☑ $d_v$ = 绝对湿度，单位是克/立方米。
- ☑ t = 当前温度，单位是摄氏度℃。
- ☑ RH = 当前相对湿度，单位是百分比（%）。
- ☑ m = 18.62。

☑ $T_n$ = 243.12℃。

☑ A = 6.112hPa。

### 14.2.2 实战演练——获取远程湿度传感器的数据

本实例将详细讲解使用湿度传感器的方法，功能是在 Android 设备中获取远程湿度传感器的数据。

| 实　例 | 功　能 | 源码路径 |
| --- | --- | --- |
| 实例 14-3 | 获取远程湿度传感器的数据 | 光盘:\daima\14\sensorEX |

（1）编写布局文件

编写布局文件 activity_thread_async_task_main.xml，在界面中分别设置 Start 和 Stop 两个按钮，在下方使用 ProgressBar 控件显示获取的湿度值。文件 activity_thread_async_task_main.xml 的具体实现代码如下：

```
<RelativeLayout xmlns:android="http://schemas.android.com/apk/res/android"
    xmlns:tools="http://schemas.android.com/tools"
    android:layout_width="match_parent"
    android:layout_height="match_parent"
    android:onClick="updateUrl"
    android:paddingBottom="@dimen/activity_vertical_margin"
    android:paddingLeft="@dimen/activity_horizontal_margin"
    android:paddingRight="@dimen/activity_horizontal_margin"
    android:paddingTop="@dimen/activity_vertical_margin"
    tools:context=".ThreadAsyncTaskMainActivity" >

    <TextView
        android:id="@+id/textView1"
        android:layout_width="wrap_content"
        android:layout_height="wrap_content"
        android:text="@string/textView1"
        android:textSize="20sp" />

    <EditText
        android:id="@+id/editText"
        android:layout_width="wrap_content"
        android:layout_height="wrap_content"
        android:visibility="invisible" />

    <TableRow
        android:id="@+id/tableRow"
        android:layout_width="fill_parent"
        android:layout_height="wrap_content"
        android:layout_below="@id/editText" >

        <Button
            android:id="@+id/startButton"
            android:layout_width="fill_parent"
            android:layout_height="wrap_content"
```

```
            android:layout_weight="1"
            android:onClick="onClickStart"
            android:text="@string/startButton" />

        <Button
            android:id="@+id/stopButton"
            android:layout_width="fill_parent"
            android:layout_height="wrap_content"
            android:layout_weight="1"
            android:onClick="onClickStop"
            android:text="@string/stopButton" />

        <Button
            android:id="@+id/setButton"
            android:layout_width="fill_parent"
            android:layout_height="wrap_content"
            android:layout_weight="1"
            android:onClick="onClickSet"
            android:text="@string/setButton"
            android:visibility="invisible" />
    </TableRow>

    <ProgressBar
        android:id="@+id/progressBar1"
        style="?android:attr/progressBarStyleHorizontal"
        android:layout_width="wrap_content"
        android:layout_height="wrap_content"
        android:layout_alignLeft="@+id/tableRow"
        android:layout_alignRight="@+id/tableRow"
        android:layout_below="@+id/tableRow"
        android:layout_marginTop="40dp" />
</RelativeLayout>
```

（2）监听用户触摸单击屏幕控件事件并处理

主 Activity 的实现文件是 ThreadAsyncTaskMainActivity.java，功能是监听用户触摸单击屏幕控件的事件，执行对应的处理方法。文件 ThreadAsyncTaskMainActivity.java 的具体实现代码如下：

```
public class ThreadAsyncTaskMainActivity extends Activity {
    private static final String URL_SENSOR = "http://lmi92.cnam.fr/ds2438/ds2438/";
    private Button startButton, stopButton, setButton;
    private ProgressBar progressBar;
    private EditText editText;
    private TextView textView;
    private String url;
    private long top, bot, acquitionTime = 3000; // 3 s en ms
    private WorkAsyncTask task;
    private HumiditySensorAbstract ds2438;

    @Override
    protected void onCreate(Bundle savedInstanceState) {
        super.onCreate(savedInstanceState);
        setContentView(R.layout.activity_thread_async_task_main);
```

```
        //监听主屏幕的控件触摸事件
        startButton = (Button) findViewById(R.id.startButton);
        stopButton = (Button) findViewById(R.id.stopButton);
        setButton = (Button) findViewById(R.id.setButton);
        editText = (EditText) findViewById(R.id.editText);
        progressBar = (ProgressBar) findViewById(R.id.progressBar1);
        textView = (TextView) findViewById(R.id.textView1);
        startButton.setEnabled(true);
        stopButton.setEnabled(false);
        url = ThreadAsyncTaskMainActivity.loadURL_SENSOR(this);

    }

    /*
     * 单击 Start 按钮后调用 WorkAsyncTask()读取湿度
     */
    public void onClickStart(View v) {
        Toast.makeText(this, "Starting...", Toast.LENGTH_SHORT).show();
        task = new WorkAsyncTask();
        task.execute();

    }
    /*
     *单击 Stop 按钮后停止读取湿度工作
     */
    public void onClickStop(View v) {
        Toast.makeText(this, "Stopping...", Toast.LENGTH_SHORT).show();
        task.cancel(true);
        stopButton.setEnabled(false);
        startButton.setEnabled(true);
        textView.setText("Acquisition... Appuyez sur start");
    }
    public void onClickSet(View v) {
        url = editText.getText().toString();
        setButton.setVisibility(View.INVISIBLE);
        startButton.setEnabled(true);
        textView.setVisibility(View.VISIBLE);
        editText.setVisibility(View.INVISIBLE);
        ThreadAsyncTaskMainActivity.saveURL_SENSOR(this, url);
    }

    /*
     * 更新远程 URL 地址
     */
    public void updateUrl(View v) {
        setButton.setVisibility(View.VISIBLE);
        textView.setVisibility(View.INVISIBLE);
        editText.setVisibility(View.VISIBLE);
        editText.setText(url);
        stopButton.setEnabled(false);
        startButton.setEnabled(false);
```

```
}

/*
 * 保存远程传感器的 URL
 */
private static void saveURL_SENSOR(Context context, String url) {
    SharedPreferences prefs = PreferenceManager
            .getDefaultSharedPreferences(context);
    Editor edit = prefs.edit();
    edit.putString("url_sensor", url);
    edit.commit();
}

/*
 *载入远程传感器的 URL
 */
private static String loadURL_SENSOR(Context context) {
    SharedPreferences prefs = PreferenceManager
            .getDefaultSharedPreferences(context);
    return prefs.getString("url_sensor", URL_SENSOR);
}

private class WorkAsyncTask extends AsyncTask<Void, Float, Void> {

    @Override
    protected Void doInBackground(Void... params) {
        while (!isCancelled()) {

            try {
                ds2438 = new HTTPHumiditySensor(url);
                top = System.currentTimeMillis();
                float taux = ds2438.value();
                bot = System.currentTimeMillis();
                long time = top - bot;
                publishProgress(taux);
                SystemClock.sleep(acquitionTime - time);
            } catch (Exception e) {

                e.printStackTrace();
            }
        }
        return null;
    }

    @Override
    protected void onProgressUpdate(Float... values) {
        Time currentTime = new Time();
        currentTime.setToNow();
        String date = currentTime.format("%d.%m.%Y %H:%M:%S");
        // String date = currentTime.format("%H:%M:%S");
        textView.setText("[" + date + "] ds2438 : " + values[0]);
```

```
                progressBar.setProgress(values[0].intValue());
            }

            @Override
            protected void onPreExecute() {
                startButton.setEnabled(false);
                stopButton.setEnabled(true);
            }

            @Override
            protected void onCancelled() {
                progressBar.setProgress(0);

            }
        }
}
```

（3）设置远程湿度传感器的初始 URL 地址

编写文件 HTTPHumiditySensor.java，功能是设置远程湿度传感器的初始 URL 地址，具体实现代码如下：

```
public class HTTPHumiditySensor extends HumiditySensorAbstract {
    // public final static String DEFAULT_HTTP_SENSOR = "http://localhost:8999/ds2438/";
    public final static String DEFAULT_HTTP_SENSOR = "http://10.0.2.2:8999/ds2438/";
    public static final long ONE_MINUTE = 60L * 1000L;

    private static String urlSensor;

    private HTTPHumiditySensor() {
        this(DEFAULT_HTTP_SENSOR);
    }

    public HTTPHumiditySensor(String urlSensor) {
        this.urlSensor = urlSensor;
    }
    public float value() throws Exception {
        StringTokenizer st = new StringTokenizer(request(), "=");
        st.nextToken();
        float f = Float.parseFloat(st.nextToken()) * 10F;
        return ((int) f) / 10F;
    }

    public long minimalPeriod() {
        if (urlSensor.startsWith("http://localhost")
                    || urlSensor.startsWith("http://10.0.2.2"))
            return 200L;
        else
            return 2000L;
    }

    public String getUrl() {
        return HTTPHumiditySensor.urlSensor;
```

```
    }

    private String request() throws Exception {
        URL url = new URL(urlSensor);
        URLConnection connection = url.openConnection();
        BufferedReader in = new BufferedReader(new InputStreamReader(
                connection.getInputStream()), 128);
        String result = new String("");
        String inputLine = in.readLine();
        while (inputLine != null) {
            result = result + inputLine;
            inputLine = in.readLine();
        }
        in.close();
        return result;
        /****/
    }

}
```

至此，整个实例介绍完毕，在真机中执行后会获取并显示远程湿度传感器的湿度。

# 第 5 篇

- 第 15 章　条形码解析技术详解
- 第 16 章　NFC 近场通信技术详解
- 第 17 章　Google Now 和 Android Wear 详解

# 第15章 条形码解析技术详解

在当前手机系统中，无论是智能手机还是普通的手机，基本上都支持手机拍照和录制视频功能，这些功能都是通过手机上的摄像头实现的。在Android系统中，上述照相机功能是通过Camera系统实现的。另外，通过摄像头可以实现条形码解析技术。条形码应用是物联网设备中的常见应用之一，为人们的生活提供了极大的便利。本章将详细讲解在Android设备中使用摄像头解析条形码的方法，为读者步入本书后面知识的学习打下基础。

## 15.1 Android拍照系统结构基础

**知识点讲解：光盘:视频\知识点\第15章\Android拍照系统结构基础.avi**

在Android系统中，Camera（照相机/拍照）系统提供了取景器、视频录制和拍摄相片等功能，并且还具有各种控制类的接口。另外，在Camera系统中还提供了Java层的接口和本地接口。其中Java框架中的Camera类实现了Java层相机接口，为照相机类和扫描类使用。而Camera的本地接口可以给本地程序调用，作为视频输入环节应用于摄像机和视频电话领域。

Android照相机系统的基本层次结构如图15-1所示。

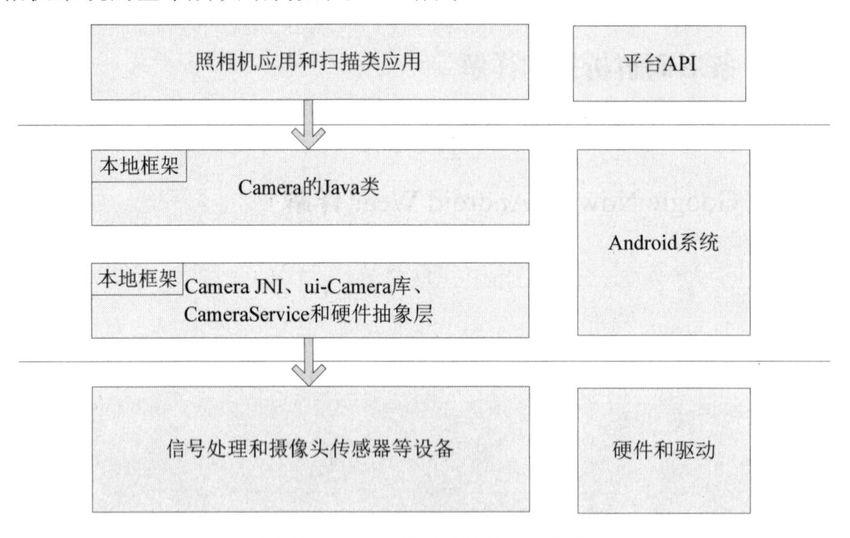


图15-1 照相机系统的层次结构

Android系统中的Camera系统包括Camera驱动程序层、Camera硬件抽象层、AudioService、Camera本地库、Camera的Java框架类和Java应用层对Camera系统的调用。Camera系统的具体结构如图15-2所示。

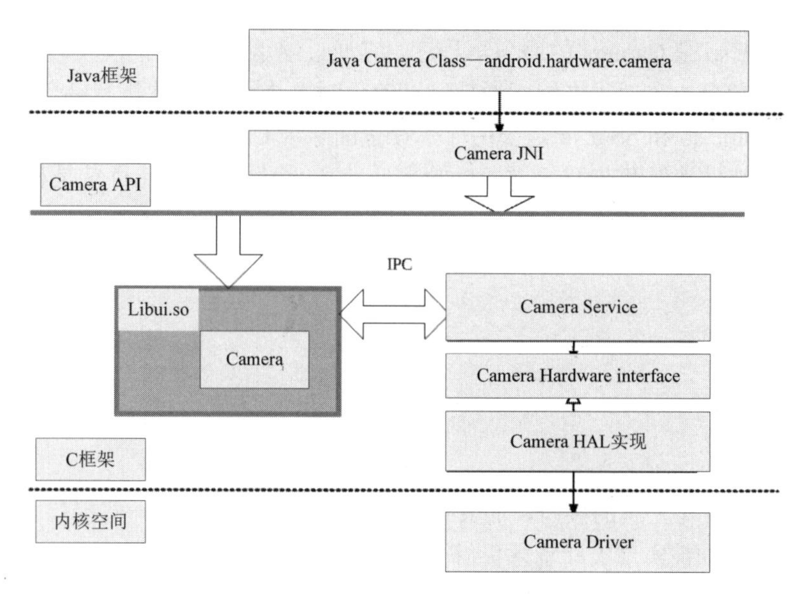


图 15-2　Camera 的系统结构

在图 15-2 所示的结构中，各个构成层次的具体说明如下所示。

（1）Camera 系统的 Java 层，代码路径是 frameworks/base/core/java/android/hardware/。

其中 Camera.java 是主要实现的文件，对应的 Java 层次的类是 android.hardware.Camera，这个类和 JNI 中定义的类是一个，有些方法通过 JNI 的方式调用本地代码得到，有些方法自己实现。

（2）Camera 系统的 Java 本地调用部分（JNI），代码路径是 frameworks/base/core/jni/android_hardware_Camera.cpp。

这部分内容编译成为目标文件 libandroid_runtime.so，主要的头文件在目录 frameworks/base/include/ui/中。

（3）Camera 本地框架，其中头文件路径是 frameworks/native/include/ui 或 frameworks/av/include/camera/。

源代码路径是 frameworks/native/libs/ui 或 frameworks/av/camera/。

这部分的内容被编译成库 libui.so 或 libcamera_client.so。

（4）Camera 服务部分，代码路径是 frameworks/av/services/camera/libcameraservice/。

这部分内容被编译成库 libcameraservice.so。

为了实现一个具体功能的 Camera 驱动程序，在最底层还需要一个硬件相关的 Camer 库（例如通过调用 video for linux 驱动程序和 Jpeg 编码程序实现）。这个库将被 Camera 的服务库 libcameraservice.so 调用。

（5）摄像头驱动程序。

此部分是基于 Linux 的 Video for Linux 视频驱动框架。

（6）硬件抽象层。

硬件抽象层中的接口代码路径是 frameworks/base/include/ui/或 frameworks/av/include/camera/。

其中的核心文件是 CameraHardwareInterface.h。

在 Camera 系统的各个库中，库 libui.so 位于核心的位置，它对上层提供的接口主要是 Camera 类，类 libandroid_runtime.so 通过调用 Camera 类提供对 Java 的接口，并且实现了 android.hardware.camera 类。

库 libcameraservice.so 是 Camera 的服务器程序，它通过继承 libui.so 的类实现服务器的功能，并且与 libui.so 中的另外一部分内容通过进程间通信（即 Binder 机制）的方式进行通信。

库 libandroid_runtime.so 和 libui.so 是公用库，在里面除了 Camera 外还有其他方面的功能。

Camera 部分的头文件被保存在 frameworks/base/include/ui/目录中，此目录是和库 libmedia.so 的源文件目录 frameworks/base/libs/ui/相对应的。

在 Camera 中主要包含如下头文件：

☑ ICameraClient.h

☑ Camera.h

☑ ICamera.h

☑ ICameraService.h

☑ CameraHardwareInterface.h

文件 Camera.h 提供了对上层的接口，而其他的几个头文件都是提供一些接口类（即包含了纯虚函数的类），这些接口类必须被实现类继承才能够使用。

当整个 Camera 在运行时，可以大致上分成 Client 和 Server 两个部分，它们分别在两个进程中运行，它们之间使用 Binder 机制实现进程间通信。这样在客户端调用接口，功能则在服务器中实现，但是在客户端中调用就好像直接调用服务器中的功能，进程间通信的部分对上层程序不可见。

从框架结构上来看，文件 ICameraService.h、ICameraClient.h 和 ICamera.h 中的 3 个类定义了 Camera 的接口和架构，ICameraService.cpp 和 Camera.cpp 两个文件用于实现 Camera 架构，Camera 的具体功能在下层调用硬件相关的接口来实现。

# 15.2 底层程序详解

**知识点讲解：光盘:视频\知识点\第 15 章\底层程序详解.avi**

在 Linux 系统中，Camera 驱动程序使用了 Linux 标准的 Video for Linux 2（V4L2）驱动程序。无论是内核空间还是用户空间，都使用 V4L2 驱动程序框架来定义数据类和控制类。所以在移植 Android 中的 Camera 系统时，也是用标准的 V4L2 驱动程序作为 Camera 的驱动程序。在 Camera 系统中，V4L2 驱动程序的任务是获得 Video 数据。

## 15.2.1 V4L2 API

V4L2 是 V4L 的升级版本，是 Linux 下视频设备程序提供的一套接口规范。包括一套数据结构和底层 V4L2 驱动接口。V4L2 驱动程序向用户空间提供字符设备，主设备号是 81。对于视频设备来说，次设备号是 0～63。次设备号在 64～127 之间的是 Radio 设备，次设备号在 192～223 之间的是 Teletext 设备，次设备号在 215～255 之间的是 VBI 设备。

V4L2 中常用的结构体在内核文件 include/linux/videodev2.h 中定义，代码如下：

```
struct v4l2_requestbuffers                //申请帧缓冲，对应命令 VIDIOC_REQBUFS
        struct v4l2_capability            //视频设备的功能，对应命令 VIDIOC_QUERYCAP
        struct v4l2_input                 //视频输入信息，对应命令 VIDIOC_ENUMINPUT
        struct v4l2_standard              //视频的制式，如 PAL 和 NTSC，对应命令 VIDIOC_ENUMSTD
```

```
        struct v4l2_format              //帧的格式，对应命令 VIDIOC_G_FMT、VIDIOC_S_FMT 等
        struct v4l2_buffer              //驱动中的一帧图像缓存，对应命令 VIDIOC_QUERYBUF
        struct v4l2_crop                //视频信号矩形边框
        v4l2_std_id                     //视频制式
```

常用的 ioctl 接口命令也在文件 include/linux/videodev2.h 中定义，代码如下：

```
VIDIOC_REQBUFS                          //分配内存
        VIDIOC_QUERYBUF                 //把 VIDIOC_REQBUFS 中分配的数据缓存转换成物理地址
        VIDIOC_QUERYCAP                 //查询驱动功能
        VIDIOC_ENUM_FMT                 //获取当前驱动支持的视频格式
        VIDIOC_S_FMT                    //设置当前驱动的视频捕获格式
        VIDIOC_G_FMT                    //读取当前驱动的视频捕获格式
        VIDIOC_TRY_FMT                  //验证当前驱动的显示格式
        VIDIOC_CROPCAP                  //查询驱动的修剪能力
        VIDIOC_S_CROP                   //设置视频信号的矩形边框
        VIDIOC_G_CROP                   //读取视频信号的矩形边框
        VIDIOC_QBUF                     //把数据从缓存中读取出来
        VIDIOC_DQBUF                    //把数据放回缓存队列
        VIDIOC_STREAMON                 //开始视频显示函数
        VIDIOC_STREAMOFF                //结束视频显示函数
        VIDIOC_QUERYSTD                 //检查当前视频设备支持的标准，例如 PAL 或 NTSC
```

## 15.2.2　操作 V4L2 的流程

V4L2 中提供了很多访问接口，可以根据具体需要选择操作方法。不过需要注意的是，很少有驱动完全实现了所有的接口功能。所以在使用时需要参考驱动源码，或仔细阅读驱动提供者的使用说明。V4L2 的操作流程如下：

（1）打开设备文件，具体代码如下：

```
int fd = open(Devicename,mode);
Devicename：/dev/video0、/dev/video1 ……
Mode：O_RDWR [| O_NONBLOCK]
```

如果需要使用非阻塞模式调用视频设备，当没有可用的视频数据时不会阻塞，而会立刻返回。

（2）获取设备的 capability，具体代码如下：

```
struct v4l2_capability capability;
int ret = ioctl(fd, VIDIOC_QUERYCAP, &capability);
```

在此需要查看设备具有什么功能，如是否具有视频输入特性。

（3）选择视频输入，代码如下：

```
struct v4l2_input input;
//……开始初始化 input
int ret = ioctl(fd, VIDIOC_QUERYCAP, &input);
```

每一个视频设备可以有多个视频输入，如果只有一路输入，则可以没有这个功能。

（4）检测视频支持的制式，具体代码如下：

```
v4l2_std_id std;
do {
                ret = ioctl(fd, VIDIOC_QUERYSTD, &std);
 } while (ret == -1 && errno == EAGAIN);
                    switch (std) {
                    case V4L2_STD_NTSC:
```

```
        //……
            case V4L2_STD_PAL:
        //……
            }
```

（5）设置视频捕获格式，具体代码如下：

```
struct v4l2_format fmt;
fmt.type = V4L2_BUF_TYPE_VIDEO_OUTPUT;
fmt.fmt.pix.pixelformat = V4L2_PIX_FMT_UYVY;
fmt.fmt.pix.height = height;
fmt.fmt.pix.width = width;
fmt.fmt.pix.field = V4L2_FIELD_INTERLACED;
ret = ioctl(fd, VIDIOC_S_FMT, &fmt);
if(ret) {
        perror("VIDIOC_S_FMT\n");
        close(fd);
        return -1;
}
```

（6）向驱动申请帧缓存，具体代码如下：

```
struct v4l2_requestbuffers req;
 if (ioctl(fd, VIDIOC_REQBUFS, &req) == -1) {
        return -1;
 }
```

在结构 v4l2_requestbuffers 中定义了缓存的数量，驱动会根据这个申请对应数量的视频缓存。通过多个缓存可以建立 FIFO，这样可以提高视频采集的效率。

（7）获取每个缓存的信息，并 MMAP（映射）到用户空间。具体代码如下：

```
typedef struct VideoBuffer {
        void *start;
        size_t length;
} VideoBuffer;
          VideoBuffer* buffers = calloc( req.count, sizeof(*buffers) );
struct v4l2_buffer buf;
          for (numBufs = 0; numBufs < req.count; numBufs++) {    //映射所有的缓存
        memset( &buf, 0, sizeof(buf) );
        buf.type = V4L2_BUF_TYPE_VIDEO_CAPTURE;
        buf.memory = V4L2_MEMORY_MMAP;
        buf.index = numBufs;
        if (ioctl(fd, VIDIOC_QUERYBUF, &buf) == -1) {              //获取到对应 index 的缓存信息，此处主要利
用 length 信息及 offset 信息来完成后面的 mmap 操作
                return -1;
        }
buffers[numBufs].length = buf.length;
        //转换成相对地址
        buffers[numBufs].start = mmap(NULL, buf.length,
                PROT_READ | PROT_WRITE,
                MAP_SHARED,
                fd, buf.m.offset);
if (buffers[numBufs].start == MAP_FAILED) {
                return -1;
        }
```

（8）开始采集视频，具体代码如下：

```
int buf_type= V4L2_BUF_TYPE_VIDEO_CAPTURE;
int ret = ioctl(fd, VIDIOC_STREAMON, &buf_type);
```

（9）取出 FIFO 缓存中已经采样的帧缓存，具体代码如下：

```
struct v4l2_buffer buf;
memset(&buf,0,sizeof(buf));
buf.type=V4L2_BUF_TYPE_VIDEO_CAPTURE;
buf.memory=V4L2_MEMORY_MMAP;
buf.index=0;//此值由下面的 ioctl 返回
if (ioctl(fd, VIDIOC_DQBUF, &buf) == -1)
{
        return -1;
}
```

通过上述代码，可以根据返回的 buf.index 找到对应的 mmap 映射好的缓存，实现取出视频数据的功能。

（10）将刚刚处理完的缓冲重新入队列尾，这样可以循环采集，具体代码如下：

```
if (ioctl(fd, VIDIOC_QBUF, &buf) == -1) {
        return -1;
}
```

（11）停止视频的采集，具体代码如下：

```
int ret = ioctl(fd, VIDIOC_STREAMOFF, &buf_type);
```

（12）关闭视频设备，具体代码如下：

```
close(fd);
```

## 15.2.3　V4L2 驱动框架

在上述使用 V4L2 的流程中，各个操作都需要有底层 V4L2 驱动的支持。在内核中有一些非常完善的例子。例如在 Linux-2.6.26 内核目录/drivers/media/video//zc301/中，文件 zc301_core.c 实现了 ZC301 视频驱动代码。

（1）V4L2 驱动注册、注销函数

在 Video 核心层文件 drivers/media/video/videodev.c 中提供了注册函数，具体代码如下：

```
int video_register_device(struct video_device *vfd, int type, int nr)
```

参数说明如下。

- ☑ video_device：要构建的核心数据结构。
- ☑ type：表示设备类型，此设备号的基地址受此变量的影响。
- ☑ nr：如果 end-base>nr>0，次设备号=base（基准值，受 type 影响）+nr；否则系统将自动分配合适的次设备号。

我们具体需要的驱动只需构建 video_device 结构，然后调用注册函数即可。例如在文件 zc301_core.c 中的实现代码如下：

```
err = video_register_device(cam->v4ldev, VFL_TYPE_GRABBER, video_nr[dev_nr]);
```

在 Video 核心层文件 drivers/media/video/videodev.c 中提供了如下注销函数：

```
void video_unregister_device(struct video_device *vfd)
```

（2）构建 struct video_device

在结构 video_device 中包含了视频设备的属性和操作方法，具体可以参考文件 zc301_core.c，代码

如下：

```
strcpy(cam->v4ldev->name, "ZC0301[P] PC Camera");
  cam->v4ldev->owner = THIS_MODULE;
  cam->v4ldev->type = VID_TYPE_CAPTURE | VID_TYPE_SCALES;
  cam->v4ldev->fops = &zc0301_fops;
  cam->v4ldev->minor = video_nr[dev_nr];
  cam->v4ldev->release = video_device_release;
  video_set_drvdata(cam->v4ldev, cam);
```

在上述 zc301 的驱动中并没有全部实现 struct video_device 中的操作函数，例如 vidioc_querycap、vidioc_g_fmt_cap，这是因为在 struct file_operations zc0301_fops 中的 zc0301_ioctl 实现了前面的所有 ioctl 操作，所以无须在 struct video_device 中再次实现 struct video_device 中的操作。

另外也可以使用下面的代码来构建 struct video_device。

```
static struct video_device camif_dev =
  {
    .name = "s3c2440 camif",
    .type = VID_TYPE_CAPTURE|VID_TYPE_SCALES|VID_TYPE_SUBCAPTURE,
    .fops = &camif_fops,
    .minor = -1,
    .release = camif_dev_release,
    .vidioc_querycap = vidioc_querycap,
    .vidioc_enum_fmt_cap = vidioc_enum_fmt_cap,
    .vidioc_g_fmt_cap = vidioc_g_fmt_cap,
    .vidioc_s_fmt_cap = vidioc_s_fmt_cap,
    .vidioc_queryctrl = vidioc_queryctrl,
    .vidioc_g_ctrl = vidioc_g_ctrl,
    .vidioc_s_ctrl = vidioc_s_ctrl,
  };
  static struct file_operations camif_fops =
  {
    .owner = THIS_MODULE,
    .open = camif_open,
    .release = camif_release,
    .read = camif_read,
    .poll = camif_poll,
    .ioctl = video_ioctl2, /* V4L2 ioctl handler */
    .mmap = camif_mmap,
    .llseek = no_llseek,
  };
```

结构 video_ioctl2 是在文件 videodev.c 中实现的，video_ioctl2 中会根据 ioctl 不同的 cmd 来调用 video_device 中的操作方法。

## 15.2.4 实现 Video 核心层

具体实现代码请参考内核文件/drivers/media/videodev.c，实现流程如下所示。

（1）注册 256 个视频设备，代码如下：

```
static int __init videodev_init(void)
        {
```

```
        int ret;
        if (register_chrdev(VIDEO_MAJOR, VIDEO_NAME, &video_fops)) {
                return -EIO;
        }
        ret = class_register(&video_class);
        ......
}
```

在上述代码中注册了 256 个视频设备和 video_class 类，video_fops 为这 256 个设备共同的操作方法。

（2）实现 V4L2 驱动的注册函数，具体代码如下：

```
int video_register_device(struct video_device *vfd, int type, int nr)
{
        int i=0;
        int base;
        int end;
        int ret;
        char *name_base;
    switch(type) //根据不同的 type 确定设备名称、次设备号
        {
                case VFL_TYPE_GRABBER:
                        base=MINOR_VFL_TYPE_GRABBER_MIN;
                        end=MINOR_VFL_TYPE_GRABBER_MAX+1;
                        name_base = "video";
                        break;
                case VFL_TYPE_VTX:
                        base=MINOR_VFL_TYPE_VTX_MIN;
                        end=MINOR_VFL_TYPE_VTX_MAX+1;
                        name_base = "vtx";
                        break;
                case VFL_TYPE_VBI:
                        base=MINOR_VFL_TYPE_VBI_MIN;
                        end=MINOR_VFL_TYPE_VBI_MAX+1;
                        name_base = "vbi";
                        break;
                case VFL_TYPE_RADIO:
                        base=MINOR_VFL_TYPE_RADIO_MIN;
                        end=MINOR_VFL_TYPE_RADIO_MAX+1;
                        name_base = "radio";
                        break;
                default:
                        printk(KERN_ERR "%s called with unknown type: %d\n",
                                __func__, type);
                        return -1;
        }
    /*计算出次设备号*/
        mutex_lock(&videodev_lock);
        if (nr >= 0 && nr < end-base) {
                /* use the one the driver asked for */
                i = base+nr;
                if (NULL != video_device[i]) {
                        mutex_unlock(&videodev_lock);
```

```
                        return -ENFILE;
                }
        } else {
                for(i=base;i<end;i++)
                        if (NULL == video_device[i])
                                break;
                if (i == end) {
                        mutex_unlock(&videodev_lock);
                        return -ENFILE;
                }
        }
        video_device[i]=vfd; //保存 video_device 结构指针到系统的结构数组中，最终的次设备号和 i 相关
        vfd->minor=i;
        mutex_unlock(&videodev_lock);
        mutex_init(&vfd->lock);
          memset(&vfd->class_dev, 0x00, sizeof(vfd->class_dev));
          if (vfd->dev)
                vfd->class_dev.parent = vfd->dev;
        vfd->class_dev.class = &video_class;
        vfd->class_dev.devt = MKDEV(VIDEO_MAJOR, vfd->minor);
        sprintf(vfd->class_dev.bus_id, "%s%d", name_base, i - base);//最后在/dev 目录下的名称
    ret = device_register(&vfd->class_dev);//结合 udev 或 mdev 可以实现自动在/dev 下创建设备节点
          ……
}
```

从上面的注册函数代码中可以看出，注册 V4L2 驱动的过程只是创建了设备节点，例如/dev/video0，并且保存了 video_device 结构指针。

（3）打开视频驱动。

当使用下面的代码在用户空间调用 open 函数打开对应的视频文件时，对应代码如下：

```
int fd = open(/dev/video0, O_RDWR);
```

对应/dev/video0 目录的文件操作结构是在文件/drivers/media/videodev.c 中定义的 video_fops，代码如下：

```
static const struct file_operations video_fops=
    {
          .owner = THIS_MODULE,
          .llseek = no_llseek,
          .open = video_open,
    };
```

上述代码只是实现了 open 操作，后面的其他操作需要使用 video_open()来实现，具体代码如下：

```
static int video_open(struct inode *inode, struct file *file)
    {
          unsigned int minor = iminor(inode);
          int err = 0;
          struct video_device *vfl;
          const struct file_operations *old_fops;
    if(minor>=VIDEO_NUM_DEVICES)
                return -ENODEV;
          mutex_lock(&videodev_lock);
          vfl=video_device[minor];
          if(vfl==NULL) {
```

```
                    mutex_unlock(&videodev_lock);
                    request_module("char-major-%d-%d", VIDEO_MAJOR, minor);
                    mutex_lock(&videodev_lock);
                    vfl=video_device[minor]; //根据次设备号取出 video_device 结构
                    if (vfl==NULL) {
                            mutex_unlock(&videodev_lock);
                            return -ENODEV;
                    }
            }
            old_fops = file->f_op;
            file->f_op = fops_get(vfl->fops);//替换此打开文件的 file_operation 结构。后面的其他针对此文
件的操作都由新的结构来负责。也就是由每个具体的 video_device 的 fops 负责
            if(file->f_op->open)
                    err = file->f_op->open(inode,file);
            if (err) {
                    fops_put(file->f_op);
                    file->f_op = fops_get(old_fops);
            }
            ……
    }
```

# 15.3 拍照系统的硬件抽象层

**知识点讲解：光盘:视频\知识点\第 15 章\拍照系统的硬件抽象层.avi**

在 Andorid 2.1 及其以前的版本中，Camera 系统的硬件抽象层的头文件保存在目录 frameworks/base/include/ui/中。

在 Andorid 2.2 及其以后的版本中，Camera 系统的硬件抽象层的头文件保存在目录 frameworks/av/include/camera/中。

在上述目录中主要包含了如下头文件。

- ☑ CameraHardwareInterface.h：在里面定义了 C++接口类，此类需要根据系统的情况来实现继承。
- ☑ CameraParameters.h：在里面定义了 Camera 系统的参数，可以在本地系统的各个层次中使用这些参数。
- ☑ Camera.h：在里面提供了 Camera 系统本地对上层的接口。

## 15.3.1 Andorid 2.1 及其以前的版本

在 Andorid 2.1 及其以前的版本中，在文件 CameraHardwareInterface.h 中首先定义了硬件抽象层接口的回调函数类型，对应的代码如下：

```
/** startPreview()使用的回调函数*/
typedef void (*preview_callback)(const sp<IMemory>& mem, void* user);

/** startRecord()使用的回调函数*/
typedef void (*recording_callback)(const sp<IMemory>& mem, void* user);

/** takePicture()使用的回调函数*/
```

```
typedef void (*shutter_callback)(void* user);

/** takePicture()使用的回调函数*/
typedef void (*raw_callback)(const sp<IMemory>& mem, void* user);

/** takePicture()使用的回调函数*/
typedef void (*jpeg_callback)(const sp<IMemory>& mem, void* user);

/** autoFocus()使用的回调函数*/
typedef void (*autofocus_callback)(bool focused, void* user);
```

然后定义类 CameraHardwareInterface，并在类中定义各个接口函数。具体代码如下：

```
class CameraHardwareInterface : public virtual RefBase {
public:
    virtual ~CameraHardwareInterface() { }
    virtual sp<IMemoryHeap>getPreviewHeap() const = 0;
    virtual sp<IMemoryHeap>getRawHeap() const = 0;
    virtual status_t startPreview(preview_callback cb, void* user) = 0;
    virtual bool useOverlay() {return false;}
    virtual status_t setOverlay(const sp<Overlay> &overlay) {return BAD_VALUE;}
    virtual void stopPreview() = 0;
    virtual bool previewEnabled() = 0;
    virtual status_t startRecording(recording_callback cb, void* user) = 0;
    virtual void stopRecording() = 0;
    virtual bool recordingEnabled() = 0;
    virtual void releaseRecordingFrame(const sp<IMemory>& mem) = 0;
    virtual status_t autoFocus(autofocus_callback,
                               void* user) = 0;
    virtual status_t takePicture(shutter_callback,
                                 raw_callback,
                                 jpeg_callback,
                                 void* user) = 0;
    virtual status_t cancelPicture(bool cancel_shutter,
                                   bool cancel_raw,
                                   bool cancel_jpeg) = 0;
    virtual CameraParameters  getParameters() const = 0;
    virtual void release() = 0;
    virtual status_t dump(int fd, const Vector<String16>& args) const = 0;
};
extern "C" sp<CameraHardwareInterface> openCameraHardware();
};
```

可以将上述代码中的接口分为如下几类。

☑ 取景预览：startPreview、stopPreview、useOverlay 和 setOverlay。

☑ 录制视频：startRecording、stopRecording、recordingEnabled 和 releaseRecordingFrame。

☑ 拍摄照片：takePicture 和 cancelPicture。

☑ 辅助功能：autoFocus（自动对焦）、setParameters 和 getParameters。

## 15.3.2 Andorid 2.2 及其以后的版本

在 Andorid 2.2 及其以前的版本中，在文件 Camera.h 中首先定义了通知信息的枚举值，对应的代码

如下：

```
enum {
    CAMERA_MSG_ERROR = 0x001,                    //错误信息
    CAMERA_MSG_SHUTTER = 0x002,                  //快门信息
    CAMERA_MSG_FOCUS = 0x004,                    //聚焦信息
    CAMERA_MSG_ZOOM = 0x008,                     //缩放信息
    CAMERA_MSG_PREVIEW_FRAME = 0x010,            //帧预览信息
    CAMERA_MSG_VIDEO_FRAME = 0x020,              //视频帧信息
    CAMERA_MSG_POSTVIEW_FRAME = 0x040,           //拍照后停止帧信息
    CAMERA_MSG_RAW_IMAGE = 0x080,                //原始数据格式照片信息
    CAMERA_MSG_COMPRESSED_IMAGE = 0x100,         //压缩格式照片信息
    CAMERA_MSG_ALL_MSGS = 0x1FF                  //所有信息
};
```

然后在文件 CameraHardwareInterface.h 中定义如下 3 个回调函数。

```
//通知回调
typedef void (*notify_callback)(int32_t msgType,
                                int32_t ext1,
                                int32_t ext2,
                                void* user);
//数据回调
typedef void (*data_callback)(int32_t msgType,
                              const sp<IMemory>& dataPtr,
                              void* user);
//带有时间戳的数据回调
typedef void (*data_callback_timestamp)(nsecs_t timestamp,
                                        int32_t msgType,
                                        const sp<IMemory>& dataPtr,
                                        void* user);
```

然后定义类 CameraHardwareInterface，在类中各个函数的具体实现和其他 Android 版本中的相同。区别是回调函数不再由各个函数分别设置，所以 startPreview 和 startRecording 缺少了回调函数的指针和 void*类型的附加参数。主要实现代码如下：

```
class CameraHardwareInterface : public virtual RefBase {
public:
    virtual ~CameraHardwareInterface() { }
    virtual sp<IMemoryHeap>getPreviewHeap() const = 0;
    virtual sp<IMemoryHeap>getRawHeap() const = 0;
    virtual void setCallbacks(notify_callback notify_cb,
                              data_callback data_cb,
                              data_callback_timestamp data_cb_timestamp,
    virtual void enableMsgType(int32_t msgType) = 0;
    virtual void disableMsgType(int32_t msgType) = 0;
    virtual bool msgTypeEnabled(int32_t msgType) = 0;
    virtual status_t startPreview() = 0;
    virtual status_t getBufferInfo(sp<IMemory>& Frame, size_t *alignedSize) = 0;
    virtual bool useOverlay() {return false;}
    virtual status_t setOverlay(const sp<Overlay> &overlay) {return BAD_VALUE;}
    virtual void stopPreview() = 0;
    virtual bool previewEnabled() = 0;
    virtual status_t startRecording() = 0;
```

```
    virtual void stopRecording() = 0;
    virtual bool recordingEnabled() = 0;
    virtual void releaseRecordingFrame(const sp<IMemory>& mem) = 0;
    virtual status_t autoFocus() = 0;
    virtual status_t cancelAutoFocus() = 0;
    virtual status_t takePicture() = 0;
    virtual status_t cancelPicture() = 0;
    virtual CameraParameters    getParameters() const = 0;
    virtual status_t sendCommand(int32_t cmd, int32_t arg1, int32_t arg2) = 0;
    virtual void release() = 0;
    virtual status_t d
}
```

因为在新版本的 Camera 系统中增加了 sendCommand()，所以需要在文件 Camera.h 中增加新命令和返回值。具体实现代码如下：

```
//函数 sendCommand 使用的命令类型
enum {
    CAMERA_CMD_START_SMOOTH_ZOOM = 1,
    CAMERA_CMD_STOP_SMOOTH_ZOOM = 2,
    CAMERA_CMD_SET_DISPLAY_ORIENTATION = 3,
};

//错误类型
enum {
    CAMERA_ERROR_UKNOWN    = 1,
    CAMERA_ERROR_SERVER_DIED = 100
};
```

### 15.3.3 实现 Camera 硬件抽象层

在函数 startPreview 的实现过程中，保存预览回调函数并建立预览线程。在预览线程的循环中，等待视频数据的到达；视频帧到达后调用预览回调函数，将视频帧送出。

取景器预览的主要步骤如下：

（1）在初始化的过程中，建立预览数据的内存队列（多种方式）。

（2）在函数 startPreview 中建立预览线程。

（3）在预览线程的循环中，等待视频数据到达。

（4）视频到达后使用预览回调机制将视频向上传送。

在此过程不需要使用预览回调函数，可以直接将视频数据输入到 Overlay 上。 如果使用 Overlay 实现取景器，则需要有以下两个变化：

☑ 在函数 setOverlay 中，从 ISurface 接口中取得 Overlay 类。

☑ 在预览线程的循环中，不是用预览回调函数直接将数据输入到 Overlay 上。

录制视频的主要步骤如下：

（1）在函数 startRecording 的实现（或者在 setCallbacks ）中保存录制视频回调函数。

（2）录制视频可以使用自己的线程，也可以使用预览线程。

（3）通过调用录制回调函数的方式将视频帧送出。

当调用函数 releaseRecordingFrame 后，表示上层通知 Camera 硬件抽象层，这一帧的内存已经用完，可以进行下一次的处理。如果在 V4L2 驱动程序中使用原始数据（RAW），则视频录制的数据和取

景器预览的数据为同一数据。当调用 releaseRecordingFrame 时，通常表示编码器已经完成了对当前视频帧的编码，对这块内存进行释放。在这个函数的实现中，可以设置标志位，标记帧内存可以再次使用。

由此可见，对于 Linux 系统来说，摄像头驱动部分大多使用 Video for Linux 2（V4L2）驱动程序，在此处主要的处理流程如下：

（1）如果使用映射内核内存的方式（V4L2_MEMORY_MMAP），构建预览的内存 MemoryHeap Base 需要从 V4L2 驱动程序中得到内存指针。

（2）如果使用用户空间内存的方式（V4L2_MEMORY_USERPTR），MemoryHeapBase 中开辟的内存是在用户空间建立的。

（3）在预览的线程中，使用 VIDIOC_DQBUF 调用阻塞等待视频帧的到来，处理完成后使用 VIDIOC_QBUF 调用将帧内存再次压入队列，然后等待下一帧的到来。

# 15.4　拍照系统的 Java 部分

**知识点讲解：光盘:视频\知识点\第 15 章\拍照系统的 Java 部分.avi**

在文件 packages/apps/Camera/src/com/android/camera/Camera.java 中，已经包含了对 Camera 的调用。在文件 Camera.java 中包含的对包的引用的代码如下：

```
import android.hardware.Camera.PictureCallback;
import android.hardware.Camera.Size;
```

然后定义类 Camera，此类继承了活动 Activity 类，在它的内部包含了一个 android.hardware.Camera。对应代码如下：

```
public class Camera extends Activity implements View.OnClickListener, SurfaceHolder.Callback{
android.hardware.Camera mCameraDevice;
}
```

调用 Camera 功能的代码如下：

```
mCameraDevice.takePicture(mShutterCallback, mRawPictureCallback, mJpegPictureCallback);
mCameraDevice.startPreview();
mCameraDevice.stopPreview();
```

startPreview、stopPreview 和 takePicture 等接口就是通过 Java 本地调用（JNI）来实现的。文件 frameworks/base/core/java/android/hardware/Camera.java 提供了一个 JAVA 类 Camera，具体代码如下：

```
public class Camera {
}
```

在类 Camera 中，大部分代码使用 JNI 调用下层得到，例如下面的代码。

```
public void setParameters(Parameters params) {
        Log.e(TAG, "setParameters()");
        //params.dump();
        native_setParameters(params.flatten());
}
```

还有下面的代码。

```
public final void setPreviewDisplay(SurfaceHolder holder) {
        setPreviewDisplay(holder.getSurface());
}
private native final void setPreviewDisplay(Surface surface);
```

在上面的两段代码中，两个 setPreviewDisplay 参数不同，后一个是本地方法，参数为 Surface 类型，前一个通过调用后一个实现，但自己的参数以 SurfaceHolder 为类型。

（1）Camera 的 Java 本地调用部分

在 Android 系统中，Camera 驱动的 Java 本地调用（JNI）部分在文件 frameworks/base/core/jni/android_hardware_Camera.cpp 中实现。

在文件 android_hardware_Camera.cpp 中定义了一个 JNINativeMethod（Java 本地调用方法）类型的数组 gMethods，具体代码如下：

```
static JNINativeMethod camMethods[] = {
{"native_setup","(Ljava/lang/Object;)V",(void*)android_hardware_Camera_native_setup },
    {"native_release","()V",(void*)android_hardware_Camera_release },
    {"setPreviewDisplay","(Landroid/view/Surface;)V",(void *)android_hardware_Camera_setPreviewDisplay },
    {"startPreview","()V",(void *)android_hardware_Camera_startPreview },
    {"stopPreview", "()V", (void *)android_hardware_Camera_stopPreview },
    {"setHasPreviewCallback","(Z)V",(void *)android_hardware_Camera_setHasPreviewCallback },
    {"native_autoFocus","()V",(void *)android_hardware_Camera_autoFocus },
    {"native_takePicture", "()V", (void *)android_hardware_Camera_takePicture },
    {"native_setParameters","(Ljava/lang/String;)V",(void *)android_hardware_Camera_setParameters },
    {"native_getParameters", "()Ljava/lang/String;",(void *)android_hardware_Camera_getParameters }
};
```

JNINativeMethod 的第一个成员是一个字符串，表示 Java 本地调用方法的名称，此名称是在 Java 程序中调用的名称；第二个成员也是一个字符串，表示 Java 本地调用方法的参数和返回值；第三个成员是 Java 本地调用方法对应的 C 语言函数。

通过函数 register_android_hardware_Camera()将 gMethods 注册为 android/media/Camera 的类，其主要实现代码如下：

```
int register_android_hardware_Camera(JNIEnv *env)
{
return AndroidRuntime::registerNativeMethods(env, "android/hardware/Camera",camMethods, NELEM(cam
Methods));
}
```

其中类 android/hardware/Camera 和 Java 类 android.hardware.Camera 相对应。

（2）Camera 的本地库 libui.so

文件 frameworks/base/libs/ui/Camera.cpp 用于实现文件 Camera.h 中提供的接口，其中最重要的代码片段如下：

```
sp<Camera> Camera::create(const sp<ICamera>& camera)
{
     ALOGV("create");
     if (camera == 0) {
          ALOGE("camera remote is a NULL pointer");
          return 0;
     }

    sp<Camera> c = new Camera(-1);
    if (camera->connect(c) == NO_ERROR) {
         c->mStatus = NO_ERROR;
         c->mCamera = camera;
         camera->asBinder()->linkToDeath(c);
```

```
            return c;
        }
        return 0;
}
```

函数 connect()的实现代码如下：

```
sp<Camera> Camera::connect(int cameraId, const String16& clientPackageName,
            int clientUid)
{
        return CameraBaseT::connect(cameraId, clientPackageName, clientUid);
}
```

函数 connect()通过调用 getCameraService 得到一个 ICameraService，再通过 ICameraService 的 cs->connect(c)得到一个 ICamera 类型的指针。调用 connect()函数会得到一个 Camera 类型的指针。在正常情况下，已经初始化完成了 Camera 的成员 mCamera。

函数 startPreview()的实现代码如下：

```
status_t Camera::startPreview()
{
        ALOGV("startPreview");
        sp <ICamera> c = mCamera;
        if (c == 0) return NO_INIT;
        return c->startPreview();
}
```

其他函数的实现过程也与函数 setDataSource 类似。在库 libmedia.so 中的其他一些文件与头文件的名称相同，分别如下。

☑ frameworks/base/libs/ui/ICameraClient.cpp

☑ frameworks/base/libs/ui/ICamera.cpp

☑ frameworks/base/libs/ui/ICameraService.cpp

此处的类 BnCameraClient 和 BnCameraService 虽然实现了 onTransact()函数，但是由于还有纯虚函数没有实现，所以不能实例化这个类。

（3）Camera 服务 libcameraservice.so

目录 frameworks/av/services/camera/libcameraservice/实现一个 Camera 的服务，此服务是继承 ICameraService 的具体实现。在此目录下和硬件抽象层“桩”实现相关的文件说明如下所示。

☑ CameraHardwareStub.cpp：Camera 硬件抽象层“桩”实现。

☑ CameraHardwareStub.h：Camera 硬件抽象层“桩”实现的接口。

☑ CannedJpeg.h：包含一块 JPEG 数据，在拍照片时作为 JPEG 数据。

☑ FakeCamera.h 和 FakeCamera.cpp：实现假的 Camera 黑白格取景器效果。

在文件 Android.mk 中，使用宏 USE_CAMERA_STUB 决定是否使用真的 Camera，如果宏为真，则使用 CameraHardwareStub.cpp 和 FakeCamera.cpp 构造一个假的 Camera；如果为假，则使用 CameraService.cpp 构造一个实际上的 Camera 服务。文件 Android.mk 的主要代码如下：

```
LOCAL_MODULE:= libcamerastub
LOCAL_SHARED_LIBRARIES:= libui
include $(BUILD_STATIC_LIBRARY)
endif # USE_CAMERA_STUB
#
# libcameraservice
```

```
#
include $(CLEAR_VARS)
LOCAL_SRC_FILES:= \
    CameraService.cpp
LOCAL_SHARED_LIBRARIES:= \
    libui \
    libutils \
    libcutils \
    libmedia
LOCAL_MODULE:= libcameraservice
LOCAL_CFLAGS+=-DLOG_TAG=\"CameraService\"
ifeq ($(USE_CAMERA_STUB), true)
LOCAL_STATIC_LIBRARIES += libcamerastub
LOCAL_CFLAGS += -include CameraHardwareStub.h
else
LOCAL_SHARED_LIBRARIES += libcamera
endif
include $(BUILD_SHARED_LIBRARY)
```

文件CameraService.cpp继承了BnCameraService的实现，在此类内部又定义了类Client，CameraService::Client继承了BnCamera。在运作的过程中，函数CameraService::connect()用于得到一个CameraService::Client。在使用过程中，主要是通过调用这个类的接口来实现完成Camera的功能。因为CameraService::Client本身继承了BnCamera类，而BnCamera类继承了ICamera，所以可以将此类当成ICamera来使用。

类CameraService和CameraService::Client的结果如下：

```
class CameraService : public BnCameraService
{
class Client : public BnCamera {};
wp<Client> mClient;
}
```

在CameraService中，静态函数instantiate()用于初始化一个Camera服务，此函数的代码如下：

```
void CameraService::instantiate() {
defaultServiceManager()->addService( String16("media.camera"), new CameraService());
}
```

其实函数CameraService::instantiate()注册了一个名称为media.camera的服务，此服务和文件Camera.cpp中调用的名称相对应。

Camera整个运作机制是：在文件Camera.cpp中调用ICameraService的接口，此时实际上调用的是BpCameraService。而BpCameraService通过Binder机制和BnCameraService实现两个进程的通信。因为BpCameraService的实现就是此处的CameraService，所以Camera.cpp虽然是在另外一个进程中运行，但是调用ICameraService的接口就像直接调用一样，从函数connect()中可以得到一个ICamera类型的指针，整个指针的实现实际上是CameraService::Client。

上述Camera功能的具体实现就是CameraService::Client所实现的，其构造函数如下：

```
CameraService::Client::Client(const sp<CameraService>& cameraService,
        const sp<ICameraClient>& cameraClient) :
mCameraService(cameraService), mCameraClient(cameraClient), mHardware(0)
{
mHardware = openCameraHardware();
mHasFrameCallback = false;
}
```

在构造函数中，通过调用 openCameraHardware()得到一个 CameraHardwareInterface 类型的指针，并作为其成员 mHardware。以后对实际的 Camera 的操作都通过这个指针进行，这是一个简单的直接调用关系。

其实真正的 Camera 功能已经通过实现 CameraHardwareInterface 类来完成。在这个库中，文件 CameraHardwareStub.h 和 CameraHardwareStub.cpp 定义了一个“桩”模块的接口，可以在没有 Camera 硬件的情况下使用。例如在仿真器的情况下使用的文件就是 CameraHardwareStub.cpp 和它依赖的文件 FakeCamera.cpp。

类 CameraHardwareStub 的结构如下：

```
class CameraHardwareStub : public CameraHardwareInterface {
class PreviewThread : public Thread {
};
};
```

在类 CameraHardwareStub 中包含了线程类 PreviewThread，此线程可以处理 PreView，即负责刷新取景器的内容。实际的 Camera 硬件接口通常可以通过对 V4L2 捕获驱动的调用来实现，同时还需要一个 JPEG 编码程序将从驱动中取出的数据编码成 JPEG 文件。

在文件 FakeCamera.h 和 FakeCamera.cpp 中实现了类 FakeCamera，用于实现一个假的摄像头输入数据的内存。定义代码如下：

```
class FakeCamera {
public:
    FakeCamera(int width, int height);
    ~FakeCamera();

    void setSize(int width, int height);
    void getNextFrameAsRgb565(uint16_t *buffer);        //获取 RGB565 格式的预览帧
    void getNextFrameAsYuv422(uint8_t *buffer);         //获取 Yuv422 格式的预览帧
    status_t dump(int fd, const Vector<String16>& args);

private:
    void drawSquare(uint16_t *buffer, int x, int y, int size, int color, int shadow);
    void drawCheckerboard(uint16_t *buffer, int size);

    static const int kRed = 0xf800;
    static const int kGreen = 0x07c0;
    static const int kBlue = 0x003e;

    int mWidth, mHeight;
    int mCounter;
    int mCheckX, mCheckY;
    uint16_t *mTmpRgb16Buffer;
};
```

当在 CameraHardwareStub 中设置参数后会调用函数 initHeapLocked()，此函数的实现代码如下：

```
void CameraHardwareStub::initHeapLocked()
{
    int picture_width, picture_height;
    mParameters.getPictureSize(&picture_width, &picture_height);
    //建立内存堆栈，创建两块内存
    mRawHeap = new MemoryHeapBase(picture_width * 2 * picture_height);
```

```
    int preview_width, preview_height;
    mParameters.getPreviewSize(&preview_width, &preview_height);
    LOGD("initHeapLocked: preview size=%dx%d", preview_width, preview_height);

    //从参数中获取信息
    int how_big = preview_width * preview_height * 2;

    if (how_big == mPreviewFrameSize)
        return;

    mPreviewFrameSize = how_big;

    mPreviewHeap = new MemoryHeapBase(mPreviewFrameSize * kBufferCount);
    //建立内存队列
    for (int i = 0; i < kBufferCount; i++) {
        mBuffers[i] = new MemoryBase(mPreviewHeap, i * mPreviewFrameSize, mPreviewFrameSize);
    }

    delete mFakeCamera;
    mFakeCamera = new FakeCamera(preview_width, preview_height);
}
```

定义函数 startPrevie()来创建一个线程，此函数的实现代码如下：

```
status_t CameraHardwareStub::startPreview(preview_callback cb, void* user)
{
    Mutex::Autolock lock(mLock);
    if (mPreviewThread != 0) {
        return INVALID_OPERATION;
    }
    mPreviewCallback = cb;
    mPreviewCallbackCookie = user;
    mPreviewThread = new PreviewThread(this);          //建立视频预览线程
    return NO_ERROR;
}
```

通过上面建立的线程可以调用预览回调机制，将预览的数据传递给上层的 CameraService。

创建预览线程函数 previewThread()，建立一个循环以得到假的摄像头输入数据的来源，并通过预览回调函数将输出传到上层中去。函数 previewThread()的主要实现代码如下：

```
int CameraHardwareStub::previewThread()
{
    mLock.lock();
        int previewFrameRate = mParameters.getPreviewFrameRate();
        //计算在当前缓冲堆之内的垂距
        ssize_t offset = mCurrentPreviewFrame * mPreviewFrameSize;
        sp<MemoryHeapBase> heap = mPreviewHeap;
            //假设照相机内部状态没有变化
        //(or is thread safe)
        FakeCamera* fakeCamera = mFakeCamera;
        sp<MemoryBase> buffer = mBuffers[mCurrentPreviewFrame];
            mLock.unlock();
    if (buffer != 0) {
        //计算在框架之间等待多久
```

```
        int delay = (int)(1000000.0f / float(previewFrameRate));
            //设置为总是合法操作，即使内存消亡仍然在操作过程中被映射
            void *base = heap->base();
            //添加模拟照相框架
        uint8_t *frame = ((uint8_t *)base) + offset;
        fakeCamera->getNextFrameAsYuv422(frame);

                //一个新客户通知框架
        mPreviewCallback(buffer, mPreviewCallbackCookie);
            //推进缓冲
        mCurrentPreviewFrame = (mCurrentPreviewFrame + 1) % kBufferCount;
        //等待它...
        usleep(delay);
    }
    return NO_ERROR;
}
```

在上述文件中还定义了其他的函数，函数的功能一看便知，在此为节省本书篇幅将不再一一进行详细讲解，请读者参考开源的代码文件。

# 15.5 开发拍照应用程序

**知识点讲解：光盘:视频\知识点\第 15 章\开发拍照应用程序.avi**

在开发 Android 应用程序的过程中，有如下两种方式可以调用系统的摄像头实现拍照功能。

☑ 通过 Intent 调用系统的照相机 Activity。

☑ 通过编码调用 Camera API。

本节将详细讲解上述两种方式的具体实现流程。

## 15.5.1 通过 Intent 调用系统的照相机 Activity

在现实中的很多应用场景中，都需要使用摄像头去拍摄照片或视频，然后在照片或视频的基础之上进行处理。但是因为 Android 系统源码是开源的，所以很多设备厂商均可使用，造成定制比较混乱的状况发生。一般来说，在需要用到摄像头拍照或摄像时，均会直接调用系统现有的相机应用去进行拍照或摄像，我们只取它拍摄的结果进行处理，这样就避免了不同设备的摄像头的一些细节问题。

当通过 Intent 直接调用系统提供的照相机功能时，复用其拍照 Activity 是最简单和最方便的办法，此时不需要考虑手机的兼容性问题，如预览拍照图片大小等。例如下面的演示代码。

```
Intent i = new Intent("android.media.action.IMAGE_CAPTURE");
startActivityForResult(i, C.REQUEST_CODE_CAMERA);
```

然后通过如下所示的代码在 onActivityResult 中获取返回的数据。

```
if(resultCode==RESULT_OK){
    Bundle extras = data.getExtras();
    if(extras!=null){
        Bitmap bitmap = (Bitmap) extras.get(C.CODE_PHOTO_BITMAP_DATA);
        …
    }
}
```

但是通过上述过程这样获取的图片是缩小过的，如果要获取原始的相片，则需要在调用时就指定相片生成的路径。例如下面的演示代码。

```
i.putExtra(MediaStore.EXTRA_OUTPUT, Uri.fromFile(new File(tempPath)));
```

然后指定相片的类型，例如下面的演示代码。

```
i.putExtra("outputFormat", Bitmap.CompressFormat.PNG.name());
```

也可以设置拍照的横竖屏，例如下面的演示代码。

```
i.putExtra(MediaStore.EXTRA_SCREEN_ORIENTATION,Configuration.ORIENTATION_LANDSCAPE);
```

## 15.5.2 调用 Camera API 拍照

除了调用内置的拍照系统之外，还可以直接调用摄像头的 API 来编写类似取景框的拍照 Activity，这种方式更加灵活。但需要考虑底层的具体结构，需要了解 Android 的版本、摄像头的分辨率组合等信息，这样才能够编写出各种屏幕分辨率组合的布局等。

另外，Android 提供的 SDK（android.hardware.Camera）不能正常地使用竖屏（Portrait Layout）加载照相机，当用竖屏模式加载照相机时会产生如下问题。

☑ 照相机成像左倾 90°。

☑ 照相机成像长宽比例不对（失比）。

造成上述问题的原因是摄像头对照物的映射是 Android 底层固定的，以 landscape 方式为正，并且产生大小为 320×480 的像。此时可以通过编写自定义代码来校正上述问题，具体编码思路如下。

（1）设置 Camera 的参数来实现转变图片预览角度，但这种方式并不是对所有的 Camera 都有效果的。例如下面的演示代码。

```
Camera.Parameters parameters = camera.getParameters();
parameters.set("orientation", "portrait");
parameters.set("rotation", 90);
camera.setParameters(parameters);
```

（2）在获取拍摄相片之后进行旋转校正，但是如果图片太大会造成内存溢出的问题。例如下面的演示代码。

```
Matrix m = new Matrix();
m.postRotate(90);
m.postScale(balance, balance);
Bitmap.createBitmap(tempBitmap, 0, 0, tempBitmap.getWidth(), tempBitmap.getHeight(),m, true);
```

在 Android 系统中，当通过 Camera API 方式实现拍照功能时，需要用到如下所示的类。

（1）Camera 类：最主要的类，用于管理 Camera 设备，常用的方法如下。

☑ open()：通过该方法获取 Camera 实例。

☑ setPreviewDisplay(SurfaceHolder)：设置预览拍照。

☑ startPreview()：开始预览。

☑ stopPreview()：停止预览。

☑ release()：释放 Camera 实例。

☑ takePicture(Camera.ShutterCallback shutter, Camera.PictureCallback raw, Camera.PictureCallback jpeg)：这是拍照要执行的方法，包含了 3 个回调参数。其中参数 shutter 是快门按下时的回调，参数 raw 是获取拍照原始数据的回调，参数 jpeg 用来获取压缩成 jpg 格式的图像数据。

☑ Camera.PictureCallback 接口：该回调接口包含了一个 onPictureTaken(byte[]data, Camera camera)

方法。在这个方法中可以保存图像数据。

（2）SurfaceView 类：用于控制预览界面。其中 SurfaceHolder.Callback 接口用于处理预览的事件，需要实现如下所示的 3 个方法。

☑ surfaceCreated(SurfaceHolderholder)：预览界面创建时调用，每次界面改变后都会重新创建，需要获取相机资源并设置 SurfaceHolder。

☑ surfaceChanged(SurfaceHolderholder, int format, int width, int height)：在预览界面发生变化时调用，每次界面发生变化之后需要重新启动预览。

☑ surfaceDestroyed(SurfaceHolderholder)：预览销毁时调用，停止预览，释放相应资源。

### 15.5.3　总结 Camera 拍照的流程

经过本节前面内容的介绍，想必大家对 Camera 拍照的基本知识有了一个全新的了解。在此根据笔者的开发经验，总结出使用 Camera 实现拍照应用的基本流程，希望为广大读者起到一个抛砖引玉的作用。

（1）如果想在自己的应用中使用摄像头，需要在 AndroidManifest.xml 中增加以下权限声明代码。

```
<uses-permissionandroid:name="android.permission.CAMERA"/>
```

（2）设定摄像头布局。

这一步是开发工作的基础，也就是说我们希望在应用程序中增加多少辅助性元素，如摄像头各种功能按钮等。这里采取最简方式，除了拍照外，没有多余摄像头功能。布局文件 camera_surface.xml 介绍如下：

```
<LinearLayoutxmlns:android="http://schemas.android.com/apk/res/android"
  android:layout_width="fill_parent"android:layout_height="fill_parent"
  androidrientation="vertical">
  <SurfaceViewandroid:id="@+id/surface_camera"
  android:layout_width="fill_parent"android:layout_height="10dip"
  android:layout_weight="1">
  </SurfaceView>
</LinearLayout>
```

在此不要在资源文件名称中使用大写字母，如果把该文件命名为 CameraSurface.xml，会带来不必要的麻烦。

上述布局非常简单，只有一个 LinearLayout 视图组，在它下面只有一个 SurfaceView 视图，也就是摄像头屏幕。

（3）编写摄像头实现代码。

在此创建一个名为 CameraView 的 Activity 类，实现 SurfaceHolder.Callback 接口。

```
publicclassCamaraViewextendsActivityimplementsSurfaceHolder.Callback
```

接口 SurfaceHolder.Callback 被用来接收摄像头预览界面变化的信息，它实现了如下 3 个方法。

☑ surfaceChanged：当预览界面的格式和大小发生改变时，该方法被调用。

☑ surfaceCreated：在初次实例化、预览界面被创建时，该方法被调用。

☑ surfaceDestroyed：当预览界面被关闭时，该方法被调用。

接下来看一下在摄像头应用中如何使用这个接口，首先看一下在 Activity 类中的 onCreate()方法，其中通过下面的代码设置摄像头预览界面将通过全屏显示，并且没有“标题（title）”，并设置屏幕格式为“半透明”。

```
getWindow().setFormat(PixelFormat.TRANSLUCENT);
requestWindowFeature(Window.FEATURE_NO_TITLE);
```

```
getWindow().setFlags(WindowManager.LayoutParams.FLAG_FULLSCREEN,
WindowManager.LayoutParams.FLAG_FULLSCREEN);
```

然后通过 setContentView 来设定 Activity 的布局为前面我们创建的 camera_surface，并创建一个 SurfaceView 对象，从 xml 文件中获得它。对应代码如下：

```
setContentView(R.layout.camera_surface);
mSurfaceView=(SurfaceView)findViewById(R.id.surface_camera);
mSurfaceHolder=mSurfaceView.getHolder();
mSurfaceHolder.addCallback(this);
mSurfaceHolder.setType(SurfaceHolder.SURFACE_TYPE_PUSH_BUFFERS);
```

通过以上代码从 surfaceview 中获得了 holder，并增加 callback 功能到 this。这意味着我们的操作（Activity）可以管理这个 surfaceview。

再看 Callback 功能的实现代码：

```
publicvoidsurfaceCreated(SurfaceHolderholder){
mCamera=Camera.open();
```

上面的 mCamera 是 Camera 类的一个对象。在 surfaceCreated()方法中“打开”摄像头。这就是启动它的方式。

接下来定义方法 surfaceChanged()让摄像头做好拍照准备，设定它的参数，并开始在 Android 屏幕中启动预览画面。在此使用了 semaphore 参数来防止冲突：当 mPreviewRunning 为 true 时，意味着摄像头处于激活状态，并未被关闭，因此我们可以使用它。

```
public void surfaceChanged(SurfaceHolderholder,intformat,intw,inth){
  if(mPreviewRunning){
 mCamera.stopPreview();
}
Camera.Parametersp=mCamera.getParameters();
  p.setPreviewSize(w,h);
  mCamera.setParameters(p);
  try{
  mCamera.setPreviewDisplay(holder);
  }catch(IOExceptione){
  e.printStackTrace();
}
mCamera.startPreview();
mPreviewRunning=true;
  }
publicvoidsurfaceDestroyed(SurfaceHolderholder){
  mCamera.stopPreview();
  mPreviewRunning=false;
  mCamera.release();
}
```

通过上述方法代码停止了摄像头，并释放相关的资源。正如大家所看到的，在此设置 mPreviewRunning 为 false 以防止在 surfaceChanged 方法中的冲突。这是因为我们已经关闭了摄像头，而且不能再设置其参数或在摄像头中启动图像预览。

最后看下面最重要的方法：

```
Camera.PictureCallbackmPictureCallback=newCamera.PictureCallback(){
  publicvoidonPictureTaken(byte[]imageData,Camerac){
}
};
```

在拍照时该方法被调用。例如可以在界面上创建一个 OnClickListener，当点击屏幕时调用 PictureCallBack()方法，此方法会提供图像的字节数组，然后使用 Android 提供的 Bitmap 和 BitmapFactory 类，将其从字节数组转换成想要的图像格式。

# 15.6　解析二维码

**知识点讲解：光盘:视频\知识点\第 15 章\解析二维码.avi**

QR Code 码是由日本 Denso 公司于 1994 年 9 月研制的一种矩阵二维码符号，它具有一维条码及其他二维条码所具有的信息容量大、可靠性高、可表示汉字及图像等多种信息、保密防伪性强等优点。本节将详细讲解使用相机解析 QR Code 码的方法。

## 15.6.1　QR Code 码的特点

从 QR Code 码的英文名称 Quick Response Code 可以看出，超高速识读特点是 QR Code 码区别于四一七条码、Data Matrix 等二维码的主要特性。由于在用 CCD 识读 QR Code 码时，整个 QR Code 码符号中信息的读取是通过 QR Code 码符号的位置探测图形，用硬件来实现，因此，信息识读过程所需时间很短，它具有超高速识读特点。用 CCD 二维条码识读设备，每秒可识读 30 个含有 100 个字符的 QR Code 码符号；对于含有相同数据信息的四一七条码符号，每秒仅能识读 3 个符号；对于 Data Martix 矩阵码，每秒仅能识读 2～3 个符号。QR Code 码的超高速识读特性，使它能够广泛应用于工业自动化生产线管理等领域。

QR Code 码具有全方位（360°）识读特点，这是 QR Code 码优于行排式二维条码如四一七条码的另一主要特点，由于四一七条码是通过将一维条码符号在行排高度上的截短来实现的，因此，它很难实现全方位识读，其识读方位角仅为±10°，能够有效地表示中国汉字和日本汉字。由于 QR Code 码用特定的数据压缩模式表示中国汉字和日本汉字，它仅用 13bit 即可表示一个汉字，而四一七条码、Data Martix 等二维码没有特定的汉字表示模式，因此仅用字节表示模式来表示汉字，在用字节模式表示汉字时，需用 16bit（两个字节）表示一个汉字，因此 QR Code 码比其他的二维条码表示汉字的效率提高了 20%。

## 15.6.2　实战演练——使用 Android 相机解析二维码

下面将通过一个具体实例来讲解使用 Android 相机解析 QR Code 二维码的方法。

| 题　目 | 目　的 | 源 码 路 径 |
|---|---|---|
| 实例 15-1 | 使用 Android 相机解析 QR Code 二维码 | 光盘:\daima\15\qrEX |

本实例的具体实现流程如下。

（1）分别创建私有 Camera 对象 mCamera01、mButton01、mButton02 和 mButton03，然后设置默认相机预览模式为 false。具体代码如下：

```
/*创建私有 Camera 对象*/
private Camera mCamera01;
```

```
private Button mButton01, mButton02, mButton03;
/*作为 review 照下来的相片之用*/
private ImageView mImageView01;
private String TAG = "HIPPO";
private SurfaceView mSurfaceView01;
private SurfaceHolder mSurfaceHolder01;

/*默认相机预览模式为 false*/
private boolean bIfPreview = false;
```

（2）设置应用程序全屏幕运行，并添加红色正方形红框 View 供用户对准条形码，然后将创建的红色方框添加至 Activity 中。具体代码如下：

```
public void onCreate(Bundle savedInstanceState)
{
  super.onCreate(savedInstanceState);
  /*使应用程序全屏幕运行，不使用 title bar*/
  requestWindowFeature(Window.FEATURE_NO_TITLE);
  setContentView(R.layout.main);
  /*添加红色正方形红框 View，供 User 对准条形码*/
  DrawCaptureRect mDraw = new DrawCaptureRect
  (
    example203.this,
    110, 10, 100, 100,
    getResources().getColor(R.drawable.lightred)
  );

  /*将创建的红色方框添加至 Activity 中*/
  addContentView
  (
    mDraw,
    new LayoutParams
    (
      LayoutParams.WRAP_CONTENT, LayoutParams.WRAP_CONTENT
    )
  );
```

（3）分别取得屏幕解析像素，绑定 SurfaceView 并设置预览大小。具体代码如下：

```
/*取得屏幕解析像素*/
DisplayMetrics dm = new DisplayMetrics();
getWindowManager().getDefaultDisplay().getMetrics(dm);

mImageView01 = (ImageView) findViewById(R.id.myImageView1);

/*以 SurfaceView 作为相机 Preview 之用*/
mSurfaceView01 = (SurfaceView) findViewById(R.id.mSurfaceView1);

/*绑定 SurfaceView，取得 SurfaceHolder 对象*/
mSurfaceHolder01 = mSurfaceView01.getHolder();

/*Activity 必须实现 SurfaceHolder.Callback*/
mSurfaceHolder01.addCallback(example203.this);
```

```
/*额外的预览大小设置，在此不使用*/
//mSurfaceHolder01.setFixedSize(320, 240);

/*
* 以 SURFACE_TYPE_PUSH_BUFFERS(3)
* 作为 SurfaceHolder 显示类型
* */
mSurfaceHolder01.setType
(SurfaceHolder.SURFACE_TYPE_PUSH_BUFFERS);

mButton01 = (Button)findViewById(R.id.myButton1);
mButton02 = (Button)findViewById(R.id.myButton2);
mButton03 = (Button)findViewById(R.id.myButton3);
```

（4）编写单击方法 mButton01 按钮的响应程序，单击后打开相机及预览二维条形码。具体代码如下：

```
/*打开相机及预览二维条形码*/
mButton01.setOnClickListener(new Button.OnClickListener()
{
  @Override
  public void onClick(View arg0)
  {

    /*自定义初始化打开相机函数*/
    initCamera();
  }
});
```

（5）编写方法单击 mButton02 按钮的响应程序，单击后停止预览。具体代码如下：

```
/*停止预览*/
mButton02.setOnClickListener(new Button.OnClickListener()
{
  @Override
  public void onClick(View arg0)
  {

    /*自定义重置相机，并关闭相机预览函数*/
    resetCamera();
  }
});
```

（6）编写单击 mButton03 按钮后的响应程序，单击后拍照处理并生成二维条形码。具体代码如下：

```
  /*拍照 QR Code 二维条形码*/
  mButton03.setOnClickListener(new Button.OnClickListener()
  {
    @Override
    public void onClick(View arg0)
    {
      /*自定义拍照函数*/
      takePicture();
    }
  });
}
```

（7）定义方法 initCamera()用于自定义初始相机函数，具体代码如下：

```
/*自定义初始相机函数*/
private void initCamera()
{
  if(!bIfPreview)
  {
    /*若相机不是在预览模式，则打开相机*/
    mCamera01 = Camera.open();
  }

  if (mCamera01 != null && !bIfPreview)
  {
    Log.i(TAG, "inside the camera");

    /*创建 Camera.Parameters 对象*/
    Camera.Parameters parameters = mCamera01.getParameters();

    /*设置相片格式为 JPEG*/
    parameters.setPictureFormat(PixelFormat.JPEG);

    /*指定 preview 的屏幕大小*/
    parameters.setPreviewSize(160, 120);

    /*设置图片分辨率大小*/
    parameters.setPictureSize(160, 120);

    /*将 Camera.Parameters 设置于 Camera*/
    mCamera01.setParameters(parameters);

    /*setPreviewDisplay 唯一的参数为 SurfaceHolder*/
    mCamera01.setPreviewDisplay(mSurfaceHolder01);

    /*立即运行 Preview*/
    mCamera01.startPreview();
    bIfPreview = true;
  }
}
```

（8）定义方法 takePicture()用于拍照并获取图像。具体代码如下：

```
/*拍照撷取图像*/
private void takePicture()
{
  if (mCamera01 != null && bIfPreview)
  {
    /*调用 takePicture()方法拍照*/
    mCamera01.takePicture
    (shutterCallback, rawCallback, jpegCallback);
  }
}
```

（9）定义方法 resetCamera()实现相机重置，然后释放 Camera 对象。具体代码如下：

```
/*相机重置*/
private void resetCamera()
```

```
{
  if (mCamera01 != null && bIfPreview)
  {
    mCamera01.stopPreview();
    /*释放 Camera 对象 */
    //mCamera01.release();
    mCamera01 = null;
    bIfPreview = false;
  }
}
private ShutterCallback shutterCallback = new ShutterCallback()
{
  public void onShutter()
  {
  }
};
private PictureCallback rawCallback = new PictureCallback()
{
  public void onPictureTaken(byte[] _data, Camera _camera)
  {
  }
};
```

（10）定义方法 onPictureTaken()，对传入的图片进行处理。首先设置 onPictureTaken 传入的第一个参数即为相片的 byte；然后使用 Matrix.postScale 方法缩小图像大小；接下来创建新的 Bitmap 对象；然后获取 4:3 图片的居中红色框部分 100×100 像素，并将拍照的图文件以 ImageView 显示出来；最后将传入的图文件译码成字符串，并定义方法 mMakeTextToast()输出提示。具体代码如下：

```
private PictureCallback jpegCallback = new PictureCallback()
{
  public void onPictureTaken(byte[] _data, Camera _camera)
  {
    try
    {
      /*onPictureTaken()传入的第一个参数即为相片的 byte*/
      Bitmap bm =
      BitmapFactory.decodeByteArray(_data, 0, _data.length);

      int resizeWidth = 160;
      int resizeHeight = 120;
      float scaleWidth = ((float) resizeWidth) / bm.getWidth();
      float scaleHeight = ((float) resizeHeight) / bm.getHeight();

      Matrix matrix = new Matrix();
      /*使用 Matrix.postScale 方法缩小 Bitmap Size*/
      matrix.postScale(scaleWidth, scaleHeight);

      /*创建新的 Bitmap 对象*/
      Bitmap resizedBitmap = Bitmap.createBitmap
      (bm, 0, 0, bm.getWidth(), bm.getHeight(), matrix, true);

      /*撷取 4:3 图片的居中红色框部分 100×100 像素*/
```

```
Bitmap resizedBitmapSquare = Bitmap.createBitmap
(resizedBitmap, 30, 10, 100, 100);

/*将拍照的图文件以 ImageView 显示出来*/
mImageView01.setImageBitmap(resizedBitmapSquare);

/*将传入的图文件译码成字符串*/
String strQR2 = decodeQRImage(resizedBitmapSquare);
if(strQR2!="")
{
  if (URLUtil.isNetworkUrl(strQR2))
  {
    /*OMIA 规范，网址条形码，打开浏览器上网*/
    mMakeTextToast(strQR2, true);
    Uri mUri = Uri.parse(strQR2);
    Intent intent = new Intent(Intent.ACTION_VIEW, mUri);
    startActivity(intent);
  }
  else if(eregi("wtai://",strQR2))
  {
    /*OMIA 规范，手机拨打电话格式*/
    String[] aryTemp01 = strQR2.split("wtai://");
    Intent myIntentDial = new Intent
    (
      "android.intent.action.CALL",
      Uri.parse("tel:"+aryTemp01[1])
    );
    startActivity(myIntentDial);
  }
  else if(eregi("TEL:",strQR2))
  {
    /*OMIA 规范，手机拨打电话格式*/
    String[] aryTemp01 = strQR2.split("TEL:");
    Intent myIntentDial = new Intent
    (
      "android.intent.action.CALL",
      Uri.parse("tel:"+aryTemp01[1])
    );
    startActivity(myIntentDial);
  }
  else
  {
    /*若仅是文字，则以 Toast 显示出来*/
    mMakeTextToast(strQR2, true);
  }
}

/*显示完图文件，立即重置相机，并关闭预览*/
resetCamera();

/*再重新启动相机继续预览*/
```

```
        initCamera();
      }
      catch (Exception e)
      {
        Log.e(TAG, e.getMessage());
      }
    }
};

public void mMakeTextToast(String str, boolean isLong)
{
  if(isLong==true)
  {
    Toast.makeText(example203.this, str, Toast.LENGTH_LONG).show();
  }
  else
  {
    Toast.makeText(example203.this, str, Toast.LENGTH_SHORT).show();
  }
}
```

（11）定义方法 checkSDCard()来判断记忆卡是否存在。具体代码如下：

```
private boolean checkSDCard()
{
  /*判断记忆卡是否存在*/
  if(android.os.Environment.getExternalStorageState().equals
  (android.os.Environment.MEDIA_MOUNTED))
  {
    return true;
  }
  else
  {
    return false;
  }
}
```

（12）定义方法 decodeQRImage(Bitmap myBmp)来解码传入的 Bitmap 图片，主要代码如下：

```
/*解码传入的 Bitmap 图片*/
public String decodeQRImage(Bitmap myBmp)
{
  String strDecodedData = "";
  try
  {
    QRCodeDecoder decoder = new QRCodeDecoder();
    strDecodedData   = new String
    (decoder.decode(new AndroidQRCodeImage(myBmp)));
  }
  catch(Exception e)
  {
    e.printStackTrace();
  }
  return strDecodedData;
}
```

（13）定义类 DrawCaptureRect 来绘制相机预览画面中的正方形方框。具体代码如下：

```
/*绘制相机预览画面中的正方形方框*/
class DrawCaptureRect extends View
{
  private int colorFill;
  private int intLeft,intTop,intWidth,intHeight;

  public DrawCaptureRect
  (
    Context context, int intX, int intY, int intWidth,
    int intHeight, int colorFill
  )
  {
    super(context);
    this.colorFill = colorFill;
    this.intLeft = intX;
    this.intTop = intY;
    this.intWidth = intWidth;
    this.intHeight = intHeight;
  }

  @Override
  protected void onDraw(Canvas canvas)
  {
    Paint mPaint01 = new Paint();
    mPaint01.setStyle(Paint.Style.FILL);
    mPaint01.setColor(colorFill);
    mPaint01.setStrokeWidth(1.0F);
    /*在画布上绘制红色的 4 条方边框作为瞄准器*/
    canvas.drawLine
    (
      this.intLeft, this.intTop,
      this.intLeft+intWidth, this.intTop, mPaint01
    );
    canvas.drawLine
    (
      this.intLeft, this.intTop,
      this.intLeft, this.intTop+intHeight, mPaint01
    );
    canvas.drawLine
    (
      this.intLeft+intWidth, this.intTop,
      this.intLeft+intWidth, this.intTop+intHeight, mPaint01
    );
    canvas.drawLine
    (
      this.intLeft, this.intTop+intHeight,
      this.intLeft+intWidth, this.intTop+intHeight, mPaint01
    );
```

```
        super.onDraw(canvas);
    }
}
```

（14）定义方法 eregi()实现自定义比较字符串处理。具体代码如下：

```
    /*自定义比较字符串函数*/
    public static boolean eregi(String strPat, String strUnknow)
    {
        String strPattern = "(?i)"+strPat;
        Pattern p = Pattern.compile(strPattern);
        Matcher m = p.matcher(strUnknow);
        return m.find();
    }

    @Override
    public void surfaceChanged
    (SurfaceHolder surfaceholder, int format, int w, int h)
    {
        Log.i(TAG, "Surface Changed");
    }

    @Override
    public void surfaceCreated(SurfaceHolder surfaceholder)
    {
        Log.i(TAG, "Surface Changed");
    }

    @Override
    public void surfaceDestroyed(SurfaceHolder surfaceholder)
    {
        Log.i(TAG, "Surface Destroyed");
    }

    @Override
    protected void onPause()
    {
        super.onPause();
    }
}
```

执行后能够通过手机拍照的方式实现二维码解析，如图 15-3 所示。

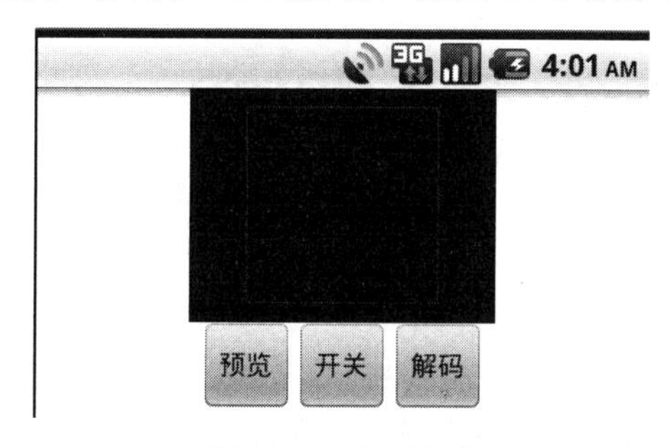


图 15-3　执行效果

# 第 16 章　NFC 近场通信技术详解

NFC 是近场通信（Near Field Communication）的缩写，此技术由非接触式射频识别（RFID）演变而来，由飞利浦半导体（现恩智浦半导体）、诺基亚和索尼共同研制开发，其基础是 RFID 及互联技术。NFC 是一种短距高频的无线电技术，在 13.56MHz 频率运行于 20 厘米距离内。其传输速度有 106kbit/s、212kbit/s 或者 424kbit/s 3 种。目前近场通信已成为 ISO/IEC IS 18092 国际标准、ECMA-340 标准和 ETSI TS 102 190 标准。NFC 采用主动和被动两种读取模式。本章将详细讲解在 Android 设备中使用近场通信技术的基本知识，为步入本书后面知识的学习打下基础。

## 16.1　近场通信技术基础

**知识点讲解：光盘:视频\知识点\第 16 章\近场通信技术基础.avi**

NFC 近场通信技术由非接触式射频识别（RFID）及互联互通技术整合演变而来，在单一芯片上结合感应式读卡器、感应式卡片和点对点的功能，能在短距离内与兼容设备进行识别和数据交换。工作频率为 13.56MHz，但是使用这种手机支付方案的用户必须更换特制的手机。目前这项技术在日、韩被广泛应用。手机用户凭着配置了支付功能的手机就可以行遍全国：他们的手机可以用作机场登机验证、大厦的门禁钥匙、交通一卡通、信用卡、支付卡等。本节将简要讲解 NFC 技术的基本知识。

### 16.1.1　NFC 技术的特点

近场通信是基于 RFID 技术发展起来的一种近距离无线通信技术。与 RFID 一样，近场通信信息也是通过频谱中无线频率部分的电磁感应耦合方式传递，但两者之间还是存在很大的区别。近场通信的传输范围比 RFID 小，RFID 的传输范围可以达到 0～1m，但由于近场通信采取了独特的信号衰减技术，相对于 RFID 来说近场通信具有成本低、带宽高、能耗低等特点。

在现实应用中，近场通信技术的主要特征如下：

☑ 用于近距离（10cm 以内）安全通信的无线通信技术。
☑ 射频频率：13.56MHz。
☑ 射频兼容：ISO 14443 和 ISO 15693，Felica 标准。
☑ 数据传输速度：106kbit/s、212kbit/s 或 424kbit/s。

### 16.1.2　NFC 的工作模式

在现实应用中，NFC 技术有如下所示的 3 种工作模式。

☑ 卡模式（Card emulation）：此模式其实就是相当于一张采用 RFID 技术的 IC 卡。可以替代现在大量的 IC 卡（包括信用卡）使用的场合，如商场刷卡、公交卡、门禁管制、车票、门票等。

此种方式下有一个极大的优点，那就是卡片通过非接触读卡器的 RF 域来供电，即便是寄主设备（如手机）没电也可以工作。

☑ 点对点模式（P2P mode）：此模式和红外线差不多，可用于数据交换，只是传输距离较短，传输创建速度较快，传输速度也快，功耗低（蓝牙也类似）。将两个具备 NFC 功能的设备链接，能实现数据点对点传输，如下载音乐、交换图片或者同步设备地址簿。因此通过 NFC，多个设备如数位相机、PDA、计算机和手机之间都可以交换资料或者服务。

☑ 读卡器模式（Reader/writer mode）：作为非接触读卡器使用，可以从海报或者展览信息电子标签上读取相关信息。

### 16.1.3　NFC 和蓝牙的对比

在现实应用中，NFC 和蓝牙（Bluetooth）都是短程通信技术，而且都被集成到移动电话。但 NFC 不需要复杂的设置程序，并且也可以简化蓝牙连接。NFC 略胜蓝牙的地方在于设置程序较短，但无法达到低功率蓝牙（Bluetooth Low Energy）的速度。在两台 NFC 设备相互连接的识别过程中，使用 NFC 来替代人工设置会使创建连接的速度大大加快，会少于十分之一秒。

NFC 的最大数据传输量是 424kbit/s，这远小于 Bluetooth V2.1（2.1Mbit/s）。虽然 NFC 在传输速度与距离方面比不上 Bluetooth，但是 NFC 技术不需要电源，对于移动电话或是移动消费性电子产品来说，NFC 的使用比较方便。NFC 的短距离通信特性正是其优点，由于耗电量低、一次只和一台机器链接，拥有较高的保密性与安全性，NFC 有利于信用卡交易时避免被盗用。NFC 的目标并非是取代蓝牙等其他无线技术，而是在不同的场合、不同的领域起到相互补充的作用。

NFC 技术和蓝牙技术主要功能参数的对比如表 16-1 所示。

表 16-1　NFC 技术和蓝牙技术主要功能参数的对比

| 说　明 | NFC | Bluetooth | Bluetooth Low Energy |
| --- | --- | --- | --- |
| RFID 兼容 | ISO 18000-3 | active | active |
| 标准化机构 | ISO/IEC | Bluetooth SIG | Bluetooth SIG |
| 网络标准 | ISO 13157 etc. | IEEE 802.15.1 | IEEE 802.15.1 |
| 网络类型 | Point-to-point | WPAN | WPAN |
| 加密 | not with RFID | available | available |
| 范围 | < 0.2m | ~10m（class 2） | ~1m（class 3） |
| 频率 | 13.56MHz | 2.4～2.5GHz | 2.4～2.5GHz |
| Bit rate | 424kbit/s | 2.1Mbit/s | ~1.0Mbit/s |
| 设置程序 | < 0.1s | < 6s | < 1s |
| 功耗 | < 15mA（read） | varies with class | < 15mA（xmit） |

## 16.2　射频识别技术详解

**知识点讲解：光盘:视频\知识点\第 16 章\射频识别技术详解.avi**

射频识别即 RFID（Radio Frequency Identification）技术，又称无线射频识别，是 NFC 技术的一个

子集。RFID 是一种通信技术，可以通过无线电信号识别特定目标并读写相关数据，而无须识别系统与特定目标之间建立机械或光学接触。在现实中常用的 RFID 技术有低频（125～134.2kHz）、高频（13.56MHz）、超高频、微波等技术。RFID 读写器也分移动式和固定式，目前 RFID 技术应用很广，例如，图书馆、门禁系统和食品安全溯源等。本节将详细讲解射频识别技术 RFID 的基本知识。

## 16.2.1 RFID 技术简介

RFID 是一种无线通信技术，可以通过无线电信号识别特定目标并读写相关数据，而无须识别系统与特定目标之间建立机械或者光学接触。从概念上来讲，RFID 类似于条码扫描，对于条码技术而言，它是将已编码的条形码附着于目标物，并使用专用的扫描读写器利用光信号将信息由条形磁传送到扫描读写器；而 RFID 则使用专用的 RFID 读写器及专门的可附着于目标物的 RFID 标签，利用频率信号将信息由 RFID 标签传送至 RFID 读写器。

无线电的信号是通过调成无线电频率的电磁场，把数据从附着在物品上的标签上传送出去，以自动辨识与追踪该物品。某些标签在识别时从识别器发出的电磁场中就可以得到能量，并不需要电池；也有标签本身拥有电源，并可以主动发出无线电波（调成无线电频率的电磁场）。标签包含了电子存储的信息，数米之内都可以识别。与条形码不同的是，射频标签不需要处在识别器视线之内，也可以嵌入被追踪物体之内。

在现实应用中，有许多行业都运用了射频识别技术。将标签附着在一辆正在生产中的汽车上，生产者可以追踪此车在生产线上的进度。仓库可以追踪药品的所在。射频标签也可以附于牲畜与宠物上，方便对牲畜与宠物的积极识别（积极识别意思是防止数只牲畜使用同一个身份）。射频识别的身份识别卡可以使员工得以进入锁住的建筑部分，汽车上的射频应答器也可以用来征收收费路段与停车场的费用。

某些射频标签附在衣物、个人财物上，甚至于植入人体内。由于这项技术可能会在未经本人许可的情况下读取个人信息，所以这项技术也会有侵犯个人隐私的忧患。

## 16.2.2 RFID 技术的组成

从结构上讲 RFID 是一种简单的无线系统，只有两个基本器件，该系统用于控制、检测和跟踪物体。系统由一个询问器和很多应答器组成。在最初的技术领域中，应答器是指能够传输信息、回复信息的电子模块。近年来，由于射频技术发展迅猛，应答器有了新的说法和含义，又被叫做智能标签或标签。RFID 电子标签的阅读器通过天线与 RFID 电子标签进行无线通信，可以实现对标签识别码和内存数据的读出或写入操作。RFID 技术可识别高速运动物体并可同时识别多个标签，操作快捷方便。

伴随着 RFID 技术的不断发展，其具体组成如下。

- ☑ 应答器：由天线、耦合元件及芯片组成，一般来说都是用标签作为应答器，每个标签具有唯一的电子编码，附着在物体上标识目标对象。
- ☑ 阅读器：由天线、耦合元件及芯片组成，读取（有时还可以写入）标签信息的设备，可设计为手持式 RFID 读写器（如 C5000W）或固定式读写器。
- ☑ 应用软件系统：是应用层软件，主要是把收集的数据进一步处理，并为人们所使用。

## 16.2.3 RFID 技术的特点

射频识别系统最重要的优点是非接触识别，它能穿透雪、雾、冰、涂料、尘垢和条形码无法使用的恶劣环境阅读标签，并且阅读速度极快，大多数情况下不到 100 毫秒。有源式射频识别系统的速写能力也是重要的优点。可用于流程跟踪和维修跟踪等交互式业务。

制约射频识别系统发展的主要问题是不兼容的标准。射频识别系统的主要厂商提供的都是专用系统，导致不同的应用和不同的行业采用不同厂商的频率和协议标准，这种混乱和割据的状况已经制约了整个射频识别行业的增长。许多欧美组织正在着手解决这个问题，并已经取得了一些成绩。标准化必将刺激射频识别技术的大幅度发展和广泛应用。

RFID 技术的主要特点如下。

☑ 快速扫描：RFID 辨识器可以同时辨识读取数个 RFID 标签。

☑ 体积小型化、形状多样化：RFID 在读取上并不受尺寸大小与形状的限制，不需为了读取精确度而配合纸张的固定尺寸和印刷品质。此外，RFID 标签更可向小型化与多样形态发展，以应用于不同产品。

☑ 抗污染能力和耐久性：传统条形码的载体是纸张，因此容易受到污染，但 RFID 对水、油和化学药品等物质具有很强的抵抗性。此外，由于条形码附于塑料袋或外包装纸箱上，所以特别容易受到折损；RFID 卷标是将数据存在芯片中，因此可以免受污损。

☑ 可重复使用：当今的条形码印刷上去之后就无法更改，RFID 标签则可以重复地新增、修改、删除 RFID 卷标内存储的数据，方便信息的更新。

☑ 穿透性和无屏障阅读：在被覆盖的情况下，RFID 能够穿透纸张、木材和塑料等非金属或非透明的材质，并能够进行穿透性通信。而条形码扫描机必须在近距离而且没有物体阻挡的情况下才可以辨读条形码。

☑ 数据的记忆容量大：一维条形码的容量是 50Bytes，二维条形码最大的容量可存储 2～3000 字符，RFID 最大的容量则有数 MegaByteso。随着记忆载体的发展，数据容量也有不断扩大的趋势。未来物品所需携带的资料量会越来越大，对卷标所能扩充容量的需求也相应增加。

☑ 安全性：由于 RFID 承载的是电子式信息，其数据内容可经由密码保护，使其内容不易被伪造及变造。

RFID 因其所具备的远距离读取、高存储量等特性而备受瞩目。它不仅可以帮助一个企业大幅提高货物、信息管理的效率，还可以让销售企业和制造企业互联，从而更加准确地接收反馈信息，控制需求信息，优化整个供应链。

## 16.2.4 RFID 技术的工作原理

RFID 技术的基本工作原理是：当标签进入磁场后，接收解读器发出的射频信号，凭借感应电流所获得的能量发送出存储在芯片中的产品信息（Passive Tag，无源标签或被动标签），或者由标签主动发送某一频率的信号（Active Tag，有源标签或主动标签），解读器读取信息并解码后，送至中央信息系统进行有关数据处理。

一套完整的 RFID 系统由阅读器（Reader）、电子标签（TAG，也就是所谓的应答器（Transponder））

及应用软件系统 3 部分组成，其工作原理是 Reader 发射一特定频率的无线电波能量给 Transponder，用以驱动 Transponder 电路将内部的数据送出，此时 Reader 便依序接收解读数据，送给应用程序做相应的处理。

以 RFID 卡片阅读器及电子标签之间的通信及能量感应方式来看，可以大致上将 RFID 分成感应耦合（Inductive Coupling）及后向散射耦合（Backscatter Coupling）两种。通常低频的 RFID 大都采用第一种方式，而较高频大多采用第二种方式。

阅读器根据使用的结构和技术不同可以是读或读/写装置，是 RFID 系统信息控制和处理中心。阅读器通常由耦合模块、收发模块、控制模块和接口单元组成。阅读器和应答器之间一般采用半双工通信方式进行信息交换，同时阅读器通过耦合给无源应答器提供能量和时序。在实际应用中，可进一步通过 Ethernet 或 WLAN 等实现对物体识别信息的采集、处理及远程传送等管理功能。应答器是 RFID 系统的信息载体，应答器大多是由耦合元件（线圈、微带天线等）和微芯片组成无源单元。

# 16.3　Android 系统中的 NFC

**知识点讲解：光盘:视频\知识点\第 16 章\Android 系统中的 NFC.avi**

NFC 通信总是由一个发起者（initiator）和一个接收者（target）组成的。通常 initiator 主动发送电磁场（RF）为被动式接收者（passive target）提供电源。其工作的基本原理和收音机类似。正是由于被动式接收者通过发起者提供电源，因此 target 可以有非常简单的形式，如标签、卡和 Sticker 的形式。另外，NFC 也支持点到点的通信（peer to peer），此时参与通信的双方都有电源支持。在 Android 系统的 NFC 模块应用中，Android 手机通常是作为通信中的发起者，也就是作为 NFC 的读写器。Android 手机也可以模拟作为 NFC 通信的接收者，并且从 Android 2.3.3 起也支持 P2P 通信。Android 系统支持如下所示的 NFC 标准。

☑ NfcANFC-A (ISO 14443-3A)
☑ NfcBNFC-B (ISO 14443-3B)
☑ NfcFNFC-F (JIS 6319-4)
☑ NfcVNFC-V (ISO 15693)
☑ IsoDepISO-DEP (ISO 14443-4)
☑ MifareClassic
☑ MifareUltralight

在 Android 系统中，NFC 模块结构从上到下的具体说明如下所示。

（1）/system/framework/framework.jar

☑ android.nfc：标准接口（NFCAdapter/NfcManager）。
☑ android.nfc.tech：标签技术。

（2）/system/Nfc.apk

☑ com.android.nfc：NFC 服务相关。
☑ .DeviceHos：底层设备接口原型。
☑ .NfcService：Nfc 服务实现 DeviceHostListener 接口。
☑ com.android.nfc.dhimpl：NFC 功能底层实现-com.android.nfc.DeviceHost (NXP)。

☑ .NativeNfcManager：implements DeviceHost。

☑ JNI-> com_android_nfc_NativeNfcManager.cpp (libnfc_jni.so)。

☑ .NativeNfcSecureElement。

☑ JNI-> com_android_nfc_NativeNfcSecureElement.cpp (libnfc_jni.so)。

（3）/system/lib/libnfc_.so

☑ libnfc-nxp => libnfc.so, libnfc_ndef.so。

☑ libnfc-nci => libnfc-nci.so。

本节将详细讲解 Android 系统中 NFC 模块源码的基本知识。

## 16.3.1　分析 Java 层

在 Android 系统中，NFC 模块的 Java 层代码位于目录\frameworks\base\core\java\android\nfc\中。

在上述目录中，包含了用来与本地 NFC 适配器进行交互的顶层类，这些类可以表示被检测到的 tags 和用 NDEF 数据格式。在 NFC 的 Java 层中，常用的顶层类如下：

（1）NfcManager

类 NfcManager 在文件/frameworks/base/core/java/android/nfc/NfcManager.java 中定义，这是一个 NFC Adapter 的管理器，可以列出所有此 Android 设备支持的 NFC Adapter，只不过大部分 Android 设备只有一个 NFC Adapter，所以在大部分情况下可以直接用静态方法 getDefaultAdapter(context)来取适配器。文件/frameworks/base/core/java/android/nfc/NfcManager.java 的具体实现代码如下：

```
public final class NfcManager {
    private final NfcAdapter mAdapter;

    /**
     * @hide
     */
    public NfcManager(Context context) {
        NfcAdapter adapter;
        context = context.getApplicationContext();
        if (context == null) {
            throw new IllegalArgumentException(
                    "context not associated with any application (using a mock context?)");
        }
        try {
            adapter = NfcAdapter.getNfcAdapter(context);
        } catch (UnsupportedOperationException e) {
            adapter = null;
        }
        mAdapter = adapter;
    }

    /**
     * Get the default NFC Adapter for this device.
     *
     * @return the default NFC Adapter
     */
```

```
    public NfcAdapter getDefaultAdapter() {
        return mAdapter;
    }
}
```

（2）NfcAdapter

类在文件/frameworks/base/core/java/android/nfc/NfcAdapter.java 中定义，此类表示本设备的 NFC Adapter，可以定义 Intent 来请求将系统检测到 tags 的提醒发送到我们的 Activity，并提供方法去注册前台 tag 提醒发布和前台 NDEF 推送。前台 NDEF 推送是当前 Android 版本唯一支持的 p2p NFC 通信方式。文件/frameworks/base/core/java/android/nfc/NfcAdapter.java 的具体实现代码如下：

```
public final class NfcAdapter {
    static final String TAG = "NFC";
    public static final String ACTION_NDEF_DISCOVERED = "android.nfc.action.NDEF_DISCOVERED";
    @SdkConstant(SdkConstantType.ACTIVITY_INTENT_ACTION)
    public static final String ACTION_TECH_DISCOVERED = "android.nfc.action.TECH_DISCOVERED";
    @SdkConstant(SdkConstantType.ACTIVITY_INTENT_ACTION)
    public static final String ACTION_TAG_DISCOVERED = "android.nfc.action.TAG_DISCOVERED";
    public static final String ACTION_TAG_LEFT_FIELD = "android.nfc.action.TAG_LOST";
    public static final String EXTRA_TAG = "android.nfc.extra.TAG";
    public static final String EXTRA_NDEF_MESSAGES = "android.nfc.extra.NDEF_MESSAGES";
    public static final String EXTRA_ID = "android.nfc.extra.ID";
    @SdkConstant(SdkConstantType.BROADCAST_INTENT_ACTION)
    public static final String ACTION_ADAPTER_STATE_CHANGED =
            "android.nfc.action.ADAPTER_STATE_CHANGED";
    public static final String EXTRA_ADAPTER_STATE = "android.nfc.extra.ADAPTER_STATE";

    public static final int STATE_OFF = 1;
    public static final int STATE_TURNING_ON = 2;
    public static final int STATE_ON = 3;
    public static final int STATE_TURNING_OFF = 4;

    /** @hide */
    public static final int FLAG_NDEF_PUSH_NO_CONFIRM = 0x1;

    /** @hide */
    public static final String ACTION_HANDOVER_TRANSFER_STARTED =
            "android.nfc.action.HANDOVER_TRANSFER_STARTED";

    /** @hide */
    public static final String ACTION_HANDOVER_TRANSFER_DONE =
            "android.nfc.action.HANDOVER_TRANSFER_DONE";

    /** @hide */
    public static final String EXTRA_HANDOVER_TRANSFER_STATUS =
            "android.nfc.extra.HANDOVER_TRANSFER_STATUS";

    /** @hide */
    public static final int HANDOVER_TRANSFER_STATUS_SUCCESS = 0;
    /** @hide */
    public static final int HANDOVER_TRANSFER_STATUS_FAILURE = 1;
```

```
/** @hide */
public static final String EXTRA_HANDOVER_TRANSFER_URI =
        "android.nfc.extra.HANDOVER_TRANSFER_URI";

static boolean sIsInitialized = false;

static INfcAdapter sService;
static INfcTag sTagService;
static HashMap<Context, NfcAdapter> sNfcAdapters = new HashMap();
static NfcAdapter sNullContextNfcAdapter;

final NfcActivityManager mNfcActivityManager;
final Context mContext;
public interface OnNdefPushCompleteCallback {
    public void onNdefPushComplete(NfcEvent event);
}
public interface CreateNdefMessageCallback {
    public NdefMessage createNdefMessage(NfcEvent event);
}
public interface CreateBeamUrisCallback {
    public Uri[] createBeamUris(NfcEvent event);
}
private static boolean hasNfcFeature() {
    IPackageManager pm = ActivityThread.getPackageManager();
    if (pm == null) {
        Log.e(TAG, "Cannot get package manager, assuming no NFC feature");
        return false;
    }
    try {
        return pm.hasSystemFeature(PackageManager.FEATURE_NFC);
    } catch (RemoteException e) {
        Log.e(TAG, "Package manager query failed, assuming no NFC feature", e);
        return false;
    }
}
public static synchronized NfcAdapter getNfcAdapter(Context context) {
    if (!sIsInitialized) {
        if (!hasNfcFeature()) {
            Log.v(TAG, "this device does not have NFC support");
            throw new UnsupportedOperationException();
        }

        sService = getServiceInterface();
        if (sService == null) {
            Log.e(TAG, "could not retrieve NFC service");
            throw new UnsupportedOperationException();
        }
        try {
            sTagService = sService.getNfcTagInterface();
        } catch (RemoteException e) {
```

```
                Log.e(TAG, "could not retrieve NFC Tag service");
                throw new UnsupportedOperationException();
            }

            sIsInitialized = true;
        }
        if (context == null) {
            if (sNullContextNfcAdapter == null) {
                sNullContextNfcAdapter = new NfcAdapter(null);
            }
            return sNullContextNfcAdapter;
        }
        NfcAdapter adapter = sNfcAdapters.get(context);
        if (adapter == null) {
            adapter = new NfcAdapter(context);
            sNfcAdapters.put(context, adapter);
        }
        return adapter;
    }

    /** get handle to NFC service interface */
    private static INfcAdapter getServiceInterface() {
        IBinder b = ServiceManager.getService("nfc");
        if (b == null) {
            return null;
        }
        return INfcAdapter.Stub.asInterface(b);
    }
    public static NfcAdapter getDefaultAdapter(Context context) {
        if (context == null) {
            throw new IllegalArgumentException("context cannot be null");
        }
        context = context.getApplicationContext();
        if (context == null) {
            throw new IllegalArgumentException(
                    "context not associated with any application (using a mock context?)");
        }
        /* use getSystemService() for consistency */
        NfcManager manager = (NfcManager) context.getSystemService(Context.NFC_SERVICE);
        if (manager == null) {
            return null;
        }
        return manager.getDefaultAdapter();
    }
    @Deprecated
    public static NfcAdapter getDefaultAdapter() {
        //introduced in API version 9 (GB 2.3)
        //deprecated in API version 10 (GB 2.3.3)
        //removed from public API in version 16 (ICS MR2)
        //should maintain as a hidden API for binary compatibility for a little longer
        Log.w(TAG, "WARNING: NfcAdapter.getDefaultAdapter() is deprecated, use " +
```

```
                "NfcAdapter.getDefaultAdapter(Context) instead", new Exception());

        return NfcAdapter.getNfcAdapter(null);
    }
    public boolean isEnabled() {
        try {
            return sService.getState() == STATE_ON;
        } catch (RemoteException e) {
            attemptDeadServiceRecovery(e);
            return false;
        }
    }
    public int getAdapterState() {
        try {
            return sService.getState();
        } catch (RemoteException e) {
            attemptDeadServiceRecovery(e);
            return NfcAdapter.STATE_OFF;
        }
    }
    public boolean enable() {
        try {
            return sService.enable();
        } catch (RemoteException e) {
            attemptDeadServiceRecovery(e);
            return false;
        }
    }
    public boolean disable() {
        try {
            return sService.disable(true);
        } catch (RemoteException e) {
            attemptDeadServiceRecovery(e);
            return false;
        }
    }

    public void setBeamPushUris(Uri[] uris, Activity activity) {
        if (activity == null) {
            throw new NullPointerException("activity cannot be null");
        }
        if (uris != null) {
            for (Uri uri : uris) {
                if (uri == null) throw new NullPointerException("Uri not " +
                        "allowed to be null");
                String scheme = uri.getScheme();
                if (scheme == null || (!scheme.equalsIgnoreCase("file") &&
                        !scheme.equalsIgnoreCase("content"))) {
                    throw new IllegalArgumentException("URI needs to have " +
                            "either scheme file or scheme content");
                }
```

```
            }
        }
        mNfcActivityManager.setNdefPushContentUri(activity, uris);
    }
    public void setBeamPushUrisCallback(CreateBeamUrisCallback callback, Activity activity) {
        if (activity == null) {
            throw new NullPointerException("activity cannot be null");
        }
        mNfcActivityManager.setNdefPushContentUriCallback(activity, callback);
    }
    public void setNdefPushMessage(NdefMessage message, Activity activity,
            Activity ... activities) {
        int targetSdkVersion = getSdkVersion();
        try {
            if (activity == null) {
                throw new NullPointerException("activity cannot be null");
            }
            mNfcActivityManager.setNdefPushMessage(activity, message, 0);
            for (Activity a : activities) {
                if (a == null) {
                    throw new NullPointerException("activities cannot contain null");
                }
                mNfcActivityManager.setNdefPushMessage(a, message, 0);
            }
        } catch (IllegalStateException e) {
            if (targetSdkVersion < android.os.Build.VERSION_CODES.JELLY_BEAN) {
                Log.e(TAG, "Cannot call API with Activity that has already " +
                        "been destroyed", e);
            } else {
                throw(e);
            }
        }
    }

    /**
     * @hide
     */
    public void setNdefPushMessage(NdefMessage message, Activity activity, int flags) {
        if (activity == null) {
            throw new NullPointerException("activity cannot be null");
        }
        mNfcActivityManager.setNdefPushMessage(activity, message, flags);
    }
    public void setNdefPushMessageCallback(CreateNdefMessageCallback callback, Activity activity,
            Activity ... activities) {
        int targetSdkVersion = getSdkVersion();
        try {
            if (activity == null) {
                throw new NullPointerException("activity cannot be null");
            }
            mNfcActivityManager.setNdefPushMessageCallback(activity, callback, 0);
```

```
            for (Activity a : activities) {
                if (a == null) {
                    throw new NullPointerException("activities cannot contain null");
                }
                mNfcActivityManager.setNdefPushMessageCallback(a, callback, 0);
            }
        } catch (IllegalStateException e) {
            if (targetSdkVersion < android.os.Build.VERSION_CODES.JELLY_BEAN) {
                Log.e(TAG, "Cannot call API with Activity that has already " +
                        "been destroyed", e);
            } else {
                throw(e);
            }
        }
    }
    public void setNdefPushMessageCallback(CreateNdefMessageCallback callback, Activity activity,
            int flags) {
        if (activity == null) {
            throw new NullPointerException("activity cannot be null");
        }
        mNfcActivityManager.setNdefPushMessageCallback(activity, callback, flags);
    }
    public void setOnNdefPushCompleteCallback(OnNdefPushCompleteCallback callback,
            Activity activity, Activity ... activities) {
        int targetSdkVersion = getSdkVersion();
        try {
            if (activity == null) {
                throw new NullPointerException("activity cannot be null");
            }
            mNfcActivityManager.setOnNdefPushCompleteCallback(activity, callback);
            for (Activity a : activities) {
                if (a == null) {
                    throw new NullPointerException("activities cannot contain null");
                }
                mNfcActivityManager.setOnNdefPushCompleteCallback(a, callback);
            }
        } catch (IllegalStateException e) {
            if (targetSdkVersion < android.os.Build.VERSION_CODES.JELLY_BEAN) {
                Log.e(TAG, "Cannot call API with Activity that has already " +
                        "been destroyed", e);
            } else {
                throw(e);
            }
        }
    }
    public void enableForegroundDispatch(Activity activity, PendingIntent intent,
            IntentFilter[] filters, String[][] techLists) {
        if (activity == null || intent == null) {
            throw new NullPointerException();
        }
        if (!activity.isResumed()) {
```

```
            throw new IllegalStateException("Foreground dispatch can only be enabled " +
                    "when your activity is resumed");
        }
        try {
            TechListParcel parcel = null;
            if (techLists != null && techLists.length > 0) {
                parcel = new TechListParcel(techLists);
            }
            ActivityThread.currentActivityThread().registerOnActivityPausedListener(activity,
                    mForegroundDispatchListener);
            sService.setForegroundDispatch(intent, filters, parcel);
        } catch (RemoteException e) {
            attemptDeadServiceRecovery(e);
        }
    }
    public void disableForegroundDispatch(Activity activity) {
        ActivityThread.currentActivityThread().unregisterOnActivityPausedListener(activity,
                mForegroundDispatchListener);
        disableForegroundDispatchInternal(activity, false);
    }
```

（3）NdefMessage 和 NdefRecord

NDEF 是 NFC 论坛定义的数据结构，用来将有效的数据存储到 NFC tags 中，如文本、URL 和其他 MIME 类型。一个 NdefMessage 扮演一个容器，这个容器存储那些发送和读到的数据。一个 NdefMessage 对象包含 0 或多个 NdefRecord，每个 NDEF record 有一个类型，如文本、URL、智慧型海报/广告，或其他 MIME 数据。在 NdefMessage 中的第一个 NfcRecord 的类型，功能是发送 tag 到一个 Android 设备上的 Activity。其中类 NdefMessage 在文件/frameworks/base/core/java/android/nfc/NdefMessage.java 中定义，具体实现代码如下：

```
public final class NdefMessage implements Parcelable {
    private final NdefRecord[] mRecords;
    public NdefMessage(byte[] data) throws FormatException {
        if (data == null) throw new NullPointerException("data is null");
        ByteBuffer buffer = ByteBuffer.wrap(data);

        mRecords = NdefRecord.parse(buffer, false);

        if (buffer.remaining() > 0) {
            throw new FormatException("trailing data");
        }
    }
    public NdefMessage(NdefRecord record, NdefRecord ... records) {
        if (record == null) throw new NullPointerException("record cannot be null");

        for (NdefRecord r : records) {
            if (r == null) {
                throw new NullPointerException("record cannot be null");
            }
        }

        mRecords = new NdefRecord[1 + records.length];
```

```
        mRecords[0] = record;
        System.arraycopy(records, 0, mRecords, 1, records.length);
    }
    public NdefMessage(NdefRecord[] records) {
        if (records.length < 1) {
            throw new IllegalArgumentException("must have at least one record");
        }
        for (NdefRecord r : records) {
            if (r == null) {
                throw new NullPointerException("records cannot contain null");
            }
        }

        mRecords = records;
    }
    public NdefRecord[] getRecords() {
        return mRecords;
    }
    public int getByteArrayLength() {
        int length = 0;
        for (NdefRecord r : mRecords) {
            length += r.getByteLength();
        }
        return length;
    }
    public byte[] toByteArray() {
        int length = getByteArrayLength();
        ByteBuffer buffer = ByteBuffer.allocate(length);

        for (int i=0; i<mRecords.length; i++) {
            boolean mb = (i == 0);
            boolean me = (i == mRecords.length - 1);
            mRecords[i].writeToByteBuffer(buffer, mb, me);
        }

        return buffer.array();
    }

    @Override
    public int describeContents() {
        return 0;
    }

    @Override
    public void writeToParcel(Parcel dest, int flags) {
        dest.writeInt(mRecords.length);
        dest.writeTypedArray(mRecords, flags);
    }

    public static final Parcelable.Creator<NdefMessage> CREATOR =
            new Parcelable.Creator<NdefMessage>() {
```

```
        @Override
        public NdefMessage createFromParcel(Parcel in) {
            int recordsLength = in.readInt();
            NdefRecord[] records = new NdefRecord[recordsLength];
            in.readTypedArray(records, NdefRecord.CREATOR);
            return new NdefMessage(records);
        }
        @Override
        public NdefMessage[] newArray(int size) {
            return new NdefMessage[size];
        }
    };

    @Override
    public int hashCode() {
        return Arrays.hashCode(mRecords);
    }
    @Override
    public boolean equals(Object obj) {
        if (this == obj) return true;
        if (obj == null) return false;
        if (getClass() != obj.getClass()) return false;
        NdefMessage other = (NdefMessage) obj;
        return Arrays.equals(mRecords, other.mRecords);
    }

    @Override
    public String toString() {
        return "NdefMessage " + Arrays.toString(mRecords);
    }
}
```

（4）Tag

类 Tag 在文件/frameworks/base/core/java/android/nfc/Tag.java 中定义，此类用于标示一个被动的 NFC 目标，如 tag、card 和钥匙挂扣，甚至是一个电话模拟的 NFC 卡。当一个 tag 被检测到时，一个 tag 对象将被创建并且封装到一个 Intent 中，然后 NFC 发布系统将这个 Intent 用 startActivity 发送到注册了接收这种 Intent 的 Activity 中。我们可以用方法 getTechList()来得到这个 tag 支持的技术细节，并创建一个 android.nfc.tech 提供的相应的 TagTechnology 对象。文件 Tag.java 的具体实现代码如下：

```
public final class Tag implements Parcelable {
    final byte[] mId;
    final int[] mTechList;
    final String[] mTechStringList;
    final Bundle[] mTechExtras;
    final int mServiceHandle;
    final INfcTag mTagService;

    int mConnectedTechnology;
    public Tag(byte[] id, int[] techList, Bundle[] techListExtras, int serviceHandle,
            INfcTag tagService) {
        if (techList == null) {
```

```
            throw new IllegalArgumentException("rawTargets cannot be null");
        }
        mId = id;
        mTechList = Arrays.copyOf(techList, techList.length);
        mTechStringList = generateTechStringList(techList);
        mTechExtras = Arrays.copyOf(techListExtras, techList.length);
        mServiceHandle = serviceHandle;
        mTagService = tagService;

        mConnectedTechnology = -1;
    }
    public static Tag createMockTag(byte[] id, int[] techList, Bundle[] techListExtras) {
        return new Tag(id, techList, techListExtras, 0, null);
    }
```

（5）tech/子目录

除此之外，在 frameworks/base/core/java/android/nfc/tech/目录中包含了查询 tag 属性和进行 I/O 操作的类。这些类分别标示一个 tag 支持的不同的 NFC 技术标准，如图 16-1 所示。

frameworks ▸ base ▸ core ▸ java ▸ android ▸ nfc ▸ tech　　搜索 tech

享 ▾　刻录　新建文件夹

| 名称 | 修改日期 | 类型 | 大小 |
|---|---|---|---|
| BasicTagTechnology.java | 2013/8/14 9:20 | JAVA 文件 | 5 KB |
| IsoDep.java | 2013/8/14 9:20 | JAVA 文件 | 8 KB |
| MifareClassic.java | 2013/8/14 9:20 | JAVA 文件 | 24 KB |
| MifareUltralight.java | 2013/8/14 9:20 | JAVA 文件 | 11 KB |
| Ndef.java | 2013/8/14 9:20 | JAVA 文件 | 15 KB |
| NdefFormatable.java | 2013/8/14 9:20 | JAVA 文件 | 7 KB |
| NfcA.java | 2013/8/14 9:20 | JAVA 文件 | 6 KB |
| NfcB.java | 2013/8/14 9:20 | JAVA 文件 | 5 KB |
| NfcBarcode.java | 2013/8/14 9:20 | JAVA 文件 | 5 KB |
| NfcF.java | 2013/8/14 9:20 | JAVA 文件 | 6 KB |
| NfcV.java | 2013/8/14 9:20 | JAVA 文件 | 5 KB |
| package.html | 2013/8/14 9:20 | HTML 文档 | 1 KB |
| TagTechnology.java | 2013/8/14 9:20 | JAVA 文件 | 9 KB |

图 16-1　frameworks/base/core/java/android/nfc/tech/目录

在图 16-1 所示的目录中，接口 TagTechnology 是表 16-2 中所有 Tag Technology 类必须实现的。

表 16-2　必须实现的 Tag Technology 类

| | |
|---|---|
| NfcA | 支持 ISO 14443-3A 标准的操作 |
| NfcB | 支持 ISO 14443-3B 标准的操作 |
| NfcF | 支持 JIS 6319-4 标准的操作 |
| NfcV | 支持 ISO 15693 标准的操作 |
| IsoDep | 支持 ISO 14443-4 标准的操作 |
| Ndef | 提供对那些被格式化为 NDEF 的 tag 的数据的访问和其他操作 |
| NdefFormatable | 对那些可以被格式化为 NDEF 的 tag 提供一个格式化的操作 |
| MifareClassic | 如果 android 设备支持 MIFARE，提供对 MIFARE Classic 目标的属性和 I/O 操作 |
| MifareUltralight | 如果 android 设备支持 MIFARE，提供对 MIFARE Ultralight 目标的属性和 I/O 操作 |

而类 NdefFormatable 对那些可以被格式化成 NDEF 格式的 tag 提供一个格式化的操作，文件 frameworks/base/core/java/android/nfc/tech/NdefFormatable.java 的具体实现代码如下：

```
public final class NdefFormatable extends BasicTagTechnology {
    private static final String TAG = "NFC";
    public static NdefFormatable get(Tag tag) {
        if (!tag.hasTech(TagTechnology.NDEF_FORMATABLE)) return null;
        try {
            return new NdefFormatable(tag);
        } catch (RemoteException e) {
            return null;
        }
    }
    /*package*/ void format(NdefMessage firstMessage, boolean makeReadOnly) throws IOException,
            FormatException {
        checkConnected();

        try {
            int serviceHandle = mTag.getServiceHandle();
            INfcTag tagService = mTag.getTagService();
            int errorCode = tagService.formatNdef(serviceHandle, MifareClassic.KEY_DEFAULT);
            switch (errorCode) {
                case ErrorCodes.SUCCESS:
                    break;
                case ErrorCodes.ERROR_IO:
                    throw new IOException();
                case ErrorCodes.ERROR_INVALID_PARAM:
                    throw new FormatException();
                default:
                    throw new IOException();
            }
            if (!tagService.isNdef(serviceHandle)) {
                throw new IOException();
            }

            if (firstMessage != null) {
                errorCode = tagService.ndefWrite(serviceHandle, firstMessage);
                switch (errorCode) {
                    case ErrorCodes.SUCCESS:
                        break;
                    case ErrorCodes.ERROR_IO:
                        throw new IOException();
                    case ErrorCodes.ERROR_INVALID_PARAM:
                        throw new FormatException();
                    default:
                        throw new IOException();
                }
            }

            if (makeReadOnly) {
                errorCode = tagService.ndefMakeReadOnly(serviceHandle);
                switch (errorCode) {
                    case ErrorCodes.SUCCESS:
                        break;
```

```
                    case ErrorCodes.ERROR_IO:
                        throw new IOException();
                    case ErrorCodes.ERROR_INVALID_PARAM:
                        throw new IOException();
                    default:
                        throw new IOException();
                }
            }
        } catch (RemoteException e) {
            Log.e(TAG, "NFC service dead", e);
        }
    }
}
```

类 MifareClassic 在文件 frameworks/base/core/java/android/nfc/tech/MifareClassic.java 中定义，如果 Android 设备支持 MIFARE，则提供对 MIFARE Classic 目标的属性和 I/O 操作。文件 MifareClassic.java 的具体实现代码如下：

```
public final class MifareClassic extends BasicTagTechnology {
    private static final String TAG = "NFC";
    public static final byte[] KEY_DEFAULT =
            {(byte)0xFF,(byte)0xFF,(byte)0xFF,(byte)0xFF,(byte)0xFF,(byte)0xFF};
    public static final byte[] KEY_MIFARE_APPLICATION_DIRECTORY =
            {(byte)0xA0,(byte)0xA1,(byte)0xA2,(byte)0xA3,(byte)0xA4,(byte)0xA5};
    public static final byte[] KEY_NFC_FORUM =
            {(byte)0xD3,(byte)0xF7,(byte)0xD3,(byte)0xF7,(byte)0xD3,(byte)0xF7};

    /** A MIFARE Classic compatible card of unknown type */
    public static final int TYPE_UNKNOWN = -1;
    /** A MIFARE Classic tag */
    public static final int TYPE_CLASSIC = 0;
    /** A MIFARE Plus tag */
    public static final int TYPE_PLUS = 1;
    /** A MIFARE Pro tag */
    public static final int TYPE_PRO = 2;

    /** Tag contains 16 sectors, each with 4 blocks */
    public static final int SIZE_1K = 1024;
    /** Tag contains 32 sectors, each with 4 blocks */
    public static final int SIZE_2K = 2048;
    /**
     * Tag contains 40 sectors. The first 32 sectors contain 4 blocks and the last 8 sectors
     * contain 16 blocks
     */
    public static final int SIZE_4K = 4096;
    /** Tag contains 5 sectors, each with 4 blocks */
    public static final int SIZE_MINI = 320;

    /** Size of a MIFARE Classic block (in bytes) */
    public static final int BLOCK_SIZE = 16;

    private static final int MAX_BLOCK_COUNT = 256;
    private static final int MAX_SECTOR_COUNT = 40;
```

```
private boolean mIsEmulated;
private int mType;
private int mSize;
public static MifareClassic get(Tag tag) {
    if (!tag.hasTech(TagTechnology.MIFARE_CLASSIC)) return null;
    try {
        return new MifareClassic(tag);
    } catch (RemoteException e) {
        return null;
    }
}
/** @hide */
public MifareClassic(Tag tag) throws RemoteException {
    super(tag, TagTechnology.MIFARE_CLASSIC);
    NfcA a = NfcA.get(tag);    //MIFARE Classic is always based on NFC a
    mIsEmulated = false;
    switch (a.getSak()) {
    case 0x01:
    case 0x08:
        mType = TYPE_CLASSIC;
        mSize = SIZE_1K;
        break;
    case 0x09:
        mType = TYPE_CLASSIC;
        mSize = SIZE_MINI;
        break;
    case 0x10:
        mType = TYPE_PLUS;
        mSize = SIZE_2K;
        //SecLevel = SL2
        break;
    case 0x11:
        mType = TYPE_PLUS;
        mSize = SIZE_4K;
        //Seclevel = SL2
        break;
    case 0x18:
        mType = TYPE_CLASSIC;
        mSize = SIZE_4K;
        break;
    case 0x28:
        mType = TYPE_CLASSIC;
        mSize = SIZE_1K;
        mIsEmulated = true;
        break;
    case 0x38:
        mType = TYPE_CLASSIC;
        mSize = SIZE_4K;
        mIsEmulated = true;
        break;
```

```
        case 0x88:
            mType = TYPE_CLASSIC;
            mSize = SIZE_1K;
            //NXP-tag: false
            break;
        case 0x98:
        case 0xB8:
            mType = TYPE_PRO;
            mSize = SIZE_4K;
            break;
        default:
            throw new RuntimeException(
                    "Tag incorrectly enumerated as MIFARE Classic, SAK = " + a.getSak());
        }
    }
```

类 MifareUltralight 在文件 frameworks/base/core/java/android/nfc/tech/MifareUltralight.java 中定义，如果 Android 设备支持 MIFARE，则提供对 MIFARE Ultralight 目标的属性和 I/O 操作。

```
public final class MifareUltralight extends BasicTagTechnology {
    private static final String TAG = "NFC";
    /** A MIFARE Ultralight compatible tag of unknown type */
    public static final int TYPE_UNKNOWN = -1;
    /** A MIFARE Ultralight tag */
    public static final int TYPE_ULTRALIGHT = 1;
    /** A MIFARE Ultralight C tag */
    public static final int TYPE_ULTRALIGHT_C = 2;
    /** Size of a MIFARE Ultralight page in bytes */
    public static final int PAGE_SIZE = 4;
    private static final int NXP_MANUFACTURER_ID = 0x04;
    private static final int MAX_PAGE_COUNT = 256;
    /** @hide */
    public static final String EXTRA_IS_UL_C = "isulc";
    private int mType;
    public static MifareUltralight get(Tag tag) {
        if (!tag.hasTech(TagTechnology.MIFARE_ULTRALIGHT)) return null;
        try {
            return new MifareUltralight(tag);
        } catch (RemoteException e) {
            return null;
        }
    }
    /** @hide */
    public MifareUltralight(Tag tag) throws RemoteException {
        super(tag, TagTechnology.MIFARE_ULTRALIGHT);
        NfcA a = NfcA.get(tag);
        mType = TYPE_UNKNOWN;
        if (a.getSak() == 0x00 && tag.getId()[0] == NXP_MANUFACTURER_ID) {
            Bundle extras = tag.getTechExtras(TagTechnology.MIFARE_ULTRALIGHT);
            if (extras.getBoolean(EXTRA_IS_UL_C)) {
                mType = TYPE_ULTRALIGHT_C;
            } else {
```

```
                mType = TYPE_ULTRALIGHT;
            }
        }
    }
    public byte[] readPages(int pageOffset) throws IOException {
        validatePageIndex(pageOffset);
        checkConnected();

        byte[] cmd = { 0x30, (byte) pageOffset};
        return transceive(cmd, false);
    }
    public void writePage(int pageOffset, byte[] data) throws IOException {
        validatePageIndex(pageOffset);
        checkConnected();
        byte[] cmd = new byte[data.length + 2];
        cmd[0] = (byte) 0xA2;
        cmd[1] = (byte) pageOffset;
        System.arraycopy(data, 0, cmd, 2, data.length);
        transceive(cmd, false);
    }
    public void setTimeout(int timeout) {
        try {
            int err = mTag.getTagService().setTimeout(
                    TagTechnology.MIFARE_ULTRALIGHT, timeout);
            if (err != ErrorCodes.SUCCESS) {
                throw new IllegalArgumentException("The supplied timeout is not valid");
            }
        } catch (RemoteException e) {
            Log.e(TAG, "NFC service dead", e);
        }
    }
    public int getTimeout() {
        try {
            return mTag.getTagService().getTimeout(TagTechnology.MIFARE_ULTRALIGHT);
        } catch (RemoteException e) {
            Log.e(TAG, "NFC service dead", e);
            return 0;
        }
    }

    private static void validatePageIndex(int pageIndex) {
        if (pageIndex < 0 || pageIndex >= MAX_PAGE_COUNT) {
            throw new IndexOutOfBoundsException("page out of bounds: " + pageIndex);
        }
    }
}
```

## 16.3.2 分析 JNI 部分

在 Android 系统中，NFC 模块的 JNI 部分代码通过目录/packages/apps/Nfc/nxp/jni 实现。

JNI 部分代码向上会和 Framework 层的 Java 代码进行交互，向下会和 libnfc 层进行交互。JNI 层的核心文件是 com_android_nfc_NativeNfcManager.cpp，功能是初始化并启动 NFC 服务，并扫描和读取 tag。文件 com_android_nfc_NativeNfcManager.cpp 的主要实现代码如下：

```
static void client_kill_deferred_call(void* arg)
{
    struct nfc_jni_native_data *nat = (struct nfc_jni_native_data *)arg;

    nat->running = FALSE;
}

static void kill_client(nfc_jni_native_data *nat)
{
    phDal4Nfc_Message_Wrapper_t   wrapper;
    phLibNfc_DeferredCall_t *pMsg;

    usleep(50000);

    ALOGD("Terminating client thread...");

    pMsg = (phLibNfc_DeferredCall_t*)malloc(sizeof(phLibNfc_DeferredCall_t));
    pMsg->pCallback = client_kill_deferred_call;
    pMsg->pParameter = (void*)nat;

    wrapper.msg.eMsgType = PH_LIBNFC_DEFERREDCALL_MSG;
    wrapper.msg.pMsgData = pMsg;
    wrapper.msg.Size = sizeof(phLibNfc_DeferredCall_t);

    phDal4Nfc_msgsnd(gDrvCfg.nClientId, (struct msgbuf *)&wrapper, sizeof(phLibNfc_Message_t), 0);
}

static void nfc_jni_ioctl_callback(void *pContext, phNfc_sData_t *pOutput, NFCSTATUS status) {
    struct nfc_jni_callback_data * pCallbackData = (struct nfc_jni_callback_data *) pContext;
    LOG_CALLBACK("nfc_jni_ioctl_callback", status);

    /* Report the callback status and wake up the caller */
    pCallbackData->status = status;
    sem_post(&pCallbackData->sem);
}

static void nfc_jni_deinit_download_callback(void *pContext, NFCSTATUS status)
{
    struct nfc_jni_callback_data * pCallbackData = (struct nfc_jni_callback_data *) pContext;
    LOG_CALLBACK("nfc_jni_deinit_download_callback", status);

    /* Report the callback status and wake up the caller */
    pCallbackData->status = status;
    sem_post(&pCallbackData->sem);
}

static int nfc_jni_download_locked(struct nfc_jni_native_data *nat, uint8_t update)
{
```

```
    uint8_t OutputBuffer[1];
    uint8_t InputBuffer[1];
    struct timespec ts;
    NFCSTATUS status = NFCSTATUS_FAILED;
    phLibNfc_StackCapabilities_t caps;
    struct nfc_jni_callback_data cb_data;
    bool result;

    /* Create the local semaphore */
    if (!nfc_cb_data_init(&cb_data, NULL))
    {
       goto clean_and_return;
    }

    if(update)
    {
        TRACE("phLibNfc_Mgt_DeInitialize() (download)");
        REENTRANCE_LOCK();
        status = phLibNfc_Mgt_DeInitialize(gHWRef, nfc_jni_deinit_download_callback, (void *)&cb_data);
        REENTRANCE_UNLOCK();
        if (status != NFCSTATUS_PENDING)
        {
            ALOGE("phLibNfc_Mgt_DeInitialize() (download) returned 0x%04x[%s]", status, nfc_jni_get_status_
name(status));
        }

        clock_gettime(CLOCK_REALTIME, &ts);
        ts.tv_sec += 5;

        /* Wait for callback response */
        if(sem_timedwait(&cb_data.sem, &ts))
        {
            ALOGW("Deinitialization timed out (download)");
        }

        if(cb_data.status != NFCSTATUS_SUCCESS)
        {
            ALOGW("Deinitialization FAILED (download)");
        }
        TRACE("Deinitialization SUCCESS (download)");
    }

    result = performDownload(nat, false);

    if (!result) {
        status = NFCSTATUS_FAILED;
        goto clean_and_return;
    }

    TRACE("phLibNfc_Mgt_Initialize()");
    REENTRANCE_LOCK();
    status = phLibNfc_Mgt_Initialize(gHWRef, nfc_jni_init_callback, (void *)&cb_data);
    REENTRANCE_UNLOCK();
```

```
    if(status != NFCSTATUS_PENDING)
    {
        ALOGE("phLibNfc_Mgt_Initialize() (download) returned 0x%04x[%s]", status, nfc_jni_get_status_name
(status));
        goto clean_and_return;
    }
    TRACE("phLibNfc_Mgt_Initialize() returned 0x%04x[%s]", status, nfc_jni_get_status_name(status));

    if(sem_wait(&cb_data.sem))
    {
       ALOGE("Failed to wait for semaphore (errno=0x%08x)", errno);
       status = NFCSTATUS_FAILED;
       goto clean_and_return;
    }

    /* Initialization Status */
    if(cb_data.status != NFCSTATUS_SUCCESS)
    {
        status = cb_data.status;
        goto clean_and_return;
    }

    /* ====== CAPABILITIES ======= */
    REENTRANCE_LOCK();
    status = phLibNfc_Mgt_GetstackCapabilities(&caps, (void*)nat);
    REENTRANCE_UNLOCK();
    if (status != NFCSTATUS_SUCCESS)
    {
       ALOGW("phLibNfc_Mgt_GetstackCapabilities returned 0x%04x[%s]", status, nfc_jni_get_status_ name
(status));
    }
    else
    {
        ALOGD("NFC capabilities: HAL = %x, FW = %x, HW = %x, Model = %x, HCI = %x, Full_FW = %d, Rev
= %d, FW Update Info = %d",
                caps.psDevCapabilities.hal_version,
                caps.psDevCapabilities.fw_version,
                caps.psDevCapabilities.hw_version,
                caps.psDevCapabilities.model_id,
                caps.psDevCapabilities.hci_version,
                caps.psDevCapabilities.full_version[NXP_FULL_VERSION_LEN-1],
                caps.psDevCapabilities.full_version[NXP_FULL_VERSION_LEN-2],
                caps.psDevCapabilities.firmware_update_info);
    }

    /*Download is successful*/
    status = NFCSTATUS_SUCCESS;

clean_and_return:
   nfc_cb_data_deinit(&cb_data);
   return status;
}
```

```
static int nfc_jni_configure_driver(struct nfc_jni_native_data *nat)
{
    char value[PROPERTY_VALUE_MAX];
    int result = FALSE;
    NFCSTATUS status;

    /* ====== CONFIGURE DRIVER ======= */
    /* Configure hardware link */
    gDrvCfg.nClientId = phDal4Nfc_msgget(0, 0600);

    TRACE("phLibNfc_Mgt_ConfigureDriver(0x%08x)", gDrvCfg.nClientId);
    REENTRANCE_LOCK();
    status = phLibNfc_Mgt_ConfigureDriver(&gDrvCfg, &gHWRef);
    REENTRANCE_UNLOCK();
    if(status == NFCSTATUS_ALREADY_INITIALISED) {
            ALOGW("phLibNfc_Mgt_ConfigureDriver() returned 0x%04x[%s]", status, nfc_jni_get_status_name
(status));
    }
    else if(status != NFCSTATUS_SUCCESS)
    {
        ALOGE("phLibNfc_Mgt_ConfigureDriver() returned 0x%04x[%s]", status, nfc_jni_get_status_name
(status));
        goto clean_and_return;
    }
    TRACE("phLibNfc_Mgt_ConfigureDriver() returned 0x%04x[%s]", status, nfc_jni_get_status_name(status));

    if(pthread_create(&(nat->thread), NULL, nfc_jni_client_thread, nat) != 0)
    {
        ALOGE("pthread_create failed");
        goto clean_and_return;
    }

    driverConfigured = TRUE;

clean_and_return:
    return result;
}

static int nfc_jni_unconfigure_driver(struct nfc_jni_native_data *nat)
{
    int result = FALSE;
    NFCSTATUS status;

    /* Unconfigure driver */
    TRACE("phLibNfc_Mgt_UnConfigureDriver()");
    REENTRANCE_LOCK();
    status = phLibNfc_Mgt_UnConfigureDriver(gHWRef);
    REENTRANCE_UNLOCK();
    if(status != NFCSTATUS_SUCCESS)
    {
        ALOGE("phLibNfc_Mgt_UnConfigureDriver() returned error 0x%04x[%s] -- this should never happen",
status, nfc_jni_get_status_name( status));
    }
```

```
    else
    {
        ALOGD("phLibNfc_Mgt_UnConfigureDriver() returned 0x%04x[%s]", status, nfc_jni_get_status_name
(status));
        result = TRUE;
    }

    driverConfigured = FALSE;

    return result;
}
```

### 16.3.3　分析底层

在 Android 系统中，NFC 模块的底层部分有驱动部分和 libnfc 库部分两大类。驱动部分在 device 目录中实现，例如 device/samsung/tuna/nfc 目录中保存了设备厂家三星电子提供的 hardware lib。而 libnfc-nci 和 libnfc-nxp 目录中的底层文件则负责 NFC 数据的读取和解析工作。

## 16.4　在 Android 系统编写 NFC APP 的方法

**知识点讲解：光盘:视频\知识点\第 16 章\在 Android 系统编写 NFC APP 的方法.avi**

当 Android 手机开启了 NFC 程序，并且检测到一个 TAG 后，TAG 分发系统会自动创建一个封装了 NFC TAG 信息的 Intent。如果多于一个应用程序能够处理这个 Intent，那么手机就会弹出一个对话框，让用户选择处理该 TAG 的 Activity。在 TAG 分发系统中定义了 3 种 Intent，按优先级从高到低排列顺序依次是 NDEF_DISCOVERED、TECH_DISCOVERED 和 TAG_DISCOVERED。

当 Android 设备检测到有 NFC Tag 靠近时，会根据 Action 声明的顺序给对应的 Activity 发送含 NFC 消息的 Intent。此处使用的 intent-filter 的 Action 类型为 TECH_DISCOVERED，从而可以处理所有类型为 ACTION_TECH_DISCOVERED 并且使用的技术为 nfc_tech_filter.xml 文件中定义的类型的 TAG。

当 Android 手机检测到一个 TAG 时，启用 Activity 的匹配过程如图 16-2 所示。

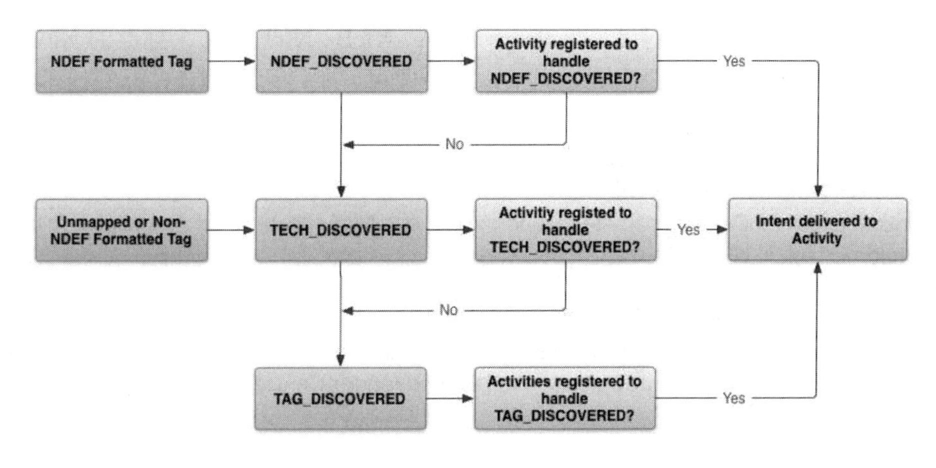


图 16-2　启用 Activity 的匹配过程

在 Android 系统中，编写 NFC APP 的基本流程如下。

（1）在相关的 androidManifest 文件中设置 NFC 权限，具体代码如下：

```
<uses-permission android:name="android.permission.NFC" />
```

（2）设置 SDK 的级别限制，例如设置为 API 10。

```
<uses-sdk android:minSdkVersion="10"/>
```

（3）声明特殊功能的限制权限，通过如下声明可以让应用程序在 Google Play 上声明使用者必须拥有 NFC 功能。

```
<uses-feature android:name="android.hardware.nfc" android:required="true" />
```

（4）实现 NFC 标签过滤。

在 Activity 的 Intent 过滤 XML 声明中，可以同时声明过滤如下 3 种 action（动作），但是需要提前知道系统在发送 Intent 时拥有的优先级。

☑ 动作 1：过滤 ACTION_TAG_DISCOVERED

```
<intent-filter>
    <action android:name="android.nfc.action.TAG_DISCOVERED"/>
    <category android:name="android.intent.category.DEFAULT"/>
</intent-filter>
```

这个最简单，也是最后一个被尝试接收 intent 的选项。

☑ 动作 2：过滤 ACTION_NDEF_DISCOVERED

```
<intent-filter>
<action android:name="android.nfc.action.NDEF_DISCOVERED"/>
<category android:name="android.intent.category.DEFAULT"/>
<data android:mimeType="text/plain" />
 </intent-filter>
```

其中最重要的是 data 的 mimeType 类型，此定义越准确，Intent 指向这个 Activity 的成功率就越高，否则系统可能不会发出用户想要的 NDEF intent。

☑ 动作 3：过滤 ACTION_TECH_DISCOVERED

首先需要在<project-path>/res/xml 下面创建一个过滤规则文件，可以任意命名，例如可以叫做 nfc_tech_filter.xml。这其中定义的是 nfc 实现的各种标准，每一个 NFC 卡都会符合多个不同的标准。可以在检测到 NFC 标签后，使用 getTechList()方法来查看所检测的 tag 究竟支持哪些 NFC 标准。在一个 nfc_tech_filter.xml 文件中可以定义多个<tech-list>结构组，每一组代表声明只接受同时满足这些标准的 NFC 标签。例如 A 组表示，只有同时满足 IsoDep、NfcA、NfcB、NfcF 这 4 个标准的 NFC 标签的 Intent 才能进入。A 与 B 组之间的关系就是只要满足其中一个即可。换句话说，NFC 标签技术满足 A 的声明也可以，满足 B 的声明也可以。

```
<resources xmlns:xliff="urn:oasis:names:tc:xliff:document:1.2">
<tech-list> --------------------------------A 组
<tech>android.nfc.tech.IsoDep</tech>  <tech>android.nfc.tech.NfcA</tech>  <tech>android.nfc.tech.NfcB</tech>
<tech>android.nfc.tech.NfcF</tech>
</tech-list>
<tech-list>------------------------------------------B 组
<tech>android.nfc.tech.NfcV</tech>  <tech>android.nfc.tech.Ndef</tech>  <tech>android.nfc.tech. NdefFormatable
</tech> <tech>android.nfc.tech.MifareClassic</tech> <tech>android.nfc.tech.MifareUltralight</tech>
 </tech-list>
</resources>
```

在 androidManifest 文件中，声明 xml 过滤的举例代码如下：

```
<activity>
 <intent-filter>
<action android:name="android.nfc.action.TECH_DISCOVERED"/>
</intent-filter>
<meta-data android:name="android.nfc.action.TECH_DISCOVERED" android:resource=" @xml/nfc_tech_filter"
/>-------------这个就是用户的资源文件名
 </activity>
```

（5）创建 NFC 标签的前台分发系统。

什么是 NFC 的前台发布系统？就是说当已经打开应用时，那么通过这个前台发布系统的设置，可以让已经启动的 Activity 拥有更高的优先级来依据在代码中定义的标准来过滤和处理 Intent，而不是让其他的声明了 Intent Filter 的 Activity 来干扰，甚至连自己声明在文件 androidManifest 中的 Intent Filter 都不会来干扰。也就是说，Foreground Dispatch 的优先级大于 Intent Filter。此时有如下所示的两种情况。

☑ 第一种情况：当 Activity 没有启动时去扫描 tag，那么系统中所有的 Intent Filter 都将一起参与过滤。

☑ 第二种情况：当 Activity 启动去扫描 tag 时，那么将直接使用在 Foreground Dispatch 中代码写入的过滤标准。如果这个标准没有命中任何 Intent，那么系统将使用所有 Activity 声明的 Intent Filter xml 来过滤。

例如在 OnCreate()中可以添加如下所示的代码。

```
   mPendingIntent = PendingIntent.getActivity(this, 0,
                new Intent(this, getClass()).addFlags(Intent.FLAG_ACTIVITY_SINGLE_TOP), 0);
        //做一个 Intent Filter 过滤 action，这里过滤的是 ndef
        IntentFilter ndef = new IntentFilter(NfcAdapter.ACTION_NDEF_DISCOVERED);
//如果对 action 的定义有更高的要求，如 data 的要求，可以使用如下的代码来定义 Intent Filter
//         try {
//              ndef.addDataType("*/*");
//         } catch (MalformedMimeTypeException e) {

//              //TODO Auto-generated catch block

//              e.printStackTrace();

//         }
     //生成 Intent Filter
        mFilters = new IntentFilter[] {
                ndef,
        };
        //做一个 tech-list。可以看到是二维数据，每一个一维数组之间的关系是或，但是一个一维数组之内的各
个项是与的关系
        mTechLists = new String[][] {
                new String[] { NfcF.class.getName()},
                new String[]{NfcA.class.getName()},
                new String[]{NfcB.class.getName()},
                new String[]{NfcV.class.getName()}
                };
```

在 onPause()和 onResume()中需要加入如下相应的代码。

```
public void onPause() {
 super.onPause();
```

```
//反注册 mAdapter.disableForegroundDispatch(this);
}
public void onResume() {
 super.onResume();
//设定 Intent Filter 和 tech-list。如果两个都为 null，就代表优先接受任何形式的 TAG action。也就是说系统会主动发 TAG intent
mAdapter.enableForegroundDispatch(this, mPendingIntent, mFilters, mTechLists);
}
```

# 16.5　实战演练——使用 NFC 发送消息

**知识点讲解：光盘:视频\知识点\第 16 章\使用 NFC 发送消息.avi**

当 Android 设备检测到有 NFC Tag 时，预期的行为是触发最合适的 Activity 来处理检测到的 Tag，这是因为 NFC 通常是在非常近的距离才起作用（<4m）。如果在此时需要用户来选择合适的应用来处理 Tag，则很容易断开与 Tag 之间的通信，因此需要选择合适的 Intent Filter 只处理读写的 Tag 类型。Android 系统支持两种 NFC 消息发送机制，分别是 Intent 发送机制和前台 Activity 消息发送机制。

☑ Intent 发送机制：当系统检测到 Tag 时，Android 系统提供 manifest 中定义的 Intent Filter 来选择合适的 Activity 来处理对应的 Tag，当有多个 Activity 可以处理对应的 Tag 类型时，则会显示 Activity 选择窗口由用户选择，如图 16-3 所示。

☑ 前台 Activity 消息发送机制：允许一个在前台运行的 Activity 拥有读写 NFC Tag 的优先权。如果系统 Android 检测到周围有 NFC Tag，且前台允许这个 Activity 可以处理该种类型的 Tag，那么这个 Activity 即可获取优先权，而不会出现选择 Activity 的窗口。

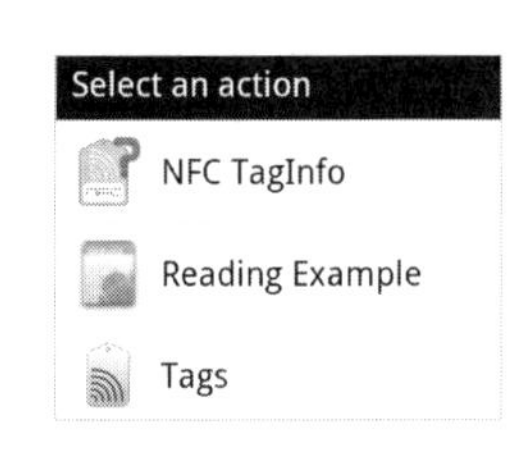


图 16-3　选择窗口

上述两种方法基本上都是使用 Intent Filter 来指明 Activity 可以处理的 Tag 类型，一个是使用 Android 的 Manifest 来说明，一个是通过代码来声明。

下面将通过一个具体实例的实现过程，来讲解在 Android 系统中使用 NFC 消息发送机制的基本方法。

| 实　例 | 功　能 | 源码路径 |
|---|---|---|
| 实例 16-1 | 演示 NFC 消息发送机制的基本用法 | 光盘:\daima\16\NFCEX |

本实例的具体实现流程如下。

（1）在文件 AndroidManifest.xml 中声明 NFC 权限，具体实现代码如下：

```
<manifest xmlns:android="http://schemas.android.com/apk/res/android"
      package="com.pstreets.nfc"
      android:versionCode="1"
      android:versionName="1.0">
    <uses-sdk android:minSdkVersion="10" />
     <uses-permission android:name="android.permission.NFC" />
     <uses-feature android:name="android.hardware.nfc" android:required="true" />
    <application android:icon="@drawable/icon" android:label="@string/app_name">
        <activity android:name=".NFCDemoActivity"
                  android:label="@string/app_name"
```

```
                android:launchMode="singleTop">
            <intent-filter>
                <action android:name="android.intent.action.MAIN" />
                <category android:name="android.intent.category.LAUNCHER" />
            </intent-filter>
            <intent-filter>
              <action android:name="android.nfc.action.NDEF_DISCOVERED"/>
              <data android:mimeType="text/plain" />
            </intent-filter>
            <intent-filter
                    >
                    <action
                        android:name="android.nfc.action.TAG_DISCOVERED"
                        >
                    </action>
                    <category
                        android:name="android.intent.category.DEFAULT"
                        >
                    </category>
              </intent-filter>
             <!-- Add a technology filter -->
            <intent-filter>
                <action android:name="android.nfc.action.TECH_DISCOVERED" />
            </intent-filter>
            <meta-data android:name="android.nfc.action.TECH_DISCOVERED"
                android:resource="@xml/filter_nfc"
            />
        </activity>
        <activity android:name=".MainActivity"
                  android:label="@string/app_name">
            <intent-filter>
                <action android:name="android.intent.action.MAIN" />
                <category android:name="android.intent.category.LAUNCHER" />
            </intent-filter>
        </activity>
    </application>
</manifest>
```

这样通过上述声明代码，当 Android 检测到有 Tag 时，会显示 Activity 选择窗口，就会显示图 16-3 所示的 Reading Example 效果。

（2）编写布局文件 main.xml，功能是通过文本控件显示当前的扫描状态，具体实现代码如下：

```
<RelativeLayout
  xmlns:android="http://schemas.android.com/apk/res/android"
  android:layout_width="match_parent"
  android:layout_height="match_parent"
  android:text="@string/title">
  <TableLayout
          android:id="@+id/purchScanTable1"
          android:layout_width="wrap_content"
          android:layout_height="wrap_content">
          <TableRow
```

```
        android:id="@+id/table1Row1"
        android:layout_width="wrap_content"
        android:layout_height="wrap_content">
        <TextView
            android:id="@+id/status_label"
            android:layout_width="wrap_content"
            android:layout_height="wrap_content"
            android:text="Current Status:    " />
        <TextView
            android:id="@+id/status_data"
            android:layout_width="wrap_content"
            android:layout_height="wrap_content"
            android:text=" Scan a Tag" />
    </TableRow>
    <View
        android:id="@+id/purchScanES1"
        android:layout_width="match_parent"
        android:layout_height="55px"
        android:layout_below="@id/status_label"
        android:background="#000000" />
    <TableRow
        android:id="@+id/table1Row2"
        android:layout_width="wrap_content"
        android:layout_height="wrap_content">
        <TextView
            android:id="@+id/block_0_label"
            android:layout_width="wrap_content"
            android:layout_height="wrap_content"
            android:text="BLOCK 0:    " />
        <TextView
            android:id="@+id/block_0_data"
            android:layout_width="wrap_content"
            android:layout_height="wrap_content"
            android:text=" " />
    </TableRow>
    <TableRow
        android:id="@+id/table1Row3"
        android:layout_width="wrap_content"
        android:layout_height="wrap_content">
        <TextView
            android:id="@+id/block_1_label"
            android:layout_width="wrap_content"
            android:layout_height="wrap_content"
            android:text="BLOCK 1:    " />
        <TextView
            android:id="@+id/block_1_data"
            android:layout_width="wrap_content"
            android:layout_height="wrap_content"
            android:text=" " />
    </TableRow>
</TableLayout>
```

```
    <View
              android:id="@+id/purchScanES1"
              android:layout_width="match_parent"
              android:layout_height="75px"
              android:layout_below="@id/purchScanTable1"
              android:background="#000000" />
    <Button
                   android:id="@+id/clear_but"
                   android:layout_width="fill_parent"
                   android:layout_height="wrap_content"
                   android:layout_below="@id/purchScanES1"
                   android:gravity="center_horizontal"
                   android:text="Clear" />
</RelativeLayout>
```

（3）编写程序文件 NFCDemoActiviy.java，当在前台运行 NFCDemoActivity 时，如果希望只有它来处理 Mifare 类型的 Tag，此时可以使用前台消息发送机制。文件 NFCDemoActivity.java 的具体实现代码如下：

```
public class NFCDemoActivity extends Activity {
    private NfcAdapter mAdapter;
    private PendingIntent mPendingIntent;
    private IntentFilter[] mFilters;
    private String[][] mTechLists;
    private TextView mText;
    private int mCount = 0;

    @Override
    public void onCreate(Bundle savedState) {
        super.onCreate(savedState);

        setContentView(R.layout.foreground_dispatch);
        mText = (TextView) findViewById(R.id.text);
        mText.setText("Scan a tag");

        mAdapter = NfcAdapter.getDefaultAdapter(this);

        mPendingIntent = PendingIntent.getActivity(this, 0,
                new Intent(this, getClass()).addFlags(Intent.FLAG_ACTIVITY_SINGLE_TOP), 0);

        IntentFilter ndef = new IntentFilter(NfcAdapter.ACTION_TECH_DISCOVERED);
        try {
            ndef.addDataType("*/*");
        } catch (MalformedMimeTypeException e) {
            throw new RuntimeException("fail", e);
        }
        mFilters = new IntentFilter[] {
                ndef,
        };

        mTechLists = new String[][] { new String[] { MifareClassic.class.getName() } };
    }
```

```
    @Override
    public void onResume() {
        super.onResume();
        mAdapter.enableForegroundDispatch(this, mPendingIntent, mFilters, mTechLists);
    }

    @Override
    public void onNewIntent(Intent intent) {
        Log.i("Foreground dispatch", "Discovered tag with intent: " + intent);
        mText.setText("Discovered tag " + ++mCount + " with intent: " + intent);
    }

    @Override
    public void onPause() {
        super.onPause();
        mAdapter.disableForegroundDispatch(this);
    }
}
```

这样在执行本实例后，每当靠近一次 Tag，计数就会增加 1。执行效果如图 16-4 所示。

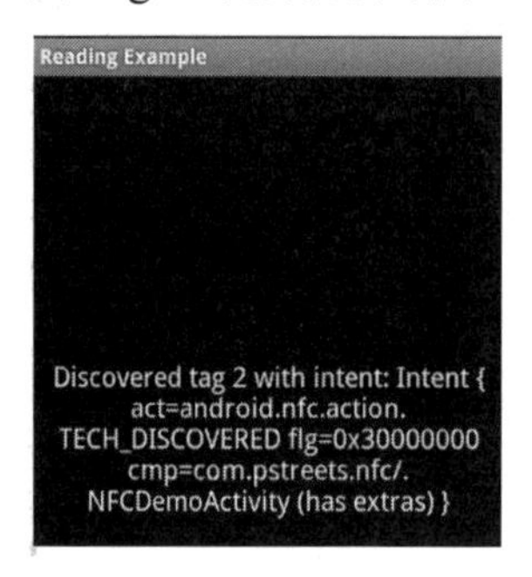


图 16-4　执行效果

# 第 17 章　Google Now 和 Android Wear 详解

Google Now 是谷歌在 I/O 开发者大会上随安卓 4.1 系统同时推出的一款应用，它会全面了解用户的各种习惯和正在进行的动作，并利用它所了解的来为用户提供相关信息。本章将详细讲解在 Android 设备中使用 Google Now 技术的基本知识，为步入本书后面知识的学习打下基础。

## 17.1　Google Now 介绍

**知识点讲解：光盘:视频\知识点\第 17 章\Google Now 介绍.avi**

Google Now 是 Google 在移动市场最重要的创新之一。通过对用户数据的挖掘，Google Now 在适当的时刻提供适当的信息，而它的卡片式推送也代表了 Google 展现信息的新方向。正如 GigaOM 的作者在某次旅行中体会到的，Google Now 成了一个有力的帮手。虽然它仍有些让人不安，但 Google Now 利大于弊。本节将详细讲解 Google Now 的基本知识，为读者步入本书后面知识的学习打下基础。

### 17.1.1　搜索引擎的升级——Google Now

Google Now 功能是 I/O 大会上的一个亮点，这个可以根据不同使用习惯来帮助用户进行多项信息的预测，虽然人机交互方面与 iOS 上的 Siri 还有很大差距，但其预测比起 Siri 更加实用。国外媒体都给了 Google Now 功能很高的评价，不过这个功能在中国受到很大的限制。

在过去的 10 年中，搜索引擎的核心是获取足够多的海量信息，搜索技术的发展过程是追赶如何更好地获取信息的过程，核心是个性化和实时信息。但是随着时代的进步和发展，现在搜索结果正在变得越来越个性化。不同的人都会看到他更感兴趣的搜索结果，提高了搜索的效率。甚至由于搜索变得过于个性化，人们获得的信息都是自己想看到的，从而让原本能够扩大人们视野的搜索变成了把人们限制在自我的世界工具。这还引发了关于搜索过分个性化可能引发的弊端的讨论。

搜索在个性化方面的努力最重要的是将搜索和社交网络结合，这样搜索引擎就能获得用户的更多信息，从而更好地帮助用户做出判断。在个性化搜索方面谷歌遇到了来自 Facebook 的挑战，拥有最多用户信息的网站是 Facebook，但它却并不向谷歌开放。从某种程度上说，谷歌推出自己的社交网络 Google+的核心也是希望获得更多的用户信息。

实时搜索更多是搜索在技术实现上的改进，当然，大部分实时信息都存在于 Twitter 和雅虎，这对谷歌也是不小的挑战。随着移动互联网的发展，位置也成为了搜索引擎提供结果的重要依据，这也是个性化的一部分。而随着位置信息的加入，围绕这一点可以打造一个生活服务的平台。

综上所述，本地搜索将是一个巨大的市场，这个时候搜索提供的已经不仅仅是信息，更应该是一种服务。正因如此，Google Now 便登上了历史舞台，接下来看一看 Google Now 能带来什么？

☑　新的应用会更加方便用户收取电子邮件，当接收到新邮件时，它就会自动弹出以便查看。

- ☑ 实现了办理登记手续的 QR CODE 终端的更新，但是这一功能目前仅限于美国联合航空公司使用。
- ☑ 具有新的镜头搜索功能，令搜索和查找更加方便准确。
- ☑ 具有步行和行车里程记录功能，这个计步器功能可通过 Android 设备的传感器来统计用户每月行驶的里程，包括步行和骑自行车的路程。
- ☑ 拥有并强化了对博物馆、电影院、餐厅等搜索帮助。
- ☑ 旅游和娱乐特色功能：包括汽车租赁、演唱会门票和通勤共享方面的卡片。公共交通和电视节目的卡片进行改善，这些卡片现在可以听音识别音乐和节目信息。用户可以为新媒体节目的开播设定搜索提醒，同时还可以接收实时 NCAA 橄榄球比分。

### 17.1.2 Google Now 的用法

其实 Google Now 并不是如同 Google Mail、Google Talk 那样的独立 App，Google Now 被 Google 集成到了 Google 搜索中。在正常情况下，开启 Google 搜索即可使用 Google Now。但是因为 Google 搜索业务已经退出中国大陆，所以 Google 也没打算让 Google Now 覆盖中国大陆用户。即使顺利安装了 Google 搜索，会依然找不到 Google Now 功能，如图 17-1 所示。

此时需要经过如下所示的步骤进行设置。

（1）登录手机设备的 Google 账户。

（2）在“设置”选项中将系统语言改为英文，如图 17-2 所示。

图 17-1 默认没有 Google Now 功能的 Google 搜索

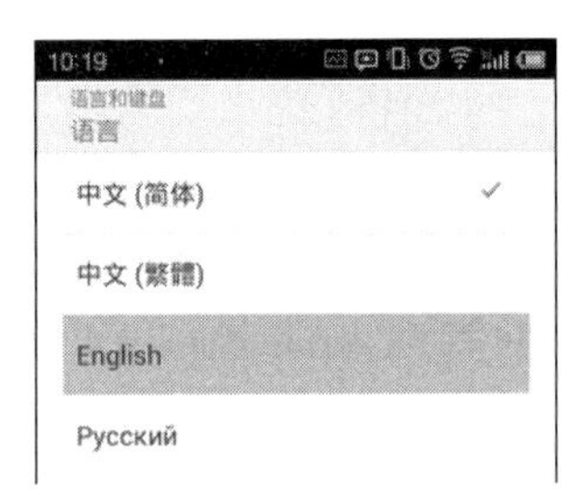


图 17-2 设置设备语言为英文

（3）再次开启 Google Search 后会发现 Google Now，如图 17-3 所示。

（4）按照提示，单击 Next 按钮即可完成 Google Now 的初始化，这时就可以使用 Google Now 了，如图 17-4 所示。

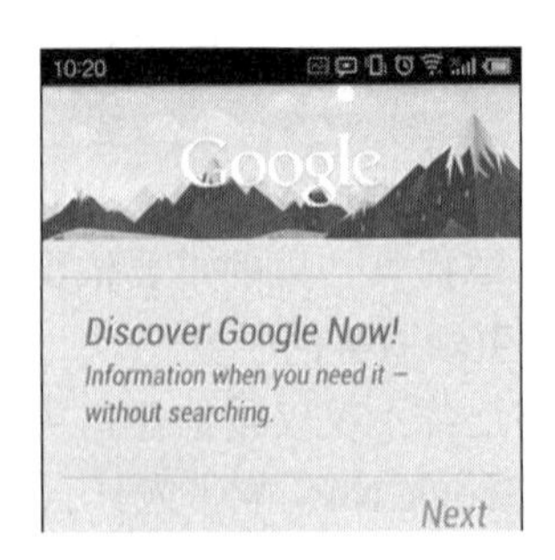


图 17-3 Google Search 中出现 Google Now

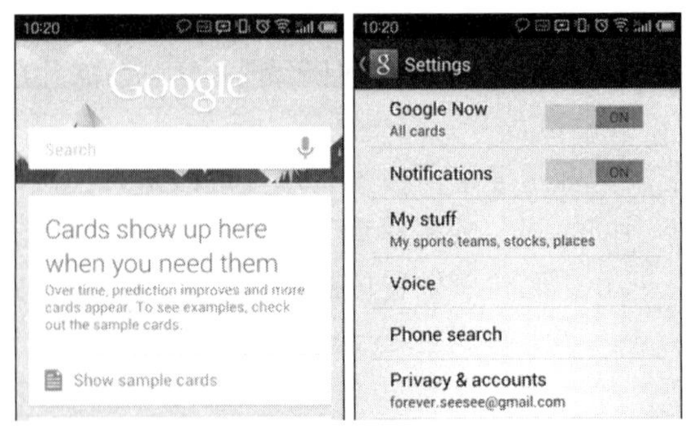


图 17-4 此时可以使用 Google Now

（5）当设置完 Google Now 后回到设置菜单，将系统语言重新设置为简体中文。设置完毕后，Google Now 不但不会被关闭，语言也变成了简体中文。这意味着 Google 本来就做好了 Google Now 的简体中文语言支持，只是没对简体中文用户开放而已，如图 17-5 所示。

（6）经过测试后会发现，虽然 Google Now 没有针对国内用户开放，但是数据依然涵盖了国内。在使用期间，公交班次、天气等信息都准确无误，连接也没遇到什么阻碍，如图 17-6 所示。

图 17-5　中文 Google Now　　　　图 17-6　使用 Google Now 的界面

**注意：** 只有在设备中登录并绑定 Google 账号后才能使用 Google Now 功能，国产行货手机没有内置添加 Google 账号功能，读者需要在获取 Root 权限后进行添加设置。

# 17.2　Android Wear 详解

**知识点讲解：光盘:视频\知识点\第 17 章\Android Wear 详解.avi**

2014 年 3 月，继谷歌眼镜之后，谷歌推出了 Android Wear 可穿戴平台，正式进军智能手表领域。与之前传闻不同的是，谷歌并未推出硬件，这意味着什么？显然，作为一个平台服务商，谷歌的目标不仅仅是一款卖得好的智能手表，而是一统整个穿戴式计算机行业。对于用户而言，Android Wear 将改变目前智能手表领域缺乏标准、各自为营的混乱状况，同时也能够与自己的 Android 手机获得更无缝化的数据共享。本节将简单看一看 Android Wear 平台给我们带来了怎样的前景和未来。

## 17.2.1　什么是 Android Wear

可以将 Android Wear 看作是一个针对智能手表等可穿戴设备优化的 Android 版本，Android Wear 界面更适合小屏幕，主要功能是面向手机与手表互联带来的新型移动体验。举个例子来说，平常乘坐公交车时难免会遇到坐过站的情况，只要在 Android Wear 手表中设定好目的地，GPS 便会开始定位，及时提醒我们“还有 1 站到达大明湖”，这样就能够避免发生坐过站的情况。

从本章前面讲解的内容可知，Google Now 应用一直致力于通过上下文联想技术提供全面、智能的搜索体验，现在 Google Now 被集成到 Android Wear 中，不需要任何按键，只需说“OK Google”以及用户想知道的内容或是进行的操作即可。

谷歌在视频中演示了相当丰富的使用场景，例如用户要去海滩冲浪，Android Wear 手表会自动弹出“海里有海蜇”的警告；在收到短信场景时，可以直接语音回复即可；在登机场景中，直接出示手表中的机票二维码就可以完成登机工作。

另外，健身应用也是 Android Wear 必备的一个功能。Android Wear 能够实时监测我们的活动状态，记录步数及热量消耗。当然，健身功能实际上还有很大的发展空间，相信谷歌和手表制造商会在日后为用户提供更多样化的健康监测形式，如手表背面内置传感器监测用户体温和心率等。

由此可见，Android Wear 是将 Android 延伸到可穿戴设备的项目。这个项目首先从智能手表开始。通过一系列的新设备和应用，Android Wear 将能够做到以下方面。

☑ 在用户最需要的时候给出有用的信息：从用户最喜欢的社交应用获取更新，使用通信应用交流，从购物应用、新闻应用那里获取通知等。

☑ 直接回答用户的问题：说一声“OK Google”来提出问题，如航班离开的时间、游戏的分数，或者完成某件事情（如呼叫出租车、发短信、预定餐厅或者设置闹钟）。

☑ 更好地监控用户的健康：通过 Android Wear 上的提醒和健康信息，达到自己健身的目标。用户最爱的健身应用能够提供实时的速度、距离和时间信息。

☑ 通向多屏世界的钥匙：Android Wear 能让用户控制其他设备。用“Ok Google”打开手机上的音乐列表，或者将最喜欢的电影投射到电视上面。在开发者的参与下，还会有更多的可能性。

目前，摩托罗拉和 LG 已经展示了概念的 Android Wear 手表，预计三星、HTC、华硕等厂商都会后续跟进。首先来看看摩托罗拉的 Moto 360 手表，它拥有一个接近传统手表的圆形金属表盘，适合在所有场合佩戴。摩托罗拉公司也承诺将使用精良的材质，保持佩戴的舒适性，如图 17-7 所示。

图 17-7　Android Wear 手表

## 17.2.2　搭建 Android Wear 开发环境

现在谷歌已经公开了 Android Wear 的预览版，只面向谷歌账号开发者用户公开。具体信息请登录 http://developer.android.com/wear/index.html，如图 17-8 所示。

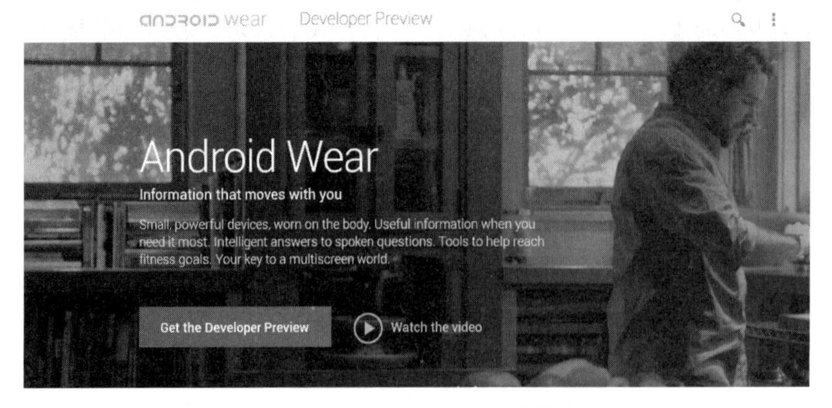


图 17-8　Android Wear 官方站点

单击图 17-8 中的 Get the Developer Preview 按钮，进入 Android Wear 开发者预览界面，在此列出了搭建开发环境的方法和开发资料，如图 17-9 所示。

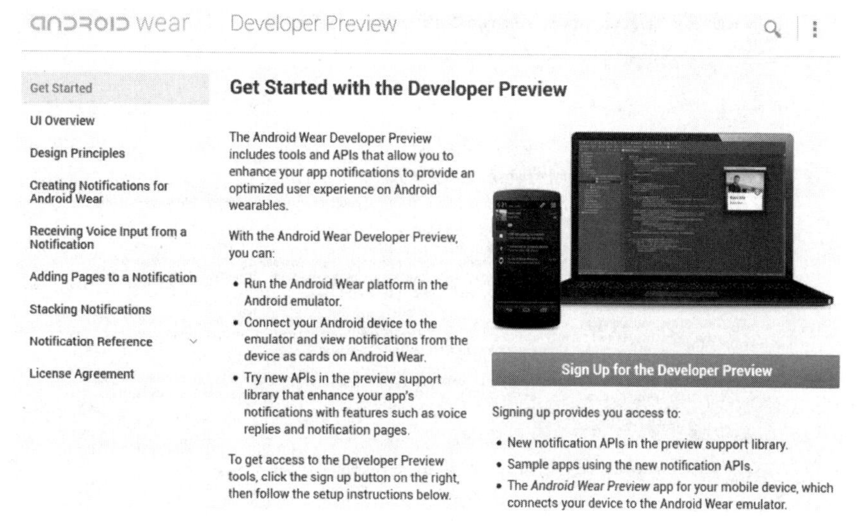


图 17-9　Android Wear 开发者预览界面

Android Wear 开发环境和 Android 应用开发环境类似，具体过程如下：

（1）根据第 2 章的内容安装 Android SDK，在 Android SDK 中包括了 Android Wear 的所有开发工具。

（2）单击图 17-9 中的按钮进入注册界面，在此界面注册为 Android Wear 预览开发者，如图 17-10 所示。

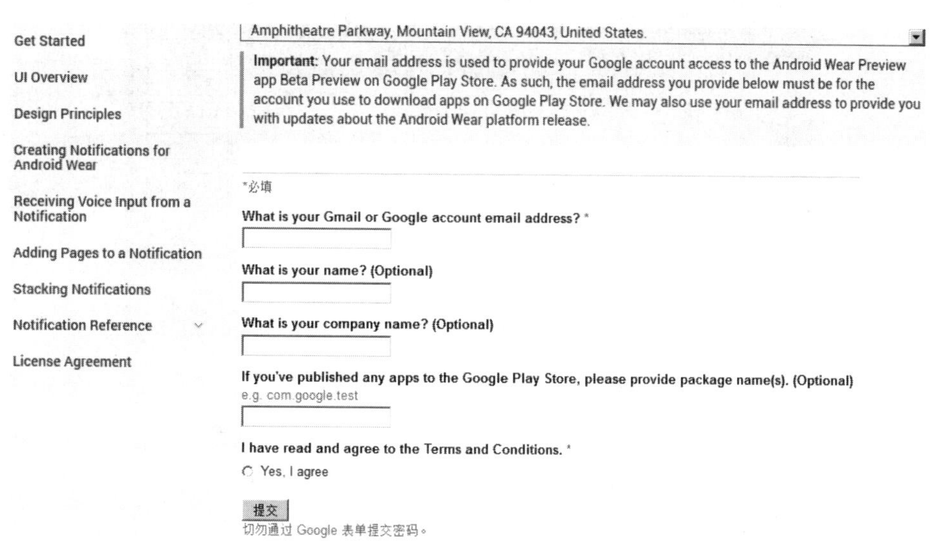


图 17-10　注册界面

（3）输入 Gmail 账户信息后单击“提交”按钮，等待谷歌发送回复的邮件信息，如图 17-11 所示。

通过邮件中的链接可以下载 Android Wear 预览版的开发程序包和演示实例，下载后的压缩包是 AndroidWearPreview.zip，解压缩后的效果如图 17-12 所示。

（4）检查 Android SDK 工具的版本 22.6 或更高，如果当前 Android SDK 工具的版本低于 22.6，则必须进行更新。

（5）在图 17-13 所示的界面中创建一个 Android 模拟器，Android Wear 要求的最低版本是 Android 4.4.2，选择 Android Wear ARM(armeabi-v7a)。

Android Wear Developer Preview - Sign Up Confirmation

noreply@google.com
发送至 我
3月19日

英文 > 中文 翻译邮件 对英文停用

Hello Developer,

Thank you for signing up for the Android Wear Developer Preview.

To begin developing on Android Wear, you'll need the Preview Support library and the Android Wear Preview app for your mobile device. Follow these steps:

- Download the Preview Support library and samples
- Opt-in to become a tester of the Android Wear Preview app in the Google Play Store. After opt-in, it could take up to 24 hours for the Android Wear Preview app to be accessible to you in Google Play. Make sure the opt-in user account is the same user signed in to Google Play.

Refer to the Android Wear Developer Get Started page for details. Since this is a preview release, please do not publicly distribute apps built with the Preview library. Also note that the APIs are potentially subject to change and you will need to modify your apps when they are released out of preview.

Share your experiences and ask questions by joining the Android Wear Developers Google+ Community. We look forward to seeing how your apps take advantage of these new APIs to provide innovative new user experiences!

— The Android Wear Team

图 17-11　谷歌回复的邮件信息

| | | | |
|---|---|---|---|
| samples | 2014/3/18 1:27 | 文件夹 | |
| LICENSE | 2014/3/18 1:27 | 文件 | 19 KB |
| README | 2014/3/18 1:27 | 文件 | 1 KB |
| wearable-preview-support.jar | 2014/3/18 1:27 | Executable Jar... | 30 KB |

图 17-12　解压缩 AndroidWearPreview.zip

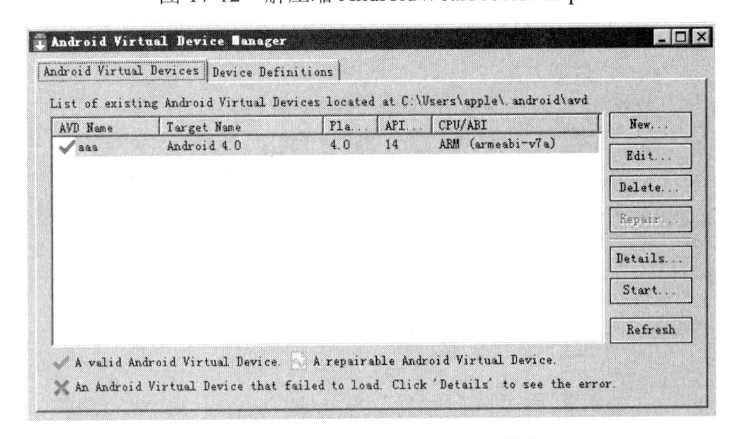


图 17-13　准备创建一个 Android 模拟器

（6）单击图 17-13 中的 New 按钮，在新界面中进行如下所示的设置。

☑　AVD name：设置创建模拟器的名字为 wear。

☑　Target：设置此值最低为 Android 4.4.2 - API Level 19。

☑　CPU/ABI：设置此值为 Android Wear ARM (armeabi-v7a)。

☑　Skin：用于设置 Android Wear 的外观，现在 Android Wear 只有两种外观，分别是方形（Android WearSquare）或圆形（AndroidWearRound）。

☑　其他选项：设置为默认值即可。

设置后的界面效果如图 17-14 所示。

单击 OK 按钮完成创建，单击 Start 按钮可以运行这个模拟器，运行后的效果如图 17-15 所示。

另外，通过谷歌回复的 Gmail 邮件可知，可以登录 https://play.google.com/apps/testing/com.google.android.wearablepreview.app 下载 Android Wear Preview，当然最简单的方法是从 Play 商店下载获取，如图 17-16 所示。

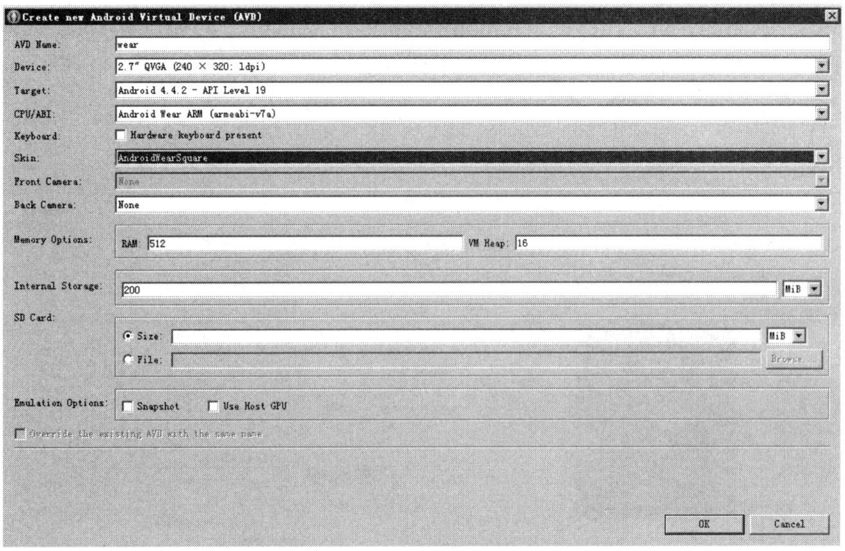

Create new Android Virtual Device (AVD)

AVD Name: wear
Device: 2.7" QVGA (240 × 320: ldpi)
Target: Android 4.4.2 - API Level 19
CPU/ABI: Android Wear ARM (armeabi-v7a)
Keyboard: Hardware keyboard present
Skin: AndroidWearSquare
Front Camera: None
Back Camera: None
Memory Options: RAM: 512　VM Heap: 16
Internal Storage: 200 MiB
SD Card:
Size: MiB
File: Browse...
Emulation Options: Snapshot　Use Host GPU
Override the existing AVD with the same name
OK　Cancel

图 17-14　创建了一个方形 Android Wear 模拟器

图 17-15　模拟器运行效果

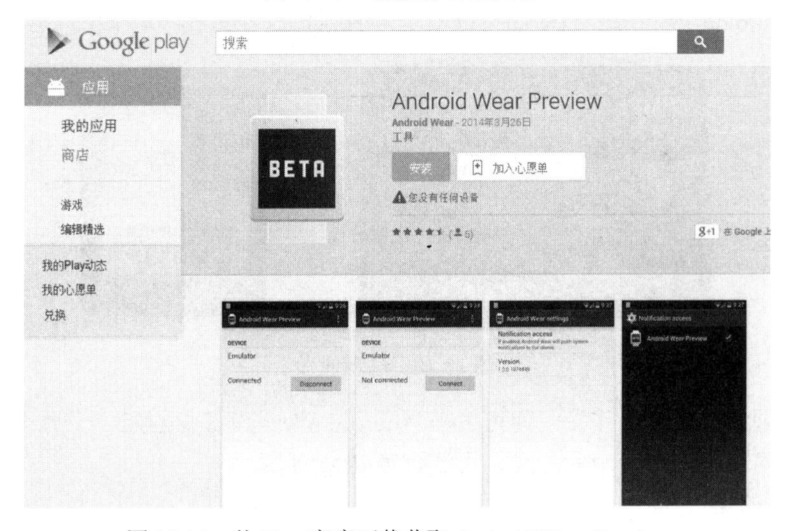


图 17-16　从 Play 商店下载获取 Android Wear Preview

另外，也可以登录 https://plus.google.com/communities/113381227473021565406 进入测试人员社区，在这里可以和一线开发人员进行交流，如图 17-17 所示。

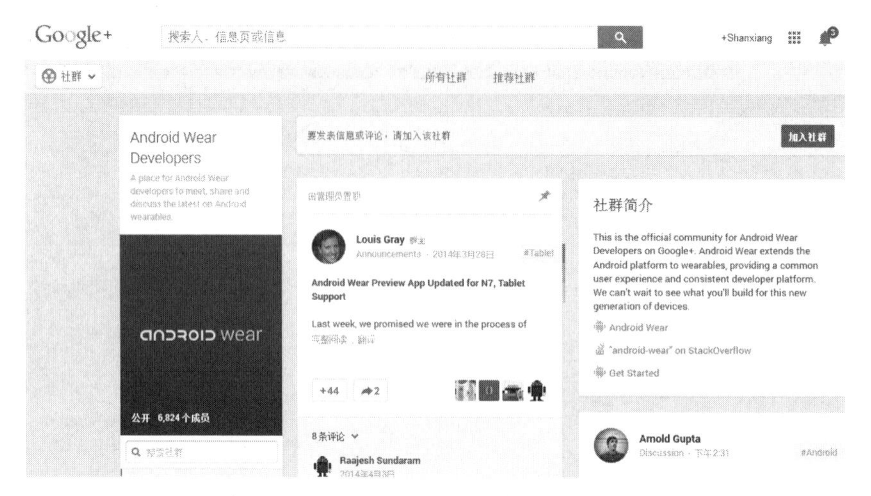


图 17-17　Android Wear 开发者交流社区

# 17.3　开发 Android Wear 程序

**知识点讲解：光盘:视频\知识点\第 17 章\开发 Android Wear 程序.avi**

在搭建完 Android Wear 开发环境之后，接下来开始讲解开发 Android Wear 程序的基本知识。本节将首先讲解开发 Android Wear 程序的知识，然后通过一个演示实例讲解具体开发过程。

## 17.3.1　创建通知

当一个手机或平板电脑等 Android 设备连接到一个 Android Wear 时，所有的通知在设备之间都是共享的。在 Android Wear 中，每个通知都会以新卡片背景流的样式出现，如图 17-18 所示。

图 17-18　出现通知

由此可见，无须经过多少工作量，便可以在 Android Wear 设备中创建一个通知应用程序。但是为了提高用户体验，当用户面对一个通知时，再通过声音来回复。

（1）引入需要的类

在开发 Android Wear 应用程序之前，必须首先详细阅读开发者预览文档。在该文档文件中提到，

Android Wear 应用程序必须包括 V4 支持库和开发者预览版支持库。所以开始的时候，应该包括项目中的代码下面的引入文件：

```
import android.preview.support.wearable.notifications.*;
import android.preview.support.v4.app.NotificationManagerCompat;
import android.support.v4.app.NotificationCompat;
```

（2）通过提醒 Builder 创建通知

在 Android Wear 中，通过用 V4 支持库可以实现最新的通知等功能，例如用操作按钮和大图标创建通知。例如在下面的演示代码中，使用 NotificationCompat API 结合新的 NotificationManagerCompat API 可以创建并发布通知。

```
int notificationId = 001;
Intent viewIntent = new Intent(this, ViewEventActivity.class);
viewIntent.putExtra(EXTRA_EVENT_ID, eventId);
PendingIntent viewPendingIntent =
        PendingIntent.getActivity(this, 0, viewIntent, 0);

NotificationCompat.Builder notificationBuilder =
        new NotificationCompat.Builder(this)
        .setSmallIcon(R.drawable.ic_event)
        .setContentTitle(eventTitle)
        .setContentText(eventLocation)
        .setContentIntent(viewPendingIntent);

NotificationManagerCompat notificationManager =
        NotificationManagerCompat.from(this);

notificationManager.notify(notificationId, notificationBuilder.build());
```

通过上述代码，当上述通知出现在手持设备中时，用户可以调用指定的 setcontentintent()方法通过触摸的方式通知 PendingIntent。当这个通知出现在 Android Wear 中时，用户也可以用通知操作来调用在手持设备上的意图。

（3）添加动作按钮

除了通过 setcontentintent()定义的主要操作外，还可以通过传递 PendingIntent 到 addaction()方法的方式添加其他操作。例如下面的代码显示了和前面类型相同的通知，但是增加了一个在地图上实现定位的事件操作。

```
Intent mapIntent = new Intent(Intent.ACTION_VIEW);
Uri geoUri = Uri.parse("geo:0,0?q=" + Uri.encode(location));
mapIntent.setData(geoUri);
PendingIntent mapPendingIntent =
        PendingIntent.getActivity(this, 0, mapIntent, 0);

NotificationCompat.Builder notificationBuilder =
        new NotificationCompat.Builder(this)
        .setSmallIcon(R.drawable.ic_event)
        .setContentTitle(eventTitle)
        .setContentText(eventLocation)
        .setContentIntent(viewPendingIntent)
        .addAction(R.drawable.ic_map,
                getString(R.string.map), mapPendingIntent);
```

（4）为通知添加一个大视图

在手持设备上，用户可以通过扩大通知卡片的方式来查看通知内容。在 Android Wear 设备中，大视图的内容是默认可见的。当在通知中添加扩展的内容后，可以调用 NotificationCompat.Builder 对象中的 setStyle()方法来实现 bigtextstyle 或 inboxstyle 样式实例。

例如在下面的代码中，添加了 NotificationCompat.bigtextstyle 实例的事件通知，这样可以包括完整的事件描述，包括可以提供比 setcontenttext()空间更多的文本内容。

```
BigTextStyle bigStyle = new NotificationCompat.BigTextStyle();
bigStyle.bigText(eventDescription);

NotificationCompat.Builder notificationBuilder =
        new NotificationCompat.Builder(this)
        .setSmallIcon(R.drawable.ic_event)
        .setLargeIcon(BitmapFractory.decodeResource(
                getResources(), R.drawable.notif_background))
        .setContentTitle(eventTitle)
        .setContentText(eventLocation)
        .setContentIntent(viewPendingIntent)
        .addAction(R.drawable.ic_map,
                getString(R.string.map), mapPendingIntent)
        .setStyle(bigStyle);
```

**注意：** 可以使用 setlargeicon()方法为任何通知添加一个背景图像。

（5）为设备添加新的功能

在 Android Wear 预览版的支持库中提供了很多新的 API，通过这些 API 可以在穿戴设备中提高通知用户体验。例如可以添加额外的页面内容，或添加用户使用语音输入文本的响应功能。通过使用这些新的 API 和实例的 NotificationCompat.Builder()构造函数可以添加新的功能。例如下面的演示代码。

```
NotificationCompat.Builder notificationBuilder =
        new NotificationCompat.Builder(mContext)
        .setContentTitle("New mail from " + sender.toString())
        .setContentText(subject)
        .setSmallIcon(R.drawable.new_mail);

Notification notification =
        new WearableNotifications.Builder(notificationBuilder)
        .setHintHideIcon(true)
        .build();
```

在上述代码中，方法 setHintHideIcon()的功能是从通知卡中移除应用程序图标。方法 setHint HideIcon()是一个新的通知功能，可以从 WearableNotifications.Builder 对象中生成。

当想要推送传递的通知时，一定要始终使用 NotificationManagerCompat API，例如下面的演示代码。

```
NotificationManagerCompat notificationManager =
        NotificationManagerCompat.from(this);

notificationManager.notify(notificationId, notification);
```

**注意：** 在笔者写作此书时，Android Wear 开发者预览版 API 只是为了开发和测试而推出的，并不是为了编写出具体应用程序。谷歌在正式公布 Android Wear SDK 之前，上述开发流程只是待定的。

## 17.3.2　创建声音

如果在创建的通知中包含了文本回复功能，例如回复一封邮件，在通常情况下会在手持设备上启动一个 Activity。当通知显示在穿戴设备上时，可以允许用户使用语音输入口述一个回复，还可以提供预先设置的文本信息让用户选择。当用户使用语音回复或者选择预设信息时，系统会发送信息到与手持设备相连的应用，该信息以一个附加品的形式与我们定义使用的通知行动的 Intent 相关联，如图 17-19 所示。

图 17-19　声音回复

**注意：** 在安卓模拟器上开发时，即使在语音输入域，也必须使用文本回复，所以要确保在 AVD 设置上已激活了 Hardware keyboard present。

（1）定义远程回复

在 Android Wear 中创建支持语音输入的行动时，首先需要使用 RemoteInput.Builder APIs 创建一个 RemoteInput 的实例。RemoteInput.Builder 构造器获取一个 String 类型的值，系统会将这个值作为一个 key 传递给 Intent extra，这个 Intent 可以将回复信息传送到手持设备中的应用程序。例如在下面的代码中创建了一个新的 RemoteInput 对象，功能是提供自定义标签给语音输入命令。

```
//传送给行动 intent 的 key 的字符串
private static final String EXTRA_VOICE_REPLY = "extra_voice_reply";
String replyLabel = getResources().getString(R.string.reply_label);
RemoteInput remoteInput = new RemoteInput.Builder(EXTRA_VOICE_REPLY)
        .setLabel(replyLabel)
        .build();
```

（2）添加预置文本进行回复

除了支持语音输入外，在 Android Wear 中还可以提供最多 5 条预置文本回复信息，以供用户进行快速回复。实现方法是调用 setChoices()方法，并将字符串数组传递给它。例如可以在资源数组中定义如下所示的回复。

```
res/values/strings.xml
<?xml version="1.0" encoding="utf-8"?>
<resources>
    <string-array name="reply_choices">
        <item>Yes</item>
        <item>No</item>
        <item>Maybe</item>
    </string-array>
</resources>
```

执行效果如图 17-20 所示。

图 17-20　添加的回复数值

然后通过如下代码释放 String 数组并将其添加到 RemoteInput 中。

```
String replyLabel = getResources().getString(R.string.reply_label);
String[] replyChoices = getResources().getStringArray(R.array.reply_choices);

RemoteInput remoteInput = new RemoteInput.Builder(EXTRA_VOICE_REPLY)
        .setLabel(replyLabel)
        .setChoices(replyChoices)
        .build();
```

（3）为主行动接收语音输入

在 Android Wear 应用中，如果 Reply 是应用程序的主行动（由 setContentIntent()方法定义），那么需要使用 addRemoteInputForContentIntent()方法将 RemoteInput 添加到主行动上。例如下面的演示代码。

```
//为回复行动创建 intent
Intent replyIntent = new Intent(this, ReplyActivity.class);
PendingIntent replyPendingIntent =
        PendingIntent.getActivity(this, 0, replyIntent, 0);
//创建通知
NotificationCompat.Builder replyNotificationBuilder =
        new NotificationCompat.Builder(this)
        .setSmallIcon(R.drawable.ic_new_message)
        .setContentTitle("Message from Travis")
        .setContentText("I love key lime pie!")
        .setContentIntent(replyPendingIntent);
//创建运程回复<语音>
RemoteInput remoteInput = new RemoteInput.Builder(EXTRA_VOICE_REPLY)
        .setLabel(replyLabel)
        .build();
//创建穿戴设备的通知并添加语音输入
Notification replyNotification =
        new WearableNotifications.Builder(replyNotificationBuilder)
        .addRemoteInputForContentIntent(remoteInput)
        .build();
```

通过使用 addRemoteInputForContentIntent()方法将 RemoteInput 对象添加到通知的主行动中后，通常 Open 按钮会显示为 Reply 按钮，当用户在 Android Wear 上选择它时，它就会启动语音输入 UI 视图界面。

（4）为次行动设置语音输入

如果 Reply 动作不是我们创建通知的主动作，而只是为次行动激活语音输入，那么可以添加 RemoteInput 到新的行动按钮（由 Action 对象定义）。通过 Action.Builder()构造器实例化 Action，它会

给行动按钮添加一个 icon 和文本标签，加上 PendingIntent，当用户选择这个行动时，系统会使用它调用应用。例如下面的演示代码。

```
//创建一个 pending intent，当用户选择这个行动时，会启用这个 intent
Intent replyIntent = new Intent(this, ReplyActivity.class);
PendingIntent pendingReplyIntent =
        PendingIntent.getActivity(this, 0, replyIntent, 0);

//创建远程输入
RemoteInput remoteInput = new RemoteInput.Builder(EXTRA_VOICE_REPLY)
        .setLabel(replyLabel)
        .build();

//创建通知行动
Action replyAction = new Action.Builder(R.drawable.ic_message,
        "Reply", pendingIntent)
        .addRemoteInput(remoteInput)
        .build();
```

然后为 Action 添加 RemoteInput.Builder，使用 addAction()方法为 WearableNotifications.Builder 添加 Action。例如下面的演示代码。

```
//创建基本的通知创建者
NotificationCompat.Builder replyNotificationBuilder =
        new NotificationCompat.Builder(this)
        .setContentTitle("New message");

// 创建通知行动并添加远程输入
Action replyAction = new Action.Builder(R.drawable.ic_message,
        "Reply", pendingIntent)
        .addRemoteInput(remoteInput)
        .build();

// 创建穿戴设备的通知并添加行动
Notification replyNotification =
        new WearableNotifications.Builder(replyNotificationBuilder)
        .addAction(replyAction)
        .build();
```

现在，当用户在 Android Wear 设备上选择 Reply 时，系统提示用户使用语音输入（如果提供了预置回复，则会显示预置列表）。当用户完成回复时，系统会调用与该行动关联的 intent，并添加作为字符值的 EXTRA_VOICE_REPLY extra（传递给 RemoteInput.Builder 构造器的字符串）到用户信息。

## 17.3.3　给通知添加页面

当想为 Android Wear 设备提供更多的信息，而且不需要用户使用手持设备打开应用时，可以在 Android Wear 上为通知添加一个或若干个页面，附加的页面会在主通知卡片的右边立即显示出来，如图 17-21 所示。

图 17-21　给通知添加页面

当创建多张页面时，第一步需要把通知显示在手机或平板设备，也就是先创建一个主通知（第一张页面），然后使用 addPage()方法每次添加一张页面，或者使用 addPages()方法从 Collection 对象添加若干页面。例如下面的演示代码。

```
//为主通知创建 Builder
NotificationCompat.Builder notificationBuilder =
        new NotificationCompat.Builder(this)
        .setSmallIcon(R.drawable.new_message)
        .setContentTitle("Page 1")
        .setContentText("Short message")
        .setContentIntent(viewPendingIntent);

//为第二张页面创建 big text 风格
BigTextStyle secondPageStyle = new NotificationCompat.BigTextStyle();
secondPageStyle.setBigContentTitle("Page 2")
               .bigText("A lot of text...");

Notification secondPageNotification =
        new NotificationCompat.Builder(this)
        .setStyle(secondPageStyle)
        .build();

Notification twoPageNotification =
        new WearableNotifications.Builder(notificationBuilder)
        .addPage(secondPageNotification)
        .build();
```

## 17.3.4　通知堆

当为手持设备创建通知时，大家可能习惯于把同类型的通知放在一个汇总通知中。例如，如果应用创建了接收信息的通知，当接收到多条信息时不会显示多条通知在手持设备上，而是使用单条通知。此时只需要提供汇总信息即可，例如“两条新信息”的提示。但是在 Android Wear 设备上，汇总信息没什么作用，因为用户无法在 Android Wear 设备上逐条阅读细节（因为他们必须在手持设备上打开应用查看更多信息）。为了 Android Wear 设备，需要把所有的通知聚集到一个堆中。通知堆以单张卡片的形式存在，用户可以展开它逐条查看，新的 setGroup()方法提供了可能，虽然在手持设备上它仍然仅提供一条汇总信息，如图 17-22 所示。

图 17-22　通知堆演示界面

（1）将通知逐条添加到 Group

为了在 Android Wear 中创建堆，需要为每条通知调用 setGroup()方法，并将唯一的 group key 传递给它们。例如下面的演示代码。

```
final static String GROUP_KEY_EMAILS = "group_key_emails";

NotificationCompat.Builder builder = new NotificationCompat.Builder(mContext)
         .setContentTitle("New mail from " + sender)
         .setContentText(subject)
         .setSmallIcon(R.drawable.new_mail);

Notification notif = new WearableNotifications.Builder(builder)
         .setGroup(GROUP_KEY_EMAILS)
         .build();
```

在默认情况下，会以添加顺序来展现通知，最新的通知显示在头部。但是也可以通过传递一个位置值给 setGroup()的第二个参数，在 group 中定义一个特殊的位置。

（2）添加一个汇总通知

在 Android Wear 应用中，提供一个汇总通知给手持设备是很重要的。除了逐条添加通知到同一个通知堆外，建议仍添加一个汇总通知，不过设置它的次序为 GROUP_ORDER_SUMMARY。例如下面的演示代码。

```
Notification summaryNotification = new WearableNotifications.Builder(builder)
         .setGroup(GROUP_KEY_EMAILS, WearableNotifications.GROUP_ORDER_SUMMARY)
         .build();
```

这条通知不会显示在 Android Wear 设备上的通知堆中，只会显示在手持设备上的唯一通知上。

## 17.3.5　通知语法介绍

（1）android.preview.support.v4.app

用 NotificationCompat.Builder 对象来设定通知的 UI 信息和行为，调用 NotificationCompat.Builder.build()来创建通知，调用 NotificationManager.notify()来发送通知。

一条通知必须包含如下信息。

☑　一个小图标：用 setSmallIcon()来设置。

☑　一个标题：用 setContentTitle()来设置。

☑　详情文字：用 setContentText()来设置。

（2）android.preview.support.wearable.notifications

这是一个提醒接口类，在里面定义了如表 17-1 所示的类。

表 17-1　提醒类

| | |
|---|---|
| RemoteInput | 远程输入类，可穿戴设备输入 |
| RemoteInput.Builder | 生成 RemoteInput 的目标 |
| WearableNotifications | 可穿戴设备类型的通知 |
| WearableNotifications.Action | 可穿戴设备类型通知的行为动作 |
| WearableNotifications.Action.Builder | 生成类 WearableNotifications.Action 对象 |
| WearableNotifications.Builder | 一个 NotificationCompat.Builder 生成器对象，为可穿戴的扩展功能提供通知方法 |

例如在下面的代码中，通过注释详细讲解并演示了各个 Android Wear 对象的基本用法。

```
int notificationId = 001; //通知 id
Intent replyIntent = new Intent(this, ReplyActivity.class); //响应 Action，可以启动 Activity、Service 或者 Broadcast
 PendingIntent pendingIntent = PendingIntent.getActivity(this, 0, replyIntent, 0);
 RemoteInput remoteInput = new RemoteInput.Builder("key")//响应输入，“key”为返回 Intent 的 Extra 的 Key 值
        .setLabel("Select")                                    //输入页标题
        .setChoices(String[])                                  //输入可选项
        .build();
Action replyAction = new Action.Builder(R.drawable, //WearableNotifications.Action.Builder 对应可穿戴设备的
Action 类
                                  "Reply", pendingIntent) //对应 pendingIntent
                                 .addRemoteInput(remoteInput)
                                 .build();

NotificationCompat.Builder notificationBuilder =   new NotificationCompat.Builder(mContext) //标准通知创建
            .setContentTitle(title).setContentText(subject).setSmallIcon(R.drawable).setStyle(style)
 .setLargeIcon(bitmap) //  设置可穿戴设备显示的背景图
 .setContentIntent(pendingIntent) //可穿戴设备左滑，有默认 Open 操作，对应手机端的点击通知
 .addAction(R.drawable, String, pendingIntent);     //增加一个操作，可加多个
Notification notification = new WearableNotifications.Builder(notificationBuilder)     //创建可穿戴类通知，为通知
增加可穿戴设备新特性，必须与兼容包中的 NotificationManager 对应，否则无效
            .setHintHideIcon(true)              //隐藏应用图标
          .addPages(notificationPages)          //增加 Notification 页
          .addAction(replyAction)               //对应上页，pendingIntent 可操作项
.addRemoteInputForContentIntent(replyAction) //可为 ContentIntent 替换默认的 Open 操作
.setGroup(GROUP_KEY, WearableNotifications.GROUP_ORDER_SUMMARY)     //为通知分组
.setLocalOnly(true) //可设置只在本地显示
.setMinPriority() //设置只在可穿戴设备上显示通知
          .build();
NotificationManagerCompat notificationManager = NotificationManagerCompat.from(this);//获得 Manager
notificationManager.notify(notificationId, notificationBuilder.build()); //发送通知
```

## 17.4　实战演练——开发一个 Android Wear 程序

**知识点讲解：光盘:视频\知识点\第 17 章\开发一个 Android Wear 程序.avi**

本节将通过一个具体实例来讲解开发一个 Android Wear 程序的方法。

| 实 例 | 功 能 | 源 码 路 径 |
| --- | --- | --- |
| 实例 17-1 | 开发一个 Android Wear 程序 | 光盘:\daima\17\wearmaster |

本实例的具体实现流程如下。

（1）编写布局文件 activity_main.xml，具体实现代码如下：

```
<RelativeLayout xmlns:android="http://schemas.android.com/apk/res/android"
    xmlns:tools="http://schemas.android.com/tools"
    android:layout_width="match_parent"
    android:layout_height="match_parent"
    android:paddingLeft="@dimen/activity_horizontal_margin"
    android:paddingRight="@dimen/activity_horizontal_margin"
    android:paddingTop="@dimen/activity_vertical_margin"
    android:paddingBottom="@dimen/activity_vertical_margin"
    tools:context="com.ezhuk.wear.MainActivity">
    <TextView
        android:text="@string/hello_world"
        android:layout_width="wrap_content"
        android:layout_height="wrap_content" />
</RelativeLayout>
```

（2）编写值文件 strings.xml，功能是设置通知的文本内容，具体实现代码如下：

```
<?xml version="1.0" encoding="utf-8"?>
<resources>
    <string name="app_name">Wear</string>
    <string name="hello_world">Test</string>
    <string name="content_title">Basic Notification</string>
    <string name="content_text">Sample text.</string>
    <string name="page1_title">Page 1</string>
    <string name="page1_text">Sample text 1.</string>
    <string name="page2_title">Page 2</string>
    <string name="page2_text">Sample text 2.</string>
    <string name="action_title">Action Title</string>
    <string name="action_text">Action text.</string>
    <string name="action_button">Action</string>
    <string name="action_label">Action</string>
    <string name="summary_title">Summary Title</string>
    <string name="summary_text">Summary text.</string>
    <string-array name="input_choices">
        <item>First item</item>
        <item>Second item</item>
        <item>Third item</item>
    </string-array>
</resources>
```

（3）编写文件 MainActivity.java 实现程序的主 Activity，功能是载入 Android Wear 的通知类 NotificationUtils，调用不同的 showNotificationXX 方法显示通知信息，具体实现代码如下：

```
package com.ezhuk.wear;

import android.app.Activity;
import android.os.Bundle;
```

```
import static com.ezhuk.wear.NotificationUtils.*;


public class MainActivity extends Activity {
    @Override
    protected void onCreate(Bundle savedInstanceState) {
        super.onCreate(savedInstanceState);
        setContentView(R.layout.activity_main);
    }

    @Override
    protected void onResume() {
        super.onResume();

        showNotification(this);
        showNotificationNoIcon(this);
        showNotificationMinPriority(this);
        showNotificationBigTextStyle(this);
        showNotificationBigPictureStyle(this);
        showNotificationInboxStyle(this);
        showNotificationWithPages(this);
        showNotificationWithAction(this);
        showNotificationWithInputForPrimaryAction(this);
        showNotificationWithInputForSecondaryAction(this);
        showGroupNotifications(this);
    }

    @Override
    protected void onPause() {
        super.onPause();
    }
}
```

（4）编写文件 NotificationUtils.java，功能是定义各种不同类型 showNotificationXX 的通知方法，具体实现代码如下：

```
import android.app.Notification;
import android.app.PendingIntent;
import android.content.Context;
import android.content.Intent;
import android.graphics.BitmapFactory;
import android.net.Uri;
import android.preview.support.v4.app.NotificationManagerCompat;
import android.preview.support.wearable.notifications.RemoteInput;
import android.preview.support.wearable.notifications.WearableNotifications;
import android.support.v4.app.NotificationCompat;


public class NotificationUtils {
    private static final String ACTION_TEST = "com.ezhuk.wear.ACTION";
    private static final String ACTION_EXTRA = "action";

    private static final String NOTIFICATION_GROUP = "notification_group";
```

```
public static void showNotification(Context context) {
    NotificationCompat.Builder builder =
            new NotificationCompat.Builder(context)
                    .setSmallIcon(R.drawable.ic_launcher)
                    .setContentTitle(context.getString(R.string.content_title))
                    .setContentText(context.getString(R.string.content_text));

    NotificationManagerCompat.from(context).notify(0,
            new WearableNotifications.Builder(builder)
                    .build());
}

public static void showNotificationNoIcon(Context context) {
    NotificationCompat.Builder builder =
            new NotificationCompat.Builder(context)
                    .setSmallIcon(R.drawable.ic_launcher)
                    .setContentTitle(context.getString(R.string.content_title))
                    .setContentText(context.getString(R.string.content_text));

    NotificationManagerCompat.from(context).notify(1,
            new WearableNotifications.Builder(builder)
                    .setHintHideIcon(true)
                    .build());
}

public static void showNotificationMinPriority(Context context) {
    NotificationCompat.Builder builder =
            new NotificationCompat.Builder(context)
                    .setSmallIcon(R.drawable.ic_launcher)
                    .setContentTitle(context.getString(R.string.content_title))
                    .setContentText(context.getString(R.string.content_text));

    NotificationManagerCompat.from(context).notify(2,
            new WearableNotifications.Builder(builder)
                    .setMinPriority()
                    .build());
}

public static void showNotificationWithStyle(Context context,
                                              int id,
                                              NotificationCompat.Style style) {
    Notification notification = new WearableNotifications.Builder(
            new NotificationCompat.Builder(context)
                    .setSmallIcon(R.drawable.ic_launcher)
                    .setStyle(style))
            .build();

    NotificationManagerCompat.from(context).notify(id, notification);
}

public static void showNotificationBigTextStyle(Context context) {
    showNotificationWithStyle(context, 3,
```

```
                new NotificationCompat.BigTextStyle()
                        .setSummaryText(context.getString(R.string.summary_text))
                        .setBigContentTitle("Big Text Style")
                        .bigText("Sample big text."));
    }

    public static void showNotificationBigPictureStyle(Context context) {
        showNotificationWithStyle(context, 4,
                new NotificationCompat.BigPictureStyle()
                        .setSummaryText(context.getString(R.string.summary_text))
                        .setBigContentTitle("Big Picture Style")
                        .bigPicture(BitmapFactory.decodeResource(
                                context.getResources(), R.drawable.background)));
    }

    public static void showNotificationInboxStyle(Context context) {
        showNotificationWithStyle(context, 5,
                new NotificationCompat.InboxStyle()
                        .setSummaryText(context.getString(R.string.summary_text))
                        .setBigContentTitle("Inbox Style")
                        .addLine("Line 1")
                        .addLine("Line 2"));
    }

    public static void showNotificationWithPages(Context context) {
        NotificationCompat.Builder builder =
                new NotificationCompat.Builder(context)
                        .setSmallIcon(R.drawable.ic_launcher)
                        .setContentTitle(context.getString(R.string.page1_title))
                        .setContentText(context.getString(R.string.page1_text));

        Notification second = new NotificationCompat.Builder(context)
                .setSmallIcon(R.drawable.ic_launcher)
                .setContentTitle(context.getString(R.string.page2_title))
                .setContentText(context.getString(R.string.page2_text))
                .build();

        NotificationManagerCompat.from(context).notify(6,
                new WearableNotifications.Builder(builder)
                        .addPage(second)
                        .build());
    }

    public static void showNotificationWithAction(Context context) {
        Intent intent = new Intent(Intent.ACTION_VIEW);
        intent.setData(Uri.parse(""));
        PendingIntent pendingIntent =
                PendingIntent.getActivity(context, 0, intent, 0);

        NotificationCompat.Builder builder =
                new NotificationCompat.Builder(context)
                        .setSmallIcon(R.drawable.ic_launcher)
```

```
                    .setContentTitle(context.getString(R.string.action_title))
                    .setContentText(context.getString(R.string.action_text))
                    .addAction(R.drawable.ic_launcher,
                            context.getString(R.string.action_button),
                            pendingIntent);

    NotificationManagerCompat.from(context).notify(7,
            new WearableNotifications.Builder(builder)
                    .build());
}

public static void showNotificationWithInputForPrimaryAction(Context context) {
    Intent intent = new Intent(ACTION_TEST);
    PendingIntent pendingIntent =
            PendingIntent.getActivity(context, 0, intent, 0);

    NotificationCompat.Builder builder =
            new NotificationCompat.Builder(context)
                    .setSmallIcon(R.drawable.ic_launcher)
                    .setContentTitle(context.getString(R.string.action_title))
                    .setContentText(context.getString(R.string.action_text))
                    .setContentIntent(pendingIntent);

    String[] choices =
            context.getResources().getStringArray(R.array.input_choices);

    RemoteInput remoteInput = new RemoteInput.Builder(ACTION_EXTRA)
            .setLabel(context.getString(R.string.action_label))
            .setChoices(choices)
            .build();

    NotificationManagerCompat.from(context).notify(8,
            new WearableNotifications.Builder(builder)
                    .addRemoteInputForContentIntent(remoteInput)
                    .build());
}

public static void showNotificationWithInputForSecondaryAction(Context context) {
    Intent intent = new Intent(ACTION_TEST);
    PendingIntent pendingIntent =
            PendingIntent.getActivity(context, 0, intent, 0);

    RemoteInput remoteInput = new RemoteInput.Builder(ACTION_EXTRA)
            .setLabel(context.getString(R.string.action_label))
            .build();

    WearableNotifications.Action action =
            new WearableNotifications.Action.Builder(
                    R.drawable.ic_launcher,
                    "Action",
                    pendingIntent)
            .addRemoteInput(remoteInput)
            .build();
```

```
        NotificationCompat.Builder builder =
                new NotificationCompat.Builder(context)
                        .setContentTitle(context.getString(R.string.action_title));

        NotificationManagerCompat.from(context).notify(9,
                new WearableNotifications.Builder(builder)
                        .addAction(action)
                        .build());
    }

    public static void showGroupNotifications(Context context) {
        Notification first = new WearableNotifications.Builder(
                new NotificationCompat.Builder(context)
                        .setSmallIcon(R.drawable.ic_launcher)
                        .setContentTitle(context.getString(R.string.page1_title))
                        .setContentText(context.getString(R.string.page1_text)))
                .setGroup(NOTIFICATION_GROUP)
                .build();

        Notification second = new WearableNotifications.Builder(
                new NotificationCompat.Builder(context)
                        .setSmallIcon(R.drawable.ic_launcher)
                        .setContentTitle(context.getString(R.string.page2_title))
                        .setContentText(context.getString(R.string.page2_text)))
                .setGroup(NOTIFICATION_GROUP)
                .build();

        Notification summary = new WearableNotifications.Builder(
                new NotificationCompat.Builder(context)
                        .setSmallIcon(R.drawable.ic_launcher)
                        .setContentTitle(context.getString(R.string.summary_title))
                        .setContentText(context.getString(R.string.summary_text)))
                .setGroup(NOTIFICATION_GROUP, WearableNotifications.GROUP_ORDER_SUMMARY)
                .build();

        NotificationManagerCompat.from(context).notify(10, first);
        NotificationManagerCompat.from(context).notify(11, second);
        NotificationManagerCompat.from(context).notify(12, summary);
    }

    public static void cancelNotification(Context context, int id) {
        NotificationManagerCompat.from(context).cancel(id);
    }

    public static void cancelAllNotifications(Context context) {
        NotificationManagerCompat.from(context).cancelAll();
    }
}
```

至此，一个简单的 Android Wear 通知程序创建完毕。执行后会实现通知功能，如图 17-23 所示。

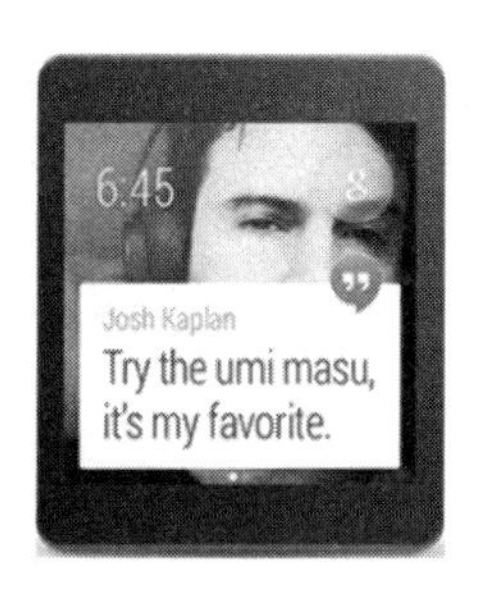


图 17-23　执行效果

有关 Android Wear 更多的演示程序，读者可以参考官方文档中的演示实例。

## 17.5　实战演练——实现手机和 Android Wear 的交互

**知识点讲解：光盘:视频\知识点\第 17 章\实现手机和 Android Wear 的交互.avi**

本节将通过一个具体实例来讲解将手机应用中的通知信息发送到 Android Wear 的方法，介绍在 Android Wear 中显示通知提示信息的过程。

| 实　例 | 功　能 | 源 码 路 径 |
|---|---|---|
| 实例 17-2 | 实现手机和 Android Wear 的交互 | 光盘:\daima\17\Android-Wear-Codelab |

本实例的具体实现流程如下。

（1）打开 Eclipse，在 AVD Manager 窗口中新建一个 Android Wear 模拟器，具体设置如图 17-24 所示。

（2）在设备中启动 Android 应用程序，并单击 Connect（连接）按钮，如图 17-25 所示。

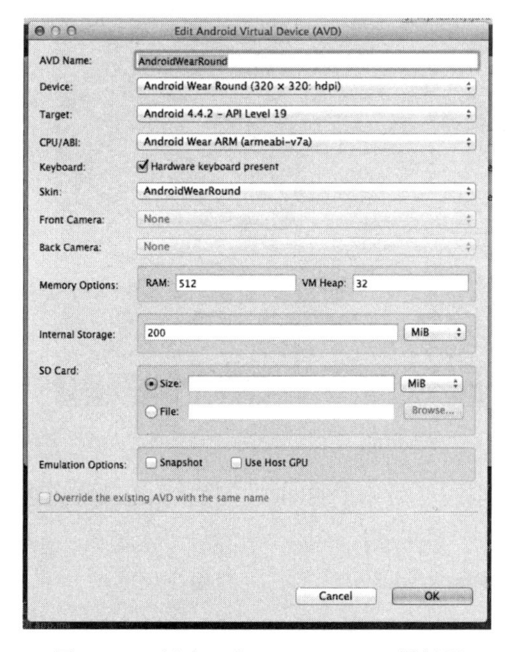


图 17-24　新建一个 Android Wear 模拟器

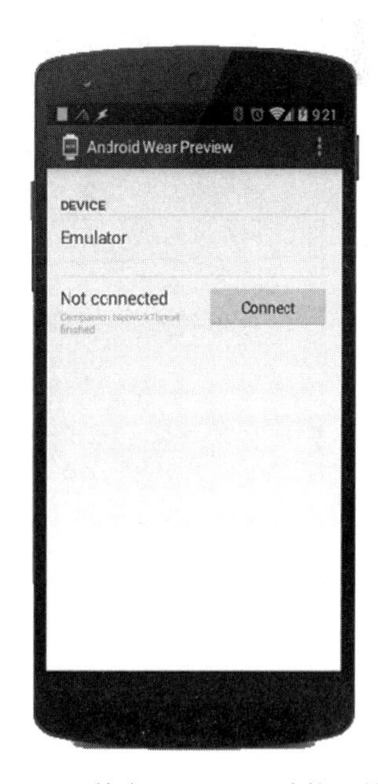


图 17-25　单击 Connect（连接）按钮

（3）导航到 android-studio/sdk/platform-tools 目录，然后使用如下所示的命令：

```
adb -d forward tcp:5601 tcp:5601
```

如果由于某种原因，仍然无法连接两个设备，尝试使用 adb devices，可以使用上述命令在每当将手机连接到电脑或每次重新启动时使用的仿真器。其中未连接时的界面如图 17-26 所示。已经连接后的界面如图 17-27 所示。

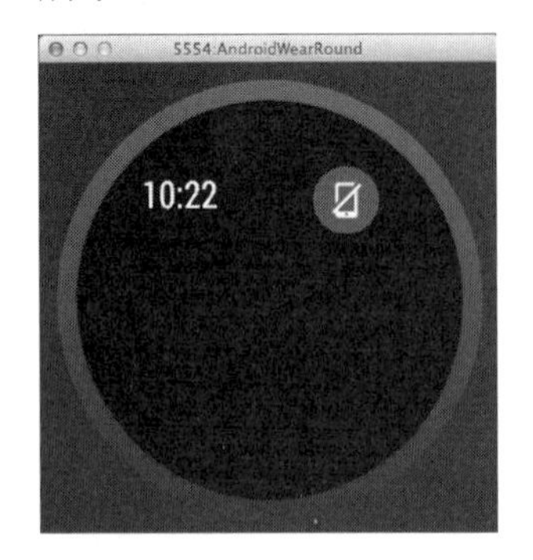


图 17-26　未连接时的界面

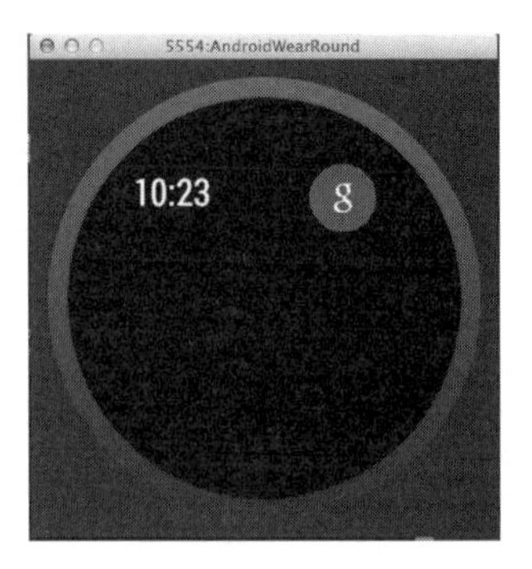


图 17-27　连接后的界面

（4）在 Eclipse 中新建一个 Android 应用程序，如图 17-28 所示。

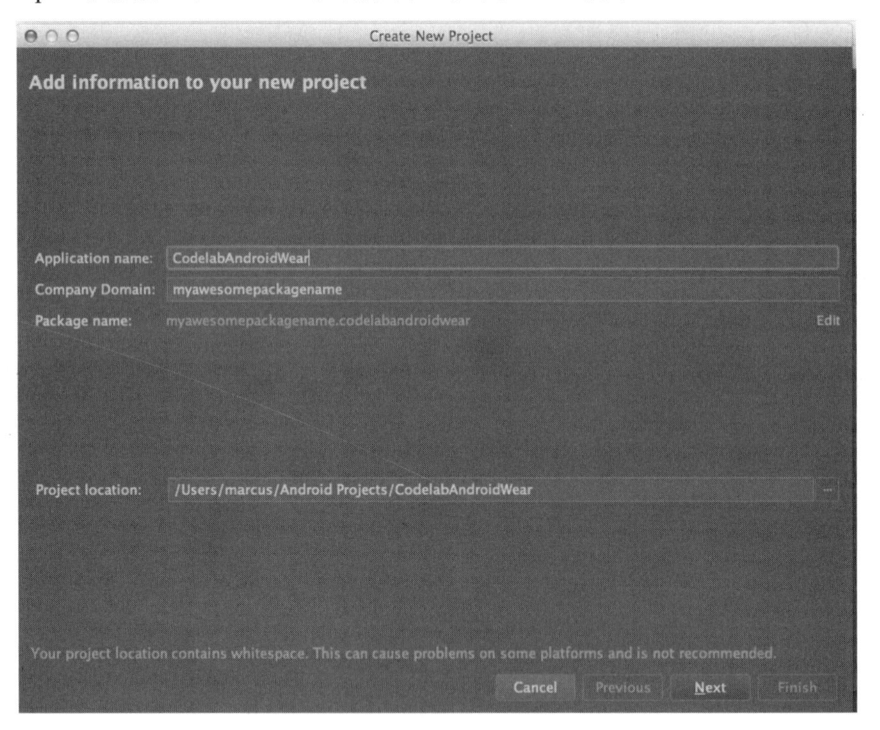


图 17-28　新建一个 Android 项目

（5）为了提高应用程序的运行速度，建议使用较低版本的 SDK，例如 Android 4.0.3，如图 17-29 所示。

（6）在 Add an activity to Mobile 界面中选择 Blank Activity 选项，如图 17-30 所示。

（7）在工程中添加需要的库文件：wearable-preview-support.jar、gradle-wrapper.jar，如图 17-31 所示。

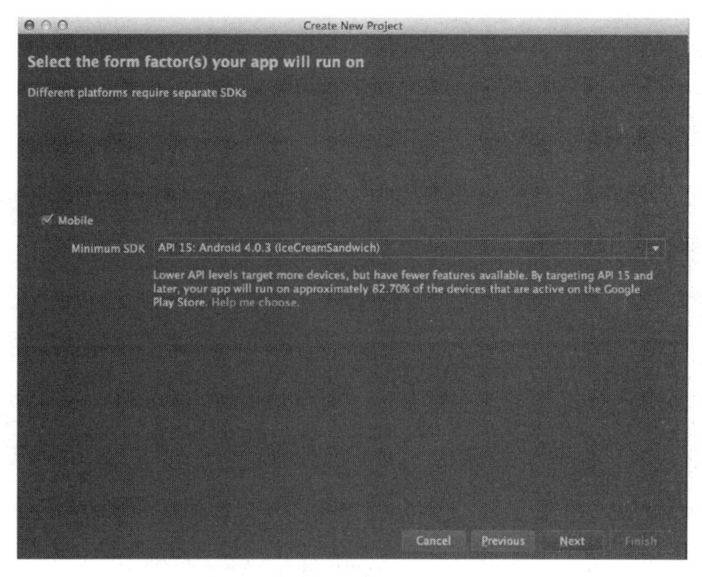


图 17-29　使用较低版本的 SDK

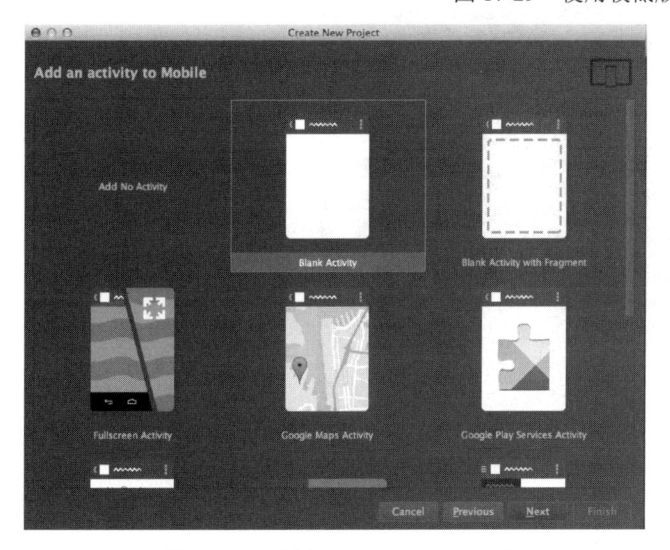


图 17-30　选择 Blank Activity 选项

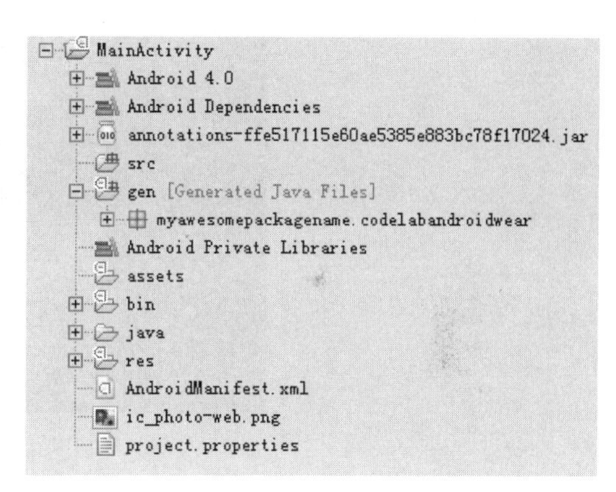


图 17-31　添加需要的库文件

（8）编写依赖文件 build.gradle，具体实现代码如下：

```
dependencies {
    compile 'com.android.support:support-v4:19.1.+'          //支持的类库
    compile fileTree(dir: 'libs', include: ['*.jar'])         //编译目录中的所有.jar 库文件
}
```

（9）开始为 Android 应用程序创建一个通知图标，在 res 文件夹上右击，并在弹出的快捷菜单中依次选择 New | Image Asset 命令，如图 17-32 所示。

（10）在弹出界面的 Asset Type 下拉列表中选择 Notification Icons，并给它命名一个资源名称，如图 17-33 所示。

（11）检查设置的姓名和目标是否正确，并单击 Finish（完成）按钮，Android 将为 Notification 自动生成所有需要尺寸的图标和文件夹，如图 17-34 所示。

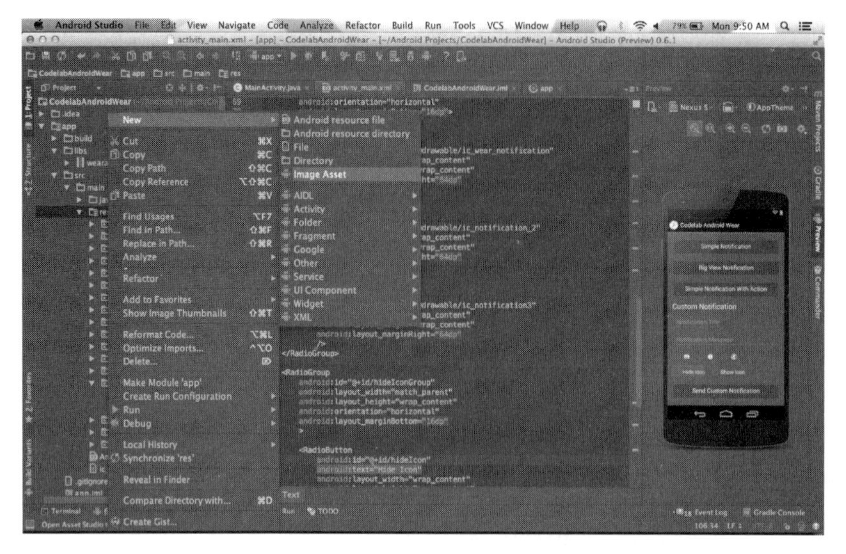


图 17-32 选择 New | Image Asset 命令

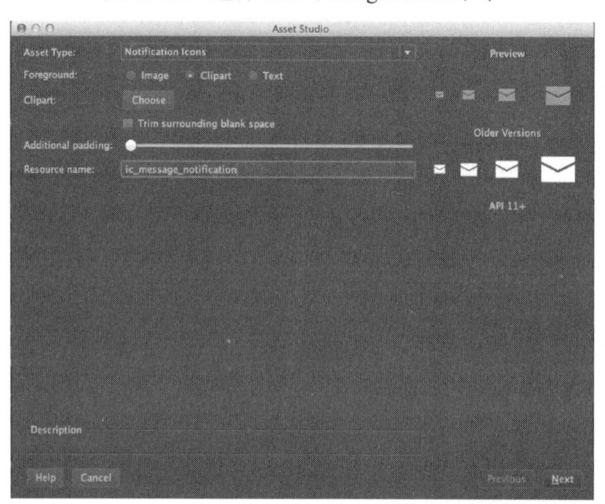


图 17-33 命名资源名称

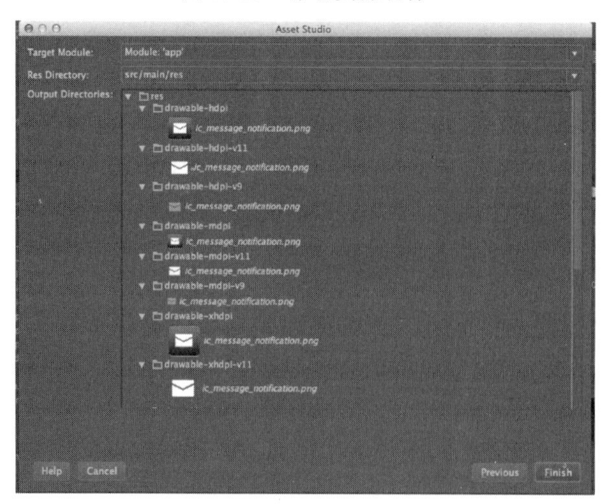


图 17-34 自动生成需要尺寸的图标和文件夹

（12）创建主 XML 布局文件 activity_main.xml，具体实现代码如下：

```
<LinearLayout
    xmlns:android="http://schemas.android.com/apk/res/android"
    xmlns:tools="http://schemas.android.com/tools"
    android:layout_width="match_parent"
    android:layout_height="match_parent"
    android:paddingLeft="@dimen/activity_horizontal_margin"
    android:paddingRight="@dimen/activity_horizontal_margin"
    android:paddingTop="@dimen/activity_vertical_margin"
    tools:context=".MainActivity"
    android:orientation="vertical"
    android:background="#34495e"
    >
    <Button
        android:id="@+id/simpleNotification"
        android:text="Simple Notification"
        android:layout_width="match_parent"
        android:layout_height="wrap_content"
        android:layout_marginBottom="@dimen/activity_vertical_margin"
        android:onClick="sendNotification"
    />
    <Button
        android:id="@+id/bigNotification"
        android:text="Big View Notification"
        android:layout_width="match_parent"
        android:layout_height="wrap_content"
        android:layout_marginBottom="@dimen/activity_vertical_margin"
        android:onClick="sendNotification"
    />
    <Button
        android:id="@+id/bigNotificationWithAction"
        android:text="Big Notification With Action"
        android:layout_width="match_parent"
        android:layout_height="wrap_content"
        android:layout_marginBottom="@dimen/activity_vertical_margin"
        android:onClick="sendNotification"
    />
    <TextView
        android:text="Custom Notification"
        android:textAppearance="?android:attr/textAppearanceLarge"
        android:layout_width="match_parent"
        android:layout_height="wrap_content"
        android:layout_marginBottom="@dimen/activity_vertical_margin"
        />
    <EditText
        android:id="@+id/notificationTitle"
        android:hint="Notification Title"
        android:layout_width="match_parent"
        android:layout_height="wrap_content"
        android:layout_marginBottom="@dimen/activity_vertical_margin"
        />
```

```
<EditText
    android:id="@+id/notificationMessage"
    android:hint="Notification Message"
    android:layout_width="match_parent"
    android:layout_height="wrap_content"
    android:layout_marginBottom="@dimen/activity_vertical_margin"
/>
<RadioGroup
    android:id="@+id/iconGroup"
    android:layout_width="match_parent"
    android:layout_height="wrap_content"
    android:orientation="horizontal"
    android:layout_marginBottom="@dimen/activity_vertical_margin">
    <RadioButton
        android:id="@+id/icon1"
        android:drawableRight="@drawable/ic_wear_notification"
        android:layout_width="wrap_content"
        android:layout_height="wrap_content"
        android:layout_marginRight="@dimen/activity_horizontal_margin"
    />
    <RadioButton
        android:id="@+id/icon2"
        android:drawableRight="@drawable/ic_notification_2"
        android:layout_width="wrap_content"
        android:layout_height="wrap_content"
        android:layout_marginRight="@dimen/activity_horizontal_margin"
    />
    <RadioButton
        android:id="@+id/icon3"
        android:drawableRight="@drawable/ic_notification3"
        android:layout_width="wrap_content"
        android:layout_height="wrap_content"
        android:layout_marginRight="@dimen/activity_horizontal_margin"
        />
</RadioGroup>
<RadioGroup
    android:id="@+id/hideIconGroup"
    android:layout_width="match_parent"
    android:layout_height="wrap_content"
    android:orientation="horizontal"
    android:layout_marginBottom="@dimen/activity_vertical_margin"
    >
    <RadioButton
        android:id="@+id/hideIcon"
        android:text="Hide Icon"
        android:layout_width="wrap_content"
        android:layout_height="wrap_content"
        android:layout_marginRight="@dimen/activity_horizontal_margin"
        />
    <RadioButton
        android:id="@+id/showIcon"
```

```
            android:text="Show Icon"
            android:layout_width="wrap_content"
            android:layout_height="wrap_content"
            android:layout_marginRight="@dimen/activity_horizontal_margin"
            />
    </RadioGroup>
    <Button
        android:id="@+id/sendCustomNotification"
        android:text="Send Custom Notification"
        android:layout_width="match_parent"
        android:layout_height="wrap_content"
        android:onClick="sendNotification"
    />
</LinearLayout>
```

上述布局文件的执行效果如图 17-35 所示。

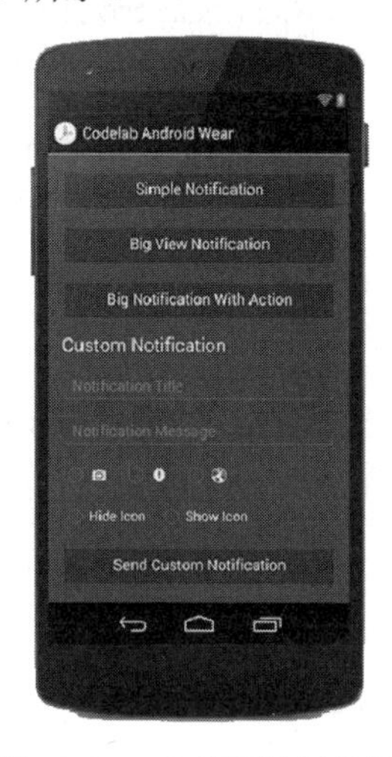


图 17-35　主界面执行效果

（13）通过使用 Android 穿戴 API，可以将手机通知信息发送到穿戴设备中。接下来首先为主界面的 android:onClick="sendNotification"按钮添加单击响应事件处理程序，具体实现代码如下：

```
public void sendNotification(View view) {
    switch(view.getId()) {
        case R.id.simpleNotification:
            break;

        case R.id.bigNotification:
            break;

        case R.id.bigNotificationWithAction:
            break;

        case R.id.sendCustomNotification:
            break;
    }
}
```

（14）将如下所示的实例变量添加到类，这些实例变量和 RadioGroup 中输入的数据相对应。

```
public class MainActivity extends Activity {

    private EditText mCustomTitle, mCustomMessage;
    private RadioGroup mCustomIconGroup, showHideIconGroup;
    private int mCustomIcon;
    private boolean showIcon = false;
    private String LOG_TAG = "WEAR";

    @Override
```

（15）实例化用户界面元素，并添加检索有关哪个元素被选中的信息。具体实现代码如下：

```
@Override
protected void onCreate(Bundle savedInstanceState) {
    super.onCreate(savedInstanceState);
    setContentView(R.layout.activity_main);

    mCustomTitle = (EditText) findViewById(R.id.notificationTitle);
    mCustomMessage = (EditText) findViewById(R.id.notificationMessage);

    mCustomIconGroup = (RadioGroup) findViewById(R.id.iconGroup);
    mCustomIconGroup.setOnCheckedChangeListener(new RadioGroup.OnCheckedChangeListener() {
        @Override
        public void onCheckedChanged(RadioGroup group, int checkedId) {
            switch (group.getCheckedRadioButtonId()) {
                case R.id.icon1:
                    mCustomIcon = R.drawable.ic_wear_notification;
                    break;
                case R.id.icon2:
                    mCustomIcon = R.drawable.ic_notification_2;
                    break;
                case R.id.icon3:
                    mCustomIcon = R.drawable.ic_notification3;
                    break;
            }
        }
    });

    showHideIconGroup = (RadioGroup) findViewById(R.id.hideIconGroup);
    showHideIconGroup.setOnCheckedChangeListener(new RadioGroup.OnCheckedChangeListener() {
        @Override
        public void onCheckedChanged(RadioGroup group, int checkedId) {
            switch (group.getCheckedRadioButtonId()) {
                case R.id.showIcon:
                    showIcon = true;
                    break;
                case R.id.hideIcon:
                    showIcon = false;
                    break;
            }
        }
    });
}
```

（16）编写发送通知信息的方法 sendNotification()，在切换之前添加所有的变量信息以便在通知之

间实现共享。具体实现代码如下：

```
public void sendNotification(View view) {

    int notificationId = 001; //id- An identifier for this notification unique within your application.
    String eventTitle = "Sample Notification"; //Title for the notificaiton
    String eventText = "Text for the notification."; //Text for the notificaiton
    String intentExtra = "This is an extra String!"; //Extra String to be passed to a intent
    String eventDescription = "This is supposed to  be a content that will not fit the normal content screen"
            + " usually a bigger text, by example a long text message or email.";

    Intent viewIntent = new Intent(this, MainActivity.class);
    PendingIntent viewPendingIntent = PendingIntent.getActivity(this, 0, viewIntent, 0);

    NotificationCompat.BigTextStyle bigStyle = new NotificationCompat.BigTextStyle();

    NotificationCompat.Builder mBuilder = null;
    Notification mNotification = null;

    NotificationManagerCompat notificationManager = NotificationManagerCompat.from(this);

    switch(view.getId()) {
        case R.id.simpleNotification:
            break;

        case R.id.bigNotification:
            break;

        case R.id.bigNotificationWithAction:
            break;

        case R.id.sendCustomNotification:
            break;
    }
}
```

（17）开始构建声明，根据用户的输入来显示通知信息。可以使用 Notification 建立和发送通知任何元素对象，如游戏得分、闹钟等。

☑ 简单通知。

实现一个简单的用 4 个元素通知的形式。第一个是可绘制的通知，使用资产管理器创建可绘制对象之一。第二个用一个标题通知 3，要显示 4 的消息。具体实现代码如下：

```
case R.id.simpleNotification:
    mBuilder = new NotificationCompat.Builder(this)
            .setSmallIcon(R.drawable.ic_wear_notification)
            .setContentTitle(eventTitle)
            .setContentText(eventText)
            .setAutoCancel(true)
            .setContentIntent(viewPendingIntent);
    break;
```

在开关后添加下面的代码来显示或通知。

```
notificationManager.notify(notificationId, mBuilder.build());
Log.d(LOG_TAG, "Normal Notification");
```

此时开始测试，运行 Android 应用程序，在手机上的运行效果如图 17-36 所示。

此时通知信息在可穿戴设备中的运行效果如图 17-37 所示。

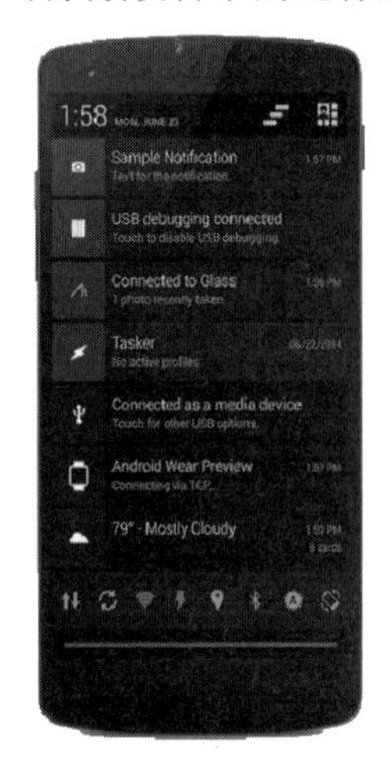

图 17-36　在手机上的运行效果

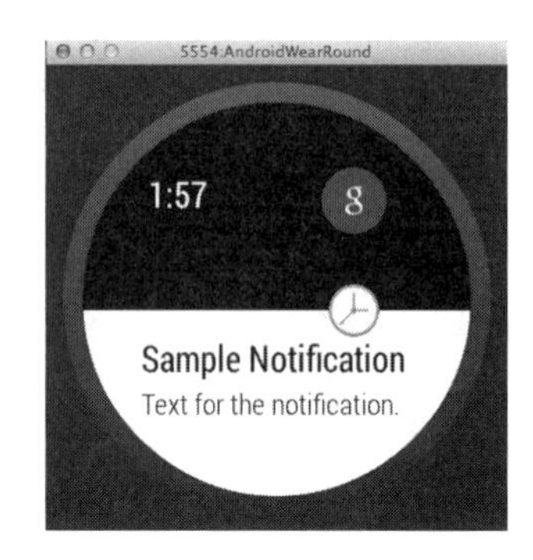


图 17-37　在可穿戴设备中的运行效果

☑ 实现大图标通知效果，大图标作为通知信息的背景显示在后面，不同的小图标将使用 BitmapFactory.decodeResource()解码 PNG 文件，此时该 setContentTitle 和 setContentText 将由 bigStyle.bigText 和 bigStyle.setBigContentTitle 覆盖。具体实现代码如下：

```
case R.id.bigNotification:
    bigStyle.bigText(eventDescription);
    bigStyle.setBigContentTitle("Override Title");
    mBuilder = new NotificationCompat.Builder(this)
        .setSmallIcon(R.drawable.ic_wear_notification)
        .setLargeIcon(BitmapFactory.decodeResource(getResources(), R.drawable.ic_sample_codelab))
        .setContentTitle(eventTitle)
        .setContentText(eventText)
        .setContentIntent(viewPendingIntent)
        .setAutoCancel(true)
        .setStyle(bigStyle);
    break;
```

大图标通知在可穿戴设备中的效果如图 17-38 所示，因为文本被设置为一个大的文本，所以提供一个允许用户向内侧滚动以阅读完整文字。

图 17-38　大图标通知

☑ 为了实现大图标通知功能，创建第二个 Activity 界面，布局文件 activity_second.xml 的具体实

现代码如下：

```
<LinearLayout xmlns:android="http://schemas.android.com/apk/res/android"
    xmlns:tools="http://schemas.android.com/tools"
    android:layout_width="match_parent"
    android:layout_height="match_parent"
    android:paddingLeft="@dimen/activity_horizontal_margin"
    android:paddingRight="@dimen/activity_horizontal_margin"
    android:paddingTop="@dimen/activity_vertical_margin"
    android:paddingBottom="@dimen/activity_vertical_margin"
    android:background="#34495e"
    android:orientation="vertical"
    tools:context="myawesomepackagename.codelabandroidwear.SecondActivity">

    <TextView
        android:id="@+id/extraMessage"
        android:text="@string/hello_world"
        android:layout_width="wrap_content"
        android:layout_height="wrap_content" />

    <ImageView
        android:id="@+id/extraPhoto"
        android:layout_width="wrap_content"
        android:layout_height="wrap_content"
        />
</LinearLayout>
```

☑ 为第二个界面设置和准备接收的一些数据，通过 PROCEDE 建立 Notificaiton 提醒信息。实例文件 SecondActivity.java 的具体实现代码如下：

```
@Override
protected void onCreate(Bundle savedInstanceState) {
    super.onCreate(savedInstanceState);
    setContentView(R.layout.activity_second);
    TextView mTextView = (TextView) findViewById(R.id.extraMessage);
    ImageView mImageView = (ImageView) findViewById(R.id.extraPhoto);
}
Intent extraIntent = getIntent();

//Get the intent information based on the names passed by your notificaiton "message" and
mTextView.setText(extraIntent.getStringExtra("message"));
mImageView.setImageResource(extraIntent.getIntExtra("photo", 0));
case R.id.bigNotificationWithAction:
    Intent photoIntent = new Intent(this, SecondActivity.class);
    photoIntent.putExtra("message", intentExtra);
    photoIntent.putExtra("photo", R.drawable.ic_sample_codelab);

    PendingIntent photoPending = PendingIntent.getActivity(this, 0, photoIntent, 0);
    bigStyle.setBigContentTitle("Mr. Flowers");
    bigStyle.bigText("Check out this picture!! :D");
    mBuilder = new NotificationCompat.Builder(this)
            .setSmallIcon(R.drawable.ic_wear_notification)
            .setLargeIcon(BitmapFactory.decodeResource(getResources(), R.drawable.ic_sample_codelab))
            .setContentIntent(viewPendingIntent)
```

```
                .addAction(R.drawable.ic_photo, "See Photo", photoPending)
                .setAutoCancel(true)
                .setStyle(bigStyle);
    break;
```

此时可以在手机中和可穿戴设备中实现大图标通知效果，如图 17-39 所示。

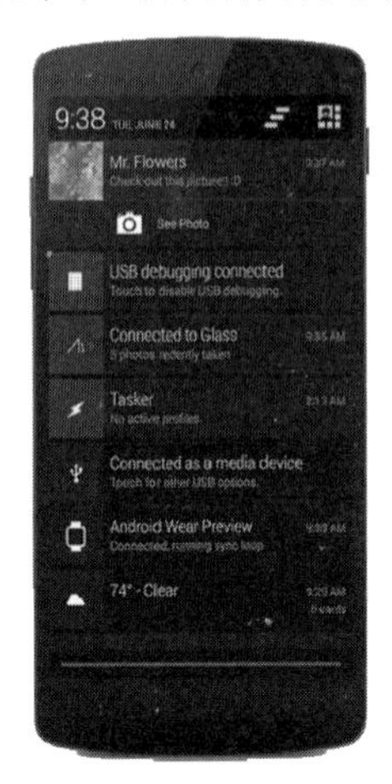

手机中的信息

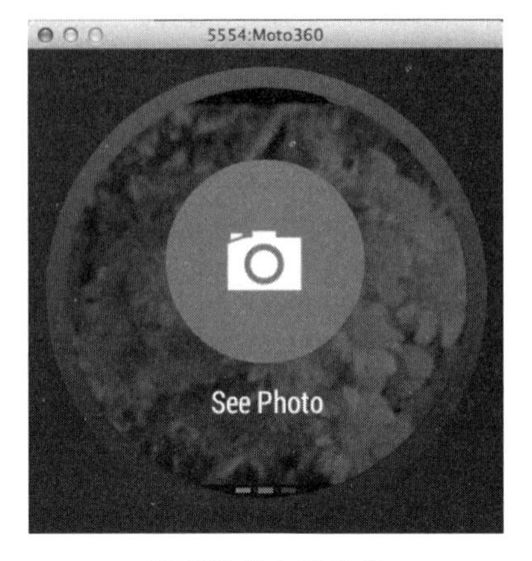


穿戴设备中的信息

图 17-39　大图标通知信息

至此，便建立了手机和智能手表之间的信息通知功能。读者可以继续编写其他类型的通知信息，例如自定义一个：让用户设置标题、一条消息，并选择图标在通知中显示，如果用户想要显示或没有应用程序图标会给出一个选项。上述功能的实现代码如下：

```
case R.id.sendCustomNotification:
    mBuilder = new NotificationCompat.Builder(this)
                .setSmallIcon(mCustomIcon)
                .setContentTitle(mCustomTitle.getText().toString())
                .setAutoCancel(true)
                .setContentText(mCustomMessage.getText().toString())
                .setContentIntent(viewPendingIntent);

    mNotification = new WearableNotifications.Builder(mBuilder)
                .setHintHideIcon(!showIcon)
                .build();
    break;
```

另外还有另一种类型的通知，以显示需要的情况下加入到电话通知功能中。在开关后替换已经存在的代码，具体实现代码如下：

```
if(view.getId() != R.id.sendCustomNotification) {
    notificationManager.notify(notificationId, mBuilder.build());
    Log.d(LOG_TAG, "Normal Notification");
} else {
    notificationManager.notify(notificationId, mNotification);
    Log.d(LOG_TAG, "Wear Notification");
}
```

读者可以书中的实例为基础，继续自由发挥自己的创作天赋。